Communications in Computer and Information Science 773

More information about this series at http://www.springer.com/series/7899

Jinfeng Yang · Qinghua Hu
Ming-Ming Cheng · Liang Wang
Qingshan Liu · Xiang Bai
Deyu Meng (Eds.)

Computer Vision

Second CCF Chinese Conference, CCCV 2017
Tianjin, China, October 11–14, 2017
Proceedings, Part III

Editors
Jinfeng Yang
Civil Aviation University of China
Tianjin
China

Qinghua Hu
Tianjin University
Tianjin
China

Ming-Ming Cheng
Nankai University
Tianjin
China

Liang Wang
Institute of Automation
Chinese Academy of Sciences
Beijing
China

Qingshan Liu
Nanjing University of Information Science and Technology
Nanjing
China

Xiang Bai
Huazhong University of Science and Technology
Wuhan
China

Deyu Meng
Xi'an Jiaotong University
Xi'an
China

ISSN 1865-0929 ISSN 1865-0937 (electronic)
Communications in Computer and Information Science
ISBN 978-981-10-7304-5 ISBN 978-981-10-7305-2 (eBook)
https://doi.org/10.1007/978-981-10-7305-2

Library of Congress Control Number: 2017960863

Printed on acid-free paper

This Springer imprint is published by Springer Nature
The registered company is Springer Nature Singapore Pte Ltd.
The registered company address is: 152 Beach Road, #21-01/04 Gateway East, Singapore 189721, Singapore

Preface

Welcome to the proceedings of the Second Chinese Conference on Computer Vision (CCCV 2017) held in Tianjin!

CCCV, hosted by the China Computer Federation (CCF) and co-organized by Computer Vision Committee of CCF, is a national conference in the field of computer vision. It aims at providing an interactive communication platform for students, faculties, and researchers from industry. It promotes not only academic exchange, but also communication between academia and industry. CCCV is one of the most important local academic activities in computer vision. In order to keep track of the frontier of academic trends and share the latest research achievements, innovative ideas, and scientific methods in the field of computer vision, international and local leading experts and professors are invited to deliver keynote speeches, introducing the latest theories and methods in the field of computer vision.

The CCCV 2017 received 465 full submissions. Each submission was reviewed by at least two reviewers selected from the Program Committee and other qualified researchers. Based on the reviewers' reports, 174 papers were finally accepted for presentation at the conference. The acceptance rate is 37.4%. The proceedings of the CCCV 2017 are published by Springer.

We are grateful to the keynote speakers, Prof. Katsushi Ikeuchi from University of Tokyo, Prof. Sven Dickinson from University of Toronto, Prof. Gérard Medioni from University of Southern California, Prof. Marc Pollefeys from ETH Zurich, Prof. Xilin Chen from Institute of Computing Technology, Chinese Academy of Science.

Thanks go to the authors of all submitted papers, the Program Committee members and the reviewers, and the Organizing Committee. Without their contributions, this conference would not be a success. Special thanks go to all of the sponsors and the organizers of the five special forums; their support made the conference a success. We are also grateful to Springer for publishing the proceedings and especially to Ms. Celine (Lanlan) Chang of Springer Asia for her efforts in coordinating the publication.

We hope you find CCCV 2017 an enjoyable and fruitful conference.

Tieniu Tan
Hongbin Zha
Jinfeng Yang
Qinhua Hu
Ming-Ming Cheng
Liang Wang
Qingshan Liu

Preface

Organization

CCCV 2017 (CCF Chinese Conference on Computer Vision 2017) was hosted by the China Computer Federation (CCF) and co-organized by the Civil Aviation University of China, Tianjin University, Nankai University, and CCF Technical Committee on Computer Vision.

Organizing Committee

International Advisory Board

Bill Freeman — Massachusetts Institute of Technology, USA
Gerard Medioni — University of Southern California, USA
Ramin Zabih — Cornell University, USA
David Forsyth — University of Illinois at Urbana-Champaign, USA
Jitendra Malik — University of California, Berkeley, USA
Fei-Fei Li — Stanford University, USA
Ikeuchi Katsushi — The University of Tokyo, Japan
Song-chun Zhu — University of California, Los Angeles, USA
Long Quan — The Hong Kong University of Science and Technology, SAR China

Steering Committee

Tieniu Tan — Institute of Automation, Chinese Academy of Sciences, China
Hongbin Zha — Peking University, China
Xinlin Chen — Institute of Computing Technology, Chinese Academy of Sciences, China
Jianhuang Lai — Sun Yat-Sen University, China
Dewen Hu — National University of Defense Technology, China
Shengyong Chen — Tianjin University of Technology, China
Yanning Zhang — Northwestern Polytechnical University, China
Tao Wang — IQIYI Inc.

General Chairs

Tieniu Tan — Institute of Automation, Chinese Academy of Sciences, China
Hongbin Zha — Peking University, China
Jinfeng Yang — Civil Aviation University of China

Program Chairs

Qinghua Hu — Tianjin University, China
Ming-Ming Cheng — Nankai University, China
Liang Wang — Institute of Automation, Chinese Academy of Sciences, China
Qingshan Liu — Nanjing University of Information Science and Technology, China

Organizing Chairs

Jufeng Yang — Nankai University, China
Pengfei Zhu — Tianjin University, China

Publicity Chairs

Qiguang Miao — Xidian University, China
Wei Jia — Hefei University of Technology, China

International Liaison Chairs

Andy Yu — ShanghaiTech University, China
Zhanyu Ma — Beijing University of Posts and Telecommunications, China

Publication Chairs

Xiang Bai — Huazhong University of Science and Technology, China
Deyu Meng — Xi'an Jiaotong University, China

Tutorial Chairs

Zhouchen Lin — Peking University, China
Xin Geng — Southeast University, China

Workshop Chairs

Zhaoxiang Zhang — Institute of Automation, Chinese Academy of Sciences, China
Rongrong Ji — Xiamen University, China

Sponsorship Chairs

Yongzhen Huang — Watrix Technology
Huchan Lu — Dalian University of Technology, China

Demo Chairs

Liang Lin — Sun Yat-Sen University, China
Chao Xu — Tianjin University, China

Competition Chairs

Wangmeng Zuo	Harbin Institute of Technology, China
Jiwen Lu	Tsinghua University, China

Website Chairs

Changqing Zhang	Tianjin University, China
Shaocheng Han	Civil Aviation University of China

Finance Chairs

Guimin Jia	Civil Aviation University of China
Ruiping Wang	Institute of Computing Technology, Chinese Academy of Sciences, China

Coordination Chairs

Lifang Wu	Beijing University of Technology, China
Shiying Li	Hunan University, China

Program Committee

Haizhou Ai	Tsinghua University, China
Xiang Bai	Huazhong University of Science and Technology, China
Yinghao Cai	Chinese Academy of Sciences, China
Xiaochun Cao	Chinese Academy of Sciences, China
Hui Ceng	University of Science and Technology Beijing, China
Hongbin Cha	Peking University, China
Zhengjun Cha	University of Science and Technology of China, China
Hongxin Chen	HISCENE
Shengyong Chen	Zhejiang University of Technology, China
Songcan Chen	Nanjing University of Aeronautics and Astronautics, China
Xilin Chen	Chinese Academy of Science, China
Yongquan Chen	HuanJing Information Technology Inc., China
Hong Cheng	University of Electronic Science and Technology of China
Jian Cheng	Chinese Academy of Science, China
Mingming Cheng	Nankai University, China
Jun Chu	Nanchang Hangkong University, China
Yang Cong	Chinese Academy of Science, China
Chaoran Cui	Shandong University of Finance and Economics, China
Cheng Deng	Xidian University, China
Weihong Deng	Beijing University of Posts and Telecommunications, China
Xiaoming Deng	Chinese Academy of Science, China

Jing Dong	Chinese Academy of Science, China
Junyu Dong	Ocean University of China
Weisheng Dong	Xidian University, China
Fuqing Duan	Beijing Normal University, China
Bin Fan	Chinese Academy of Science, China
Xin Fan	Dalian University of Technology, China
Yuchun Fang	Shanghai University, China
Jufu Feng	Peking University, China
Jianjiang Feng	Tsinghua University, China
Xianping Fu	Dalian Maritime University, China
Shenghua Gao	ShanghaiTech University, China
Yong Gao	JieShang Visual technology Inc., China
Shiming Ge	Chinese Academy of Science, China
Xin Geng	Southeast University, China
Guanghua Gu	Yanshan University, China
Jie Gui	Chinese Academy of Science, China
Yulan Guo	National University of Defense Technology, China
Zhenhua Guo	Tsinghua University, China
Aili Han	Shandong University, China
Junwei Han	Northwestern Polytechnical University, China
Huiguang He	Chinese Academy of Science, China
Lianghua He	Tongji University, China
Ran He	Chinese Academy of Science, China
Yutao Hou	NVIDIA Semiconductor Technical Services Inc.
Zhiqiang Hou	Air Force Engineering University
Haifeng Hu	Sun Yat-sen University, China
Hua Huang	Beijing Institute of Technology, China
Di Huang	Beijing University of Aeronautics and Astronautics
Kaiqi Huang	Chinese Academy of Science, China
Qingming Huang	University of Chinese Academy of Sciences
Yongzhen Huang	Chinese Academy of Science, China
Yanli Ji	University of Electronic Science and Technology of China
Rongrong Ji	Xiamen University, China
Wei Jia	Chinese Academy of Science, China
Tong Jia	Northeastern University, China
Yunde Jia	Beijing Institute of Technology, China
Muwei Jian	Ocean University of China
Yugang Jiang	Fudan University, China
Shuqiang Jiang	Chinese Academy of Science, China
Cheng Jin	Fudan University, China
Lianwen Jin	South China University of Technology, China
Xiaoyuan Jing	Wuhan University, China
Xiangwei Kong	Dalian University of Technology, China
Jianhuang Lai	Sun Yat-sen University, China
Zhen Lei	Chinese Academy of Science

Hua Li	Chinese Academy of Science, China
Jian Li	National University of Defense Technology
Jia Li	Beijing University of Aeronautics and Astronautics
Teng Li	Anhui University, China
Xi Li	Zhejiang University, China
Ce Li	Lanzhou University of Technology, China
Chaofeng Li	Jiangnan University, China
Chunguang Li	Beijing University of Posts and Telecommunications, China
Chunming Li	University of Electronic Science and Technology of China
Jianguo Li	Intel
Jinping Li	University of Jinan, China
Lingling Li	Zhengzhou Institute of Aeronautical Industry Management, China
Peihua Li	Dalian University of Technology, China
Shiying Li	Hunan University, China
Yongjie Li	University of Electronic Science and Technology of China, China
Zechao Li	Nanjing University of Science and Technology, China
Ziyang Li	Chinese Academy of Science, China
Zongmin Li	China University of Petroleum, China
Shengcai Liao	Chinese Academy of Science, China
Liang Lin	Sun Yat-sen University, China
Weiyao Lin	Shanghai Jiao Tong University, China
Zhouchen Lin	Peking University, China
Chenglin Liu	Chinese Academy of Science, China
Haibo Liu	Harbin Engineering University, China
Heng Liu	Anhui University of Technology, China
Hong Liu	Peking University, China
Huaping Liu	Tsinghua University, China
Jing Liu	Chinese Academy of Science, China
Li Liu	National University of Defense Technology, China
Qingshan Liu	Nanjing University of Information Science and Technology, China
Shigang Liu	Shanxi Normal University, China
Weifeng Liu	China University of Petroleum, China
Yiguang Liu	Sichuan University, China
Yue Liu	Beijing Institute of Technology, China
Guoliang Lu	Shandong University, China
Huchuan Lu	Dalian University of Technology
Xiaoqiang Lu	Xi'an Institute of Optics and Precision Mechanics, Chinese Academy of Sciences, China
Jiwen Lu	Tsinghua University, China
Feng Lu	Beijing University of Aeronautics and Astronautics, China

Yao Lu	Beijing Institute of Technology, China
Bin Luo	Anhui University, China
Ke Lv	University of Chinese Academy of Sciences, China
Bingpeng Ma	University of Chinese Academy of Sciences, China
Huimin Ma	Tsinghua University, China
Wei Ma	Beijing University of Technology, China
Zhanyu Ma	Beijing University of Posts and Telecommunications, China
Lin Mei	Third Institute of Public Security
Deyu Meng	Xi'an Jiaotong University, China
Qiguang Miao	Xidian University, China
Rongrong Ni	Beijing Jiaotong University, China
Xiushan Nie	Shandong University of Finance and Economics
Jifeng Ning	Northwest A&F University, China
Jianquan Ouyang	Xiangtan University, China
Yuxin Peng	Peking University, China
Yu Qiao	Shenzhen Institutes of Advanced Technology, Chinese Academy of Sciences, China
Lei Qin	Chinese Academy of Sciences, China
Pinle Qin	North University of China
Jianhua Qiu	Beijing Wisdom Eye Technology, China
Chuanxian Ren	Sun Yat-sen University, China
Qiuqi Ruan	Beijing Jiaotong University, China
Nong Sang	Huazhong University of Science and Technology, China
Shiguang Shan	Chinese Academy of Science, China
Shuhan Shen	Chinese Academy of Science, China
Jianbing Shen	Beijing Institute of Technology, China
Linlin Shen	Shenzhen University, China
Peiyi Shen	Xidian University, China
Wei Shen	Shanghai University, China
Mingli Song	Zhejiang University, China
Fei Su	Beijing University of Posts and Telecommunications, China
Hang Su	Tsinghua University, China
Dongmei Sun	Beijing Jiaotong University, China
Jian Sun	Xi'an Jiaotong University, China
Taizhe Tan	Guangdong University of Technology, China
Tieniu Tan	Chinese Academy of Science, China
Xiaoyang Tan	Nanjing University of Aeronautics and Astronautics, China
Jin Tang	Anhui University, China
Jinhui Tang	Nanjing University of Science and Technology, China
Yandong Tang	Chinese Academy of Science, China
Zengfu Wang	Chinese Academy of Science, China
Hanzi Wang	Xiamen University, China

Hanli Wang	Tongji University, China
Hongyuan Wang	Changzhou University, China
Hongpeng Wang	Harbin Institute of Technology, China
Jinjia Wang	Yanshan University, China
Jingdong Wang	Microsoft Research
Liang Wang	Chinese Academy of Science, China
Qi Wang	Northwestern Polytechnical University
Qing Wang	Northwestern Polytechnical University
RuiPing Wang	Chinese Academy of Science, China
Shengke Wang	Ocean University of China
Shiquan Wang	Philips Research Institute of China
Sujing Wang	Chinese Academy of Science, China
Tao Wang	IQIYI Inc.
Wei Wang	Chinese Academy of Science, China
Yuanquan Wang	Hebei University of Technology, China
Yuehuan Wang	Huazhong University of Science and Technology, China
Yunhong Wang	Beijing University of Aeronautics and Astronautics, China
Shikui Wei	Beijing Jiaotong University, China
Gongjian Wen	National University of Defense Technology, China
Xiangqian Wu	Harbin Institute of Technology, China
Lifang Wu	Beijing University of Technology, China
Jianxin Wu	Nanjing University, China
Jun Wu	Chongqing Kaiser Technology, China
Yadong Wu	Southwest University of Science and Technology, China
Yihong Wu	Chinese Academy of Sciences, China
Yongxian Wu	South China University of Technology, China
Guisong Xia	Wuhan University, China
Shiming Xiang	Chinese Academy of Science, China
Xiaohua Xie	Sun Yat-sen University, China
Junliang Xing	Chinese Academy of Sciences, China
Hongkai Xiong	Shanghai Jiao Tong University, China
Chao Xu	Tianjin University, China
Mingliang Xu	Zhengzhou University, China
Yong Xu	Harbin Institute of Technology, China
Yong Xu	South China University of Technology, China
Xinshun Xu	Shandong University, China
Feng Xue	HeFei University of Technology, China
Jianru Xue	Xi'an Jiaotong University, China
Yan Yan	Xiamen University, China
Dongming Yan	Chinese Academy of Sciences, China
Junchi Yan	IBM China Research Institute, China
Bo Yan	Fudan University, China
Chenhui Yang	Xiamen University, China

Dong Yang	Beijing Wisdom Eye Technology Inc., China
Gongping Yang	Shandong University, China
Jian Yang	Beijing Institute of Technology, China
Jie Yang	Shanghai Jiaotong University, China
Jinfeng Yang	Civil Aviation University of China
Jufeng Yang	Nankai University, China
Lu Yang	University of Electronic Science and Technology of China
Wankou Yang	Southeast University, China
Jian Yao	Wuhan University, China
Mao Ye	University of Electronic Science and Technology of China
Xucheng Yin	University of Science and Technology Beijing, China
Yilong Yin	Shandong University, China
Xianghua Ying	Peking University, China
Xingang You	Beijing Institute of Electronics Technology and Application, China
Xinge You	Huazhong University of Science and Technology, China
Jian Yu	Beijing Jiaotong University, China
Shiqi Yu	Shenzhen University, China
Xiaoyi Yu	Peking University, China
Ye Yu	HeFei University of Technology, China
Zhiwen Yu	South China University of Technology, China
Jun Yu	Hanzhou Electronic Science and Technology University, China
Jingyi Yu	ShanghaiTech University, China
Xiaotong Yuan	Nanjing University of Information Science and Technology, China
Di Zang	Tongji University, China
Honggang Zhang	Beijing University of Posts and Telecommunications, China
Junping Zhang	Fudan University, China
Junge Zhang	Chinese Academy of Sciences, China
Lin Zhang	Tongji University, China
Shihui Zhang	Yanshan University, China
Wei Zhang	Fudan University, China
Wei Zhang	Shandong University, China
Wenqiang Zhang	Fudan University, China
Xiaoyu Zhang	Chinese Academy of Sciences, China
Yan Zhang	Nanjing University, China
Yanning Zhang	Northwestern Polytechnical University, China
Yifan Zhang	Chinese Academy of Sciences, China
Yimin Zhang	Intel China Research Center, China
Yongdong Zhang	Chinese Academy of Science, China
Yunzhou Zhang	Northeastern University

Zhang Zhang	Chinese Academy of Science, China
Zhaoxiang Zhang	Chinese Academy of Science, China
Guofeng Zhang	Zhejiang University, China
Yao Zhao	Beijing Jiaotong University, China
Cairong Zhao	Tongji University, China
Yang Zhao	HeFei University of Technology, China
Yuqian Zhao	Central South University
Weishi Zheng	Sun Yat-sen University, China
Wenming Zheng	Southeast University
Bineng Zhong	Huaqiao University, China
Dexing Zhong	Xi’an Jiaotong University, China
Hanning Zhou	Beijing Gourd Software Technology Inc., China
Rigui Zhou	Shanghai Maritime University, China
Zhenfeng Zhu	Beijing Jiaotong University
Liansheng Zhuang	University of Science and Technology of China
Beiji Zou	Central South University
Wangmeng Zuo	Harbin Institute of Technology, China

Sponsoring Institutions

megvii | Face++

SENSETIME
商汤科技

艾思科尔科技
Isecure Technology

平安科技
PINGAN TECHNOLOGY

VRVIEW賓景

水滴科技
WATRIX TECHNOLOGY

极视角

上海中科智谷人工智能工业研究院
Shanghai Academy Of Artificial Intelligence

SEGWAY
ROBOTICS

Contents – Part III

Object Detection and Classification

Object Identification

Photography and Video

Robot Vision

Shape Representation and Matching

Statistical Methods and Learning

Object Detection and Classification

Classification of Foreign Object Debris Using Integrated Visual Features and Extreme Learning Machine

Kai Hu[1(✉)], Dongshun Cui[2,3], Yuan Zhang[1], Chunhong Cao[1], Fen Xiao[1], and Guangbin Huang[2]

[1] The MOE Key Laboratory of Intelligent Computing and Information Processing and the College of Information Engineering, Xiangtan University, Xiangtan 411105, China
{kaihu,yuanz,caoch,xiaof}@xtu.edu.cn

[2] School of Electrical and Electronic Engineering, Nanyang Technological University, Singapore 639798, Singapore
{DCUI002,EGBHuang}@ntu.edu.sg

[3] Energy Research Institute @NTU (ERI@N), Interdisciplinary Graduate School, Nanyang Technological University, Singapore 639798, Singapore

Abstract. In this paper, we develop a novel algorithm for classifying foreign object debris (FOD) based on the integrated visual features and extreme learning machine (ELM). After image preprocessing, various types of characteristics of the FOD image such as the color names, the scale-invariant feature transform (SIFT), and the histograms of oriented gradient (HOG) features are extracted in the proposed algorithm. These features are then combined into an integrated visual feature vector to characterize foreign objects in the image. Further, according to the extracted integrated visual features, classification is carried out using the ELM classifier. The experimental results show that the proposed classification algorithm outperforms other state-of-the-art methods. Furthermore, we also demonstrate the effectiveness of the use of the integrated visual features and the ELM classifier.

Keywords: Foreign object debris · Integrated visual features
Feature extraction · Extreme learning machine · Classification

1 Introduction

Foreign objects (as shown in Fig. 1) such as wrenches, rubber pieces of tires, screws, stones, and metal, are named foreign object debris (FOD) which may cause foreign object damage and seriously threaten flight safety [7]. On July 25, 2000, a concord flight operated by Air France was crashed by a metal strip and one hundred and thirteen people lost their lives in this accident [4]. Therefore, FOD detection and classification are very important for preventing such concord disaster.

J. Yang et al. (Eds.): CCCV 2017, Part III, CCIS 773, pp. 3–13, 2017.
https://doi.org/10.1007/978-981-10-7305-2_1

Fig. 1. Examples of FOD on the airport (collected from the internet). (a) Wrench; (b) Tire debris; (c) Screw; (d) Stone; (e) Metal; and (f) Others.

Since 2000, a variety of countries began to develop technology for FOD detection. So far there are four detection systems have been successfully applied in airports in field worldwide. They are Tarsier Radar system developed by United Kingdom (UK) [2], FODetect system by Israel [3], FOD Finder system by United States (US) [1], and iFerret system by Singapore [26]. These systems employed different methods and sensors for detecting FOD and giving promising results on the detection of FOD, but they cannot directly classify FOD. Even though some systems can obtain the FOD images, they will not automatically complete the FOD classification. If the characteristics of different types of foreign objects are used to help the classification of FOD, it will show great significance on the prevention of FOD damage and can provide the utilization rate of the airport runway.

In recent years, many methods for the classification of FOD on airports have been developed. As we know, a detection or classification algorithm is often influenced by the feature extraction algorithm and the classifier performance [14,15]. Therefore, most of the FOD classification approaches mainly focus the study on the novel feature extraction algorithm or the efficient classifier. Wang et al. [29] proposed a foreign object debris detection and identification method, in which the Gabor features are extracted and the nearest neighbor (NN) classifier is used to obtain the final classification results. Niu et al. [25] proposed a FOD detection system which based on Gabor wavelets and support vector machine (SVM). They firstly used Gabor wavelets to extract useful features for describing FOD images and then employed the SVM for the classification.

Next, they proposed an improved version using Garbor wavelets and SVM for FOD classification [24]. Xiang et al. [31] proposed a FOD recognition method based on depth features and Adaboost classifier. Han et al. [11,12] proposed novel FOD classification algorithms based on low-level features and subspace features. Cao et al. [5] proposed a FOD detection framework using region proposal technique and convolution neural networks (CNN).

Although the above methods have achieved promising results, there is much room to improve on the classification accuracy, especially, the features to characterize many different types of FOD are remains to be further investigated. In this paper, a novel FOD classification framework based on the integrated visual features and extreme learning machine (ELM) classifier is developed. The proposed algorithm not only considers color and texture features for the feature extraction of FOD images, but also employs ELM for an efficient classification. The experimental results on both various feature extraction methods and different classifiers demonstrate the performance of our proposed FOD classification algorithm.

The organization of this paper are presented as follows. In Sect. 2, we present the proposed classification algorithm, including the integrated visual feature extraction and the classification using ELM. In Sect. 3, we evaluate the proposed classification algorithm on the FOD dataset from different perspectives. We also compare the proposed algorithm with other state-of-the-art FOD classification methods. Finally, the conclusions of this paper are provided in Sect. 4.

2 Methods

A novel FOD classification algorithm based on the integrated visual features and extreme learning machine is proposed in this paper. The overview of the proposed FOD classification algorithm is shown in Fig. 2.

2.1 Image Preprocessing

In this paper, the resolution of original FOD images obtained by the FOD detection system is 1024×1024 pixels. Actually, foreign object only occupied a small area on the whole image. To obtain stable and accurate classification result, we perform image preprocessing on original FOD images. First, an edge detection method is used to find the edge of the foreign objects and then the FOD images are normalized to 64×64 for the classification [11,12].

2.2 Integrated Visual Feature Extraction

Color feature. Color is an effective tool for describing human visual properties and it is often used to represent objects in image segmentation and classification. Due to the influences of angle, illumination and shadow, it usually produces variations in the color descriptions and thus leads to the color feature extraction is difficult [21]. However, color has been applied in image classification with high

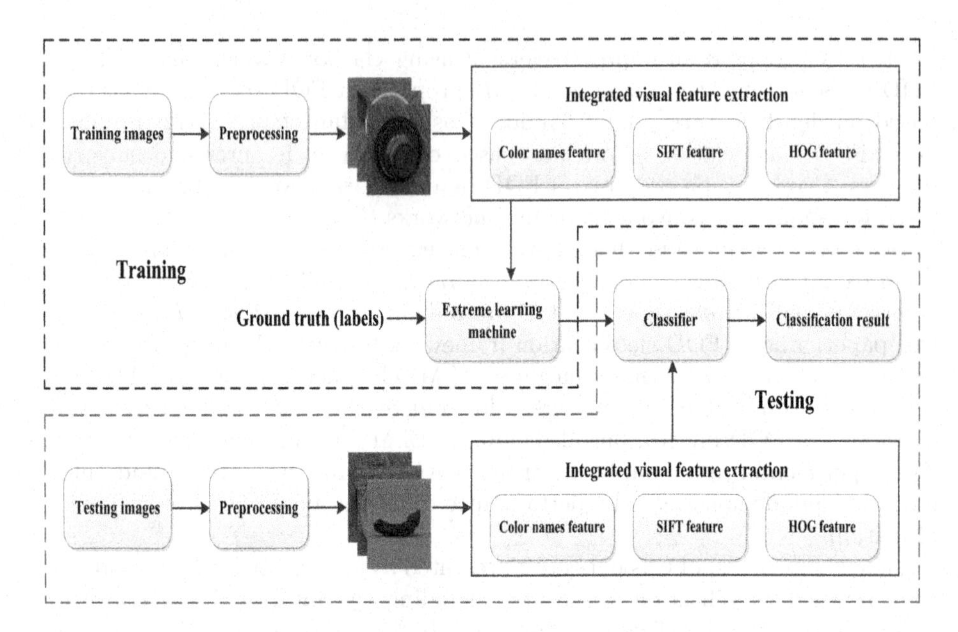

Fig. 2. The overview of the proposed FOD classification algorithm.

robustness since color do not depend on the deformation and scale variation of the image. In 2007, Weijer et al. [30] proposed a color naming feature extraction method in which the contribution is assigning linguistic color labels to image pixels.

Texture feature. Scale-invariant feature transform (SIFT) proposed by David Lowe [23] is an effective algorithm to characterize local texture features of images. It has been demonstrated by Lowe that the SIFT keypoint descriptors exhibit high performance for local feature representation with invariance of image scale and rotation, illumination change, and image noise. The SIFT feature extraction algorithm contains four steps which are detection of scale-space extrema; localization of keypoints; assignment of orientation; and calculation of keypoint descriptors.

Another useful local texture feature extraction algorithm is histograms of oriented gradient (HOG) which is proposed by Dalal and Triggs [9]. In their method, the image is firstly divided into many small spatial regions (cells). Secondly, the histogram of gradient orientation for each larger spatial regions (blocks) is calculated. The combined histograms are then formed the representations. Finally, after the contrast normalization on all of the cells in the block, the HOG descriptors are obtained.

Feature extraction. In the proposed algorithm, the scale invariant feature transform descriptors [23] and the histograms of oriented gradient descriptors [9] for texture feature as well as the color names feature descriptors [30] are extracted as visual features to characterize the FOD image.

In the actual extraction of color names feature, SIFT feature and HOG feature, we firstly sample the feature at multiple scales with a grid spacing of 4. Secondly, a dictionary with size 100 has been learned. Further, the locality-cnstrained linear coding (LLC) at 2-level spatial pyramid is employed to calculate the feature descriptor for each region [21]. Finally, we combine the color names feature, SIFT feature and HOG feature into an integrated visual feature vector to characterize the FOD image.

2.3 Classification Using ELM

Extreme learning machine proposed by Huang et al. [18] is a single-hidden layer feedforward neural network (SLFN), which has several new features when compared to the traditional neural networks: it has only one hidden layer; the hidden layer need not be tuned; input weights and hidden layer biases of ELM can be chosen randomly; and the output weights of ELM are determined analytically. These special features make the learning of ELM simple and efficient, and this leads to it has both universal approximation [17] and classification capabilities. Therefore, up to now ELM has been widely applied in both academia and industry such as to solve the clustering [19], regression [13], classification [15], detection [28], and feature learning problems [16,20].

For M samples $(\mathbf{x}_i, \mathbf{t}_i)$, L hidden nodes and active function $g(x)$, the output of standard SLFN can be formulated as

$$\sum_{i=1}^{L} \beta_i g_i(\mathbf{x}_j) = \sum_{i=1}^{L} \beta_i g(\mathbf{w}_i \cdot \mathbf{x}_j + b_i) = \mathbf{o}_j, j = 1, \ldots, M, \tag{1}$$

where $\mathbf{w}_i$ is the weight vector connecting the ith hidden node and the input nodes, while β_i is the weight vector between the ith hidden node and the output nodes. b_i is the threshold of ith hidden node. The standard SLFN can approximate the M samples with zero error, that is to say there exist β_i, $\mathbf{w}_i$ and b_i satisfy

$$\sum_{i=1}^{L} \beta_i g(\mathbf{w}_i \cdot \mathbf{x}_j + b_i) = \mathbf{t}_j, j = 1, \ldots, M. \tag{2}$$

These equations can be rewritten as

$$\mathbf{H}\beta = \mathbf{T}, \tag{3}$$

where $\mathbf{H}$ is the output matrix of hidden layer. We can easily get

$$\beta = \mathbf{H}^{\dagger}\mathbf{T}, \tag{4}$$

where $\mathbf{H}^{\dagger}$ is the Moore-Penrose generalized inverse of $\mathbf{H}$.

Training. Before the training, the feature matrix F for the training FOD images is constructed based on the integrated visual feature extraction. The procedure of training using ELM is presented as follows. Firstly, we randomly choose the parameters of hidden layer nodes $\mathbf{w}_i$ and b_i. Secondly, according to $\mathbf{w}_i$ and b_i, the output matrix $\mathbf{H}$ is obtained using the feature matrix F. Thirdly, β is obtained with the usage of $\mathbf{H}$ and the ground truth (labels) of the training FOD images and finally to finish the ELM training.

Testing. After the ELM training, the parameters $\mathbf{w}_i$, b_i and β are obtained that means the ELM classifier for FOD classification has been constructed. In the testing, the integrated visual features of the testing FOD images are firstly extracted. Then we compute the output matrix $\mathbf{H}$ according to $\mathbf{w}_i$ and b_i. Furthermore, the classification of FOD is carried out using β.

3 Experimental Results and Discussion

In this section, the performance of the proposed FOD classification algorithm is evaluated using a FOD dataset of which all images are collected by the FOD detection subsystem [11]. The FOD dataset contains five categories for a total of 320 images. It includes 81 images of wrench, 74 images of plastic pip, 64 images of tire debris, 50 images of fuel-tank cap, and 51 images of metal bar. As mentioned before, an edge detection method is used to find the edge of the foreign objects and then the FOD images are normalized to 64×64. Examples of the FOD images used in the experiments are shown in Fig. 3.

As presented by Han et al. in [11], it is difficult to perform classification on this FOD dataset because that there exist many artefacts such as large intra-class variability, small interclass dissimilarity, occlusions, scale variation, and background clutter in the images. Therefore, in this paper we extract various of types of object characteristics to effectively represent the foreign objects. In the proposed FOD classification algorithm, the color names, SIFT, and HOG features are extracted for each FOD image and then all features are combined into an integrated visual feature vector to characterize the FOD image. Subsequently, we use ELM to classify each image into different FOD categories. Conveniently, we note the proposed classification algorithm as IVF-ELM.

To evaluate the performance of the proposed classification algorithm, a series of experiments are conducted using different feature extraction methods and different classifiers in terms of classification accuracy. In the experiments, we randomly select only 33.3% of images as the testing set and the remainders are used for the training set. The experiments are repeated for 10 times and the average result is computed as the final classification accuracy.

In order to evaluate the effectiveness of the integrated visual features, the classification accuracy of the proposed classification algorithm using integrated visual features as well as using only color names feature, SIFT feature or HOG feature are summarized in Table 1. All the results are obtained using the ELM classifier. From the results, it can be seen that the classification accuracy using

Fig. 3. Examples of FOD images for the classification. (a) Wrench; (b) Plastic pip; (c) Tire debris; (d) Fual-tank cap; and (e) Metal bar.

the integrated visual features outperforms all that using only the single feature, although the result of wrench is slightly lower than that of HOG. This is because the proposed classification algorithm which uses the integrated visual features to consider the color names, SIFT, and HOG of the FOD, provides a more comprehensive classification performance.

In addition, taking the results of color names, SIFT, and HOG in Table 1 for comparison, it also can be seen that different features show different classification results on various foreign objects. For example, although color names feature is an effective technique to characterize object visual properties, it obtained the lowest result on metal bar which has variations on intra-class description and illumination condition. Since SIFT and HOG features play important roles on local texture feature description, HOG obtained the highest accuracy on wrench and tire debris, and SIFT obtained the highest accuracy on plastic pip and fuel-tank cap. Whereas the integrated visual features obtained the optimal classification result almost in all FOD classes.

Furthermore, to evaluate the performance of the ELM classifier, we conduct another experiment using the proposed classification algorithm with different classifiers such as the nearest neighbor, SVM, and ELM. Figure 4 shows the comparison of the proposed algorithm using the different classifiers mentioned above.

Table 1. Comparative results of the proposed algorithm using different feature extraction methods in terms of the classification accuracy.

FOD	Classification accuracy (%)			
	Color names	SIFT	HOG	Integrated
Wrench	92.22	97.78	100	98.15
Plastic pip	93.95	96.37	93.13	97.12
Tire debris	97.21	91.19	97.64	100
Fual-tank cap	90.33	100	96.36	100
Mental bar	81.18	91.76	91.76	91.76

Table 2. Comparative results of the proposed algorithm using different classifiers in terms of the classification accuracy.

Classifier	Classification accuracy (%)			
	Color names	SIFT	HOG	Integrated
NN	87.95	91.16	85.96	93.03
SVM	86.33	90.93	86.29	92.60
ELM	91.56	95.52	96.06	97.56

The quantitative results are given in Table 2. From the results, it can be seen that no matter what the feature extract method is used, ELM performs better than NN and SVM by providing the best classification accuracy. It also can be seen that no matter what the classifier is used, the results based on the integrated visual features higher than those of based on the single feature. The results demonstrate the effectiveness of the use of the integrated visual features and the ELM classifier.

Moreover, in order to evaluate the overall performance of the proposed classification algorithm, we compare it with other state-of-the-art methods, e.g., the FOD classification method based on low-level features proposed by Han et al. [11] and the classification and regression method proposed by Liaw and Wiener [22]. For convenience, we refer above compared methods as LLFM and CRM, respectively. Comparison of the different classification methods are presented in Table 3. The best result is highlighted using bold font. The results of the LLFM and CRM methods are extracted from the paper of [11]. As seen from Table 3, the proposed IVF-ELM algorithm outperforms the competing methods in terms of the classification accuracy. The accuracy of the proposed IVF-ELM algorithm increases at least 5.04% compared with other state-of-the-art methods.

Actually, in this paper, we propose a simple and efficient FOD classification algorithm. The validity of the extracted integrated visual features (including color names, SIFT, and HOG) and the ELM classifier have been demonstrated through the experiments presented above. However, there exist some limitations which can be further improved. Firstly, the integrated visual features can be

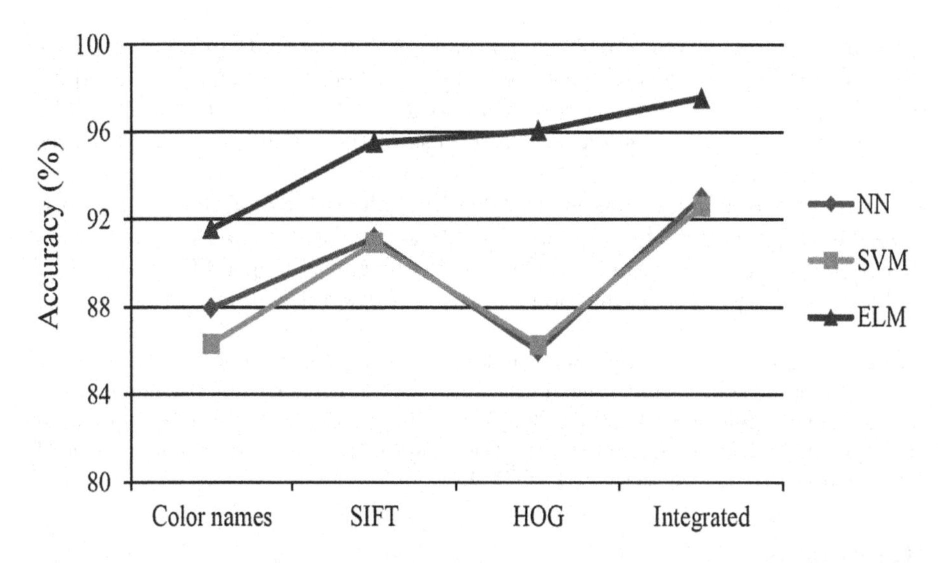

Fig. 4. Comparison of the proposed classification algorithm using the different feature extraction methods and the different classifiers.

Table 3. Comparison between the proposed algorithm and other state-of-the-art methods in terms of the classification accuracy.

Method	Source	Feature	Classifier	Accuracy (%)
LLFM	Han et al. [11]	Color and SIFT	NN	92.52
CRM	Liaw and Wiener [22]	SIFT	Random forest	85.05
IVF-ELM	Proposed	Color names, SIFT and HOG	ELM	**97.56**

optimized by the feature dimension reduction or the feature selection [6,10] to search the most efficient features for characterizing FOD. Secondly, in the experiments we find that other classifiers, e.g., SVM, exhibits the same efficient classification ability in a certain type of FOD, and therefore, an ensemble classifier [8,27], which combines various classifier properties, can be considered to improve the performance of the proposed FOD classification algorithm on a large dataset.

4 Conclusion

In this paper, a novel FOD classification algorithm using integrated visual features and extreme learning machine is proposed. The integrated visual features, which contribute from color names, SIFT, and HOG, are extracted in the proposed algorithm for the FOD images after the image preprocessing. Next, extreme learning machine, an efficient classifier is employed to classify the FOD images. In the experiment section, we conduct a series of experiments using different feature extraction methods and different classifiers to evaluate the performance of the proposed classification algorithm. The experimental results show that the proposed algorithm outperforms other state-of-the-art methods. The classification accuracy is enhanced by the use of the integrated visual features and the ELM classifier.

Next, we will employ some feature learning algorithms for the feature optimization and therefore to promote the FOD classification performance. Furthermore, we will plan to use a ensemble classifier on a bigger FOD dataset to evaluate our proposed classification algorithm in the future research.

Acknowledgements. The authors would like to thank Zhenqi Han for providing the FOD dataset and some useful discussions. This work was supported by the China Scholarship Council under Grant 201608430008, the National Natural Science Foundation of China under Grant 61401386, and the Scientific Research Fund of Hunan Provincial Education Department under Grant 16C1545.

References

1. FOD Finder system. http://www.fodfinder.com
2. Tarsier-QinetiQ Automatic Runway Debris Detection (2009). http://www2.QinetiQ.com/Tarsier
3. Xsight-Advanced Radar and Optic Sensors for FOD Detection and Homeland Security (2009). http://www.xsightsys.com/fodetect.htm
4. Air France Flight 4590, August 2015. http://en.wikipedia.org/wiki/Air_France_Flight_4590
5. Cao, X., Gong, G., Liu, M., Qi, J.: Foreign object debris detection on airfield pavement using region based convolution neural network. In: 2016 International Conference on Digital Image Computing: Techniques and Applications (DICTA), pp. 1–6. IEEE (2016)
6. Chandrashekar, G., Sahin, F.: A survey on feature selection methods. Comput. Electr.Eng. **40**(1), 16–28 (2014)
7. Advisory Circular: 150/5210-24: Airport Foreign Object Debris (FOD) Management (2010). http://www.fodfinder.com
8. Cogranne, R., Fridrich, J.: Modeling and extending the ensemble classifier for steganalysis of digital images using hypothesis testing theory. IEEE Trans. Inf. Forensics Secur. **10**(12), 2627–2642 (2015)
9. Dalal, N., Triggs, B.: Histograms of oriented gradients for human detection. In: IEEE Computer Society Conference on Computer Vision and Pattern Recognition, CVPR 2005, vol. 1, pp. 886–893. IEEE (2005)
10. Guyon, I., Elisseeff, A.: An introduction to variable and feature selection. J. Mach. Learn. Res. **3**(Mar), 1157–1182 (2003)
11. Han, Z., Fang, Y., Xu, H.: Fusion of low-level feature for FOD classification. In: 2015 10th International Conference on Communications and Networking in China (ChinaCom), pp. 465–469. IEEE (2015)
12. Han, Z., Fang, Y., Xu, H., Zheng, Y.: A novel FOD classification system based on visual features. In: Zhang, Y.J. (ed.) ICIG 2015. LNCS, vol. 9217, pp. 288–296. Springer, Cham (2015). https://doi.org/10.1007/978-3-319-21978-3_26
13. He, Q., Shang, T., Zhuang, F., Shi, Z.: Parallel extreme learning machine for regression based on mapreduce. Neurocomputing **102**, 52–58 (2013)

14. Hu, K., Gao, X., Li, F.: Detection of suspicious lesions by adaptive thresholding based on multiresolution analysis in mammograms. IEEE Trans. Instrum. Meas. **60**(2), 462–472 (2011)
15. Hu, K., Yang, W., Gao, X.: Microcalcification diagnosis in digital mammography using extreme learning machine based on hidden Markov tree model of dual-tree complex wavelet transform. Expert Syst. Appl. **86**, 135–144 (2017)
16. Huang, G.B.: What are extreme learning machines? Filling the gap between Frank Rosenblatts dream and John von Neumanns puzzle. Cogn. Comput. **7**(3), 263–278 (2015)
17. Huang, G.B., Zhou, H., Ding, X., Zhang, R.: Extreme learning machine for regression and multiclass classification. IEEE Trans. Syst. Man Cybern. Part B (Cybern.) **42**(2), 513–529 (2012)
18. Huang, G.B., Zhu, Q.Y., Siew, C.K.: Extreme learning machine: theory and applications. Neurocomputing **70**(1), 489–501 (2006)
19. Javed, K., Gouriveau, R., Zerhouni, N.: A new multivariate approach for prognostics based on extreme learning machine and fuzzy clustering. IEEE Trans. Cybern. **45**(12), 2626–2639 (2015)
20. Kasun, L.L.C., Zhou, H., Huang, G.B., Vong, C.M.: Representational learning with elms for big data. IEEE Intell. Syst. **28**(6), 31–34 (2013)
21. Khosla, A., Xiao, J., Torralba, A., Oliva, A.: Memorability of image regions. In: International Conference on Neural Information Processing Systems, vol. 25, pp. 296–304 (2012)
22. Liaw, A., Wiener, M.: Classification and regression by randomforest. R News **2**(3), 18–22 (2002)
23. Lowe, D.G.: Distinctive image features from scale-invariant keypoints. Int. J. Comput. Vis. **60**(2), 91–110 (2004)
24. Niu, B., Gu, H., Gao, Z.: A novel foreign object debris classification method for runway security. In: Electronics, Information Technology and Intellectualization: Proceedings of the International Conference EITI 2014, Shenzhen, China, 16–17 August 2014, p. 73. CRC Press (2015)
25. Niu, B., Gu, H., Sun, J., Chen, N.: Research of fod recognition based on gabor wavelets and SVM classification. J. Inf. Comput. Sci. **10**(6), 1633–1640 (2013)
26. Stratech: iFerret system. http://www.stratechsystems.com
27. Su, Y., Shan, S., Chen, X., Gao, W.: Hierarchical ensemble of global and local classifiers for face recognition. IEEE Trans. Image Process. **18**(8), 1885–1896 (2009)
28. Tang, J., Deng, C., Huang, G.B., Zhao, B.: Compressed-domain ship detection on spaceborne optical image using deep neural network and extreme learning machine. IEEE Trans. Geosci. Remote Sens. **53**(3), 1174–1185 (2015)
29. Wang, Y., Wu, W., Zhang, D.Y.: Foreign object debris detection and identification system based on computer vision. Video Eng. **5**(34), 102–104 (2010)
30. Van de Weijer, J., Schmid, C., Verbeek, J.: Learning color names from real-world images. In: IEEE Conference on Computer Vision and Pattern Recognition, CVPR 2007, pp. 1–8. IEEE (2007)
31. Xiang, Y., Cao, X.G.: Recognition algorithm of fod based on the depth feature and adaboost. Electron. Des. Eng. **23**, 183–186 (2015)

Two Novel Image-Based CAPTCHA Schemes Based on Visual Effects

Ping Zhang, Haichang Gao(✉), Zhouhang Cheng, and Fang Cao

Institute of Software Engineering, Xidian University,
Xi'an 710071, Shaanxi, People's Republic of China
hchgao@xidian.edu.cn

Abstract. CAPTCHA is a security mechanism designed to differentiate between computers and humans, and is used to defend against malicious bot programs. Text-based and image-based CAPTCHAs are two of the most widely deployed schemes, but most of the existing CAPTCHA schemes are either too difficult for humans to solve or not safe enough since a lot of researchers have attacked them successfully. These CAPTCHAs are also language-dependent and they cannot be automatically generated. So it is urgent to explore new possible schemes of CAPTCHA. In this paper, we mainly made two contributions to CAPTCHA. First, we propose two novel image-based CAPTCHA schemes based on visual effects. DeRection is one CAPTCHA scheme that takes advantage of human ability on capturing deformed regions of an image in the case of contrast. CONSCHEME is another one that capitalizes on human ability of understanding the content in a three-dimensional space. We conducted preliminary experiments over more than 110 users to verify the usability and security of two schemes. Second, by analyzing the characteristics of these two schemes proposed in this paper and comparing them with the existing image-based ones, we propose a set of new guidelines for the design of image-based CAPTCHA.

Keywords: CAPTCHA · Language independent · Web service
Visual effects · Image-based

1 Introduction

CAPTCHA (Completely Automated Public Turing Test to Tell Computers and Humans Apart) [1] is a program which is used to tell computers and humans apart. It can protect the websites from cracking password, automated voting, receiving spam mails, automated registering accounts and attacking password systems.

Currently, most websites adopt text-based CAPTCHA that using sophisticated distortion, rotation or noise interferences to prevent from machines recognition. Users need to decipher characters within an image. However, the distorted characters really reduce human accuracy, and such schemes were broken by [2–8]. Therefore, image-based CAPTCHA becomes a promising alternative because

J. Yang et al. (Eds.): CCCV 2017, Part III, CCIS 773, pp. 14–25, 2017.
https://doi.org/10.1007/978-981-10-7305-2_2

these schemes have many advantages, for example, being not vulnerable to segmentation attacks, and being user-friendly by mouse-based interaction. But it also suffers from various drawbacks, the images adopted by many image-based CAPTCHAs need to be manually tagged which means these schemes are not automatically generated, and most image-based CAPTCHAs are still dependent on language. Furthermore, with the rapid development of the deep learning algorithm, most text-based and image-based schemes that based on classification can be well attacked. In addition, audio and video-based CAPTCHAs have lower utilization rate and require higher bandwidth, many websites don't support these types of CAPTCHAs. Therefore, it is urgent and difficult to design a widely usable CAPTCHA.

In this paper, we present two schemes that can overcome the issues above. In the DeRection (Deformed Regions Detection in a GIF image) scheme, we present a GIF image which contains ten frames and every frame randomly has 2 to 6 deformed regions. Users need two steps to pass the challenge. First, click the GIF image to focus on one frame. Second, click all the deformed regions in this frame. In the CONSCHEME (Counting the Number of Stacking Cubes in a Three-dimensional scene) scheme, we use Unity to create a three-dimensional space with several cubes stacking together. Interacting with mouse and keyboard, users can get any perspectives of the scene. This challenge requires users to count the quantity of the cubes.

The two mechanisms have many features in common. Firstly, both of them have low language dependence since users only need to click on the deformed regions within the image in the DeRection and enter a number of cubes in the CONSCHEME. Besides, they can be generated online automatically. The images of DeRection can be obtained from websites and all we need from the CONSCHEME is a Unity plugin that can be embedded into web browser and reused once generated. In contrast, many existing image-based CAPTCHAs need a database of manually tagged images. By definition, they are not real CAPTCHAs which must be a completely automated public Turing test. We have demonstrated experiments to evaluate these two schemes and the results indicate that they both have good robustness and usability. The limitations of the proposed schemes are that they require higher bandwidth compared with traditional image-based CAPTCHAs. And users need to set up the Unity plugin when using CONSCHEME for the first time.

The rest of this paper is organized as follows. In Sect. 2, we introduce the related works. In Sect. 3, we provide the details of design and implement of the schemes. In Sect. 4, we present the experiments and analysis of the proposed scheme. And in Sect. 5, we propose a set of image-based CAPTCHA design guidelines. Finally, we make a brief conclusion in Sect. 6.

2 Related Works

CAPTCHA relies on the capability gap between humans and computers in solving Artificial Intelligence problems. Existing CAPTCHA systems can be broadly grouped into three classes: (1) text-based, (2) image-based, and (3) audio or video-based.

Text-based CAPTCHAs are the most widely used schemes [2]. Designers usually use sophisticated distortion, rotation, noise and complex background on English letters and Arabic numerals to prevent automatic attack from achieving a high success rate.

Security is the most worrying factor of text-based CAPTCHAs. A large number of attack methods have been proposed. In 2003, Mori and Malik [2] broke EZ-Gimpy and Gimpy by analyzing the shape context with a success rate of 92% and 33% respectively. In 2008, Yan and El Ahmad [3] proposed a new segmentation method to attack Microsoft MSN with 92% segmentation and 60% overall success rate. In 2013, Gao et al. [4] broke several hollow CAPTCHAs with a generic method with the success rate range from 36% to 89%. In 2014, Goodfellow et al. [5] using deep CNN solving reCAPTCHA with 99.8% success rate. In 2015, Karthik and Recasens [6] applied CNN-based method on Microsoft CAPTCHAs with 57.05% CAPTCHA suc-cess rate of. More recently, in NDSS 2016, Gao [7] found a simple generic attack that firstly employed Gabor filter on attacking a wide range of CAPTCHAs.

With each failed CAPTCHA scheme, CAPTCHA designers accumulate experiences and then design better schemes, with increased friendly usability and improved security. Image-based CAPTCHAs have been proposed as an alternative for the text-based CAPTCHA systems. They can be broadly divided into two types.

First type is to find a corresponding word to describe images, such as Naming CAPTCHA [8] and IMAGINATION [9]. Naming CAPTCHA requires user to type a word to describe the common object in six images. IMAGINATION asks user to click an image's geometry center from a synthetic image then select a correct label from a given list. However, typing a word may cause misspelling and polysemy, and the random guess rate of selecting from a list is relatively high. This type also has strong language dependence. IMAGINATION was attacked by Zhu et al. [10].

Second type is to classify the given images. For example, Asirra [11] shows 12 images of cats and dogs, asking user to select all the cats. SEMAGE [12] shows 8 real or cartoon images of animals, allowing user to choose images of same animals. 12306.cn scheme [13] shows 8 images, users need to recognize the text above the images and pick out all the targets to match the text. This type requires a database of correctly tagged images, which causes a waste of resources. And once the dataset is obtained, it will be much easier to break them. Zhu et al. [10] have broken Asirra with a success rate of 10.3%.

There are some other types of image-based CAPTCHAs. "What's up" [14] asks user to adjust the orientation of three images. ARTiFACIAL [15] asks user to identify a face and click its six points from a synthetic image and it has been broken by Li [16]. FR-CAPTCHA [17] and FaceDCAPTCHA [18] are two face-based schemes, the former asks user to select two face images of the same person while the latter is to distinguish human faces from cartoon-face images. Our team has broken these two CAPTCHAs with 42% and 48% success rate respectively [19].

3 Design and Implement

3.1 DeRection

In DeRection scheme, the users are given a GIF image, every frame of this image has 2 to 6 deformed regions. Users need to find out all of the deformed regions in the frame to pass the challenge. Users can click the "Change an Image" button to change the GIF image and click the "Change a Frame" button to change a frame of the GIF image.

Every GIF image is made up of 10 images with randomly 2 to 6 deformed regions. We download a picture from the Internet and judge if it is appropriate to be deformed by calculating its texture complexity. If it is, we use it to generate 10 images with different deformed regions and the position to be deformed should also be suitable. So the whole problem relies on finding a method to calculate texture complexity of an image.

There are many methods to extract features from an image and measure the texture complexity. Among these methods, GLCM (Gray-level co-occurrence matrix) [20] is a widely-used and well-performed one, we also adopted it in this paper. GLCM is the statistical method of examining the textures that considers the spatial relationship of the pixels. The GLCM characterizes the texture of an image by calculating how often pairs of pixel with specific values and in a specified spatial relationship occur in an image, creating a GLCM, and then extracting statistical measures from this matrix.

According to [20], we calculated the ASM (Angular Second Moment), Entropy, Contrast, IDM (Inverse Different Moment) and Correlation values and set weights to them based on the GLCM of three directions, they are 0°, 45° and 90°. The texture complexity of a region in an image is measured by the formula:

$$R = \sum_{i=1}^{3} (a_1 \cdot J_i + a_2 \cdot H_i + a_3 \cdot G_i + a_4 \cdot Q_i + a_5 \cdot Conv_i) \tag{1}$$

J, *H*, *G*, *Q* and *Cov* represents ASM, Entropy, Contrast, IDM and Correlation respectively. *a1*, *a2*, *a3*, *a4* and *a5* are constant parameters and their values are 0.2012, 0.2673, 0.2814, 0.1517 and 0.1126.

However, in our experiments, we cannot find a suitable threshold to accurately divide the regions. To solve this, we tested each of the feature parameters separately and there are many discoveries. For example, the complexity of the image is not strictly related to the value of Correlation. Through a series of experiments, finally we adopted Entropy and Contrast and introduced Variance, a new feature parameter to measure the texture complexity. Variance describes the discrete degree of the value of the variables to their mathematical expectation. We used Variance to measure the degree of discrete of the grayscale value of an image. Finally, we calculated according to the formula:

$$R' = \sum_{i=1}^{3} (0.5 \cdot E_i + 0.5 \cdot Conv_i) + \frac{v}{10} \tag{2}$$

We calculated the GLCM of three directions, and we set the final threshold of R' to 14. This method is acceptable for classifying effect and time. Then, we compared the value of R and R' and we found that the images got with R has a large ambiguous area, which made it hard to divide the regions into two parts. We testd 50 different regions in more than 450 different images and got the R and R', and we found that ambiguous area got from R is about 62.4%. So we finally choose R' to determine whether a region is suitable to be deformed and got a relatively satisfactory result. If the region is selected as deformed region, then we need to perform deformation. Deformation is all about moving pixels from one coordinate to another coordinate. We use the formulas below:

$$newX = \frac{(x - cenX)\sqrt{(x - cenX)^2 + (y - cenY)^2}}{radius} + cenX \tag{3}$$

$$newY = \frac{(y - cenY)\sqrt{(x - cenX)^2 + (y - cenY)^2}}{radius} + cenY \tag{4}$$

$(newX, newY)$ is the coordinate of the pixel in the original image and (x, y) is the coordinate in the processed image. $(cenX, cenY)$ and radius are the coordinates of the center and the radius of the deformation region. Figure 2 shows the effect of convex lens deformation. The origin of the deformation is at the origin of all four different color concentric circles. While calculating mapping, the mapping point of the original point within the deformation area, which is a circle, moves along the direction away from the center of the circle.

The original version of DeRection is presenting users with a static image. But the results did not reach our expectations. Through analysis we found that the deformation regions are hard to find in a static image, but in GIF image, they can capture people's eyes. So we generated many GIF images and users need to click the GIF image to get one frame of the GIF image. The verification mode remains the same but it is much easier for users to pass the challenge. But the problem came that crackers can solve our scheme by comparing the pixels between frames to find out the deformation regions. To overcome this, we modified every pixel value after generating deformed regions. Figure 2 shows one frame of final version of DeRection.

3.2 CONSCHEME

CONSCHEME is another image-based CAPTCHA scheme we proposed, which is an interactive three-dimensional CAPTCHA system created by Unity.

We use Unity to create a three-dimensional room on the floor of which are a lot of cubes stacked. The walls, ceiling and floor are labeled the same stickers as the cubes. An example of CONSCHEME and its three views are shown in Fig. 3. The original version of this scheme had no floor or walls and the cubes just float in the air with no texture, and people felt dazzled when rotate the cubes. With texture and walls, people can focus on the cubes in proper perspective. Benefit from Unity's good interactivity, users can scroll the mouse wheel to zoom in or zoom out the cubes, click the arrow keys on the keyboard to rotate the object and drag with the left mouse button to change the perspective of the cubes. When a unity-plugin is produced, we export it into web format and embed it into the web page. It can produce different number of cubes stacked together in different ways. Users are asked to count out the number of cubes and input it to pass the challenge.

4 Experiment and Analysis

To verify the design idea of our schemes, we carried out an experiment and asked human users to use our proposed CAPTCHAs and record relevant data such as accuracy and recognition time. Participants are more than 110, mostly sophomores in Xidian University. We built a website to present users with the challenge of our schemes to collect data. Users can submit their homework only if they have passed the challenge. It took us months to collect and analyze the data of all sets. We then improved our schemes according the data and the feedback information from the users and got the final versions of both schemes. The legal data of each scheme are shown in Table 1.

Table 1. The accuracy rate and recognition time of each set.

Scheme	The right number	Total number	Accuracy rate	Recognition time
DeRection	911	1054	86.43%	12.77s
CONSHEME	832	978	85.07%	15.12s

As shown in Table 1, the accuracy rate of DeRection and CONSCHEME are roughly the same and higher than 85%. We can draw a conclusion from the data that for both of the schemes, users can pass approximately eight to nine challenges when they are given ten of our CAPTCHAs. The Table 1 also shows the recognition time of each scheme. It takes 12.77 s and 15.12 s to complete the DeRecion and CONSCHEME. It is widely accepted that CAPTCHA should be completed no more than 30 s [1], so the two schemes are satisfying under this principle.

4.1 Design Analysis

Usability and robustness are the most important features that all the CAPTCHA schemes focused on. We tried but failed to find a CAPTCHA similar to ours from the existing CAPTCHAs. Finally, we decided to take the Asirra to compare with ours for its complete data. Asirra, as showed in Fig. 1, shows 12 images of cats and dogs and asks users to pick out all the cats. Similar to our first scheme, users need to click some regions in an image. And the solution space is finite for the limited pictures in the image, which makes it comparable to our CONSCHEME.

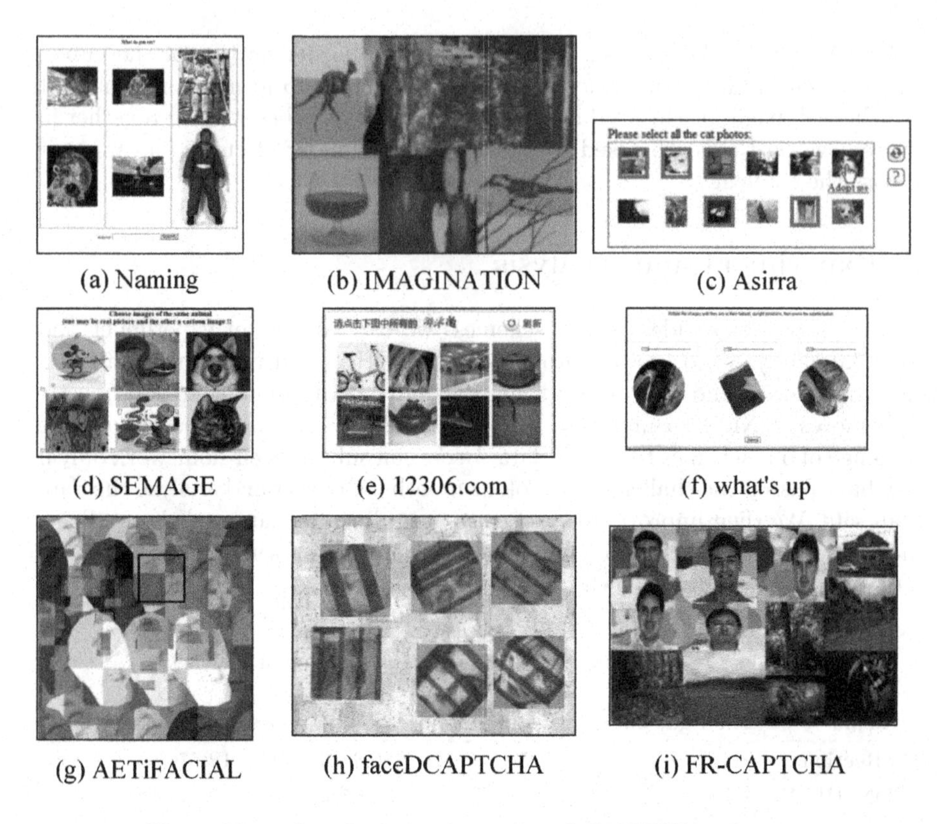

Fig. 1. Examples of existing image-based CAPTCHA schemes.

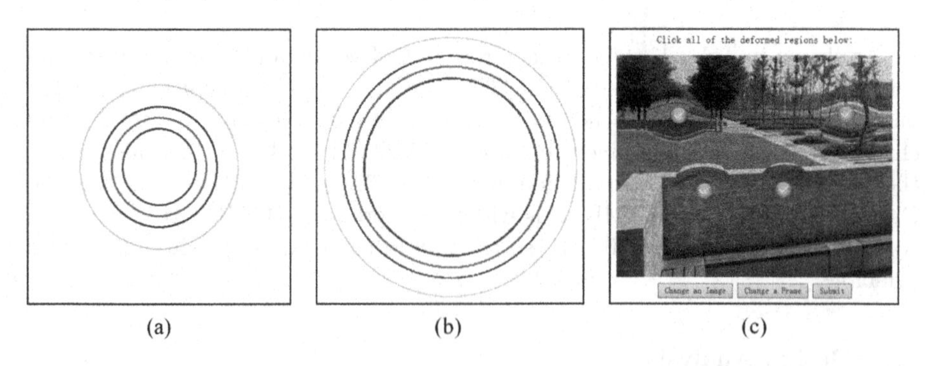

Fig. 2. The effect of convex deformation and an example of DeRection: (a) Original image; (b) After deformed; (c) An example of DeRection.

Usability analysis. Good usability indicates that users can quickly and accurately pass the test. However, in pursuit of robustness, many schemes usually sacrifice the usability too much. Some image-based CAPTCHA schemes carry out excessive deformation, noise and rotation. For example, the CAPTCHA scheme

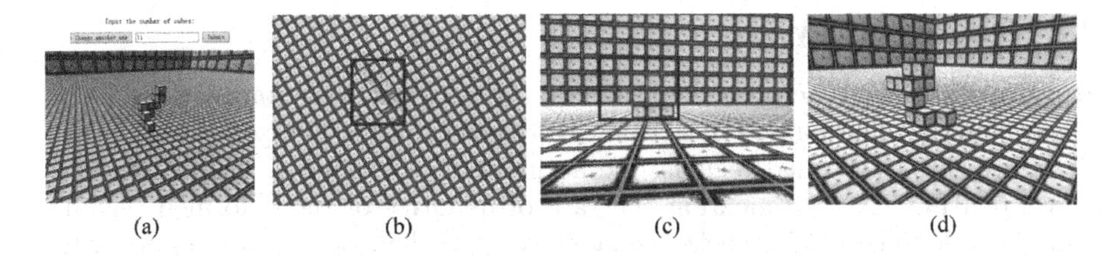

Fig. 3. Three views of the stacked cubes in CONSCHEME: (a) Example of CONSCHEME; (b) Top view; (c) Main view and left view; (d) General view.

adopted by 12306.cn [13] mainly adopt 8 low-resolution images thus it is very difficult for users to recognize and distinguish all of the objects, so the usability of it is reduced. In fact, when firstly put into use, the 12306.cn scheme got a huge number of complaints. Asirra has good usability because the small images are clearly and orderly placed and users only need to click the cats. The schemes of ours are also user friendly. In DeRection, users can capture the different regions in GIF image easily. The CONSCHEME utilizes human's ability to recognize objects in three-dimensional space, which is a capability that human have been trained and used since their birth. What's more, the feedback messages say this scheme is attractive and interesting.

Generation analysis. According to the definition of CAPTCHA, can be automatically generated and evaluated is the basic criterion for CAPTCHA design. However, due to the difficulties in generation, most of existing image-based CAPTCHA schemes adopt a tagged images database that already existed or established during the generation process of the designers. While generating, they just use tagged images from the database and put them together in some ways. Asirra adopted a tagged images database to generate their challenges. On the contrary, both of our proposed CAPTCHAs can be generated online and do not adopt any tagged images database. In the DeRection scheme, all the original images we adopt can be downloaded from the Internet real time, so it does not adopt any image database. As with the CONSCHEME scheme, once the unity plugin is finished, it can generate different number of stacked cube while using.

Language independence. Our designs are language independent which can be used by people all over the world. As mentioned above, users using our schemes only need to count out how many cubes there are in the space or pick out all of the deformed regions in the image, thus it is not necessary for user to master English or other specific language. Asirra is also language independent. Unfortunately, some image-based CAPTCHA systems are strongly language dependent, such as IMAGINATION [9] asks users to select a correct label from a given list and Naming CAPTCHA [8] requires users to understand the objects and type a word that describe the object of images in the textbox.

4.2 Security Analysis

Random guess attack. In the DeRection, the size of every image is $450 * 350$ pixels, and the maximum radius of the deformed regions is 50 pixels. Considering there are three deformed regions, the accuracy rate of random guess attack is about 0.0046%. Not to mention the rate of quantity of the deformed regions and the tolerance of the clicked zone is smaller than $50 * 50$ pixels. As with CONSCHEME scheme, the number of stacked cubes is between 5 and 17 so the random success rate is 1/13. Nevertheless, we can set multiple rounds of challenge. Such technique is already in use in Assira [11] and SEMAGE [12]. Furthermore, web service providers using our scheme can adopt CAPTCHA Token Buckets [11] to make it stronger and more secure that attackers cannot break our schemes without efforts.

Other attacks. During our design, we considered a lot of issues to achieve better robustness. In DeRection, after generating deformed regions, we modified every pixel value to resist attack through simple comparison between pixels. In CONSCHEME, the cubes' color and texture are the same as backgrounds, and the angles to get the three views are missed so bots cannot easily attack it by mathematical means. However, Asirra is broken at 10.3% by support vector machine classifiers trained on color and texture features extracted from images. Furthermore, some traditional image-based CAPTCHA schemes adopt a limited tagged image database and the labels and images are both limited, once we get the whole database by constantly refreshing and saving, we can solve out them without too much price. For example, the 12306.cn scheme contains about 580 categories and not more than one million images through a rough estimate. If the time is long enough, people can get the whole database thus break this scheme.

5 Image-Based CAPTCHA Design Guidelines

We analyzed the design process and the characteristics of our proposed CAPTCHA schemes and summarized a series of guidelines as follows:

- **Automate generate (AG):** The challenge should be generated automatically. But many existing schemes need human labor to gather raw material or tag images. For example, the images adopted "what's up" [14] CAPTCHA should be preprocessed by removing bad images by human.
- **Machine attacks (MA):** The challenge should be strong enough and the best existing techniques are far from solving the problem or there is no existing specific ways to attack the test successfully.
- **Resistance to no-effort attacks (RTNA):** The challenge should survive no-effort attacks. No-effort attacks are those can solve a CAPTCHA challenge without solving the hard AI problem. For example, "What's up" CAPTCHA was broken by random guess attack with success rate of 4.48% [10].
- **Secret database (SD):** Whether the adopted database is open or not.

- **Get the database (GTD):** Whether the attackers are able to access the secret database. Some databases are available while others are not. For example, Asirra's image database is provided by a novel, mutually beneficial partnership with Petfinder.com and attackers may get the database from the website. And images adopted by 12306.cn are kept secret and crackers need to keep refreshing the login interface to get the whole database.
- **Easy to human (ETH):** The challenge should be quickly and easily used by a human user. Any test that requires longer than 30 s becomes less useful in practice. And the average success rate of the user to pass this challenge should be as large as possible. The CAPTCHA adopted by the 12306.cn scheme got a huge number of complaints when firstly put into use.
- **Input modality (IM):** The challenge should be easy for users to interact with. The basic interactive forms include typing the keyboard, clicking the mouse and touching screen. The clicking is better than typing in to the specified textbox.
- **Language dependence (LD):** The challenge is related to a changing keyword. After users have acquired verification mode of the CAPTCHA, there are still some variations of the keywords in many tests, and users need to understand what these keywords mean to pass the challenge. For example, in every 12306.cn challenge, there is a changing distorted keyword that representing in the upper images for users to recognize.
- **Educational difference (ED):** Whether educational differences have an impact on passing this challenge. Some categories of objects need to be known by learning, and people may know something in different ways due to cultural differences.

According to the factors above, we can summarize guidelines about evaluating a CAPTCHA scheme. Firstly, a good CAPTCHA scheme should be generated automatically without a database of tagged images or with a database accessible to all of us. Besides, a good scheme can resist to common machine attacks and easy no-effort attacks such as random guess algorithm. What's more, to reach a good usability, the higher the success rate and the faster the pass time, the better. Generally it is acceptable that the accuracy rate should not be lower than 80% and the recognition time should not be longer than 30 s. It is better but not necessary to offer an interesting CAPTCHA. Then, the operation method should be as simple as possible, it is best for users to just use the mouse. But typing with keyboard is also generally accepted since nearly all of the text-based and many image-based CAPTCHAs require users to input something. Last but not least, the CAPTCHA scheme is best to be language independent and the users should not be required with any educational background.

As is shown in Table 2, the two schemes of ours are promising. They can be generated automatically with no database and there are no existing machine attacks or no-effort attacks against them. And users only need mouth and keyboard to use. The accuracy rate and recognition time are both acceptable but might be worse than the others in Table 2, but we have to take different

Table 2. The guidelines and evaluation of existing CAPTCHAs. ("—" means no data available or required, K&M means Keyboard and Mouse).

Schemes	AG	MA	RTNA	SD	GTD	ETH	IM	LD	ED
Naming	No	98.44% [21]	Yes	Yes	Easy	74%/24s	Keyboard	Yes	Yes
What's up	No	—	No [10]	Yes	Hard	84%	Mouse	No	No
Asirra	No	10.3% [10]	Yes	Yes	Easy	83.4%/15s	Mouse	No	No
ARTiFACIAL	Yes	18%/1.47s [16]	Yes	No	—	99.7%/14s	Mouse	No	No
SEMAGE	No	—	Yes	Yes	Hard	94.8%/11.64s	Mouse	No	No
IMAGINATION	No	4.95%/0.96s [10]	Yes	Yes	Easy	85%	Mouse	Yes	Yes
12306.cn	No	—	Yes	Yes	Hard	—	Mouse	Yes	Yes
FaceDCAPTCHA	No	48%/6.2s [19]	Yes	Yes	Easy	92.47%	Mouse	No	No
FR-CAPTCHA	No	42%/<14s [19]	Yes	Yes	Easy	94%	Mouse	Yes	No
DeRection	Yes	—	Yes	No	—	86.43%/12.77s	Mouse	No	No
CONSCHEME	Yes	—	Yes	No	—	85.07%/15.12s	K&M	No	No

experimental backgrounds into consideration. For example, the results of the Naming CAPTCHA are not that convincing because they came from 20 users who were paid $10–$15 for completing 100 rounds challenge.

6 Conclusion

In this paper, we propose two image-based CAPTCHA schemes both of which are based on visual effects. The DeRection asks users to find all of the deformed regions in one frame of a GIF image. And the CONSCHEME CAPTCHA asks users to count the number of stacking cubes in a three-dimensional scene.

By analyzing the results obtained using demographically diverse group of volunteers, we can assert that both the DeRection scheme and the CONSCHEME scheme are convenient to use and easy to solve. Moreover, we summarized a series of guidelines for design good image-based CAPTCHAs. We believe that the proposed DeRection and CONSCHEME schemes facilitate security against bots in online services without compromising user convenience. And the guidelines can be widely adopted by researchers on further researches on image-based CAPTCHAs.

Acknowledgments. The authors thank the reviewers for their careful reading of this paper and for their helpful and constructive comments. This project is supported by the National Natural Science Foundation of China (61472311) and the Fundamental Research Funds for the Central Universities.

References

1. Von Ahn, L., Blum, M., Langford, J.: Telling humans and computers apart automatically. Commun. ACM **47**(2), 56–60 (2004)
2. Mori, G., Malik, J.: Recognizing objects in adversarial clutter: breaking a visual CAPTCHA. In: Proceedings of the 2003 IEEE Computer Society Conference on Computer Vision and Pattern Recognition, vol. 1, p. I. IEEE (2003)

3. Yan, J., El Ahmad, A.S.: A low-cost attack on a Microsoft CAPTCHA. In: Proceedings of the 15th ACM Conference on Computer and Communications Security, pp. 543–554. ACM (2008)
4. Gao, H., Wang, W., Qi, J., Wang, X., Liu, X., Yan, J.: The robustness of hollow CAPTCHAs. In: Proceedings of the 2013 ACM SIGSAC Conference on Computer and Communications Security, pp. 1075–1086. ACM (2013)
5. Goodfellow, I.J., Bulatov, Y., Ibarz, J., Arnoud, S., Shet, V.: Multi-digit number recognition from street view imagery using deep convolutional neural networks. arXiv preprint arXiv:1312.6082 (2013)
6. Karthik, C.P., Recasens, R.A.: Breaking Microsoft's CAPTCHA. Technical report (2015)
7. Yan, J.: A simple generic attack on text CAPTCHAs (2016)
8. Chew, M., Tygar, J.D.: Image recognition CAPTCHAs. In: Zhang, K., Zheng, Y. (eds.) ISC 2004. LNCS, vol. 3225, pp. 268–279. Springer, Heidelberg (2004). https://doi.org/10.1007/978-3-540-30144-8_23
9. Datta, R., Li, J., Wang, J.Z.: Imagination: a robust image-based CAPTCHA generation system. In: Proceedings of the 13th Annual ACM International Conference on Multimedia, pp. 331–334. ACM (2005)
10. Zhu, B.B., Yan, J., Li, Q., Yang, C., Liu, J., Xu, N., Yi, M., Cai, K.: Attacks and design of image recognition CAPTCHAs. In: Proceedings of the 17th ACM Conference on Computer and Communications Security, pp. 187–200. ACM (2010)
11. Elson, J., Douceur, J.R., Howell, J., Saul, J.: Asirra: a CAPTCHA that exploits interest-aligned manual image categorization. In: ACM Conference on Computer and Communications Security, vol. 7, pp. 366–374. Citeseer (2007)
12. Vikram, S., Fan, Y., Gu, G.: SEMAGE: a new image-based two-factor CAPTCHA. In: Proceedings of the 27th Annual Computer Security Applications Conference, pp. 237–246. ACM (2011)
13. China Railway Customer Service Website. 12306 CAPTCHA (2017). http://www.12306.cn/mormhweb/
14. Gossweiler, R., Kamvar, M., Baluja, S.: What's up CAPTCHA? A CAPTCHA based on image orientation. In: Proceedings of the 18th International Conference on World Wide Web, pp. 841–850. ACM (2009)
15. Rui, Y., Liu, Z.: Artifacial: automated reverse turing test using facial features. Multimedia Syst. **9**(6), 493–502 (2004)
16. Li, Q.: A computer vision attack on the artifacial CAPTCHA. Multimedia Tools Appl. **74**(13), 4583–4597 (2015)
17. Goswami, G., Powell, B.M., Vatsa, M., Singh, R., Noore, A.: FR-CAPTCHA: CAPTCHA based on recognizing human faces. PloS ONE **9**(4), e91708 (2014)
18. Goswami, G., Powell, B.M., Vatsa, M., Singh, R., Noore, A.: FaceDCAPTCHA: face detection based color image CAPTCHA. Future Gener. Comput. Syst. **31**, 59–68 (2014)
19. Gao, H., Lei, L., Zhou, X., Li, J., Liu, X.: The robustness of face-based CAPTCHAs. In: 2015 IEEE International Conference on Computer and Information Technology; Ubiquitous Computing and Communications; Dependable, Autonomic and Secure Computing; Pervasive Intelligence and Computing (CIT/IUCC/DASC/PICOM), pp. 2248–2255. IEEE (2015)
20. Zulpe, N., Pawar, V.: GLCM textural features for brain tumor classification. IJCSI Int. J. Comput. Sci. Issues **9**(3), 354–359 (2012)
21. Raj, A., Jain, A., Pahwa, T., Jain, A.: Picture CAPTCHAs with sequencing: their types and analysis. Int. J. of Digit. Soc. **1**(3), 208–220 (2010)

An Unsupervised Domain Adaptation Algorithm Based on Canonical Correlation Analysis

Pan Xiao[1], Bo Du[1(✉)], and Xue Li[2]

[1] School of Computer, Wuhan University, Wuhan, Hubei, China
remoteking@whu.edu.cn
[2] LIESMARS, Wuhan University, Wuhan, Hubei, China

Abstract. This paper addresses the unsupervised domain adaptation problem, which is especially challenging as the target domain does not provide explicitly label information. To solve this problem, we develop a new algorithm based on canonical correlation analysis (CCA). Specifically, we first use CCA to project both domain data onto the correlation subspace. To exploit the target domain data further, we train an SVM classifier by the source domain to obtain the pre-label of the target domain. Considering that the label space between the source and target domain may be different or even disjoint, we introduce a class adaptation matrix to adapt them. An objective function taking all factors mentioned above into consideration is designed. Finally, we learn a classification matrix by iterative optimization. Empirical studies on benchmark tasks of action recognition demonstrate that our algorithm can improve classification accuracy significantly.

Keywords: Domain adaptation · Canonical correlation analysis
Unsupervised learning

1 Introduction

In computer vision, domain adaptation (DA) has become a very popular topic. It addresses the problem that we need to solve the same learning tasks across different domains [2,20]. Generally, we can divide domain adaptation into two parts: unsupervised DA in which target domain data are completely unlabeled, and semi-supervised DA where a small number of instances in the target domain are labeled. We focus on the unsupervised scenario, which is especially challenging as the target domain does not provide explicitly any information on how to optimize classifiers. The goal of unsupervised domain adaptation is to derive a classifier for the unlabeled target domain data by extracting the information that is invariant across source and target domains.

Canonical correlation analysis (CCA) is often used to deal with DA problems since it can obtain two projection matrices to maximize the correlation between two different domains [9]. The derived correlation subspace can preserve common features of both domains very well.

J. Yang et al. (Eds.): CCCV 2017, Part III, CCIS 773, pp. 26–37, 2017.
https://doi.org/10.1007/978-981-10-7305-2_3

In our work, an efficient unsupervised domain adaptation algorithm based on CCA is developed. Specifically, we first make use of CCA to derive the correlation subspace. In order to explore the target domain data further, we use the source domain data to train a SVM classifier and then obtain the pre-label of the target domain. Considering that the label space between source and target domain may be different or even disjoint, we introduce a class adaptation matrix to adapt them. Taking all factors mentioned above into consideration, we design an objective function. Finally, a fine classifier can be obtained by iterative optimization.

The rest of the paper is organized as follows. Section 2 first introduces the related work of DA and CCA. In Sect. 3, we discusses our proposed unsupervised domain adaptation algorithm based on canonical correlation analysis in detail. Section 4 shows the experimental results in a cross-domain action recognition dataset. The last section gives some conclusive discussions.

2 Related Work

We now review some state-of-the-art domain adaptation methods and the recent works related with deep learning are also discussed. Finally, we introduce the main idea of CCA.

2.1 Domain Adaptation Methods

Generally speaking, domain adaptation problems can be solved by instance-based and feature-based approaches.

The goal of instance-based approaches is to re-weight the source domain instances by making full use of the information of target domain. For example, Dai et al. [3] proposed an algorithm based on Adaboost, which can iteratively reinforce useful samples to help train classifiers. Shi et al. [21] attempted to find a new representation for the source domain, which can reduce the negative effect of misleading samples. In [11], a heuristic algorithm was developed to remove misleading instances of the source domain. Li et al. [13] proposed a framework that can iteratively learn a common space for both domain. Several methods [15,16,26–28] proposed by Wu et al. and Liu et al. can also help us solve the domain adaptation problem effectively.

The purpose of feature-based approaches is to discover common latent features. For instance, a method integrating subspaces on the Grassman manifold was developed to learn a feature projection matrix for both domains in [7]. Zhang et al. [34] introduced a novel feature extraction algorithm, which can efficiently encode the discriminative information from limited training data and the sample distribution information from unlimited test data. In [5], a projection aligning subspaces of both domains was designed. The distributions of the feature space and the label space are considered in [8] to learn conditional transferable components. In [22–24], three subspaces extraction methods were proposed, which provides the new way to find the common subspaces of both domains. The method in

[19] attempted to project both domains into a Reproducing Kernel Hilbert Space (RKHS) and then obtain some transfer components based on Maximum Mean Discrepancy (MMD). In [30], the independence between the samples learned features and domain features is maximized to reduce the domains' discrepancy.

The discrepancies between domains [32] can be reduced through deep networks, which learn feature representation disentangling the factors of variations behind data [1]. Recent works have demonstrated that deep neural networks are powerful for learning transferable features [6,17,18,25]. Specifically, these methods embeds DA modules into deep networks to improve the performance, which mainly correct the shifts in marginal distributions, assuming conditional distributions remain unchanged after the marginal distribution adaptation. However, the recent research also finds that the features extracted in higher layers need to depend on the specific dataset [33].

2.2 Canonical Correlation Analysis

We briefly review canonical correlation analysis (CCA) as follows.

Suppose that $X^s = \{x_1^s, \ldots, x_n^s\} \in \mathbb{R}^{d_s \times n}$ and $X^t = \{x_1^t, \ldots, x_n^t\} \in \mathbb{R}^{d_t \times n}$ are source and target domain dataset respectively. n denotes the number of samples. CCA can obtain two projection vectors $u^s \in \mathbb{R}^{d_s}$ and $u^t \in \mathbb{R}^{d_t}$ to maximize the correlation coefficient ρ:

$$\max_{u^s,u^t} \rho = \frac{{u^s}^\top \sum_{st} u^t}{\sqrt{{u^s}^\top \sum_{ss} u^s}\sqrt{{u^t}^\top \sum_{tt} u^t}}, \tag{1}$$

where $\sum_{st} = X^s {X^t}^\top$, $\sum_{ss} = X^s {X^s}^\top$, $\sum_{tt} = X^t {X^t}^\top$, and $\rho \in [0,1]$. According to [9], we can regard (1) as a generalized eigenvalue decomposition problem, there is

$$\sum_{st}\left(\sum_{tt}\right)^{-1}\sum_{st}^\top u^s = \eta \sum_{ss} u^s \tag{2}$$

Then, u^t can be calculated by $\sum_{tt}^{-1}\sum_{st}^\top u^s/\eta$ after u^s is obtained. To avoid overfitting and singularity problems, two terms $\lambda_s I$ and $\lambda_t I$ are added into $\sum_{ss}$ and $\sum_{tt}$ respectively. We have

$$\sum_{st}\left(\sum_{tt} + \lambda_t I\right)^{-1}\sum_{st}^\top u^s = \eta\left(\sum_{ss} + \lambda_s I\right) u^s \tag{3}$$

Generally speaking, we can obtain more than one pair of projection vectors $\{u_i^s\}_{i=1}^L$ and $\{u_i^t\}_{i=1}^L$. L denotes the dimensions of the CCA subspace. CCA can determine projection matrices $P_s = \{u_1^s, \ldots, u_d^s\} \in \mathbb{R}^{d_s \times L}$ and $P_t = \{u_1^t, \ldots, u_d^t\} \in \mathbb{R}^{d_t \times L}$, which can project the source and target domain data (X^s and X^t) onto the correlation subspace. Once the correlation subspace spanned by $\left\{u_i^{s,t}\right\}_{i=1}^L$ is derived, we can recognize the target domain data by the model trained from the source domain data.

3 Our Method

Our approach mainly consists of four steps. Firstly, we use the CCA to find the source and target domain's projection matrices and then project both domain data onto the correlation subspace. The second step is to train a SVM classifier to obtain the pre-label matrix of the target domain data. Then, we introduce a sigmoid function to process dataset on the correlation subspace. Finally, by minimizing the norm of classification errors, we obtain a class adaptation matrix and a classification matrix simultaneously.

3.1 The Correlation Subspace

We denote $X_S = (x_1, x_2, \ldots, x_{N_S})^\top, x_i \in \mathbb{R}^d$ as the source domain data and $X_{Tu} = (x_1, x_2, \ldots, x_{N_{Tu}})^\top, x_i \in \mathbb{R}^d$ as the target domain data. Then we can use CCA mentioned above to find the projection matrices $P_S \in \mathbb{R}^{d\times L}$ and $P_{Tu} \in \mathbb{R}^{d\times L}$ for labeled source domain and unlabeled target domain data respectively. L denotes the dimension of the correlation subspace. Moreover, we denote $X_S^P \in \mathbb{R}^{N_S\times L}$ and $X_{Tu}^P \in \mathbb{R}^{N_{Tu}\times L}$ as data matrix of source and target domain projected onto the correlation subspace. Then, we have

$$X_S^P = X_S P_S \tag{4}$$

$$X_{Tu}^P = X_{Tu} P_{Tu} \tag{5}$$

3.2 The Pre-label of Target Domain

Let $Y_S = (y_1, y_2, \ldots, y_{N_S})^T \in \mathbb{R}^{N_S\times c}$ be the label matrix of source domain with c classes. In our algorithm, we propose to obtain the pre-label of target domain by training a SVM classifier on the CCA correlation subspace. And we denote $Y_{Tu} = (y_1, y_2, \ldots, y_{N_{Tu}})^T \in \mathbb{R}^{N_{Tu}\times c}$ as the pre-label matrix.

3.3 The Sigmoid Function

What's more, a sigmoid function $G(\cdot)$ is introduced to process both domain dataset on the correlation subspace. The role of $G(\cdot)$ is to preform a non-linear mapping, which can improve the generalization ability of our model further. Specifically, we have

$$R_S = G(X_S^P) = G(X_S P_S) \tag{6}$$

$$R_{Tu} = G(X_{Tu}^P) = G(X_{Tu} P_{Tu}) \tag{7}$$

3.4 The Classification Matrix and Class Adaptation Matrix

We first define a classification matrix $\beta \in \mathbb{R}^{L\times c}$. It aims to classify both domain data onto the right class as accurate as possible. That is to say, $R_S\beta$ and $R_{Tu}\beta$

should be similar to Y_S and Y_{Tu} respectively. Specifically, we define the objective function as

$$\min_{\beta} F(\beta) = \|\beta\|_{q,p} + C_S \|R_S\beta - Y_S\|_F^2 + C_{Tu} \|R_{Tu}\beta - Y_{Tu}\|_F^2 \quad (8)$$

where $\|\cdot\|_{q,p}$ and $\|\cdot\|_F^2$ are the $l_{q,p}$-norm and Frobenius norm respectively. C_S and C_{Tu} are the penalty coefficient for both domain data. Specifically, $\|\beta\|_{q,p}$ can be written as

$$\|\beta\|_{q,p} = (\sum_{i=1}^{m}(\sum_{j=1}^{n} |\beta_{ij}|^q)^{p/q})^{1/p} \quad (9)$$

$q \geq 2$ and $0 \leq p \leq 2$ are set to impose sparsity on β. It's difficult to solve the objective function when $p = 0$, therefore, we let $p = 1$. The classification accuracy will not be improved with larger q [10], so we set $q = 2$. Finally, the objective function can be described as

$$\min_{\beta} F(\beta) = \|\beta\|_{2,1} + C_S \|R_S\beta - Y_S\|_F^2 + C_{Tu} \|R_{Tu}\beta - Y_{Tu}\|_F^2 \quad (10)$$

We also introduce a class adaptation matrix $\Theta \in \mathbb{R}^{c\times c}$ to adapt in label space. This is because the label space between source and target domains may be different [29]. So label adaptation may help obtain a better classification model. To incorporate label adaptation into our method, we can redefine the objective function as

$$\begin{aligned} \min_{\beta,\Theta} \ F(\beta,\Theta) = \|\beta\|_{2,1} + C_S \|R_S\beta - Y_S\|_F^2 + \\ C_{Tu} \|R_{Tu}\beta - Y_{Tu} \circ \Theta\|_F^2 + \gamma \|\Theta - I\|_F^2 \end{aligned} \quad (11)$$

$\|\Theta - I\|_F^2$ is a term to control the class distortion. And γ is the trade-off parameter. The symbol $\circ$ denotes a multiplication operator, which can perform label adaptation between domains. In [4], the importance of unlabeled data has been emphasized. It's believed that a large number of unlabeled target domain data containing meaningful information for classification may not be fully explored. We minimize the error between the $R_{Tu}\beta$ and $Y_{Tu} \circ \Theta$ to explore the unlabeled data further.

The problem in our method turns out how to find the optimal classification matrix β and class adaptation matrix Θ simultaneously.

3.5 Optimization Algorithm

We can obtain the solution for the objective function (11) easily since β and Θ is differentiable.

Firstly, by fixing $\Theta = I$, we can get the derivative of (11) with respect to β. And there is

$$\frac{\partial F(\beta,\Theta)}{\partial \beta} = 2Q\beta + 2C_S R_S^T(R_S\beta - Y_S) + 2C_{Tu}R_{Tu}^T(R_{Tu}\beta - Y_{Tu} \circ \Theta) \quad (12)$$

in which $Q \in \mathbb{R}^{L \times L}$ is a diagonal matrix. We can regard the i-th element of Q as

$$Q_{ii} = \frac{1}{2\left\|\beta_i\right\|_2} \tag{13}$$

in which β_i can be seen as the i-th row of β.

In our algorithm, to avoid $\beta_i = 0$, we incorporate a very small value $\epsilon > 0$ into (13). Specifically, we use $\left\|\beta_i\right\|_2 + \epsilon$ to update Q. So the Eq. (13) can be rewritten as follows

$$Q_{ii} = \frac{1}{2(\left\|\beta_i\right\|_2 + \epsilon)}, \epsilon > 0 \tag{14}$$

We can let the Eq. (12) be zero, namely $\frac{\partial F(\beta,\Theta)}{\partial \beta} = 0$, then the optimal β can be obtained, there is

$$\beta = (Q + C_S R_S^T R_S + C_{Tu} R_{Tu}^T R_{Tu})^{-1}(C_S R_S^T Y_S + C_{Tu} R_{Tu}^T Y_{Tu} \circ \Theta) \tag{15}$$

Second, according to the formula (15), we substitute the fixed β value into the objective function. The optimization problem (11) becomes

$$\min_{\Theta} F(\Theta) = C_{Tu}\left\|R_{Tu}\beta - Y_{Tu} \circ \Theta\right\|_F^2 + \gamma\left\|\Theta - I\right\|_F^2 \tag{16}$$

Then, we can obtain the derivative of (16) with respect to Θ. Specifically, we have

$$\frac{\partial F(\beta,\Theta)}{\partial \Theta} = -2C_{Tu}Y_{Tu}^T(R_{Tu}\beta - Y_{Tu} \circ \Theta) + 2\gamma(\Theta - I) \tag{17}$$

Similarly, by setting (17) to be zero, we have

$$\Theta = (C_{Tu}Y_{Tu}^T Y_{Tu} + \gamma I)^{-1}(C_{Tu}Y_{Tu}^T R_{Tu}\beta + \gamma I) \tag{18}$$

The result can be obtained by iteratively optimizing β and Θ. The optimization procedure of our model is summarized in Algorithm 1. T_{max} denotes the number of maximum iteration. In this paper, we set T_{max} to be 50. Once the number of iteration reach T_{max}, the iterative update procedure would be terminated.

4 Experimental Results

4.1 Experimental Setting

Dataset. The Inria Xmas Motion Acquisition Sequences (IXMAS)[1] records 11 actions. Each action can be seen as a category. There are 12 actors involved in this action shooting and they perform each action three times. Therefore, 396 instances are captured by one camera in total. As seen from Fig. 1, five cameras (domains) are used to capture the actions simultaneously. To extract features from each image, we follow the procedure in [14]. Finally, each image can be regarded as a vector of 1000 dimensions. This dataset aims to set a standard for human action recognition.

[1] http://4drepository.inrialpes.fr/public/viewgroup/6.

Algorithm 1. Domain Adaptation Based on Canonical Correlation Analysis

Input: Source domain data X_S, Target domain data X_{Tu}, Source domain label Y_S, The number of maximum iteration T_{max}
Output: Classification matrix β^t, Class adaptation matrix Θ^t
1: Calculate P_S and P_{Tu} based on canonical correlation analysis;
2: Calculate X_S^P and X_{Tu}^P using (4) and (5);
3: Obtain the pre-label Y_{Tu} of target domain data by training a SVM classifier on the CCA correlation subspace;
4: Calculate R_S and R_{Tu} using (6) and (7);
5: $t \leftarrow 1$
6: $Q^t \leftarrow I_{L \times L}$
7: $\Theta^t \leftarrow I_{c \times c}$
8: **while** not converged ($t < T_{max}$) **do**
9: Calculate the classification matrix β^t using (15);
10: Update Θ^{t+1} using (18);
11: Update Q^{t+1} using (14);
12: $t \leftarrow t + 1$
13: **end while**

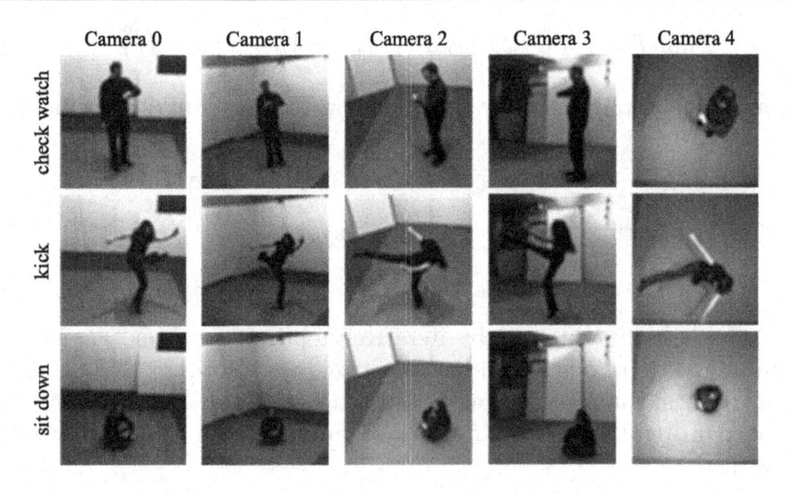

Fig. 1. Example actions of the IXMAS dataset. Each row represents an action at five different views.

Implementation Details. We follow the operation in [31] to obtain the CCA projection matrices for both domains. Specifically, two thirds of domains' samples in each catagory are selected. And the training set consists of 30 labeled samples per category in source domain and all unlabeled samples in target domain. The test set consists of all unlabeled target domain data. Then we follow the procedure mentioned in Sect. 3 to train a classifier and get the classification accuracies. The above procedure is repeated ten times. We give the average classification accuracy in Table 1.

Table 1. The classification accuracies and standard errors (%) for all methods on the IXMAS dataset

Domains	SVM	SA	TCA	GFK	MIDA	Ours
cam0→cam1	15.3±1.1	63.8±1.0	48.3±1.3	49.4±1.4	59.8±0.5	**81.3±0.8**
cam0→cam2	12.1±0.9	72.1±1.3	54.1±4.4	55.2±6.8	58.0±1.3	**86.4±1.3**
cam0→cam3	7.3±0.8	72.6±1.3	55.4±4.7	55.5±9.0	57.7±1.4	**91.9±1.0**
cam0→cam4	10.7±0.7	73.0±1.1	45.3±3.8	51.2±6.0	54.3±2.8	**92.5±1.3**
cam1→cam0	7.5±0.8	71.6±0.9	26.6±2.0	59.0±1.7	53.5±1.6	**89.0±0.8**
cam1→cam2	13.3±1.3	68.6±1.1	66.7±2.1	53.5±1.5	67.6±1.7	**84.9±1.3**
cam1→cam3	7.1±0.8	72.1±0.9	59.5±2.8	60.7±11.3	57.1±1.7	**90.9±0.9**
cam1→cam4	10.9±0.6	70.5±2.1	44.1±4.7	61.9±3.2	51.8±2.1	**92.0±1.0**
cam2→cam0	16.8±2.1	72.3±1.2	47.8±6.7	61.7±5.2	60.0±1.0	**88.7±1.4**
cam2→cam1	12.3±0.6	71.8±0.6	29.3±3.1	59.1±7.6	57.2±1.8	**84.7±0.8**
cam2→cam3	11.1±0.7	68.8±1.1	68.7±1.2	50.5±11.4	67.2±1.6	**91.6±0.8**
cam2→cam4	8.0±0.7	72.9±1.5	51.2±5.1	60.6±3.7	51.1±3.7	**88.1±1.1**
cam3→cam0	5.3±1.0	74.4±2.0	68.3±3.6	56.3±2.0	59.5±1.2	**84.0±1.1**
cam3→cam1	6.9±1.0	72.8±1.0	63.3±2.5	52.1±7.1	58.8±1.3	**90.6±1.0**
cam3→cam2	12.4±0.8	76.5±2.6	66.8±3.1	65.5±2.7	59.6±0.9	**82.5±0.9**
cam3→cam4	10.4±0.9	68.6±1.6	47.2±3.2	47.3±7.8	67.7±2.0	**87.5±1.2**
cam4→cam0	18.5±1.3	63.9±0.8	60.7±1.2	55.8±3.8	63.9±1.3	**86.5±0.7**
cam4→cam1	12.5±1.4	71.0±1.6	42.0±4.6	66.1±0.9	57.5±1.3	**89.4±1.0**
cam4→cam2	6.7±0.8	74.4±1.4	65.6±3.2	59.0±6.0	58.1±1.7	**90.4±0.9**
cam4→cam3	9.6±1.3	76.5±1.9	65.4±3.2	65.4±12.8	59.5±0.9	**89.0±0.6**
Average	10.7±1.0	71.4±1.4	53.8±3.3	57.3±5.6	59.0±1.6	**88.1±1.0**

4.2 Comparison Methods

We compare our framework with a baseline and several classic unsupervised domain adaptation methods.

SVM [12]. We regard SVM as the baseline. SVM has become a classic method to solve classification problems. To solve the DA problem, we use the original features in both domains directly. Specifically, We build a prediction model based on the source domain data and then classify instances in the target domain. Since SVM is not developed for DA problem, the final result on target domain may be the worst when compared with other methods.

Subspace Alignment (SA) [5]. This algorithm is very simple. It learns PCA subspaces of both domains at first. And then a linear mapping aligning the PCA subspaces is derived. After that, we can build models based on the source domain to classify the target domain data on the common subspace.

Transfer Component Analysis (TCA) [19]. This algorithm is designed according to maximum mean discrepancy (MMD), which can measure the distance between two distributions. By minimizing MMD, a projection matrix narrowing the distance between both domains can be obtain. This method can also map both domain data into a kernel space. In our experiments, Gaussian RBF kernels are taken.

Geodesic Flow Subspaces (GFK) [7]. This method applies the Grassman manifold to solve DA problems. First of all, the PCA or PLSA subspaces of both domains are computed. Then the subspaces are embedded into the Grassman manifold. And we can use the subspaces to obtain super-vertors by transforming the original features. Finally, low dimensional feature vectors are derived and we can train a prediction model on them.

Maximum Independence Domain Adaptation (MIDA) [30]. MIDA introduces Hilbert-Schmidt independence criterion to adapt different domains. Specifically, in order to reduce the difference across domains, we can try to obtain the maximum of the independence between the learned features and the sample features.

4.3 Parameter Tuning

In our method, there are totally four parameters including C_S, C_{Tu}, ϵ and γ. Generally speaking, it is not appropriate for an algorithm to tune the four parameters at the same time. Actually, there is no need to tune all of them. We can find the optimal solution by freezing two parameters. To be specific, we set $\epsilon = 1$ and $\gamma = 0.1$. Then we search for the best values of Cs and C_{Tu} within the ranges $\{4^0, 4^1, 4^2, 4^3, 4^4, 4^5, 4^6\}$ and $\{10^{-3}, 10^{-2}, 10^{-1}, 10^0, 10^1, 10^2, 10^3\}$ respectively. Finally, the best performance of our model is reported.

For SVM and other four state-of-the-art DA methods, we follow the procedures in corresponding paper to tune parameters and then report the best classification results.

4.4 Experimental Results and Comparisons

The classification accuracies and standard errors are summarized in Table 1. Cam0-cam5 represent different domains. Specifically, the form A→B states that A is the source domain and B is the target domain. For example, cam0→cam1 represents that images captured by cam0 are used as the source domain and images captured by cam1 are regarded as the target domain. The classification accuracy of SVM can be seen from the second column of Table 1. And the results of the classic unsupervised DA methods are shown in the third to the sixth column. The last column is the result of our proposed method. Totally, 20 domain pairs are given and we bold the best results for each pair. From Table 1, we can conclude that

- The classification model trained by SVM doesn't perform well. As can be seen from the table, average accuracy is around 11% and most of the results are no more than 15%. In real applications, such a model is useless.

- We can obtain better prediction models by training classifiers based on those classic unsupervised DA methods (SA, TCA, GFK, MIDA). The average classification accuracy for each method is above 50%. It's worth noting that the result of SA is highest (71.4%) compared to TCA, GFK and MIDA. That is to say, SA is more suitable to deal with IXMAS dataset.
- The classification result can be improved further by our model. Specifically, the average accuracy of our proposed algorithm is 88.1%. The result is good enough since it is improved around 77% points compared with SVM.

5 Conclusion

A new unsupervised domain adaptation algorithm based on canonical correlation analysis is proposed in this paper. Our method shows competitive performance when compared with some state-of-the-art methods, e.g. SVM, SA, TCA, GFK, MIDA.

Acknowledgement. This work was supported in part by the National Natural Science Foundation of China under Grants U1536204, 60473023, 61471274, and 41431175, China Postdoctoral Science Foundation under Grant No. 2015M580753.

References

1. Bengio, Y., Courville, A., Vincent, P.: Representation learning: a review and new perspectives. IEEE Trans. Pattern Anal. Mach. Intell. **35**(8), 1798–1828 (2013)
2. Blitzer, J., Kakade, S., Foster, D.P.: Domain adaptation with coupled subspaces, pp. 173–181 (2011)
3. Dai, W., Yang, Q., Xue, G.R., Yu, Y.: Boosting for transfer learning. In: International Conference on Machine Learning, pp. 193–200 (2007)
4. Duan, L., Xu, D., Tsang, I.W.: Domain adaptation from multiple sources: a domain-dependent regularization approach. IEEE Trans. Neural Netw. Learn. Syst. **23**(3), 504 (2012)
5. Fernando, B., Habrard, A., Sebban, M., Tuytelaars, T.: Unsupervised visual domain adaptation using subspace alignment. In: International Conference on Computer Vision, pp. 2960–2967 (2013)
6. Ganin, Y., Lempitsky, V.: Unsupervised domain adaptation by backpropagation. In: International Conference on Machine Learning, pp. 1180–1189 (2015)
7. Gong, B., Shi, Y., Sha, F., Grauman, K.: Geodesic flow kernel for unsupervised domain adaptation. In: Computer Vision and Pattern Recognition, pp. 2066–2073 (2012)
8. Gong, M., Zhang, K., Liu, T., Tao, D., Glymour, C., Schölkopf, B.: Domain adaptation with conditional transferable components, pp. 2839–2848 (2016)
9. Hardoon, D.R., Szedmak, S.R., Shawe-Taylor, J.R.: Canonical correlation analysis: an overview with application to learning methods. Neural Comput. **16**(12), 2639–2664 (2004)
10. Hou, C., Nie, F., Li, X., Yi, D., Wu, Y.: Joint embedding learning and sparse regression: a framework for unsupervised feature selection. IEEE Trans. Cybern. **44**(6), 793–804 (2014)

11. Jiang, J., Zhai, C.X.: Instance weighting for domain adaptation in NLP. In: Meeting of the Association of Computational Linguistics, pp. 264–271 (2007)
12. Joachims, T.: Making large-scale SVM learning practical, pp. 499–526 (1998)
13. Li, X., Zhang, L., Du, B., Zhang, L., Shi, Q.: Iterative reweighting heterogeneous transfer learning framework for supervised remote sensing image classification. IEEE J. Sel. Topics Appl. Earth Obs. Remote Sens. **PP**(99), 1–14 (2017)
14. Liu, J., Shah, M., Kuipers, B., Savarese, S.: Cross-view action recognition via view knowledge transfer. In: Computer Vision and Pattern Recognition, pp. 3209–3216 (2011)
15. Liu, T., Tao, D., Song, M., Maybank, S.J.: Algorithm-dependent generalization bounds for multi-task learning. IEEE Trans. Pattern Anal. Mach. Intell. **39**(2), 227–241 (2017)
16. Liu, T., Tao, D.: Classification with noisy labels by importance reweighting. IEEE Trans. Pattern Anal. Mach. Intell. **38**(3), 447–461 (2016)
17. Long, M., Cao, Y., Wang, J., Jordan, M.: Learning transferable features with deep adaptation networks. In: International Conference on Machine Learning, pp. 97–105 (2015)
18. Long, M., Zhu, H., Wang, J., Jordan, M.I.: Unsupervised domain adaptation with residual transfer networks. In: Advances in Neural Information Processing Systems, pp. 136–144 (2016)
19. Pan, S.J., Tsang, I.W., Kwok, J.T., Yang, Q.: Domain adaptation via transfer component analysis. IEEE Trans. Neural Netw. **22**(2), 199–210 (2011)
20. Saenko, K., Kulis, B., Fritz, M., Darrell, T.: Adapting visual category models to new domains. In: Daniilidis, K., Maragos, P., Paragios, N. (eds.) ECCV 2010. LNCS, vol. 6314, pp. 213–226. Springer, Heidelberg (2010). https://doi.org/10.1007/978-3-642-15561-1_16
21. Shi, Q., Du, B., Zhang, L.: Domain adaptation for remote sensing image classification: a low-rank reconstruction and instance weighting label propagation inspired algorithm. IEEE Trans. Geosci. Remote Sens. **53**(10), 5677–5689 (2015)
22. Tao, D., Li, X., Wu, X., Maybank, S.J.: General tensor discriminant analysis and Gabor features for gait recognition. IEEE Trans. Pattern Anal. Mach. Intell. **29**(10), 1700 (2007)
23. Tao, D., Li, X., Wu, X., Maybank, S.J.: Geometric mean for subspace selection. IEEE Trans. Pattern Anal. Mach. Intell. **31**(2), 260–274 (2009)
24. Tao, D., Tang, X., Li, X., Wu, X.: Asymmetric bagging and random subspace for support vector machines-based relevance feedback in image retrieval. IEEE Trans. Pattern Anal. Mach. Intell. **28**(7), 1088–1099 (2006)
25. Tzeng, E., Hoffman, J., Zhang, N., Saenko, K., Darrell, T.: Deep domain confusion: maximizing for domain invariance (2014). arXiv preprint arXiv:1412.3474
26. Wu, J., Cai, Z., Zeng, S., Zhu, X.: Artificial immune system for attribute weighted naive Bayes classification. In: The 2013 International Joint Conference on Neural Networks (IJCNN), pp. 1–8. IEEE (2013)
27. Wu, J., Hong, Z., Pan, S., Zhu, X., Cai, Z., Zhang, C.: Multi-graph-view learning for graph classification. In: 2014 IEEE International Conference on Data Mining (ICDM), pp. 590–599. IEEE (2014)
28. Wu, J., Pan, S., Zhu, X., Zhang, C., Wu, X.: Positive and unlabeled multi-graph learning. IEEE Trans. Cybern. **47**(4), 818–829 (2017)
29. Xiang, E.W., Pan, S.J., Pan, W., Su, J., Yang, Q.: Source-selection-free transfer learning. In: Proceedings of the International Joint Conference on Artificial Intelligence, IJCAI 2011, Barcelona, Catalonia, Spain, July, pp. 2355–2360 (2011)

30. Yan, K., Kou, L., Zhang, D.: Domain adaptation via maximum independence of domain features (2016)
31. Yeh, Y.R., Huang, C.H., Wang, Y.C.: Heterogeneous domain adaptation and classification by exploiting the correlation subspace. IEEE Trans. Image Process. **23**(5), 2009–2018 (2014). A Publication of the IEEE Signal Processing Society
32. Yosinski, J., Clune, J., Bengio, Y., Lipson, H.: How transferable are features in deep neural networks? In: Advances in neural information processing systems, pp. 3320–3328 (2014)
33. Yosinski, J., Clune, J., Bengio, Y., Lipson, H.: How transferable are features in deep neural networks? Eprint Arxiv, vol. 27, pp. 3320–3328 (2014)
34. Zhang, L., Zhang, L., Tao, D., Huang, X.: Sparse transfer manifold embedding for hyperspectral target detection. IEEE Trans. Geosci. Remote Sens. **52**(2), 1030–1043 (2014)

UAV-Based Vehicle Detection by Multi-source Images

Shangjie Jiang(✉), Bin Luo, Jun Liu, Yun Zhang, and LiangPei Zhang

State Key Laboratory of Information Engineering in Surveying, Mapping and Remote Sensing, Wuhan University, Wuhan 430072, China
shangjiejiang@whu.edu.cn

Abstract. Detecting vehicles from autonomous unmanned aerial vehicle (UAV) systems is attracting the attention of more and more researchers. This technique has also been widely applied in traffic monitoring and management. Differing from the other object detection frameworks which just use data from a single source (usually visible images), we adopt multi-source data (visible and thermal infrared images) for a robust detection performance. Since deep learning techniques have shown great performance in object detection, we utilize "You only look once" (YOLO), which is a state-of-the-art real-time object detection framework for automatic vehicle detection. The main contributions of this paper are as follows. (1) Through integrating a thermal infrared imaging sensor and a visible-light imaging sensor on the UAV, we build a multi-source data acquisition system. (2) The rich information from the multi-source data is fully exploited in the proposed detection framework to further improve the accuracy of the detection result.

Keywords: UAV · Vehicle detection · Deep learning
Thermal infrared image

1 Introduction

Unmanned aerial vehicles (UAVs) were first used primarily in military applications. More recently, UAVs have been used as remote sensing tools to present aerial views in both scientific research fields and the civilian domain [1,6]. UAVs are able to fly at lower altitudes and collect images at much higher resolutions than the conventional platforms such as satellites and manned aircraft [16]. Furthermore, UAVs can be used in a variety of scenarios because of the high maneuverability, simple control, security, and reliability. Of course, the wide application is also down to the sensors equipped on the UAVs. Traditional UAVs equipped with visual sensors are widely used in surveying, mapping, and inspection, but both the type and quality of the data have been tremendously enhanced due to the recent advances in sensor technology [7,10]. In particular, infrared sensors have excellent imaging capabilities as the wavelength of infrared sensors is beyond the visible spectrum. Furthermore, the cost of thermal sensors has

J. Yang et al. (Eds.): CCCV 2017, Part III, CCIS 773, pp. 38–49, 2017.
https://doi.org/10.1007/978-981-10-7305-2_4

decreased dramatically [4]. As a result, UAVs equipped with both visible-light cameras and thermal infrared cameras have proved useful in many applications, such as power line inspection, solar panel inspection, search and rescue, precision agriculture, fire fighting, etc.

Vehicle detection is an important object detection application [2,19], and detection from aerial platforms has become a key aspect of autonomous UAV systems in rescue or surveillance missions [3]. Autonomous UAV systems have gained increased popularity in civil and military applications due to their convenience and effectiveness. With the rapid growth in the number of traffic vehicles in recent years, traffic regulation is facing a huge challenge [15], and autonomous location reporting for detected vehicles can alleviate the need for manual image analysis [14]. The methods of vehicle detection are mainly divided into background modeling and methods based on apparent features, but the main difficulty is that the images are influenced by light, viewing angles, and occlusions [5]. In view of the above difficulties, scholars have attempted to use the traditional machine learning methods, but the results, to date, have not matched researchers expectations [3]. In this paper, to address these issues, we focus on a deep learning method based on multi-source data.

In 2013, the region-based convolutional neural network (R-CNN) algorithm took detection accuracy to a new height. Since then, a number of new deep learning methods have been proposed, including Fast-R-CNN, Faster-R-CNN, the single shot multi-box detector (SSD) [11], and You only look once (YOLO). YOLO frames object detection as a regression problem to spatially separated bounding boxes and associated class probabilities [12,13]. The improved YOLOv2 is state-of-the-art on standard detection tasks like PASCAL VOC and COCO. Using a novel, multi-scale training method. YOLOv2 in the mean average precision (mAP) outperforms methods such as Faster-R-CNN with ResNet and SSD, while running significantly faster [13].

This paper makes the following main contributions: (1) we propose a novel multi-source data acquisition system based on the UAV platform; (2) we attempt to detect vehicles based on deep learning using multi-source data obtained by the UAV; and (3) we take the multi-source data into consideration to improve the accuracy of the detection result.

The rest of this paper is organized as follows. An overview of the approach to data acquisition and processing is presented in Sect. 2. Section 3 describes the experimental process and the experimental results. Finally, we conclude the paper and discuss possible future work in Sect. 4.

2 Approach

In the following, we present an overview of the proposed approach for vehicle detection using visible and thermal infrared images. The whole process consists of four main parts: (1) obtain multi-source data through the UAV system; (2) conduct image correction and registration through feature point extraction and homography matrix; (3) integrate the multi-source data by image fusion and band combination; and (4) train the data and detect the vehicles using the YOLO model. A flowchart of the proposed approach is shown in Fig. 1.

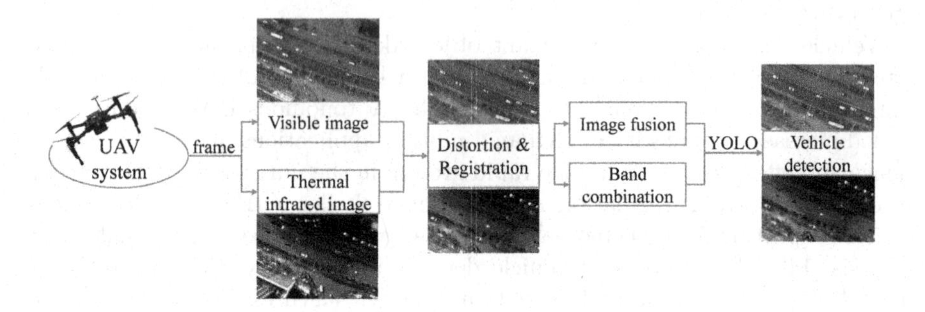

Fig. 1. Flowchart of the proposed approach.

2.1 Multi-source Data Acquisition System

The UAV platform consists of a flight controller, propulsion system, GPS, rechargeable battery, and cameras with high-definition image transmission [9]. The use of both visible and thermal infrared cameras is essential for multi-source data. However, very few aircraft systems support the two kinds of cameras at the same time, and there are problems with the synchronization and transmission of multi-source data. To address these problems, the UAV system for obtaining multi-source data needs two extra cameras and a PC motherboard, a 4G module, a base station, and a lithium battery. The framework of the multi-source acquisition system is shown in Fig. 2.

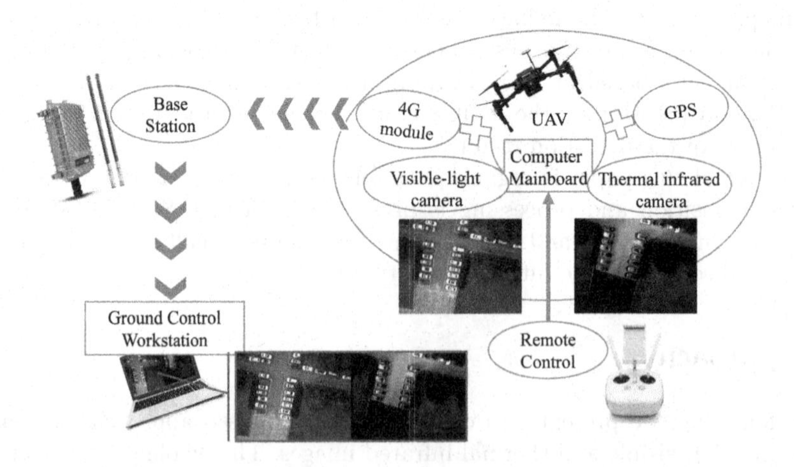

Fig. 2. Multi-source data acquisition system flow diagram.

2.2 Image Correction and Registration

Modern camera lenses always suffer from distortion. In order to alleviate the influence on the image registration caused by image distortion, distortion correction is essential. We adopt a simple checkerboard to correct the distortion of the visible-light camera, but the common checkerboard is ineffective for distortion correction of the images captured by the thermal infrared camera. The reason for this is that the black and white grids appear almost the same in the thermal infrared image. Fortunately, we find a square as shown in Fig. 3(a) which is made of different colors of marbles. Different colors of marbles have different temperatures under sunlight owing to the different reflectivity and absorptivity. Higher-temperature marbles appear as lighter areas in the thermal infrared image, and lower-temperature marbles are the opposite. The square in thermal infrared image is shown in Fig. 3(b). We employ the maximally stable extremal regions (MSER) algorithm to locate the black regions of the calibration area through appropriate thresholds. The thresholds include the size of the area and the average of the region pixels, due to the black points in the image being small. Then, based on the black regions, we calculate the centers of the regions, and we consider the centers as the corner points in the checkerboard. The points extracted are shown in Fig. 3(c). After obtaining the checkerboards for the two kinds of camera, we used the camera calibrator tool in MATLAB to correct the distortion.

Fig. 3. (a) The calibration area in visible image. (b) The calibration area in thermal infrared image. (c) Detecting the points in the calibration area in thermal infrared image.

In order to process the visible image and thermal infrared image, the thermal infrared image should be registered based on the visible image, because the visible image has a higher resolution than the thermal infrared image. In this framework, the registration requires translation, rotation, and scaling. The images have similar patterns, but the pixel intensities are quite different or even inverse in the same region, which contributes to the complexity and difficulty of the feature matching. Therefore, we select the ground control points (GCPs) manually. And the proposed framework matches the multi-source data by homography matrix [17].

2.3 Vehicle Detection

In the proposed detection framework, we detect the vehicles with the multi-source data, so the key point of this paper is how to use the multi-source data to increase the accuracy of detection compared with the result of single-source data. We propose two ways of combining the visible and thermal infrared data for a better detection result: one is image fusion based on image weighted fusion and image band combination; the other is decision fusion based on the result of the detection in the visible and thermal infrared images. The vehicle detection method used in this framework is YOLO.

YOLO. YOLO runs a single convolutional network to predict multiple bounding boxes and class probabilities for those boxes. The network of YOLO uses features from the entire image to predict each bounding box, and predicts all the bounding boxes across all the classes. The system models the detection as the following process [12]:

1. Divide the input image into an $S \times S$ grid.
2. Predict B bounding boxes and confidence scores for those boxes in each grid cell. The confidence is defined as $Pr(Object) * IOU_{\text{pred}}^{\text{truth}}$(intersection over union), and each bounding box contains five predictions: x,y,w,h, and $confidence$, where (x, y, w, h) represent the center of the box and the width and height of the box.
3. Predict C conditional class probabilities in each grid cell, $Pr(Class_i|Object)$, which are conditioned on the grid cells containing an object.
4. Multiply the class probabilities and confidence at the test time, which indicates the specific class confidence scores for each box [8,12].

Through the above process, YOLO learns generalizable representations of objects from the entire image, and is extremely fast. We therefore use YOLO as the vehicle detection method in the proposed framework. In addition, we enhance YOLO by expanding the types of input image, making it possible to support not only natural images but also multi-band images.

Image Fusion

Weighted fusion. In order to add the information of the thermal infrared image to the information of the visible image, we adopt simple weighted fusion, which is called the weighted averaging (WA) method. This is the simplest and most direct image fusion method. In the thermal infrared image, the vehicle and the background have great differences in pixel values. Fusing the visible image with the pixel values of the thermal infrared image can restrain the complex background to a certain extent, so the weighted fusion we apply is straightforward and efficient. The implementation of weighted fusion is simple, as shown in the following formula:

$$B_i = w * V_i + (1 - w) * I, i = 1, 2, 3 \tag{1}$$

where B, V, and I represent the new image with three bands, the visible image, and the thermal infrared image, respectively, w represents the weights, and i represents the band number.

Band combination. We combine the image bands in order to learn more image features in the deep learning framework. The visible image has three bands, and the thermal infrared image has one band. By combining the bands of the two images, the new image has four bands, which provide richer image information.

Decision Fusion. The detection results of specific vehicles in the multi-source data are usually different; in other words, the detection boxes in the visible and thermal infrared images are different, and the redundant or complementary results provided by the multi-source data can actually be aggregated. Therefore, we adopt decision fusion to combine the detection results based on two boxes: $box_1 = (x_1, y_1, w_1, h_1, p_1)^T$; $box_2 = (x_2, y_2, w_2, h_2, p_2)^T$. Where (x, y) is the center coordinate of the box, (w, h) are the width and height of the box, and p represents the confidence score of the box. The new box ($box_{new} = (x_{new}, y_{new}, w_{new}, h_{new})^T$) is calculated as the formula (2). An example of the detection box fusion is shown in Fig. 4. In addition, we need to take the overlapping areas of the detection boxes into consideration. We therefore employ a threshold to judge whether to combine the two boxes or not.

$$\begin{bmatrix} x_{new} \\ y_{new} \\ w_{new} \\ h_{new} \end{bmatrix} = \frac{p_1}{p_1 + p_2} * \begin{bmatrix} x_1 \\ y_1 \\ w_1 \\ h_1 \end{bmatrix} + \frac{p_1}{p_1 + p_2} * \begin{bmatrix} x_2 \\ y_2 \\ w_2 \\ h_2 \end{bmatrix} \quad (2)$$

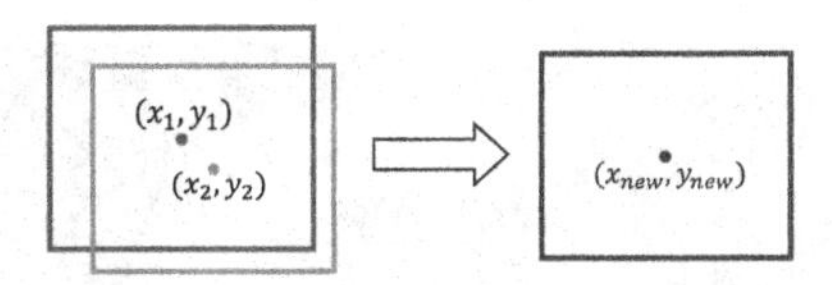

Fig. 4. Detection boxes fusion.

3 Experiments and Analysis

We obtained the multi-source data sets from the UAV acquisition system. The data sets were then preprocessed and divided into training data sets and test data sets. We then trained the data in the deep learning framework. Finally, we used the trained model to detect the vehicles in the test data sets, and we compared the vehicle detection results in the different kinds of data sets.

3.1 Data Collection

We used a DJI MATRICE 100 quadcopter as the UAV, for its user-friendly control system and flexible platform. We equipped the UAV with two cameras for the multi-source data. One camera was a USB-connected industrial digital camera with a pixel size of 5.0 μm × 5.2 μm. The other was a card-type infrared camera with the wavelength range 8–14 μm. The developed image preservation system saved the multi-source data simultaneously in video mode.

After obtaining the video from the UAV system, we selected some frames at intervals to ensure the diversity in the data set. We then obtained the visible and thermal infrared images from the video frames by cropping (the size of the video frame was 1280 × 960, and the size of each image was 640 × 480).

3.2 Data Pre-processing

The cameras carried on the UAV system suffered from distortion. We therefore used a big checkerboard to correct the visible-light camera, and the square in previous introduction to correct the distortion of the thermal infrared camera. By selecting the GCPs manually, we used the homography matrix to complete the registration. We obtained the transformation relationship for every pixel between the two images based on the registration result. Thanks to the fixed cameras and the invariant model, we employed the transformation relationship for all the images.

The size of all the images was 640 × 480. After the registration, the thermal infrared images had invalid areas, so we cropped the images based on the infrared images. The size of the cropped images was then 500 × 280. A set of images is shown as an example in Fig. 5, where it can be seen that the registration result basically meets the processing requirements.

Fig. 5. (a) Visible image. (b) Registered thermal infrared image.

Based on the visible image data set and the thermal infrared image data set, we undertook the weighted image fusion with different weight values: $w_1 = 0.7$; $w_2 = 0.8$; $w_3 = 0.9$. We then prepared the band combination data, which contained three bands of the visible image and one band of the infrared image. We thus obtained the six data sets.

The next step was to select the training data and the test data. We labeled the vehicles in the images. The labels contained five parameters $(class, x, y, w, h)$. The first parameter expressed the class, (x, y) was the top-left corner of the vehicle, and (w, h) was the width and height of the vehicle. In the experiment, we labeled 1000 vehicles for the training data and 673 vehicles for the test data in every data set. Some typical samples from the visible data set and the thermal infrared data set are shown in Fig. 6.

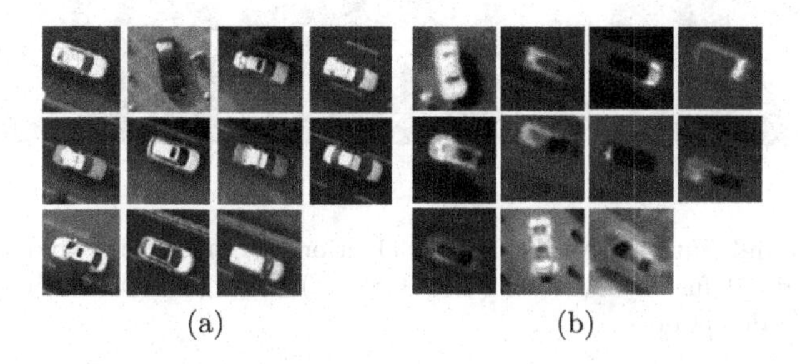

Fig. 6. (a) Tagged samples from the visible image. (b) Tagged samples from the thermal infrared image.

3.3 Model Training for Vehicle Detection

We used YOLO to train the six data sets, with a batch size of 15, a momentum of 0.9, a weight decay of 0.0005, a learning rate of 0.00005, and the input images resized to 448×448. For the six data sets, we tried to make sure that every data set was convergent and saved the training results of every data set.

3.4 Vehicle Detection

Based on the training results, we used the training modules to detect the vehicles in the six test data sets, with each data set having its own training module. We used the same threshold ($threshold = 0.5$) for each data set, which is defined as the confidence. This gave us the vehicle confidence scores for each box.

To visualize the detection performance, we take two test images from the six data sets as an example. Specifically, the first image has 6 vehicles in the ground truth shown in Fig. 7(a), and the second image has 20 vehicles in the ground truth shown in Fig. 8(a). The results of the vehicle detection are shown in Figs. 7(b)–(f) and 8(b)–(f), and the detection result of the decision fusion image is shown in three bands of the four bands.

As shown in the Figs. 7 and 8, the vehicles labelled with 3, 4 in Fig. 7(a) are not detected vehicles in visible image, But they are detected vehicles in Fig. 7(c). And the vehicles labelled with 1,2, 3, 4, 5, 6, 7 in Fig. 8(a) are not detected vehicles in visible image. But some of these had been detected with other strategies. For example, the vehicles labelled with 2, 3, 6 are detected vehicles in weighted fusion image.

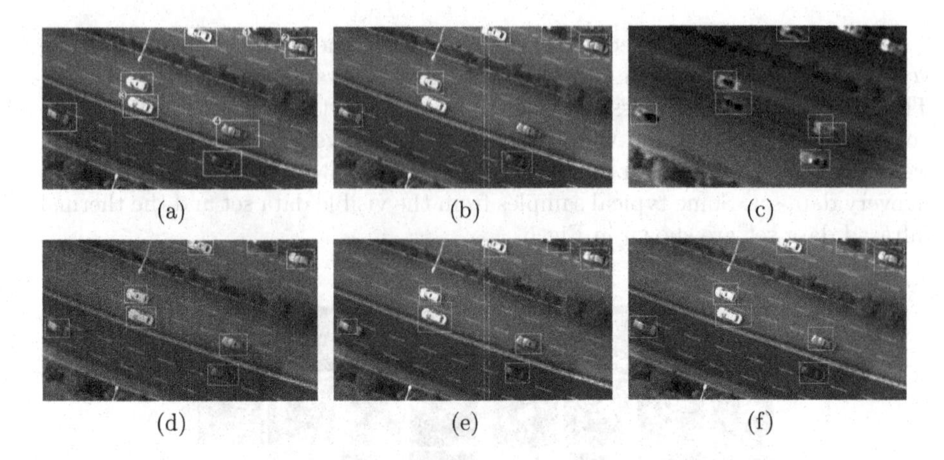

Fig. 7. (a) Ground truth. (b) Visible image detection (c) Thermal infrared image detection. (d) Weighted fusion detection ($w = 0.8$). (e) Band combination detection. (f) Decision fusion detection.

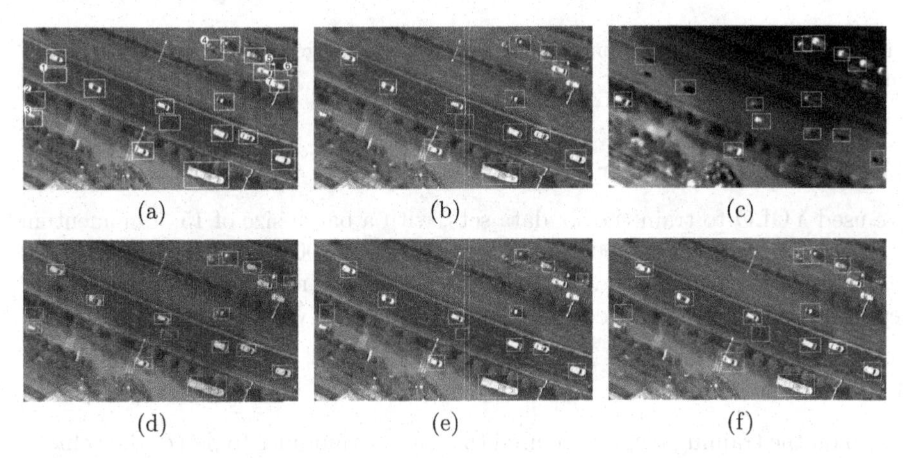

Fig. 8. (a) Ground truth. (b) Visible image detection (c) Thermal infrared image detection. (d) Weighted fusion detection ($w = 0.8$). (e) Band combination detection. (f) Decision fusion detection.

3.5 Comparative Experiment

To evaluate the vehicle detection performance, four commonly used criteria were computed: the false positive rate (FPR), the missing ratio (MR), accuracy (AC), and error ratio (ER). These criteria are defined as follows:

$$FPR = \frac{Number\ of falsely\ detected\ vehicle}{Number\ of\ detected\ vehicle} \times 100\% \tag{3}$$

$$MR = \frac{Number\ of\ missing\ vehicle}{Number\ of\ vehicle} \times 100\% \tag{4}$$

$$AC = \frac{Number\ of\ detected\ vehicle}{Number\ of\ vehicle} \times 100\% \tag{5}$$

$$ER = FPR + MR \tag{6}$$

For every detected vehicle, if it had an intersection overlap with a test vehicle of greater than 0.5, we considered it to be a true detected vehicle. If a test vehicle was not found within the detected vehicle with an intersection overlap of greater than 0.5, we considered it to be a missing vehicle. Table 1 shows the detection performance of the different data sets [18].

Table 1. Comparison of the different data sets (test vehicle sample sets: 673)

Data	FPR	MR	AC	ER
Visible image	3.29%	14.26%	82.91%	17.56%
Thermal infrared image	1.57%	33.73%	65.23%	35.30%
Weighted fusion ($w = 0.8$)	4.24%	5.35%	90.64%	9.59%
Band combination	1.54%	12.93%	85.74%	14.46%
Decision fusion	3.65%	6.39%	90.19%	10.04%

Based on the experimental results, we can conclude that the proposed detection framework shows a good vehicle detection performance. The detection is less effective in the thermal infrared data set due to the lower resolution. Not only can the weighted fusion and band combination data sets improve the accuracy of the vehicle detection result, to a certain degree, but the decision fusion using two base image data sets can also increase the accuracy of the detection. The weighted fusion was best able to improve the accuracy of the vehicle detection in these data sets when the weight was set to 0.8.

4 Conclusions

In this study, our main aim was to design a small UAV system with two types of cameras to obtain multi-source data. By using a deep learning framework to detect vehicles in the visible and thermal infrared images which have been corrected and registered, the results show that the framework is an efficient way to detect vehicles. The proposed framework adopts weighted fusion, band combination, and decision fusion methods to use the multi-source data. The experimental results show that the addition of the thermal infrared image data set can improve the accuracy of the vehicle detection.

In the future, we will try to obtain extreme weather data, and find more efficient methods to increase the accuracy of the image registration. Moreover, we will explore more possibilities of using multi-source data to further improve the accuracy of vehicle detection.

References

1. Breckon, T.P., Barnes, S.E., Eichner, M.L., Wahren, K.: Autonomous real-time vehicle detection from a medium-level UAV. In: Proceedings of 24th International Conference on Unmanned Air Vehicle Systems, p. 29–1. sn (2009)
2. Chen, X., Xiang, S., Liu, C.L., Pan, C.H.: Vehicle detection in satellite images by parallel deep convolutional neural networks. In: 2013 2nd IAPR Asian Conference on Pattern Recognition (ACPR), pp. 181–185. IEEE (2013)
3. Chen, X., Xiang, S., Liu, C.L., Pan, C.H.: Vehicle detection in satellite images by hybrid deep convolutional neural networks. IEEE Geosci. Remote Sens. Lett. **11**(10), 1797–1801 (2014)
4. Dai, C., Zheng, Y., Li, X.: Layered representation for pedestrian detection and tracking in infrared imagery. In: IEEE Computer Society Conference on Computer Vision and Pattern Recognition-Workshops 2005, CVPR Workshops, p. 13. IEEE (2005)
5. Gaszczak, A., Breckon, T.P., Han, J.: Real-time people and vehicle detection from UAV imagery (2011)
6. Gleason, J., Nefian, A.V., Bouyssounousse, X., Fong, T., Bebis, G.: Vehicle detection from aerial imagery. In: 2011 IEEE International Conference on Robotics and Automation (ICRA), pp. 2065–2070. IEEE (2011)
7. Han, J., Zhang, D., Cheng, G., Guo, L., Ren, J.: Object detection in optical remote sensing images based on weakly supervised learning and high-level feature learning. IEEE Trans. Geosci. Remote Sens. **53**(6), 3325–3337 (2015)
8. Redmon, J.: YOLO: real-time object detection. https://pjreddie.com/darknet/yolo/
9. Kaaniche, K., Champion, B., Pégard, C., Vasseur, P.: A vision algorithm for dynamic detection of moving vehicles with a UAV. In: Proceedings of the 2005 IEEE International Conference on Robotics and Automation, 2005, ICRA 2005, pp. 1878–1883. IEEE (2005)
10. Li, Z., Liu, Y., Hayward, R., Zhang, J., Cai, J.: Knowledge-based power line detection for UAV surveillance and inspection systems. In: 23rd International Conference on Image and Vision Computing New Zealand 2008, IVCNZ 2008, pp. 1–6. IEEE (2008)
11. Liu, W., Anguelov, D., Erhan, D., Szegedy, C., Reed, S., Fu, C.-Y., Berg, A.C.: SSD: single shot multibox detector. In: Leibe, B., Matas, J., Sebe, N., Welling, M. (eds.) ECCV 2016. LNCS, vol. 9905, pp. 21–37. Springer, Cham (2016). https://doi.org/10.1007/978-3-319-46448-0_2
12. Redmon, J., Divvala, S., Girshick, R., Farhadi, A.: You only look once: unified, real-time object detection. In: Proceedings of the IEEE Conference on Computer Vision and Pattern Recognition, pp. 779–788 (2016)
13. Redmon, J., Farhadi, A.: YOLO9000: better, faster, stronger. arXiv preprint arXiv:1612.08242 (2016)
14. Rodríguez-Canosa, G.R., Thomas, S., del Cerro, J., Barrientos, A., MacDonald, B.: A real-time method to detect and track moving objects (DATMO) from unmanned aerial vehicles (UAVs) using a single camera. Remote Sens. **4**(4), 1090–1111 (2012)
15. Sun, Z., Bebis, G., Miller, R.: On-road vehicle detection: a review. IEEE Trans. Pattern Anal. Mach. Intell. **28**(5), 694–711 (2006)
16. Turner, D., Lucieer, A., Watson, C.: Development of an unmanned aerial vehicle (UAV) for hyper resolution vineyard mapping based on visible, multispectral, and thermal imagery. In: Proceedings of 34th International Symposium on Remote Sensing of Environment, p. 4 (2011)

17. Ueshiba, T., Tomita, F.: Plane-based calibration algorithm for multi-camera systems via factorization of homography matrices. In: null, p. 966. IEEE (2003)
18. Zhang, F., Du, B., Zhang, L., Xu, M.: Weakly supervised learning based on coupled convolutional neural networks for aircraft detection. IEEE Trans. Geosci. Remote Sens. **54**(9), 5553–5563 (2016)
19. Zhao, T., Nevatia, R.: Car detection in low resolution aerial images. Image Vis. Comput. **21**(8), 693–703 (2003)

Moving Object Detection via Integrating Spatial Compactness and Appearance Consistency in the Low-Rank Representation

Minghe Xu, Chenglong Li, Hanqin Shi, Jin Tang, and Aihua Zheng(✉)

School of Computer Science and Technology, Anhui University,
No. 111 Jiulong Road, Hefei 230601, China
{xuminghe001,lcl1314,shihq726}@foxmail.com, {tj,ahzheng214}@ahu.edu.cn

Abstract. Low-rank and sparse separation models have been successfully applied to background modeling and achieved promising results on moving object detection. It is still a challenging task in complex environment. In this paper, we propose to enforce the spatial compactness and appearance consistency in the low-rank and sparse separation framework. Given the data matrix that accumulates sequential frames from the input video, our model detects the moving objects as sparse outliers against the low-rank structure background. Furthermore, we explore the spatial compactness by enforcing the consistency among the pixels within the same superpixel. This strategy can simultaneously promote the appearance consistency since the superpixel is defined as the pixels with homogenous appearance nearby the neighborhood. The extensive experiments on public GTD dataset suggest that, our model can better preserve the boundary information of the objects and achieves superior performance against other state-of-the-arts.

Keywords: Low-rank representation · Smoothness constraint
Spatial compactness · Appearance consistency

1 Introduction

Moving object detection is a fundamental problem in video analysis, and plays a critical role in numerous vision applications, such as intelligent transportation [1], vehicle navigation [25] and scene understanding [17]. Over the years, many approaches have been proposed for moving object detection while background subtraction has been recognized as one of the most competitive approaches.

Conventional background modelling methods include single Gaussian distribution [23], Mixture of Gaussian [7,21], and their variations, VIBE [2] and fuzzy concepts based methods [3]. However, these methods model the background for each pixel independently and lack of the relations between the consecutive frames, thus they are very sensitive to noises and occlusions.

M. Xu—Master student.

J. Yang et al. (Eds.): CCCV 2017, Part III, CCIS 773, pp. 50–60, 2017.
https://doi.org/10.1007/978-981-10-7305-2_5

Recently, the low-rank and sparse separation framework has emerged by decomposing the video sequence into low-rank background and sparse foregrounds (moving objects). One pioneering work is Robust Principal Component Analysis (RPCA) [12,14,22], which decomposes a given matrix/frames into a low-rank background matrix and sparse foreground matrix. Candès et al. [6] proposed to recover the low-rank and sparse components individually by solving a convenient convex program called Principal Component Pursuit (PCP). Zhou et al. [27] proposed to handle both small entrywise noises and gross sparse errors. Dou et al. [10] proposed a incremental learning based LRR model using K-SVD for dictionary learning. Zhou et al. [26] proposed to relax the requirement of sparse and random distribution of corruption by preserving l_0-penalty and modeling the spatial contiguity of the sequence. In order to enforce the appearance consistency onto the spatial neighboring relationship, Xin et al. [24] introduced the intensity similarities to the neighboring pixels via regularization terms for both the foreground and background matrices. However, these methods constructed the graph only based on pixel level which ignored the spatial compactness. Recently, Javed et al. [11] proposed a superpixel-based online matrix decomposition method which separate the low-rank background and sparse foregrounds on the superpixel level. However, the performance may excessively rely on the superpixel prior which often produces unfaithful segmentation.

As we observed, the objects are generally spatially compact and consistent in appearance which means the pixels in the same concept of spatial region with close appearance tend to belong to the same pattern (foreground/background). Based on this observation, our main effort is to explore the spatial compactness and the appearance consistency of the objects based on the general framework of low-rank and sparse separation. Specifically, we first encourage the appearance consistency for the object by weighting the neighboring pixel pairs with the appearance similarity. Furthermore, we enforce the global spatial compactness on the superpixel level by constructing the informative graphs for the pixels within the same superpixel. Noted that the superpixel strategy can also promote the appearance consistency since a superpixel is defined as the perceptually consistent unit in appearance.

2 Our Approach

In this section, we will present our model by elaborating the enforcement of spatial compactness and appearance consistency in the low-rank and sparse separation framework, followed by the alternating optimization algorithm.

2.1 Problem Formulation

In this paper, we formulate the problem of foreground detection as a low-rank and sparse separation model. A video sequence $\mathbf{D} = [\mathbf{f}_1, \mathbf{f}_2, \ldots, \mathbf{f}_n] \in \mathbb{R}^{m \times n}$ is composed of n frames by of m pixels per frame. $\mathbf{B} \in \mathbb{R}^{m \times n}$ is a background matrix, which denotes the underlying background images. Our goal is to discover the object mask $\mathbf{S}$ from data matrices $\mathbf{D}$, where $\mathbf{S}_{ij}$ is a binary matrix:

$$\mathbf{S}_{ij} = \begin{cases} 0, & \text{if } ij \text{ is background,} \\ 1, & \text{if } ij \text{ is foreground.} \end{cases} \tag{1}$$

We assume that the underlying background images are linearly correlated and the foregrounds are sparse and contiguous, which has been successfully applied in background modeling [16,26]. Furthermore, for the background region where $\mathbf{S}_{ij} = 0$, we assume that $\mathbf{D}_{ij} = \mathbf{B}_{ij} + \epsilon_{ij}$, where ϵ_{ij} denotes i.i.d. Gaussian noise. Based on the above assumptions, we have:

$$\begin{aligned} &\min_{\mathbf{B}, \mathbf{S}_{ij} \in \{0,1\}} \quad \alpha \| vec(\mathbf{S}) \|_0 \\ &s.t. \ \mathbf{S}_\perp \circ \mathbf{D} = \mathbf{S}_\perp \circ (\mathbf{B} + \epsilon), \ rank(\mathbf{B}) \leq r, \end{aligned} \tag{2}$$

where α is a penalized factor, and $||\mathbf{X}||_0$ indicates the l_0 norm of a vector. The operator "$\circ$" denotes element-wise multiplication of two matrices, $\mathbf{S}_\perp$ denotes the region of $\mathbf{S}_{ij} = 0$, and r is a constant that suppresses the complexity of the background model.

Appearance consistency. Due to the non-convexity of l_0 norm of the matrix $\mathbf{S}$, a common practice is to introduce a contiguous constraint to form a MRF [8] model which can be solved by graph cuts [4,15]. In order to preserve the spatial smoothness of the objects, [16,26] constructed the graph based on the neighboring pixels. However, it is necessary to enforce the appearance similarity onto the neighboring pixels for the informative graphs [11,24]. Therefore, we construct the smoothness by:

$$||\mathbf{C}\ vec(\mathbf{S})||_1 = \sum_{(ij,kl)\in\varepsilon} w_{ij,kl}\ |\mathbf{S}_{ij} - \mathbf{S}_{kl}|, \tag{3}$$

where, $||\mathbf{X}||_1 = \sum_{\mathbf{ij}} |\mathbf{X_{ij}}|$ denotes the l_1-norm, ε denotes the edge set connecting spatially neighboring pixels, $(ij, kl) \in \varepsilon$ when pixel ij and kl are spatially connected. $\mathbf{C}$ is the node-edge incidence matrix denoting the connecting relationship among pixels, and $vec(\mathbf{S})$ is a vectorize operator on matrix $\mathbf{S}$. Among them, consider the first term $||\mathbf{C}\ vec(\mathbf{S})||_1$ in Eq. (7) represents the difference between the adjacent pixels $w_{ij,kl}$ indicates the adaptive weighting factor between the pixels and is defined as:

$$w_{ij,kl} = \exp\frac{-||d_{ij} - d_{kl}||_2^2}{2\sigma^2} \tag{4}$$

where d_{ij} and d_{kl} represent the intensity of pixel ij and kl respectively and σ is a tunning parameter. Based on this construction, as shown in Fig. 1(a), the higher probability that a pair of pixels belongs to the same segment (with close intensity), the stronger correlation between this pair, which can further enforce the appearance consistency between neighboring pixels.

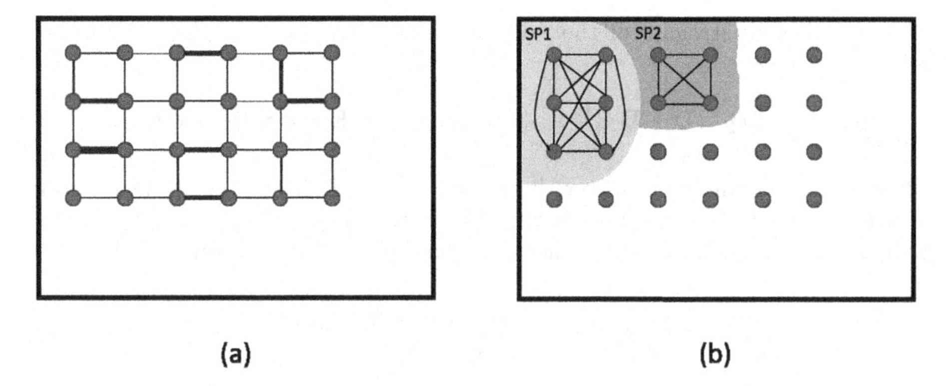

Fig. 1. Illustration of generating the informative graphs. (a) Constructing the weighted graphs for the neighboring pixels where the thicker links between pixel pairs indicate the higher appearance similarity. (b) Constructing graphs between the pixel pairs within the same superpixel. (Color figure online)

Spatial compactness. It is observed that, the pixels from the same superpixel, which is a perceptually consistent unit in color and texture, are basically derived from the same concept (background/foreground). In order to enforce this spatial compactness, we further construct the fully connected graph between the pixels within each superpixel (as shown in Fig. 1(b)) generated by the lazy random walks (LRW) [20] and introduce the spatial compactness into the model via:

$$||\mathbf{A}\ vec(\mathbf{S})||_1 = \sum_{(ij,pq)\in\mathcal{N}} |\mathbf{S}_{ij} - \mathbf{S}_{pq}|, \tag{5}$$

where, $\mathcal{N}$ indicates edge set connecting all the pixel pairs within each superpixel and $\mathbf{A}$ is the node-edge incidence matrix denoting the connecting relationship among pixels. It can also promote the appearance consistency since the superpixel consists of the consistent unit in color and texture.

As concluded, we can integrate our formulation by enforcing the spatial compactness and appearance consistency into Eq. (2) as:

$$\begin{aligned} &\min_{\mathbf{B},\mathbf{S}_{ij}\in\{0,1\}} \alpha\|vec(\mathbf{S})\|_0 + \mu||\mathbf{E}\ vec(\mathbf{S})||_1, \\ &s.t.\ \mathbf{S}_\perp \circ \mathbf{D} = \mathbf{S}_\perp \circ (\mathbf{B} + \epsilon),\ rank(\mathbf{B}) \le r, \end{aligned} \tag{6}$$

with:

$$||\mathbf{E}\ vec(\mathbf{S})||_1 = \beta||\mathbf{C}\ vec(\mathbf{S})||_1 + \gamma||\mathbf{A}\ vec(\mathbf{S})||_1, \tag{7}$$

where μ, β and γ are tuning parameters.

2.2 Model Optimization

Equation (6) is a NP-hard problem, to make Eq. (6) tractable, we relax the rank operator on $\mathbf{B}$ with the nuclear norm, the nuclear norm has proven to be an effective convex surrogate of the rank operator [19]. Therefore, Eq. (6) can be reformulated as:

$$\min_{\mathbf{B},\mathbf{S}_{ij}\in\{0,1\}} \frac{1}{2} ||P_{\mathbf{S}_\perp}(\mathbf{D}-\mathbf{B})||_F^2 + \alpha\|vec(\mathbf{S})\|_0 + \mu||\mathbf{E}\ vec(\mathbf{S})||_1 + \lambda\|\mathbf{B}\|_*, \quad (8)$$

where λ is a balance parameter. $||\cdot||_*$ and $||\cdot||_F$ indicate the nuclear norm of a matrix and the Frobenius norm of a matrix, respectively. $P_{\mathbf{S}_\perp}(\mathbf{X})$ is the complement to $P_{\mathbf{S}}(\mathbf{X})$ which is the orthogonal projection of matrix $\mathbf{X}$ denoted by:

$$P_{\mathbf{S}}(\mathbf{X})(i,j) = \begin{cases} 0, & if\ \mathbf{S}_{ij} = 0, \\ \mathbf{X}_{ij}, & if\ \mathbf{S}_{ij} = 1, \end{cases} \quad (9)$$

Therefore, we adopt an alternating algorithm by separating Eq. (8) over $\mathbf{B}$ and $\mathbf{S}$ in the following two steps.

B− ***subproblem.*** Given an current estimate of the foreground mask $\hat{\mathbf{S}}$, estimating $\mathbf{B}$ by minimizing Eq. (8) turns out to be the matrix completion problem. This is to learn a low-rank background matrix from partial observations.

$$\min_{\mathbf{B}} \frac{1}{2} ||P_{\hat{\mathbf{S}}_\perp}(\mathbf{D}-\mathbf{B})||_F^2 + \lambda\|\mathbf{B}\|_*, \quad (10)$$

The optimal $\mathbf{B}$ in Eq. (13) can be computed by the SOFT-IMPUTE [18] algorithm. Which based on the following Lemma [5]:

Lemma 1. Given a matrix $\mathbf{Z}$, the solution to the optimization problem

$$\min_{\mathbf{X}} \frac{1}{2} ||\mathbf{Z}-\mathbf{X}||_F^2 + \lambda\|\mathbf{X}\|_*, \quad (11)$$

is given by $\hat{\mathbf{X}} = \Theta_\lambda(\mathbf{Z})$, where Θ_λ means the singular value thresholding

$$\Theta_\lambda(\mathbf{Z}) = \mathbf{U}\Sigma_\lambda\mathbf{V}^T, \quad (12)$$

Here, $\Sigma_\lambda = diag[(d_1-\lambda)_+, \cdots, (d_r-\lambda)_+]$, $\mathbf{U}\Sigma_\lambda\mathbf{V}^T$ is the SVD of $\mathbf{Z}$, $\Sigma = diag[d_1 - d_r]$ and $t_+ = max(t, 0)$. Rewriting Eq. (10), we have:

$$\begin{aligned} &\min_{\mathbf{B}} \frac{1}{2} ||P_{\hat{\mathbf{S}}_\perp}(\mathbf{D}-\mathbf{B})||_F^2 + \lambda\|\mathbf{B}\|_*, \\ &= \min_{\mathbf{B}} \frac{1}{2} ||[P_{\hat{\mathbf{S}}_\perp}(\mathbf{D}) + P_{\hat{\mathbf{S}}}(\mathbf{B})] - \mathbf{B})||_F^2 + \lambda\|\mathbf{B}\|_*, \end{aligned} \quad (13)$$

According to Lemma 1, given an arbitrary initialization $\hat{\mathbf{B}}$, the optimal solution can be obtained by iteratively using Eq. (14):

$$\hat{\mathbf{B}} \longleftarrow \Theta_\lambda(P_{\hat{\mathbf{S}}_\perp}(\mathbf{D}) + P_{\hat{\mathbf{S}}}(\hat{\mathbf{B}})), \quad (14)$$

S− ***subproblem.*** Given an current estimate of the background position matrix $\hat{\mathbf{B}}$, Eq. (8) can be transferred into following optimization functions:

$$\min_{\mathbf{S}} \frac{1}{2} ||P_{\mathbf{S}_\perp}(\mathbf{D}-\hat{\mathbf{B}})||_F^2 + \alpha\|vec(\mathbf{S})\|_0 + \mu||\mathbf{E}\ vec(\mathbf{S})||_1, \quad (15)$$

Algorithm 1. Optimization Algorithm to Eq. 8

Require: $\mathbf{D} = [\mathbf{l}_1, \mathbf{l}_2, \ldots, \mathbf{l}_n] \in \mathbb{R}^{m \times n}$.
Set $\mathbf{B} = \mathbf{D}$, $\mathbf{S} = \mathbf{0}$, $\tau = 1e-4$, $maxIter = 20$.
Ensure: $\hat{\mathbf{S}}, \hat{\mathbf{B}}$.
1: Using SOFT-IMPUTE algorithm to optimize energy function Eq. (10), by computing $\hat{\mathbf{B}}$: $\hat{\mathbf{B}} \longleftarrow \Theta_\lambda(P_{\hat{\mathbf{S}}_\perp}(\mathbf{D}) + P_{\hat{\mathbf{S}}}(\hat{\mathbf{B}}))$
2: **if** $rank(\hat{\mathbf{B}}) \leq K$ **then**
3: tuning parameters λ, returns run to step 1.
4: **end if**
5: Using graph cuts algorithm to optimize energy function Eq. (15) by computing $\hat{\mathbf{S}}$: $\hat{\mathbf{S}} = arg\ min_{\mathbf{S}} \sum_{i,j}(\alpha - \frac{1}{2}(\mathbf{D}_{ij} - \hat{\mathbf{B}}_{ij}))^2 \mathbf{S}_{ij} + \mu||\mathbf{E}\ vec(\mathbf{S})||_1$
6: Check the convergence condition: if the maximum objective change between two consecutive iterations is less than τ or the maximum number of iterations reaches $maxIter$, then terminate the loop.

The energy function Eq. (15) can be rewritten in line with the standard form of a first-order Markov Random Fields [8] as:

$$
\begin{aligned}
&\frac{1}{2}\ ||P_{\mathbf{S}_\perp}(\mathbf{D} - \hat{\mathbf{B}})||_F^2 + \alpha||vec(\mathbf{S})||_0 + \mu||\mathbf{E}\ vec(\mathbf{S})||_1 \\
&= \frac{1}{2}\sum_{i,j}(\mathbf{D}_{ij} - \hat{\mathbf{B}}_{ij})^2(1 - \mathbf{S}_{ij}) + \alpha\sum_{i,j}\mathbf{S}_{ij} + \mu||\mathbf{E}\ vec(\mathbf{S})||_1, \\
&= \sum_{i,j}(\alpha - \frac{1}{2}(\mathbf{D}_{ij} - \hat{\mathbf{B}}_{ij}))^2\mathbf{S}_{ij} + \mu||\mathbf{E}\ vec(\mathbf{S})||_1 + \frac{1}{2}\sum_{i,j}(\mathbf{D}_{ij} - \hat{\mathbf{B}}_{ij})^2.
\end{aligned}
\tag{16}
$$

When $\hat{\mathbf{B}}$ is fixed, $\frac{1}{2}\sum_{i,j}(\mathbf{D}_{ij} - \hat{\mathbf{B}}_{ij})^2$ is constant. Meanwhile, $\mathbf{S}_{ij}$ beside the $(\beta - \frac{1}{2}(\mathbf{D}_{ij} - \hat{\mathbf{B}}_{ij}))^2$ is also constant. Known Markov unary term and pairwise smoothing term, one can easily obtain the optimal foreground matrix though graph cuts method [4,15] since $\mathbf{S}_{ij} \in \{0, 1\}$ is discrete.

A sub-optimal solution can be obtained by alternating optimizing $\mathbf{B}$ and $\mathbf{S}$ and the algorithm is summarised in Algorithm 1.

3 Experiments

We evaluate our method against the state-of-the-arts on the public challenging GTD dataset [16]. It consists of 25 video sequence pairs in both visual and thermal modality. In this paper, we evaluate the proposed method on visual modality videos. The GTD dataset [16] contains fifteen different scenes and various challenges including intermittent motion, low illumination, bad weather, intense shadow, dynamic scene and background clutter etc.

3.1 Evaluation Settings

Parameters. In our model of Eq. (8), the parameter λ controls the complexity of the background model which is first roughly estimated by the rank of the background model. The parameter α which controls the sparsity of the foreground masks is set as $\alpha = 16.2\sigma^2$, where σ^2 is estimated online by the mean variance of $\{\mathbf{D_{ij}} - \hat{\mathbf{B_{ij}}}\}$. The parameter μ controls spatial smoothness between pixels that satisfies the constructed informative graphs, and is set as $\mu = 0.205$. The parameter β and γ control the relative contribution of each term in Eq. (7), respectively. We determine β and γ by adjusting its ratio to α, and empirically set as $\{\beta, \gamma\} = \{2.7\alpha, 0.13\alpha\}$. Moreover, we set $\sigma = 25$ in Eq. (3), and set the number of superpixel patches $\mathcal{A} = 650$.

Evaluation Criterion. The Precision, Recall, F-measure are first comprehensively evaluated, which are defined as following:

$$\begin{aligned} \text{Precision} &= \frac{\text{TP}}{\text{TP} + \text{FP}}, \\ \text{Recall} &= \frac{\text{TP}}{\text{TP} + \text{FN}}, \\ \text{F-measure} &= 2\frac{\text{Precision} \times \text{Recall}}{\text{Precision} + \text{Recall}}. \end{aligned} \tag{17}$$

where TP = True Positives, indicating the foreground pixels correctly labeled as foreground. FP = False Positives, referring the background pixels incorrectly labeled as foreground. TN = True Negatives, corresponding to background pixels correctly labeled as background. FN = False Negatives, referring to foreground pixels incorrectly labeled as background [9]. F-measure is a comprehensive measurement to balance the argument between precision and recall.

Furthermore, the Mean Absolute Error (MAE) is evaluated to measure the disagreement between the detected results and the groundtruth:

$$MAE = \frac{1}{N \times \mathcal{F}} \sum_{i=1}^{\mathcal{F}} \sum_{p \in DR, \acute{p} \in GT} XOR(p, \acute{p}) \tag{18}$$

where N denotes and resolution of the frame and $\mathcal{F}$ denotes the number of the frames in the video clip. DR and GT indicate the "Detection Result" and the "Ground Truth" respectively. $XOR(*)$ denotes the logic operator "*exclusive OR*". $p, \acute{p} \in \{0, 1\}$ denotes the background/foreground pixels.

3.2 Comparison Results

We compare our approach with four state-of-the-art moving object detection algorithms including DECOLOR [26], GMM [13], VIBE [2] and PCP [6]. To keep things fair, we choose the default parameters released by the authors for corresponding methods.

Qualitative Results. Figure 2 demonstrates the detected results on a certain frame of six video clips from GTD dataset [16]. From which we can see, our method can produce finer boundary information and better suppress the influence of the noise.

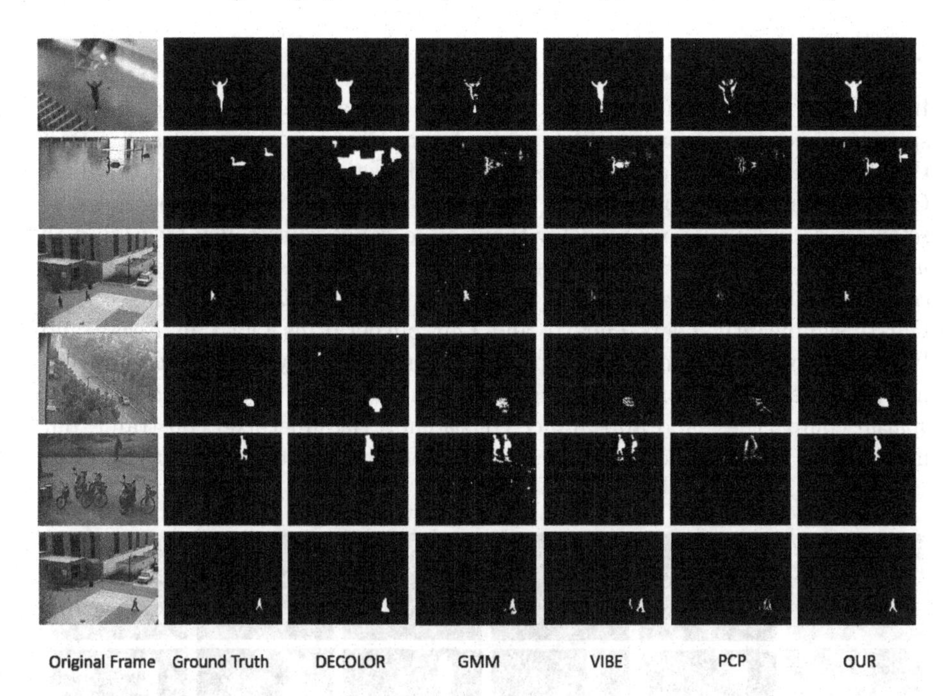

Fig. 2. Sample results of our method against the state-of-the-arts on six video sequences from GTD dataset.

Quantitative Results. Table 1 reports precision, recall, F-measure, and MAE on public GTD dataset [16]. We can see our method significantly outperforms the state-of-the-arts in precision, F-measure, and MAE. Although the recall of our method looks lower than DECOLOR [26], from Fig. 2 we can see, DECOLOR [26] tends to produce coarse boundary which always leads to high recall. The F-measure which is the comprehensive criteria between precision and recall together with the MAE verify the promising performance of our method.

Table 1. The precision, recall, F-measure and MAE values on GTD public dataset, where the bold fonts of results indicate the best performance.

Algorithm	DECOLOR	GMM	VIBE	PCP	OUR
Precision	0.54	0.51	0.40	0.29	**0.62**
Recall	**0.82**	0.64	0.47	0.18	0.79
F-measure	0.59	0.51	0.39	0.22	**0.67**
MAE	0.006	0.0125	0.0169	0.0155	**0.005**

3.3 Component Analysis

In order to validate the spatial compactness and appearance consistency via superpixel constraint, we evaluate several variations of our model and report the results on Table 2 and visualize several detection results on Fig. 3, where Ours: the proposed model; Our-I: our model without spatial compactness by setting γ to 0; Our-II: our model without appearance consistancy by setting all $w_{ij,kl}$ to 1; Our-III: our model without spatial compactness and the appearance consistency by setting all $w_{ij,kl} = 1$ and $\gamma = 0$. From Table 2 we can see that: Our-II significantly beats Our-III and Our outperforms Our-I in Recall and F-measure, which suggest that superpixel constraint plays importance role on moving object detection. From Fig. 3 we can see that: After introducing the spatial compactness and appearance consistency via superpixel constraint, our method can better preserve the boundary information and suppress the noise.

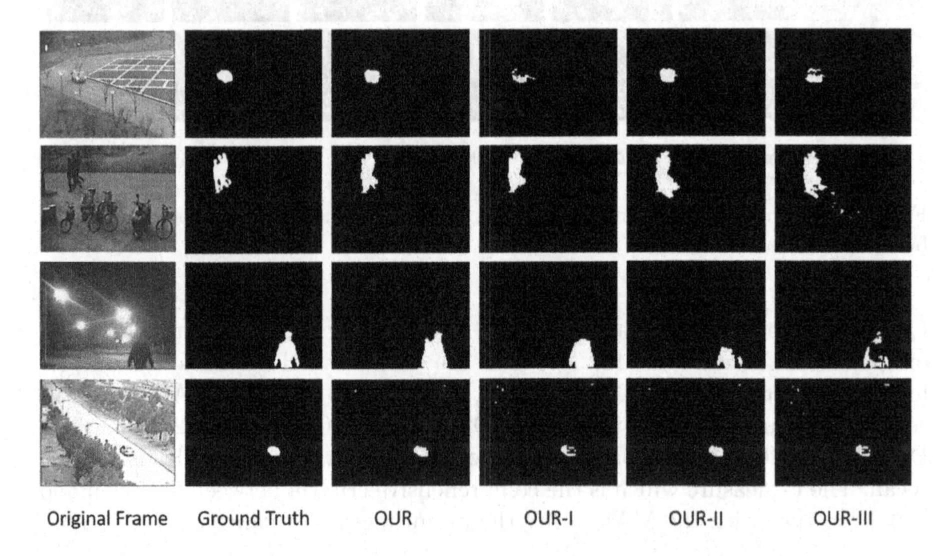

Fig. 3. Example results of our method and its variants on four video sequences from GTD dataset.

Table 2. Average precision, recall, and F-measure of our method and its variants on the GTD dataset. The bold fonts of results indicate the best performance.

Algorithm	Precision	Recall	F-measure
Ours	**0.62**	**0.79**	**0.67**
Ours-I	**0.62**	0.77	0.64
Ours-II	0.57	0.78	0.62
Ours-III	0.51	0.66	0.55

4 Conclusion

In this paper, we have proposed a novel method for moving object detection under the low-rank and sparse separation framework. We have first emphasized the neighboring pixels with close appearance. We have further explored the spatial compactness and appearance consistency between the pixels within the same superpixel. Extensive experiments against state-of-the-arts on the public video sequences suggest that, the proposed method can better preserve the boundary of the objects and robust to the noise. In future work, we will focus on extending our model to online or streaming fashion for real-life applications.

Acknowledgement. This study was funded by the National Nature Science Foundation of China (61502006, 61671018), the Natural Science Foundation of Anhui Province (1508085QF127), the Natural Science Foundation of Anhui Higher Education Institutions of China (KJ2017A017) and Co-Innovation Center for Information Supply & Assurance Technology, Anhui University.

References

1. Al-Sultan, S., Al-Bayatti, A.H., Zedan, H.: Context-aware driver behavior detection system in intelligent transportation systems. IEEE Trans. Veh. Technol. **62**(9), 4264–4275 (2013)
2. Barnich, O., Van Droogenbroeck, M.: ViBe: a universal background subtraction algorithm for video sequences. IEEE Trans. Image Process. **20**(6), 1709–1724 (2011)
3. Bouwmans, T.: Background subtraction for visual surveillance: a fuzzy approach. In: Handbook on Soft Computing for Video Surveillance, pp. 103–134 (2012)
4. Boykov, Y., Veksler, O., Zabih, R.: Fast approximate energy minimization via graph cuts. IEEE Trans. Pattern Anal. Mach. Intell. **23**(11), 1222–1239 (2001)
5. Cai, J.F., Candès, E.J., Shen, Z.: A singular value thresholding algorithm for matrix completion. SIAM J. Optim. **20**(4), 1956–1982 (2010)
6. Candès, E.J., Li, X., Ma, Y., Wright, J.: Robust principal component analysis? J. ACM (JACM) **58**(3), 1–36 (2011)
7. Chauhan, A.K., Krishan, P.: Moving object tracking using Gaussian mixture model and optical flow. Int. J. Adv. Res. Comput. Sci. Softw. Eng. **3**(4), 243–246 (2013)
8. Darbon, J.: Global optimization for first order Markov random fields with submodular priors. Discret. Appl. Math. **157**(16), 3412–3423 (2009)
9. Davis, J., Goadrich, M.: The relationship between precision-recall and ROC curves. In: International Conference on Machine Learning, pp. 233–240 (2006)
10. Dou, J., Li, J., Qin, Q., Tu, Z.: Moving object detection based on incremental learning low rank representation and spatial constraint. Neurocomputing **168**(C), 382–400 (2015)
11. Javed, S., Oh, S.H., Sobral, A., Bouwmans, T., Jung, S.K.: Background subtraction via superpixel-based online matrix decomposition with structured foreground constraints. In: IEEE International Conference on Computer Vision Workshop, pp. 930–938 (2016)
12. Jiang, B., Ding, C., Tang, J.: Graph-Laplacian PCA: closed-form solution and robustness. In: Proceedings of the IEEE Conference on Computer Vision and Pattern Recognition, pp. 3492–3498 (2013)

13. KaewTraKulPong, P., Bowden, R.: An improved adaptive background mixture model for real-time tracking with shadow detection. In: Remagnino, P., Jones, G.A., Paragios, N., Regazzoni, C.S. (eds.) Video-Based Surveillance Systems, pp. 135–144. Springer, Heidelberg (2002). https://doi.org/10.1007/978-1-4615-0913-4_11
14. Ke, Q., Kanade, T.: Robust L1 norm factorization in the presence of outliers and missing data by alternative convex programming. In: 2005 Proceedings of the IEEE Conference on Computer Vision and Pattern Recognition, vol. 1, pp. 739–746. IEEE (2005)
15. Kolmogorov, V., Zabin, R.: What energy functions can be minimized via graph cuts? IEEE Trans. Pattern Anal. Mach. Intell. **26**(2), 147–159 (2004)
16. Li, C., Wang, X., Zhang, I., Tang, J., Wu, H., Lin, L.: WELD: Weighted low-rank decomposition for robust grayscale-thermal foreground detection. IEEE Trans. Circuits Syst. Video Tech. **1**(1), 1–14 (2016)
17. Lin, D., Fidler, S., Urtasun, R.: Holistic scene understanding for 3D object detection with RGBD cameras. In: Proceedings of the IEEE International Conference on Computer Vision, pp. 1417–1424 (2013)
18. Mazumder, R., Hastie, T., Tibshirani, R.: Spectral regularization algorithms for learning large incomplete matrices. J. Mach. Learn. Res. **11**, 2287–2322 (2010)
19. Recht, B., Fazel, M., Parrilo, P.A.: Guaranteed minimum-rank solutions of linear matrix equations via nuclear norm minimization. SIAM Rev. **52**(3), 471–501 (2010)
20. Shen, J., Du, Y., Wang, W., Li, X.: Lazy random walks for superpixel segmentation. IEEE Trans. Image Process. **23**(4), 1451 (2014)
21. Stauffer, C., Grimson, W.E.L.: Adaptive background mixture models for real-time tracking. In: 1999 Proceedings of the IEEE International Conference on Computer Vision, vol. 2, pp. 246–252 (1999)
22. Torre, F.D.L., Black, M.J.: A framework for robust subspace learning. Int. J. Comput. Vis. **54**(1–3), 117–142 (2003)
23. Wren, C.R., Azarbayejani, A., Darrell, T., Pentland, A.P.: Pfinder: real-time tracking of the human body. IEEE Trans. Pattern Anal. Mach. Intell. **19**(7), 780–785 (1997)
24. Xin, B., Tian, Y., Wang, Y., Gao, W.: Background subtraction via generalized fused Lasso foreground modeling. In: 2015 Proceedings of the IEEE Conference on Computer Vision and Pattern Recognition, pp. 4676–4684 (2015)
25. Zhou, H., Kong, H., Wei, L., Creighton, D., Nahavandi, S.: Efficient road detection and tracking for unmanned aerial vehicle. IEEE Trans. Intell. Transp. Syst. **16**(1), 297–309 (2015)
26. Zhou, X., Yang, C., Yu, W.: Moving object detection by detecting contiguous outliers in the low-rank representation. IEEE Trans. Pattern Anal. Mach. Intell. **35**(3), 597–610 (2013)
27. Zhou, Z., Li, X., Wright, J., Candes, E., Ma, Y.: Stable principal component pursuit. In: 2010 IEEE International Symposium on Information Theory, pp. 1518–1522. IEEE (2010)

MFRPN: Towards High-Quality Region Proposal Generation in Object Detection

Dingqian Zhang(✉), Hui Zhang, Wanling Zeng, Zhongxing Han, and Xiaohui Hu

Institute of Software Chinese Academy of Sciences, Beijing, China
dingqian2015@iscas.ac.cn

Abstract. Most state-of-the-art object detection networks need region proposals in their two-step framework. Popular region proposal networks can provide hundred proposals with acceptable accuracy. In this paper, we introduce a Multiple Filters Region Proposal Network (MFRPN) that can change its structure with dataset. We calculate the suitable sizes of filters and use multiple filters with appropriate reference boxes to make the regression of coordinates of proposals more accurate. To illustrate the proposed MFRPN, we adopt the framework of Faster R-CNN [1] and replace the RPN with the MFRPN. As a result, we get 0.98% improvement in mean AP on PASCAL VOC 2007 and 1.45% on PASCAL VOC 2012.

Keywords: Object detection · Multiple filters · Reference box
Region proposal

1 Introduction

Object detection is to detect specific objects in images. Generally, it consists two steps: finding where the objects are (proposals generation), then giving these objects category labels and confidence scores (objects classification). This two-step division matches to visual mechanism of human beings. We first give a scan of the whole image to get the region we really care about. Then we observe carefully for more details to identify what we look at. Although one-step object detection algorithms (e.g., YOLO [12] and SSD [10]) exist, their prediction accuracy is lower than two-step algorithms. In this paper, we focus on improving two-step object detection algorithms. Based on the difference of one-step and two-step algorithms, we can draw a conclusion that proposals are useful for object detection asking for high accuracy. Our algorithm Multiple Filters Region Proposal Network (MFRPN) settles down to generating high-quality proposals. As we will

D. Zhang—Student of ISCAS.

This work is supported by the Natural Science Foundation of China (U1435220) (61503365).

J. Yang et al. (Eds.): CCCV 2017, Part III, CCIS 773, pp. 61–72, 2017.
https://doi.org/10.1007/978-981-10-7305-2_6

show in following sections, detection network with MFRPN shows advantages in detection accuracy.

In early time, proposals are generated by some classical algorithms (e.g., Selective Search [16] and EdgeBox [17]) which provide about thousands of proposals per image to insure covering all the possible objects. However, these solutions are summaried by experts. They have their weaknesses in some situations. Thanks to the development of deep learning, the work of extracting features from pictures is done perfectly by Convolutional Neural Networks (CNN). CNN has rich representation capacity and powerful generalization ability and it can take advantage of computing ability of GPU. Therefore, we can get more accurate features of objects easily.

Nowadays, most state-of-the-art object detection networks use the Region Proposal Network (RPN) [14] to generate proposals for object detection (e.g., Faster R-CNN [14] and R-FCN [3]). RPN has several convolutional layers, one regression layer and one classification layer. To effectively generate different scales or aspect ratios proposals, RPN uses multiple references for every predicted boxes regression work (typically, 9 different reference boxes). But in this method, many hyper-parameters are set directly, such as, the sizes of reference boxes. The construction of a nice network model should be data-driven. If parameters of a model are changed with dataset, we believe it has universality and stability. The original RPN is not suitable for images which include small (Fig. 1a) or dense (Fig. 1b) objects. Because proposals of these images need more accurate predicted coordinates and a confined network cannot deal with it well. We also notice that the sizes of receptive fields are important to generate proposals. Sizes of receptive fields should be close to the sizes of reference boxes.

If receptive fields are too small, it may cause under-fitting problem. Because the information is too little to make correct decision. Oppositely, If receptive fields are too large, it may cause over-fitting problem. Because much information is redundant. For example, a network may misunderstand that chairs must be put next to a table. Our MFRPN has the ability to deal with details and all multiple filters (kernel sizes are different) on the last layer of CNN has suitable reference boxes. We believe MFRPN can sovle the problems above.

(a)

(b)

Fig. 1. Example images with complex contents: (a) full of small objects, (b) dense similar objects.

Figure 2 shows our idea clearly. The size of receptive field determine how much information a RPN can get in one time regression. If receptive fields are too large or too small, the features extracted by a RPN will be more or less then the object itself. So we choose suitable sizes of reference boxes in our object detection network. Since different categories of objects have their own standard sizes. We should use different filters for different objects. And training networks with multiple filters simultaneously can make regression smoother. In conclusion, our main contributions are:

1. We emphasis appropriate mount of information is quite important for regression. The information here is a part of an image in receptive field. So the size of the filter on the last convolution layer should be changed with dataset and match the object's size.
2. We summarize the phenomenon that appropriate sizes of reference boxes are important for boxes regression. It means the sizes of reference boxes should match their filters receptive fields rather than the object itself.
3. We design multiple filters region proposal network. For small and dense object detection, we can use multiple filters with similar sizes. For detecting object with various sizes, we can use multiple filters with various sizes.

Fig. 2. Our idea: Small receptive field (blue) cannot get the whole features of object. Large receptive filed (green) will bring wrong features. Appropriate receptive filed (red) make regression more accurate. (Color figure online)

2 Related Work

There are diverse methods for generating object proposals. The most widely used unsupervised method is Selective Search [16]. It is a clustering method. Other popular clustering method are EdgeBox [17] and MCG [1]. The advantage of clustering methods is they can provide proposals of multiple scales and sizes spontaneously. BING [2], MultiBox [4] are typical in supervised methods. BING uses the binary feature which can be promoted computing speed by registers. MultiBox uses CNN to generate proposals. But recently proposed promising solution, Region Proposal Network (RPN) [14], has more excellent performance. It uses multi-task loss function to combine proposals generation and coordinates regression together in one step. RPN can reduce the number of proposals to less than 300 with higher recall rate. In our experiments, we uses RPN as the baseline algorithm.

The RPN in Faster RCNN adopts 3 aspect ratios (1:1, 1:2, 2:1) and 3 scales (128^2, 256^2, 512^2) anchor boxes as reference boxes basing on the Simonyan and Zisserman model (VGG-16) [15]. Different aspect ratios anchor boxes are convenient to generate different shape proposals. Different scales anchor boxes are used for generating multiple scales proposal more accurately. The size of 3 scales are designed elaborately which are smaller than the receptive field, nearly equal to the receptive field and larger than the receptive field. But their information are all come from the same piece of feature map. That is to say if the predicted field matches the size of receptive field, the predicted result will be more reliable. So we claim that reference boxes should not much larger than receptive field of filter. And we should pay more attention to details in receptive field.

YOLO 9000 [13] clusters images by size and changes the reference boxes' sizes with dataset. But it doesn't change the filters' sizes. We believe that changing dataset should bring changes in filters' sizes. And changing filters should bring changes in reference boxes, too. Appropriate filters are more important than reference boxes.

As for multiple filters train simultaneously, SPP [6] and Grid Loss [11] does the similar but different job. SPP proposes a spatial pyramid pooling layer which uses multiple kernel sizes to get a fixed-length representation. Grid Loss proposes a novel loss layer for CNN which minimizes error rate on both sub-blocks and the whole feature map. These algorithms focus on use the relationship of part and whole to detect objects. Because we expect to focus the whole object itself, we use kernels of multiple sizes separately.

In recent years, many novel methods have been proposed to improve prediction accuracy given by RPN. HyperNet [8] and FPN [9] are typical two of these methods. They try to use both high level and low level feature together to predict proposals. But we put attention to improve accuracy of coordinates regression. MFRPN is a new path to get better proposals and can cooperate with their networks.

3 Our Approach

In this section, we introduce Multiple Filter Region Proposal Network (MFRPN). We will explain how we design our network, how it works, and how to use MFRPN in an object detection network in detail.

3.1 Confirming Sizes of Filters

In this step, we will cluster all bounding boxes in our training set. And try to find appropriate filters which receptive fields are the nearest top K to clustering result. Figure 3 shows the relationship between reference box and bounding box. In this paper, we use k-means clustering method to solve this problem. There are three steps:

1. Run k-means clustering on dataset to divide sizes of objects into K categories.
2. According to the used network, calculate the corresponding relationship between filters and receptive fields.
3. Confirm sizes of filters on the last convolution layer which receptive fields are the nearest top K to bounding boxes clustering result.

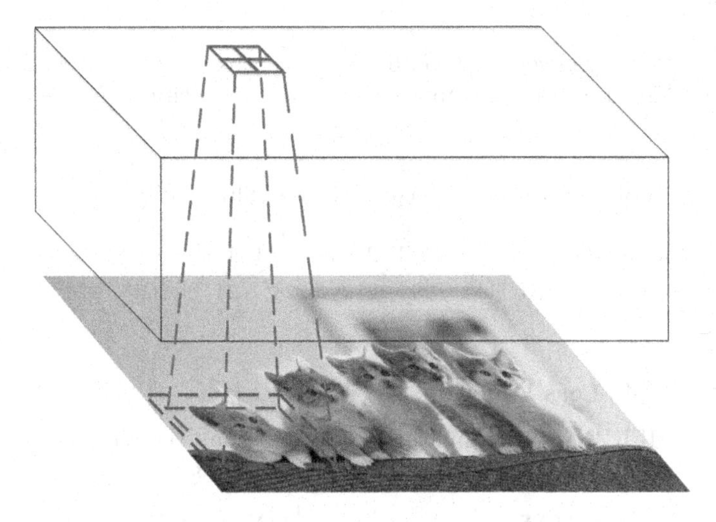

Fig. 3. The relationship between receptive field (blue) and bounding box (red) in our network. We try to use filters with appropriate receptive fields. (Color figure online)

3.2 Multiple Filters Region Proposal Networks

After calculating filters' sizes, We will build our region proposal network. The MFRPN is several convolutional layers (Fig. 4). We use K small networks to slide over the feature map output by the last convolutional layer [14]. Note that the K here is equal to K in k-means clustering. These small networks take $n_1^2 \sim n_k^2$ part of the last feature map. That means the kernel sizes of the networks' filters are $n_1 \sim n_k$. Although the sizes of filters are different, we equal the numbers of outputs of filters. In this paper, the number of outputs is 512. So every filter generates 512-d lower-dimensional features. Every independent convolutional layer is followed by two 1×1 convolutional layers. One for classification, one for coordinates regression.

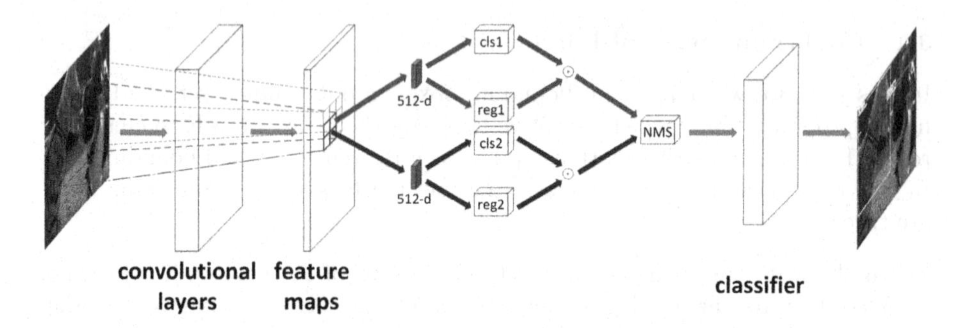

Fig. 4. MFRPN object detection architecture. This architecture can be extended to include more than two filters to generate high-quality proposals. Our experiments just use 2 filters for demonstration. Different filters provide different proposals which are merged by NMS.

In the adopted network (VGG-16), the receptive fields of filters is showed in Table 1. We will choose the appropriate filter for each category clustered in Sect. 3.1.

Table 1. The size of receptive fields of filters in VGG-16.

Size of filter	1×1	2×2	3×3	4×4	5×5	6×6
Size of receptive field	196^2	212^2	228^2	244^2	260^2	276^2

3.3 Extensible Loss Function

For training MFRPNs, we use multi-task loss function. And all filters contribute to the loss. The loss function for an image is defined as:

$$L(p_{ki}, t_{ki}) = \frac{1}{N_{cls}} \sum_k \sum_i L_{cls}(p_{ki}, p_i^*) + \frac{1}{N_{reg}} \sum_k \lambda_k \sum_i p_i^* L_{reg}(t_{ki}, t_i^*). \quad (1)$$

In this equation, i is the index of a reference box in a mini-batch, k is the index of multiple filters. p_{ki} is the probability of reference box being an object predicted by filter k. But the ground-truth label p_i^* is independent of filters. If reference box is positive, p_i^* is 1. And if reference box is negative, p_i^* is 0. In second part of the equation, t_{ki} (t_i^*) is 4-d feature stands for coordinates of the center of a predicted box (reference box), height and width normalized in method proposed in [5]. L_{cls} is log loss, L_{reg} is robust loss [5], defined as:

$$L_{reg}(t_{ki}, t_i^*) = \sum_{i \in x,y,w,h} smooth_{L_1}(t_{ki} - t_i^*). \quad (2)$$

in which

$$smooth_{L_1} = \begin{cases} 0.5x^2 & if\ |x| < 1 \\ |x| - 0.5 & otherwise. \end{cases} \quad (3)$$

The two part loss functions are normalized by N_{cls} and N_{reg}. In this paper N_{cls} (=256) and N_{reg} (≈2400) are balanced by λ_k, and λ_k are all 10 in this paper. As we can see, the closer to ground truth, the more accurate regression is. Because the loss is quadratic, when the difference value is in $(-1, 1)$. But MFRPN uses multiple filters to train convolutional layers together. In this case, there are more difference values in $(-1, 1)$. So our algorithm makes the regression smoother.

3.4 Training Object Detection Network with MFRPN

For training MFRPN, we follow the training method in [14]. But we use different labels and box-targets for different filters. In each mini-batch, we use 256 different reference boxes for each filter.

In this paper, we use the Fast R-CNN [5] as our classification network. For traning object detection network with MFRPN, we use 4-Step Alternating Training [14] algorithm.

4 Experiments

To compare with Faster R-CNN we replace the RPN of Faster R-CNN with our MFRPN. It is worth mentioning that our convolution layers can be replace by other CNNs like ResNet [7] and our classification network can be replaced, too. We use PASCAL VOC 2007 and 2012 dataset in training and testing phase.

4.1 Experiments on Choosing Filters

We choose k-means as our clustering method. The distance between too objects is defined as:

$$distance(a, b) = -IOU(a, b) \tag{4}$$

In this step, the position of object is useless. So we move all the bounding boxes to the top left corner and then calculate the distance. Table 2 shows some clustering results on PASCAL VOC 2007 and PASCAL VOC 2012 dataset.

Table 2. Part of clustering results on PASCAL VOC 2007 training set. We change the number of clustering categories for each experiment.

	K = 2	K = 3
VOC 2007	70 × 93 263 × 253	54 × 74 147 × 183 340 × 287
VOC 2012	71 × 90 277 × 262	53 × 67 150 × 188 352 × 294

We will use $K = 2$ in this paper. Combined with Table 1 can be seen, the appropriate sizes of filters are 1×1 and 5×5 for PASCAL VOC 2007, 1×1 and 6×6 for PASCAL VOC 2012.

4.2 Experiments on Changing Reference Boxes

We claim that reference box must match the receptive field. We use the Faster R-CNN [14] as the basic experimental method. Faster R-CNN use the RPN to generate proposals. On the last convolutional layer, the size of filter is 3. From the Table 1 we can find the receptive field of this filter is 218. Faster R-CNN uses multiple references. The sizes of the references are $128^2, 256^2, 512^2$. As we can see, 512^2 is much larger than the filter's receptive field, so we change the references to $64^2, 128^2, 256^2$. The results are showed in Table 3.

Table 3. Detection results on PASCAL VOC 2007. We use different sizes of reference boxes, noting that the receptive fields are all close to 256.

	Size of filter	Reference boxes	Mean AP
1	3	$128^2, 256^2, 512^2$	69.94
2	3	$64^2, 128^2, 256^2$	70.62

From these results, we find the original network gets the lowest mean AP. Because the original reference boxes are much larger than the receptive field. It means, in a mini-batch, there are too many unknown factors to predict. The reference boxes experimental results meet our idea that appropriate sizes of reference boxes are good for generating high-quality proposals.

In our next experiments, we chose $1 \times 1, 5 \times 5$ and 6×6 filters to extract feature vectors. So we choose $64^2, 128^2, 256^2$ as the sizes of our reference boxes.

4.3 Experiments on Using Multiple Filters

We believe multiple filters can improve detection accuracy. Since images of PASCAL VOC are small. So our filters are similar sizes. Our theory is dividing receptive field more accurately is necessary. So in our experiments, multiple filters are smaller than single filter. We use multiple filters on the last convolutional layer of RPN and keep other parameters the same as Faster R-CNN. The results are showed in Table 4. When using multiple filters, the mean AP increases by over 0.15%. Although the improvement is little, it proves that multiple filters have a beneficial effect on generating high-quality proposals.

Table 4. Detection results on PASCAL VOC 2007. We use multiple filters. In this experiment, sizes of filters are 2 and 3.

	Size of filter	Reference box	Mean AP
1	3	$128^2, 256^2, 512^2$	69.94
2	2 & 3	$128^2, 256^2, 512^2$	70.02

4.4 Object Detection Networks with MFRPNs

So far, we have proved most of our ideas are useful for improving proposals' quality. We train object detection network (Faster R-CNN) with MFRPN. Since our dataset is PASCAL VOC, we set the sizes of multiple filters are 1 and 5 for 2007 dataset. And the sizes are changed to 1 and 6, when use 2012 training set. And the reference boxes are $64^2, 128^2, 256^2$. If the images are bigger, we will change the number and sizes of filters and reference boxes as well. We keep other parameters the same as Faster R-CNN. But the mean AP is 69.7698%, lower than we expect.

Because we get more proposals and most of these have more accurate vertex coordinates, mixing some proposals together is important in our experiments. NMS is a typical solution to proposals fusion. In Faster R-CNN, NMS reduces the number of proposals to 300 before they are sent to classification network in the final testing phase. But in our experiments, MFRPN needs more. Further, several experiments have been performed to find the appropriate number of proposals that should be kept and the results are shown in Table 5. In order to indicate our improvement is not the result of more number of proposals. We give classification network more proposals from original RPN, too. After these experiments, we keep 1300 proposals left after NMS and finish training the object detection network with MFRPN. The detection results are showed in Table 6. The mean AP of proposed MFRPN is higher than our baseline. To show the stability of our method, we keep the parameters same as what we set on PASCAL VOC 2007 and change the dataset to PASCAL VOC 2012. The detection results are showed in Table 7. Our method increases mean AP to 68.44%. The increment proves our method is stable. Figure 5 shows some results on the VOC test-dev set.

Table 5. Detection results on PASCAL VOC 2007. We use MFRPN or original RPN to generate proposals and detectors of Faster R-CNN to classify objects. In these experiments, numbers of proposals kept after NMS are different.

Number of proposals	300	400	500	600	700	800	900	1000	1100	1200	1300	1400
Mean AP (MFRPN) (%)	69.77	69.94	70.23	70.32	70.58	70.61	70.63	70.75	70.84	70.91	70.91	70.78
Mean AP (RPN) (%)	69.94	70.03	70.04	70.00	69.98	70.01	70.02	70.02	70.02	70.02	70.02	70.02

Table 6. Results on PASCAL VOC 2007 test set (trained on VOC 2007 trainval) with detectors of Faster R-CNN and VGG-16. The proposals are generated by different methods and MFRPN provide 1300 proposals for the detector.

	mean AP	areo	bike	bird	boat	bottle	bus	car	cat
RPN	69.94	68.55	78.20	67.28	57.66	51.23	79.57	**79.36**	**85.15**
MFRPN	**70.92**	**71.66**	**80.06**	**69.55**	**60.63**	**52.46**	**81.21**	78.96	84.69

chair	cow	table	dog	horse	mbike	person	plant	sheep	sofa	train	tv
49.44	75.49	64.60	82.53	83.16	**77.64**	76.20	36.83	**72.95**	**67.30**	**78.03**	**67.87**
51.30	**78.87**	**66.26**	**83.55**	**84.99**	75.54	**76.84**	**40.07**	71.87	66.03	77.46	66.46

Table 7. Results on PASCAL VOC 2012 test set (trained on VOC 2012 trainval) with detectors of Faster R-CNN and VGG-16. The proposals are generated by different methods. To show the stability of our network, we keep 1300 proposals after NMS in MFRPN experiment, too.

	mean AP	areo	bike	bird	boat	bottle	bus	car	cat
RPN	66.99	82.33	76.43	71.02	48.37	45.20	72.08	72.27	87.25
MFRPN	**68.44**	**83.92**	**78.17**	**71.17**	**51.69**	**46.80**	**77.24**	**72.63**	**88.14**

chair	cow	table	dog	horse	mbike	person	plant	sheep	sofa	train	tv
42.18	**73.72**	50.03	86.76	78.68	78.36	77.35	34.50	**70.11**	**57.08**	77.14	58.93
43.47	73.04	**51.81**	**86.96**	**80.24**	**81.54**	**77.86**	**36.02**	69.01	56.64	**81.35**	**61.19**

Fig. 5. Examples of our detection results.

We also compare the running time between baseline and our method. Because we use multi-task loss to implement our method, the training time is almost the same as the baseline. But during test phase, our method spend more time. In our experiment (NVIDIA TITAN X (Pascal)), the average test time is 0.350 s. Although the time is higher than the baseline which is 0.236 s, the speed is acceptable.

5 Conclusion

In this paper, we propose the Multiple Filters Region Proposal Network (MFRPN) for generating high-quality region proposals. The proposed MFRPN can change its structure with dataset. According to the general classification of objects' sizes, MFRPN can choose nice filters to cover objects automatically.

Then MFRPN adopts appropriate reference boxes and multiple filters to get more accurate proposals.

It can cooperate with most two-step object detection networks. And it is compatible with other improved methods of RPN. In conclusion, our method improves state-of-the-art object detection not only accuracy but also stability.

References

1. Arbeláez, P., Pont-Tuset, J., Barron, J.T., Marques, F., Malik, J.: Multiscale combinatorial grouping. In: Proceedings of the IEEE Conference on Computer Vision and Pattern Recognition, pp. 328–335 (2014)
2. Cheng, M.M., Zhang, Z., Lin, W.Y., Torr, P.: Bing: binarized normed gradients for objectness estimation at 300fps. In: Proceedings of the IEEE Conference on Computer Vision and Pattern Recognition, pp. 3286–3293 (2014)
3. Dai, J., Li, Y., He, K., Sun, J.: R-FCN: object detection via region-based fully convolutional networks. In: Advances in Neural Information Processing Systems, pp. 379–387 (2016)
4. Erhan, D., Szegedy, C., Toshev, A., Anguelov, D.: Scalable object detection using deep neural networks. In: Proceedings of the IEEE Conference on Computer Vision and Pattern Recognition, pp. 2147–2154 (2014)
5. Girshick, R.: Fast R-CNN. In: Proceedings of the IEEE International Conference on Computer Vision, pp. 1440–1448 (2015)
6. He, K., Zhang, X., Ren, S., Sun, J.: Spatial pyramid pooling in deep convolutional networks for visual recognition. In: Fleet, D., Pajdla, T., Schiele, B., Tuytelaars, T. (eds.) ECCV 2014. LNCS, vol. 8691, pp. 346–361. Springer, Cham (2014). https://doi.org/10.1007/978-3-319-10578-9_23
7. He, K., Zhang, X., Ren, S., Sun, J.: Deep residual learning for image recognition. In: Proceedings of the IEEE Conference on Computer Vision and Pattern Recognition, pp. 770–778 (2016)
8. Kong, T., Yao, A., Chen, Y., Sun, F.: HyperNet: towards accurate region proposal generation and joint object detection. In: Proceedings of the IEEE Conference on Computer Vision and Pattern Recognition, pp. 845–853 (2016)
9. Lin, T.Y., Dollár, P., Girshick, R., He, K., Hariharan, B., Belongie, S.: Feature pyramid networks for object detection. arXiv preprint arXiv:1612.03144 (2016)
10. Liu, W., Anguelov, D., Erhan, D., Szegedy, C., Reed, S., Fu, C.-Y., Berg, A.C.: SSD: single shot multibox detector. In: Leibe, B., Matas, J., Sebe, N., Welling, M. (eds.) ECCV 2016. LNCS, vol. 9905, pp. 21–37. Springer, Cham (2016). https://doi.org/10.1007/978-3-319-46448-0_2
11. Opitz, M., Waltner, G., Poier, G., Possegger, H., Bischof, H.: Grid loss: detecting occluded faces. In: Leibe, B., Matas, J., Sebe, N., Welling, M. (eds.) ECCV 2016. LNCS, vol. 9907, pp. 386–402. Springer, Cham (2016). https://doi.org/10.1007/978-3-319-46487-9_24
12. Redmon, J., Divvala, S., Girshick, R., Farhadi, A.: You only look once: unified, real-time object detection. In: Proceedings of the IEEE Conference on Computer Vision and Pattern Recognition, pp. 779–788 (2016)
13. Redmon, J., Farhadi, A.: Yolo9000: better, faster, stronger. arXiv preprint arXiv:1612.08242 (2016)
14. Ren, S., He, K., Girshick, R., Sun, J.: Faster R-CNN: towards real-time object detection with region proposal networks. In: Advances in Neural Information Processing Systems, pp. 91–99 (2015)

15. Simonyan, K., Zisserman, A.: Very deep convolutional networks for large-scale image recognition. arXiv preprint arXiv:1409.1556 (2014)
16. Uijlings, J.R., Van De Sande, K.E., Gevers, T., Smeulders, A.W.: Selective search for object recognition. Int. J. Comput. Vis. **104**(2), 154–171 (2013)
17. Zitnick, C.L., Dollár, P.: Edge boxes: locating object proposals from edges. In: Fleet, D., Pajdla, T., Schiele, B., Tuytelaars, T. (eds.) ECCV 2014. LNCS, vol. 8693, pp. 391–405. Springer, Cham (2014). https://doi.org/10.1007/978-3-319-10602-1_26

Pedestrian Detection by Using CNN Features with Skip Connection

Peng Zhang[1,2] and Zengfu Wang[1,2](✉)

[1] Institute of Intelligent Machines, Chinese Academy of Sciences, Hefei 230031, China
[2] University of Science and Technology of China, Hefei 230026, China
hizhangp@mail.ustc.edu.cn, zfwang@ustc.edu.cn

Abstract. The CNN based pedestrian detection is developing rapidly in recent years. Compared to the features used in former pedestrian detection models, the features from deep CNN have outperformed in many aspects. In this paper, we focus on the problem of pedestrian detection by using CNN features with skip connections and mainly address the corresponding issues in urban scenes: different spatial scales and optical blur of pedestrian. We propose an effective end-to-end pedestrian detector, which fuses different features from multi-layers to recover the coarse features of small scale pedestrians and optical blur pedestrians. The experimental results show that our method achieves state-of the-art performance on Caltech Pedestrian Detection Benchmark.

Keywords: Pedestrian detection · Skip connection · Normalization

1 Introduction

Recently, pedestrian detection which aims to locate pedestrian instances in an image has drawn much attention beyond general object detection, because of its attractive applications in automatic driving, video surveillance, person re-identification and robotics.

Extensive research on pedestrian detection have been put forward [18,30], these papers achieved state-of-the-art performance on well-established benchmark dataset, such as Caltech Pedestrian Detection Benchmark [7] and KITTI Vision Benchmark [8]. However, with the decelerated improvement on detection quality (miss rate), it seems that we are reaching the bottleneck of pedestrian detection.

Currently, there remains two main challenging points of pedestrian detection in urban scenes: (1) different spatial scales and (2) optical blur. The scale of pedestrian instance in Caltech dataset ranges from 16 pixels to about 128 pixels, and nearly 70% pedestrian instances lies in between 30 to 80 pixels [7]. Compared to large-scale pedestrian instances, small-scale ones suffer from blurred boundaries and breezing appearance as a result of low resolution, which makes it difficult to distinguish them from background and similar objects (such as

J. Yang et al. (Eds.): CCCV 2017, Part III, CCIS 773, pp. 73–83, 2017.
https://doi.org/10.1007/978-981-10-7305-2_7

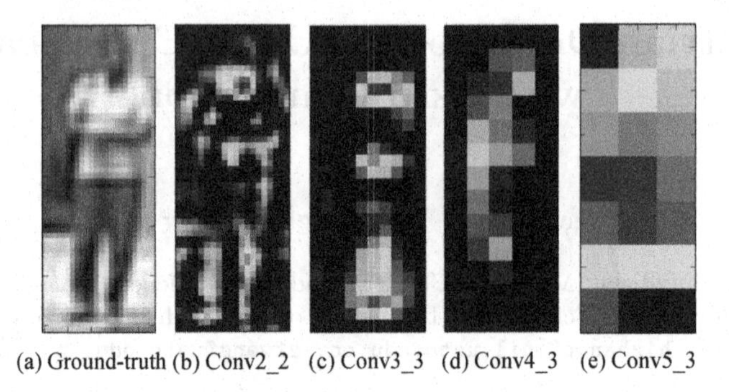

Fig. 1. The spatial feature distribution of pedestrian instance in each convolutional layer. For a typical ground-truth input, fine, detailed feature evolves to coarse, abstract feature from low layer to high layer.

street lamp and trash can). Meanwhile, observing the feature maps from the same convolutional layer, pedestrian instances with large-scale can provide more detailed information than small-scale ones. These difference will make the model less sensitive to small-scale pedestrian instances. Optical blur mainly results from the movement of the camera in the car. Fast relative displacement between camera and pedestrian instances will fuzzy the optical features, only coarse feature maps are obtained compared to other ones with similar scale.

Faster R-CNN [22] is a quite successful method in the field of general object detection, and has been the base of many recent popular pedestrian detection methods [5,18,27,30]. Instead of using hand-craft features, such as HOG+SVM rigid and deformable part detector (DPM), Faster R-CNN is based on deep CNN. Faster R-CNN consists of two parts: a fully convolutional region proposal network (RPN) and followed by an R-CNN [10] classifier. Convolutional neural network feature generated by RPN shows strong robustness to pedestrian instances in pedestrian detection models. However, native Faster R-CNN architecture cannot handle small-scale pedestrian instances and optical blur problems. Following the architecture of Faster R-CNN, we propose a skip connection feature model for pedestrian detection to overcome above two challenges. As shown in Fig. 1, lower convolutional layers provide discriminative features to capture intra edge variations, higher layers encode semantic concepts for object category classification. With appropriate combination of feature maps from these convolutional layers, synthetic feature maps can provide more expressive and distinguishable features for small-scale or blurry pedestrian instances. Using region proposals generated by RPN from conv5_3 layer in VGG16 [24], we extract specific convolutional feature maps in conv2_2, conv3_3, conv4_3 and conv5_3 layer for classifier. Unlike prior works using multi-stage pipeline for detection, we construct an end-to-end pedestrian detection model for consistent learning. Our model achieves 8.91% miss rate on Caltech Pedestrian Detection Benchmark, at the speed of 0.24 s/image with Titan X (Maxwell).

Our method has three contributions: (1) We introduce skip connection feature for pedestrian detection which combines different feature maps from different convolutional layers. (2) We prove that batch normalization is more suitable for skip connection feature combination. (3) We show that an end-to-end detection system is much easier to train compared to multi-stage system.

2 Related Work

2.1 Pedestrian Detection

Considering the prior works on pedestrian detection, we can divide them into three families [3]: DPM variants (MT-DPM [28]), Deep Neural Network (CompACT-Deep [5], SAF R-CNN [18]) and Decision forests (ChnFtrs [6], Katamari [3]).

Feature extraction is a very important step for pedestrian detection. Hand-craft features such as Integral Channel Feature (ICF) [6] and Aggregated Channel Feature (ACF) [29] are extracted from the combination of three types of 6 channels (LUV color channels, normalized gradient magnitude, and histogram of oriented gradients), and then used to generate features of different levels [6], local sum of squares [2], or decorated LDCF (Linear Discriminant Channel Feature) [21]. These are very polular feature extraction methods before region-based convolutional neural network (R-CNN) achieved the top performance for general object detection based on AlexNet model. Deep convolutional neural network (DCNN) has been proved to be a powerful feature extractor on a wide variety of computer vision, SCF+AlexNet [13], DeepParts [25] and many other tasks [5,18,27,31] have significantly out-performed hand-crafted feature models using deep convolutional feature.

The most recent surveys [5,30] indicate that Convolutional Neural Network (CNN) performs a good feature extractor, but downstream classifier (the *fc* layers) degrades the results due to the low resolution of feature maps and lack of bootstrapping strategy. To overcome above drawbacks they learn boosted classifiers on top of hybrid feature maps which extracted from several layers, these improvements build an expressive feature map and powerful classifier, which help to achieve top performance. For example, RPN+BF [30] uses RPN in Faster R-CNN as a class-agnostic detector to generate region proposals, then adopt RoI Pooling to extract fixed-length features from regions, finally, these features are used to train a cascaded boosted forest classifier. This multi-stage strategy divides the detection process into several modules, making it more complex, time-consuming and hard to train, while our method is much simpler and easier for consistent learning.

2.2 Skip-Layer Connections

Skip-layer connections come from Fully Convolution Network (FCN) [20], which has an excellent performance on semantic segmentation. In this paper, information in the coarse, high layer is combined with that in the fine, low layer to generate a impressive feature map for semantic segmentation.

For a pre-trained CNN model, CNN features at different layers have different properties for an object. Low convolutional layer provides detailed local edge features which is helpful to distinguish targets from the background and similar object, but they are not robust to the change of appearance, such as deformation and blur. High convolutional layer captures more abstract and semantic information, the are more robust to different appearance and easy to classify each object into different classes, however, they suffer from the low resolution of feature maps, restricting their spatial distinguishing ability.

Recent HyperNet [17] extracts feature maps from several layers, applies max pooling or deconvolution to each feature maps to obtain a single output cube which is call Hyper Feature and leads to a more accurate object detection result.

3 Skip Connections Feature

In this section, we introduce our skip connection features (illustrated in Fig. 2), which are extracted from different convolutional layers of the neural network model. With the proper fusion of these normalized features, the model is able to achieve the state-of-the-art result on Caltech Pedestrian Detection Benchmark.

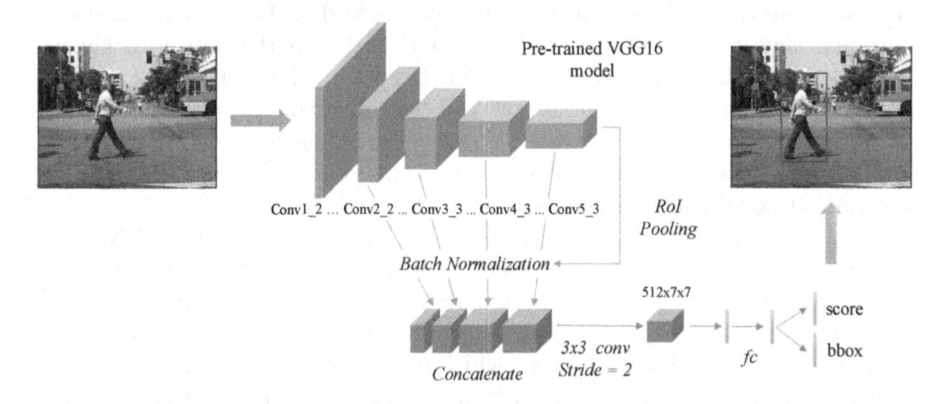

Fig. 2. Skip connections pedestrian detection architecture. Our model takes in an input image, uses pretrained VGG16 model to extract feature maps and evaluates 128 RoI generated by RPN, then RoI Pooling is applied to several layers to extract multiple level's abstraction, after batch-normalizing each layer's feature map, we apply a convolution layer to the concatenated feature maps, classification and adjustment are predicted for each proposal.

3.1 Feature Extraction

Exploring recent successful object detectors such as SPPnet, Faster R-CNN [22] and R-FCN [16], they all extract feature maps from the last convolution layer (conv5_3 in VGG16 [24] and res5c in ResNet-101 [12]), then feed feature maps into the *fc* layers for classification scores and adjusted bounding boxes.

Given an image, we keep the image's aspect and resize the short side into 800 pixels. As a result of 4 pooling layers in VGG16 model, size of feature maps is 50 pixels. And for a typical pedestrian instance with a median height of 48 pixels in the image [7], size of its feature maps is only about 6 pixels (after the input image resized into 800 pixels). RPN, which is based on Fixed-size sliding windows, can avoid collapsing bins [30]. However, RoI pooling [9] may suffer from the low resolution and insufficient information.

Inspired by the design of HyperNet [17] and Inside-Outside Net [1], using region proposals generated by RPN, we adopt RoI pooling to several layers (conv2_2, conv3_3, conv4_3 and conv5_3) to extract feature maps with different resolution. Thanks to the design of RoI pooling, original *fc* layers will have no limitations to the size of feature maps. For example, we can extract feature maps from RoIs on conv2_2 (of a stride = 2), then pool the feature maps into a fixed-size of $C \times 13 \times 13$, where C depends on the number of channels in each layer.

We downsample or upsample feature maps from each feature maps to keep a $C \times 13 \times 13$ fixed-size before fed into the *fc* layers, it is also needed for feature fusion in the next phase. It's worth noting that we leave the number of feature maps' channels C unchanged, because accuracy of classification is much more important than that of location in pedestrian detection. We keep feature maps in higher layers with more channels and less channels in lower layers, aiming to allow higher layers weigh bigger influence than lower ones.

3.2 Layer Fusion

Feature maps extracted from different layers have coarse-to-fine information across CNN model, through a careful fusion method, we can provide more expressive spatial information for the *fc* layers.

Since feature maps after RoI Pooling are down sampled or up sampled to the same size, a straightforward idea would be to concatenate these feature maps together and reduce the dimensionality using convolutional layer. However, feature maps at different layers may have very different amplitudes, feature map with high amplitude can suppress one with low amplitude and result in unstable learning. To combine coarse-to-fine information efficiently, normalizing the amplitude of different feature maps is needed such that each layer has similar distribution.

Instead of using L2 normalization in [1], we apply Batch Normalization [14] to each feature map. Batch Normalization makes training easier and brings in several advantages:

1. Faster training: Learning rate can be accelerated by Batch Normalization compared to other pedestrian detection models, and leads to faster convergence.
2. Amplitude Normalization: Batch Normalization layer can efficiently normalize data in feature maps (zeros means, unit variances, and decorrelated), and take care of backward propagation. On the contrast, feature normalization needs to be carefully addressed when applied to the *fc* layers using L2 normalization [19].

3. More accurate: In a batch-normalized model, each feature map has similar amplitude, we can apply unified learning rate to each layer and lead to a better learning, hence accuracy of the model.

In order to gain high resolution of feature maps for R-CNN classifier, we set the size of RoI Pooling kernel to 13×13. For the feature maps whose size less than 13×13, we add a deconvolution operation to perform upsampling and a convolutional operation is applied to the other. Then we concatenate them together to create a skip-connections-feature cube.

To match the *fc* layers in VGG16, a convolution layer with 3×3 kernel and stride of 2 is applied at the top. This operation not only reduces 13×13 feature maps into 7×7, but also selects class independent and location independent information from skip-connections-feature cube.

4 Experiments

We train and evaluate our model on the popular Caltech Pedestrian Dataset [7], and compare our results with other algorithms.

4.1 Datasets

Caltech Pedestrian Dataset and its detection benchmark [7] is one of the most popular pedestrian detection dataset. The dataset contains about 10 h of 640×480 30 Hz video of urban traffic collected from a vehicle camera. There are about 250,000 frames, which consist of 350,000 bounding boxes and 2300 unique annotated pedestrian instances.

To avoid the weakness of lacking training data, we prepare our training images one out of each 3 frames in original training data (set00–set05), instead of one out of each 30 frames for standard training set and test set. Then training images are augmented by horizontal flip with probability 0.5 and images with empty bounding boxes or occluded pedestrian are removed.

The standard testing data (set06–set10) are evaluated under the "reasonable" setting (on 50-pixel or taller, unoccluded or partially occluded pedestrians). We compare our performance using log-average miss rate (MR) which is calculated by averaging each miss rate on False Positive Per Image (FPPI) in $[10^{-2}, 10^{0}]$.

4.2 Implementation Details

We use the VGG16 model weights pre-trained on ImageNet [23] dataset to extract feature maps for detection. Since we need to extract feature maps from conv2_2 layer to conv5_3 layer, it is necessary to fine-tune starting from conv2_1 layer and keep previous layers frozen.

Following the setting of Faster R-CNN [22], the pool5 layer is removed and feature maps extracted by conv5_3 layer are feed into RPN directly to generate region proposals. As mentioned above, to get more spacial information for RCNN layer, we set the size of RoI Pooling kernel to 13×13, and a 3×3 kernel convolutional layer with the stride of 2 is followed to match the size of the *fc* layers.

One of the key points in improvement of pedestrian detection is enlarging the resolution, so an image is resized such that its short side has 800 pixels before feed into neural network. For RPN training, we set an anchor to be positive if it has an Intersection-over-Union (IoU) ratio greater than 0.5 with one of the ground-truth bounding box, we also adjust the anchor according to the dataset [7], the aspect ratios of anchor are set to 1.0, 2.0 and 3.0, and the scales are starting from 20 pixels height with a scaling stride of 2x to 160 pixels. The usage of a wide range of scales specific to pedestrian can avoid many false positive instances and make it easy for detecting pedestrian of different size.

Our model is implemented on the public available platform caffe [15]. We fine-tune our model with a weight decay of 0.0005 and a momentum of 0.9 and use single-scale training by default. We keep the first two layers frozen and update the parameters of other layers with a learning rate of 0.001 for 30 k iterations and 0.0001 for next 30 k iterations on Caltech Pedestrian Dataset.

4.3 Result Evaluation

To investigate the influence of our model on pedestrian detection, we conduct several groups of experiments on various factors and architectures.

Table 1. Comparison of detection performance on different layers' feature maps on the Caltech Reasonable test set. Note that all region proposals are generated from conv5_3 layer.

Convolutional layer	Channels	Ratio	Miss rate (%)
Conv3_3	256	4	13.84
Conv4_3	512	8	**11.63**
Conv5_1	512	16	13.16
Conv5_2	512	16	12.07
Conv5_3	512	16	11.65

Different Convolutional Layers. Fusion of multi-layers' feature maps has been proven beneficial in detection [17,30] and other computer vision applications [11]. In our model, we combine several feature maps from selected convolution layers. To verify the various performance of feature maps in different convolution layers, we adopt RoI Pooling on different convolutional layers and generate region proposals from the same convolutional layers (conv5_3) for fair comparisons. Table 1 shows the results of using different RoI pooling feature maps in our method. Conv4_3 and conv5_3 perform good results (11.63% and 11.65%), indicating that high resolution (conv4_3) and semantic feature (conv5_3) are the main effect factors of the final performance. The decline trend after an initial ascent from conv4_3 to conv5_3 means that intermediate convolutional layers may suffer from low resolution and insufficient semantic features (Fig. 3).

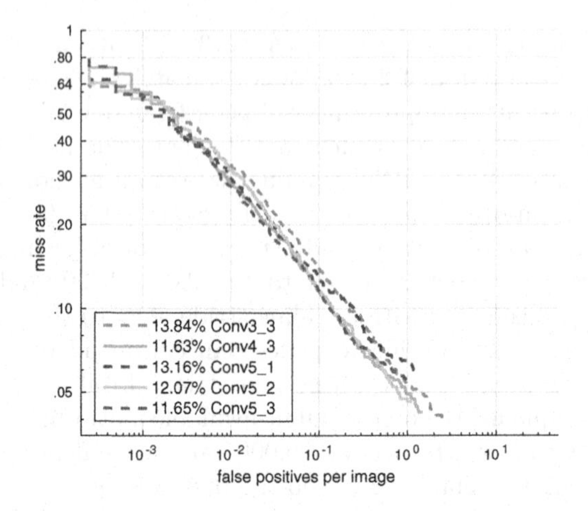

Fig. 3. Comparison on the Caltech Reasonable test set, the performance of each layer varies from 10^{-2} to 10^{0} on False Positive Per Image (FPPI).

Normalization. Normalizing the amplitude of different feature maps is an effective way to combine coarse-to-fine information from different layers. To analysis the performance of this strategy, we compare the results under different normalization method in Table 2. It can be observed that unnormalized concatenation with each feature map leads to unstable learning and not very good result (13.68%), LRN normalization decreases the miss rate by 3.68% and Batch Normalization gains the state-of-the-art performance (8.91%), showing that normalization are benificial for stable learning in skip-layer feature fusion and batch normalization is the key to get better performance.

Comparison with state-of-the-art. Figure 4 shows the result of our model trained on the Caltech training set and evaluated on the Caltech Reasonable test set, which achieves a miss rate of 8.91%. We compare our result with several former state-of-the-art models on Caltech Pedestrian Benchmark, including RPN+BF [30], SAF R-CNN [18], CompACT-Deep [5], DeepParts [25], Checkerboards+ [32], MS-RCNN [4], and TA-CNN [26]. It can be observed that our model outperforms other models and achieves state-of-the-art performance.

We also compare our model with Faster R-CNN [22], CompACT-Deep [5] and RPN+BF [30] in Table 3. With the design of RPN, Faster R-CNN is capable of

Table 2. Comparison of detection performance on unnormalized, LRN and BatchNorm concatenation on the Caltech Reasonable test set.

Convolutional layer	Normalization	Miss rate (%)
Conv2_2, ..., Conv5_3	None	13.68
Conv2_2, ..., Conv5_3	LRN	10.00
Conv2_2, ..., Conv5_3	BatchNorm	**8.91**

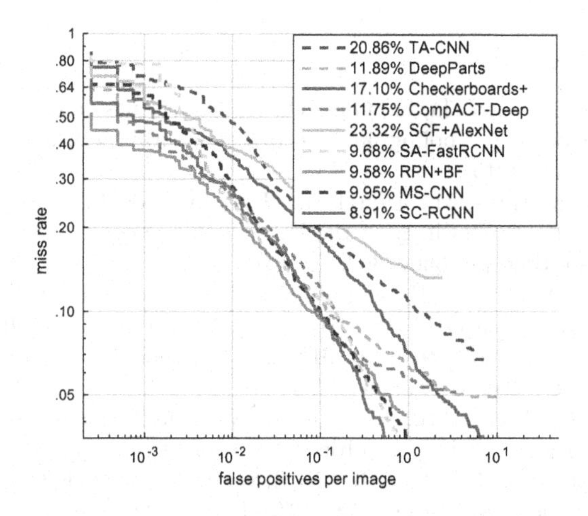

Fig. 4. Log-average miss rate on the Caltech Reasonable test set.

Table 3. Comparison of detection performance of our method with several prior state-of-the-art methods on the Caltech Reasonable test set, including test speed and miss rate.

Method	Test rate (s/image)	Miss rate (%)
Faster R-CNN [22]	**0.23**	12.65
CompACT-Deep [5]	0.37	9.68
RPN+BF [30]	0.50	9.60
Proposed	0.24	**8.91**

handling different size of pedestrian instances, but the feature maps generated by RoI Pooling is not expressive enough for efficient discrimination. Our model and RPN+BF both try to fuse feature maps from different convolutional layers, unlike RPN+BF, they trained a cascaded boosted forest as classifier, we directly feed these normalized feature map into the *fc* layers for classification and regression. The result shows that training an end-to-end system leads to smaller miss rate with faster test speed (0.24 s/image in our model compared to 0.50 s/image in RPN+BF).

5 Conclusion

In this paper, we propose an effective end-to-end pedestrian detector, which combines feature maps from different convolutional layers generated by RPN in Faster R-CNN. By fusing coarse features from high layers with fine features from low layers, we produce a more expressive feature map for the *fc* layers classifier. Our model is capable of handling pedestrian instances with different spatial scales and optical blur. Evaluation result in Caltech Pedestrian Dataset shows that our model achieves state-of-the-art performance on its benchmark and also has fastest test rate.

References

1. Bell, S., Zitnick, C.L., Bala, K., Girshick, R.: Inside-outside net: detecting objects in context with skip pooling and recurrent neural networks. arXiv preprint arXiv:1512.04143 (2015)
2. Benenson, R., Mathias, M., Tuytelaars, T., Van Gool, L.: Seeking the strongest rigid detector. In: Proceedings of the IEEE Conference on Computer Vision and Pattern Recognition, pp. 3666–3673 (2013)
3. Benenson, R., Omran, M., Hosang, J., Schiele, B.: Ten years of pedestrian detection, what have we learned? In: Agapito, L., Bronstein, M.M., Rother, C. (eds.) ECCV 2014. LNCS, vol. 8926, pp. 613–627. Springer, Cham (2015). https://doi.org/10.1007/978-3-319-16181-5_47
4. Cai, Z., Fan, Q., Feris, R.S., Vasconcelos, N.: A unified multi-scale deep convolutional neural network for fast object detection. In: Leibe, B., Matas, J., Sebe, N., Welling, M. (eds.) ECCV 2016. LNCS, vol. 9908, pp. 354–370. Springer, Cham (2016). https://doi.org/10.1007/978-3-319-46493-0_22
5. Cai, Z., Saberian, M., Vasconcelos, N.: Learning complexity-aware cascades for deep pedestrian detection. In: Proceedings of the IEEE International Conference on Computer Vision, pp. 3361–3369 (2015)
6. Dollár, P., Tu, Z., Perona, P., Belongie, S.: Integral channel features (2009)
7. Dollar, P., Wojek, C., Schiele, B., Perona, P.: Pedestrian detection: an evaluation of the state of the art. IEEE Trans. Pattern Anal. Mach. Intell. **34**(4), 743–761 (2012)
8. Geiger, A., Lenz, P., Urtasun, R.: Are we ready for autonomous driving? The KITTI vision benchmark suite. In: 2012 IEEE Conference on Computer Vision and Pattern Recognition (CVPR), pp. 3354–3361. IEEE (2012)
9. Girshick, R.: Fast R-CNN. In: Proceedings of the IEEE International Conference on Computer Vision, pp. 1440–1448 (2015)
10. Girshick, R., Donahue, J., Darrell, T., Malik, J.: Rich feature hierarchies for accurate object detection and semantic segmentation. In: Proceedings of the IEEE Conference on Computer Vision and Pattern Recognition, pp. 580–587 (2014)
11. He, K., Zhang, X., Ren, S., Sun, J.: Spatial pyramid pooling in deep convolutional networks for visual recognition. In: Fleet, D., Pajdla, T., Schiele, B., Tuytelaars, T. (eds.) ECCV 2014. LNCS, vol. 8691, pp. 346–361. Springer, Cham (2014). https://doi.org/10.1007/978-3-319-10578-9_23
12. He, K., Zhang, X., Ren, S., Sun, J.: Deep residual learning for image recognition. In: Proceedings of the IEEE Conference on Computer Vision and Pattern Recognition, pp. 770–778 (2016)
13. Hosang, J., Omran, M., Benenson, R., Schiele, B.: Taking a deeper look at pedestrians. In: Proceedings of the IEEE Conference on Computer Vision and Pattern Recognition, pp. 4073–4082 (2015)
14. Ioffe, S., Szegedy, C.: Batch normalization: accelerating deep network training by reducing internal covariate shift. arXiv preprint arXiv:1502.03167 (2015)
15. Jia, Y., Shelhamer, E., Donahue, J., Karayev, S., Long, J., Girshick, R., Guadarrama, S., Darrell, T.: Caffe: convolutional architecture for fast feature embedding. In: Proceedings of the 22nd ACM International Conference on Multimedia, pp. 675–678. ACM (2014)
16. Dai, J., Li, Y., He, K., Sun, J.: R-FCN: object detection via region-based fully convolutional networks. arXiv preprint arXiv:1605.06409 (2016)

17. Kong, T., Yao, A., Chen, Y., Sun, F.: HyperNet: towards accurate region proposal generation and joint object detection. arXiv preprint arXiv:1604.00600 (2016)
18. Li, J., Liang, X., Shen, S., Xu, T., Yan, S.: Scale-aware fast R-CNN for pedestrian detection. arXiv preprint arXiv:1510.08160 (2015)
19. Liu, W., Rabinovich, A., Berg, A.C.: ParseNet: looking wider to see better. arXiv preprint arXiv:1506.04579 (2015)
20. Long, J., Shelhamer, E., Darrell, T.: Fully convolutional networks for semantic segmentation. In: Proceedings of the IEEE Conference on Computer Vision and Pattern Recognition, pp. 3431–3440 (2015)
21. Nam, W., Dollár, P., Han, J.H.: Local decorrelation for improved pedestrian detection. In: Advances in Neural Information Processing Systems, pp. 424–432 (2014)
22. Ren, S., He, K., Girshick, R., Sun, J.: Faster R-CNN: towards real-time object detection with region proposal networks. In: Advances in Neural Information Processing Systems, pp. 91–99 (2015)
23. Russakovsky, O., Deng, J., Su, H., Krause, J., Satheesh, S., Ma, S., Huang, Z., Karpathy, A., Khosla, A., Bernstein, M., et al.: ImageNet large scale visual recognition challenge. Int. J. Comput. Vis. **115**(3), 211–252 (2015)
24. Simonyan, K., Zisserman, A.: Very deep convolutional networks for large-scale image recognition. arXiv preprint arXiv:1409.1556 (2014)
25. Tian, Y., Luo, P., Wang, X., Tang, X.: Deep learning strong parts for pedestrian detection. In: Proceedings of the IEEE International Conference on Computer Vision, pp. 1904–1912 (2015)
26. Tian, Y., Luo, P., Wang, X., Tang, X.: Pedestrian detection aided by deep learning semantic tasks. In: Proceedings of the IEEE Conference on Computer Vision and Pattern Recognition, pp. 5079–5087 (2015)
27. Xiang, Y., Choi, W., Lin, Y., Savarese, S.: Subcategory-aware convolutional neural networks for object proposals and detection. In: 2017 IEEE Winter Conference on Applications of Computer Vision (WACV), pp. 924–933. IEEE (2017)
28. Yan, J., Zhang, X., Lei, Z., Liao, S., Li, S.Z.: Robust multi-resolution pedestrian detection in traffic scenes. In: Proceedings of the IEEE Conference on Computer Vision and Pattern Recognition, pp. 3033–3040 (2013)
29. Yang, B., Yan, J., Lei, Z., Li, S.Z.: Aggregate channel features for multi-view face detection. In: 2014 IEEE International Joint Conference on Biometrics (IJCB), pp. 1–8. IEEE (2014)
30. Zhang, L., Lin, L., Liang, X., He, K.: Is faster R-CNN doing well for pedestrian detection? arXiv preprint arXiv:1607.07032 (2016)
31. Zhang, S., Benenson, R., Omran, M., Hosang, J., Schiele, B.: How far are we from solving pedestrian detection? arXiv preprint arXiv:1602.01237 (2016)
32. Zhang, S., Benenson, R., Schiele, B.: Filtered channel features for pedestrian detection. In: 2015 IEEE Conference on Computer Vision and Pattern Recognition (CVPR), pp. 1751–1760. IEEE (2015)

Edge-Preserving Background Estimation Using Most Similar Neighbor Patch for Small Target Detection

Yuehuan Wang[1,2](✉), Xueping Xu[1], Nuoning Yue[1], and Jie Chen[3]

[1] School of Automation, Huazhong University of Science and Technology, Wuhan, People's Republic of China
yuehwang@hust.edu.cn

[2] National Key Laboratory of Science and Technology on Multi-spectral Information Processing, Wuhan, People's Republic of China

[3] The 9th Designing of China Aerospace Science Industry Corp., Beijing, China

Abstract. Infrared small targets can easily be submerged in complex backgrounds in single frame infrared small target detection, the edges usually cause high false alarms and lead to erroneous detection results. In this paper, a novel edge-preserving background estimation method based on most similar neighbor patch is proposed to attenuate this problem. First, we make the best of structural features in infrared image and introduce an improved local adaptive contrast measure (ILACM) to measure the patch similarity. Then most similar neighbor patch with maximum patch similarity can be utilized to realize edge-preserving background estimation model. At last, we can obtain target image by eliminating estimated background from original image. Experiments and comparisons with state-of-the-art methods show that our method has better background estimation performance in diverse infrared images and improves SCR values of the images significantly.

Keywords: Local adaptive contrast measure
Most similar neighbor patch · Edge-preserving background estimation
Small target detection

1 Introduction

Small target detection is widely applied in remote sensing, industry, military, precision-guided munitions and many other fields. Due to the complicated backgrounds in practical application scenarios, most existing detection algorithms usually lead to many false alarms and cannot detect the target accurately. So there are still many problems needed to be further researched in the small target detection.

At present, researchers detect the small target in infrared image based on two main theoretical foundations. For one thing, the size of the small target is relatively small in the image, usually containing only a few pixels [5,6]; For the other

J. Yang et al. (Eds.): CCCV 2017, Part III, CCIS 773, pp. 84–95, 2017.
https://doi.org/10.1007/978-981-10-7305-2_8

thing, the gray value distribution is different from surrounding local back-ground, that is, the target has no spatial correlation with the surroundings [4,11–13].

Thus, a class of background estimation methods that utilize gray value differences between the small target and its surroundings is proposed, such as Local Contrast Method (LCM) [4] and kernel-based nonparametric regression method [7]. However, the edges also have a high level of contrast with their surroundings, these methods will have poor performance when edge exists and could only segregate target from the flat background image. Therefore, it is necessary to propose a method to protect the edges in the background image.

Addressing this problem, the edge-preserving background estimation methods are proposed to reduce false alarms and increase the signal clutter ratio (SCR) [9], such as bilateral filter methods [1,2]. Unfortunately, they are easy to smooth out the edges in the process of filtering. Some methods dedicate to detect the direction of edges and use it to accomplish background estimation. For example, Kim proposed Modified-Mean Subtraction Filter (M-MSF) [10], but the method could only preserve boundaries with horizontal directions. Yuan et al. proposed two-dimensional least mean square (TDLMS) filter to detect the small target [14], but owing to its sensitivity to outliers, the background estimation method usually produces unsatisfying results in junction area of different edges [3], and usually results in high false alarm rate in subsequent target detection.

In this paper, we propose a more accurate background estimation method, which utilizes structure features between patches instead of gray value contrast between pixels. First, by analyzing the structural features of flat regions, edge regions, and small target in infrared images, we find that the patch in the flat region or in a complete edge always contains same structure with its surroundings. So we can obtain an improved local adaptive contrast measure (ILACM) by calculating the summation of the Euclidean Distance between a center patch and its surrounding areas. Then, ILACM is used to measure the patch similarity among different patches. At last, we utilize the patch with maximum patch similarity to estimate the background image in order to preserve more edges in the background image.

This paper is organized as following: an improved local adaptive contrast measure (ILACM) and patch similarity based on ILACM are proposed in Sect. 2. In Sect. 3, edge-preserving background estimation method using most similar neighbor patch is introduced. Section 4 presents experimental results on the collected dataset. The conclusion will be presented in the last section.

2 Patch Similarity Based on ILACM

A typical infrared image can be divided into three parts: flat regions, edge regions and small target. Here, a typical infrared image and its 3D Surface Map are shown in Fig. 1, which consists of target (in the red box), edges (as in the blue box), and flat regions (as in the yellow box). Through the Fig. 1, we can get the following conclusions. One scene with homogeneous gray value usually has the same infrared radiation, so the patch in it always has a high similarity to the surrounding areas, such as patch b_i and its neighbor patches $b_{(i-1)}, b_{(i+1)}$. Similarly, Edge, as a complete structure, although it has different gray values with some of its neighborhoods, we can still find there are some similar regions locating along or around the edge on account of the structural integrity, such as edge patch c_i and $c_{(i-1)}, c_{(i+1)}$. While as an isolated object, the target gray value is totally different from its surrounding neighborhood pixels. So it is difficult to find its similar patch.

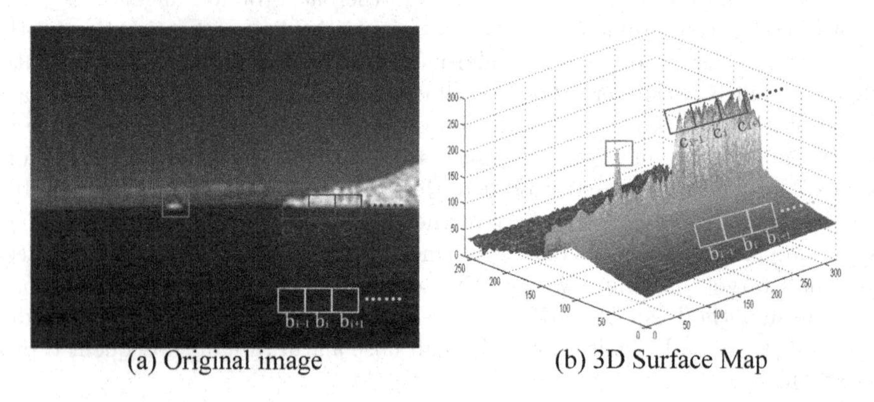

Fig. 1. One representative infrared small target image and its 3D surface map. (Color figure online)

Thus, we first propose an improved local adaptive contrast measure (ILACM). Then we use it to calculate patch similarity.

Different from LCM that only uses gray value contrast between pixels, the proposed ILACM calculates the summation of the Euclidean Distance between the center patch and its surrounding areas. The structural feature and the varied sizes of the small target are adequately considered.

Let V_0 be a patch in the image, and its size is $(2v+1) \times (2v+1)$, which can be changed with the size of target. n areas with same size around it are denoted as $V_k (k = 1, 2, \ldots n)$. Let (x_0, y_0) and (x, y) as the center pixel position in the V_0 and V_k. $I(x_0, y_0)$ and $I_k(x, y)$ denote the gray value of them. The number of pixels between (x_0, y_0) and (x, y) is depended on search scope. Thus, the kth local difference contrast is expressed by the following formula:

$$ILACM_k(x_0, y_0) = \sum_{i=-v}^{v} \sum_{j=-v}^{v} |I(x_0 + i, y_0 + j) - I_k(x + i, y + j)|^2 \quad (1)$$

Thus, once the $ILACM_k$ about a center patch and one of its surrounding areas k is obtained, the patch similarity can be calculated using Eq. (2).

$$S_k = \frac{1}{ILACM_k + 1} = \frac{1}{\sum_{i=-v}^{v} \sum_{j=-v}^{v} |I(x_0 + i, y_0 + j) - I_k(x + i, y + j)|^2 + 1} \quad (2)$$

Obviously, the dynamic range of S_k is $(0, 1]$, which should be inverse proportion to Euclidean Distance between the gray value of $I(x_0 + i, y_0 + j)$ and $I_k(x + i, y + j)$. The maximum value 1 is obtained only if all corresponding elements between V_0 and V_k are the same. The greater the difference between center patch V_0 and its neighbor V_k, the smaller patch similarity S_k will be.

Thus, when we want to estimate the gray value of the center pixel of a patch, we can first utilize Eq. (2) to find the most similar neighbor patch in its surrounding areas, and then use it to estimate the gray value of the center pixel of the patch. If an edge is included in the patch, we can also find a patch that has a high similarity with it and preserve them well in our estimates. Furthermore, the target does not have similar patches around and cant be estimated using patch similarity, so the effect of small target on background estimation can be precluded effectively.

3 Background Estimation Model Using Most Similar Neighbor Patch

In small target detection applications, the original infrared image generally consist of three components: target image, background image and noise image, which can be indicated by the following Eq. (3):

$$F(x, y) = F_T(x, y) + F_B(x, y) + F_N(x, y) \tag{3}$$

where, (x, y) is pixel coordinates in the original image. $F(x, y)$ indicates the original infrared image. $F_T(x, y)$ is target image. $F_B(x, y)$ is background image, which can be divided into uniform gray-scale regions and edge regions. The noise image $F_N(x, y)$ includes stochastic noises. Generally, clutter background has much more influence on small target detection than noise image. Thus, the challenge of small target detection is to get a fidelity background image estimation, which means that the edges in original image should be preserved in estimated background.

As mentioned above, we have proposed patch similarity based on ILACM, which is possible to search for a patch with maximum similarity to estimate the edge and preserves the edge in the background image successfully.

Here, based on patch similarity, we will propose an edge-preserving background estimation model based on most similar neighbor patch. In the Sect. 2, we have introduced the computation of patch similarity. Consequently, we should firstly search the most similar neighbor patch V_s for background estimation in its neighboring patches. Considering the center patch has the highest similarity with itself, a preserving band should be established to exclude its own impact on the estimated results. In addition, the central patch should have a higher level of similarity to the closer patches than remote ones, so the search region does not need to be too large. This search strategy not only ensures a more accurate background estimation model but also reduces computational complexity.

As depicted in Fig. 2, the search region $G(x, y)$ is decomposed into multiple small patches, the red rectangle denotes the center patch V_0, whose size is

$(2v+1)\times(2v+1)$. The blue rectangle serving as a preserving band to prevent the effect of center patch, its size is $(2p+1)\times(2p+1)$. The orange region between green and blue is effective search area, if $2v+1=\epsilon$, the coordinate range of $G(x,y)$ is $[x_0-p-\epsilon : x_0+p+\epsilon, y_0-p-\epsilon : y_0+p+\epsilon]$. Hence, the most similar neighbor patch V_s can be obtained by following equation:

$$i = \arg\max_i \, (S_1, S_2, \ldots, S_n) \tag{4}$$

$$V_s = V_i \tag{5}$$

where i represents the ith image patch in the search region, S_i indicates the patch similarity of the central patch and its ith surrounding patch in the search area.

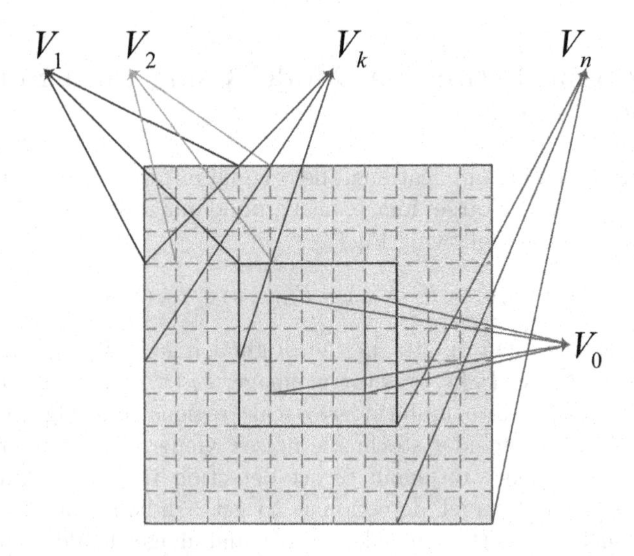

Fig. 2. Illustration of most similar neighbor patch search strategy. (Color figure online)

Furthermore, if the center pixel $I(x_0,y_0)$ and the pixel $I(x_s,y_s)$ in V_s are in the same structure, the probability of the same gray value they share will be higher. So we can estimate the center pixel using Gradient Inverse Weight [8], and it can be formulated using following equation:

$$\hat{I}(x_0,y_0) = \frac{1}{W}\sum_{x_s,y_s\in V_s}\frac{1}{|I(x_0,y_0)-I(x_s,y_s)|+1}\times I(x_s,y_s) \tag{6}$$

where $\hat{I}_{(x_0,y_0)}$ is estimated background gray value of (x_0,y_0), and W is normalization constant, which can be expressed as Eq. (7).

$$W = \sum_{x_s,y_s\in V_s}\frac{1}{|I(x_0,y_0)-I(x_s,y_s)|+1} \tag{7}$$

Based on the above analysis, we verify the proposed method using some typical real infrared images. The algorithm flow is shown in Fig. 3, the input of our model is a representative small target image against complicated background. The corresponding 3-D surface map is shown in the first column, and the small target is completely submerged in edges. Then, we utilize the proposed model to obtain the background estimation image in the second column. At last, we can get target image by eliminating the background from original image, and its 3-D surface map is displayed in the last column. As we can see, the clutters and noise residual are well eliminated, so our proposed method can achieve the purpose on edge protection.

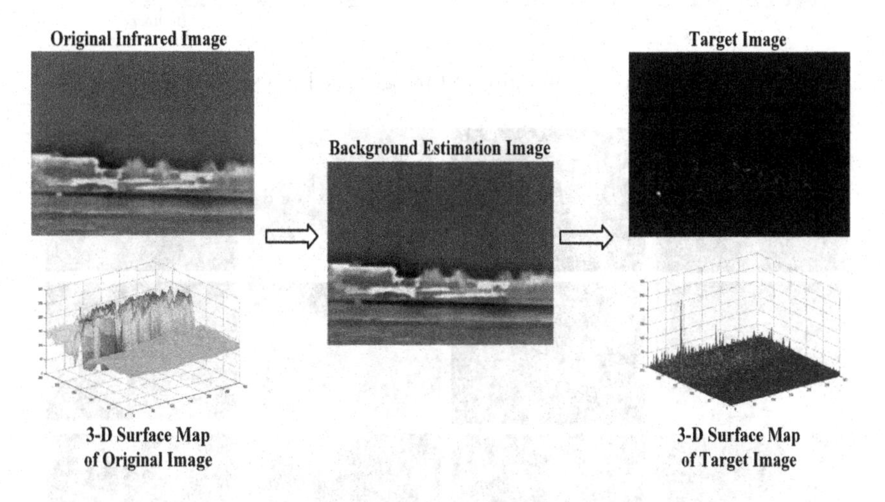

Fig. 3. Algorithmic framework of the proposed small infrared target detection system.

4 Experiments and Analysis

In this section, the edge-preserving capability of our edge-preserving background estimation method is tested in real infrared images. Then, the effectiveness and practicality of the proposed method are demonstrated by comparing with some of the arts. As shown in Fig. 4, we select eight typical infrared images. In these eight images, different edges such as buildings, sea-sky, clouds, and lands are sufficient to appraising the edge-preserving ability of different background estimation methods. In experiments, the size of center patch is 3×3 and the size of the search area is set as 11×11.

The results of the proposed method are displayed in Fig. 5, the first column (a) and third column (c) are the background estimation results. Obviously, the edge information among different scenes has been effectively and accurately preserved. As shown in column (b) and column (d), the target images are obtained by eliminating the background images from the original images. Since the edge clutter is effectively retained in the background image, it is effectively suppressed in the target image.

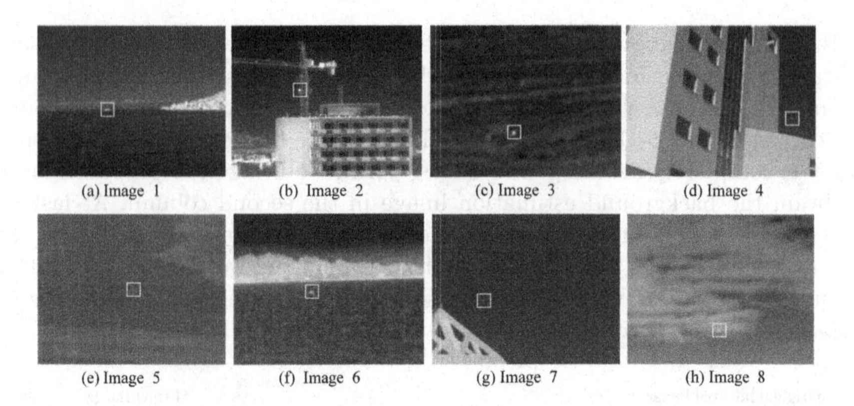

Fig. 4. Eight original infrared images used in experiments.

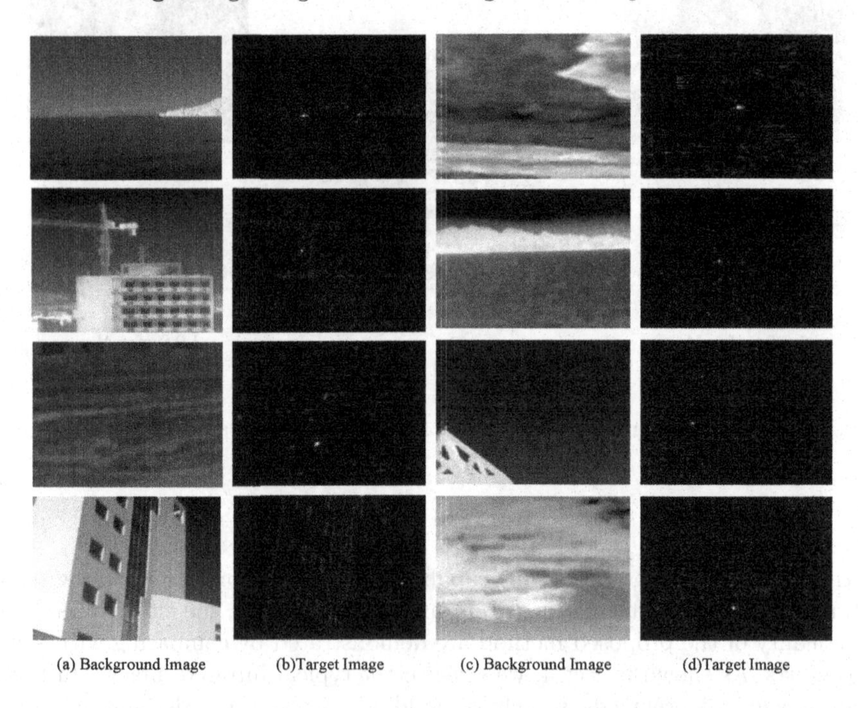

Fig. 5. Background images and target images for eight real infrared images mentioned above.

In the above experiment, we can easily find that the proposed method preserves the complicated edge in the estimated background image exactly and could implement the detection of small targets accurately. So the interference of the edges on the infrared small target detection can be effectively suppressed in our background estimation process.

Here, in order to show the superiority of our method, some baseline methods, including Bilateral Filter (BF) [1], TDLMS [14], kernel-based nonparametric

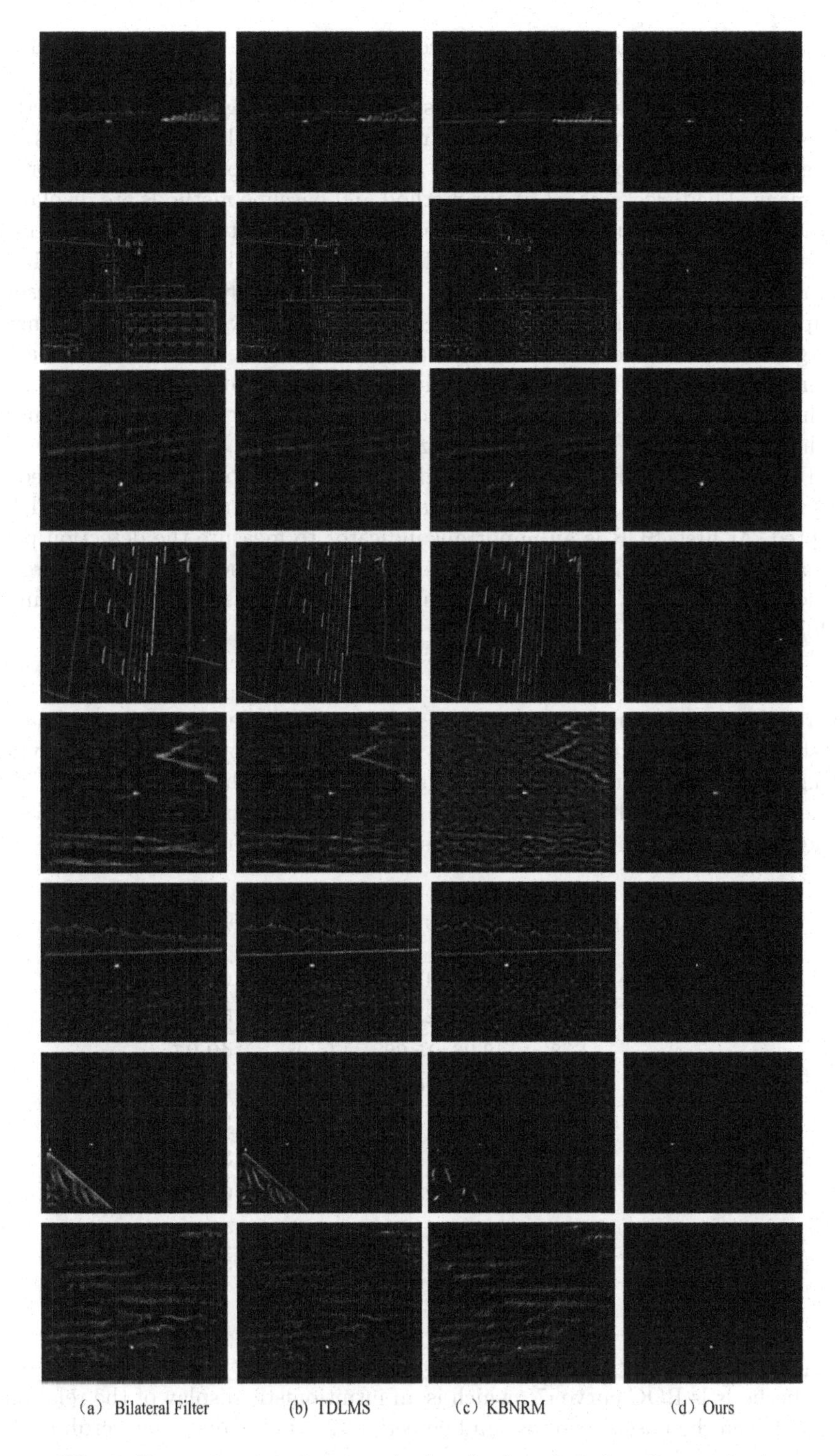

Fig. 6. Target images of our method and other three baseline methods.

regression method (KBNRM) [7] are compared with the proposed method on small target images with different complex and noisy backgrounds.

As well known that the fewer edges contained in target images, the better edge-preserving background estimation results we obtain. In the experiments, we can get target image by eliminating the estimated background image from original image. The target images of our method and baseline methods are displayed in Fig. 6, the first column is target image results of bilateral filter, the second column is results of TDLMS method, the third column is results of KBNRM, and the last column is the results of our method. Obviously, by subjective visual comparison, the filtering results in Bilateral Filter, TDLMS and KBNRM remain a large amount of edges, which mean that they are unable to estimate the background correctly. Then we can obviously observe that the target images of our method contain the least edges. It demonstrates that our background estimate method preserves the most accurate edges.

Then, in order to further illustrate the superiority of our algorithms in edge protection and target detection, qualitative analysis from two aspects will be adopted. At first, SCR is an important indicator to measure the detection performance of diverse methods, the higher the SCR, the more prominent target, the less edge clutters, and the better the detection results. The SCR is defined as follows:

$$SCR = \frac{S}{C} \tag{8}$$

where, S is the intensity of target, C is standard deviation of target image. As shown in Table 1, the target images which obtained by our edge-preserving background estimation method has highest SCR of all this four methods. The highest SCR implies that least edges are left in the target image, so our proposed background estimation method can preserve edges efficiently.

Table 1. SCRs of target images obtained by different methods.

Image	Original	BF	TDLMS	KBNRM	Ours
1	7.01	22.80	23.83	26.46	**58.22**
2	3.78	13.26	15.00	16.08	**40.97**
3	14.21	17.47	26.69	25.29	**47.55**
4	1.91	9.37	11.01	8.37	**50.89**
5	14.60	11.63	14.33	11.78	**31.62**
6	3.96	16.58	23.59	18.75	**62.39**
7	3.61	13.94	21.50	9.67	**42.84**
8	10.96	14.41	18.55	15.03	**37.58**

Another important indicator to demonstrate the effect of small target detection methods is ROC curve [7], which is an intuitionistic display of the relationship between the false alarm rate and detection rate in the detection results. The

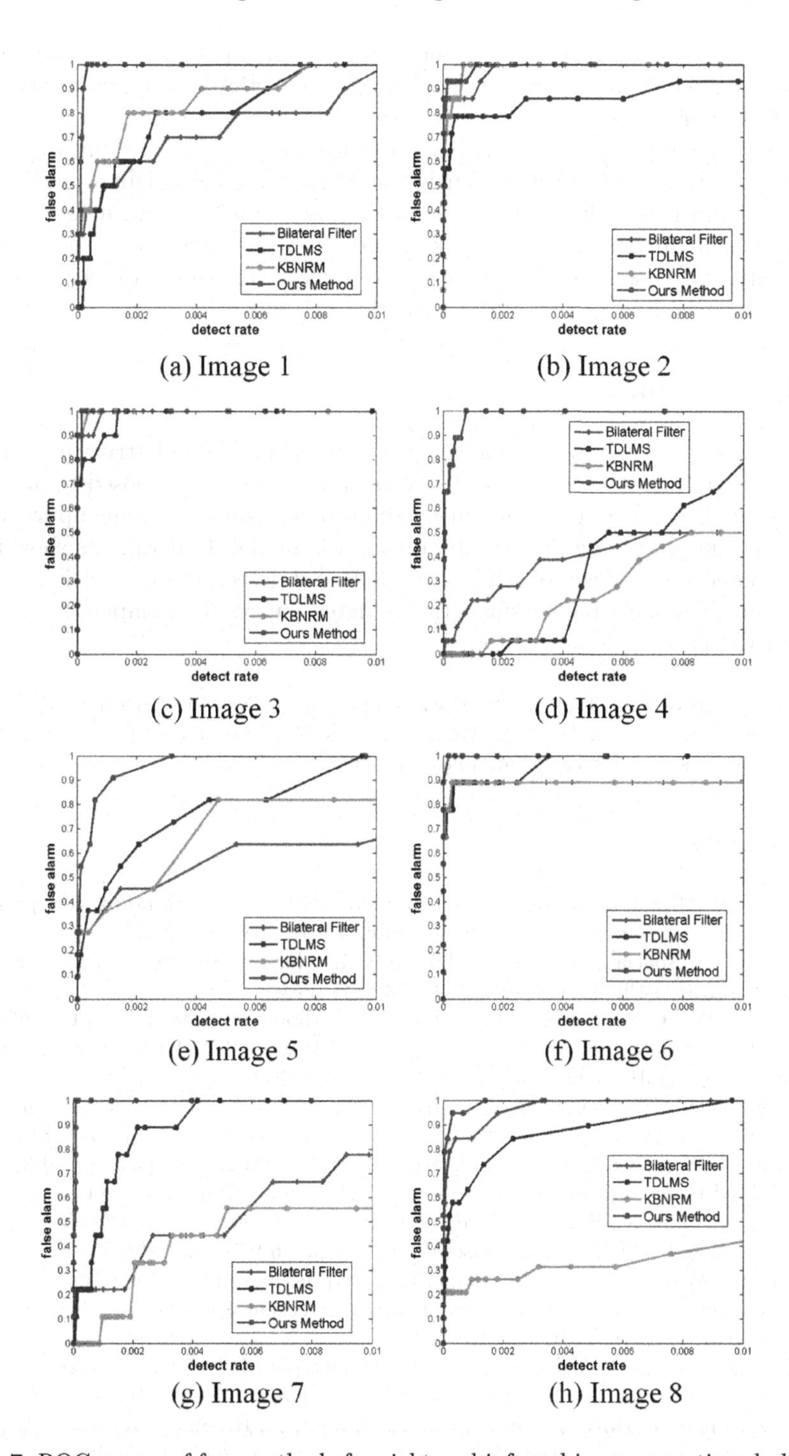

Fig. 7. ROC curves of four methods for eight real infrared images mentioned above.

ROC often determined by the clutter in the target image. If the target image contains less edge clutters, we can get a higher detection rate under a relatively lower false alarm rate.

The ROCs of comparison methods are displayed in Fig. 7. From Fig. 7, we can see that the ROCs based on our proposed most similar neighbor patch have the maximum target detection rate and the minimum false alarm rate in each image, which implies that our background estimation method contains least edge clutters. In the above experiments, the effectiveness and practicality of the proposed method have been authenticated simultaneously.

5 Conclusion

In this paper, by utilizing the local gray value contrast and structure feature of images, an improved local adaptive contrast measure is proposed to measure patch similarity. Then, most similar neighbor patch is used to come up with our effective edge-preserving background estimation model. Experiments show that our proposed method can obtain higher SCR and detection rate, get lower false alarm rate and have better small target detection results compared with the state-of-the-arts.

Acknowledgements. This research was supported by National Advanced Research Project of China No. 41415020402. We also gratefully acknowledge the editors and the anonymous reviewers for their valuable comments.

References

1. Bae, T.W.: Small target detection using bilateral filter and temporal cross product in infrared images. Infrared Phys. Technol. **54**(5), 403–411 (2011)
2. Bae, T.W.: Spatial and temporal bilateral filter for infrared small target enhancement. Infrared Phys. Technol. **63**, 42–53 (2014)
3. Bai, K., Wang, Y.: A least trimmed square method for clutter removal in infrared small target detection. In: Proceedings of the SPIE 8918, MIPPR 2013: Automatic Target Recognition and Navigation, 26 October 2013
4. Chen, C., Li, H., Wei, Y., Xia, T., Tang, Y.Y.: A local contrast method for small infrared target detection. IEEE Trans. Geosci. Remote Sens. **52**(1), 574–581 (2014)
5. Deng, H., Sun, X.: Infrared small-target detection using multiscale gray difference weighted image entropy. IEEE Trans. Geosci. Remote Sens. **52**(1), 60–72 (2016)
6. Deng, H., Sun, X.: Small infrared target detection based on weighted local difference measure. IEEE Trans. Geosci. Remote Sens. **54**(7), 4204–4215 (2016)
7. Gu, Y., Wang, C., Liu, B.: A kernel-based nonparametric regression method for clutter removal in infrared small-target detection applications. IEEE Geosci. Remote Sens. Lett. **7**(3), 469–473 (2010)
8. Kim, J., Jeong, J.: A new adaptive linear interpolation algorithm using pattern weight based on inverse gradient. In: IEEE Conference, Publications, p. 58 (2009)
9. Li, Z., Chen, J., Hou, Q., Fu, H., Dai, Z., Jin, G., Li, R.: Sparse representation for infrared dim target detection via a discriminative over-complete dictionary learned online. Sensors **14**(6), 9451–9470 (2014)

10. Kim, S., Yang, Y., Lee, J.: Horizontal small target detection with cooperative background estimation and removal filter. In: International Conference on Acoustics, Speech, and Signal Processing, pp. 1761–1764 (2011)
11. Sun, S.G., Kwak, D.M., Jang, W.B., Jong Kim, D.: Small target detection using center-surround difference with locally adaptive threshold. In: Proceedings of the 4th International Symposium on Image and Signal Processing and Analysis, pp. 402–407 (2014)
12. Wang, B., Liu, S., Li, Q., Lei, R.: Blind-pixel correction algorithm for an infrared focal plane array based on moving-scene analysis. Opt. Eng. **45**(3), 364–367 (2006)
13. Xie, K., Fu, K., Zhou, T., Zhang, J., Yang, J., Wu, Q.: Small target detection based on accumulated center-surround difference measure. Infrared Phys. Technol. **67**, 234–778 (2014)
14. Yuan, C., Liu, R., Yang, J.: Small target detection using two-dimensional least mean square (TDLMS) filter based on neighborhood analysis. Sensors **29**(2), 188–200 (2008)

Space Target Detection in Video Satellite Image via Prior Information

Xueyang Zhang(✉) and Junhua Xiang

College of Aerospace Science and Engineering,
National University of Defense Technology, Changsha 410073, China
zxy1135@qq.com, xiangjunhua@126.com

Abstract. Compared to ground-based observation, space-based observation is an effective approach to catalog and monitor increasing space targets. In this paper space target detection in video satellite image with star image background is studied. An adaptive space target detector based on prior information is proposed. Firstly, bilateral filter is used to decrease noise. Then a single frame image is segmented using adaptive thresholding. Considering the continuity of target motion and brightness change, adaptive thresholding is based on local image properties and prior information of previous frames detection and Kalman filter. Then the algorithm uses the correlation of target motion in multi-frame to detect the target from stars. Experimental results with video image from Tiantuo-2 satellite show that this algorithm provides a good way for space target detection.

Keywords: Space target detection · Small target detection
Video satellite · Adaptive thresholding · Kalman filter
Tiantuo-2 satellite

1 Introduction

In September 8, 2014, Tiantuo-2 (TT-2) designed by National University of Defense Technology independently was successfully launched into orbit, which is the first Chinese interactive earth observation microsatellite using video imaging system. The mass was 67 kg, and the altitude was 490 km. An experiment based on interactive control strategies with human in the loop was carried out to realize continuous tracking and monitoring of moving targets.

Now video satellite has been widely concerned by domestic and foreign researchers, and several video satellites have been launched into orbits, such as Skysat-1 and 2 by Skybox Imaging, TUBSAT series satellites by Technical University of Berlin [15], and the video satellite by Chang Guang Satellite Technology. They can obtain video image with different on-orbit performance.

Increasing space objects have a greater impact on human space activities, and cataloging and monitoring them become a hot issue in the field of space environment [4,10,13,16]. Compared to ground-based observation, space-based

J. Yang et al. (Eds.): CCCV 2017, Part III, CCIS 773, pp. 96–107, 2017.
https://doi.org/10.1007/978-981-10-7305-2_9

observation is not restricted by weather or geographical location, and avoids the disturbance of the atmosphere to the objects signal, which has a unique advantage [4,10,19]. To use video satellite to observe space objects is an effective approach.

Target detection and tracking in satellite video image is an important part of space-based observation. For general problem of target detection and tracking in video, Optical-flow [6], block-matching [8], template detection [3,12] and so on have been proposed. But most of these methods are based on gray-level and enough texture information is highly required [26]. In fact, small target detection in optical images with star background mainly has following difficulties: (1) because the target occupies only one or a few pixels in the image, the shape of the target is not available. (2) Due to background stars and noise introduced by space environment detecting equipment, the target almost submerges in the complex background bright spots, which increases the difficulty of target detection greatly. (3) Attitude motion of the target leads to changing brightness, even losing the target in several frames.

Aiming at these difficulties, many scholars have proposed various algorithms, mainly including Track before Detection (TBD) and Detect before Tracking (DBT). Multistage Hypothesizing Testing (MHT) [1] and dynamic programming based algorithm [9,18,25] can be classified into TBD, which is effective when the image Signal to Noise Ratio is very low. But high computation complexity and hard threshold always follow them [14]. Actually, DBT is usually adopted for target detection in star images [2,7,22]. A space target detection algorithm in video based on motion information was proposed by [24], which solved the first and second difficulty partly. But the third difficulty remains unsolved and may lead to target losing.

Aiming at the third difficulty, an adaptive space target detector based on prior information is proposed in this paper. Considering the continuity of target motion and brightness change, adaptive thresholding based on local image properties and prior information of previous frames detection and Kalman filter is used to segment a single frame image. Experimental results about video image from TT-2 demonstrate the effectiveness of the algorithm.

2 Characteristic Analysis of Satellite Video Image

Video satellite image of space contains deep space background, stars, targets, and noise introduced by imaging devices and cosmic rays. The mathematical model is given by [20]

$$f(x,y,k) = f_B(x,y,k) + f_s(x,y,k) + f_T(x,y,k) + n(x,y,k) \quad (1)$$

where $f_B(x,y,k)$ is gray level of deep space background, $f_s(x,y,k)$ is gray level of stars, $f_T(x,y,k)$ is gray level of targets and $n(x,y,k)$ is gray level of noise. (x,y) is pixel coordinate in the image. k is the number of the frame.

In video image, stars and weak small targets occupy only one or a few pixels. It is difficult to distinguish the target from background stars by morphological characteristics or photometric features. Besides, attitude motion of the target leads to changing brightness, even losing the target in several frames. Therefore its almost impossible to detect the target in a single frame image, and necessary to use the continuity of target motion and brightness change in multi-frame images.

Figure 1 is local image of video from TT-2, where (a) is star and (b) is target (debris). Its impossible to distinguish them by morphological characteristics or photometric features.

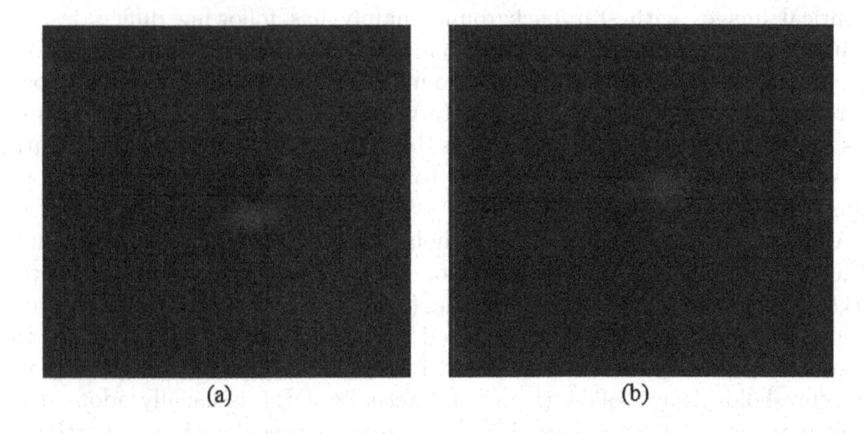

Fig. 1. Images of star and target

3 Target Detection via Prior Information

When attitude of video satellite is stabilized, background stars are moving extremely slowly in the video and can be considered static in several frames. At the same time, noise is random (dead pixels appear in a fixed position) and only targets are moving continuously. This is the most important distinction on motion characteristics of their image. Because of platform jitter, targets cant be detected by simple frame difference method. An adaptive space target detector based on prior information is proposed. The prior information comes from previous frames detection and Kalman filter. The procedure of the target detector is shown in Fig. 2.

3.1 Image Denoising

Noise in video image mainly includes the space radiation noise, the space background noise, and the CCD dark current noise and so on. Bilateral filter can be used to denoise, which is a simple, non-iterative scheme for edge-preserving smoothing [17]. The weights of the filter have two components, the first of which is the same weighting used by the Gaussian filter. The second component takes

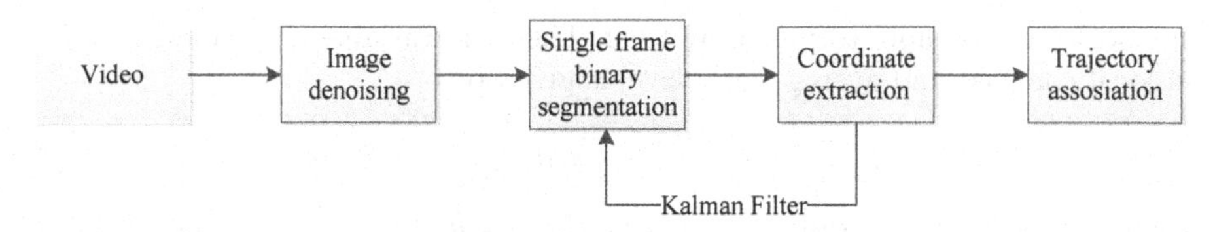

Fig. 2. Procedure of target detector

into account the difference in intensity between the neighboring pixels and the evaluated one. The diameter of the filter is set to 5, and weights are given by

$$w_{ij} = \frac{1}{K} \exp^{-(x_i-x_c)^2+(y_i-y_c)^2/2\sigma_s^2} \exp^{-(f(x_i,y_i)-f(x_c,y_c))^2/2\sigma_r^2} \quad (2)$$

where K is the normalization constant, (x_c, y_c) is the center of the filter, $f(x, y)$ is the gray level at (x, y), σ_s is set to 10 and σ_r is set to 75.

3.2 Single Frame Binary Segmentation

In order to distinguish between stars and targets, it is necessary to remove the background in each frame image, but stray light leads to uneven gray level distribution of background. Figure 3(a) is a single frame of video image and Fig. 3(b) is its gray level histogram. The single peak shape of the gray histogram shows that classical global threshold method, which is traditionally used to segment star image [5,20–23,25], cant be used to segment the video image.

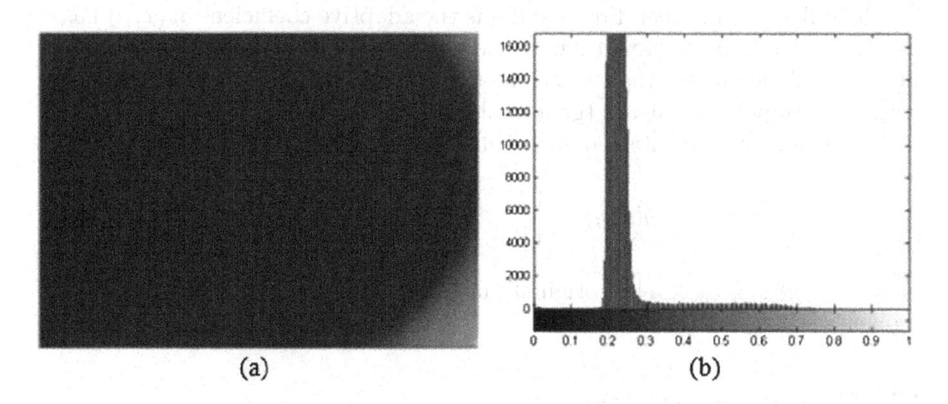

Fig. 3. Single frame image and its gray level histogram

Whether it is a star or a target, its gray level is greater than the pixels in its neighborhood. Consider using variable thresholding based on local image properties to segment image. Calculate standard deviation σ_{xy} and mean value m_{xy} for the neighborhood of every point (x, y) in image, which are descriptors of the local contrast and average gray level. Then the variable thresholding based on local contrast and average gray level is given by

$$T_{xy} = a\sigma_{xy} + bm_{xy} \tag{3}$$

where a and b are constant and greater than 0. b is the contribution of local average gray level to the thresholding and can be set to 1. a is the contribution of local contrast to the thresholding and is the main parameter to set according to the target characteristic.

But on the other hand, for space moving targets, their brightness is sometimes changing according to the changing attitude. If a was set to be constant, the target would be lost in some frames. Considering the continuity of target motion, if in kth frame (x, y) is detected as probable target coordinate, the target detection probability is much greater in the 5×5 window of (x, y) in $(k+1)$th frame. Integrated with the continuity of target brightness change, if there is no probable target detected in the 5×5 window of (x, y) in $(k+1)$th frame, a_k can be reduced with a factor, i.e. $a_k\rho(\rho < 1)$. So the adaptive thresholding based on local image properties and prior information of previous frames detection and Kalman filter is given by

$$\begin{aligned} T_k(x, y) &= a_k(x, y)\sigma_{xy} + bm_{xy} \\ a_k(x, y) &= a_{k-1}(x, y)\rho^{P(x,y,k|k-1)} \end{aligned} \tag{4}$$

where $P(x, y, k|k-1)$ is the probability that (x, y) in kth frame is in the 5×5 window of the predicted coordinate of probable target detected in $(k-1)$th frame, equals 0 or 1. The predicted coordinate is derived from Kalman filter, as shown in next section.

The difference between Eqs. 3 and 4 is the adaptive coefficient $a_k(x, y)$ based on prior information of previous frames detection and Kalman filter. a is initially set to be a little greater than 1 and reset to the initial value when the gray level at (x, y) becomes large again (greater than 150).

Image binary segmentation algorithm is given by

$$g(x, y) = \begin{cases} 1 & \text{if } f(x, y) > T_{xy} \\ 0 & \text{if } f(x, y) \le T_{xy} \end{cases} \tag{5}$$

where $f(x, y)$ is gray level of original image at (x, y) and $g(x, y)$ is gray level of segmented image at (x, y).

3.3 Coordinate Extraction

In the ideal optical system, the point target in the CCD focal plane occupies one pixel, but in practical imaging condition, circular aperture diffraction results in the target in the focal plane diffusing multi-pixel. So the coordinate of the target in the pixel frame is determined by the position of the center of gray level. Simple gray weighted centroid algorithm is used to calculate the coordinates, with positioning accuracy up to 0.1–0.3 pixels [11], by

$$(x_S, y_S) = \frac{\sum\limits_{(x,y)\in S} f(x,y)\cdot(x,y)}{\sum\limits_{(x,y)\in S} f(x,y)} \tag{6}$$

where S is target area after segmentation, $f(x, y)$ is gray level of original image at (x, y), and (x_S, y_S) is the coordinate of S.

Kalman filter is an efficient recursive filter that estimates the internal state of a linear dynamic system from a series of noisy measurements, which can be used to predict probable targets coordinate in next frame.

Assuming the state vector of the target at the kth frame is $\mathbf{x}_k = (x_k, y_k, vx_k, vy_k)^T$, i.e. its coordinate and velocity (unit: pixel/Δt) in pixel frame, and the system equation is

$$\begin{cases} \mathbf{x}_k = F\mathbf{x}_{k-1} + \mathbf{w} = \begin{bmatrix} 1\,0\,1\,0 \\ 0\,1\,0\,1 \\ 0\,0\,1\,0 \\ 0\,0\,0\,1 \end{bmatrix} \mathbf{x}_{k-1} + \mathbf{w} \\ \mathbf{z}_k = H\mathbf{x}_k + \mathbf{v} = \begin{bmatrix} 1\,0\,0\,0 \\ 0\,1\,0\,0 \end{bmatrix} \mathbf{x}_k + \mathbf{v} \end{cases} \tag{7}$$

where F is the state transition matrix, H is the measurement matrix, $\mathbf{z}_k$ is the measurement vector, i.e. the coordinate at the kth frame, $\mathbf{w}$ is the process noise which is assumed to be zero mean Gaussian white noise with covariance Q, denoted as $\mathbf{w} \sim N(0, Q)$, and $\mathbf{v}$ is the measurement noise which is assumed to be zero mean Gaussian white noise with covariance R, denoted as $\mathbf{v} \sim N(0, R)$.

Let $\hat{\mathbf{x}}_{k|l}$ be the estimate of $\mathbf{x}_k$ given measurements up to and including at the lth frame, where $l \le k$, $P_{k|k}$ be the posteriori error covariance matrix to measure the estimated accuracy of the state estimate. Then $\hat{\mathbf{x}}_{k|k}$ and $P_{k|k}$ represent the state of the filter. The procedure of Kalman filter is as following:

Initialization. Initialize $\hat{\mathbf{x}}_{0|0}$ and $P_{0|0}$. For $\hat{\mathbf{x}}_{0|0} = (x_{0|0}, y_{0|0}, vx_{0|0}, vy_{0|0})^T$, $x_{0|0}, y_{0|0}$ are obtained by single frame binary segmentation and gray weighted centroid algorithm, and $vx_{0|0}, vy_{0|0}$ is set to zero.

Prediction. The prediction phase uses the state estimate from the previous frame to produce an estimate of the state at the current frame. The predicted coordinates will be used in single frame binary segmentation and trajectory association.

$$\begin{aligned} \hat{\mathbf{x}}_{k|k-1} &= F\hat{\mathbf{x}}_{k-1|k-1} \\ P_{k|k-1} &= FP_{k-1|k-1}F^T + Q \end{aligned} \tag{8}$$

Update. The update phase combined the prediction with the current measurement to refine the state estimate.

$$\begin{aligned} K_k &= P_{k|k-1}H^T(HP_{k|k-1}H^T + R)^{-1} \\ \hat{\mathbf{x}}_{k|k} &= \hat{\mathbf{x}}_{k|k-1} + K_k(\mathbf{z}_k - H\hat{\mathbf{x}}_{k|k-1}) \\ P_{k|k} &= (I - K_kH)P_{k|k-1} \end{aligned} \tag{9}$$

where K_k is an intermediate variable called the Kalman gain.

3.4 Trajectory Association

After processing a frame image, some potential target coordinates are obtained. Associate them with existing trajectories, or generate new trajectories, by nearest neighborhood filter. The radius of the neighborhood is determined by target characteristic, which needs to be greater than the moving distance of target image in a frame and endure losing the target in several frames.

When a trajectory has 20 points, need to judge whether its a target. Velocity in the state vector is used. If the norm of sum of the velocity of the points in the trajectory is greater than a given thresholding, its a target; otherwise its not. That is to judge whether the point is moving. The thresholding here is mainly to remove image motion caused by the instability of the satellite platform and other noise. The thresholding is usually set to 2.

Thus targets are detected based on the continuity of motion in multi frame and trajectories are updated.

4 Experimental Results

The adaptive space target detector is verified using video image from TT-2. TT-2 has 4 space video sensors and video image used in this paper is from the high-resolution camera, whose focal length is 1000 mm, FOV is $2°30'$, and $d_x = d_y = 8.33\,\mu\text{m}$. Video image is 25 frames per second and the resolution is 960×576.

Continuous 1000 frame images taken in attitude stabilization are processed. In Eq. 4, a_{xy} is initially set to 1.1 and b is set to 1. The reduced factor is set to 0.8. The neighborhood in image segmentation is set to 7×7. In Eq. 7, Q is set to $10^{-4}I_4$ and R is set to $0.2I_2$, where I_n is the $n \times n$ identity matrix. In the initialization of Kalman filter, $P_{0|0}$ is set to $\begin{bmatrix} 0.2 & 0 & 0 & 0 \\ 0 & 0.2 & 0 & 0 \\ 0 & 0 & 0 & 0 \\ 0 & 0 & 0 & 0 \end{bmatrix}$. The radius of neighborhood in trajectory association is set to 5.

In Fig. 4, 3 target areas are obtained after segmenting the 9th frame (as shown in (a)) and 5 target areas are obtained after segmenting the 30th frame (as shown in (b)). These areas include targets, stars and noise, which cant be distinguished in a single frame. Figure 4 also shows the brightness change of the target.

Figure 5 is derived by overlaying the 1000 frame images. The trajectory of the target can be found in the white box of Fig. 5. Figure 5 shows that the brightness of the target varies considerably.

The adaptive space target detector detects the target well as shown in Fig. 6. For image of 960×576 is too large, Fig. 6 is interested in local image and a white box is used to identify the target every 50 frames. Figure 7 gives the trajectory of the target.

Fig. 4. Result of segmenting the 9th and 30th frame image

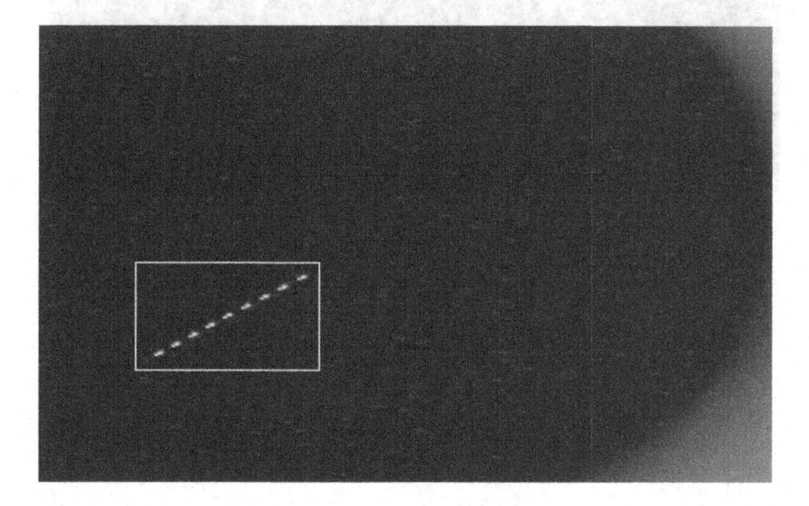

Fig. 5. Overlay of the 1000 frame images

Besides, for these 1000 frames, if a_{xy} in Eq. 4 is set to constant, the target will be lost in 421 frames whereas Eq. 4 with adaptive coefficient detected the target in 947 frames. Probability of detection improved a lot.

Another example is to use the adaptive space target detector to process an overexposed video of 187 frames taken in attitude stabilization. Overlaying the 187 frame images gives Fig. 8. The trajectory of the target in the black box of Fig. 8 is easily neglected by naked eyes.

The adaptive space target detector detects the target well as shown in Fig. 9. Figure 9 are local images of the 30th, 60th, 90th, 120th, 150th, and 180th frame and a white box is used to identify the target at each frame. Figure 10 gives the trajectory of the target.

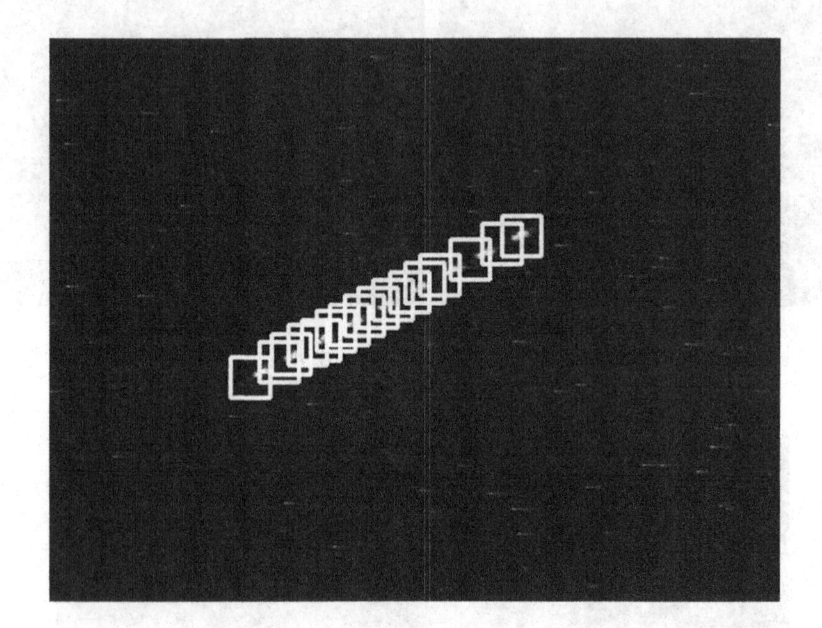

Fig. 6. Detecting the target in the image

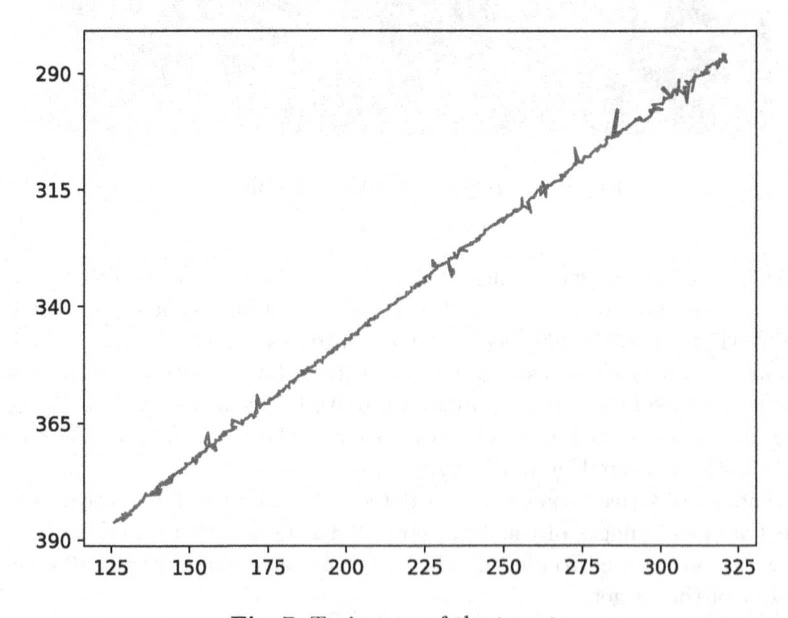

Fig. 7. Trajectory of the target

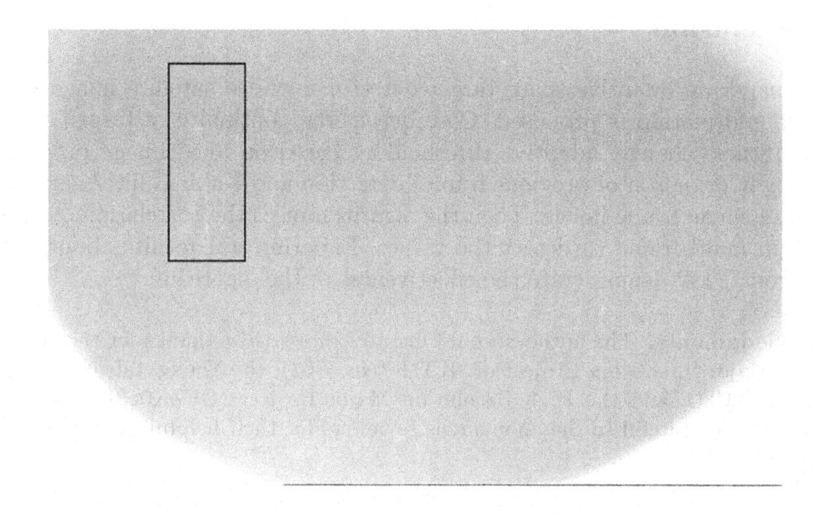

Fig. 8. Overlay of the 187 frame overexposed images

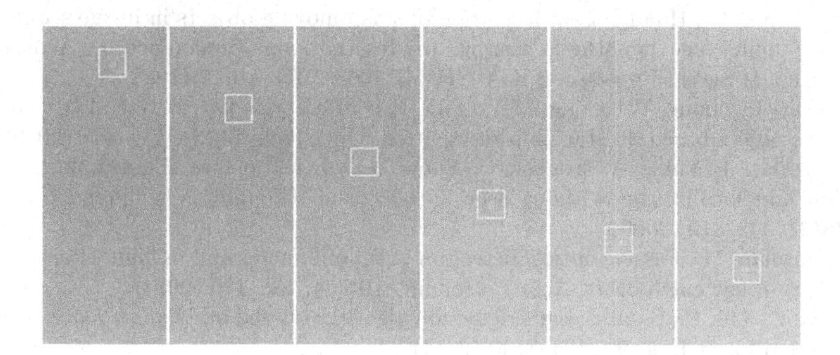

Fig. 9. Detecting the target in the overexposed images

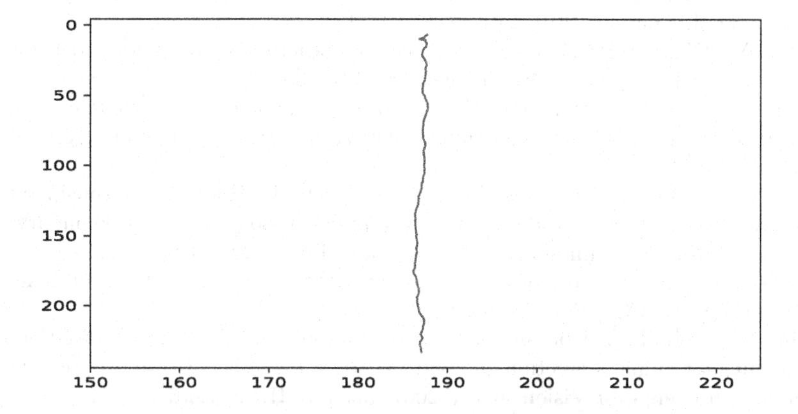

Fig. 10. Trajectory of the target in an overexposed video

5 Conclusion

In this paper an adaptive space target detector in video satellite image based on prior information is proposed. Considering the continuity of target motion and brightness change, adaptive thresholding based on local image properties and prior information of previous frames detection and Kalman filter is used to segment a single frame image. Then the algorithm uses the correlation of target motion in multi-frame to detect the target. Experimental results about video image from TT-2 demonstrate the effectiveness of the algorithm.

Acknowledgments. The authors would like to express their thanks for the support from the Major Innovation Project of NUDT (No. 7-03), the Young Talents Training Project of NUDT and the High Resolution Major Project (GFZX04010801). Also, the authors are grateful to the anonymous reviewers for their helpful suggestions and remarks.

References

1. Blostein, S.D., Huang, T.S.: Detection of small moving objects in image sequences using multistage hypothesis testing. In: International Conference on Acoustics, Speech, & Signal Processing, ICASSP, pp. 1068–1071, vol. 2 (1988)
2. Cheng, J., Zhang, W., Cong, M.Y., Pan, H.B.: Research of detecting algorithm for space object based on star map recognition. Opt. Tech. **36**(3), 439–444 (2010)
3. Coughlan, J., Yuille, A., English, C., Snow, D.: Efficient deformable template detection and localization without user initialization. Comput. Vis. Image Underst. **78**(3), 303–319 (2000)
4. Gruntman, M.: Passive optical detection of submillimeter and millimeter size space debris in low earth orbit. Acta Astronaut. **105**(1), 156–170 (2014)
5. Han, Y., Liu, F.: Small targets detection algorithm based on triangle match space. Infrared Laser Eng. **9**, 3134–3140 (2014)
6. Horn, B.K.P., Schunck, B.G.: Determining optical flow. Artif. Intell. **17**(81), 185–203 (1980)
7. Huang, T., Xiong, Y., Li, Z., Zhou, Y., Li, Y.: Space target tracking by variance detection. J. Comput. **9**(9), 2107–2115 (2014)
8. Jing, X., Chau, L.P.: An efficient three-step search algorithm for block motion estimation. IEEE Trans. Multimedia **6**(3), 435–438 (2004)
9. Johnston, L.A., Krishnamurthy, V.: Performance analysis of a dynamic programming track before detect algorithm. IEEE Trans. Aerosp. Electron. Syst. **38**(1), 228–242 (2002)
10. Krutz, U., Jahn, H., Kuhrt, E., Mottola, S., Spietz, P.: Radiometric considerations for the detection of space debris with an optical sensor in leo as a secondary goal of the asteroidfinder mission. Acta Astronaut. **69**(5), 297–306 (2011)
11. Liebe, C.C.: Accuracy performance of star trackers-a tutorial. IEEE Trans. Aerosp. Electron. Syst. **38**(2), 587–599 (2002)
12. Lin, Z., Davis, L.S., Doermann, D., DeMenthon, D.: Hierarchical part-template matching for human detection and segmentation. In: IEEE 11th International Conference on Computer Vision, ICCV 2007, pp. 1–8. IEEE (2007)
13. Phipps, C.: A laser-optical system to re-enter or lower low earth orbit space debris. Acta Astronaut. **93**, 418–429 (2014)

14. Punithakumar, K., Kirubarajan, T., Sinha, A.: A sequential monte carlo probability hypothesis density algorithm for multitarget track-before-detect. In: Proceedings of SPIE, vol. 59131S, p.59131S-8 (2005)
15. Steckling, M., Renner, U., Röser, H.P.: DLR-TUBSAT, qualification of high precision attitude control in orbit. Acta Astronaut. **39**(9), 951–960 (1996)
16. Sun, R., Zhan, J., Zhao, C., Zhang, X.: Algorithms and applications for detecting faint space debris in GEO. Acta Astronaut. **110**, 9–17 (2015)
17. Tomasi, C., Manduchi, R.: Bilateral filtering for gray and color images. In: International Conference on Computer Vision, pp. 839–846 (1998)
18. Tonissen, S.M., Evans, R.J.: Performance of dynamic programming techniques for track-before-detect. IEEE Trans. Aerosp. Electron. Syst. **32**(4), 1440–1451 (1996)
19. Wang, Y., Mitrovic Minic, S., Leitch, R., Punnen, A.P.: A grasp for next generation sapphire image acquisition scheduling. Int. J. Aerosp. Eng. **2016**, 7 (2016). https://doi.org/10.1155/2016/3518537
20. Wang, Z., Zhang, Y.: Algorithm for CCD star image rapid locating. Chin. J. Space Sci. **26**(3), 209–214 (2006)
21. Yao, R., Zhang, Y.N., Yang, T., Duan, F.: Detection of small space target based on iterative distance classification and trajectory association. Opt. Precis. Eng. **20**(1), 179–189 (2012)
22. Zhang, C.H., Chen, B., Zhou, X.D.: Small target trace acquisition algorithm for sequence star images with moving background. Opt. Precis. Eng. **16**(3), 524–530 (2008)
23. Zhang, J., Xi, X., Zhou, X.: Space target detection in star image based on motion information. In: ISPDI 2013 - Fifth International Symposium on Photoelectronic Detection and Imaging, vol. 8913, p. 891307 (2013)
24. Zhang, X., Xiang, J.: Space target detection in video based on motion information. J. Nav. Aeronaut. Astronaut. Univ. **31**(2), 113–116 (2016)
25. Zhang, Y.Y., Wang, C.X.: Space small targets detection based on improved DPA. Acta Electronica Sinica **38**(3), 556–560 (2010)
26. Zhu, Y., Hu, W., Zhou, J., Duan, F., Sun, J., Jiang, L.: A new starry images matching method in dim and small space target detection. In: IEEE Computer Society Fifth International Conference on Image and Graphics, pp. 447–450 (2009)

Scene Recognition with Sequential Object Context

Yuelian Wang and Wei Pan(✉)

College of Information Engineering, Capital Normal University, Beijing 100048, China
wangyuelian2355@gmail.com, bjpanwei@163.com

Abstract. Convolutional Neural Networks (CNNs) have been widely used for many computer vision tasks and produce discriminative and rich representations for images or regions of an image. Recognizing scenes requires both local object features and global semantic information as a scene image is usually composed of multiple objects which are organized with specific spatial distribution. To address these problems, in this paper, we propose a deep network architecture which models the sequential object context of scenes to capture object level information. We first detect a set of obejcts in a scene image, and then apply a pre-trained CNN to extract discriminative features for these objects. Then we use a Long Short-Term Memory (LSTM) network to get the context features by progressively receiving all contextual objects. The learned sequential object context incorporates object-object relationship and object-scene relationship in an end-to-end trainable manner. We evaluate our model on two benchmark datasets and achieve promising results compared to state-of-the-art methods.

Keywords: Convolutional neural network · Long short-term memory Scene recognition · Sequential object context

1 Introduction

Scene recognition is a critical task in the computer vision community. It has a wide range of applications such as assistive human companions, robotic agent path planning, monitoring systems and so on. State-of-the-art approaches in scene recognition are based on the successful combination of deep representations and large-scale datasets. Specifically, deep convolutional neural networks (CNNs) trained on ImageNet [24] and Places [37] have shown significant improvement in performance over methods using hand-engineered features and have been used to set baseline performance for visual recognition.

As a scene images is usually composed of multiple objects which are organized with specific spatial distribution, classifying it requires not only the holistic features of the whole image, but also the local features of objects in the image. However, CNNs learn image features in a layer-wise manner where low layers capture general features that resemble either Gabor filters or color blobs and

J. Yang et al. (Eds.): CCCV 2017, Part III, CCIS 773, pp. 108–119, 2017.
https://doi.org/10.1007/978-981-10-7305-2_10

high layers learn specific features which are semantic and representative even though they greatly depend on the chosen dataset and task [35]. The low layer general features are gradually transformed into high layer powerful features with multiple convolutional layers and pooling layers. This feature learning mechanism of CNNs suggests that they might not be the best suited architectures for classifying scene images where local object features follow a complex distribution in the spatial space. The reason is that the spatial aggregation implementation of pooling layers in a CNN is simple in some extent, and does not retain much information about local feature distributions. When crucial inference happens in the fully connected layers near the top of the CNN, aggregated features fed into these layers are in fact global features that neglect local feature distributions. The global CNN features are not efficient enough to capture contextual knowledge like the complex interaction of objects in a scene.

In addition to the entire image, it has been demonstrated that an image representation based on objects can be very useful in visual recognition tasks for scenes. Li *et al.* [14] propose a high-level image representation where an image is represented as a scale-invariant response map of a large number of pre-trained generic object detectors. The object-based representation carries rich semantic level image information and achieves superior performance on many high level visual recognition tasks. Li *et al.* [15] propose a hierarchical probabilistic graphical model to perform scene classification with the contextual information in form of object co-occurrence is explicitly represented by a probabilistic chain structure. Liao *et al.* [19] propose a architecture which encourages deep neural networks to incorporate object-level information with a regularization of semantic segmentation for scene recognition. Wu *et al.* [31] use a region proposal technique to generate a set of high-quality patches potentially containing objects and then a scene image representation is obtained by pooling the feature response maps of all the learned meta objects at multiple spatial scales to retain more information about their local spatial distribution.

In this paper, we propose a architecture to learn sequential object context which encodes rich object-level context using a LSTM network on top of a set of discriminative objects, as shown in Fig. 1. This architecture attempts to learn powerful semantic representations in scenes by modeling object-object and scene-object relationships within a single system. The intuition is that we human beings first scan objects in an image and then reason the relationships between these objects to decide what scene category the image belongs to. The joint existence of a set of objects in a scene highly influences the final scene category. Additionally, the LSTM units are capable of modeling the relationship between the objects by progressively taking in object features at each time step.

In our framework, we first use a region proposal network to detect a set of objects for each scene image. These objects are sorted by their locations in the image to form the object-based context sequence. And then, we use scene-centric Places CNN to extract features for the whole image to capture global scene information. At the same time, we use object-centric ImageNet CNN to extract features for the detected objects. After this, the representations of the

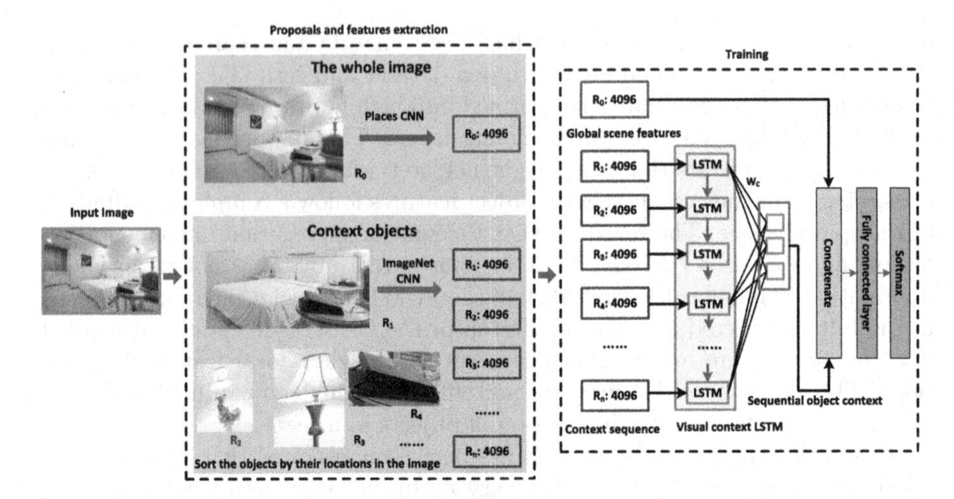

Fig. 1. The framework of our method. Firstly, a set of objects are detected in a scene image. The objects are arranged by their locations in the image. And then we use Places CNN and ImageNet CNN to extract features for the whole image and these objects respectively. The features of these objects are put into the visual context LSTM to form the sequential object context. At last, the global scene features of the whole image and the learned context are concatenated and the combination of them are put into a sub network to classify the image.

object context sequence are put into a LSTM network to form the discriminative and representative sequential object context. At last, the global scene features of the whole image and the learned context are concatenated and the combination of them are put into a sub network to classify the image. In this way, the network can learn about the scene class probability distribution given it has seen a specific set of objects through time. In summary, the main contributions of our paper are as follows:

- We firstly use an LSTM network to explicitly learn sequential object context for scene recognition. The learned discriminative and representative context contains information from all of the objects in the image.
- We empirically show that the sequential object context is complementary to the global scene information extracted form the whole image. Leveraging both global scene features and local sequential object context, our method achieves promising results compared to state-of-the-art methods on many challenging benchmarks.

The rest of the paper is organized as follows. We give a brief overview of related work in Sect. 2. Section 3 describes the proposed method. Section 4 describes the experiments and Sect. 5 concludes the paper.

2 Related Work

Scene recognition. Earlier work in scene recognition focuses on carefully hand crafted representations such as image contours, high contrast points, histogram of oriented gradients and so on [28]. Recently, with the great success of deep convolutional networks, features extracted from CNNs have been the primary candidate in most visual recognition tasks [4,24,37]. As CNNs trained on ImageNet [24] achieve impressive performance in object recognition, CNNs trained on Places [37] get significant performance in scene recognition. In order to get better performance, there are two primary ways to take full advantage of CNN features. The first way is to extract abundant features from local patches and aggregate them into effective scene representations [6,8,31,34]. Usually, these approaches combine multiple local patches and multiple scales features, and these features are pooled using VLAD [8] or Fisher vector [6] encoding. The second way is to leverage features which are extracted from complementary CNNs [10,16]. Thus, these different features can have complementary characteristics. Li *et al.* [16] have demonstrated that the combination of features from deep neural networks with various architectures can significant improve the performance as features obtained from heterogeneous CNNs have different characteristics since each network has a different architecture with different depth and the design of receptive fields. Herranz *et al.* [10] have improved that the concatenation of features extracted from object-oriented and scene-oriented networks results in significant recognition gains. In this paper, we assume that knowledge about objects in a scene image is helpful in scene recognition since objects are main components of scenes. We propose a framework to explore object context for scene recognition.

Context modeling. The utilization of context information for computer vision has attracted a lot of attention. Choi *et al.* [5] propose a graphical model to exploit co-occurrence, position, scale and global context which together is used to identify out-of-context objects in a scene. Torrala *et al.* [27] have shown how to exploit visual context to perform robust place recognition, categorization of novel places, and object priming. Izadinia *et al.* [12] have proposed a method to learn scene structures that can encode three main interlacing components of a scene: the scene category, the context-specific appearance of objects, and their layouts. Recently, RNN-based architectures have been widely used to model context for a lot of visual tasks, such as object detection [2], segmentation [11,18], scene labeling [3,25], human re-identification [29] and so on. The primary mechanism behind RNN is that the connections with previous states enables the network to memorize information from past inputs and thereby capture the contextual dependency of the sequential data. In the same spirit, we use a LSTM network to model context to boost the classification performance for scene images.

CNN-LSTM Models. CNN have been widely used to learn discriminative features for a wide range of visual tasks [18,23,24,30] and Recurrent Neural Networks (RNNs) have been widely used for sequence learning. Recently, a lot of deep architectures use the joint learning of CNN and RNN to get feature representations as well as their dependencies. Tasks combining visual and language like image captioning [17,30,33] and visual question answering [1,9] use CNN to get image features while use LSTM to generate natural language expressions as image descriptions or answers. As the architectures which are only composed of CNN can get features from large receptive fields but can not allow for finer pixel-level label assignment. Architectures with LSTM components can learn dependencies between pixels and improve agreement among their labels. A lot of work have used CNN-LSTM architectures for scene labeling [3] and semantic segmentation [18]. In this work, we use a CNN-LSTM based network to learn informative and discriminative for scene recognition.

3 Our Method

Overview: As a scene image is usually composed of multiple objects, the context of a scene image encapsulates rich information about how scenes and objects are related to each other. Such contextual information has the potential to enable a coherent understanding of scene images. In order to leverage such informative information, we propose sequential object context incorporating global scene information for scene recognition, as shown in Fig. 1. We first detect a set of objects for each scene image. To model the distribution of these objects, they are sorted by their locations in the image. And then, we use deep neural networks to extract features for the whole image and the detected objects. The representations of the object context sequence are put into a LSTM network to form the discriminative and representative sequential object context. At last, the global scene features of the whole image and the learned context are concatenated and the combination of them are put into a sub network to classify the image. The goal of our method is to complement the deep CNN features extracted from the whole image with local context from objects within a scene. The following sections provide the details of our method and its training procedure.

3.1 Object Proposal Extraction

As we aim to incorporate better local visual context for scene recognition, we first need to detect objects in scene images. We train a Faster R-CNN [23] detector and build our system on top of the detections, as shown in Fig. 2. Because a Region Proposal Network (RPN) in the Faster R-CNN takes an image as input and outputs a set of rectangular object proposals, each with an objectness score, we can select discriminative visual objects depending on the output of the RPN. Specifically, to get local proposals, we select top-n detected objects to represent important local objects according to their class confidence scores obtained from Faster R-CNN. We train our Faster R-CNN model using the VGG-16 convolutional architecture [26]. The model is first pre-trained on ImageNet [24] dataset and then fine-tuned on the training set of MS COCO [20] dataset, as MS COCO contains a lot of images which are composed of multiple objects.

After this, these objects are sorted by their locations in the image with the order of from left to right and top to down. Specially, the entire image is also considered as a special object and is denoted as R_0. We use $seq(I)$ to denote the initial sequential representations of the local objects, which contains a sequence of representations $seq(I) = \{R_1, R_2, \ldots R_N\}$, where R_1 to R_N are the local objects. R_0 is the corresponding global representation of the entire image.

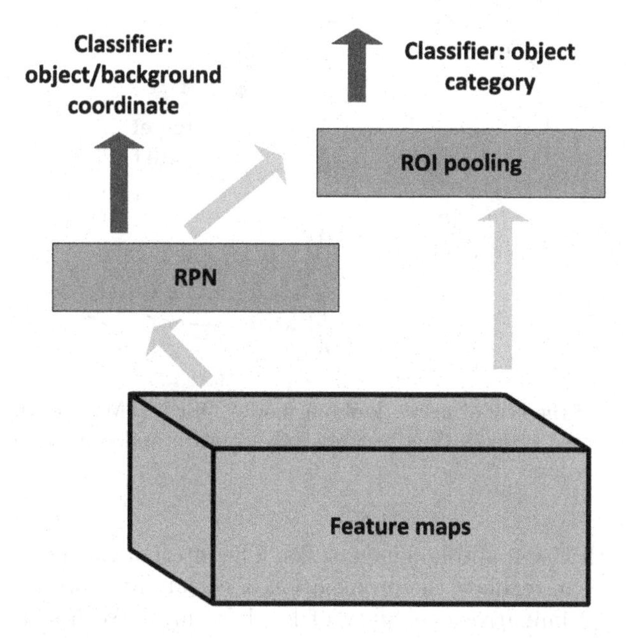

Fig. 2. The object detector in our model. We select top-n detected objects according to their class confidence scores.

3.2 Features Extraction

Convolutional Neural Networks (CNNs) have been widely used for many visual tasks due to its powerful representation ability. In our work, we also use CNNs to extract features for the whole image R_0 and the sequential representations of the local objects $seq(I) = \{R_1, R_2, \ldots R_N\}$. It has been demonstrated that scene-centric knowledge (Places) and object-centric knowledge (ImageNet) are complementary, and the combination of these two can significantly improve the performance [10]. So we use a Palces CNN to extract the 'fc7' layer features for R_0 and use an ImageNet CNN to extract the 'fc7' layer features for sequential local objects $seq(I)$. The features of the entire image is denoted as V_0, where $V_0 = CNN_P(R_0)$. The features of the $i-th$ object is denoted as V_i, where $V_i = CNN_I(R_i)$. The context sequence can be denoted as $seq_V(I) = \{V_1, V_2, \ldots V_N\}$.

3.3 Sequential Object Context Modeling

The core idea of our work is motivated by the previous works [7,21,29] which have demonstrated that the LSTM architectures can model abundant context

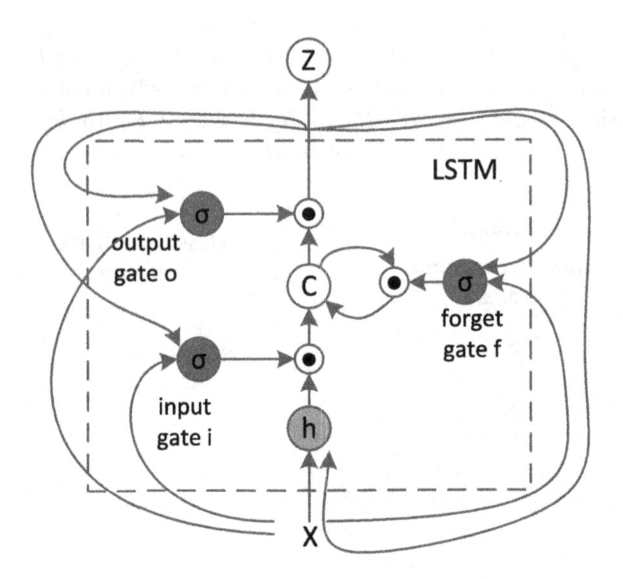

Fig. 3. Diagram of the LSTM network of our model. Our LSTM network continuously receives the detected objects thus progressively capture and aggregate the relevant contextual information.

features for both visual and language tasks. The internal gating mechanisms in the LSTM cells can regulate the propagation of certain relevant contexts, which enhance the discriminative capability of local features. We first introduce the LSTM network which is used in our method and then present the input and output of the LSTM module.

The architectural of the visual context LSTM is illustrated in Fig. 3. It receives the output of the previous time step, as well as the input at the current time step, as the inputs of the current unit. Mathematically, the update equations at time l can be formulated as:

$$i_l = \sigma(\mathbf{W}_{ix}x_l + \mathbf{W}_{im}m_{l-1}) \tag{1}$$

$$f_l = \sigma(\mathbf{W}_{fx}x_l + \mathbf{W}_{fm}m_{l-1}) \tag{2}$$

$$o_l = \sigma(\mathbf{W}_{ox}x_l + \mathbf{W}_{om}m_{l-1}) \tag{3}$$

$$c_l = f_l \odot c_{l-1} + i_l \odot \phi(\mathbf{W}_{cx}x_l + \mathbf{W}_{cm}m_{l-1}) \tag{4}$$

$$m_l = o_l \odot \phi(c_l) \tag{5}$$

where i_l, f_l and o_l represent the input gate, forget gate, and output gate at time step l respectively; c_l is the state of the memory cell and m_l is the hidden state; $\odot$ represents the element-wise multiplication, $\sigma(\cdot)$ represents the sigmoid function and $\phi(\cdot)$ represents the hyperbolic tangent function; $W_{[\cdot][\cdot]}$ denote the parameters of the model.

The LSTM takes in $seq_V(I) = \{V_1, V_2, \dots V_N\}$ and encodes each object into a fixed length vector. Thus, we have encoding hidden states computed from:

$$h_t = LSTM(seq_V(I)_t, h_{t-1}),\ t = 1, 2, \dots N \tag{6}$$

Once the hidden representations from all the context objects are obtained, they are combined to obtain the sequential object context $conV$ as shown below:

$$conV = \mathbf{W}_{\mathbf{C}}^T[(h_1)^T, (h_2)^T, \dots (h_r)^T, \dots, (h_N)^T],\ r = 1, 2, \dots, N \tag{7}$$

where $\mathbf{W_C}$ is the transformation matrix we need to learn and $[\cdot]^T$ indicates the transpose operation.

The final features $V(I)$ used to recognize a scene image are obtained from the combination of the global scene features V_0 and the sequential object context features $conV$. $V(I)$ are put into a sub network which is mainly composed of a fully-connected layer and a softmax layer to classify the image.

$$V(I) = [V_0,\ conV] \tag{8}$$

3.4 Training Details

We train our model on the framework of Caffe [13]. The visual context LSTM and the sub classification network are optimized in a end-to-end manner. For each scene image, we detect $n = 10$ objects to form the consequential object context. We use the mini-batch stochastic gradient descent method with the batch size of 20. The hidden state size of the visual context LSTM is set to 512, and the size of the fully-connected layer of the classification network is 4096.

4 Experiments

4.1 Dataset

To verify the effectiveness of our method, we evaluate the performance of our method on two benchmark datasets: MIT 67 [22] and SUN 397 [32].

MIT 67: MIT Indoor 67 [22] contains 67 categories of indoor images, with 80 images per category available for training as well as $20 * 67$ images for test. Indoor scenes tend to be rich in objects compared to object-centric images, which in general makes the task more challenging.

SUN 397: SUN 397 [32] is a scene benchmark containing 397 categories, including indoor, man-made and natural categories. This dataset is very challenging, not only because of the large number of categories, but also because the more limited amount of training data with 50 images per category for training and 50 images per category for test.

4.2 Quantitative Results

We conduct experiments on two benchmark datasets to qualitatively verify the effectiveness of our method. Our experiments mainly aim to demonstrate the usefulness of the consequential context not to get the best performance, so we just use the Alexnet networks to extract features. Table 1 shows the performance comparison of the proposed algorithm with the baseline algorithm. Alexnet 205 denotes that we use the Alexnet which is trained on Places 205 to extract fc7 features for the entire image and then feed the features to the sub classification network. Alexnet 205 & SOC denotes the combination of the features of the entire image and the learned sequential object context (SOC). It can be seen that the combination of global and local features of the proposed architecture outperforms the global scene information of the baseline algorithm for all the datasets. We get the same results when we use the network trained on Places 365 to extract the features. The comparison of our method with state-of-the-art methods is also show in Table 1. It shows that our method achieves promising results compared to state-of-the-art methods.

Table 1. The recognition performance on MIT 67 and SUN 397 datasets. * indicates that the performance are got with our own implementation. SOC is short for sequential object context.

Method	MIT 67	SUN 397
Alexnet 205*	68.25	54.36
Alexnet 205 & SOC*	69.36	55.78
Alexnet 365*	70.22	56.02
Alexnet 365 & SOC*	**71.86**	**57.72**
ImageNet Alexnet [37]	56.79	42.61
Places 205 Alexnet [37]	68.24	54.32
Places 365 Alexnet [36]	70.72	56.12
MS Orderless Pooling [8]	68.88	51.98

5 Conclusion

In this paper, we propose a deep model to model sequential object context for scene recognition. As scene images are rich in objects and the global scene information extracted from the entire image with CNNs neglect local object distributions, the learned sequential object context features are strong complementary representations which contain full information obtained from local objects. Experimental results show that the combination of the global scene information and the learned local sequential object context significantly improves the recognition performance. By using the LSTM module, our network can selectively propagate relevant contextual information and thus enhance the discriminative capacity of the local features.

In future work, we will use deeper networks as our feature extractors to get more powerful features. We will also incorporate multi-scale CNN features to our network to get better performance.

Acknowledgements. This work was supported in part by the National Natural Science Foundation of China under Grant No. 61202027. This work was also funded by the Project of Construction of Innovative Teams and Teacher Career Development for Universities and Colleges Under Beijing Municipality under Grant No. IDHT20150507.

References

1. Agrawal, A., Lu, J., Antol, S., Mitchell, M., Zitnick, C.L., Parikh, D., Batra, D.: VQA: visual question answering. Int. J. Comput. Vis. **123**(1), 4–31 (2017)
2. Bell, S., Zitnick, C.L., Bala, K., Girshick, R.: Inside-outside net: detecting objects in context with skip pooling. In: CVPR (2016)
3. Byeon, W., Breuel, T.M., Raue, F., Liwicki, M.R.: Scene labeling with LSTM recurrent neural networks. In: CVPR (2015)
4. Chatfield, K., Simonyan, K., Vedaldi, A., Zisserman, A.: Return of the devil in the details: delving deep into convolutional nets. In: Proceedings of the British Machine Vision Conference, BMVC 2014 (2014)
5. Choi, M.J., Torralba, A., Willsky, A.S.: Context models and out-of-context objects. Pattern Recogn. Lett. **33**(7), 853–862 (2012)
6. Dixit, M., Chen, S., Gao, D., Rasiwasia, N., Vasconcelos, N.: Scene classification with semantic fisher vectors. In: Proceedings of the IEEE Conference on Computer Vision and Pattern Recognition, CVPR 2015, pp. 2974–2983 (2015)
7. Fernández, S., Graves, A., Schmidhuber, J.: An application of recurrent neural networks to discriminative keyword spotting. In: de Sá, J.M., Alexandre, L.A., Duch, W., Mandic, D. (eds.) ICANN 2007. LNCS, vol. 4669, pp. 220–229. Springer, Heidelberg (2007). https://doi.org/10.1007/978-3-540-74695-9_23
8. Gong, Y., Wang, L., Guo, R., Lazebnik, S.: Multi-scale orderless pooling of deep convolutional activation features. In: Fleet, D., Pajdla, T., Schiele, B., Tuytelaars, T. (eds.) ECCV 2014. LNCS, vol. 8695, pp. 392–407. Springer, Cham (2014). https://doi.org/10.1007/978-3-319-10584-0_26
9. Goyal, Y., Khot, T., Summers-Stay, D., Batra, D., Parikh, D.: Making the V in VQA matter: elevating the role of image understanding in Visual Question Answering. In: Conference on Computer Vision and Pattern Recognition (CVPR) (2017)
10. Herranz, L., Jiang, S., Li, X.: Scene recognition with CNNs: objects, scales and dataset bias. In: CVPR (2016)
11. Hu, R., Rohrbach, M., Darrell, T.: Segmentation from natural language expressions. In: Leibe, B., Matas, J., Sebe, N., Welling, M. (eds.) ECCV 2016. LNCS, vol. 9905, pp. 108–124. Springer, Cham (2016). https://doi.org/10.1007/978-3-319-46448-0_7
12. Izadinia, H., Sadeghi, F., Farhadi, A.: Incorporating scene context and object layout into appearance modeling. In: CVPR (2014)
13. Jia, Y., Shelhamer, E., Donahue, J., Karayev, S., Long, J., Girshick, R., Guadarrama, S., Darrell, T.: Caffe: convolutional architecture for fast feature embedding. In: Proceedings of the 22nd ACM International Conference on Multimedia, MM 2014, pp. 675–678. ACM, New York (2014)

14. Li, L., Su, H., Xing, E., Fei-Fei, L.: Object bank: a high-level image representation for scene classification and semantic feature sparsification. In: Advances in Neural Information Processing Systems (2010)
15. Li, X., Guo, Y.: An object co-occurrence assisted hierarchical model for scene understanding. In: Proceedings of the British Machine Vision Conference (2012)
16. Li, X., Herranz, L., Jiang, S.: Heterogeneous convolutional neural networks for visual recognition. In: Chen, E., Gong, Y., Tie, Y. (eds.) PCM 2016. LNCS, vol. 9917, pp. 262–274. Springer, Cham (2016). https://doi.org/10.1007/978-3-319-48896-7_26
17. Li, X., Song, X., Herranz, L., Zhu, Y., Jiang, S.: Image captioning with both object and scene information. In: Proceedings of the 2016 ACM on Multimedia Conference, MM 2016, pp. 1107–1110. ACM, New York (2016)
18. Liang, X., Shen, X., Feng, J., Lin, L., Yan, S.: Semantic object parsing with graph LSTM. In: Leibe, B., Matas, J., Sebe, N., Welling, M. (eds.) ECCV 2016. LNCS, vol. 9905, pp. 125–143. Springer, Cham (2016). https://doi.org/10.1007/978-3-319-46448-0_8
19. Liao, Y., Kodagoda, S., Wang, Y., Shi, L., Liu, Y.: Understand scene categories by objects: a semantic regularized scene classifier using convolutional neural networks. In: IEEE International Conference on Robotics and Automation (ICRA) (2016)
20. Lin, T.Y., et al.: Microsoft COCO: common objects in context. In: Fleet, D., Pajdla, T., Schiele, B., Tuytelaars, T. (eds.) ECCV 2014. LNCS, vol. 8693, pp. 740–755. Springer, Cham (2014). https://doi.org/10.1007/978-3-319-10602-1_48
21. Palangi, H., Deng, L., Shen, Y., Gao, J., He, X., Chen, J., Song, X., Ward, R.: Deep sentence embedding using long short-term memory networks: analysis and application to information retrieval. IEEE/ACM Trans. Audio Speech Lang. Process. **24**, 694–707 (2016)
22. Quattoni, A., Torralba, A.: Recognizing indoor scenes. In: Proceedings of the IEEE Conference on Computer Vision and Pattern Recognition Workshops, CVPR Workshops 2009, pp. 413–420 (2009)
23. Ren, S., He, K., Girshick, R., Sun, J.: Faster R-CNN: towards real-time object detection with region proposal networks. In: NIPS (2015)
24. Russakvovsky, O., Deng, J., Su, H., Krause, J., Satheesh, S., Ma, S., Huang, Z., Karpathy, A., Kholsa, A., Bernstein, M., Berg, A., Fei-Fei, L.: Imagenet large scale visual recognition challenge. Int. J. Comput. Vis. **115**(3), 211–252 (2015)
25. Shuai, B., Zuo, Z., Wang, G., Wang, B.: DAG-Recurrent neural networks for scene labeling. In: CVPR (2016)
26. Simonyan, K., Zisserman, A.: Very deep convolutional networks for large-scale image recognition. In: ICLR (2015)
27. Torralba, A., Murphy, K.P., Freeman, W.T., Rubin, M.A.: Context-based vision system for place and object recognition. In: ICCV (2003)
28. Tuytelaars, T., Mikolajczyk, K.: Local invariant feature detectors: a survey. Found. Trends. Comput. Graph. Vis. **3**(3), 177–280 (2008)
29. Varior, R.R., Shuai, B., Lu, J., Xu, D., Wang, G.: A siamese long short-term memory architecture for human re-identification. In: Leibe, B., Matas, J., Sebe, N., Welling, M. (eds.) ECCV 2016. LNCS, vol. 9911, pp. 135–153. Springer, Cham (2016). https://doi.org/10.1007/978-3-319-46478-7_9
30. Vinyals, O., Toshev, A., Bengio, S., Erhan, D.: Show and tell: a neural image caption generator. In: CVPR (2015)
31. Wu, R., Wang, B., Wang, W., Yus, Y.: Harvesting discriminative meta objects with deep CNN features for scene classification. In: ICCV (2015)

32. Xiao, J., Hays, J., Ehinger, K.A., Oliva, A., Torralba, A.: Sun database: large-scale scene recognition from abbey to zoo. In: Proceedings of the IEEE Conference on Computer Vision and Pattern Recognitions, CVPR 2010, pp. 3485–3492 (2010)
33. Xu, K., Ba, J., Kiros, R., Cho, K., Courville, A., Salakhutdinov, R., Zemel, R., Bengio, Y.: Show, attend and tell: neural image caption generation with visual attention. In: ICML (2015)
34. Yoo, D., Park, S., Lee, J.Y., Kweon, I.S.: Multi-scale pyramid pooling for deep convolutional representation. In: Computer Vision and Pattern Recognition Workshops (CVPRW) (2015)
35. Yosinski, J., Clune, J., Bengio, Y., Lipson, H.: How transferable are features in deep neural networks? In: NIPS (2014)
36. Zhou, B., Khosla, A., Lapedriza, A., Torralba, A., Oliva, A.: Places: an image database for deep scene understanding. arXiv preprint arXiv:1610.02055 (2016)
37. Zhou, B., Lapedriza, A., Xiao, J., Torralba, A., Oliva, A.: Learning deep features for scene recognition using places database. In: Proceedings of the 28th Annual Conference on Neural Information Processing Systems 2014, NIPS 2014, vol. 1, pp. 487–495 (2014)

Associated Metric Coding Network for Pedestrian Detection

Shuai Chen and Bo Ma(✉)

Beijing Laboratory of Intelligent Information Technology,
Beijing Institute of Technology, Beijing, China
{chenshuai,bma000}@bit.edu.cn

Abstract. Convolutional neural networks (CNNs) have played a significant role in pedestrian detection, owing to their capacity of learning deep features from original image. It is noteworthy that most of the existing generalized objection detection networks must crop or warp the inputs to fixed-size which leads to the low performance on multifarious input sizes. Moreover, the lacking of hard negatives mining constrains the ability of recognition. To alleviate the problems, an associated work network which contains a metric coding net (MC-net) and a weighted association CNN (WA-CNN), is introduced. With region proposal net in low layer, MC-net is introduced to strengthen the difference of intra-class. WA-CNN can be regarded as a network to reinforce the distance of inter-class and it associates the MC-net to accomplish the detection task by a weighted strategy. Extensive evaluations show that our approach outperforms the state-of-the-art methods on the Caltech and INRIA datasets.

Keywords: Pedestrian detection · Region proposal net
Weight association CNN · Metric coding net

1 Introduction

Detecting the pedestrians from original images which contain a wealth of object information, such as car, tree and the sky, is a very challenging work. The significant advances in traditional model [4,26,28] and deep model [11,15–17,22] in this area have been witnessed in recent years.

The traditional model which extracts the low-level features (*e.g.* HoG [4], Haar [26], HoG-LBP [28]) from images and then selects rich representations to train the classifier(*e.g.* SVM [4], boosting classifiers [6]), is a widely used strategy, but it is hard to be optimized unitedly for decreasing the error rate.

In the deep model, CNNs have played a significant role in the pedestrian detection, owing to their capacity of learning representative and discriminative features from the original images. For example, as the number of negatives is significantly larger than that of positive ones in one database, Tian et al. [24]

B. Ma—This work was supported in part by the National Natural Science Foundation of China (No. 61472036).

J. Yang et al. (Eds.): CCCV 2017, Part III, CCIS 773, pp. 120–131, 2017.
https://doi.org/10.1007/978-981-10-7305-2_11

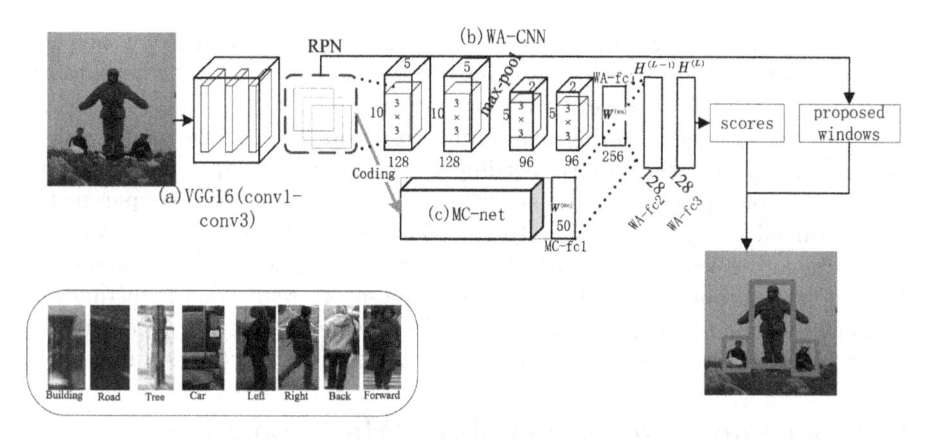

Fig. 1. The associated work network contain three parts: (a) the RPN in low layer. (b) weighted association CNN. (c) metric coding net

transfered scene attribute information from existing background scene segmentation databases to the pedestrian dataset for learning representative features.

However, most of the previous deep models must crop or warp the images to fixed-size (*e.g.* 224×224 in VGG16 [23]) which leads to the low performance on multifarious input sizes [9]. To solve this problem, spatial pyramid pooling in deep convolutional networks (SPP-net) [9] has been proposed to pool the feature maps with arbitrary size before the full-connected layers using spatial pyramid pooling. Further, [21] proposed a Region Proposal Network (RPN) that it shares full-image convolutional features with the detection network, thus enabling nearly cost-free region proposals. But these strategies are not suitable for little size inputs (the size of pedestrian) with a very deep CNN (*e.g.* VGG16, VGG19 [23]), meanwhile lacking of any strategy for hard negatives mining constrains the ability of recognition in these methods. The problems have been attracting increasing attention for accurate, yet efficient, pedestrian detection. [32] redesign RPN for pedestrian size and unite Boosted Forests (BF) to mine the hard negatives. But using RPN united the BF in low layer simply will lost many rich features for large sizes and they are hard to be optimized unitedly.

Driven by these observations and considered the excellent performance of VGG16 and RPN. We propose an effective baseline for pedestrian detection, apply RPN in low layer for generating the proposal windows with arbitrary size. Furthermore, as the mining of negative samples is significant, an associated work network feeding with the labeled multi-class negative and positive samples, is introduced in our work. It contains two networks: metric coding net (MC-net) and weighted association CNN (WA-CNN). MC-net which is based on metric learning theory, is devised to reinforce the intra-class distance. With the strategy, the network will encode the feature by the template parameter, and the generated codes can be seen as the comparability determination between the feature and the template parameter. Finally, WA-CNN which is designed to strengthen the inter-class difference by a deep model, associates the metric codes to accomplish the detection task using a weighted loss function.

This work has the following main **contributions**. (1) MC-net is devised to reinforce the intra-class distance. Feeding with the feature maps extracted from labeled viewpoint pedestrian and no-pedestrian images, the template parameter will be trained. After the training, the metric codes are encoded by the net, and the codes can be seen as a comparability measurement between the inputs and the template parameter. (2) WA-CNN is proposed to reinforce the distance of inter-class in our network with a deep CNN network, and it will associate the metric codes to accomplish the detection task with a weighted loss function.

2 From Supervised Generalized Max Pooling to the Template of Multi-class

Our goal is to propose a template learning mechanism that we attempt to represent multi-class by vectors. Motivated by a property proposed in Generalized Max Pooling (GMP) [12] that the dot-product similarity between the max-pooling representation (a vector) and a feature matrix is a constant value:

$$\boldsymbol{\psi}'\boldsymbol{\phi} = \boldsymbol{\alpha}, \tag{1}$$

where $\boldsymbol{\psi}'$ is the feature matrix, $\boldsymbol{\phi}$ is max-pooling representation, $\boldsymbol{\alpha}$ is a vector with all elements being a constant value and the value of the constant has no influence [12].

We generalize the max-pooling representation to be a template vector (representing one class). Because of the randomicity of $\boldsymbol{\alpha}$, we enforce the dot-product similarity between the feature and the template vector to be a generalized max pooling vector which is the mean value of all max-pooling vectors belonging to one class. The primal formulation is

$$\sum_{i=1}^{\tau} \boldsymbol{\psi}'_i\boldsymbol{\phi} = \boldsymbol{\alpha}, \tag{2}$$

where $\boldsymbol{\psi}'_i$ is the i-th labeled feature matrix of τ images belonging to one class. The vector $\boldsymbol{\alpha}$ can be seen as a supervised method to calculate the max pooling, and it can be seen as a comparability determination between the template vector and the feature matrix as well.

In order to learn the template vector, we can turn Eq. (2) into a least square regression problem

$$\Gamma = \frac{1}{2}||\sum_{i=1}^{\tau} \boldsymbol{\psi}'_i\boldsymbol{\phi} - \boldsymbol{\alpha}||^2. \tag{3}$$

To the problem, we must calculate several template vectors to represent multi-class in this system. Thus, we introduce metric coding net (MC-net) to generalize vectors $\boldsymbol{\phi}$ as a template parameter which is learned by a neural network, to represent multi-class template.

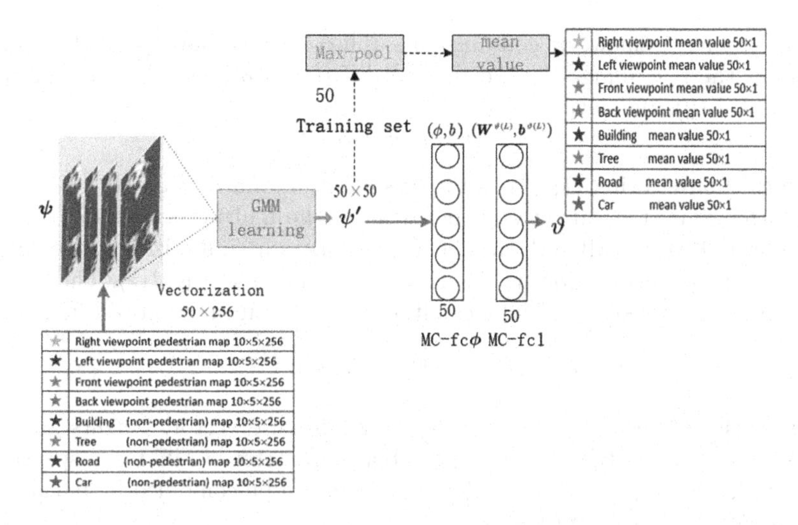

Fig. 2. Metric coding net with GMM learning and two full-connected layers.

2.1 Formulation of MC-Net

Because the low layers focus on the local features and they encode more discriminative features to capture intra-class variations [27]. Inspired by this property, MC-net is introduced to reinforce the intra-class distance by the low layer feature maps.

The net employs the feature maps $\psi \in \mathbb{R}^{10\times5\times256}$ as input and all feature maps are reshaped into a matrix ($\psi'' \in \mathbb{R}^{50\times256}$) with 50-dimensional vector and 256 feature maps by vectorization.

To derive a more compact and discriminative representation, we utilize Gaussian Mixture Model (GMM) to model the generation process of feature maps. Assume that the feature maps are subject to parametric distribution $\mathcal{P}_\lambda(\psi'')$. Then, $\mathcal{P}_\lambda(\psi'')$ can be written as

$$\mathcal{P}_\lambda(\psi'') = \sum_{t=1}^{T} \omega_t p_t(\psi''), \tag{4}$$

where p_t is the t-th component of GMM with

$$p_t(\psi'') = \frac{1}{(2\pi)^{d/2}|\boldsymbol{\Sigma}_t|^{1/2}} e^{\left(-\frac{1}{2}(\psi''-\boldsymbol{\mu}_t)^T \boldsymbol{\Sigma}_t{}^{-1}(\psi''-\boldsymbol{\mu}_t)\right)}, \tag{5}$$

and $\lambda = \{\omega_t, \boldsymbol{\mu}_t, \boldsymbol{\Sigma}_t\}_{t=1,\cdots,T}$ (T is 25 in our work) denotes the parameters of GMM training by ψ''. Because the weight parameters bring little additional information, we use $\psi' = \{\boldsymbol{\mu}_t, \Sigma_t\}_1^T \in \mathbb{R}^{50\times50}$ to represent the feature maps.

The coding framework contains one full-connected layer (MC-fcϕ, input map 1) at the beginning. The weight ϕ of MC-fcϕ can be seen as the template vector in Eq. (2), the output

$$\boldsymbol{\vartheta} = \boldsymbol{\psi}'\boldsymbol{\phi} + \boldsymbol{b} \tag{6}$$

will be calculated using Eq. (3) as the loss function and $\boldsymbol{\alpha} = \boldsymbol{\vartheta} - \boldsymbol{b}$.

To strengthen the capacity of representing the multi-class, we increase one full-connected layer (MC-fc1) in the framework and initialize the weights of MC-fc1 and MC-fc$\boldsymbol{\phi}$ randomly. The detailed setting is in Fig. 1(c). The forward propagation is passed from MC-fc$\boldsymbol{\phi}$ without activation function to MC-fc1 by

$$\boldsymbol{\vartheta} = \boldsymbol{W}^{\vartheta(L)}\left(\boldsymbol{\psi}'\boldsymbol{\phi} + \boldsymbol{b}\right) + \boldsymbol{b}^{\vartheta(L)}, \tag{7}$$

where $\boldsymbol{\vartheta}$, $\boldsymbol{W}^{\vartheta(L)}$, $\boldsymbol{b}^{\vartheta(L)}$ indicate the top-layer feature vector, weights and bias respectively, $\boldsymbol{\phi}$, $\boldsymbol{b}$ are the weight and the bias parameter of MC-fc$\boldsymbol{\phi}$ respectively. Without the activation function, the two layers can be combined by linear combination, Eq. (7) can be written as

$$\boldsymbol{\vartheta} = \boldsymbol{\psi}'\left(\boldsymbol{W}^{\vartheta(L)} \oplus \boldsymbol{\phi}\right) + \left(\boldsymbol{b} \oplus \boldsymbol{b}^{\vartheta(L)}\right) \tag{8}$$

where $\oplus$ is the operation of linear combination. This method is equivalent to increase the dimension of $\boldsymbol{\phi}$ simply. We use the non-linear activation function to reconstitute the network

$$\boldsymbol{\vartheta} = \boldsymbol{W}^{\vartheta(L)}\left(ReLu\left(\boldsymbol{\psi}'\boldsymbol{\phi} + \boldsymbol{b}\right)\right) + \boldsymbol{b}^{\vartheta(L)}, \tag{9}$$

where $ReLu$ is the rectified linear function [13]. The output

$$\boldsymbol{\vartheta} = \boldsymbol{\psi}'\left(\boldsymbol{W}^{\vartheta(L)} \bowtie \boldsymbol{\phi}\right) + \left(\boldsymbol{b} \bowtie \boldsymbol{b}^{\vartheta(L)}\right) \tag{10}$$

can be seen as the comparability determination between the input feature $\boldsymbol{\psi}'$ and the multi-class template parameter, $\bowtie$ is a generalized symbol of non-linear combination by $ReLu$.

2.2 The Training of the Net

Let $\boldsymbol{B} = \{(\boldsymbol{\psi}, \boldsymbol{\alpha}_i)\}_{i=1}^{K}$ be the training set, K is the number of training images. Specifically, corresponding to the max pooling vector in Eq. (2), $\boldsymbol{\alpha}_i$ denotes eight labeled mean values, where we pool $\boldsymbol{\psi}'$ to $\boldsymbol{\psi}^{\max} \in \mathbb{R}^{50\times1}$ by max-pooling and calculate the mean value $\boldsymbol{\alpha}_i$ in each labeled class.

Corresponding to Eq. (3),

$$E^{(MC)} = \frac{1}{2}||\boldsymbol{\vartheta} - \boldsymbol{\alpha}||^2, \tag{11}$$

is used as the loss function, where $\boldsymbol{\vartheta}$ is the output of the net, $\boldsymbol{\alpha}$ is the labeled mean value, as show in Table 1.

During the training, we set a maximum epoch number (2000) and the training process will be terminated when the objective converges on a relative little value.

Table 1. The labeled mean value.

	Labeled class ($\boldsymbol{\alpha}$)	Symbol
Pedestrian viewpoint	Front viewpoint	α_{front}
	Left viewpoint	α_{left}
	Right viewpoint	α_{right}
	Back viewpoint	α_{back}
Non-pedestrian	Building	$\alpha_{building}$
	Tree	α_{front}
	Road	α_{road}
	Car	α_{car}

Therefore, after the training, the parameter $\left[\left(\boldsymbol{W}^{\vartheta(L)} \bowtie \boldsymbol{\phi}\right), \left(\boldsymbol{b}^{\vartheta(L)} \bowtie \boldsymbol{b}\right)\right]$ can be seen as a generalized template. After inputting an arbitrary image, the output will be the metric code $\boldsymbol{\vartheta}$, it can be used directly as the comparability metric between the image and template.

3 Weighted Association CNN

As analyzed in [27], different layers encode different types of features and higher layers semantic concepts on object categories. Motivated by the property, WA-CNN is proposed to reinforce the distance of inter-class by a deep model.

3.1 The Formulation of WA-CNN

Let $\boldsymbol{D} = \{(\psi, y'_i)\}_{i=1}^{K}$ be the training feature maps set, where $y'_i = (y_i, \varrho_i^p, \varrho_i^n)$ is a three-tuple. y_i indicates whether a feature map is pedestrian or not. Binary labels $\varrho_i^p = \{\varrho_i^{pj}\}_{j=1}^4$, $\varrho_i^p = \{\varrho_i^{nj}\}_{j=1}^4$ represent the viewpoint pedestrian and non-pedestrian, and the labels are shown in Fig. 1.

As shown in Fig. 1(b), WA-CNN employs feature maps $\psi \in \mathbb{R}^{10\times5\times256}$ as input by stacking four convolutional layers (WA-conv1 to WA-conv4), one max-pool and three full-connected layers (WA-fc1 to WA-fc3), and the detailed setting is shown in Fig. 1(b). For all these layers, we utilize the rectified linear function [13] as the activation function.

As shown in Fig. 1(b), to strengthen the discriminant validity of intra-class, WA-CNN associates the metric codes which are generated by MC-net,

$$\begin{aligned} \boldsymbol{H}^{(L-1)} &= ReLu(\boldsymbol{W}^{(wa)}\boldsymbol{H}^{(L-2)} + \boldsymbol{b}^{(wa)} \\ &+ \boldsymbol{W}^{(mc)}\boldsymbol{\vartheta} + \boldsymbol{b}^{(mc)}), \end{aligned} \quad (12)$$

where $\boldsymbol{H}^{(L)}$ is the top-layer feature vector of WA-CNN, $\boldsymbol{W}^{(wa)}$, $\boldsymbol{b}^{(wa)}$ and $\boldsymbol{W}^{(mc)}$, $\boldsymbol{b}^{(mc)}$ are the parameter matrices corresponding to the two networks respectively.

We use

$$
\begin{aligned}
E^{(WA)} &= -\sum_{i=1}^{K} \log p\left(y_i, \varrho^p, \varrho^n | \psi, \vartheta\right) \\
&= -y \log p\left(y|\psi, \vartheta\right) - \sum_{i=1}^{4} \varrho^{pi} \log p\left(\varrho^p|\psi, \vartheta\right) \\
&\quad - \sum_{j=1}^{4} \varrho^{nj} \log p\left(\varrho^n|\psi, \vartheta\right),
\end{aligned}
\tag{13}
$$

as the loss function and the loss function is expand to three parts, the main pedestrian, the viewpoint pedestrian and the non-pedestrian. The main task is to predict the pedestrian label y. ϱ^{pi}, ϱ^{nj} are the i-th pedestrian estimations and the j-th non-pedestrian estimations. $p\left(y|\psi, \vartheta\right)$, $p\left(\varrho^p|\psi, \vartheta\right)$, $p\left(\varrho^n|\psi, \vartheta\right)$ are modeled by softmax functions.

3.2 The Training of WA-CNN

Because the main task is to predict the pedestrian label, the others are the auxiliary tasks. Thus, in the phase of training, we reformulate Eq. (11) using ω and ε to associate multiple tasks by a weighted strategy as the following

$$
\begin{aligned}
E^{(WA)} &= -y \log p\left(y|\psi, \vartheta\right) - \sum_{i=1}^{4} \omega_i \varrho^{pi} \log p\left(\varrho^p|\psi, \vartheta\right) \\
&\quad - \sum_{j=1}^{4} \varepsilon_j \varrho^{nj} \log p\left(\varrho^n|\psi, \vartheta\right).
\end{aligned}
\tag{14}
$$

In our work, ω and ε can be values between zero and one. we set $\forall \omega_i = 0.1,\ i = 1, 2, ...4$, $\forall \varepsilon_j = 0.1,\ j = 1, 2, ...4$ simply.

With the training set $\boldsymbol{D} = \{(\psi, {y'}_i)\}_{i=1}^{K}$, WA-CNN is trained to reinforce the distance of inter-class, further.

4 Overview on Our Method

Figure 1 shows our pipeline of pedestrian detection, where VGG16 (conv1–conv3) with RPN extract the candidate regions from the images with arbitrary image size. The generated feature maps of candidate regions $\psi \in \mathbb{R}^{10\times5\times256}$ will be reconstituted to the set $\boldsymbol{B} = \{(\psi, \boldsymbol{\alpha}_i)\}_{i=1}^{K}$ and $\boldsymbol{D} = \{(\psi, {y'}_i)\}_{i=1}^{K}$ with the labels, and they will be the training sets for our associated work network.

Our associated work network contains two network, MC-net and WA-CNN. WA-CNN can be seen a network to reinforce the distance of inter-class, on the contrary, the Mc-net plays the role in enhancing the instance difference of intra-class.

As show in Fig. 2, with the training set $\boldsymbol{B} = \{(\boldsymbol{\psi}, \boldsymbol{\alpha}_i)\}_{i=1}^{K}$ which contain the labeled viewpoint pedestrian and non-pedestrian images, MC-net learn the template parameter $\left[\left(\boldsymbol{W}^{\vartheta(L)} \bowtie \boldsymbol{\phi}\right), \left(\boldsymbol{b}^{\vartheta(L)} \bowtie \boldsymbol{b}\right)\right]$. After the training, it codes the feature maps $\boldsymbol{\psi}$ with template parameter, and the output $\boldsymbol{\vartheta}$ is a comparability determination between the input map and the template parameter.

Finally, the outputs of MC-net and WA-CNN are jointly learned by the two full-connected layers of our network. And the weighted loss function is designed to accomplish the detection task with the joint features.

5 Experiments

To evaluate the performance of our detector on Caltech-Test [7] and INRIA [8] datasets, the evaluation protocol is following with [7] strictly.

The training data generated by transferring scene attribute information from existing background scene segmentation databases to seventeen attributes in pedestrian dataset by TA-CNN [24]. We only use eight attributes (showing in Table 1) as the training data, and the data are reconstituted into two parts: the viewpoint pedestrian (left, right, front, back) and the non-pedestrian (tree, car, road and building). Note that our network does not employ any motion and context information.

For Caltech-test reasonable subset, all results of our network are obtained by training on the reconstituted training data and evaluating on Caltech-Test (set06–set10). And, to evaluate the generalization capacity of the our network, we report overall results on INRIA-test in this section. All results of our network are obtained by training on reconstituted training data and evaluating on INRIA-test.

Hereinafter, RPN will be fine-turning in our network and WA-CNN and MC-net which are the brand-new network, will be evaluated on the performance and effectiveness (Table 2).

Table 2. The runtime and performance on Caltech.

Method	Window process	Hardware	Time/img(s)	MR (%)
LDCF	-	CPU	0.6	24.8
CCF	-	Titan Z GPU	13	17.3
Ours	RPN	GeForce GTX 1080 × 2	**0.9**	**13.74**

5.1 Effectiveness of Different Components in WA-CNN

Under the framework of deep neural networks, we compare the result of our network (WA-CNN+non-linear MC-net) with WA-CNN+linear MC-net and WA-CNN without MC-net to verify the capacity of MC-net in representing multi-class.

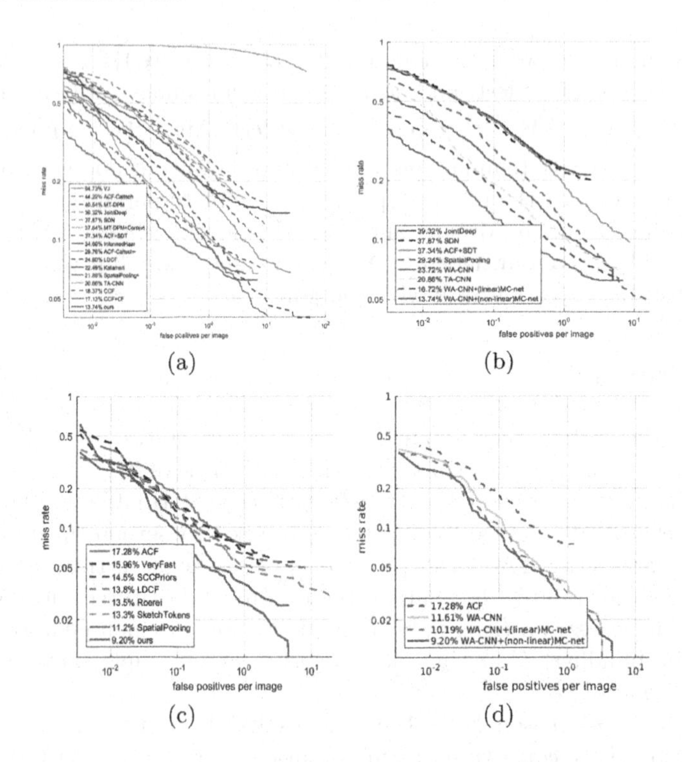

Fig. 3. Results on Caltech-test reasonable subset and INRIA dataset: (a), (c): Overall performance (b), (d): Log-average miss rate reduction procedure.

Caltech-Test reasonable subset: we systematically study the effectiveness of different components on our network. After the training, WA-CNN without MC-net gets **23.71%** miss rate. With this baseline, we implement WA-CNN+linear MC-net. As shown in Fig. 3(b), WA-CNN+linear MC-net gets **16.72%** miss rate, and it gets **7 %** improvement. To verify the capacity of the non-linear MC-net in our network, we re-train MC-net with non-linear activation function. The result shows in Fig. 3(b), MC-net (non-linear) achieves **13.74%** performance, and improves **3.02%** than linear MC-net. The result is also compared with the other deep models: JointDeep, SDN, ACF+SDT, SpatialPooling, TA-CNN, and our method gets the lowest miss rate.

INRIA: WA-CNN without MC-net gets **11.61%** miss rate and WA-CNN+linear MC-net gets **10.19%** miss rate. MC-net (non-linear) improves **9.20%** performance, as shown in Fig. 3(d). The results show that our network has a good capacity on the generalization.

5.2 Comparisons with State-of-the-Art Methods

Finally, the results of our network with existing best-performing methods which contain handcrafted features and deep neural networks are evaluated.

Caltech-Test reasonable subset: we compare the result of our network with existing best-performing methods, including VJ [25], HOG [4], ACF-Caltech [5], MT-DPM [29], MTDPM+Context [29], JointDeep [16], SDN [11], ACF+SDT [20], InformedHaar [33], ACF-Caltech+ [14], SpatialPooling [19], LDCF [14], Katamari [3], SpatialPooling+ [18],TA-CNN [24],CCF [30], CCF+CF [30]. Figure 3(a) reports the results, and our method achieves the smallest miss rate (**13.74**%) compared to all existing methods.

INRIA: We compare the result of our network with existing best-performing methods, including ACF, VeryFast [1], SCCpriors [31], LDCF, Roerei Z [2], SketchTokens [10], SpatialPooling, and parts method mentioned in Sect. 4.1. As shown in Fig. 3(c), our method achieves the lowest miss rate.

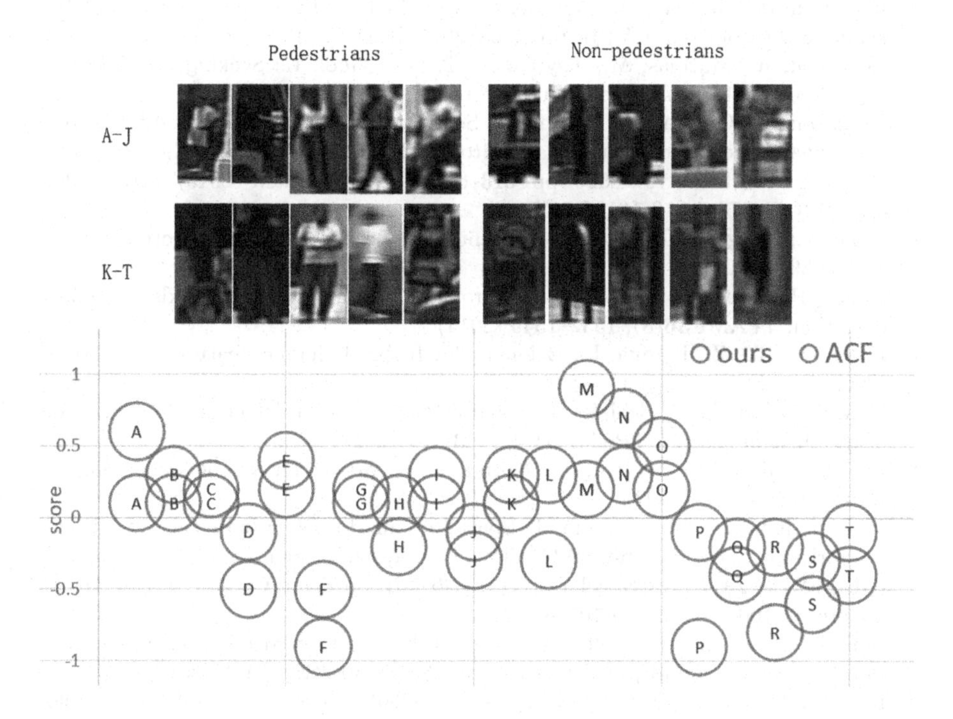

Fig. 4. Compares with ACF, our network show a good performance in hard negatives mining, the scores in hard negatives is discriminative than ACF.

5.3 Evaluation on Hard Negatives Mining

In order to evaluate our method on hard negatives mining intuitively, we chose twenty images (hard negatives) which are difficult to other methods in Caltech dataset. The scores of ACF are as our baseline in the evaluation, and the scores of our network are calculated by softmax functions. To be fair, the scores of ACF are normalized to [−1,1]. Figure 4 shows that our method have a good performance to mine the hard negatives.

6 Conclusions

In this paper, with the plenty negative and positive samples, MC-net and WA-CNN are introduced to associated work for mining the hard negatives in pedestrian detection. They enforce the intra-class and inter-class differences using the properties of low-level and high-level features in a CNN model. Under the network, the problem of input size was alleviated by a flexible using on RPN. Extensive experiments demonstrate the effectiveness of the proposed method.

References

1. Benenson, R., Mathias, M., Timofte, R., Van Gool, L.: Pedestrian detection at 100 frames per second. In: CVPR, pp. 2903–2910. IEEE (2012)
2. Benenson, R., Mathias, M., Tuytelaars, T., Van Gool, L.: Seeking the strongest rigid detector. In: CVPR, pp. 3666–3673 (2013)
3. Benenson, R., Omran, M., Hosang, J., Schiele, B.: Ten years of pedestrian detection, what have we learned? In: Agapito, L., Bronstein, M.M., Rother, C. (eds.) ECCV 2014. LNCS, vol. 8926, pp. 613–627. Springer, Cham (2015). https://doi.org/10.1007/978-3-319-16181-5_47
4. Dalal, N., Triggs, B.: Histograms of oriented gradients for human detection. CVPR **1**, 886–893 (2005)
5. Dollár, P., Appel, R., Belongie, S., Perona, P.: Fast feature pyramids for object detection. TPAMI **36**(8), 1532–1545 (2014)
6. Dollar, P., Tu, Z., Perona, P., Belongie, S.: Integral channel features. In: BMVC (2009)
7. Dollar, P., Wojek, C., Schiele, B., Perona, P.: Pedestrian detection: an evaluation of the state of the art. TPAMI **34**(4), 743–761 (2012)
8. Ess, A., Leibe, B., Van Gool, L.: Depth and appearance for mobile scene analysis. In: ICCV, pp. 1–8 (2007)
9. He, K., Zhang, X., Ren, S., Sun, J.: Spatial pyramid pooling in deep convolutional networks for visual recognition. In: Fleet, D., Pajdla, T., Schiele, B., Tuytelaars, T. (eds.) ECCV 2014. LNCS, vol. 8691, pp. 346–361. Springer, Cham (2014). https://doi.org/10.1007/978-3-319-10578-9_23
10. Lim, J.J., Zitnick, C.L., Dollár, P.: Sketch tokens: a learned mid-level representation for contour and object detection. In: CVPR, pp. 3158–3165 (2013)
11. Luo, P., Tian, Y., Wang, X., Tang, X.: Switchable deep network for pedestrian detection. In: CVPR, pp. 899–906 (2014)
12. Murray, N., Perronnin, F.: Generalized max pooling. In: CVPR, pp. 2473–2480 (2014)
13. Nair, V., Hinton, G.E.: Rectified linear units improve restricted Boltzmann machines. In: ICML, pp. 807–814 (2010)
14. Nam, W., Dollár, P., Han, J.H.: Local decorrelation for improved pedestrian detection. In: NIPS, pp. 424–432 (2014)
15. Ouyang, W., Wang, X.: A discriminative deep model for pedestrian detection with occlusion handling. In: CVPR, pp. 3258–3265 (2012)
16. Ouyang, W., Wang, X.: Joint deep learning for pedestrian detection. In: ICCV, pp. 2056–2063 (2013)
17. Ouyang, W., Zeng, X., Wang, X.: Modeling mutual visibility relationship in pedestrian detection. In: CVPR, pp. 3222–3229 (2013)

18. Paisitkriangkrai, S., Shen, C.: Pedestrian detection with spatially pooled features and structured ensemble learning. In: TPAMI, pp. 1243–1257 (2016)
19. Paisitkriangkrai, S., Shen, C., van den Hengel, A.: Strengthening the effectiveness of pedestrian detection with spatially pooled features. In: Fleet, D., Pajdla, T., Schiele, B., Tuytelaars, T. (eds.) ECCV 2014. LNCS, vol. 8692, pp. 546–561. Springer, Cham (2014). https://doi.org/10.1007/978-3-319-10593-2_36
20. Park, D., Zitnick, C.L., Ramanan, D., Dollár, P.: Exploring weak stabilization for motion feature extraction. In: CVPR, pp. 2882–2889 (2013)
21. Ren, S., He, K., Girshick, R., Sun, J.: Faster R-CNN: towards real-time object detection with region proposal networks. In: TPAMI, p. 1 (2016)
22. Sermanet, P., Kavukcuoglu, K., Chintala, S., LeCun, Y.: Pedestrian detection with unsupervised multi-stage feature learning. In: CVPR, pp. 3626–3633 (2013)
23. Simonyan, K., Zisserman, A.: Very deep convolutional networks for large-scale image recognition. arXiv preprint arXiv:1409.1556 (2014)
24. Tian, Y., Luo, P., Wang, X., Tang, X.: Pedestrian detection aided by deep learning semantic tasks. In: CVPR, pp. 5079–5087 (2015)
25. Viola, P., Jones, M.J.: Robust real-time face detection. IJCV **57**(2), 137–154 (2004)
26. Viola, P., Jones, M.J., Snow, D.: Detecting pedestrians using patterns of motion and appearance. IJCV **63**(2), 153–161 (2005)
27. Wang, L., Ouyang, W., Wang, X., Lu, H.: Visual tracking with fully convolutional networks. In: ICCV, pp. 3119–3127 (2015)
28. Wang, X., Han, T.X., Yan, S.: An HOG-LBP human detector with partial occlusion handling. In: ICCV, pp. 32–39. IEEE (2009)
29. Yan, J., Zhang, X., Lei, Z., Liao, S., Li, S.Z.: Robust multi-resolution pedestrian detection in traffic scenes. In: CVPR, pp. 3033–3040 (2013)
30. Yang, B., Yan, J., Lei, Z., Li, S.Z.: Convolutional channel features. In: ICCV, pp. 82–90 (2015)
31. Yang, Y., Wang, Z., Wu, F.: Exploring prior knowledge for pedestrian detection. In: BMVC, pp. 1–12 (2015)
32. Zhang, L., Lin, L., Liang, X., He, K.: Is faster R-CNN doing well for pedestrian detection? In: Leibe, B., Matas, J., Sebe, N., Welling, M. (eds.) ECCV 2016. LNCS, vol. 9906, pp. 443–457. Springer, Cham (2016). https://doi.org/10.1007/978-3-319-46475-6_28
33. Zhang, S., Bauckhage, C., Cremers, A.B.: Informed Haar-like features improve pedestrian detection. In: CVPR, pp. 947–954 (2014)

Large-Scale Slow Feature Analysis Using Spark for Visual Object Recognition

Da Li[1,2], Zhang Zhang[1,2(✉)], and Tieniu Tan[1]

[1] CRIPAC, Institute of Automation, Chinese Academy of Sciences, Beijing, China
da.li@cripac.ia.ac.cn, {zzhang,tnt}@nlpr.ia.ac.cn
[2] University of Chinese Academy of Sciences (UCAS), Beijing, China

Abstract. Data-driven feature learning has achieved great success in various visual recognition tasks. However, to handle millions of training image/video data efficiently, a high performance parallel computing platform combining with powerful machine learning algorithms plays a fundamental role in large-scale feature learning. In this paper, we present a novel large-scale feature learning architecture based on Slow Feature Analysis (SFA) and Apache Spark, where the slowness learning principle is implemented to learn invariant visual features from millions of local image patches. To validate the effectiveness of the proposed architecture, extensive experiments on pedestrian recognition have been performed on the INRIA pedestrian dataset. Experimental results show that the performance on pedestrian recognition can be promoted significantly with the growth of training patches, which demonstrates the necessity of large scale feature learning clearly. Furthermore, in comparisons with classical Histogram of Oriented Gradients (HOG) and Convolutional Neural Network (CNN) features, the slow features learnt by large-scale training patches can also achieve comparable performance.

Keywords: Large-scale feature learning · Slow feature analysis · Spark Parallel computing · Pedestrian recognition

1 Introduction

Extracting robust visual feature is an essential step towards achieving high performance in real-world visual recognition tasks. Recently, large-scale feature learning has achieved great success in various visual recognition tasks, in which powerful machine learning algorithms are integrated closely with high performance parallel computing platforms, clusters or GPU, so as to support data intensive computing from millions of training images. In this paper, we propose a novel large-scale feature learning architecture based on Slow Feature Analysis (SFA) [15] and Apache Spark [2], where the slowness learning principle is implemented to learn invariant visual features from millions of local image patches.

SFA is an unsupervised feature learning method, which intends to learn the invariant and slowly varying features from input signals. It has been successfully

J. Yang et al. (Eds.): CCCV 2017, Part III, CCIS 773, pp. 132–142, 2017.
https://doi.org/10.1007/978-981-10-7305-2_12

used for action recognition [12,21], trajectory clustering [20], handwriting digital recognition [7] and dynamic scenes classification [14]. Recently, Escalante-B and Wiskott [16] extend the SFA to supervised dimensionality reduction and the notion of slowness is generalized from temporal sequences to training graphs (a.k.a GSFA). Different with previous work, e.g., Sparse coding [9] and Autoencoder [6], most of which learn features to minimize the reconstruction error, we utilize the GSFA to explore the neighbor relationships in large-scale local patches from a view of manifold learning, where the learnt features will encode the intrinsic local geometric structures existing in a large number of training local patches.

Feature learning by the GSFA is to construct a feature space, in which the nearby points in original space will have similar locations. Thinking about a gray image with 16×16 pixels, and the value of each pixel ranges from 0 to 255. It means that there may exits 256^{256} possibilities [19] in the input space. Thus, the feature space learnt with limited data may have two issues: (1) Losing mush useful information makes visual samples belong to different patterns locate nearly in the feature space, i.e., inaccuracy. (2) Incomplete distribution of the training samples makes the learnt features are not effective in practical application, i.e., overfitting. On the other hand, it is the most important step of GSFA to construct the local neighbor relation graph in input space. Such graph is usually constructed through finding the top k nearest neighbors (k-NN). The hypothesis, near patches have similar patterns, is established only with large-scale data. While the inaccurate k-NN results will result in inaccurate slow feature functions.

As mentioned above, it is necessary to learn robust features with large-scale data. Meanwhile, relation graph construction in the input space is the key step of GSFA. However, it is a $O\left(n^2\right)$ problem [13] to construct it with k-NN searching. So it is a computational challenge with such large-scale data (about 10 million in this paper). The available of high performance hardware and parallel computing platform make it possible to alleviate this problem. Now, Hadoop [1] and Spark [2] attract much attention in big data tasks. The advantages of Spark, i.e., inmemory computation, sufficient APIs and good compatibility, make it faster and more convenient than MapReduce in Hadoop. Recently, some projects on deep learning with Spark are released, such as CaffeOnSpark [17] and SparkNet [10].

Motivated by above reasons, in this paper we construct a parallel architecture for large-scale slow feature learning using Spark. In summary, the efforts of this paper includes:

(1) We present a new large-scale feature learning architecture based on SFA and Spark, which is used to train invariant features with more than 10 million of local patches. The task can be completed efficiently within 140 h.
(2) We study the effect of the quantity of training patches for the visual recognition task on pedestrian recognition based on this architecture. The experimental results demonstrate that the performance is indeed improved significantly with the growth of training patches consistently.

(3) The learnt slow features can achieve superior performance than that of classical hand-crafted feature, i.e., Histogram of Oriented Gradients (HOG) [3] feature, which validate the effectiveness of large-scale slow feature learning.

2 Method

In this section, we first give a brief introduction on SFA, mainly about its essential problem and mathematical representation. Then, we explain the details about the parallel computing architecture for large-scale SFA with Spark. Finally, we will discuss how to use the learnt slow feature functions for an instance of visual recognition tasks, i.e., pedestrian recognition.

2.1 Slow Feature Analysis

Thinking about an object move through one's visual field, the signals fall on his retina change rapidly. However, the relevant abstract information, e.g. identity, changes slowly in a timescale [16]. It is known as slowness principle which first formulated by Hinton in 1989 [5]. Based on the slowness principle, SFA is to learn a group of slow feature functions to transform the input signal to a new feature space, in which the transformed signal varies as slow as possible. It is first proposed by Wiskott [15]. GSFA [16] is the extension of SFA, which generalizes the slowness principle from sequences to a training graph. Because the lack of temporal information in the training local patches, we adopt the GSFA to learn slow features in this paper. Mathematically, GSFA is formulated as follows:

Given a training graph $G = (V, E)$ with a set of nodes $V = \{\boldsymbol{x}(1), .., \boldsymbol{x}(N)\}$ and a set of edges $E := (\boldsymbol{x}(n), \boldsymbol{x}(n'))$, where $\langle n, n' \rangle$ denotes a pair of samples with $1 \leq n, n' \leq N$. GSFA aims at learning a group of input-output functions $\{g_j\}, 1 \leq j \leq J$ such that the J-dimensional output signals $y_j(n) = g_j(\boldsymbol{x}(n))$ have the minimum difference with their nearest neighbors,

$$minimize\ \Delta_j = \frac{1}{R} \sum_{n,n'} \gamma_{n,n'} \left(y_j\left(n'\right) - y_j\left(n\right) \right)^2 \tag{1}$$

s.t.

$$\frac{1}{Q} \sum_{n} \nu_n y_j(n) = 0 \ \ weighted\ zero\ mean\ , \tag{2}$$

$$\frac{1}{Q} \sum_{n} \nu_n \left(y_j\left(n\right) \right)^2 = 1 \ \ weighted\ unit\ variance\ , \tag{3}$$

$$\frac{1}{Q} \sum_{n} \nu_n y_j(n) y_{j'}(n) = 0 \ \ weighted\ decorrelation\ , \tag{4}$$

with

$$R = \sum_{n,n'} \gamma_{n,n'}\ and\ Q = \sum_{n} \nu_n\ , \tag{5}$$

where ν and γ denote the weight of node and the weight of edge respectively. And $\gamma_{n,n'} = \gamma_{n',n}$ means that the edges are undirected. Constraint (2) is used to make constraint (3) and (4) with concise form. Constraint (3) avoids the trivial solution. And constraint (4) ensures that different functions g_j take different properties of the input signal. Meanwhile constraint (4) enforces the functions g_j are ordered based on their slowness, in which the first is the slowest one.

For the linear GSFA, $g_j(\boldsymbol{x}) = \boldsymbol{w}_j^T\boldsymbol{x}$. This optimization problem equals to solve a generalized eigenvalue problem,

$$AW = BW\Lambda, \tag{6}$$

with

$$A = \frac{1}{R}\sum_{n,n'} \gamma_{n,n'}(\boldsymbol{x}(n') - \boldsymbol{x}(n))(\boldsymbol{x}(n') - \boldsymbol{x}(n))^T \tag{7}$$

and

$$B = \frac{1}{Q}\sum_{n} \nu_n(\boldsymbol{x}(n) - \widehat{\boldsymbol{x}})(\boldsymbol{x}(n) - \widehat{\boldsymbol{x}})^T \tag{8}$$

where $\widehat{\boldsymbol{x}} = (1/Q)\sum_n \nu_n \boldsymbol{x}(n)$ is the weighted mean of all the samples. For the nonlinear GSFA, the inputs can be firstly mapped to a nonlinear expansion space and then solve the problem with the steps in the linear case. The nonlinear expansion function is defined by

$$\boldsymbol{h}(\boldsymbol{x}) := [h_1(\boldsymbol{x}), ..., h_M(\boldsymbol{x})] \ . \tag{9}$$

So the slow feature functions of GSFA can be calculated with three steps:

(1) Construct the undirected graph $G = (V, E)$; and confirm the weights of nodes and edges. In this paper, $\nu_n = 1$ and $\gamma_{n,n'} = 1/N_{edges}$.
(2) Map the inputs to a nonlinear space with a nonlinear function and centralize them. In this paper, we use the quadratic expansion. For a I-dimensional input signal, $\boldsymbol{h}(\boldsymbol{x}) = [x_1, ..., x_I, x_1x_1, x_1x_2, ..., x_1x_I]$.
(3) Solve the generalized eigenvalue problem $AW = BW\Lambda$. The eigenvectors corresponding to the first J smallest eigenvalues are selected as slow feature functions.

2.2 Large-Scale SFA in Spark

The proposed parallel computing architecture for large-scale SFA runs on a cluster consists of 10 machines with 96 CPU cores, as shown in Fig. 1, in which one for master node and the other nines for worker nodes. And the total memory is 510 GB. Since the advantages of Spark in big data processing, we choose it as our computational platform to implement SFA with large-scale data. For a Spark platform, the core concept is Resilient Distributed Datasets (RDD) [18], which is a distributed memory abstraction for in-memory computations in cluster. Furthermore, the application running on Spark consists of a driver program and many executors, in which driver program runs the main function and executors run different tasks. The driver program launches different tasks to workers with a variance of RDD transformations and actions (shown in the brace of Fig. 1). The launched tasks will be implemented by the executors in each worker, and the results will return to driver program or produce new RDDs. Our large-scale SFA parallel architecture based on above principle as well.

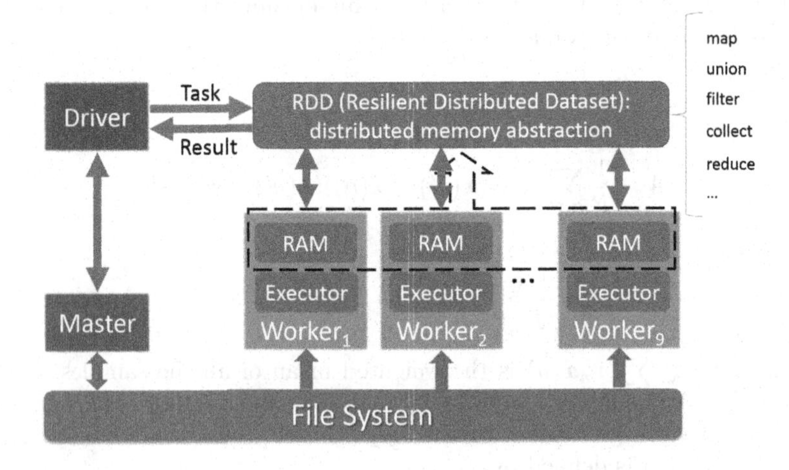

Fig. 1. An overview of the computing platform with Spark.

An overview of the architecture for large-scale SFA is shown in Fig. 2. We may summarize the large-scale SFA with Spark as five tasks:

(1) Data loading and patches extraction. Load the images data from hard disks to Spark RDD. Then extract patches from each image by dense sampling in parallel. The extracted patches will be stored as new Spark RDDs.
(2) Pre-processing. Implement PCA whitening on extracted patches to remove the redundant information and make the patches with zero mean and unit variance. It will also bring convenience for the following operations. Its result (a transformation matrix) will act on the input of nonlinear expansion.
(3) Graph construction. It is the core step in GSFA. We construct the graph with k-NN searching in parallel, in which the broadcast variables are used to improve the efficiency and the patches are divided into multiple RDDs (see Fig. 2) to satisfy the limitations of memory resources. FLANN[1] is used for k-NN searching locally (more details in Algorithm 1). The k-NN results and their source patches are used to compute the differences matrix after nonlinear expansion.
(4) Nonlinear expansion (Eq. (9)). Implement a quadratic expansion to the pre-processed patches data in parallel.

[1] http://www.cs.ubc.ca/research/flann/.

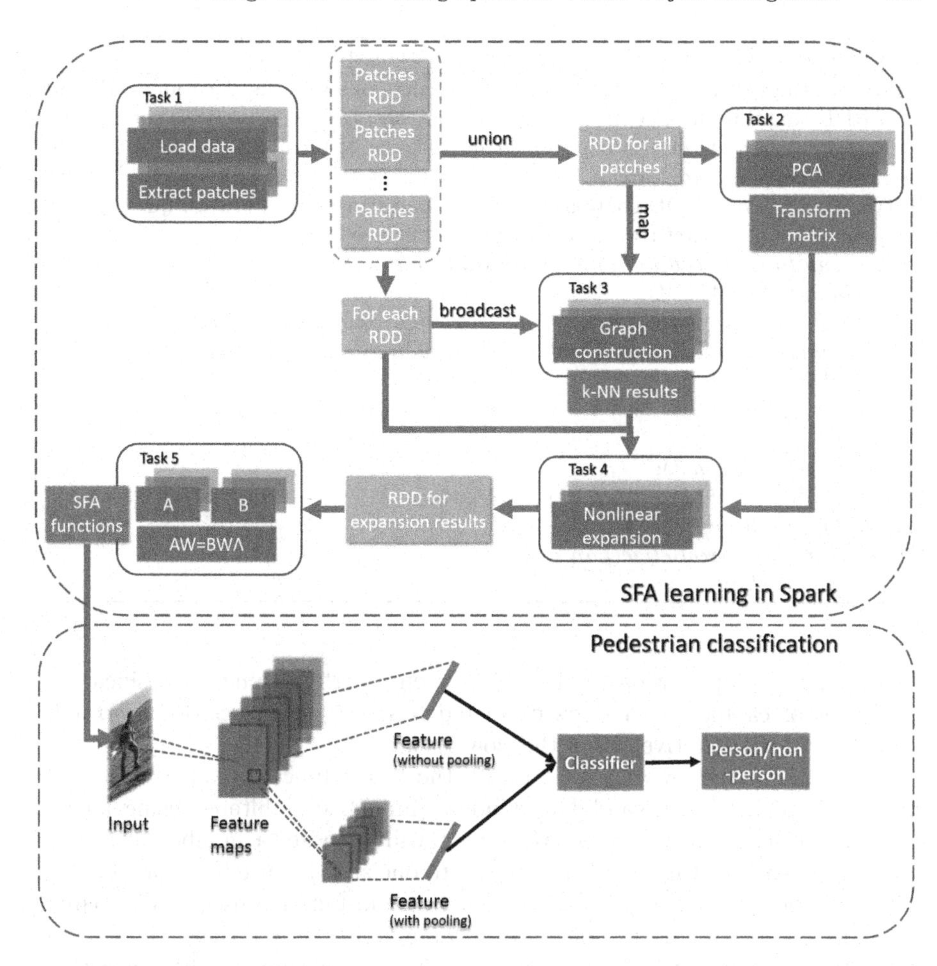

Fig. 2. Diagram of large-scale SFA in Spark for pedestrian classification. The upper part is the architecture for large-scale SFA in Spark. The lower part shows how to extract feature using slow feature functions in pedestrian classification. Two methods (with max pooling or not) are tested in experiment. (Color figure online)

(5) Solve the generalized eigenvalue problem. Firstly, compute the covariance matrices (Eqs. (7) and (8)) in parallel. Then collect the matrices from Spark RDD to the memory of master node to calculate the eigenvalues and eigenvectors.

2.3 Pedestrian Recognition Using Slow Feature Functions

Pedestrian recognition is one of the essential tasks in visual object recognition. Instead of detecting pedestrian over whole images where the performance depends not only on the feature itself, but also on the post-processing step, e.g.,

Algorithm 1. Graph construction: k-NN searching in Spark

Input: *sc*: the spark context which defines the entry of a spark application; *rdd_total*: a RDD contains all the extracted training patches; *rdd_query*: a RDD contains the query patches (its size is determined by the size of memory); *k*: the number of nearest neighbors to search.

Output: *rdd_knn*: a RDD contains the top k nearest neighbors of each query patch.

```
1: query_data_broadcast := sc.broadcast(rdd_query.collect());
2: rdd_knn_local := rdd_total.map(labmda element : {
3:   flann := FLANN();
4:   if isComingFromDifferentImages(element, query_data_broadcast)
5:     knn_local := flann.nn(element, query_data_broadcast, k, kdtree);
6:   end if
7:   return knn_local;
8: });
9: rdd_knn_local.cache();
10: knn_global := rdd_knn_local.treeAggregate();
11: rdd_knn_local.unpersist();
12: rdd_knn := sc.parallelize(knn_global);
13: return rdd_knn
```

Non-Maximum Suppression (NMS) [11], we only perform binary classifications over a set of candidate windows extracted from whole images, for a straight evaluation of the effectiveness of the slow features.

As shown in the lower part of Fig. 2, the learnt function can be seen as a filter (red block). For a candidate window (input), it is filtered using all the learnt functions with a specific stride that will generate a number of feature maps where each feature map corresponds to one slow feature function. Then, a nonlinear normalization function, e.g., the Sigmoid function used in this paper, is performed to clip the responses with large values. After that, we test two methods to generate the final feature vector: (1) concatenate the normalized feature maps directly; (2) execute max-pooling operation over a local region on the feature maps before the concatenation. To predict whether there is a person in the input window or not, a linear Support Vector Machine (SVM) is trained as the classifier.

3 Experimental Results

In this paper, we select the INRIA Person dataset [3] as the training and test dataset. It totally includes 2416 positive training windows and 1126 positive test windows, where each window corresponding to one person annotated in the dataset. To generate negative windows, we randomly extract ten windows per training negative image (12180 windows) and five windows per test negative image (2265 windows). The size of each window is 96×160 pixels.

We perform the slow feature learning with two strategies, i.e., *unsupervised SFA* and *supervised SFA*. For the supervised SFA, three kinds of methods are implemented as shown in the lower part of Table 1. The key parameters in training SFA are chosen as follows: patch size is 16×16 pixels, k in k-NN searching is 3 and dense sampling stride is 4. When generate final features, we set filtering stride to 8 and window size to 64×128 pixels. For different numbers of selected functions (as shown in Table 1), the dimensions of final features are 2100, 2100, 13000 and 3150, respectively. Figure 3 shows the results of the visualization of the first 25 slow feature functions learnt by supervised and unsupervised learning strategies.

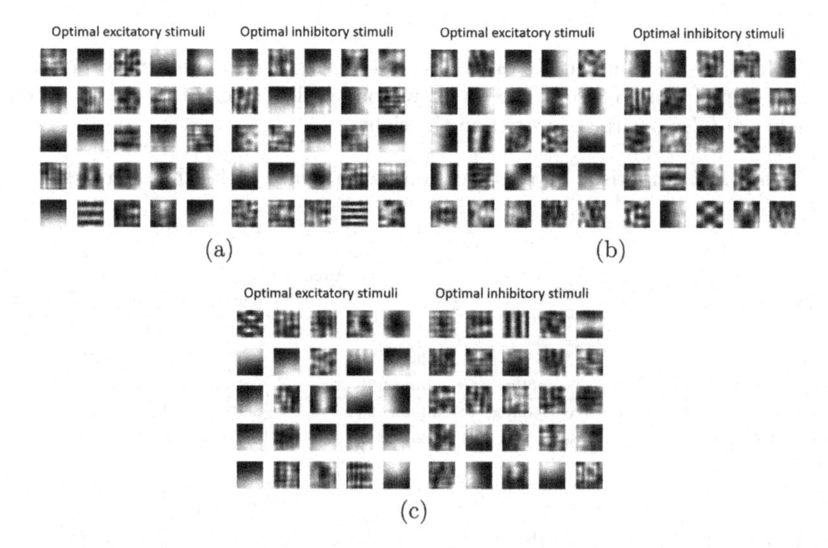

Fig. 3. Visualization of learnt slow feature functions.(a) The slow feature functions learnt by negative samples. (b) The slow feature functions learnt by positive samples. (c) The slow feature functions learnt by unsupervised SFA.

We study the relationship between the populations of training patches and the classification performance, under the case of unsupervised SFA. From Fig. 4, it clearly shows that the classification performance can be improved (precision is increased and miss rate is decreased) with the growth of training patches, which indicates that the large-scale training samples enhance the ability of the learnt slow features for pedestrian recognition. Meanwhile, compare the elapsed time of SFA learning between single-core machine and our proposed SFA+SPARK architecture, the speed-up ratio ranges from 3.56 to 14.67 when the number of training patches increases from 6k to 120k. It indicates that the efficiency has been improved significantly, especially when the data size is very large. And it can handle more than 10 million local patches within 140 h efficiently.

We compare both unsupervised and supervised SFA with the HOG feature and CNN feature. In experiment, we use the scikit-image[2] library to extract

[2] https://github.com/scikit-image/scikit-image.

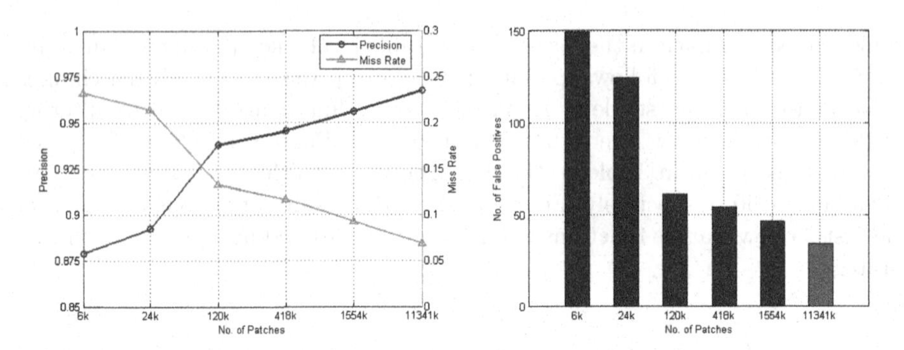

Fig. 4. Effect of the number of training samples on classification performance. The variations of precision and miss rate is shown on the left side, while the variation of the number of false positives located on the right.

Table 1. Comparison on classification performance with other features.

Methods	#Patches	#Slow feature functions	Precision	Miss rate	#False Positives (FP)
HOG [3]	-	-	0.9773	0.0521	18
CNN [8]	-	-	**0.9862**	**0.0415**	**0**
Unsupervised SFA	11 millions	20	0.9673	0.0680	34
Supervised SFA (positives w/o max pooling)	1.8 millions	20	0.9650	0.0777	31
Supervised SFA (positives with max pooling)	1.8 millions	200	0.9703	0.0583	35
Supervised SFA (positives & negtives)	11 millions	30	**0.9800**	**0.0477**	**14**

HOG feature, in which the cells per block is 2×2, pixels per cell is 8×8, number of orientations is 9 and the stride is 8. So the dimension of HOG feature is 3780. While the CNN feature is trained on the INRIA dataset using Caffe[3]. We select the output of fc7 as the final features which dimension is 4096.

The classification results of different methods are list in Table 1. The performance of our method trained with both the positive and negative samples under supervised learning strategy is superior to the performance of the HOG feature. However, it is worse than the CNN feature. It may because the CNN extracts features with multiple layers that can capture much high-level information. The hierarchical SFA [4] learning architecture may be introduced into our future work to improve the performance. From the table, we can also conclude that, the performance of slow functions trained with both the positive and negative samples is superior to that of the slow features trained only using the positive samples. Furthermore, along with the increasing of the number of slow feature functions, the max pooling may achieve more superior performance, compared to the feature representation obtained by directly concatenation. In summary, compared with the HOG feature and CNN feature, the comparable performance validates the effectiveness of our large-scale SFA+SPARK learning architecture.

[3] https://github.com/BVLC/caffe.

4 Conclusion

In this paper, we present a large-scale feature learning architecture based on SFA and Apache Spark. It is used to learn invariant visual features from more than 10 million of local patches under supervised and unsupervised learning strategies. Then, we extract features using the learnt slow feature functions to do pedestrian classification on the INRIA person dataset. The experiment results indicate that the efficiency of the SFA learning using large-scale data can be improved greatly with our proposed parallel computing architecture. It also demonstrates that the classification performance can be improved along with the increasing of the number of training patches. Furthermore, the performance of learned slow feature is comparable to the classic HOG feature and CNN feature in pedestrian classification.

Acknowledgments. This work is jointly supported by the National Key Research and Development Program of China (2016YFB1001005), the National Natural Science Foundation of China (Grant No. 61473290), the National High Technology Research and Development Program of China (863 Program) under Grant 2015AA042307.

References

1. Apache: Apache hadoop. https://hadoop.apache.org/
2. Apache: Apache spark. https://spark.apache.org/
3. Dalal, N., Triggs, B.: Histograms of oriented gradients for human detection. In: CVPR 2005, vol. 1, pp. 886–893. IEEE (2005)
4. Escalante-B, A.N., Wiskott, L.: Heuristic evaluation of expansions for non-linear hierarchical slow feature analysis. In: Machine Learning and Applications and Workshops (ICMLA), vol. 1, pp. 133–138. IEEE (2011)
5. Hinton, G.E.: Connectionist learning procedures. Artif. Intell. **40**(1), 185–234 (1989)
6. Hinton, G.E., Salakhutdinov, R.R.: Reducing the dimensionality of data with neural networks. Science **313**(5786), 504–507 (2006)
7. Huang, Y., Zhao, J., Tian, M., Zou, Q., Luo, S.: Slow feature discriminant analysis and its application on handwritten digit recognition. In: IJCNN 2009, pp. 1294–1297. IEEE (2009)
8. Jia, Y., Shelhamer, E., Donahue, J., Karayev, S., Long, J., Girshick, R., Guadarrama, S., Darrell, T.: Caffe: convolutional architecture for fast feature embedding. In: Proceedings of the ACM International Conference on Multimedia, pp. 675–678. ACM (2014)
9. Lee, H., Battle, A., Raina, R., Ng, A.Y.: Efficient sparse coding algorithms. In: Advances in Neural Information Processing Systems, pp. 801–808 (2006)
10. Moritz, P., Nishihara, R., Stoica, I., Jordan, M.I.: SparkNet: training deep networks in spark. arXiv preprint arXiv:1511.06051 (2015)
11. Sermanet, P., Kavukcuoglu, K., Chintala, S., LeCun, Y.: Pedestrian detection with unsupervised multi-stage feature learning. CVPR **2013**, 3626–3633 (2013)
12. Sun, L., Jia, K., Chan, T.H., Fang, Y., Wang, G., Yan, S.: DL-SFA: deeply-learned slow feature analysis for action recognition. In: Proceedings of the IEEE Conference on Computer Vision and Pattern Recognition, pp. 2625–2632 (2014)

13. Talwalkar, A., Kumar, S., Rowley, H.: Large-scale manifold learning. In: CVPR 2008, pp. 1–8. IEEE (2008)
14. Theriault, C., Thome, N., Cord, M.: Dynamic scene classification: learning motion descriptors with slow features analysis. In: CVPR 2013, pp. 2603–2610. IEEE (2013)
15. Wiskott, L., Sejnowski, T.J.: Slow feature analysis: unsupervised learning of invariances. Neural Comput. **14**(4), 715–770 (2002)
16. Wiskott, L., et al.: How to solve classification and regression problems on high-dimensional data with a supervised extension of slow feature analysis. J. Mach. Learn. Res. **14**(1), 3683–3719 (2013)
17. Yahoo: Caffe on spark (2016). https://github.com/yahoo/CaffeOnSpark
18. Zaharia, M., Chowdhury, M., Das, T., Dave, A., Ma, J., McCauley, M., Franklin, M.J., Shenker, S., Stoica, I.: Resilient distributed datasets: a fault-tolerant abstraction for in-memory cluster computing. In: Proceedings of the 9th USENIX Conference on Networked Systems Design and Implementation. USENIX Association (2012)
19. Zhang, L.: Big data for visual recognition: potentials and challenges. Technical report, Microsoft Research Asia (2012)
20. Zhang, Z., Huang, K., Tan, T., Yang, P., Li, J.: RED-SFA: relation discovery based slow feature analysis for trajectory clustering. In: Proceedings of the IEEE Conference on Computer Vision and Pattern Recognition, pp. 752–760 (2016)
21. Zhang, Z., Tao, D.: Slow feature analysis for human action recognition. IEEE Trans. Pattern Anal. Mach. Intell. **34**(3), 436–450 (2012)

Fine-Grained Visual Classification Based on Image Foreground and Sub-category Similarity

Xianjin Jiang, Xin Lin, Yi Ji, Jianyu Yang, and Chunping Liu(✉)

School of Computer Science and Technology, Soochow University, Suzhou, Jiangsu, China
cpliu@suda.edu.cn

Abstract. We propose a fine-grained visual classification algorithm based on image foreground and sub-category similarity. In the processing of feature extracting, our model calculates the gradient of image pixels in a classification network to obtain the foreground of the image. Then input the foreground image and the original image into the bilinear convolution network to obtain the feature of the image. At the classification stage, we propose an improved SD-SVM algorithm, which takes the advantages of the similarities among sub-categories and the differences among the similarities of sub-category. Experimental results manifest that our algorithm can achieve 85.12% accuracy on the CUB-2011 dataset and 85.21% accuracy on the FGVC-aircrafts dataset even with only the category labels, which outperforms state-of-the-art fine-grained categorization methods.

1 Introduction

Fine-grained categorization, also known as sub-category image classification, has attracted wide attention in recent years. Unlike Pascal VOC's tasks for classifying boats, bicycles and cars, fine-grained visual classification algorithm distinguishes sub-categories with high similarity. Therefore, this task is more difficult than most image classification tasks.

In general, fine-grained visual classification algorithm contains two steps: feature extraction and classification. At the stage of feature extraction, the annotations of object level and part level are useful to improve the accuracy of classification. Some of existing classification algorithms [1,6,9] use manual annotation information. Because the cost of manual annotation information is expensive, some algorithms [2,11,12] only use the category label to extract the features. At the classification stage, previous fine-grained categorization algorithms are directly use Multi-classification SVM. Besides, Lin et al. [2] pointed out that the experiment using multi-classification SVM performs better than using softmax.

In this paper, a fine-grained visual classification algorithm is designed with only the category labels. In our model, we compute the gradient of CNN to obtain the foreground, refer to 3.1. Besides, we use the bilinear CNN to extract features, refer to 3.2. We found that the similarity between categories can improve the accuracy at the classification stage. The probability of classifying the image into

J. Yang et al. (Eds.): CCCV 2017, Part III, CCIS 773, pp. 143–154, 2017.
https://doi.org/10.1007/978-981-10-7305-2_13

similar category is higher than that of classifying the image into not similar category. So we modify the multi-classification SVM for that refer to Sect. 3.2.

2 Related Work

In this subsection, we introduce some researches of fine-grained visual classification in term of feature extraction and multi-classification in recent years.

2.1 Feature Extraction

Feature extraction of fine-grained visual classification mainly include two categories. One is fusing features of object and part level with manual annotation information. The other is automatic extracting feature by deep learning. Additional manual annotation information about object level and part level, can play an important role in fine-grained visual classification tasks. Zhang et al. [6] proposed the part R-CNN algorithm. The part R-CNN model uses the manual annotations of object level and part level to train R-CNN [7] network. The R-CNN network is used to detect the object and part during the test stage. Finally, we obtain the final feature combining the object features and the part feature. Branson et al. [8] also proposed a Pose Normalized CNN algorithm. The algorithm uses the key points of manual annotation to obtain the object and part, and the localization of object and part is normalized. A final feature is obtained by combining the two normalized features of object and part.

In recent years, more and more studies that tend to not use the annotation information of object and part level achieved a very good effect. To replace the effect of annotation information, Simon and Rodner [13] designed a constellation algorithm that uses convolutional network features. A Final feature is extablished by fusing the extracting features of object and part using key points. Different from the above method, Lin et al. [2] designed a novel Bilinear CNN network model, it uses the original image as the input of the classification network, and achieves 84.1% accuracy on the CUB200-2011 data set. Zhang et al. [3] construct complex features using deep filters, and achieve the highest accuracy for the database only with the category label.

Bilinear CNN has achieved great success on fine-grained problems, but this method takes the original image as input and is greatly influenced by image background. To solve this problem, our model obtains the bounding box of the image by calculating gradient of image pixels in the classification network. Then obtain features form the original image and the image cutting by bounding box.

2.2 Multi-classification SVM

SVM was originally designed for binary classification problem. When dealing with multiple classification problem, we need to construct the appropriate multi-class classifier. At present, there are two methods to construct SVM classifier:

(1) Direct method: the kind of methods directly modify the objective function, and merge the parameters of multiple classifications into an optimization problem. This method has high computational complexity, and is only suitable for small problems. (2) Indirect method: This kind of methods achieve the framework of multi-classifier through the combination of multiple two classifier. These methods can be classified two categories: one-to-rest and one-to-one.

One-to-rest [17] is one of the earliest and most widely used methods. This method firstly constructs k classifiers (k is the total category), then a object is classified to the kth category or the remaining categories using the kth classifier. One-to-one [16] method trains a classifier between two sub-categories, so there will be $k(k-1)/2$ classifiers for a k-type problem.

In the fine-grained classification problem, there is a stronger similarity between sub-categories. Tradition classification models not use the similarity between sub-category and the difference between the similarities of sub-category. This paper establishes SD-SVM classification model that learns these information at the training stage to advance the accuracy of fine-grained classification.

3 Methods

In this subsection, we will introduce proposed fine-grained visual recognition algorithm. We first introduce how to get the bounding box of object in the image. Then we introduce the feature extraction using the bilinear CNN. Finally, we introduce improved SD-SVM multi-classifier.

3.1 Generating Bounding Box

We see a classification network as a mapping $y = f(x)$, where y is the final score vector and x is the input image. The mapping of the input layer to the conv1 layer is expressed as $f^{(1)}$, the mapping from the nth layer to the (n+1)th layer is $f^{(n)}$. So the whole network can be expressed as $f(x) = f^{(n)}f^{(n-1)}\cdots f^{(1)}(x)$. We defined $g^{(k)} = f^{(k)}f^{(k-1)}\cdots f^{(1)}(x)$. The output of the kth layer is expressed as $x^{(k)} = g^{(k-1)}(x)$.We can compute the gradient of y for input layer x:

$$\frac{\partial f}{\partial x} = \frac{\partial f}{\partial g^{(n-1)}}\frac{\partial g^{(n-1)}}{\partial g^{(n-2)}}\cdots\frac{\partial g^{(1)}}{\partial x} = \frac{\partial g_i^{(n)}}{\partial x^{(n)}}\frac{\partial g^{(n-1)}}{\partial x^{(n-1)}}\cdots\frac{\partial g^{(1)}}{\partial x} \tag{1}$$

The input image consists of three channels. For each pixel of the input image, we calculate the gradient of the three channels of y, and take the average of the three gradients as the gradient value of the pixel. We got a gradient image of the same size as the original image. As shown in Fig. 1(b). Since the gradient map obtains the most relevant part of the object, it may not have the information of the whole object. So we use the GraphCut [18] algorithm to obtain the mask of object segmentation. The advantage of the algorithm is taking advantage of some of the foreground information and the continuity of color. Then we obtain the enclose mask with the smallest rectangular border, which is the

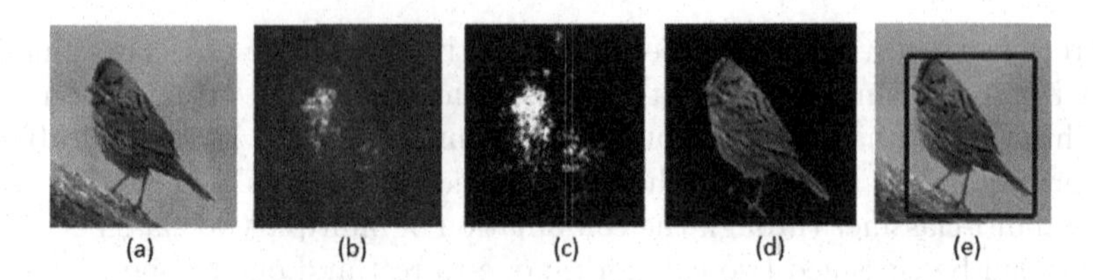

Fig. 1. An example of calculating gradient of pixel. (a) is the input image, (b) is the gradient maps, (c) is the foreground image, (d) is the result of GrapCut, (e) is the result of bounding box.

bounding box we need. We acquire the foreground information of image using the threshold gradient map. In our experiments, we use the region which gradient is greater than 95% as foreground. The foreground information is shown in Fig. 1(c). The mask of image and bounding boxes are showing in Fig. 1(d) and Fig. 1(e) respectively.

3.2 Feature Extraction

We extract the features by bilinear CNN model. A bilinear model M consists of a quaternion: $M = (f_A, f_B, P, C)$. f_A and f_B represents feature function that extract from bilinear CNN network A and B respectively. P is a pooling function and C is a classification function. The output of two networks is transformed into the final feature by bilinear operation and pooling function. For more details about bilinear CNN, please refer this paper [2].

3.3 Improved SD-SVM Multi-classifier

At the training stage, we extract the features of the training image set using B-CNN, and use the one-to-rest strategy to obtain 200 SVM classifiers. At the testing stage, the feature of testing image I_i is expressed as f_i. Using the return values of 200 trained SVM classifiers, we can link two hundred return values into a new feature vector f_s^i for image$i$, s means the feature come from SVM, and $f_s^i(u)$ is the return value that the image is classified into u class. The forms of our feature vector is different from the previous classification methods that select the maximum value of 200 return values. The category of image is assigned according to the loss function in traditional SVM classifier. The objective function is as follows:

$$\begin{aligned} & v = \arg\min\nolimits_u \ loss(f_s^i, u) \\ & s.t. \\ & loss(f_s^i, u) = -f_s^i(u) \end{aligned} \tag{2}$$

The return value of SVM represents the distance between the vector and the optimal decision surface in the vector space. This method receives good results when the maximum return value is large. However, with the decrease of the maximum return value, the accuracy of classification becomes worse, as shown in Table 1. The test set has 5794 pictures, and we count the accuracy

Table 1. The performance of different maximum return value

$\max(f_s^i)$	Correct number	Total	Accuracy
$<inf$	4872	5794	84.09%
<0.8	2601	3489	74.55%
<0.6	921	1632	56.43%
<0.4	87	291	29.9%

with pictures whose $\max(f_s^i)$ are respectively lower than inf, 0.8, 0.6, 0.4. Total represents the images within range respectively, and we give the corresponding accuracy.

For this problem, we set a threshold ε_1 for the maximum return value. We proposed SD-SVM model which is a combination of two correct methods S-SVM (SVM Modified by similarity of categories) and D-SVM (SVM Modified by difference between the similarities of sub-categories) to correct the classification model when $\max(f_s^i) < \varepsilon_1$. The threshold is derived from the distribution of the return value of the training set. We let ε_1 fit: $p(f_s^i(u) > \varepsilon_1 | I_i \notin u) = 0.03$. The formula means the probability of the u-th return value of image i is greater than ε_1 equal 0.3 when image i is not belong to u. We estimate the probability as follow:

$$\widehat{p}(f_s^i(u) > \varepsilon_1 | I_i \notin u) = \frac{\sum_{f_s^i(u)>\varepsilon_1, I_i \notin u} 1}{\sum_{I_i \notin u} 1} \tag{3}$$

Modified by similarity of categories. In fine-grained classification, the distinct between different categories is smaller than other classification problems, many sub-categories shares strong similarity. Using the SVM return value of the training set, we can create a category similarity matrix: $M_{uv} = E(f_s^i(u)), I_i \in v$. $E(\cdot)$ is the expect function. Let $f_p^v = [M_{1v}, M_{2v}, \cdots M_{200v}]^T$ represent the priori feature of category v. When the category u and v are similar, the value of M_{uv} is relatively large. So we build the S-SVM model:

$$\begin{aligned} & v = \arg\min\nolimits_v \ loss(f_s^i, f_p^v, v) \\ & s.t. \\ & loss(f_s^i, f_p^v, v) = -f_s^i(v) + 1/n * \sum\nolimits_{f_p^v(u)>\varepsilon_2} (f_s^i(u) - f_p^v(u))^2 \end{aligned} \tag{4}$$

where n is the number satisfied $f_p^v(u) > \varepsilon_2$. We judge the condition of $f_p^v(u) > \varepsilon_2$ to determine whether the two categories are similar enough. ε_2 is decided from the distribution of the return value of the training set. For example, we assume that there are three similar categories for each category on average, so we need to get a similar relationship of 1.5% for two hundred categories of bird datasets. We have to calculate ε_2 satisfying: $p(f_p^v(u) > \varepsilon_2 | u \neq v) = 0.015$. We estimate the probability as follow:

$$\widehat{p}(f_p^v(u) > \varepsilon_2 | u \neq v) = \frac{\sum_{f_p^v(u)>\varepsilon_2, u \neq v} 1}{\sum_{u \neq v} 1} \tag{5}$$

In order to showing the results of classification using our first kind of correction model(S-SVM). Figure 2 shows an example of the successful classification. Figure 2(e) is a test image which belongs to category 1 (Black footed Albatross). Figure 2(a) is an example of category 1, and Fig. 2(b) to (d) is a image that is similar to the category 1. Figure 2(f) is an example of category 45 (Northern Fulmar). Figure 2(g) to (i) is a image that is similar to the category 45. The return value of Fig. 2(e) for category 1 is 0.3923, and the return value of Fig. 2(e) for category 45 is 0.4462. Using the traditional classification method, this image is classified as category 45 incorrectly. Since the overall similarity of the image in the first line is higher than that of the third line, the image is classified correctly in our model.

Fig. 2. A successful example for classification with S-SVM. (e) is the image to be classified. (a) and (f) are the two categories which own the highest score. (b), (c), (d) are the closest three categories to (a). (g), (h), (i) are the closest three categories to (f). Since the similarity between (e) and (a)–(d) is greater than that of (f)–(i), we correctly classify (e) as category (a).

Modified by difference between the similarities of sub-categories. Since the similarity between categories is affected by many factors, the similarity is not transitive. For example, category a is similar to category b as black mouth, and category b is similar to category c as pointed mouth, however, category a is not similar to category c. When it is difficult to distinguish between category a and category b for given image, we can judge by the similarity information between the image and category c.

Based on this idea, we create a strongly discriminant category set $\omega(v)$ for each category v. We use $S(u, v)$ to express the ability of category u to distinguish category v with other categories similar to category v. $f_p^v(w) > \varepsilon_2$ means category w is similar to category v. We define $max3(\cdot)$ as a function return three of the maximum in the input. We selected the largest of the three in $S(u,v)$ into $\omega(v)$. So we build the D-SVM model:

$$
\begin{aligned}
& v = \arg\min_v \; loss(f_s^i, f_p^v, v) \\
& s.t. \\
& loss(f_s^i, f_p^v, v) = -f_s^i(v) + 1/n * \textstyle\sum_{u\in\omega(v)}(f_s^i(u) - f_p^v(u))^2 \\
& \omega(v) = max3(S(u,v)) \\
& S(u,v) = \textstyle\sum_{f_p^v(w)>\varepsilon_2}(f_p^v(u) - f_p^w(u))^2
\end{aligned} \tag{6}
$$

In order to showing the results of classification using our second kind of correction model(D-SVM). Figure 3 shows an example. Figure 3(d) is an image in the test set, and it's category is category 41 (Scissor tailed Flycatcher). Figure 3(a) to (c) are sample images of categories 41, category 189 (Red bellied Woodpecker), category 36 (Northern Flicker), category 189 is similar to category 36 and category 41, while category 41 is not similar to category 36. The return value of Fig. 3(d) for category 41 is 0.4254, and the return value for category 189 is 0.4649. In the traditional classification method, this image is misclassified as category 189. Since Fig. 3(d) is not similar to category 36, the image is classified as category 41 correctly in our model.

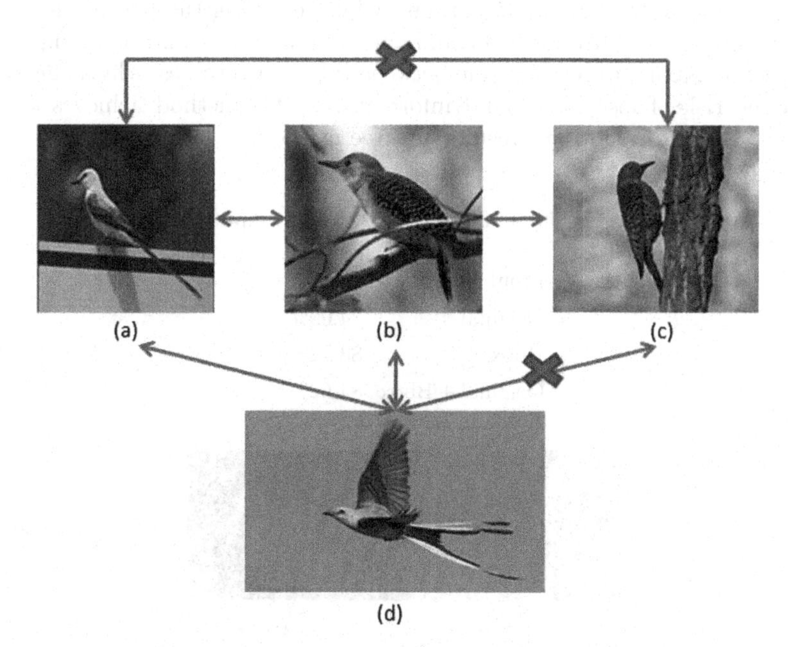

Fig. 3. A successful example for classification with D-SVM. (d) is the image to be classified. (a) and (b) are the two categories which own the highest score. (a) is not similar with (c). Meanwhile (b) is similar with (c). Since result shows (d) is not similar with (c). So we correctly classify (d) as category (a).

4 Experimental Results and Analysis

In order to verify that the proposed method is effective to improve the accuracy of fine-grained visual recognition, experiments are done in two fine-grained visual identification datasets (Caltech-UCSD Birds-200-2011 and FGVC-aircraft). At the stage of getting the bounding box, we use the DeCAF using the CNN framework that is provided by [1]. The DeCAF is trained on the ILSVRC 2012 dataset.

4.1 CUB-2011

The CUB-2011 dataset consists of 11,788 bird images belonging to 200 subcategories. We train the B-CNN [D, M] network based on the training set as [2]. We get $\varepsilon_1 = 0.58$, $\varepsilon_2 = 0.18$ in the training stage based on the method in Sect. 3.3.

Use Bounding box. We separately use the original image and the image intercepted by bounding box as the input. The experimental results are shown in Table 2. The method using the original image achieves 84.09% accuracy. The method using Bounding box achieves 83.74% accuracy. The accuracy is slightly lower than the accuracy of the original image. While the correct Bounding box is ful for removing the background of the image, the method of acquired Bounding box can not achieve the effect of manual calibration. Some images also lose important information of the foreground while removing the background, as is show in Fig. 4. The third method combining original image and bounding box is to select the maximum return value between two return values, which effectively avoids the risk of loss foreground information. This method achieves 84.62% accuracy. Table 2 is show the results of three method.

Table 2. The performance of different input image

Input	Accuracy
Original	84.09%
Bbox	83.74%
Original + Bbox	84.62%

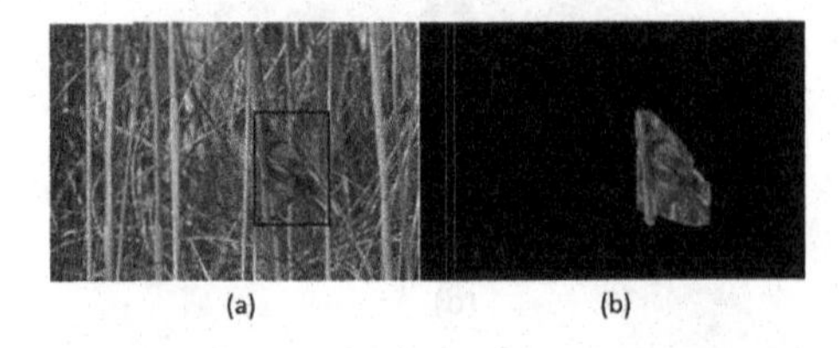

Fig. 4. An example of incorrect bounding box. (a) shows the wrong bounding box, (b) shows the defective foreground image obtained by wrong bounding box.

We also tried to calculate the bounding box for the training image during the training stage, but the result is so bad because the training stage could not avoid the wrong bounding box as the test stage. The network will train the wrong image as the correct category.

Correct SVM multi-classification. Table 3 shows the results of classification using our proposed S-SVM, D-SVM, SD-SVM. S-SVM classification method improved the accuracy by 0.3%. D-SVM classification method also increased the accuracy by 0.3%. While we adopt loss function using both correction methods to modify loss function, the accuracy of our model achieves 84.55%. Combining with the bounding box and SD-SVM classification algorithm, our proposed method can further improve the accuracy of fine-grained classification. Furthermore, we firstly extract the features of bounding box image and the original image, and then calculate the loss of classification using the extracted feature and SD-SVM algorithm. The final category is determined by the minimum loss function value. The final result is up to 85.12%.

Table 3. The performance on variants of our method

Algorithm	Accuracy
B-CNN	84.09%
B-CNN + S-SVM	84.40%
B-CNN + D-SVM	84.40%
B-CNN + SD-SVM	84.55%
B-CNN + Bbox + SD-SVM	85.12%

Comparison of threshold ε_1. The threshold ε_1 is a parameter that is used to determine the dividing line of the improved algorithm. This paper gets ε_1 value from the distribution of training set. Figure 5 shows the experimental results through manual changing ε_1. We can see that all three methods get higher accuracy opposite to traditional methods. When the threshold equals 0, our method is equivalent to the traditional method. The accuracy rise along with the threshold, and it falls until coming to the point about 0.5 or 0.6. The bigger of the return value, the more reliable the value is. So, our method get little improvement in this case. Besides, we found that SD-SVM usually outperforms S-SVM and D-SVM. We can see that when the maximum of SVM return value is less than a certain value, our method is always better than the unmodified method. While using the distributed of training set to obtain the threshold can also achieves a good result.

Comparison with previous works. Table 4 shows the results of existing best algorithms and our proposed method. Our approach is superior to all current fine-grained recognition algorithm with only category labels, and is better than

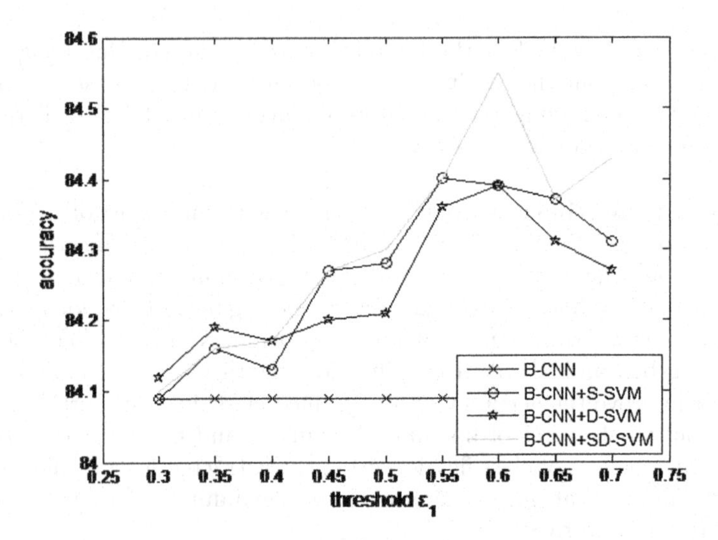

Fig. 5. The performance of different values of ε_1.

Table 4. Comparison of different methods on CUB-2011

Method	Train label	Test label	Accuracy
ours	n/a	n/a	85.12%
Two attention(2015)[12]	n/a	n/a	77.9%
STN(2015)[5]	n/a	n/a	84.1%
B-CNN(2015)[2]	n/a	n/a	84.09%
Pick filter(2016)[3]	n/a	n/a	84.54%
No part(2015)[9]	Bbox	n/a	82.0%
SPDA(2016)[4]	Bbox+Parts	Bbox	85.14%
FOAF(2014)[15]	Bbox+Parts	Bbox+Parts	81.2%
PN-CNN(2014)[8]	Bbox+Parts	Bbox+parts	85.4%

most algorithms of the fine-grained classification with the expensive manual annotation. The accuracy of our method is not as good as PN-CNN [8] and SPDA [4], which uses manual annotation of the Bbox level and part level during the training and testing stages.

4.2 FGVC-Aircraft

The FGVC-aircraft dataset consists of 10,000 aircraft images belonging to 100 subcategories. We use the B-CNN [D, D] network to train the training set as [2]. We get $\varepsilon_1 = 0.49$, $\varepsilon_2 = 0.20$ in the training stage based on the method in Sect. 3.3.

Comparison with previous work. Table 5 shows the results of different methods. Since the FGVC-aircraft data set does not provide Bbox-level annotation, many existing fine-grained classification algorithms can not be realized. As we can see from the table, our method achieves the highest accuracy.

Table 5. Comparison of different methods on FGVC-aircraft

Algorithm	Accuracy
Symbiotic segmentation(2013)[10]	72.5%
FV-SIFT(2014)[14]	80.7%
B-CNN(2015)[2]	84.1%
ours	85.21%

5 Conclusion

In this paper We propose a fine-grained visual classification algorithm based on image foreground and sub-category Similarity. Our algorithm combines the advantages of unsupervised object detection algorithm, feature extraction of bilinear CNN, as well as the similarity of sub-category. Experimental results show that our method outperforms state-of-the-art fine-grained categorization methods with only the category labels.

Acknowledgement. This work is supported by Provincial Natural Science Foundation of Jiangsu (Grant Nos. BK20151260, BK20151254), Six talent peaks project in Jiangsu Province (DZXX-027), National Natural Science Foundation of China (NSFC Grant Nos. 61272258, 61170124, 61301299, 61272005), Key Laboratory of Symbolic Computation and Knowledge Engineering of Ministry of Education, Jilin University (Grant No. 93K172016K08), and Collaborative Innovation Center of Novel Software Technology and Industrialization.

References

1. Donahue, J., Jia, Y., Vinyals, O., et al.: DeCAF: a deep convolutional activation feature for generic visual recognition. In: International Conference on Machine Learning, pp. 647–655 (2014)
2. Lin, T.Y., RoyChowdhury, A., Maji, S.: Bilinear cnn models for fine-grained visual recognition. In: Proceedings of the IEEE International Conference on Computer Vision, pp. 1449–1457 (2015)
3. Zhang, X., Xiong, H., Zhou, W., et al.: Picking deep filter responses for fine-grained image recognition. In: Proceedings of the IEEE Conference on Computer Vision and Pattern Recognition, pp. 1134–1142 (2016)
4. Zhang, H., Xu, T., Elhoseiny, M., et al.: SPDA-CNN: unifying semantic part detection and abstraction for fine-grained recognition. In: Proceedings of the IEEE Conference on Computer Vision and Pattern Recognition, pp. 1143–1152 (2016)

5. Jaderberg, M., Simonyan, K., Zisserman, A.: Spatial transformer networks. In: Advances in Neural Information Processing Systems, pp. 2017–2025 (2015)
6. Zhang, N., Donahue, J., Girshick, R., Darrell, T.: Part-based R-CNNs for fine-grained category detection. In: Fleet, D., Pajdla, T., Schiele, B., Tuytelaars, T. (eds.) ECCV 2014. LNCS, vol. 8689, pp. 834–849. Springer, Cham (2014). https://doi.org/10.1007/978-3-319-10590-1_54
7. Girshick, R., Donahue, J., Darrell, T., et al.: Rich feature hierarchies for accurate object detection and semantic segmentation. In: Proceedings of the IEEE Conference on Computer Vision and Pattern Recognition, pp. 580–587 (2014)
8. Branson, S., Van Horn, G., Belongie, S., et al.: Bird species categorization using pose normalized deep convolutional nets. arXiv preprint arXiv. 1406.2952 (2014)
9. Krause, J., Jin, H., Yang, J., et al.: Fine-grained recognition without part annotations. In: Proceedings of the IEEE Conference on Computer Vision and Pattern Recognition, pp. 5546–5555 (2015)
10. Chai, Y., Lempitsky, V., Zisserman, A.: Symbiotic segmentation and part localization for fine-grained categorization. In: Proceedings of the IEEE International Conference on Computer Vision, pp. 321–328 (2013)
11. Jaderberg, M., Simonyan, K., Zisserman, A.: Spatial transformer networks. In: Advances in Neural Information Processing Systems, pp. 2017–2025 (2015)
12. Xiao, T., Xu, Y., Yang, K., et al.: The application of two-level attention models in deep convolutional neural network for fine-grained image classification. In: Proceedings of the IEEE Conference on Computer Vision and Pattern Recognition, pp. 842–850 (2015)
13. Simon, M., Rodner, E.: Neural activation constellations: unsupervised part model discovery with convolutional networks. In: Proceedings of the IEEE International Conference on Computer Vision, pp. 1143–1151 (2015)
14. Gosselin, P.H., Murray, N., Jgou, H., et al.: Revisiting the fisher vector for fine-grained classification. Pattern Recognit. Lett. **49**, 92–98 (2014)
15. Zhang, X., Xiong, H., Zhou, W., et al.: Fused one-vs-all mid-level features for fine-grained visual categorization. In: Proceedings of the 22nd ACM International Conference on Multimedia, pp. 287–296. ACM (2014)
16. Li, H., Qi, F., Wang, S.: A comparison of model selection methods for multi-class support vector machines. In: Gervasi, O., Gavrilova, M.L., Kumar, V., Laganá, A., Lee, H.P., Mun, Y., Taniar, D., Tan, C.J.K. (eds.) ICCSA 2005. LNCS, vol. 3483, pp. 1140–1148. Springer, Heidelberg (2005). https://doi.org/10.1007/11424925_119
17. Kreßel, U.H.G.: Pairwise classification and support vector machines. In: Advances in Kernel Methods, pp. 255–268. MIT Press (1999)
18. Boykov, Y.Y., Jolly, M.P.: Interactive graph cuts for optimal boundary & region segmentation of objects in ND images. In: Proceedings of the Eighth IEEE International Conference on Computer Vision, ICCV 2001, vol. 1, pp. 105–112. IEEE (2001)

Faster R-CNN for Small Traffic Sign Detection

Zhuo Zhang[1], Xiaolong Zhou[1(✉)], Sixian Chan[1], Shengyong Chen[1], and Honghai Liu[2]

[1] College of Computer Science and Technology, Zhejiang University of Technology, Hangzhou, China
zxl@zjut.edu.cn

[2] School of Computing, University of Portsmouth, Portsmouth, UK

Abstract. Traffic sign detection is essential in autonomous driving. It is challenging especially when large proportion of instance to be detected are in small size. Directly applying state-of-the-art object detection algorithm Faster R-CNN for small traffic sign detection renders unsatisfactory detection rate, while a higher accuracy will be performed if the input images are upsampled. In this paper, we first investigate Faster R-CNN's network architecture, and regard its weak performance on small instances as improper receptive field. Then we augment its architecture according to receptive field with a higher accuracy achieved and no obvious incremental computational cost. Experiments are conducted to validate the effectiveness of proposed method and give an comparison to the state-of-the-art detection algorithms on both accuracy and computational cost. The experimental results demonstrate an improved detection accuracy and an competitive computing speed of the proposed method.

Keywords: Traffic sign detection · Convolutional Neural Network
Receptive field

1 Introduction

Traffic signs such as traffic lights and road signs play an important role in driving scene. They are designed to inform drivers of the current traffic situation, and their location information bridges the detection and recognition procedures. In driving assistant system, finding their location and determine their category can help drivers further reduce accident happening rate. We are interested in determining traffic signs' positions, and in this paper we focus on small traffic signs detection, since most instances only occupy a small relative area. Being a concrete case of object detection, we consider applying existing object detection algorithm on this task.

In recent years, deep Convolutional Neural Network (CNN) based methods on object detection have a fairly good performance. They use large amounts of data for training, or take advantage of transfer learing, and with the help of GPU computation, high accuracy is gained.

J. Yang et al. (Eds.): CCCV 2017, Part III, CCIS 773, pp. 155–165, 2017.
https://doi.org/10.1007/978-981-10-7305-2_14

Faster R-CNN [16] is one of the state-of-the-art object detection algorithms. It employs convolution network to generate region proposals and further refine their locations and categories. This approach achieves impressive performance on the PASCAL VOC [6] benchmark. The dataset we use is released from CCF BDCI 2016 competition. As shown in Fig. 1, this dataset contains large amounts of traffic signs that cover a small region of whole image, which is very different from the VOC2007 and pushes us in a dilemma when applying Faster R-CNN to achieve a high accuracy. One way to solve this problem is to upsample input images such that the effective objects' sizes would be similar to those in VOC2007. However, this will lead to a larger computational cost. In contrast, if we use the same resize configuration on input images or don't resize for fairly small objects, the computational cost would be better but the accuracy would be undesirable.

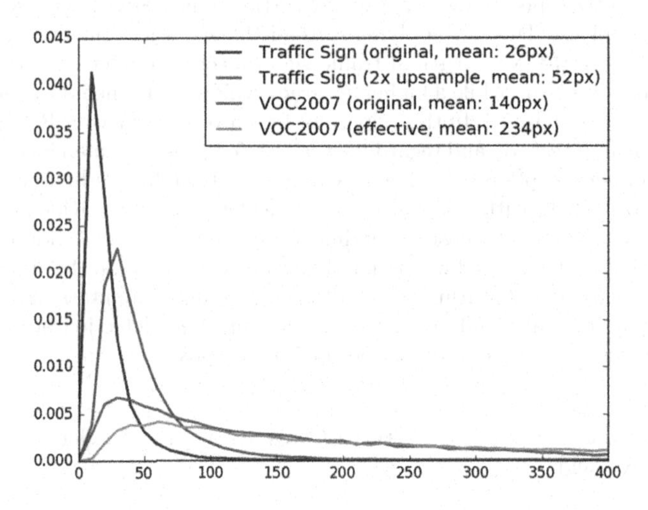

Fig. 1. Square area distribution of VOC2007 dataset and small traffic sign dataset. Obviously, even with 2x upsampling, traffic signs are still smaller than VOC2007 objects, which make it harder to detect.

In this paper, we try to propose an effective method to balance the computational cost and detection accuracy. Actually, upsampling input images is to decrease the network's receptive field. To decrease the receptive field, we can modify the network architecture instead of resizing the input images. For example, we can change convolution filer's size and stride, or change feature map resolution. If there is no input images resizing, we could gain fairly good performance by modifying the network architecture to keep a receptive field similar to upsampling scheme's. Comparing to the strategy of no input images resizing and original network, the accuracy would be guaranteed. Comparing to input upsampling with orignal network, the accuracy may be slightly lower, but the computational speed in both training and inference is fast and model size is also decreased. Meanwhile, we replace Faster R-CNN detection subnetwork's fully connection layers to convolution layers, which will further reduce the model size.

In this paper, we focus on proposing an improved Region Proposal Network (RPN) network architecture (RF + DilatedConv) to apply the Faster R-CNN on the challenging task of small traffic sign detection. Although the proposed method follows the Faster R-CNN, there are at least three major differences.

(1) We extend the Faster R-CNN in a new application to detect the challenging (small) traffic signs.
(2) We propose a receptive field guided Region Proposal Network (RPN) which boosts proposal quality.
(3) In the R-CNN detection subnetwork, we use fully convolution network to replace fully connected network, which keeps the accuracy and reduces the model size.

The rest of this paper is organized as follows. Section 2 briefly reviews the related work in both traffic sign and generic object detection. Section 3 presents analysis on RPN, gives computation basics of receptive field, and details the modification of the RPN architecture based on receptive field. In Sect. 4, extensive experiments are reported, presenting the correctness of the proposed method and competitive computing speed. Section 5 concludes this work.

2 Related Work

Early traffic sign detections are mainly in ideal conditions, where target objects occupy a large or medium proportion of the image. Most of them are clear and less occlusion. Researchers combine color and geometry characteristics to tackle this problem. For example, Fleyeh [7] detects traffic sign based on the color segmentation. Xu et al. [14] take advantage of shape symmetry for judging traffic signs. Later, more practical benchmark GTSDB [11] is proposed. Encouraged by the success of HOG feature and SVM classifier in human and generic object detection [4], this algorithm and its variants renders good accuracy on corresponding datasets [5,11,19]. However, the GTSDB benchmark is still not representative of that encountered in real tasks.

After Convolutional Neural Network (CNN) is rekindled in image classification [12,17], many CNN based object detection algorithms are proposed [3,8–10,13,15,16], essentially based on the rich representation of deep layers and additional adaptive subnetworks. Among them, RCNN [9], SPPNET [10], and Fast RCNN [8] first use existing region proposal method to generate candidate regions, then a DCNN model learns feature representation from all of them and gives a trained model, which is used for candidate region classification in inference time. All of them consists of two stage separate modules. Differ from that, Faster R-CNN [16], YOLO [15], SSD [13], and R-FCN [3] use only one network for the whole object detection task, thus the features are all learned rather than designed or partly designed via manually designing. These DCNN based algorithms can also be divided into two types: Faster R-CNN and R-FCN. Similarly, they also have two stages: learn a region proposal generator and then classify these proposals via following network with predicted location refined. While on the other hand, YOLO [15] and SSD [13] consist of only one single network, and they generate predicated object region with class labels directly.

Based on the work of Faster R-CNN, some algorithms are proposed for concrete purposed detection task. MSCNN [1] extends RPN [16] to multi-scale so that receptive field can match objects of different scales. RPN-BF [20] adopts RPN to generate high-resolution feature map to detect small pedestrian instances. Both MSCNN and RPN-BF deal with small object detection, and it is equivalent to gain smaller receptive field by obtaining high-resolution RPN last layer feature map. However, they conclude the bad performance of RPN or Faster R-CNN on these objects to improper receptive field, and their improvements are based on this discovery. Our method is based on them, but gives a formal receptive field calculation and thus provides a helpful reference for small object detection.

3 Proposed Method

3.1 Analysis on RPN

Faster-RCNN is a two-stage object detector, consisting of RPN and Fast RCNN subnetworks. RPN generates candidate object regions and Fast RCNN network classifies them and refines their locations. Region proposals' qualities determine the final detection performance on a large scale. They are found in low quality in the task of small traffic signs' detection. This subsection explores the cause for this deficiency and proposes three improvement requirements.

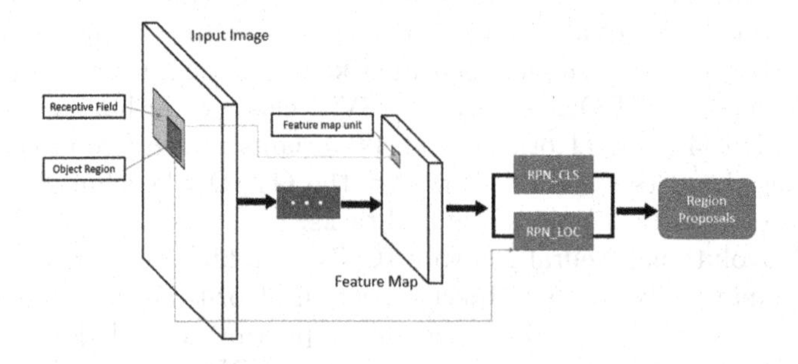

Fig. 2. Architecture of RPN.

The architecture of RPN is shown in Fig. 2. It has several convolution and pooling layers, followed by data manipulation and loss layers, which joins both classification and location regression tasks and generates region proposals finally. Feeding images to the network, feature maps are generated after each convolution and pooling layer. The resolution of the last layer's feature maps is much smaller than input images. Each unit on the last feature map depends on a region of pixels of the input image, i.e. its receptive field (NRF). The size of NRF should be close to the target objects, since oversized NRF fuses too much background information and makes features less discriminative. Decreasing the network's receptive field is the first requirement in our experiment.

In RPN, it generates a set of windows with specified scale and aspect ratios. This set of windows share the same center, and the center can be any receptive field's center, thus all these windows consist of the network's reference windows. An anchor is labeled positive if it has an Intersection-over-Union (IoU) greater than 0.5 with one ground truth box, and otherwise negative. Each anchor's corresponding unit on the last feature map for classification task is with the same label. For small object detection, oversized reference windows generate less positive samples, which harms the classification and region proposal's quality. Thus, shrinking reference windows' size is the second improvement requirement. Note that the RoI pooling layer bridges the gap between RPN and Fast RCNN, candidate objects' corresponding region on the last feature map serves as the input of Fast RCNN detection sub-network. Small target objects give birth to small region proposals, thus corresponding feature map region has low-resolution. This gives rise to less discriminative features because of collapsing bins, and thus degrades the downstream classifier. Therefore, the third improvement requirement comes that candidate objects' corresponding feature map region should be big enough for RoI pooling.

We can use simple tricks to meet the three requirements. By dropping out layer, modifying filter's stride or size, and using dilated convolution, the receptive field gets smaller. By decreasing anchor's base size, the reference window shrinks. By decreasing filter's stride, the feature map's resolution becomes larger. We give the fundamental for receptive field's calculation in the next subsection.

3.2 Calculation of Receptive Field

In convolutional networks, each unit's value depends on a region of the input. That region in the input is used as the receptive field for that unit. The input layer filters' size and stride determine how big the receptive field can be. Actually, each unit in the input region also depends on another region of the more previous layers (if any). We generalize this concept by replacing one unit with multiple units that consist of a square region. Thus, a unit's receptive field can be determined layer by layer. Since RPN receptive field is employed to match ground truth objects, we care how big the receptive field is on the input image. We treat the input image as the 0th layer, and assume that the network consists of n convolution layers (including pooling layer). Let the receptive filed of one unit of the last convolution layer's feature map be RF, it can be calculated via:

$$RF = F_n(1) \tag{1}$$

$$F_n(m) = F_{n-1}(k_n + (m-1)s_n) \tag{2}$$

$$F_0(m) = m \tag{3}$$

where $F_n(m)$ denotes the receptive field of m units on layer n, k_n and s_n represent the size and stride of nth layer's kernel respectively. The 0th layer is the input image itself.

3.3 Dilated Convolution

Dilated convolution is a general form of convolution operation [18]. It sums activation of signals with equal distances, which looks like using a filter with holes. We call it p-dilated convolution when the distance is p, and the equal distance of nearby pixels is $p + 1$. Obviously, when the distance equals to 1, it is the ordinary convolution operation. For 2D images, the p-dilated convolution operation can be defined as:

$$(W *_p I)(x, y) = \sum_{s=-a}^{a} \sum_{t=-a}^{a} W(s, t) I(x - ps, y - pt) \tag{4}$$

where I represents the image, W is the convolutional filter, (x, y) is the center point for filter and the filter's length is $2a + 1$, $*_p$ means the distance of nearby pixels selected for computation is p (on horizontal or vertical orientation).

Comparing to vanilla convolution operation, p-dilated convolution generates the sized feature map but with larger receptive field. Specifically, for p-dilated convolution, the filter's length is updated as:

$$k' = pk - 1 \tag{5}$$

With the updated kernel size k', the receptive field increases. For example, assuming all the kernels' length is 3 and stride is 1, the first layer and second layer do 1-dilated and 2-dilated convolution respectively, then the second feature maps' each unit's receptive field size is 7 instead of 5.

4 Experiments

4.1 Dataset and Algorithm Parameters Setting

In order to verity the effectiveness of our proposed method, we use same set of training and test data and choose 3 methods to compare: Faster R-CNN, our proposed method and SSD, and evaluated their performance with Average Precision (AP) at intersection and union area overlap ratio of 0.5.

The dataset is from the preliminary contest of CCF BDCI 2016 Traffic sign detection in self-driving scenario, which contains 4000 images with 720 height and 1280 width. These images are picked up from taxi's driving recorder and vary in illumination and angles. About 28000 traffic signs (mainly traffic lights and road signs) are labeled. Each image contains about 7 sign instances on average. We randomly choose 3000 images for training, the rest for validation. Five concrete methods are performed on this dataset with corresponding Average Precision evaluated for comparison.

Roughly we compare three algorithms: Faster R-CNN, our proposed Receptive Field net (RFnet), and SSD. For the purpose of comparing different receptive field, Faster R-CNN is with original size image and 2x up-sampled image as input. Our RFnet is designed to compete the latter in accuracy, and being a common RFnet and a RFnet with dilated convolution. We choose ZF-net [2] to fine-tune them on, since this backbone network consumes a video ram that a NVIDIA GTX970 GPU can handle, especially for 2x up-sampled Faster R-CNN. Being competitive in both accuracy and speed in common object detection, we also train and test a SSD model with VGG16 backbone network, to compare with Faster R-CNN and our RFnet on small traffic sign detection. Hence, there are 5 experiment schemes in total. Scheme names and corresponding measurement are listed on Table 1, and Precision-Recall curve and AP value plotted on Fig. 3.

Table 1. Details of experiment schemes.

Scheme	Receptive field	LCS resolution	Model size	Speed	AP (70000 iteration)
FRCNN-ZF-1x-input	171	16	255M	0.14 s	30.7%
FRCNN-ZF-2x-input	85	16	255M	0.37 s	44.6%
SSD-VGG16-512	-	-	91M	0.13 s	41.9%
RFnet-ZF (ours)	83	8	34M	0.41 s	50.0%
RFnet-ZF-dilated (ours)	85	8	29M	0.39 s	48.2%

4.2 Experiment Schemes and Results

For Faster R-CNN and our RFnet, we fine-tune them on ZF-net with 70000 training iteration. For SSD we fine-tune it on VGG16 with 50000 training iteration. All these schemes are trained on 3000 images and evaluated on 1000 images with AP@0.5 as the measurement. Their performance is plotted on Fig. 3. For a fixed recall, the higher precision the better accuracy. For different receptive field, original input images and 2x up-sampled images are fed into Faster R-CNN, denoted as FRCNN-ZF-1x-input and FRCNN-ZF-2x-input, which generate receptive fields of 171 and 85 respectively. Since the traffic sign detection dataset contains large number of small size instance (see Fig. 1), and Faster R-CNN expects not too small object sizes, it is not surprising that FRCNN-ZF-2x-input performs better than FRCNN-ZF-1x-input.

Our RFnet also gives a smaller receptive field that nearly the same as FRCNN-ZF-2x-input, without input image re-scaling. Based on Faster R-CNN network architecture, we decrease the last shared convolution layer's receptive field by means of altering convolution or pooling layer's kernel size and stride, and dilated convolution trick if possible. We give two concrete networks. One is

RFnet-ZF, which drops conv5 layer and decreases the second pooling layer's size from 3 to 2 and stride from 2 to 1. The other is RF-net-dilated, which drops conv4 and conv5 and decreases pool2's size from 3 to 2, stride from 2 to 1, and pool1's size from 3 to 4. Both of them shrink the receptive field to nearly half of before, without re-scaling of input. The anchor window's basic sizes also decrease, for the sake of better IoU of anchor box and ground truth box. In experiments, we use 2, 4, 8 as basic sizes. Since our RFnet is based on Faster R-CNN, we also perform 70000 iterations of back-propagation. As illustrated in Fig. 3, our RFnet performs the best, which achieves 50% AP value, and RFnet-dilated performs the second best, which achieves 48.2% AP. Comparing to FRCNN-ZF-1x-input, anchor boxes and RPN training samples are generated on larger resolution feature map of last shared convolution layer (denote as LSC resolution), while they have same input sizes. Therefore, RFnet uses half receptive field and outperforms Faster R-CNN a large margin. Comparing with FRCNN-ZF-2x-input, our RFnets have better accuracy while they keep the same receptive field. This is mainly due to the modified anchor size, which makes the network consumes longer time both in training and test phase. As illustrated on Table 1, RFnet-ZF is with a 5.4% higher accuracy than FRCNN-ZF-2x-input, with the cost of 0.04 s longer for each image inference.

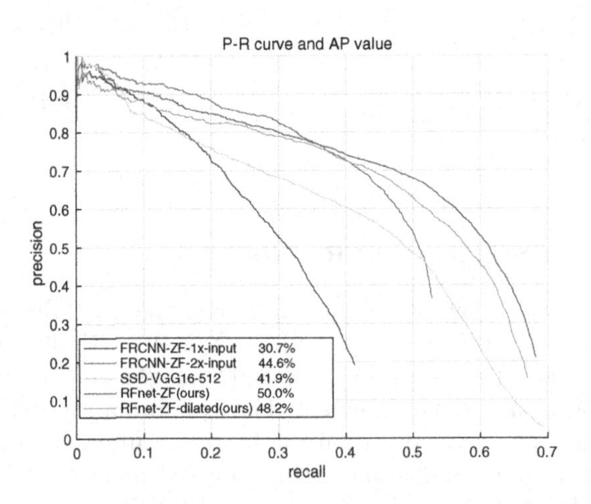

Fig. 3. Performance on the traffic sign detection dataset

We also do comparison with other state-of-the-art CNN based object detection algorithms to show the superior performance of the proposed method. We choose SSD, the representative of only one stage conv-net based method, known for its accuracy and fast inference speed. Since the limitation of time, we only pick SSD with VGG16 as backbone network and input image size of 512. Obviously, this scheme should have better accuracy than ZF-net based SSD. We use batch size of 4 and learing rate of 0.00025, and consider the batch size of Faster R-CNN and RFnet as 1. We perform 50000 iterations for SSD. We shrink the input to 512 × 512 and denote this net as SSD-VGG16-512. For accuracy, as illustrated on Fig. 3, SSD-VGG16-512 has higher precision than FRCNN-ZF-1x-input at most time, while lower than FRCNN-ZF-2x-input at low recall rate, but higher precision for higher recall. Thus, SSD is with 41.9% AP, smaller but near the performance of FRCNN-ZF-2x-input with 44.6% AP. Considering the number of parameters, i.e. the model size, SSD-VGG16-512 is 91M, which is only 35% of FRCNN with ZFnet, and also runs faster than FRCNN, SSD is competitive to it. Meanwhile, our RFnets surpass SSD on nearly all possible recall rate, and reaches 48.2% and 50.0% AP, which is much higher than SSD. This accuracy difference shows the effectiveness of our proposed RFnet. Since RFnet also use convolution layers to replace fully connected layers, the model size also shrinks. Luckily, RFnets use around 1/3 number of parameters than SSD and have better accuracy. The shortcoming of RFnets to SSD is the inference speed. Our RFnets consumes 2 more times than SSD, with nearly 0.4 s for each image.

Figure 4 demonstrates the detection results of five algorithms. From the first row (r1) to the last row (r5), the detection results are obtained by the

Fig. 4. Detection results of five detection algorithms on same set of images.

FRCNN-ZF-1x-input, FRCNN-ZF-2x-input, SSD-VGG16-512, RFnet-ZF, and RFnet-ZF-dilated, respectively. The results show that the proposed RFnet detects more correct small traffic signs.

5 Conclusion

In this paper, we have presented a simple but effective baseline that adopted Faster-RCNN's network architecture for small traffic sign detection. An analysis on the RPN network has showed that there needs a proper match among receptive field, reference window and traffic sign. We have proposed to modify the convolution network's architecture. The specific method included increasing anchors' density and feature maps' resolution, and decreasing receptive field and reference window size. These improvements brought accurate candidate regions, and kept smoothing connection with the following detection sub-network. The proposed method has gained remarkable performance boost. Meanwhile, the network generated smaller sized model due to less layers in use, and ran faster at test stage. The experimental results demonstrated the good performance of the proposed method in detecting small traffic signs, which could be further employed to multi-scale object detection.

Acknowledgments. This work was supported in part by National Natural Science Foundation of China (61403342, U1509207, 61325019, 61673329, 61603341).

References

1. Cai, Z., Fan, Q., Feris, R.S., Vasconcelos, N.: A unified multi-scale deep convolutional neural network for fast object detection. In: Leibe, B., Matas, J., Sebe, N., Welling, M. (eds.) ECCV 2016. LNCS, vol. 9908, pp. 354–370. Springer, Cham (2016). https://doi.org/10.1007/978-3-319-46493-0_22
2. Chatfield, K., Simonyan, K., Vedaldi, A., Zisserman, A.: Return of the devil in the details: delving deep into convolutional nets. arXiv preprint arXiv:1405.3531 (2014)
3. Dai, J., Li, Y., He, K., Sun, J.: R-FCN: object detection via region-based fully convolutional networks. In: Advances in Neural Information Processing Systems, pp. 379–387 (2016)
4. Dalal, N., Triggs, B.: Histograms of oriented gradients for human detection. In: IEEE Computer Society Conference on Computer Vision and Pattern Recognition, CVPR 2005, vol. 1, pp. 886–893. IEEE (2005)
5. El Margae, S., Sanae, B., Mounir, A.K., Youssef, F.: Traffic sign recognition based on multi-block LBP features using SVM with normalization. In: 2014 9th International Conference on Intelligent Systems: Theories and Applications (SITA-14), pp. 1–7. IEEE (2014)
6. Everingham, M., Van Gool, L., Williams, C.K., Winn, J., Zisserman, A.: The pascal visual object classes (VOC) challenge. Int. J. Comput. Vision **88**(2), 303–338 (2010)
7. Fleyeh, H.: Shadow and highlight invariant colour segmentation algorithm for traffic signs. In: 2006 IEEE Conference on Cybernetics and Intelligent Systems, pp. 1–7. IEEE (2006)

8. Girshick, R.: Fast R-CNN. In: Proceedings of the IEEE International Conference on Computer Vision, pp. 1440–1448 (2015)
9. Girshick, R., Donahue, J., Darrell, T., Malik, J.: Rich feature hierarchies for accurate object detection and semantic segmentation. In: Proceedings of the IEEE Conference on Computer Vision and Pattern Recognition, pp. 580–587 (2014)
10. He, K., Zhang, X., Ren, S., Sun, J.: Spatial pyramid pooling in deep convolutional networks for visual recognition. In: Fleet, D., Pajdla, T., Schiele, B., Tuytelaars, T. (eds.) ECCV 2014. LNCS, vol. 8691, pp. 346–361. Springer, Cham (2014). https://doi.org/10.1007/978-3-319-10578-9_23
11. Houben, S., Stallkamp, J., Salmen, J., Schlipsing, M., Igel, C.: Detection of traffic signs in real-world images: the German traffic sign detection benchmark. In: The 2013 International Joint Conference on Neural Networks (IJCNN), pp. 1–8. IEEE (2013)
12. Krizhevsky, A., Sutskever, I., Hinton, G.E.: Imagenet classification with deep convolutional neural networks. In: Advances in Neural Information Processing Systems, pp. 1097–1105 (2012)
13. Liu, W., Anguelov, D., Erhan, D., Szegedy, C., Reed, S., Fu, C.-Y., Berg, A.C.: SSD: single shot MultiBox detector. In: Leibe, B., Matas, J., Sebe, N., Welling, M. (eds.) ECCV 2016. LNCS, vol. 9905, pp. 21–37. Springer, Cham (2016). https://doi.org/10.1007/978-3-319-46448-0_2
14. Qingsong, X., Juan, S., Tiantian, L.: A detection and recognition method for prohibition traffic signs. In: 2010 International Conference on Image Analysis and Signal Processing (IASP), pp. 583–586. IEEE (2010)
15. Redmon, J., Divvala, S., Girshick, R., Farhadi, A.: You only look once: unified, real-time object detection. In: Proceedings of the IEEE Conference on Computer Vision and Pattern Recognition, pp. 779–788 (2016)
16. Ren, S., He, K., Girshick, R., Sun, J.: Faster R-CNN: towards real-time object detection with region proposal networks. In: Advances in Neural Information Processing Systems, pp. 91–99 (2015)
17. Russakovsky, O., Deng, J., Su, H., Krause, J., Satheesh, S., Ma, S., Huang, Z., Karpathy, A., Khosla, A., Bernstein, M., et al.: Imagenet large scale visual recognition challenge. Int. J. Comput. Vis. **115**(3), 211–252 (2015)
18. Yu, F., Koltun, V.: Multi-scale context aggregation by dilated convolutions. In: ICLR (2016)
19. Zaklouta, F., Stanciulescu, B.: Real-time traffic sign recognition using spatially weighted HOG trees. In: 2011 15th International Conference on Advanced Robotics (ICAR), pp. 61–66. IEEE (2011)
20. Zhang, L., Lin, L., Liang, X., He, K.: Is faster R-CNN doing well for pedestrian detection? In: Leibe, B., Matas, J., Sebe, N., Welling, M. (eds.) ECCV 2016. LNCS, vol. 9906, pp. 443–457. Springer, Cham (2016). https://doi.org/10.1007/978-3-319-46475-6_28

S-OHEM: Stratified Online Hard Example Mining for Object Detection

Minne Li, Zhaoning Zhang(✉), Hao Yu, Xinyuan Chen, and Dongsheng Li

National Laboratory for Parallel and Distributed Processing,
National University of Defense Technology, Changsha 410073, China
zzningxp@gmail.com

Abstract. One of the major challenges in object detection is to propose detectors with highly accurate localization of objects. The online sampling of high-loss region proposals (hard examples) uses the multitask loss with equal weight settings across all loss types (e.g., classification and localization, rigid and non-rigid categories) and ignores the influence of different loss distributions throughout the training process, which we find essential to the training efficacy. In this paper, we present the *Stratified Online Hard Example Mining (S-OHEM)* algorithm for training higher efficiency and accuracy detectors. S-OHEM exploits OHEM with stratified sampling, a widely-adopted sampling technique, to choose the training examples according to this influence during hard example mining, and thus enhance the performance of object detectors. We show through systematic experiments that S-OHEM yields an average precision (AP) improvement of 0.5% on *rigid categories* of PASCAL VOC 2007 for both the IoU threshold of 0.6 and 0.7. For KITTI 2012, both results of the same metric are 1.6%. Regarding the mean average precision (mAP), a relative increase of 0.3% and 0.5% (1% and 0.5%) is observed for VOC07 (KITTI12) using the same set of IoU threshold. Also, S-OHEM is easy to integrate with existing region-based detectors and is capable of acting with post-recognition level regressors.

1 Introduction

One of the major and fundamental challenges in object detection is to increase localization accuracy, which indicates the detector's ability to predict correct regions of target objects. The metric is typically measured by the bounding box overlap, i.e., the intersection over union (IoU) of the ground truth and predicted bounding boxes. While previous challenges (e.g., PASCAL VOC [3] and KITTI [5]) normally requires an IoU threshold of 0.5 to be considered a correct detection, real-world applications usually call for a higher accuracy (e.g., IoU ≥ 0.7). For example, the vehicle and pedestrian detection in autonomous driving need an accurate measurement of distance through real-time road traffic captures.

This work was done by Minne Li as a student.

J. Yang et al. (Eds.): CCCV 2017, Part III, CCIS 773, pp. 166–177, 2017.
https://doi.org/10.1007/978-981-10-7305-2_15

Recent literature has focused on the modification of region-based detection models at the post-recognition level to boost the localization accuracy [4,6,7]. However, limited work has addressed the problem from a data perspective. Data is important. The rapid advancement in the data collection, storage, and processing technology has made machine learning, especially deep learning, much easier by lightening the burden of generalizing well to unseen data with a limited number of training data [10].

However, the challenge of learning from imbalanced data [12] still exists. Within the "Recognition Using Regions" paradigm [11], the training set of object detection is divided into two distinct groups of annotated objects and background regions, and the number of examples in these groups experience a huge imbalance. Online Hard Example Mining (OHEM) [27] is proposed to overcome the data imbalance by integrating bootstrapping technique [30] with region-based detectors, and can be effortlessly implemented on most of the region-based detectors.

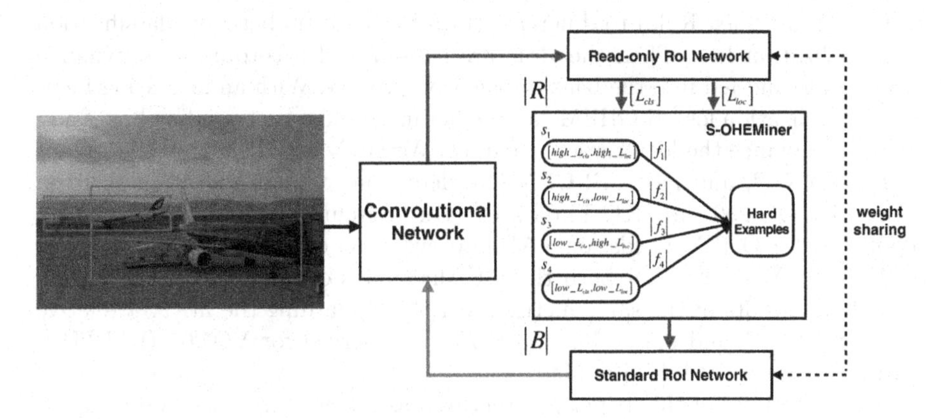

Fig. 1. Architecture of the Stratified Online Hard Example Mining algorithm (S-OHEM). We use the parameter denotation from [27]. In each mini-batch iteration, N is the number of images sampled from the dataset, R is the number of forward-propagated RoIs, and B is the number of subsampled RoIs to be fed into backpropagation. We denote classification loss by L_{cls} and localization loss by L_{loc}. S-OHEMiner conducts stratified sampling over R region proposals according to the sampling distribution at current training stage and produces B RoIs to be fed into backpropagation. We maintain a read-only RoI network and a standard RoI network with sharing weights for efficient memory allocation, derived from [27]. The blue solid stream indicates the process of forward-propagation and the green dashed stream shows the backpropagation process. More details are described in Sect. 3.3.

In this paper, we propose S-OHEM, the *Stratified Online Hard Example Mining* algorithm for training region-based deep convolutional network detectors to enhance localization accuracy, as shown in Fig. 1. The intuition of our method is that feeding hard examples to the backpropagation process could overcome the dilemma of unbalanced data, resulting in a more efficient and effective training process [27]. In the field of object detection, the hard example is defined as region proposal with higher training loss. Thus, previous hard example mining method (e.g., OHEM) is conducted by sampling region proposals according to a distribution that favors high loss instances. However, the training loss defined in previous work is the multitask loss with equal weight settings across all loss types (e.g., classification, localization, mask [13], or rigid categories and non-rigid categories). This approach ignores the influence of different loss types throughout the training process, which we found essential to the training efficacy (e.g., localization loss is more important during the latter part of the training period). Therefore, maintaining a sampling distribution according to this influence during hard example mining should enhance the performance of object detectors.

S-OHEM exploits *stratified sampling*, a sampling method involving the division of a population into distinct groups known as *strata* [21] (homogeneous subgroups, in which the inner items are similar to each other). During each mini-batch iteration, S-OHEM firstly assigns candidate examples (in the form of Region of Interests, RoIs) to different strata by the ratio between classification and localization loss. Then the RoIs are subsampled according to a dynamic distribution and fed into the backpropagation process. With an increasing focus on the localization loss, S-OHEM can predict more accurate bounding boxes and therefore enhance the localization accuracy. We apply S-OHEM to the standard Fast R-CNN [8] and Faster R-CNN [26] detection method and evaluate it on PASCAL VOC 2007 and KITTI datasets. Our systematic experimental analysis reports that S-OHEM yields some AP improvements of 0.5% on *rigid categories* of PASCAL VOC 2007 for both the IoU thresholds of 0.6 and 0.7. For KITTI 2012, both results of the same metric are 1.6%. Regarding the mAP, a relative increase of 0.3% and 0.5% (1% and 0.5%) is observed for VOC07 (KITTI12) with the same set of IoU threshold.

The remainder of this paper is structured as follows. In Sect. 2, we compare our work with related research with a focus on the improvement of localization accuracy and the use of data in object detection. In Sect. 3, we describe the design of the algorithm. In Sect. 4, we show the experimental results, and in Sect. 5, we conclude this work.

2 Related Work

Object detection has significantly benefited from the advancement of image classification task. The remarkable feature extraction ability of Deep Convolutional Networks [16,20,28,31,32] has equipped us with abundant information for the classification of region proposals. In addition, the continuously developing practical strategies (e.g., activation functions [15,24,34], regularization [18,29,30], and optimization [2,17,19]) further contribute to the efficacy of deep neural networks.

Several region-based detectors depend on the strong classification capability of deep convolutional networks to evaluate generated RoIs. R-CNN is the first to adopt this approach by evaluating each RoI separately. Fast R-CNN [8] improved this method by allowing computation sharing through projecting RoIs to a shared feature map (called RoIPool layer, derived from SPPnets [14]), resulting in better speed and accuracy. It was then integrated with the region proposal module (the Region Proposal Network, RPN) by sharing their convolutional features and extended to a unified network with "attention" [1] mechanism, leading to further speedup and accuracy enhancement. R-FCN [22] eliminates the fully-connected layers of region-based detectors and turns the whole model fully convolutional with the backbones of state-of-the-art image classifiers [16,32] to fully share computation, contributing to a significant speedup. Mask R-CNN [13], which adds a small Fully Convolutional Network (FCN) [23] as a parallel branch to standard Faster R-CNN and replaces the RoIPool layer with the RoIAlign layer, is the latest descendant of this stream and achieves significant advancement in several benchmarks of both the detection and segmentation tasks. However, most of these models use the multitask loss with equal weight settings without considering the influence of different loss type throughout the training process.

Recent work has focused on the post-recognition level of region-based detection models to boost the localization accuracy. Gidaris and Komodakis [6] proposed a CNN-model for bounding box regression, which is used with iterative localization and bounding box voting. LocNet [7] aims to enhance the localization accuracy by assigning a probability to each border of a loosely localized search region for being related to the object's bounding box. It's different from the bounding box regression approaches [4] adopted by most of the aforementioned region-based detectors and can be served as an effective alternative.

However, little work has focused on the advancement of region-based detectors from a data perspective. Online Hard Example Mining (OHEM) [27] integrates bootstrapping [30] (or *hard example mining*) with region-based detectors for a small extra computational cost, but still lacks enough focus on the localization accuracy because of the derived multi-task loss imbalance. Further discussion is available in Sect. 3.

3 Model Design

In this section, we argue that the current way of choosing hard examples lacks enough focus on localization accuracy and is suboptimal, and we will show that our approach results in better training (lower training loss), higher localization performance, and higher average precision. Firstly, we discuss the design motivation. Then we give a brief introduction of stratified sampling and definition of stratified constraint in this work. Finally, we present the design and implementation of our *Stratified Online Hard Example Mining* algorithm (S-OHEM).

3.1 Motivation

Most of the region-based detectors derive the multitask learning from Fast R-CNN, and assume equal contributions of classification loss and localization loss throughout the training process. However, this assumption is not often the case. We apply the original OHEM on standard Fast R-CNN and Faster R-CNN, then report the classification and localization loss throughout the training process on PASCAL VOC and KITTI datasets separately.

As is illustrated in Fig. 2, the classification loss is consistently larger than the localization loss (more than double in average). But this could result in a problem. Let's consider a situation where we have two region proposals RoI A and RoI B and are asked to choose one as the hard example for backpropagation. Based on the preliminary experiment result shown in Fig. 2, we make a common assumption that the training loss for RoI A and RoI B is $L_{cls}(A) = 0.21$, $L_{loc}(A) = 0.11$, and $L_{cls}(B) = 0.19$, $L_{loc}(B) = 0.12$ respectively. Recall that the classification loss is defined as log loss $L_{cls}(p, u) = -\log p_u$ for true class u [8], and thus the probability for the true class is 61.5% and 64.5% for RoI A and RoI B respectively. It's not a significant gap of the class prediction probability between these two RoIs, and we can believe they have similar performance for the classification task.

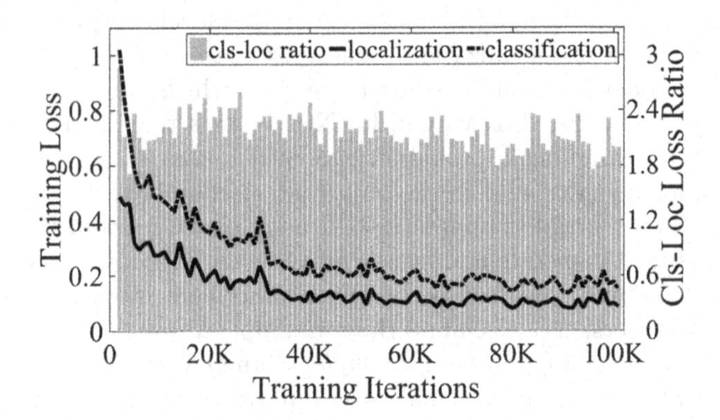

Fig. 2. Influence of different loss types throughout the training process. For better visualization, we average out the training loss of every 1000 iterations.

Regarding the localization loss, the gap between RoI A and RoI B is 0.01 ($L_{loc}(B) - L_{loc}(A) = 0.12 - 0.11 = 0.01$). Within the smooth L_1 loss settings [8], this gap turns to a 0.14 difference between the bounding boxes of ground truth and prediction. Note that this gap is quite significant when we use the parameterization for bounding box offsets given in [9], and therefore we are supposed to choose RoI B as the hard example for better localization accuracy and prediction quality. However, within the equal-weight multitask loss settings, RoI A will be chosen as the hard one. Thus, the previous hard example mining approach lacks focus on localization accuracy.

3.2 Stratified Sampling

Stratified sampling is a sampling method involving the division of a population into distinct groups known as *strata* [21]. These strata are homogeneous subgroups of the original data with similar inner items. Stratified sampling can get higher statistical precision because the variability within subgroups sharing the same properties is lower than that of the entire population [33]. Therefore. stratified sampling improves the representativeness by reducing sampling error.

Each stratum constraint s_k is denoted by $s_k = (p_k, f_k)$, where p_k is a propositional formula and f_k is the required sample size. In this work, the four stratum constraint is defined by the ratio between classification loss (L_{cls}) and localization loss: s_1 = (high L_{cls} and high L_{loc}, f_1), s_2 = (high L_{cls} and low L_{loc}, f_2), s_3 = (low L_{cls} and high L_{loc}, f_3), and s_4 = (low L_{cls} and low L_{loc}, f_4). The required sample size and threshold of high loss (hard examples) change dynamically throughout the training process.

3.3 Stratified Online Hard Example Mining Algorithm

Given the observation that the previous hard example mining approach ignores the influence of different loss types throughout the training process and lacks focus on localization accuracy, we now demonstrate our approach of *Stratified Online Hard Example Mining* (S-OHEM).

The architecture of S-OHEM is shown in Fig. 1. In each mini-batch iteration, S-OHEM firstly generates region proposals of the input images, forward-propagates all of them across the region-based detector, and gathers the training loss of each RoI. Then each RoI is assigned to one of the four strata defined in Sect. 3.2. Different loss type combinations represent how well the current detector performs in classification and localization tasks on each RoI respectively. Inside each stratum, hard examples are chosen by sorting the RoIs by loss. After that, all RoIs are subsampled according to a dynamic distribution, and a total number of B hard examples are fed into the backpropagation process. The sampling distribution of RoIs from each stratum changes dynamically throughout the learning process, as each loss type maintains different contribution to the detector model at different training stages. Specifically, the effect of classification loss is more important in the beginning, while the localization loss contributes more at later training stages.

For implementation, we keep a record of history training loss and start to change the sampling distribution when the loss becomes stable (e.g., after 40K iterations shown in Fig. 2). At the beginning of training, we only sample the first B RoIs with high L_{cls} (i.e., sample from s_{12}, the union of strata s_1 and s_2). When loss becomes stable, we gradually focus on choosing the RoIs with high L_{loc} (i.e., sample from the union of s_2 and s_3, denoted by s_{23}) by increasing the sampling ratio between s_{23} and s_1. Because of the gradually increasing focus on the localization loss, S-OHEM can predict more accurate bounding boxes and thus enhance the localization accuracy.

An equivalent alternative is available. To make it simple, we denote the contribution coefficient of L_{cls} and L_{loc} to hard example selection by α and β respectively. And our approach aims to find the optimal value of α and β in Formula (1) at different training stages. L_{select} is only for hard example mining, and the actual loss backpropagated across the network will not be affected.

$$L_{select} = \alpha L_{cls} + \beta L_{loc} \quad (1)$$

When training begins, we only sample the first B RoIs with high L_{cls} by setting α and β in Formula (1) to 1 and 0 respectively. When loss becomes stable, we gradually focus on choosing the RoIs with high L_{loc} by gradually decreasing the value of α and increasing β in Formula (1).

S-OHEM will not have a significant influence on the training time because most of the forward computation is shared between RoIs [8], and the number of backpropagated examples is much smaller than that of all region proposals of the input images. To overcome co-located RoIs and loss double counting, we follow the solution of [27] and apply non-maximum suppression (NMS) [25] to perform deduplication before the sampling procedure. NMS works by finding the highest loss RoI, and eliminating all other RoIs with lower loss and high overlap with the selected region. Besides, we derive their method of maintaining a read-only RoI network and a standard RoI network with sharing weights for efficient memory allocation. It is also worth noting that S-OHEM can be combined with any post-recognition regressors introduced in Sect. 2, because it focuses on enhancing the localization accuracy from the data perspective.

4 Experiments and Results

In this section, we conduct systematic experiments to evaluate the proposed S-OHEM and compare it with original OHEM. We describe the experimental setup in Sect. 4.1, and demonstrate the efficiency and accuracy of the algorithm by examining the training loss and average precision.

4.1 Experimental Setup

We use the standard and popular CNN architecture VGG16 from [28], and evaluate the algorithms on the PASCAL VOC 2007 and KITTI Object Detection Evaluation 2012 dataset. In the PASCAL VOC experiment, training is done on the trainval set and testing on the test set. In the KITTI 2012 experiment, we use the first 5000 images to form the training set and the remaining 2481 images for testing. All models are trained with SGD for 80k mini-batch iterations and followed the same setup from Sect. 4.1. For average precision, we report the results with IoU thresholds of 0.5, 0.6, and 0.7, to evaluate the localization accuracy in a wider range of IoU thresholds. We use Fast R-CNN [8] as the detector base for our PASCAL VOC experiment, and Faster R-CNN [26] for the KITTI 2012 experiment, to prove the usability of our approach. The initial learning rate is set to 0.001 and dropped in "steps" by a factor of 0.1 every 30K iterations. We process 2 images in each mini-batch iteration, and subsample 128 RoIs to feed them into backpropagation. Note that the baseline of OHEM reported in Table 2 (row 1–2) were reproduced and are slightly higher than the ones reported in [27].

For both experiments, we follow the procedure described in Sect. 3.3 to control the contribution coefficient of L_{cls} and L_{loc}. In the beginning, α and β are set to 1 and 0 when training starts. Then we gradually increase β to the ratio between classification and localization loss when the loss becomes stable. Specifically, β will increase to 1.9 and 2.3 for the VOC07 and KITTI12 experiment respectively.

4.2 Results and Analysis

Training Convergence. We firstly analyze the training loss for both methods by logging the average training loss every 10K steps. Figure 3 shows the average loss per RoI for VGG16 with settings presented in Sect. 4.1. We see that S-OHEM yields lower loss in both classification and localization than the original OHEM, validating our claims that S-OHEM leads to better training than OHEM. Also, the results indicate that S-OHEM is better in classification confidence and localization accuracy during the training process.

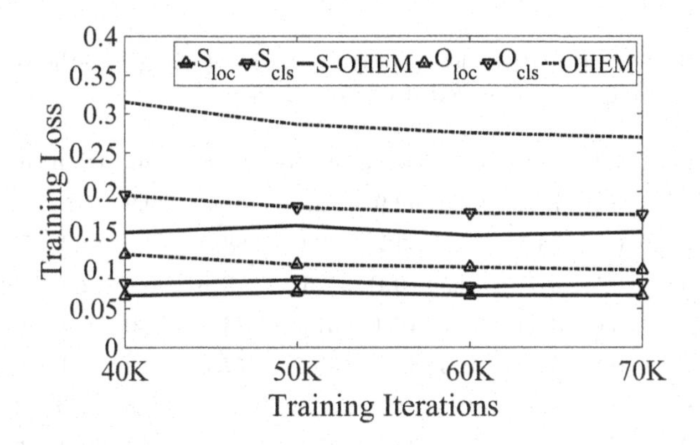

Fig. 3. Training loss for S-OHEM and OHEM. We show the average loss per RoI for VGG16. These results indicate that S-OHEM is better in classification confidence and localization accuracy during the training process.

VOC 2007. Table 1 shows that on VOC07, S-OHEM improves the mAP of OHEM from 71% to 71.1% for an IoU threshold of 0.5, and an improvement of 0.4% and 0.3% for IoU 0.6 and 0.7 respectively. For category-specific improvements, S-OHEM performs well in most of the rigid categories (bold categories in Table 1) across all three IoU thresholds, especially for IoU 0.7.

As is listed on Table 3(a), we compute the mAP among rigid categories and show increase of 0.1%, 0.5%, and 0.5% for IoU 0.5, 0.6, and 0.7 respectively. It's also interesting to find that S-OHEM performs quite well in detecting *cats* for IoU threshold 0.6, which indicates the better bounding boxes generated by S-OHEM in this environment.

Table 1. VOC 2007 test detection average precision (%). All methods use VGG16 and bounding-box regression. Legend: IoU: IoU threshold.

method	IoU	mAP	**aero**	**bike**	bird	**boat**	**bottle**	**bus**	**car**	cat	**chair**	cow	**table**	dog	horse	**mbike**	persn	plant	sheep	**sofa**	**train**	**tv**
OHEM	0.5	71.0	72.1	80.4	68.9	60.5	47.1	81.5	79.6	82.8	54.1	77.3	70.7	81.7	81.4	76.7	74.4	41.6	70.0	69.6	76.5	73.6
S-OHEM	0.5	71.1	72.8	80.9	69.2	60.2	47.9	81.4	79.5	82.5	53.8	76.6	70.3	81.9	81.5	77.5	74.5	41.6	70.1	70.2	76.0	73.6
improv	0.5	0.1	**0.7**	**0.5**	0.3	-0.3	**0.8**	-0.1	-0.1	-0.3	-0.3	-0.7	-0.4	0.2	0.1	**0.8**	0.1	0	0.1	**0.6**	-0.5	0
OHEM	0.6	62.2	63.5	74.5	57.7	47.1	38.0	76.0	74.4	70.9	42.0	70.8	61.7	72.7	74.9	68.1	62.6	30.7	59.3	63.5	66.5	68.2
S-OHEM	0.6	62.7	64.9	74.4	58.5	48.1	38.5	76.6	73.9	75.3	42.0	71.8	60.3	72.8	74.7	69.2	62.0	30.8	59.1	65.0	67.6	68.5
improv	0.6	0.5	**1.4**	-0.1	**0.8**	**1.0**	**0.5**	**0.6**	-0.5	**4.4**	0	**1.0**	-1.4	0.1	-0.2	**1.1**	-0.6	0.1	-0.2	**1.5**	**1.1**	0.3
OHEM	0.7	48.3	52.8	58.2	42.1	32.0	27.3	68.6	63.0	56.5	31.0	56.3	44.9	50.0	55.6	55.6	44.0	16.6	49.2	48.9	55.5	58.5
S-OHEM	0.7	48.6	55.2	57.8	41.4	32.5	27.9	69.0	63.8	56.9	30.4	58.2	44.9	50.0	54.1	56.1	44.2	16.4	48.6	49.9	56.2	58.9
improv	0.7	0.3	**2.4**	-0.4	-0.7	**0.5**	**0.6**	**0.4**	**0.8**	**0.4**	-0.6	**1.9**	0	0	-1.5	**0.5**	0.2	-0.2	-0.6	**1.0**	**0.7**	**0.4**

KITTI 2012. The evaluation results on KITTI 2012 is shown in Table 2. S-OHEM improves the mAP of OHEM from 63.9% to 64.9% for an IoU threshold of 0.6, and an improvement of 0.5% for IoU 0.7. We also compute the mAP among rigid categories and list results in Table 3(b). Note that the Note that the *misc* category is classified as *rigid* based on our observation of the dataset. We show some increase of 1.6% for both IoU thresholds 0.6 and 0.7.

Table 2. KITTI 2012 test detection average precision (%). All methods use VGG16 and bounding-box regression. Legend: IoU: IoU threshold.

Method	IoU	mAP	car	persn	cyclist	truck	van	tram	misc
OHEM	0.5	78.5	78.5	62.9	72.4	87.5	89.9	87.6	71.0
S-OHEM	0.5	78.5	78.2	63.7	73.9	88.3	89.0	85.3	70.8
improv	0.5	0	−0.3	**0.8**	**1.5**	**0.8**	−0.9	−2.3	−0.2
OHEM	0.6	63.9	68.7	47.4	57.7	73.8	79.0	70.8	49.6
S-OHEM	0.6	64.9	68.3	48.8	55.7	77.7	78.2	73.4	52.4
improv	0.6	1	−0.4	**1.4**	−2	**3.9**	−0.8	**2.6**	**2.8**
OHEM	0.7	42.9	50.5	29.1	37.3	52.7	59.8	42.0	28.8
S-OHEM	0.7	43.4	49.8	28.8	33.3	61.2	60.1	39.3	31.7
improv	0.7	0.5	−0.7	−0.3	−4	**8.5**	0.3	−2.7	**2.9**

Rigid and Non-rigid Category. Our experimental results have shown that S-OHEM performs quite well on rigid categories of both the VOC07 and KITTI12 dataset. The reason is that rigid bodies can reach better classification accuracy on pre-trained deep convolutional networks ascribed to its strong resistance to deformation. Therefore, the influence of different loss distribution throughout the training process (as described in Sect. 3.1) is more likely to happen on rigid bodies. Also, the border distribution of rigid bodies is more similar to each other and is thus easier to learn.

Table 3. Category specific mean average precision (%). All methods use VGG16 and bounding-box regression. Legend: **IoU:** IoU threshold. (a) On VOC 2007 test set. (b) On KITTI 2012 test set.

method	IoU	rigid	non-rigid
OHEM	0.5	70.2	72.2
S-OHEM	0.5	70.3	72.2
improv	0.5	0.1	0
OHEM	0.6	62.0	62.4
S-OHEM	0.6	62.5	63.1
improv	0.6	**0.5**	**0.7**
OHEM	0.7	49.7	46.2
S-OHEM	0.7	50.2	46.2
improv	0.7	**0.5**	0

(a)

method	IoU	rigid	non-rigid
OHEM	0.5	82.9	67.7
S-OHEM	0.5	82.3	68.8
improv	0.5	-0.6	**1.1**
OHEM	0.6	68.4	52.6
S-OHEM	0.6	70.0	52.3
improv	0.6	**1.6**	-0.3
OHEM	0.7	46.8	33.2
S-OHEM	0.7	48.4	31
improv	0.7	**1.6**	-2.2

(b)

5 Conclusion

In this paper, we proposed Stratified Online Hard Example Mining (S-OHEM) algorithm, a simple and effective method for training region-based deep convolutional network detectors to enhance localization accuracy. During hard example mining, S-OHEM exploits stratified sampling and focuses on the influence of different loss types throughout the training process. Experimental analysis shows that S-OHEM outperforms OHEM regarding training convergence and localization accuracy, and achieves some AP improvements on *rigid categories* of PASCAL VOC 2007 and KITTI 2012. Besides, S-OHEM addresses the localization enhancing problem merely from the data perspective and can be easily plugged into existing region-based detectors. Furthermore, the state-of-the-art Mask R-CNN [13] also derives the equal-weight multi-task loss with an addition task of semantic segmentation, which is improvable through S-OHEM. S-OHEM can also be applied to other multi-task loss, including the loss of semantic segmentation, key-point detection, etc.

References

1. Bahdanau, D., Cho, K., Bengio, Y.: Neural machine translation by jointly learning to align and translate. CoRR abs/1409.0473 (2014)
2. Duchi, J., Hazan, E., Singer, Y.: Adaptive subgradient methods for online learning and stochastic optimization. J. Mach. Learn. Res. **12**(Jul), 2121–2159 (2011)
3. Everingham, M., Van Gool, L., Williams, C.K., Winn, J., Zisserman, A.: The pascal visual object classes (VOC) challenge. Int. J. Comput. Vis. **88**(2), 303–338 (2010)
4. Felzenszwalb, P.F., Girshick, R.B., McAllester, D., Ramanan, D.: Object detection with discriminatively trained part-based models. IEEE Trans. Pattern Anal. Mach. Intell. **32**(9), 1627–1645 (2010)
5. Geiger, A., Lenz, P., Urtasun, R.: Are we ready for autonomous driving? The KITTI vision benchmark suite. In: Conference on Computer Vision and Pattern Recognition (CVPR) (2012)
6. Gidaris, S., Komodakis, N.: Object detection via a multi-region and semantic segmentation-aware CNN model. In: Proceedings of the IEEE International Conference on Computer Vision, pp. 1134–1142 (2015)

7. Gidaris, S., Komodakis, N.: Locnet: improving localization accuracy for object detection. In: Proceedings of the IEEE Conference on Computer Vision and Pattern Recognition, pp. 789–798 (2016)
8. Girshick, R.: Fast R-CNN. In: Proceedings of the IEEE International Conference on Computer Vision, pp. 1440–1448 (2015)
9. Girshick, R., Donahue, J., Darrell, T., Malik, J.: Rich feature hierarchies for accurate object detection and semantic segmentation. In: Proceedings of the IEEE Conference on Computer Vision and Pattern Recognition, pp. 580–587 (2014)
10. Goodfellow, I., Bengio, Y., Courville, A.: Deep Learning. MIT Press (2016). http://www.deeplearningbook.org
11. Gu, C., Lim, J.J., Arbeláez, P., Malik, J.: Recognition using regions. In: IEEE Conference on Computer Vision and Pattern Recognition, CVPR 2009, pp. 1030–1037. IEEE (2009)
12. He, H., Garcia, E.A.: Learning from imbalanced data. IEEE Trans. Knowl. Data Eng. **21**(9), 1263–1284 (2009)
13. He, K., Gkioxari, G., Dollár, P., Girshick, R.: Mask R-CNN. arXiv preprint arXiv:1703.06870 (2017)
14. He, K., Zhang, X., Ren, S., Sun, J.: Spatial pyramid pooling in deep convolutional networks for visual recognition. In: Fleet, D., Pajdla, T., Schiele, B., Tuytelaars, T. (eds.) ECCV 2014. LNCS, vol. 8691, pp. 346–361. Springer, Cham (2014). https://doi.org/10.1007/978-3-319-10578-9_23
15. He, K., Zhang, X., Ren, S., Sun, J.: Delving deep into rectifiers: surpassing human-level performance on imagenet classification. In: Proceedings of the IEEE International Conference on Computer Vision, pp. 1026–1034 (2015)
16. He, K., Zhang, X., Ren, S., Sun, J.: Deep residual learning for image recognition. In: Proceedings of the IEEE Conference on Computer Vision and Pattern Recognition, pp. 770–778 (2016)
17. Hinton, G., Srivastava, N., Swersky, K.: Neural networks for machine learning lecture 6a overview of mini-batch gradient descent (2012)
18. Ioffe, S., Szegedy, C.: Batch normalization: accelerating deep network training by reducing internal covariate shift. In: ICML, JMLR Workshop and Conference Proceedings, vol. 37, pp. 448–456. JMLR.org (2015)
19. Kingma, D.P., Ba, J.: Adam: a method for stochastic optimization. CoRR abs/1412.6980 (2014)
20. Krizhevsky, A., Sutskever, I., Hinton, G.E.: Imagenet classification with deep convolutional neural networks. In: Advances in Neural Information Processing Systems, pp. 1097–1105 (2012)
21. Li, M., Li, D., Shen, S., Zhang, Z., Lu, X.: DSS: a scalable and efficient stratified sampling algorithm for large-scale datasets. In: Gao, G.R., Qian, D., Gao, X., Chapman, B., Chen, W. (eds.) NPC 2016. LNCS, vol. 9966, pp. 133–146. Springer, Cham (2016). https://doi.org/10.1007/978-3-319-47099-3_11
22. Li, Y., He, K., Sun, J., et al.: R-FCN: object detection via region-based fully convolutional networks. In: Advances in Neural Information Processing Systems, pp. 379–387 (2016)
23. Long, J., Shelhamer, E., Darrell, T.: Fully convolutional networks for semantic segmentation. In: Proceedings of the IEEE Conference on Computer Vision and Pattern Recognition, pp. 3431–3440 (2015)
24. Nair, V., Hinton, G.E.: Rectified linear units improve restricted Boltzmann machines. In: Proceedings of the 27th International Conference on Machine Learning (ICML 2010), pp. 807–814 (2010)

25. Neubeck, A., Van Gool, L.: Efficient non-maximum suppression. In: 18th International Conference on Pattern Recognition, ICPR 2006, vol. 3, pp. 850–855. IEEE (2006)
26. Ren, S., He, K., Girshick, R., Sun, J.: Faster R-CNN: towards real-time object detection with region proposal networks. In: Advances in Neural Information Processing Systems, pp. 91–99 (2015)
27. Shrivastava, A., Gupta, A., Girshick, R.: Training region-based object detectors with online hard example mining. In: Proceedings of the IEEE Conference on Computer Vision and Pattern Recognition, pp. 761–769 (2016)
28. Simonyan, K., Zisserman, A.: Very deep convolutional networks for large-scale image recognition. CoRR abs/1409.1556 (2014). http://arxiv.org/abs/1409.1556
29. Srivastava, N., Hinton, G.E., Krizhevsky, A., Sutskever, I., Salakhutdinov, R.: Dropout: a simple way to prevent neural networks from overfitting. J. Mach. Learn. Res. **15**(1), 1929–1958 (2014)
30. Sung, K.K.: Learning and example selection for object and pattern detection. Ph.D. thesis, Cambridge, MA, USA (1996). aAI0800657
31. Szegedy, C., Liu, W., Jia, Y., Sermanet, P., Reed, S., Anguelov, D., Erhan, D., Vanhoucke, V., Rabinovich, A.: Going deeper with convolutions. In: Proceedings of the IEEE Conference on Computer Vision and Pattern Recognition, pp. 1–9 (2015)
32. Szegedy, C., Vanhoucke, V., Ioffe, S., Shlens, J., Wojna, Z.: Rethinking the inception architecture for computer vision. In: Proceedings of the IEEE Conference on Computer Vision and Pattern Recognition, pp. 2818–2826 (2016)
33. Thompson, S.K.: Stratified Sampling, pp. 139–156. Wiley (2012). http://dx.doi.org/10.1002/9781118162934.ch11
34. Xu, B., Wang, N., Chen, T., Li, M.: Empirical evaluation of rectified activations in convolutional network. CoRR abs/1505.00853. http://arxiv.org/abs/1505.00853

ScratchNet: Detecting the Scratches on Cellphone Screen

Zhao Luo[1,2], Xiaobing Xiao[3], Shiming Ge[1,2](✉), Qiting Ye[1,2], Shengwei Zhao[1,2], and Xin Jin[4]

[1] Institute of Information Engineering, Chinese Academy of Sciences, Beijing, China

[2] School of Cyber Security, University of Chinese Academy of Sciences, Beijing, China

geshiming@iie.ac.cn

[3] School of Software and Microelectronics, Peking University, Beijing, China

[4] Department of Computer Science and Technology, Beijing Electronic Science and Technology Institute, Beijing, China

Abstract. In the process of cellphone screen manufacture, equipment failures and human errors may lead to screen scratches. Traditional manual ways check scratches by human eyes, which often costs large manpower and time, but with poor effectiveness. To address this issue, we proposes an automated scratches detection method by cascading two main modules. First, the scratch filtering module detects big scratches and localizes small scratches candidates with a serial of low-level stages. Then, the scratches classification module applies a lightweight CNN model, ScratchNet, to identify each small scratch candidate whether it is real scratch or not. To train the ScratchNet, we build a Scratches on Cellphone Screen (SCS) dataset with 50K samples in 5 categories. The experimental results on SCS testing set show that the proposed method achieves an accuracy of 96.35% on classifying small scratches, which outperforms the LeNet model and the other classifiers.

Keywords: Scratch detection · Product quality inspection Cellphone screen · Convolutional neural networks

1 Introduction

Cellphones are increasingly common. Every mobile phone brand rolls out new products and it is more frequently that people replace their own mobile phones. At the same time, handset firms have continued to pursue innovation in the material and shape of phone screen. It sets a higher request to the technology of defecting scratches on the cellphone screen, so how we do that quickly and well? Conventional manual detecting methods can't meet the production requirements. Along with the development of industrialization and the arrival of the Industry 4.0 and AI 2.0, machine vision is more and more prosperous, which is

J. Yang et al. (Eds.): CCCV 2017, Part III, CCIS 773, pp. 178–186, 2017.
https://doi.org/10.1007/978-981-10-7305-2_16

used in all walks of life. For example, rail track flaw detection [3], glass defects inspection [14,15,22], PCB coating detection [4,12], LED inspection [16] and printing quality checking [17,21]. It brings about huge improvement to production speed and product quality. It also reduces the cost of labor and enhances enterprise competitive power. From a macro view, it increases the using rate of social source and frees people from manual labor.

The main contributions of this paper are three folds:

- We propose a scratches inspection method to detect the scratches on the cellphone screen by cascading low-level image processing and a lightweight convolutional neural network (CNN) classifier, which can provide fast and accurate scratch detection performance.
- We build a new screen scratch dataset, SCS, to facilitate the scratch detection model.
- We design a lightweight CNN model, ScratchNet, which achieves an accuracy of 96.35% on classifying small scratches, which outperforms the other CNN models and other classifiers.

2 Related Work

In this part, we introduce some related works, mainly including surface defects inspection and convolutional neural networks.

2.1 Surface Scratches Inspection

In [15], Peng uses downward threshold to remove the background pattern like stripes firstly. Then he segment defections by fixed threshold method and OTSU. The method work well in common defects, but it gets poor results when inspecting small and slight optical defects. In detecting PU-packings, Choiu [1] combines radius inspection method and projection, which means it can detect seven major types of flaw. However, the speed of inspection isn't enough stability and accuracy rate only reach 90.1%. Zhao and Kong [22] take Canny, Binary Feature Histogram (BFH) and AdaBoost when analyzing the glass quality. However, this method is limited to inspect two types of defects and it's measuring precision is about 25–100 pixels. With the development of machine vision, the way of detecting defects emerge one after another. Liang [11] applys PCA and Redundant Dictionary to inspect flaws on touch screens. Choi [2] adopt the Phase Only Transform (PHOT) in detecting of surface defects. Soukup [19] takes CNN to inspect metal surface defects. What's more, some people capture surface defects by other ways, for example, McGrail [13] uses standard medical ultrasound scanners to collect information of insulating material flaws.

2.2 Convolutional Neural Networks

LeCun and Bengio recognize handwritten digits [10] by LeNet model and got better performance. Goodfellow designs CNN to identify Google StreeView house number [7]. Ciresan applys CNN to traffic sign recognition benchmark [5]. Besides, CNN combined with other methods make a unprecedented breakthrough in driverless cars and faces recognition. Not only the extension of its application, but also the innovation of itself, such as Krizhevsky' AlexNet [9], Simonyan' VGGNet [18], Szegedy' GoogleNet [20] and He' ResNet [8] are more and more powerful.

3 Dataset and Method

3.1 Dataset

Data collection is very meaningful and important in facilitating scratch detection. In this paper, we build a Scratch on Cellphone Screen (SCS) dataset. First, we take a lot of cellphone screen images in various lighting environments and capturing poses. Then, through human checking, we crop a large scale of small scratches as well as background patches from screen images. In addition, according to the scratch shape in blob or line as well as saliency degree in salient or weak, we group the examples into five categories: background examples, salient blob examples, weak blob examples, salient line examples, and weak line examples. Then, we clean data and ask 3 persons to perform cross-check in order to ensure the quality of the dataset. Finally, the examples are normalized into a unified size of 32 × 32 pixels, and divided into training set and testing set. In total, the dataset has 5K images including 4.6K images for collecting training

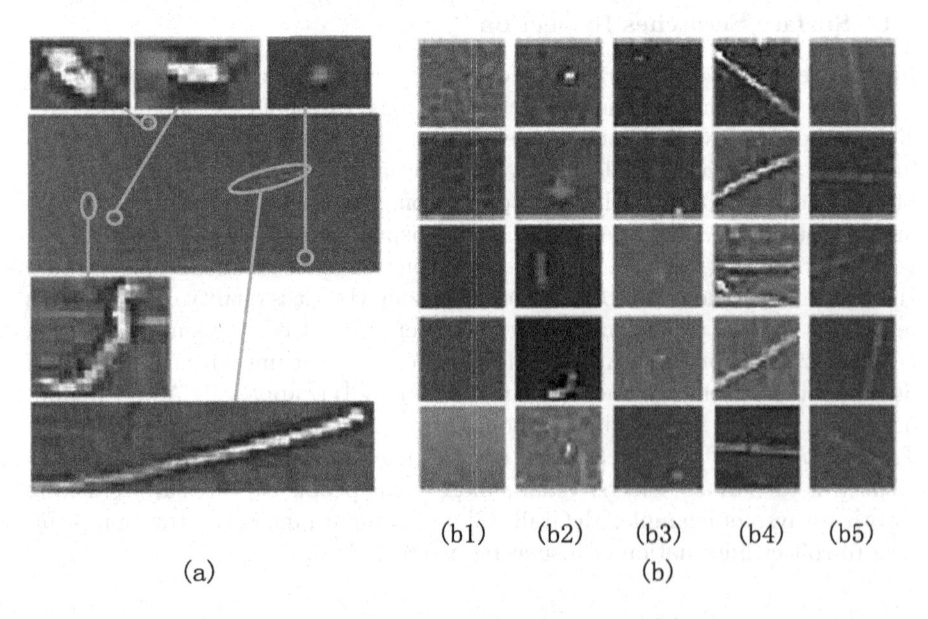

Fig. 1. Examples of scratches on the cellphone screen: (a) different categories of scratches are collected. (b) some examples of five scratch categories: (b1) background example, (b2) salient blob example, (b3) weak blob example, (b4) salient line example, (b5) weak line example.

set and the rest 4K images for testing. The training set includes five categories of examples: 12K background examples, 9K salient blob examples, 12K weak blob examples, 8K salient line examples, and 8K weak line examples. The testing images include five categories of scratches.

As shown in Fig. 1, scratches are diversity reflected on two main aspects: (1) the scratches have discrepant shapes, such as line-shape and blob-shape, and (2) their appearances are also very different due to diversified collection environments. Therefore, it will bring a great challenge to the scratch defection algorithms.

3.2 Scratch Filtering

To speed up detection, we start with scratch filtering processing, which can first filter out the big scratches and make the processing focus on the more difficult small scratches. As shown in Fig. 2, we process the captured images via four main stages, including color transformation, binarization, morphologic processing and connected component analysis. After that, we can get some big scratches and small scratch candidates. In this manner, the big scratches are filtered out in the following processing, which greatly reduces the detection complexity.

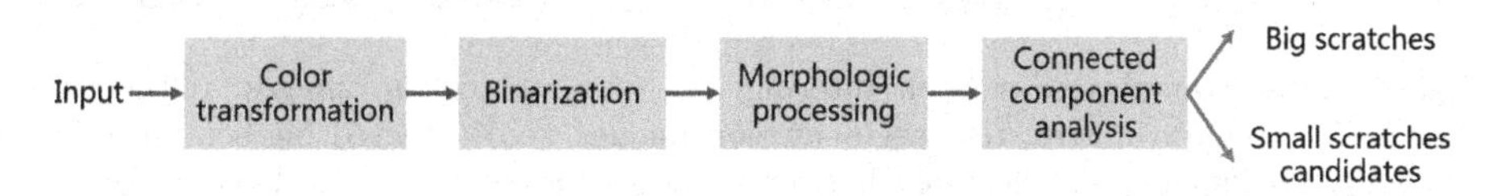

Fig. 2. Flowchart of scratch filtering. It could filter out the big scratches and make the processing focus on more difficult small scratch candidates.

Color transformation. Since detecting scratches directly on the colour image may have liter effects, we simplify the processing in the gray level image by investigating that the gray image has much litter liter effects. Therefore, the captured colour image could be transformed into gray image.

Binarization. In general, the cellphone screen is very smooth. If the scratches exist, it will generate gray level change in the regions of scratches. According to this fact, simply binarization by using a certain threshold will coarsely expose the scratches. Toward this end, we apply a threshold value to binarize the gray image, which segment the gray image into background and foreground (scratches). Suppose that $f(x, y)$ is gray value at point (x, y) and $g(x, y)$ is the value after segmentation, then

$$g(x, y) = \begin{cases} 255 & f(x, y) > T \\ 0 & f(x, y) \leq T \end{cases} \tag{1}$$

Morphologic processing. After binarization stage, the segmentation image often has discontinuous big scratches and very fractional small foregrounds. To address these issues, we adopt erosion and dilation operators in morphologic processing, which could ensure the continuity of big scratches while efficiently eliminate very fragmentary scratch candidates.

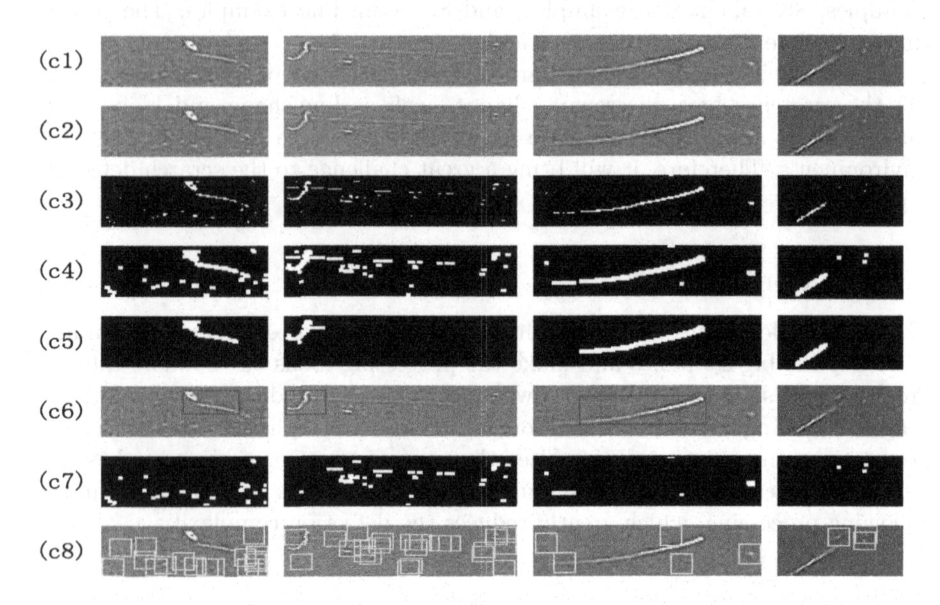

Fig. 3. Some examples of scratch filtering. (c1) original colour images, (c2) gray images, (c3) binary images, (c4) segmentation images, (c5) detected big scratches in segmentation images, (c6) detected big scratches in gray images, (c7) detected small scratch candidates in segmentation images, and (c8) detected small scratch candidates in gray images.

Connected component analysis. Then, we apply connected component analysis (CCA) to the filtered segmentation image. By using 8-connected region algorithm, we can label the connected regions and get their information, such as area and location. Suppose that there are n connected regions and their areas are represented as $\{S_i, i = 1, 2, ..., n\}$. Accordingly, we could distinguish connected region by thresholding its area and classifying it as big scratch or small scratch candidate:

$$h(i) = \begin{cases} 1 & S_i > s \\ 0 & S_i \leq s. \end{cases} \tag{2}$$

where, s is area threshold value, h is a binary classification function. If $h(i) = 1$ the region S_i is considered as big scratch and otherwise small scratch candidate.

In actual production process, if some big scratches are detected on cellphone screens, the testing examples will be considered as unqualified. This process will accelerate inspecting process without further detection. The method of detecting small scratches sounds like Girshick's RCNN [6], but we can assume the rule that the candidate box of small scratch is fixed according to the raw image patches of small scratch candidates. Figure 3 illuminates some examples of the scratch filtering. It shows the processing results in each stage. Note that scratch filtering shifts the task to classification of small scratch candidates, as shown in Fig. 3(c8).

3.3 Scratch Classification

After scratch filtering, the remaining small scratch candidates need to be identified. As shown in Fig. 3(c8), it is a challenging task due to the various appearance of small scratches candidates in weak lighting, shape deformation, and so on. To address these issues, we design a lightweight CNN, ScratchNet (see Fig. 4) to obtain robust representation of these small scratches, which can overcome these complex environment better than traditional methods. In details, our ScratchNet combines the structure of LeNet [10] and convolutional layers of VGGNet [18]. In this manner, our ScratchNet is not only as lightweight as LeNet [10], but also as strong as VGGNet [18]. As shown in Fig. 4, ScratchNet has eight layers, where the first six layers is convolution and each is a stack of two 3×3 conv to extract robust feature representation. After every convolutional layer, we apply a max pooling operation to lower dimensions of features. The last two layers respectively include one fully connection layers and one softmax layer to classify small scratches.

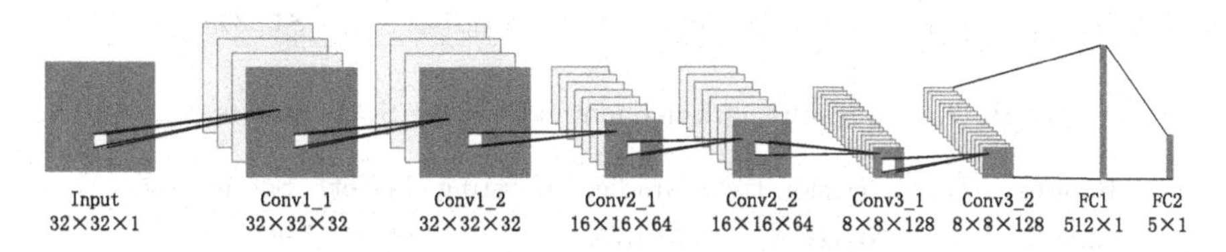

Fig. 4. The structure of ScratchNet. It includes three convolution layers, two fully connection layers and one softmax layer. Besides, the output of each layer is shown.

4 Experiments

We evaluate our methods by two experiments to demonstrate its effectiveness.

Firstly, we compare our ScratchNet with classic convolutional neural network, LeNet [10] on SCS. As shown in Fig. 5, not only ScratchNet is slightly better than LeNet in accuracy, but also ScratchNet has faster convergence speed than LeNet. The results on SCS by ScratchNet (See Table 1) demonstrate the effectiveness of our methods.

Secondly, we apply some machine learning methods on the SCS dataset. The results are given in Table 2. From the results, we can see that different methods have different performance. Obviously, other machine learning methods give lower classification performance than CNN. In addition, our ScratchNet has best results.

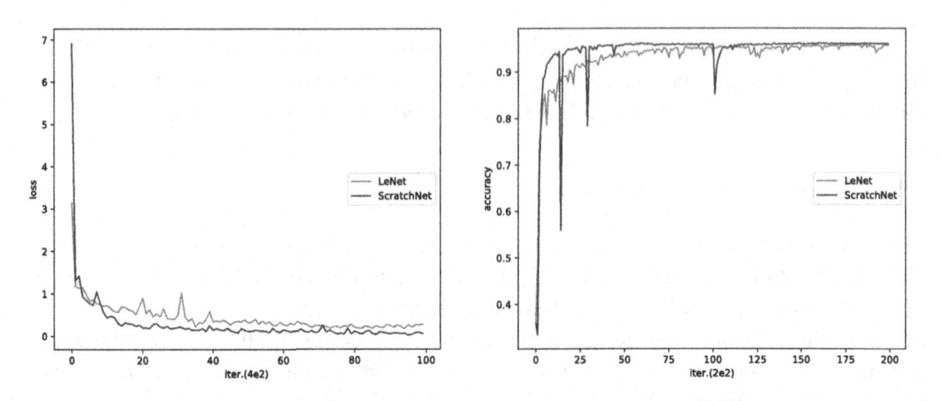

Fig. 5. The training loss (left) and accuracy (right) between ScratchNet and LeNet.

Table 1. Classification performance on SCS test set with ScratchNet.

Background	Notable batch	No notable batch	Notable line	No notable line
99.32%	0%	0.67%	0%	0%
0%	96.45%	2.26%	0.97%	0.32%
1.60%	1.04%	96.46%	0%	0.90%
0%	4.67%	0.33%	92.00%	3%
1.0%	0.67%	12.33%	1.67%	84.33%

Table 2. The results with five methods on the SCS datasets.

Random forest	Decision tree	Gradient boosting	LeNet	ScratchNet
83.89%	70.52%	73.15%	95.97%	96.35%

5 Conclusion

Our method is very practical in detecting scratches of cellphone screens. At first, if testing screens has big scratches, we can regard it as substandard screen and stop. Then we analyze candidates of small scratches as required. It improves the speed of detection, as well enhance the precision, cut down on hardware costs and advance quality. However, many existing CNNs are focus on relatively big objects and there is some room for improvement in small object inspection, so we can optimize our ScratchNet and apply ScatchNet in other issues. Except that, although we can find fixed candidates box quickly by some laws, our solution is not end-to-end just like Faster RCNN [6]. So, we can do further researches on the issue that how to introduce some useful rules into the CNN model.

Acknowledgments. This work is supported in part by the National Key Research and Development Plan (Grant No. 2016YFC0801005), the National Natural Science Foundation of China (Grant No. 61402463) and Open Foundation Project of Robot Technology Used for Special Environment Key Laboratory of Sichuan Province in China (Grant No. 16kftk01).

References

1. Chiou, Y.C., Li, W.C.: Flaw detection of cylindrical surfaces in PU-packing by using machine vision technique. Measurement **42**(7), 989–1000 (2009)
2. Choi, J., Kim, C.: Unsupervised detection of surface defects: a two-step approach. In: IEEE International Conference on Image Processing, pp. 1037–1040 (2012)
3. Clark, R.: Rail flaw detection: overview and needs for future developments. NDT E Int. **37**(2), 111–118 (2004)
4. Coulombe, A., Cantin, M., Bérard, L., Gauthier, J.: Method and system for detecting defects on a printed circuit board (2004)
5. Dan, C., Meier, U., Masci, J., Schmidhuber, J.: A committee of neural networks for traffic sign classification. **42**(4), 1918–1921 (1921)
6. Girshick, R., Donahue, J., Darrell, T., Malik, J.: Rich feature hierarchies for accurate object detection and semantic segmentation, pp. 580–587 (2014)
7. Goodfellow, I.J., Bulatov, Y., Ibarz, J., Arnoud, S., Shet, V.: Multi-digit number recognition from street view imagery using deep convolutional neural networks. Comput. Sci (2013)
8. He, K., Zhang, X., Ren, S., Sun, J.: Deep residual learning for image recognition. In: Computer Vision and Pattern Recognition, pp. 770–778 (2016)
9. Krizhevsky, A., Sutskever, I., Hinton, G.E.: Imagenet classification with deep convolutional neural networks. In: International Conference on Neural Information Processing Systems, pp. 1097–1105 (2012)
10. LeCan, Y., Bottou, L., Bengio, Y., Haffner, P.: Gradient-based learning applied to document recognition. Proc. IEEE **86**(11), 2278–2324 (1998)
11. Liang, L.Q., Li, D., Fu, X., Zhang, W.J.: Touch screen defect inspection based on sparse representation in low resolution images. Multimedia Tools Appl. **75**(5), 2655–2666 (2016)
12. Loh, H.H., Lu, M.S.: Printed circuit board inspection using image analysis. IEEE Trans. Ind. Appl. **35**(2), 426–432 (1999)
13. Mcgrail, A., Risino, A., Auckland, D.W., Varlow, B.R.: Use of a medical ultrasonic scanner for the inspection of high voltage insulation. IEEE Electr. Insul. Mag. **9**(6), 5–10 (2002)
14. Pei, K.: Study of on-line glass defect inspection system based on embedded image processing. Electron. Meas. Technol (2009)
15. Peng, X., Chen, Y., Yu, W., Zhou, Z., Sun, G.: An online defects inspection method for float glass fabrication based on machine vision. Int. J. Adv. Manuf. Technol. **39**(11–12), 1180–1189 (2008)
16. Perng, D.B., Liu, H.W., Chang, C.C.: Automated SMD LED inspection using machine vision. Int. J. Adv. Manuf. Technol. **57**(9–12), 1065–1077 (2011)
17. Shang, H.C., Chen, Y.P., Yu, W.Y., Zhou, Z.D.: Online auto-detection method and system of presswork quality. Int. J. Adv. Manuf. Technol. **33**(7–8), 756–765 (2006)
18. Simonyan, K., Zisserman, A.: Very deep convolutional networks for large-scale image recognition. Comput. Sci (2014)
19. Soukup, D., Huber-Mörk, R.: Convolutional neural networks for steel surface defect detection from photometric stereo images. In: Bebis, G., Boyle, R., Parvin, B., Koracin, D., McMahan, R., Jerald, J., Zhang, H., Drucker, S.M., Kambhamettu, C., El Choubassi, M., Deng, Z., Carlson, M. (eds.) ISVC 2014. LNCS, vol. 8887, pp. 668–677. Springer, Cham (2014). https://doi.org/10.1007/978-3-319-14249-4_64

20. Szegedy, C., Liu, W., Jia, Y., Sermanet, P., Reed, S., Anguelov, D., Erhan, D., Vanhoucke, V., Rabinovich, A.: Going deeper with convolutions. In: Computer Vision and Pattern Recognition, pp. 1–9 (2015)
21. Verikas, A., Lundström, J., Bacauskiene, M., Gelzinis, A.: Advances in computational intelligence-based print quality assessment and control in offset colour printing. Expert Sys. Appl. **38**(10), 13441–13447 (2011)
22. Zhao, J., Kong, Q.J., Zhao, X., Liu, J., Liu, Y.: A method for detection and classification of glass defects in low resolution images. In: International Conference on Image and Graphics, pp. 642–647 (2011)

Small UAV Detection in Videos from a Single Moving Camera

Lian Du(✉), Chenqiang Gao, Qi Feng, Can Wang, and Jiang Liu

Chongqing Key Laboratory of Signal and Information Processing,
Chongqing University of Posts and Telecommunications, Chongqing 400065, China
dulian006@163.com

Abstract. The rapid application of Unmamned Aerial Vehicles (UAV) has triggered serious threats to public security, individual privacy, military security, etc. Thus, discovering unknown UAVs fast and reliably becomes more and more important. Among UAV detection techniques, the vision-based method is almost the lowest cost and the most easily-configured one. In this paper, we propose a UAV detection method based on a single moving camera to handle the problem for UAVs with fast moving speed. Firstly, we employ a motion estimation method to stabilize videos. Then, a low-rank based model is adopted to obtain the object proposals and finally, a CNN-SVM approach is used to further confirm real UAV objects. Two real UAV datasets are used to evaluate the proposed method and experimental results show that our method outperforms the baseline methods.

Keywords: Unmanned Aerial Vehicle · Small object detection
Machine intelligence · Low-rank analysis

1 Introduction

In recent years, Unmanned Aerial Vehicles (UAVs) have been rapidly applied to different aspects of our society. While UAVs benefit our society, they also have posed serious threats to public security, individual privacy, military security, etc. Thus, effective UAV detection technology becomes very important to handle the problem of UAV threat. Currently, commonly-used techniques are often based on acoustics [6,13], radio frequency [17,18] and radar detection [14,19]. While these techniques usually require expensive equipment and strict configuration, vision-based techniques are inexpensive and easily-configured. Thus, vision-based UAV detection techniques have good prospect.

As a new and urgent task, there have not been many research works to handle it, up to now. Rozantsev [21] proposes a regression-based method for UAV detection in videos. Two boosted-tree regressors are employed for motion stabilization. Later, spatio-temporal cubes with fixed size length are used to detect UAVs. This method could achieve good performance when the size of flying objects changing gradually, while it tends to fail to detect objects with fast

J. Yang et al. (Eds.): CCCV 2017, Part III, CCIS 773, pp. 187–197, 2017.
https://doi.org/10.1007/978-981-10-7305-2_17

Fig. 1. Some examples of small UAVs flying in challenging backgrounds.

varying sizes. Other works [15,22] aim to detect UAVs based on fixed cameras by reconstructing the 3D trajectories of the UAVs. Nevertheless, the monitoring scope for UAV surveillance is limited due to the fixed viewpoint of cameras.

In fact, UAVs usually move irregularly and they easily move out from the sensing scope of fixed cameras. To continuously capture the UAV, one of the methods is to keep the camera moving with the UAV. However, this moving camera would bring many more challenges to the UAV detection task. For example, the image backgrounds would quickly vary from ground to sky or ground-sky. As UAVs sometimes fly in low sky, they are often "engulfed" by the confusing backgrounds, such as trees or some similar man-made things, as shown in Fig. 1. Specifically, it is hard, even for humans, to discriminate UAVs from the background in poor imaging conditions, specifically when their sizes are small. These challenges usually make the current existing methods failed with low detection precisions.

In this paper, we propose a new approach to detect UAVs from a single moving camera. To handle the problem of fast changing backgrounds resulted from the moving camera, a feature-based video stabilization method is firstly employed. Then we use a two-stage framework to complete the UAV detection task. In the first stage, we use a low-rank based method to get the UAV proposals, considering small sizes, motion property of UAVs and complex backgrounds, similar to [9,27]. After obtaining the proposals, we adopt a Support Vector Machine (SVM) based object classification method to delete the false objects and confirm the real UAV objects in the second stage. In order to obtain robust feature representation, a deep Convolutional Neural Network (CNN) is adopted to extract the features from object candidates, like [4,23]. We have evaluated our method on public RLP dataset [21] and our own DROM dataset. The results show that the proposed method has better performance, compared to the state-of-the-art methods.

The rest of this paper is organized as follows: In the Sect. 2, we introduce the proposed method in detail, including the video stabilization, candidate object detection and UAV classification modules. Section 3 explains the employed two datasets and evaluation results on the two datasets. The conclusion is drawn in Sect. 4.

2 The Proposed Method

2.1 Overview of Our Approach

Figure 2 shows the flowchart of the proposed approach. Our method mainly consists of three modules: video stabilization, low-rank based candidate object region detection, and SVM-based UAV classification. Firstly, we apply a robust feature-based stabilization method to the original video image sequence to reduce the influence of motion background. Secondly, we employ low-rank based motion detection method [9,27] to generate initial set of small motion regions which are future used to extract the UAV proposals. In order to improve the reliability of following CNN-SVM classification, we densely sample each proposal to produce multiple object candidates, as shown as step 5 in Fig. 2. Finally, each candidate is fed into a CNN-SVM framework for classification.

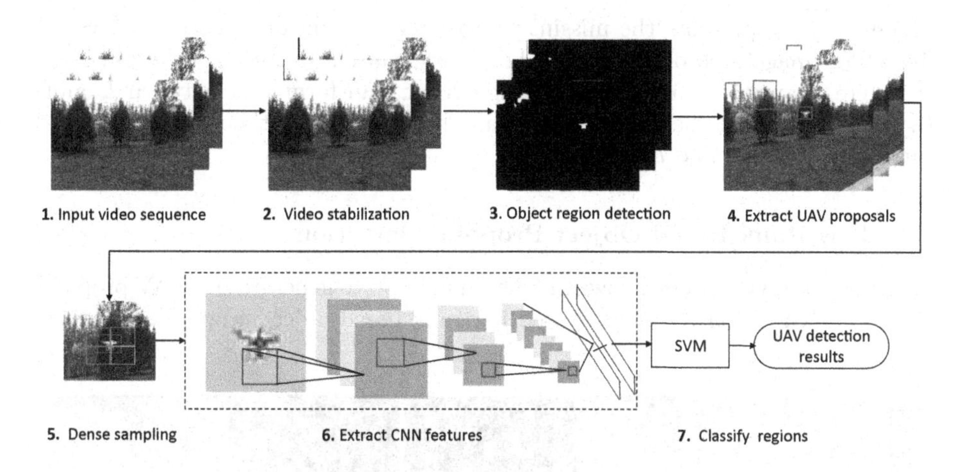

Fig. 2. Flowchart of our approach.

2.2 Video Stabilization

In this section, we introduce a robust video stabilization method to remove background motion resulted from moving camera. The commonly-used video stabilization methods tend to fail when the objects in the video sequence move fast, which is often the case in our applications. Our method deals with the problem by constantly updating reference frames online. This stabilization algorithm consists of three stages, transformation model estimation, image warping and reference frames updating.

In this paper, given a set of video frames $V = \{I_i\}_{i=1}^{N}$, we denote the I_r as current reference frame for frame $I_t \in V$. In the first stage, we employ a feature-based approach to estimate the transformation model H_r^t between frame I_r and I_t. With the assumption that if the background plane has not moved or changed significantly between frames, a common motion estimation approach for video stabilization is to estimate a homography or affine transformation. This transform is actually capturing the camera motion [2,3]. Similar to this idea, we also model camera motion as a 2D translation model. We firstly choose SURF [5] descriptor for key-points detection and description. Then, the transformation model is estimated from the key-points matches by adopting the MLESAC algorithm [1].

In the second stage, we obtain the compensated frames I_t' by warping frames I_t with translation model H_r^t. However, the homography estimation will fail if the change between two frames is too large. At last, we deal with this problem by dynamically updating the reference frames. As defined in Eq. 1, we define the transformation degree between I_t' and its reference frame I_r as α_t.

$$\alpha_t = \frac{S_t^*}{S_t}, \tag{1}$$

where the S_t^* represents the missing image area of transformed frame I_t', S_t is the whole image area of frame I_t'. The transformation degree α_t belongs to [0, 1], where a larger α_t value implies a significant transformation between I_r and I_t. In this paper, if the α_t is larger than the threshold θ, then we update the current reference frame I_r with I_t.

2.3 Low-Rank Based Object Proposal Detection

In this section, we employ a low-rank based method to generate the UAV proposals. As shown in Fig. 3, our proposed low-rank based object proposal detection

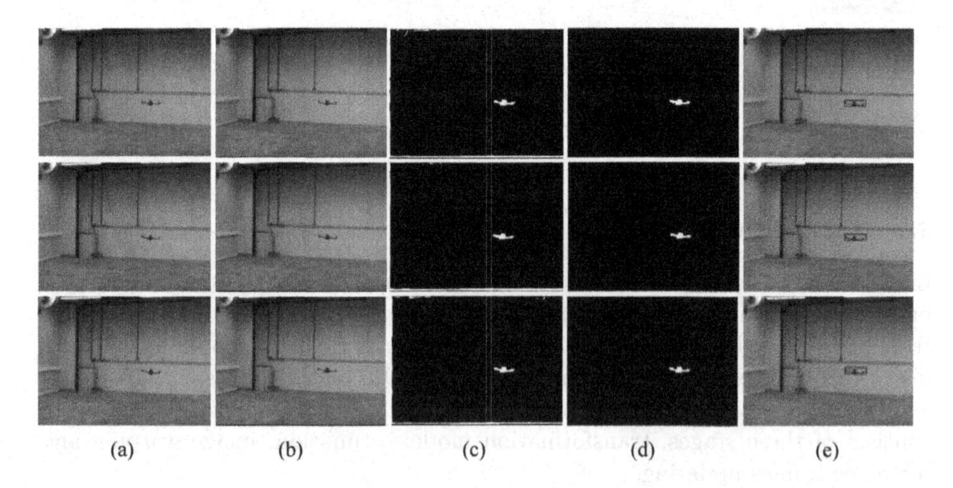

Fig. 3. Illustration of the candidate object detection processes. (a) the input video sequence. (b) the results after video stabilization. (c) and (d) the results after low-rank based detection and post-process. (e) Mapping proposals which are extracted from results of (d). (Color figure online)

methodology mainly includes three stages, low-rank detection, post-processing and proposals mapping. Figure 3(b) represents the motion compensated frames computed by method in Sect. 2. Figure 3(c) shows the initial mask obtained by low-rank detection method. As shown in Fig. 3(d), to remove isolated points in the whole image, we adopt the morphological opening operation, in which the structural element is defined as 3×3. Then, we can generate bounding boxes from the mask images, which are the UAV proposals. At last, since there is shift between the motion compensated frames and original frames, we map the region proposals to original frames with transformation model. As shown in Fig. 3(e), the 'blue' bounding boxes represent the position of proposals in motion compensated frames, and the mapping bounding boxes with colour 'red' are the proposals in original frames.

The main task of low-rank methods is to decompose the input matrix C composed of the observed vectorized image frames into a low-rank matrix L representing the background, and a sparse matrix S representing the moving objects. Thus, the problem formulation can be defined as:

$$\min_{L,S} rank(L) + \lambda \|S\|_0, \quad s.t. \ C = L + S, \tag{2}$$

where λ is a regularizing parameter. Equation 2 represents a highly non-convex optimization problem, and it can be approximately solved by its relaxing convex envelope via replacing the L0 regularization with the L1 regularization:

$$\min_{L,S} \|L\|_* + \lambda \|S\|_1, \quad s.t. \ C = L + S, \tag{3}$$

Here $\|\cdot\|_*$ indicates the nuclear norm of L. The problem is also known as the Robust Principal Component Analysis (RPCA) problem [7]. To solve the problem, we adopt the low-rank matrix estimation algorithm in [27] to estimate L and S from the input video sequence.

Furthermore, our method improves the limitation that the traditional low-rank method can only work in a batch mode because it can only work on video sequence blocks with high correlation. In this paper, we combine video stabilization and low-rank based object detection method to solve the problem by updating the reference frames. Based on the discussion in Sect. 2.2, the compensated frames that computed with the same reference frames are highly correlated. Therefore, we can use low-rank method to detecting moving objects in these frames. The algorithm of object region detection method is shown in Algorithm 1.

2.4 UAV Classification Based on CNN-SVM

The region proposals generated by above steps may contain non-uav objects such as pedestrians, flying birds and other moving objects. To distinguish between UAVs and other non-uav objects, we adopt the SVM-based classification method to cope with this task. According to experiments, objects that occupy less than

Algorithm 1. Low-rank based Object Proposal Detection.

Input: a set of video frames $V = \{I_i\}_{i=1}^{N}$, λ;
Output: detection masks M, region proposals B;
1: Initialize: reference frame $I_r \leftarrow I_1$, $t \leftarrow 1$
2: **while** $t \leq$ N of I_i **do**
3: Calculate α_t as Eq. 1;
4: **if** $\alpha_t \geq \lambda$ **then**
5: Take out transformed frames $\{C_p\} = \{I_i'\}_{i=r}^{t-1}$
6: Estimate sparse matrix S_p with $\{C_p\}$ by using method in [27], $p \in [r, t-1]$
7: Generate object proposals B_p' from S_p
8: Mapping proposals B_p from B_p' with $H_r^k, k \in [r, t-1]$
9: $I_r \leftarrow I_t$
10: Save image masks and proposals: M_p, B_p
11: **else**
12: continue;
13: **end if**
14: **end while**
15: $B = \sum B_p$, $M = \sum M_p$;

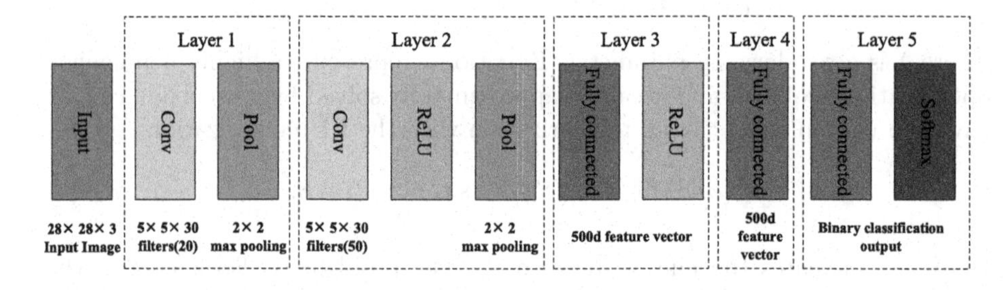

Fig. 4. Our CNN architecture.

10 pixels in the image are hardly to classify by features, our SVM-based classification method only deals with proposals whose size is larger than 10 pixels. As shown in Fig. 2, the classification method mainly includes two stages: CNN feature extraction and SVM-based classification.

Most CNN architecture such as ResNet [11], VGG [24], GoogLeNet [25] has a very deep structure which can cause time-consuming problem for feature extraction. Thus, as shown in Fig. 4, we use a simple CNN architecture to extract features for our approach. We firstly train the CNN model on our datasets using the Caffe implementation [12] of the CNN. Then, we extract features from each region proposal based on the trained CNN model. We use the output of the second fully-connected layer as the feature description which represents the highest level feature of the reference model. Then, we train a linear SVM classifier using the LIBSVM library [8] based on the CNN features to classify the object candidates.

In the online classification stage, we firstly follow the dense sampling strategy to generate k object candidates at the location of each proposals. This ensures the detected UAV is in the center of the patch so as to improve the detection precision. Secondly, we use the offline trained classifier to recognize these candidates and score them. In this case, some bounding boxes may highly overlap with each other. To reduce redundant bounding boxes, we adopt the non-maximum suppression (NMS) algorithm [10,16] on the bounding boxes based on their scores obtained by SVM.

3 Experiments

In this section, we firstly introduce implementation details of our experiments and two datasets used in this paper. Then, we exhibit our evaluation results on the two datasets. The state-of-the-art methods, including Selective Search [26] and HBT-Regression [21] are selected as the baseline methods for comparison.

3.1 Datasets

In this paper, we evaluate our method on two datasets, the RLP [21] and our DROM dataset. They are captured in real scenes and include many real-world challenges such as fast illumination changes and complex backgrounds. Figure 5 shows some examples of the variety of UAV types in the two datasets. The size of UAVs in these two datasets varies from about 6 to 120 pixels.

The RLP dateset consists of 14 video clips and is publicly available in [21]. These videos are recorded in various indoor and outdoor environments and different lighting and weather conditions, which are acquired by a camera mounted on drone.

The DROM dataset comprises of 10 video sequences captured with 800 $\times$ 600 resolution. They were acquired by us using a hand-held camera in different outdoor environments.

Fig. 5. Samples containing UAVs from two datasets. The left ones are examples of UAVs from the RLP dataset [21], and the right are from our DROM dataset.

3.2 Metrics

Similar to [21], we use precision-recall curves and the Average Precision (AP) measures as the evaluation metric. Precision is computed as the number of true positives detected by the algorithm, divided by the total number of detections. Recall is the number of true positives divided by the number of the positively labeled test samples.

3.3 Implementation Details

For both datasets, we use 40% of the data to train all the classification model of all methods, and the rest 60% are used to test. To ensure the compensated frames are highly correlated in Sect. 2, we empirically set the parameter θ in Eq. 1 as small as 0.05 for candidate object detection. Table 1 summarizes the influence of the parameter k on the average precision of our method on two datasets. As shown in Table 1, we set k as 8 for the dense sampling step so as to achieve best performance. Similar to [20], the predicted bounding box is considered as correct if its overlap ratio with the ground truth is larger than 0.5.

Table 1. The influence of the parameter k on the average precision (AP) (%) of our method on two datasets.

DROM dataset	k	2	4	6	8	10
	AP	61.49	66.28	68.22	71.03	70.06
RLP dataset	k	2	4	6	8	10
	AP	65.74	70.89	74.45	79.08	78.05

3.4 Results and Discussion

Similar to [21], we have evaluated our method on 8000 bounding boxes on the RLP and DROM datasets. We compare our method with the Selective Search [26] and HBT-Regression [21] in Fig. 6 with precision-recall curves. Compared to other methods, our method achieves better performance than the baseline method for both datasets. We also find that the Selective Search method performs worst. It happens most probably that the Selective Search method depends on colour, texture and size information to generate object locations. However, these information are usually hard to be acquired for small UAVs in fast changing frames. On the contrary, our method find it easier to find UAVs by exploiting the motion information of UAVs.

Table 2 summarizes the average precision of different methods on two datasets. We can achieve about 10% average precision improvement over the HBT-Regression method on average for both datasets. The HBT-Regression method uses spatio-temporal cubes with fixed size length, which often fails to detect UAVs with fast changing sizes. However, our low-rank based approach can find UAVs even in the cases where the UAV size varies from 30 to 100 pixels in a short time.

Figure 7 gives some challenging detection results of our method. These UAVs are "engulfed" by the confusing backgrounds specially when their sizes are small, which are almost invisible to the human eye. Through experimental comparison, as shown in Fig. 7, UAVs that are mixed with similar backgrounds cannot be detected by [21,26] but they are found by our algorithm.

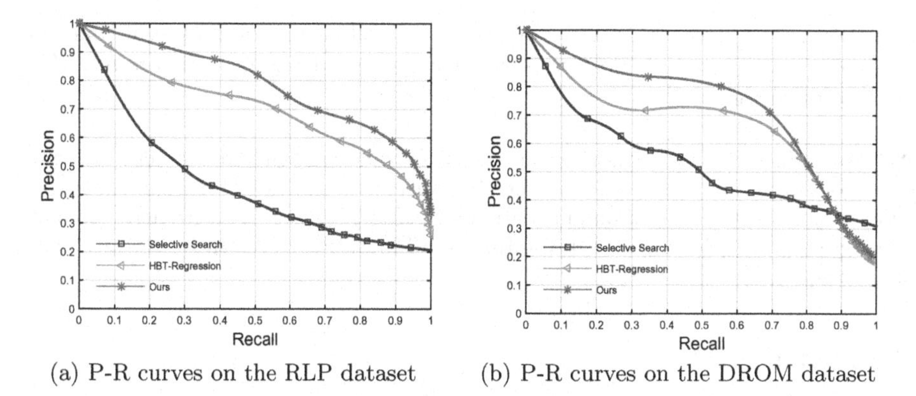

(a) P-R curves on the RLP dataset (b) P-R curves on the DROM dataset

Fig. 6. Comparing various methods on two datasets with precision-recall curves. Our approach achieves better performance compared to other methods.

Table 2. The average precision (%) of different detection methods on two datasets.

Method	Average precision	
	RLP dataset	DROM dataset
Selective search [26]	42.74	54.01
HBT-Regression [21]	68.89	61.99
Ours	**79.08**	**71.03**

(a) Results on the RLP dataset (b) Results on the DROM dataset

Fig. 7. Some challenging detection results of our method on two datasets.

4 Conclusion

In this paper, we propose a new UAV detection approach from a single moving camera. We introduce a video stabilization method and combine it with low-rank method to improve the limitation that traditional low-rank method cannot update the low-rank model online. In addition, our approach combines the appearance features and motion clues of objects so that it works well on small moving targets such as UAVs. We have evaluated our method against the state-of-the-art methods on two datasets. The evaluation results reveal that our method can achieve the best performance on two datasets with around 75% average precision.

Acknowledgments. This work is supported by the National Natural Science Foundation of China (No. 61571071), Wenfeng innovationand start-up project of Chongqing University of Posts and Telecommunications (No. WF201404), Undergraduate research training program (No. A2014-39), the Research Innovation Program for Postgraduate of Chongqing (No. CYS17222).

References

1. MLESAC: a new robust estimator with application to estimating image geometry. Comput. Vis. Image Underst. **78**(1), 138–156 (2000)
2. Fast and accurate global motion compensation: Pattern Recogn. **44**(12), 2887–2901 (2011)
3. HEASK: robust homography estimation based on appearance similarity and keypoint correspondences. Pattern Recogn. **47**(1), 368–387 (2014)
4. People counting based on head detection combining adaboost and CNN in crowded surveillance environment. Neurocomputing **208**, 108–116 (2016). SI: BridgingSemantic
5. Bay, H., Tuytelaars, T., Van Gool, L.: SURF: speeded up robust features. In: Leonardis, A., Bischof, H., Pinz, A. (eds.) ECCV 2006. LNCS, vol. 3951, pp. 404–417. Springer, Heidelberg (2006). https://doi.org/10.1007/11744023_32
6. Busset, J., Perrodin, F., Wellig, P., Ott, B., Heutschi, K., Rhl, T., Nussbaumer, T.: Detection and tracking of drones using advanced acoustic cameras (2015)
7. Candès, E.J., Li, X., Ma, Y., Wright, J.: Robust principal component analysis? J. ACM **58**(3), 11:1–11:37 (2011)
8. Chang, C.C., Lin, C.J.: LIBSVM: a library for support vector machines. ACM Trans. Intell. Syst. Technol. **2**(3), 27:1–27:27 (2011)
9. Gao, C., Meng, D., Yang, Y., Wang, Y., Zhou, X., Hauptmann, A.G.: Infrared patch-image model for small target detection in a single image. IEEE Trans. Image Process. **22**(12), 4996–5009 (2013)
10. Girshick, R., Donahue, J., Darrell, T., Malik, J.: Rich feature hierarchies for accurate object detection and semantic segmentation. In: The IEEE Conference on Computer Vision and Pattern Recognition (CVPR), June 2014
11. He, K., Zhang, X., Ren, S., Sun, J.: Deep residual learning for image recognition. In: The IEEE Conference on Computer Vision and Pattern Recognition (CVPR), June 2016
12. Jia, Y., Shelhamer, E., Donahue, J., Karayev, S., Long, J., Girshick, R., Guadarrama, S., Darrell, T.: Caffe: convolutional architecture for fast feature embedding. arXiv preprint arXiv:1408.5093 (2014)
13. Kim, J., Park, C., Ahn, J., Ko, Y., Park, J., Gallagher, J.C.: Real-time UAV sound detection and analysis system. In: 2017 IEEE Sensors Applications Symposium (SAS), pp. 1–5, March 2017
14. Klare, J., Biallawons, O., Cerutti-Maori, D.: Detection of UAVs using the MIMO radar MIRA-CLE Ka. In: Proceedings of EUSAR 2016: 11th European Conference on Synthetic Aperture Radar, pp. 1–4, June 2016

15. Martínez, C., Mondragón, I.F., Olivares-Méndez, M.A., Campoy, P.: On-board and ground visual pose estimation techniques for UAV control. J. Intell. Robot. Syst. **61**(1), 301–320 (2011)
16. Neubeck, A., Gool, L.V.: Efficient non-maximum suppression. In: 18th International Conference on Pattern Recognition (ICPR 2006), vol. 3, pp. 850–855 (2006)
17. Nguyen, P., Ravindranatha, M., Nguyen, A., Han, R., Vu, T.: Investigating cost-effective RF-based detection of drones. In: Proceedings of the 2nd Workshop on Micro Aerial Vehicle Networks, Systems, and Applications for Civilian Use, pp. 17–22. DroNet 2016. ACM, New York (2016)
18. Nguyen, P., Truong, H., Ravindranathan, M., Nguyen, A., Han, R., Vu, T.: Matthan: drone presence detection by identifying physical signatures in the drone's RF communication. In: Proceedings of the 15th Annual International Conference on Mobile Systems, Applications, and Services, MobiSys 2017, pp. 211–224. ACM, New York (2017). http://doi.acm.org/10.1145/3081333.3081354
19. Ren, J., Jiang, X.: Regularized 2-D complex-log spectral analysis and subspace reliability analysis of micro-doppler signature for UAV detection. Pattern Recogn. **69**, 225–237 (2017)
20. Ren, S., He, K., Girshick, R., Sun, J.: Faster R-CNN: towards real-time object detection with region proposal networks. In: Cortes, C., Lawrence, N.D., Lee, D.D., Sugiyama, M., Garnett, R. (eds.) Advances in Neural Information Processing Systems, vol. 28, pp. 91–99. Curran Associates, Inc., New York (2015)
21. Rozantsev, A., Lepetit, V., Fua, P.: Flying objects detection from a single moving camera. In: The IEEE Conference on Computer Vision and Pattern Recognition (CVPR), June 2015
22. Rozantsev, A., Sinha, S.N., Dey, D., Fua, P.: Flight dynamics-based recovery of a UAV trajectory using ground cameras. CoRR abs/1612.00192 (2016)
23. Sharif Razavian, A., Azizpour, H., Sullivan, J., Carlsson, S.: Cnn features off-the-shelf: an astounding baseline for recognition. In: The IEEE Conference on Computer Vision and Pattern Recognition (CVPR) Workshops, June 2014
24. Simonyan, K., Zisserman, A.: Very deep convolutional networks for large-scale image recognition. CoRR abs/1409.1556 (2014)
25. Szegedy, C., Liu, W., Jia, Y., Sermanet, P., Reed, S., Anguelov, D., Erhan, D., Vanhoucke, V., Rabinovich, A.: Going deeper with convolutions. In: The IEEE Conference on Computer Vision and Pattern Recognition (CVPR), June 2015
26. Uijlings, J.R.R., van de Sande, K.E.A., Gevers, T., Smeulders, A.W.M.: Selective search for object recognition. Int. J. Comput. Vis. **104**(2), 154–171 (2013)
27. Zhou, X., Yang, C., Yu, W.: Moving object detection by detecting contiguous outliers in the low-rank representation. IEEE Trans. Pattern Anal. Mach. Intell. **35**(3), 597–610 (2013)

A More Efficient CNN Architecture for Plankton Classification

Jiangpeng Yan[1,2(✉)], Xiu Li[1,2], and Zuoying Cui[1,2]

[1] Department of Automation, Tsinghua University, Beijing 100084, China
yanjump@outlook.com, czy15@mails.tsinghua.edu.cn
[2] Graduate School at Shenzhen, Tsinghua University, Shenzhen 518055, China
li.xiu@sz.tsinghua.edu.cn

Abstract. With mass data collected by seafloor observation networks, an autonomous system which helps to annotate these pictures are in great demand. In this paper, we study the relationship between the network architecture and the classification accuracy for the Plankton Dataset collected by Oregon State University's Hatfield Marine Science Center. We use multiple classic deep convolutional neural networks (CNN) models to compare the benefit and cost of deeper models which have performed quite well in ImageNet Large Scale Visual Recognition Challenge (ILSVRC) (http://www.image-net.org/challenges/LSVRC) competitions and we discover a hidden degeneration phenomenon. Then we conclude some skills to make CNN smaller and finally propose a more efficient network architecture whose model is much smaller (only 1.5 MB), faster (32.2 fps) and achieve a top-5 accuracy of 96% in the Plankton Dataset. This model can be actually deployed in the seafloor observation network system with its advantages.

Keywords: Plankton classification · Neural networks
Network architecture

1 Introduction

With the rapid development of the seafloor observation networks, lots of visual resources are captured by underwater camera for marine research. Take Oregon State University's Hatfield Marine Science Center for example. Hatfield scientists got 50 million plankton pictures over an 18-day period, which was more than 80 TB. 30,336 images of 121 classes in these pictures are labeled as the Plankton Dataset to hold a Kaggle data science competition[1] in 2015. Faced with such a big number of pictures, an autonomous system which helps to classify and annotate these pictures are in great demand. It is very important for this system to balance classification accuracy and speed properly.

In the competition we mentioned above, classification accuracy matters more than speed. The No.1 team called "Deep Sea", used a parallel CNN model and

[1] https://www.kaggle.com/c/datasciencebowl/data.

J. Yang et al. (Eds.): CCCV 2017, Part III, CCIS 773, pp. 198–208, 2017.
https://doi.org/10.1007/978-981-10-7305-2_18

achieved a top-5 accuracy over 98% and a frame rate of 1.4 fps, which is far from the general video frame rate 25–30 fps. In order to get the high accuracy, this team used one network to learn pixel features and the other to learn additional image features including image size in pixels, Haralick texture features, etc. The final ensemble network model is more than 300M bytes, and quite difficult to train.

In this paper, we try to find out the relationship between the network architecture and the classification accuracy of the Plankton Dataset. We use multiple classic deep CNN models and compare the benefit and cost of deeper models which have performed quite well in ILSVRC competitions and we discover a hid-den degeneration phenomenon for the Plankton Dataset. Then we conclude some skills to make CNN lighter and finally propose a simple but efficient network architecture whose model is much smaller, faster and achieve a top-5 accuracy of 96% in the Plankton Dataset, which can be actually deployed in the seafloor observation network system.

2 Related Work

The deep CNN technology have come into vogue since 2012, when Krizhevsky et al. [10] achieved a record-breaking top-1 and top-5 error rates of 37.5% and 17.0% in ILSVRC competition object classification task [1] using a 7-layer CNN architecture called AlexNet. After that, more and more researchers became interested in CNN method and a series of breakthroughs for object classification was made [5,15–17].

2.1 Going Deeper

As early as 1990s, Lecun et al. [11,12] applied CNN in digital handwritten number and document classification. Due to limitations of computation devices, shallow CNN architecture did not make much progress than other traditional methods. Obviously, the depth of CNN plays an important role in the significant results, but its also harder for networks with tens of layers to converge because of vanishing gradients [2,4]. Many researchers' works about normalization [8,14] help to solve this obstacle. In 2014, researchers from the Visual Geometry Group of Oxford University used VggNet [15] to get better performance in ImageNet. According to [15], the 19-layer VggNet performs best, and we cannot get better accuracy by increasing more layers. Whats worse, stacking more layers may lead to degradation. To solve this problem, He et al. [5] proposed a residual network (ResNet) frame using residual shortcuts, show in Fig. 1. In [5], He compares CNN models with 50, 101, and 152 layers. The 152-layer network gets the state-of-the-art top-5 error accuracy of 3.57% in ImageNet. With the help of residual learning, it seems that we can stack conventional layers as more as we want to get better performance.

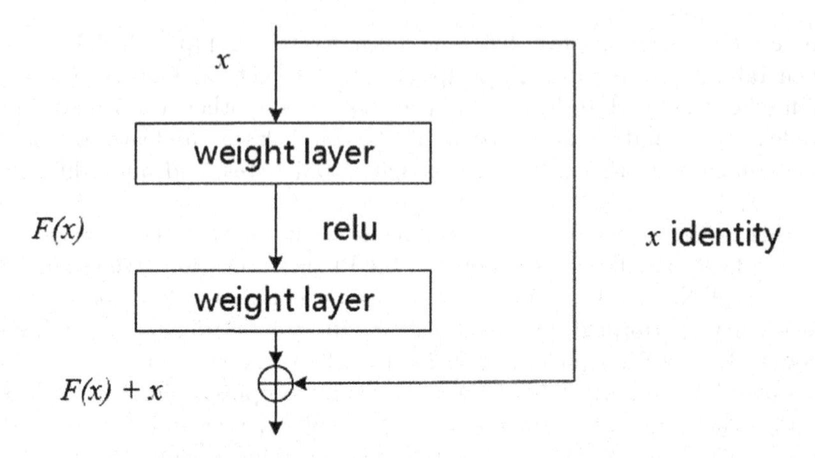

Fig. 1. A basic building block where every few stacked layers learn the residual function with feed-forward shortcut connection.

2.2 Going Smaller

As a result of deeper architecture, the size of CNN grows larger and larger. This means we need not only more expensive computing devices like GPU but also more time to train and deploy the CNN models. There are different ways to make CNN models smaller, such as pruning [3], binarized netural network [6], etc. This paper focuses on how to design the architecture more efficient at beginning. In 2014, Lin, et al. [13] proposed Network-in-Network structure using 1×1 filters to reduce computation. Szegedy et al. [17] created GoogLeNet family by using inception structure that contains 1×1, 3×3 filters layers, show in Fig. 2, to build more efficient and deeper models. Researchers from Berkley proposed SqueezeNet [7] using the cascaded structure that is named with "Fire" layer based on inception. SqueezeNet achieves the same accuracy of AlexNet with the model size squeezed to only 4.8M.

Motived by the development of CNN architecture, we try to find out which architecture can perform best in the Plankton Dataset, so that we can deploy the model in seafloor observation networks to deal with tons of data collected by undersea cameras. We need a model with not only high classification accuracy but also high speed.

3 The Hidden Degeneration

As mentioned above, the past few years have witnessed the break-throughs created by multiple classic CNN architectures with more and more convolutional layers. Before we introduce our final model, we show some experiments we did on the Plankton Dataset. In order to find out how deeper CNN models affect the accuracy in the Plankton Dataset, the models we have tested include CaffeNet[2],

[2] https://github.com/BVLC/caffe/tree/master/models/bvlc_reference_caffenet.

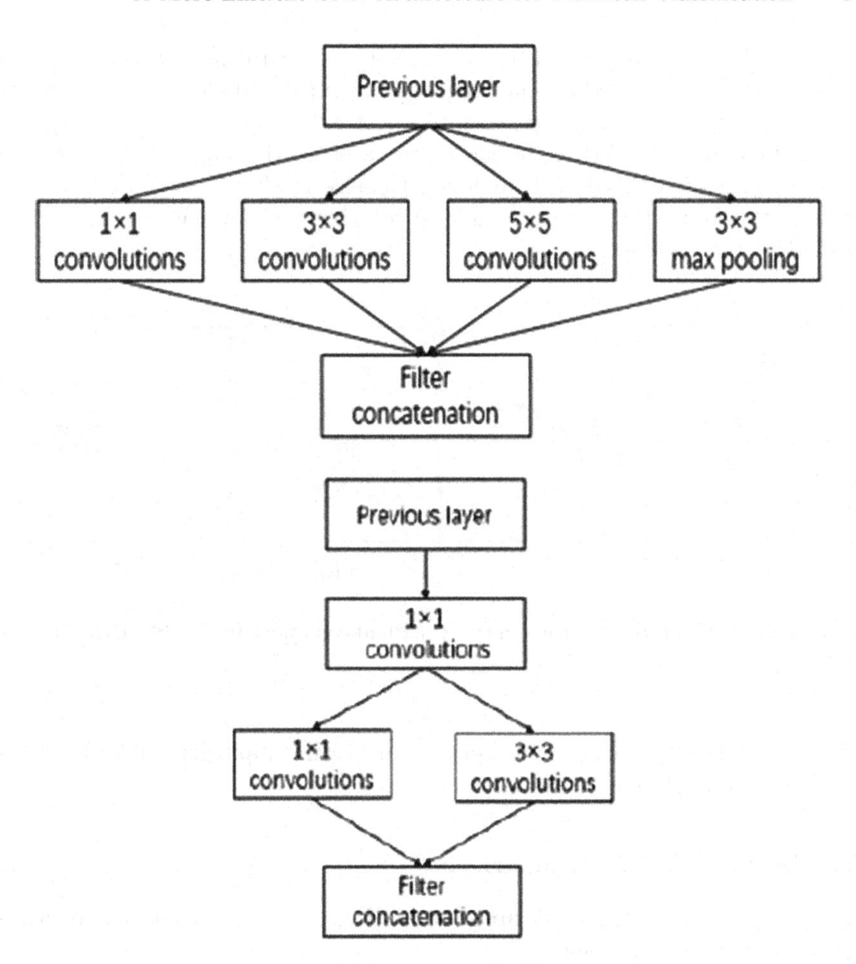

Fig. 2. A typical inception layer structure (up) contains convolutional layers with filters of different size, and the "Fire" layer of SqueezeNet (down) are designed with similar micro structure.

VggNet-19 (Vgg-19) [15], ResNet [5] with different layer numbers of 19, 50, 101. At first we held the view that advanced model with more layers may performed better, but we discovered that the basic CaffeNet has a quite good result and more layers may not only lead to the loss on Top-1 accuracy but also cannot help to improve the Top-5 accuracy. We named this phenomenon as the hidden degeneration, details are described below.

3.1 Data Augmentation and Experiment Environment

The Plankton dataset consists of 30,336 grayscale training images of varying size, divided unevenly into 121 classes which correspond to different species of plankton. This dataset was used for the National Data Science Bowl, a data science competition hosted on the Kaggle platform. We divide the dataset into separate validation, testing and training set of 3,037, 3,037 and 24,262 images respectively and rescaled them to 256×256 based on the length of their longest side. We augmented the data to increase the size of the dataset which can be useful for the prevention of overfitting by rotating the original images 0°, 90°, 180°, 270°, show in Fig. 3.

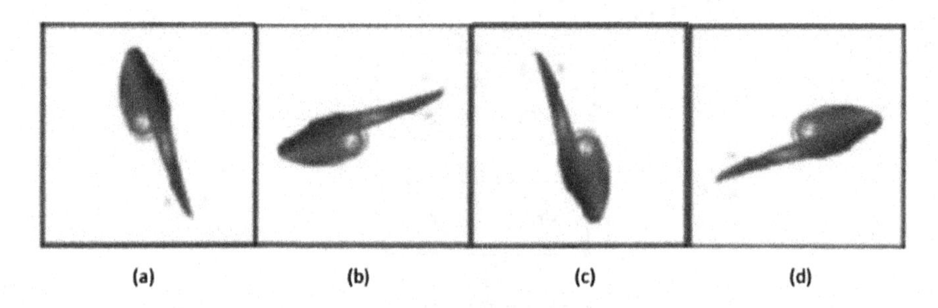

Fig. 3. Data augmentation by rotating target Plankton image 0° (a), 90° (b), 180° (c), and 270° (d).

We use Caffe [9] as experiment environment and implement different CNN models using one piece of GTX1080Ti GPU.

3.2 Multiple Models Comparison

The architecture of these models are shown in Fig. 4. ResNet has too many layers so we only show a single block.

CaffeNet is a replication of AlexNet with pooling layer before normalization by researchers from Berkley. It has 5 convolutional layers and 3 fully-connected layers. 11×11 shown in Fig. 4 means the size of filters and 96 means the number of kernels in the specific layer. Vgg-19 can be regarded as a AlexNet stacking more layers. ResNet is much different from other models with residual shortcuts and less fully-connected layers.

All the nets are trained with stochastic gradient descent (SGD) with back-propagation [11], having the same solver parameters with a learning rate of 0.001, a momentum of 0.9, and a weight decay of 0.0005.

We use the Plankton Dataset as train source directly at first, the results are shown in Table 1.

According to the results, we can find that unexpectedly although CaffeNet is not as deep as Vgg-19, Res-19, or Res-50, it has higher classification accuracy and speed with less storage space. The deepest model ResNet with 101 layers only makes little progress comparing to CaffeNet with a great setback in classification

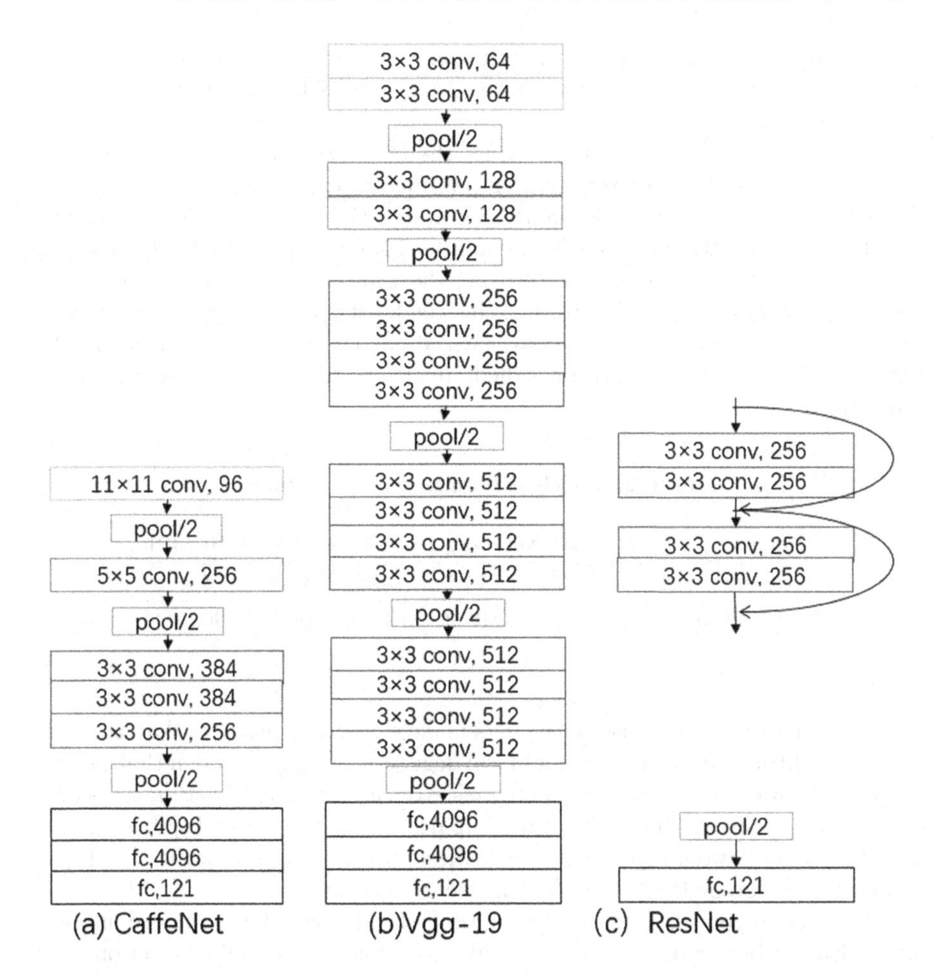

Fig. 4. There are mutiple CNN models used in the Plankton Dataset.

Table 1. Models performance on the Plankton Dataset

Model	CaffeNet	Vgg-19	Res-19	Res-50	Res-101
Top-1 accuracy (%)	74.5	60.6	70.7	72.9	74.7
Top-5 accuracy (%)	94.6	89.7	93.6	95.7	95.9
Model size (MB)	224.0	547.2	44.0	93.0	173.4
Frame rate (fps)	16.9	2.9	12.4	2.7	0.9

rate. As we clarified above, CaffeNet and Vgg-19 have nearly the same micro structure, while ResNet is much more different from them. It is shown that Vgg-19's result is worse than CaffeNet's with more convolutional layers. As for ResNet, the more convolutional layers there are, the higher the accuracy we get.

As a result, we cannot get final conclusion whether more conventional layers help to improve the classification accuracy for the Plankton Dataset by this experiment.

We hold the view that this is because the Plankton Dataset is a small scale dataset. As a result, CaffeNet has less parameters than other nets but can be trained better. To prove our inference, we design the research process in below.

The ILSVRC2012 dataset which contains 141 GB images of 1,000 classes is used to train all the nets instead of using the Plankton Dataset directly. After the models hit limits on ILSVRC2012, the weighs of previous layers in these nets that we finally saved are adopted to learn about the features of images in the Plankton Dataset. This method is called "finetune" in common. The results are shown in Table 2.

Table 2. Models performance after finetune

Model	CaffeNet	Vgg-19	Res-19	Res-50	Res-101
Top-1 accuracy (%)	77.7	74.1	78.5	76.2	76.8
Top-5 accuracy (%)	96.1	95.2	95.4	96.1	96.8

After finetune, we can see all the nets make significant progress than training on the Plankton Dataset directly, and the degeneration phenomenon that deeper models get lower accuracy is more obvious. We can see that Vgg-19 has a worse result but more layers than CaffeNet, comparing both Top-1 and Top-5 accuracy. Res-19 has the fewest layers but the highest Top-1 accuracy, while the Top-5 accuracy of different ResNet models are nearly the same.

The comparison experiment shows that the deeper and advanced models do not lead to better results on the Plankton Dataset. Residual connection does not make significant progress than CaffeNet in the experiment. And models with proper layers can help to get a better result.

Comprehensively speaking, the mass data collected by underwater camera calls for more efficient model which takes less storage space, runs more rapidly and keeps a proper accuracy. The new model should take the hidden degeneration into consideration.

4 Propose Framework

4.1 Key Points to Make CNN Smaller

In fact, when researchers try to make CNN architecture deeper with the limited storage of GPU, lots of tricks have been applied in different models to make them smaller. Here we conclude some key points on designing efficent architecture [15,17]:

First, use smaller filter instead of 5×5, 11×11 convolutional filter. Because a 5×5 convolutional layer has the same sense field as two cascaded 3×3 convolution-al layers, and the latter method reduce computation because $3 \times 3 + 3 \times 3 < 5 \times 5$. So we can find that Vgg-19 only uses 3×3 filter rather than AlexNet with 5×5, 11×11 filters.

Second, avoid adding fully-connected layers. Fully-connected layer has much more parameters than convolutional layer. One of the alternative methods is to use convolutional layer with global average pooling, but this can lead to loss of accuracy. This trick is also adopted by He in [5], but he adds one fully connected layer at last.

Third, using inception to enrich nonlinear representation of feature maps. The former two ideas are decreasing as more parameters as possible, but this will make accuracy decrease as well. Motived by SqueezeNet [7], we try to use more 1×1 inception structure to keep the accuracy with least cost.

Based on key points listed above, we proposed a more efficient framework for the Plankton Dataset.

4.2 Framework Architecture

We now introduce the CNN architecture we proposed in Fig. 5. The Model contains 14 convolutional layers in total. To squeeze the size, we do not use any fully-connected layers. All the convolution layers are designed with filters smaller than 3×3.

However, this may increase classification error. So 1×1 inception structure are adopted to maintain the accuracy.

According to the hidden degeneration we discover, the depth has an optimal value for the Plankton Dataset. We increase the net layer by layer to and finally choose the 14-layer model. Some of experiments are shown in Table 3.

Table 3. The framework with different layer numbers

Model	12 layers	14 layers	16 layers	18 layers
Top-1 accuracy (%)	74.6	76.4	76.3	76.0
Top-5 accuracy (%)	94.8	96.2	96.0	96.1

As mentioned above, SqueezeNet [7] proposed by Berkley researchers also use the 1×1 inception structure and they call two cascaded layers as Fire module, but they also use late downsample and different convolutional filter size compar-ing with our architecture. We conclude the results of different models in Table 4.

Our architecture is only 1.5 MB, and can reach real time operation rate of 32.2 fps with nearly no loss in classification accuracy.

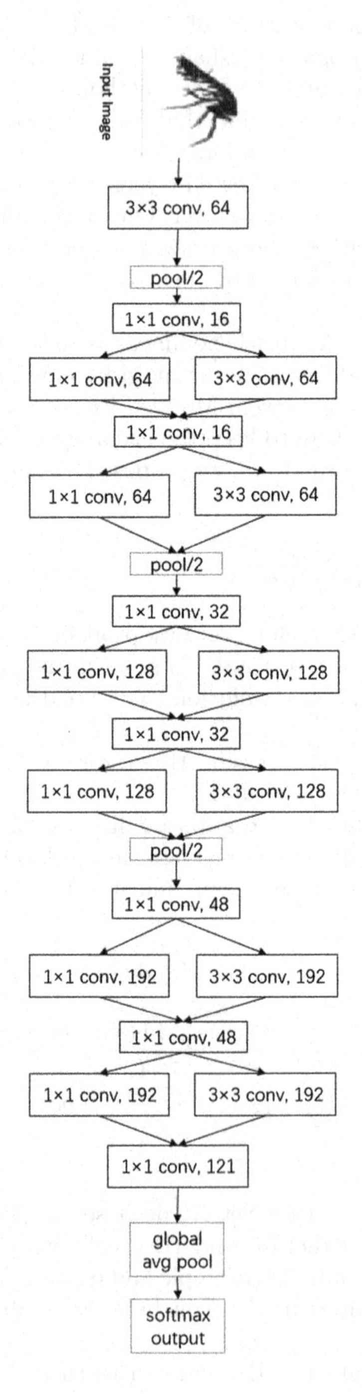

Fig. 5. There are mutiple CNN models used in the Plankton Dataset.

Table 4. Plankton classification comparision

Model	Ours	CaffeNet	Res-19	Res-50	Res-101	SqueezeNet
Top-1 accuracy (%)	76.4	77.7	78.5	76.2	76.8	74.9
Top-5 accuracy (%)	96.1	96.1	95.4	96.1	96.8	95.8
Model size (MB)	1.5	224.0	44.0	93.0	173.4	3.1
Frame rate (fps)	32.2	16.9	12.4	2.7	0.9	18.4

5 Conclusion

The development of CNN has been quite rapidly recently, however, for the specific problem such the Plankton Dataset, advanced models generlize not so well. Lots of work remains to be done for researchers.

In this paper, we compare multiple models to classify planktons. We find that the deeper and advanced models do not lead to better results on the Plankton Dataset. What's worse, deeper nets take up more storage space and run more slowly, so that cannot be actually deployed in the seafloor observation network. To solve this problem, we proposed a framework inspired by concluding the common skills to make CNN smaller, our architecture archive a real time process speed and nearly no loss of classification accuracy.

References

1. Deng, J., Dong, W., Socher, R., Li, L.J., Li, K., Fei-Fei, L.: Imagenet: a large-scale hierarchical image database. In: IEEE Conference on Computer Vision and Pattern Recognition, pp. 248–255 (2009)
2. Glorot, X., Bengio, Y.: Understanding the difficulty of training deep feedforward neural networks. In: Proceedings of the Thirteenth International Conference on Artificial Intelligence and Statistics, pp. 249–256 (2010)
3. Han, S., Mao, H., Dally, W.J.: Deep compression: compressing deep neural networks with pruning, trained quantization and Huffman coding. In: International Conference on Learning Representation (2016)
4. He, K., Zhang, X., Ren, S., Sun, J.: Delving deep into rectifiers: surpassing human-level performance on imagenet classification. In: IEEE International Conference on Computer Vision, pp. 1026–1034 (2015)
5. He, K., Zhang, X., Ren, S., Sun, J.: Deep residual learning for image recognition. In: IEEE Conference on Computer Vision and Pattern Recognition, pp. 770–778 (2016)
6. Hubara, I., Courbariaux, M., Soudry, D., El-Yaniv, R., Bengio, Y.: Binarized neural networks. In: International Conference on Neural Information Processing Systems, pp. 4107–4115 (2016)
7. Iandola, F.N., Han, S., Moskewicz, M.W., Ashraf, K., Dally, W.J., Keutzer, K.: Squeezenet: alexnet-level accuracy with 50x fewer parameters and <0.5 MB model size. arXiv preprint arXiv:1602.07360 (2016)

8. Ioffe, S., Szegedy, C.: Batch normalization: accelerating deep network training by reducing internal covariate shift. In: International Conference on Machine Learning, pp. 448–456 (2015)
9. Jia, Y., Shelhamer, E., Donahue, J., Karayev, S., Long, J., Girshick, R., Guadarrama, S., Darrell, T.: Caffe: convolutional architecture for fast feature embedding. In: ACM International Conference on Multimedia, pp. 675–678 (2014)
10. Krizhevsky, A., Sutskever, I., Hinton, G.E.: Imagenet classification with deep convolutional neural networks. In: International Conference on Neural Information Processing Systems, pp. 1097–1105 (2012)
11. LeCun, Y., Boser, B., Denker, J.S., Henderson, D., Howard, R.E., Hubbard, W., Jackel, L.D.: Backpropagation applied to handwritten zip code recognition. Neural Comput. **1**(4), 541–551 (1989)
12. LeCun, Y., Bottou, L., Bengio, Y., Haffner, P.: Gradient-based learning applied to document recognition. Proc. IEEE **86**(11), 2278–2324 (1998)
13. Lin, M., Chen, Q., Yan, S.: Network in network. arXiv preprint arXiv:1312.4400 (2013)
14. Saxe, A.M., McClelland, J.L., Ganguli, S.: Exact solutions to the nonlinear dynamics of learning in deep linear neural networks. arXiv preprint arXiv:1312.6120 (2013)
15. Simonyan, K., Zisserman, A.: Very deep convolutional networks for large-scale image recognition. In: International Conference on Learning Representation (2015)
16. Srivastava, R.K., Greff, K., Schmidhuber, J.: Training very deep networks. In: International Conference on Neural Information Processing Systems (2015)
17. Szegedy, C., Liu, W., Jia, Y., Sermanet, P., Reed, S., Anguelov, D., Erhan, D., Vanhoucke, V., Rabinovich, A.: Going deeper with convolutions. In: IEEE Conference on Computer Vision and Pattern Recognition, pp. 1–9 (2015)

Zero-Shot Learning with Deep Canonical Correlation Analysis

Zhong Ji(✉), Xuejie Yu, and Yanwei Pang

School of Electrical and Information Engineering, Tianjin University, Tianjin 300072, China
jizhong@tju.edu.cn

Abstract. Zero-shot learning (ZSL) improves the scalability of conventional image classification systems by allowing some testing categories having no training data. One key component is to learn a shared embedding space where both side information of object categories and visual representation of object images can be projected to for nearest neighbor search. Although great progress has been made, existing approaches mainly focus on shallow models, which cannot learn a strong generalized embedding space. To this end, this paper proposes a novel deep embedding model for ZSL, which formulates the embedding space with Deep Canonical Correlation Analysis (DCCA). Specifically, the side information and the visual representation are transformed via two independent deep neural networks, and then they are highly linearly correlated in the final output layer. Extensive experiments on two publicly popular datasets demonstrated the effectiveness and superiority the proposed approach.

Keywords: Zero-shot learning · Image classification
Canonical Correlation Analysis · Deep neural network

1 Introduction

Image classification technique has made great progress owing to the prosperous progress of Convolutional Neural Networks (CNN) and large scale annotated datasets. However, one obstacle it must face is that there must be enough annotation data for training. It is impractical to provide the labels for all categories since there are so many categories in the real world. Moreover, labeling all the classes for training is expensive and time-consuming.

In contrast, we human beings have the ability to recognize new categories without having to look at actual visual samples. Imaging that a child has never seen a zebra, however, he/she has been told that the zebra looks like a horse but with black and white stripes. When this child sees a zebra in the first time, he/she may recognize the zebra with his/her own inferential capability. To imitate this ability, Zero-Shot Learning (ZSL) technique is proposed and received increasingly attention recently [1–5].

J. Yang et al. (Eds.): CCCV 2017, Part III, CCIS 773, pp. 209–219, 2017.
https://doi.org/10.1007/978-981-10-7305-2_19

ZSL predicts new unseen categories by transferring the knowledge obtained from seen categories with labeled data and side information such as attributes [6] and word vectors [7]. Specifically, attributes define a few properties of an object, such as the color, the shape, and the presence or absence of a certain body part. Word vectors represent category names as vectors with a distributed language representation, which is constructed by a linguistic base, such as Wikipedia. Both attributes and word vectors are shared across both seen and unseen categories. In this way, both categories are associated with semantic vectors in the category embedding space spanned by the side information. In another words, the side information acts as a bridge to make the knowledge transferred from the seen categories to the unseen categories.

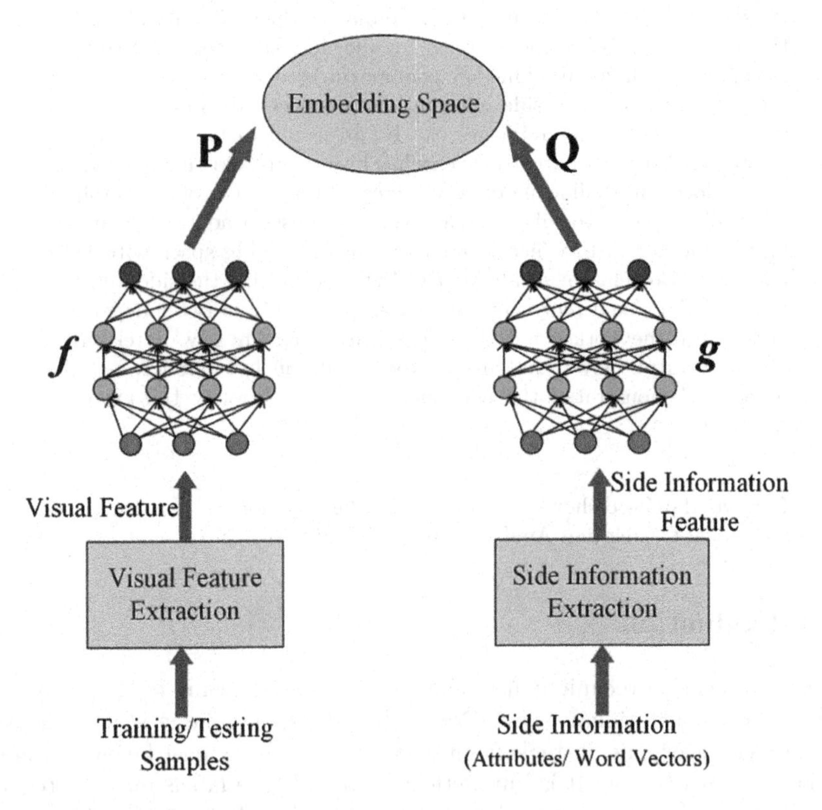

Fig. 1. The pipeline of the proposed DCCA-ZSL approach

Since the visual features and the side information are different modalities, a shared embedding space is required to compute their interaction. Actually, the construction of the embedding space is one of the key components in ZSL. Many attempts have been made to build an effective embedding model, such as CCA [8], LatEm [2], MCME [3], ReCMR [4] and SJE [9]. However, these approaches are based on shallow models, which cannot learn the strong generalized embedding models. To this end, this paper proposes a novel deep embedding model based on Deep Canonical Correlation Analysis (DCCA), and the proposed approach is called DCCA-ZSL. Figure 1 shows its pipeline.

This paper is structured as follows. Section 2 introduces the related work of ZSL. Section 3 presents the detail of the proposed DCCA-ZSL approach. Experimental results and analyses are given in Sects. 4 and 5 concludes this paper.

2 Related Work

From the perspective of embedding space, most existing ZSL approaches can be grouped into linear-based, bilinear-based, and nonlinear-based methods. Specifically, the linear-based methods generally embed the visual space to the side information space, or vice versa. For example, Linear Regression (LR) is one of the representative methods [10]. It is a straightforward cross-modality method using L2-regularized least-square loss to build an objective function to map knowledge from the one space to the other space. The authors in [10] also demonstrated that the embedding from the visual space to side information space is more effective than the inverse embedding.

Bilinear-based methods are most commonly used in ZSL. In general, a bilinear embedding is a function combining elements of two vector spaces to yield an element of a third vector space, and is linear in each of its arguments. One representative method in this group is Canonical Correlation Analysis (CCA). Lazaridou et al. [8] employs CCA to maximize the correlation between the visual feature and the semantic feature. Motivated by CCA, [3] presents a manifold regularized cross-modal embedding approach for ZSL by formulating the manifold constraint for intrinsic structure of the visual features as well as aligning pairwise consistency. ESZSL method in [1] constructs a general framework to model the relationships between visual features, class attributes and class labels with a bilinear model, and the closed-form solution makes it efficient. SJE method in [9] relates the input embedding and output embedding through a compatibility function, and implements ZSL by finding the label corresponding to the highest joint compatibility score. Recently, Yu et al. [4] developed a bilinear embedding model by employing the hinge ranking loss to exploit the structures among different modalities and devise efficient regularizers to constrain the variation of the samples in the identical modality. Xian et al. [2] presented a novel discriminative bilinear embedding model by applying multiple linear compatibility units and allowing each image to choose one of them. The model is trained with a ranking based objective function that penalizes incorrect rankings of the true class for a given image.

Nonlinear-based methods are receiving more attentions in recent years. DAP [11] and IAP [11] are the earlier attempt toward this line. They build a network to formulate the embedding relations. Specifically, DAP utilizes the class attributes as the middle layer between the input instances and the output category labels, while IAP takes the seen categories as the middle layer. Socher et al. [5] used a two-layer network to embed the visual space into a side information space. It is a simple embedding model and is not a shared space. Recently, Yang and Hospedales [12] proposed a neural network based ZSL method to build a shared space. However, the objective function is different from ours, and the network is also a two-layer one.

3 Proposed DCCA-ZSL Method

In this section, we elaborate our proposed DCCA-ZSL method. We first introduce the notations, then depict the details of DCCA-ZSL and finally analyze the optimization process briefly.

3.1 Notations

We denote the data matrix of visual feature as $X = [x_1, \cdots, x_N] \in \mathbb{R}^{D_x \times N}$and that of side information feature as $Y = [y_1, \cdots, y_N] \in \mathbb{R}^{D_y \times N}$,where N is the sample size, D_x and D_y are the dimensionalities. The f and g are used to denote the nonlinear mapping implemented by deep networks. A deep network f of depth m implements a nested mapping with the form $f(x) = f_m((\cdots f_1(x; W_1, b_1)\cdots); W_m, b_m)$, where W_i are the weight parameters of layer $i(i = 1, \cdots, m)$ and f_i is the mapping of layer i that can be represented as $f_i(x) = s(W_i^T x + b_i)$. And s is the activation function. The typical choices of which are sigmoid, tanh, ReLU, etc. We use θ_f to denote the vector of all parameters W_m and b_m for visual feature X, and similarly for θ_g to denote the parameters of semantic feature Y. The dimensionality of the embedding space, which is equal to the number of output units in the two deep networks, is denoted as d.

3.2 Canonical Correlation Analysis

Canonical Correlation Analysis (CCA) is a kind of classical statistical method. The aim of CCA is to find pairs of linear projections, $w_x \in \mathbb{R}^{D^x}$ and $w_y \in \mathbb{R}^{D^y}$, that are maximally correlated for the input random vectors X and Y. At the same time, the different dimensions are constrained to be uncorrelated. And if the noise in either view is uncorrelated with the other view, the learned representations should not contain the noise in the uncorrelated dimensions. The objective function can be formularized as follows:

$$\rho = \max_{w_x, w_y} \frac{w_x^T C_{xy} w_y}{\sqrt{w_x^T C_{xx} w_x \cdot w_y^T C_{yy} w_y}} \tag{1}$$

where $C_{xx} = XX^T \in R^{D^x \times D^x}$, $C_{yy} = YY^T \in R^{D^y \times D^y}$ are the within-set covariance matrices and $C_{xy} = XY^T \in R^{D^x \times D^y}$ is the between-sets covariance matrix and $C_{xy} = C_{yx}^T$.

3.3 The DCCA-ZSL Model

The target of DCCA is to make the output functions $f(X)$ and $g(Y)$ are maximally related in the embedding space (see Fig. 1). Therefore, the objective function can be represented as

$$(\theta_f, \theta_g) = arg \max_{\theta_f, \theta_g} corr(f(X, \theta_f), g(Y, \theta_g)). \tag{2}$$

Rewrite the function in the form of matrix operations and then we have:

$$\max_{\theta_f, \theta_g, P, Q} \quad \frac{1}{N} tr(P^T f(X) g(Y)^T Q), \tag{3}$$

$$s.t. \quad P^T(\frac{1}{N} f(X) f(X)^T + r_x I) P = I, \tag{4}$$

$$Q^T(\frac{1}{N} g(Y) g(Y)^T + r_y I) Q = I, \tag{5}$$

$$P_i^T f(X) g(Y)^T q_j = 0, for\ i \neq j. \tag{6}$$

where P and Q are the CCA directions that project the transformed features, r_x and r_y are the regularization parameters, $P^T f(\cdot)$ and $Q^T g(\cdot)$ are the final learned mapping for the two modalities.

Similarity metric is important in kNN search in the testing phase [16]. In our work, we select the cosine distance is used as the similarity metric to seek the nearest side information feature for each visual sample according to:

$$\max \quad cos(P^T f(X), g(Y)^T Q). \tag{7}$$

And then the corresponding category of the side information feature is chosen as the predicted result.

In addition, we make a brief introduction to the solution and optimization of DCCA-ZSL method. Let H_1 and H_2 be the representation matrices produced by the deep models on the two different features. $\overline{H_1}$ and $\overline{H_2}$ are the corresponding centered matrices. And Σ_{12} is the between-sets covariance matrix. Σ_{11} and Σ_{22} are within-set covariance matrices.

$$\Sigma_{12} = \frac{1}{n-1} \overline{H_1 H_2}^T, \tag{8}$$

$$\Sigma_{11} = \frac{1}{n-1} \overline{H_1 H_1}^T + r_1 I, \tag{9}$$

$$\Sigma_{22} = \frac{1}{n-1} \overline{H_2 H_2}^T + r_2 I, \tag{10}$$

where n is the size of training samples. Then the total correlation of H_1 and H_2 can be calculated by the sum of the top k singular values of matrix

T= $\Sigma_{11}^{-1/2}\Sigma_{12}\Sigma_{22}^{1/2}$. And if $k = d$, the correlation is exactly the matrix trace norm of T.

$$corr(H_1, H_2) = \|T\|_t r = tr(T^T T)^{1/2}, \tag{11}$$

Stochastic Gradient Descent (SGD) [13] algorithm is used to optimize the model. We first randomly pick a minibatch of p training pairs and feed them forward to compute the transformed matrices H_1, H_2 and the correlation matrix T. Then the gradient of T with respect to H_1 and H_2 is computed. Finally, we use back propagation algorithm to adjust the network parameters according to the gradient descent direction until the model converges.

4 Experimental Results and Analysis

4.1 Datasets and Settings

Datasets. To evaluate the effectiveness of the proposed DCCA-ZSL approach, extensive experiments are conducted on two benchmark datasets, AwA [11] and CUB [14]. Specifically, **Animal with Attribute (AwA)** [11] consists of 30,475 animal images, which belong to 50 categories. Each category is annotated with 85 attributes. **Caltech-UCSD Bird2011 (CUB)** [14] includes 11,788 images from 200 bird subspecies. And each category is annotated with 312 attributes. Moreover, CUB is a fine-grained dataset which is more challenging for image classification. For AwA and CUB datasets, we use the standard 40/10 and 150/50 split settings, respectively.

Features. VGG-verydeep-19 [15] (denoted as VGGNet for short) is used as the visual feature extraction model, and the 4096-dimensional feature of the second fully connected layer of this model is selected as the visual features of input images. Attributes and word vectors are used as side information, respectively. Particularly, we choose the Word2Vec model trained on the Wikipedia corpus as text feature extraction model, from which 1000-dimentional and 400-dimentional word vectors are used in AwA and CUB datasets, respectively.

Evaluation Metric. The average per-class top-1 accuracy [9] on the test sets is reported over 10 trials.

4.2 Results on AwA and CUB Datasets

CCA is chosen as the baseline embedding model, whose corresponding ZSL method is called CCA-ZSL. Besides, four state-of-the-art methods are selected for comparison, that is SJE [9], LatEm [10], DAP [11], and ESZSL [1]. Among these, CCA-ZSL and ESZSL are implemented by ourselves, while the results of SJE, LatEm and DAP are from the orignal papers or website. The results are summarized in Table 1.

Table 1. Comparison results on different datasets (in %)

Methods	AwA		CUB	
	T	A	T	A
SJE [9]	51.2	66.7	28.4	50.1
LatEm [10]	61.1	71.9	31.8	45.5
DAP [11]	-	57.5	-	-
ESZSL [1]	67.4	75.3	30.4	46.8
CCA-ZSL	65.6	72.5	30.4	46.2
DCCA-ZSL	71.8	75.6	35.1	47.3

From Table 1, we can observe that:

(1) The proposed DCCA-ZSL method is competitive with the state-of-the-art methods. It beats all the copetitors in AwA dataset in different side information spaces and gains best performance in CUB when using word vectors. Compared with the CCA-ZSL, DCCA-ZSL achieves an improvement of 6.2% and 4.7% in AwA and CUB with word vectors, respectively. As for attribute feature, it has a 3.1% and 1.1% improvements, respectively.
(2) The performance improvement of DCCA-ZSL with word vectors is more obvious than that with attributes. This is due to the fact that word vectors are extracted from corpus by unsupervised learning. Therefore, the learned features contain more noise than the human labeled attributes. Using deep network can further remove redundant information and achieve feature fusion, thereby improving classification performance.
(3) The classification accuracy is much higher for those with attributes than those with word vectors. This is because that attributes are annotated by experts, which can better reflect class correlation than word vectors.
(4) We also notice that the performance in AwA dataset is much higher than those in CUB dataset. The reason is that CUB is a fine-grained dataset, which is more challenging for image classification.

Additionally, Fig. 2 provide the confusion matrices of DCCA-ZSL and CCA-ZSL, since they can distinctly and intuitively shown the classification performance. Each column of the matrix represents the instances in a predicted class while each row represents the instances in an actual class. All correctly predicted images are located in the reverse diagonal of the matrix. And the darker the color, the more the numbers of correctly classified images. From Fig. 2, we can observe that the overall performance of DCCA-ZSL is better than that of CCA-ZSL. The darker colors in our model are more concentrated in the reverse diagonal than in CCA-ZSL. And the classification results of four categories, i.e., chimpanzee, leopard, Persian cat, and humpback whale, are with fairly high accuracy.

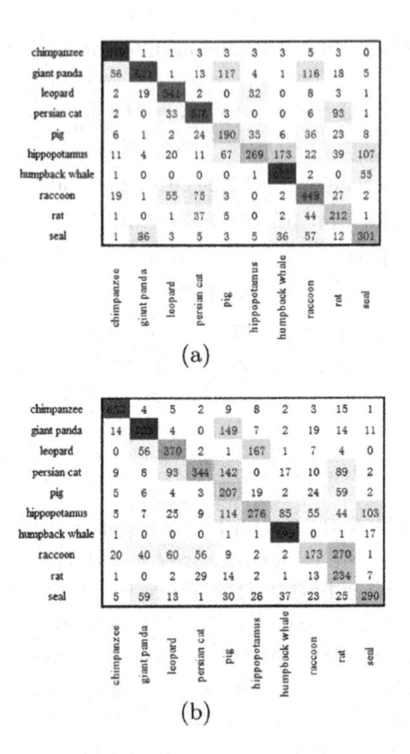

Fig. 2. The confusion matrices of DCCA-ZSL (a) and CCA-ZSL (b).

4.3 Parameter Sensitivity

This subsection analyzes the following four types of parameters: common space dimensionality, number of network layers, batchsize and hyper-parameters.

Without loss of generality, we take the experiments in CUB dataset with word vectors as examples. Fix other parameters, we first evaluate the impact of the common space dimensionality (denoted as d), and the result is shown in Fig. 3.

At the beginning, the accuracy increases with d until reaching a peak. This is because that after the mapping to the common space, the redundant information is removed and the effective feature information is preserved. Thus the classification accurate will rise. Then the increasing of d will take a risk of introducing noise, resulting in decreased classification performance. This paper finally selects $d = 40$ for both datasets.

Then, we evaluate the impact of number of network layers. In the experiments, we notice that increasing the layers of the visual feature mapping network (denoted as f) and keep the word vectors mapping network (denoted as g) unchanged, the classification performance does not improve significantly or even decline. This may be due to the fact that visual feature has been sufficiently extracted by CNN so that the performance improvement is not obvious with more layers. Thus, we set the layers of the visual transformation network

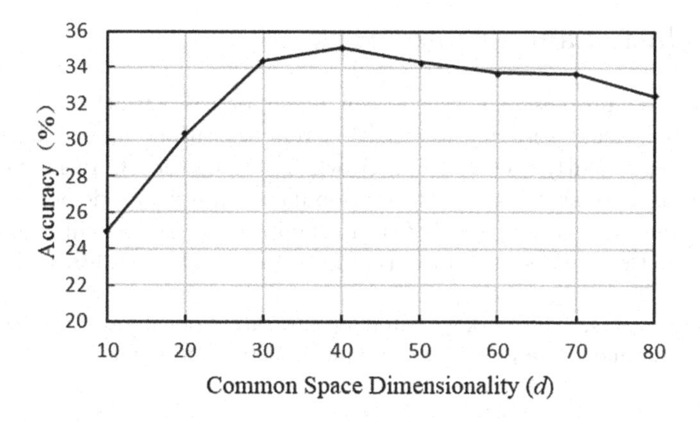

Fig. 3. Impact of common space dimensionality on CUB dataset with word vectors.

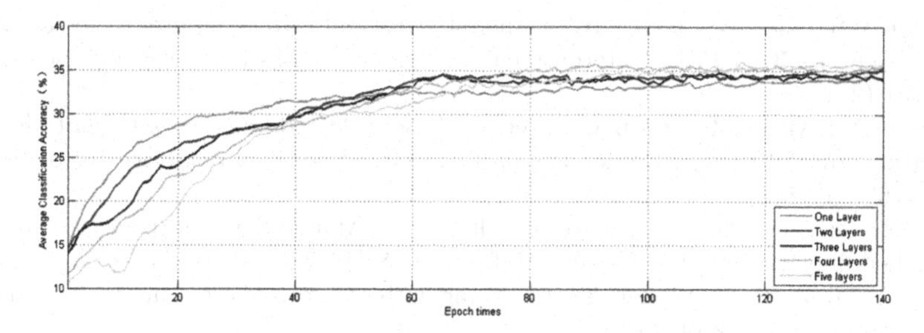

Fig. 4. Impact of the word vector network depth on CUB dataset.

to be 2. Under this setting, we observe the impact of word vector network depth, as shown in Fig. 3. It can be seen that the classification performance improves with the raise of the layers. Meanwhile, the convergence rate of the model slows down and the training time becomes longer. Making a compromise among these indexes, we choose 5 layers for word vector network.

Finally, we evaluate the impact of batchsize and hyper-parameters. The proposed DCCA-ZSL model is optimized by Stochastic Gradient Descent (SGD) algorithm. It randomly picks a minibatch of n training pairs, where n represents the batchsize. With the increase of batchsize, the model converges much faster and the learned gradient descent direction is more accurate. Besides, larger batchsize reduces the times of iterations and the network shock. However, oversized batchsize may make the algorithm get into local minimum values. Considering all of these factors, the batchsize in this paper is set to 200 and 150 for AwA and CUB, respectively.

It is also found that the two hyper-parameters rcov1 and rcov2 for regulation have great effects on the results. The parameters are selected from 0.0001, 0.001, 0.01, 0.1, 1, 10, 100 according to the corresponding experiments.

5 Conclusion and Future Work

In this paper, we propose an effective deep model for ZSL with the idea of DCCA. It embeds the visual and side information representations into a shared space with two independently deep neural network, and takes the CCA as the objective function. Extensive experiments on two popular datasets have demonstrated the superiority and promise of DCCA-ZSL. In the future work, we will explore other objective functions and side information in a deep model to address ZSL task.

Acknowledgments. This work was supported by the National Natural Science Foundation of China under Grant 61472273, Grant 61632018 and Grant 61771329.

References

1. Romeraparedes, B., Torr, P.H.S.: An embarrassingly simple approach to zero-shot learning. In: 32th JMLR International Conference on Machine Learning, pp. 2152–2161 (2015)
2. Xian, Y., Akata, Z., Sharma, G., et al.: Latent embeddings for zero-shot classification. In: 29th IEEE Conference on Computer Vision and Pattern Recognition, pp. 69–77 (2016)
3. Ji, Z., Yu, Y., Pang, Y., Guo, J., Zhang, Z.: Manifold regularized cross-modal embedding for zero-shot learning. Inf. Sci. **378**, 48–58 (2017)
4. Yu, Y., Ji, Z., Pang, Y.: Zero-shot learning with regularized cross-modality ranking. Neurocomputing **259**, 14–20 (2017)
5. Socher, R., Ganjoo, M., Sridhar, H., et al.: Zero-shot learning through cross-modal transfer. In: Advances in Neural Information Processing Systems, pp. 935–943 (2013)
6. Farhadi, A., Endres, I., Hoiem, D., et al.: Describing objects by their attributes. In: the 19th IEEE Conference on Computer Vision and Pattern Recognition, pp. 1778–1785 (2009)
7. Mikolov, T., Sutskever, I., Chen, K., et al.: Distributed representations of words and phrases and their compositionality. Adv. Neural Inf. Process. Syst. **26**, 3111–3119 (2013)
8. Lazaridou, A., Bruni, E., Baroni, M.: Is this a wampimuk? Cross-modal mapping between distributional semantics and the visual world. In: Proceedings of the Association for Computational Linguistics, pp. 1403–1414 (2014)
9. Akata, Z., Reed, S., Walter, D., et al.: Evaluation of output embeddings for fine-grained image classification. In: 27th IEEE Conference on Computer Vision and Pattern Recognition, pp. 2927–2936 (2014)
10. Shigeto, Y., Suzuki, I., Hara, K., Shimbo, M., Matsumoto, Y.: Ridge regression, hubness, and zero-shot learning. In: Appice, A., Rodrigues, P.P., Santos Costa, V., Soares, C., Gama, J., Jorge, A. (eds.) ECML PKDD 2015. LNCS (LNAI), vol. 9284, pp. 135–151. Springer, Cham (2015). https://doi.org/10.1007/978-3-319-23528-8_9
11. Lampert, C.H., Nickisch, H., Harmeling, S.: Learning to detect unseen object classes by between-class attribute transfer. In: 19th IEEE Conference on Computer Vision and Pattern Recognition, pp. 951–958 (2009)
12. Yang, Y., Hospedales, T.M.: A unified perspective on multi-domain and multi-task learning. In: International Conference on Learning Representations, pp. 1–9 (2015)

13. Wang, W., Arora, R., Livescu K., et al.: On deep multi-view representation learning objectives and optimization. arXiv preprint arXiv:1602.01024 (2016)
14. Wah, C., Branson, S., Welinder, P., Perona, P., Belongie, S.: The Caltech UCSD Birds-200-2011 Dataset, Technical report (2011)
15. Simonyan, K., Zisserman, A.: Very deep convolutional networks for large-scale image recognition. In: International Conference on Learning Representations (2014)
16. Guo, Y., Ding, G., Han, J., et al.: Robust iterative quantization for efficient lp-norm similarity search. In: International Joint Conference on Artificial Intelligence, pp. 3382–3388. AAAI Press (2016)

A Novel Level Set Model for Image Segmentation with Interactive Label Regularization Term

Huang Tan, Ming Chen(✉), Qiaoliang Li, and Shaojun Qu

Key Laboratory of High Performance Computing and Stochastic Information Processing (HPCSIP) (Ministry of Education of China),
College of Mathematics and Computer Science, Hunan Normal University, Changsha 410081, Hunan, People's Republic of China
{tanhuang,chenming,liqiaoliang,qshj}@hunnu.edu.cn

Abstract. We propose an interactive level set segmentation method with a novel user's label regularization term. This new edge-based model can force the evolution of level set function to follow the hard constraints given by the user and effectively solve the problem that user can't extract some specified object in the image due to the local optimum. The new regularization term is constructed by multiplication of the hard constraints and the level set function. Since the regularization term we defined only works on the pixels labeled by the user, the evolution of level set function can be affected accurately by user's label for foreground and background. Experimental results are provided to demonstrate the efficiency and accuracy of the new model. Our method can make the segmentation accurately reflect the user's label. The new method supports the real-time feedback in the segmentation process. We also analyzed the weight of the regularization term, and the best weight of the regularization term is provided.

Keywords: Interactive image segmentation · Real-time feedback
Regularization term · Level set

1 Introduction

Image segmentation refers to partitioning an image into several disjoint subsets such that each corresponds to a meaningful part of the image. It is a classical and

---

H. Tan—Supported by the Construct Program of the Key Discipline in Hunan Province.

M. Chen—Supported by the National Natural Science Foundation of China (NSFC), No.11471110.

Q. Li—Supported by the National Natural Science Foundation of China (NSFC), No. 11471002; Hunan Provincial Science and Technology Plan, No. 2013FJ4052.

J. Yang et al. (Eds.): CCCV 2017, Part III, CCIS 773, pp. 220–232, 2017.
https://doi.org/10.1007/978-981-10-7305-2_20

fundamental problem in computer vision. Despite many years of research, general purpose image segmentation as an ill-posed problem is still a very challenging task.

Active contour model [4,6] is one of the successful methods of segmentation. It uses the theory of dynamics model that an initial curve is derived to target contour under the internal force of the curve itself and the external force of image data. Level-set-based active contour models [15,16] expresses implicitly the contour as the zero level of a level set function, and evolves the curve based on an upgrade equation, finally, get smooth, closed, and high-precision segmentation curves. According to the properties of its energy function, active contour models are classified into region-based model [7,12,18] and edge-based model [3]. The region-based models utilize local information to guide contour curve move to the boundary of object approximately, while the edge-based on utilizes a stopping function to attract contours to the desired boundaries.

Geodesic active contour (GAC) [3], which was independently introduced by Caselles et al. [2] and Malladi et al. [14], is a very popular edge-based model. Its basic idea is to represent contours as the zero level set, and to evolve the level set function according to a partial differential equation (PDE). GAC has several advantages over the traditional parametric active contours. Firstly, the contour represented by the level set function can break or merge naturally during the evolution. Secondly, the level set function always remains the function on a fixed grid, which allows efficient numerical schemes. However, it has been proved to be locally optimal, it is possible that the contour would stop in or out of the object when it encounters an obvious boundary during the evolution.

Using interactive segmentation algorithm, user can make some foreground and background labels to give some prior knowledge before segmentation executing. With the prior knowledge, interactive segmentation algorithm can extract objects more accurately. At present, there are many works on the research of interactive segmentation. Regularization term is often used to incorporate the hard constraints of user's label into the cost function, for example using L2-norm distance to construct regularization term [17]. It is also a popular approach to use probability model such as Gauss Mixed Model, Support Vector Machine and Geodesic which are built with user's label to assist the segmentation [13,19]. Graph-based approaches use user's label to represent the information of must link or can't link as hard or soft constraints [1,5].

There are some works on interactive level set segmentation, which use regularization term of hard constraints and probability model. [20] used belief propagation to minimize a global cost function according to local level sets. The propagation starts with one user marked point, and iteratively extends the user information from the labeled pixel to its neighborhood by calculating the beliefs of the pixels in the same level as the marked pixel. This method was designed for medical image, but has not good result for object of heterogeneous images. [10] used user's label to integrates discriminative classification models and distance transforms with the level set method, and the terms of energy function are based on a probabilistic classifier and an unsigned distance transform of salient edges.

This method is effective for heterogeneous image, but it can't support user to feedback in interactive process.

In this paper, we propose an interactive level set segmentation method for edge-based model. We use the multiplication of the hard constraints which record the user's label and the level set function as a regularization term and add it into the energy function. Each element of the hard constraints represents the label of the corresponding pixels in the image. If the pixel is not marked, the value of corresponding element in the hard constrains is 0; if marked as foreground, the value is -1; if marked as background, the value is 1. With the new regularization term, the evolution of level set function can only be impacted at the location of user's label, so that the evolution process can be accurately carried out under the influence of the user's label. Usually, the initial segmentation may not be perfect. According to the observed results, user can mark new foreground and background labels for the next evolution. By these new labels, the algorithm can efficiently evolve the current level set function without recomputing from initial contour. Satisfactory results is obtained by repeating the process of mark-evolution. The experiment results show the efficiency of our method.

2 Background

Our interactive label regularizer is tested under GAC model. Hence, we briefly introduce it.

2.1 Level Set Segmentation Based on GAC Model

Given an image with size $m * n$, we denote pixels set by $\Omega = \{(x, y); x = 1, \ldots, m, y = 1, \ldots, n\}$, and the feature $I = \{I(x, y); (x, y) \in \Omega\}$. In the method of level set, Ω is taken as a continuous region.Active contours is denoted by C, and represented by the zero level set $C(t) = \{(x, y)|\phi(t, x, y) = 0\}$. Here, we define

$$\phi(t, x, y) = \begin{cases} < 0 \text{ if } (x, y) \in inside(C(t)) \\ > 0 \text{ if } (x, y) \in outside(C(t)). \end{cases} \tag{1}$$

The evolution equation of the level set function ϕ can be written in the following general form:

$$\frac{\partial \phi}{\partial t} + F|\nabla \phi| = 0 \tag{2}$$

The function F is called speed function. For image segmentation, the F depends on the image data and the level set function.

In [8], g is edge detection function defined by $g = \frac{1}{1+|\nabla G_\delta * I|^2}$, where G_δ is the Gaussian kernel with standard deviation δ. Based on g, it constructs an energy term, called the external term that drives the motion of the zero level set toward the desired image features, such as object boundaries. The external energy based on $\phi(x, y)$ is as follow:

$$E_{g,\lambda,v}(\phi) = \lambda L_g(\phi) + v A_g(\phi) \tag{3}$$

where $\lambda > 0$, v is a constant. The definitions of $L_g(\phi)$ and $A_g(\phi)$ are as follow:

$$L_g(\phi) = \int_{\Omega} g\delta(\phi)|\nabla\phi|dxdy$$

$$A_g(\phi) = \int_{\Omega} gH(-\phi)dxdy$$

L_g represents the length of zero level set and A_g represents the area inside the contour that control the speed of evolution. H represents Heaviside function defined as follows:

$$H(z) = \begin{cases} 1 \text{ if } z \geq 0 \\ 0 \text{ if } z < 0. \end{cases} \tag{4}$$

In practical applications, H is generally replaced by one relaxing version. This paper uses a substitute as follows:

$$H_\varepsilon(z) = \begin{cases} 1 & \text{if } z > \varepsilon \\ 0 & \text{if } z < -\varepsilon \\ \frac{1}{2}(1 + \frac{z}{\varepsilon} + \frac{1}{\pi}sin(\frac{\pi z}{\varepsilon}) & \text{if } |z| < \varepsilon. \end{cases} \tag{5}$$

$\delta(z) = \frac{dH(z)}{dz}$ is called Dirac measure. Here, $\delta(z)$ is derived from the relaxed H.

According to the Euler-Lagrange equation

$$\frac{\partial\phi}{\partial t} = -\frac{\partial E}{\partial\phi} \tag{6}$$

we can get the following variational formulation of $\frac{\partial\phi}{\partial t}$ in (3):

$$L(\phi) = \frac{\partial\phi}{\partial t} = \lambda\delta(\phi)div(g\frac{\nabla\phi}{|\nabla\phi|}) + vg\delta(\phi) \tag{7}$$

which is the gradient flow that minimizes the energy function. By using (7) and according to

$$\phi_{i,j}^{k+1} = \phi_{i,j}^{k} + \tau L(\phi_{i,j}^{k}) \tag{8}$$

we can process the evolution of level set function.

2.2 The Drawback of GAC Model

It is crucial to keep the evolving level set function as an approximate signed distance function during the evolution, especially in a neighborhood around the zero level set. Naturally, Li et al. [8] proposed the following integral:

$$P(\phi) = \int_{\Omega} \frac{1}{2}(|\nabla\phi| - 1)^2 dxdy. \tag{9}$$

The energy function of [8] is defined as follow:

$$E(\phi) = \mu P(\phi) + E_{g,\lambda,v}(\phi). \tag{10}$$

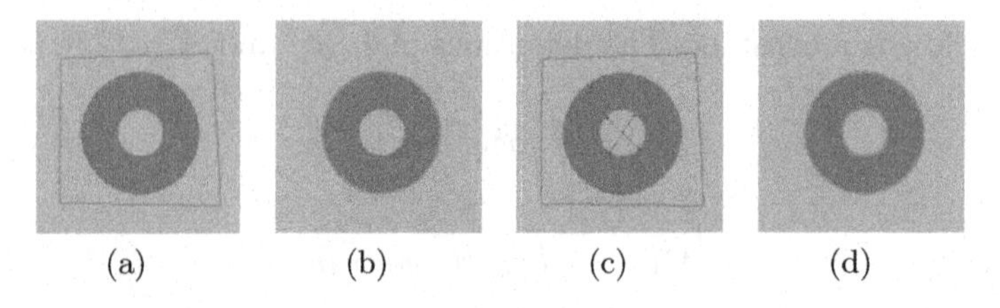

Fig. 1. Extract the black ring in the figure using the general GAC model and our method. (a) is the initial contour used by GAC model, (b) is the result of general GAC model, it can't extract the black ring due to the local optimum, (c) is the initial contour with user's label, where red scribbles provide hints for objectives to be avoided. (d) is the result of our method.

The new $L(\phi)$ is:

$$L(\phi) = \mu[\Delta\phi - div(\frac{\nabla\phi}{|\nabla\phi|})] + \lambda\delta(\phi)div(g\frac{\nabla\phi}{|\nabla\phi|}) + vg\delta(\phi) \tag{11}$$

We use this energy function to extract the black ring in Fig. 1(a). The evolution of the contour stopped at the outside boundary of the black ring which is shown in Fig. 1(b), and the segmentation result is the black ring and the gray circle inside, which is not what we want. It shows that GAC model tends to get a local optimal solution. But in Fig. 1(c) and (d), our proposed method can get a satisfactory result with additional user's label. That means, the active contour would stop evolution when the level set meets boundary. Moreover, some special object can not be extracted.

In addition, [9] proposed a level set energy function that added a penalty term to tradition level set energy function to force the level set function to be close to a signed distance function, which completely eliminated the need of the costly re-initialization procedure. [9] added a distance regularization term to the energy function of [8], so as to restrict the evolution of the zero level set in a give range rather than the whole level set function. This method can reduce the number of iterations of evolution and improve the accuracy of segmentation. However, it still cannot avoid above local solution.

3 The Level Set Model with an Interactive Label Regularization Term

Given the user scribble seeds that belong to the set F and set B, which repetitively provide hints of sub-regions to be extracted and to be avoided. The input image will be classified into these two sets F and B. u represents the matrix of user's label, defined as follows:

$$u(x,y) = \begin{cases} -1 \text{ if } (x,y) \in F(\text{blue scribbles}) \\ 0 \text{ if } (x,y) \text{ is not marked} \\ 1 \text{ if } (x,y) \in B(\text{red scribbles}) \end{cases} \tag{12}$$

In this paper, we propose an interactive level set segmentation method for edge-based model. We add a regulation term that is constructed by user's label to the energy function (10), the new energy function is as follow:

$$E(\phi) = \mu P(\phi) + E_{g,\lambda,v}(\phi) + kE_u(\phi, u) \tag{13}$$

where ϕ is level set function, $k > 0$ is the weight of E_u.

3.1 User's Label Regularization Term Based on L2-Norm Distance

Firstly, we test the L2-norm distance to construct the user's label regularization term, which was also used together with graph-based methods in Shen et al. [17]. They defined a likelihood function $l(x, y)$ for each pixel (x, y), which represents the likelihood of the pixel belonging to the corresponding set:

$$(x, y) \in \begin{cases} F & \text{if } l(x, y) < 0 \\ B & \text{if } l(x, y) > 0 \\ unknown & \text{if } l(x, y) = 0 \end{cases} \tag{14}$$

$$E_u(l, u) = \sum_{x,y} \|l(x, y) - u(x, y)\|^2. \tag{15}$$

This regularization term is effective for normalized cut and graph cut. However, experimental results show that using (15) to calculate E_u in the level-set-based energy function (13) is not effective. From Fig. 2(b) we can see that the user's mark is of no use and there are a lot of scattered contours in the background. This is mainly because, $u(x, y) = 0$ means the label of (x, y) is unknown,while $\phi(x, y) = 0$ means (x, y) locates at current contour. The different hints give arise to a penalty when $u(x, y) = 0$ and $\phi(x, y) \neq 0$, and then disturbs the evolution process of level set function and mislead to another goal.

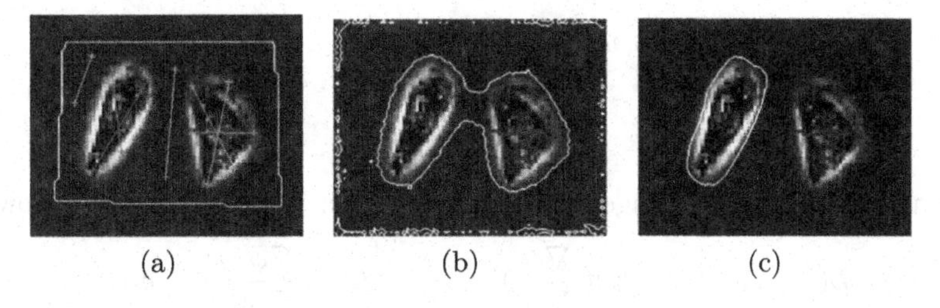

(a) (b) (c)

Fig. 2. The result of calculating E_u with L2-norm form and our method. (a) is the initial contours, where blue and red scribbles respectively provide hints for objectives to be extracted and to be avoided. (b) is the segmentation result by using L2-norm form. (c) is the segmentation result by using our user's label regularization term. (Color figure online)

3.2 Proposed Regularization Term

In order to avoid these scattered contours in the background and lead to the real goal, we should delete the side effect from the unmarked part. Here, we proposed to use the multiplication of the hard constraints from user's label and the level set function. We use the multiplication form to calculate $E_u(\phi, u)$, the value of the locations of the result matrix correspond to the unmarked part is zero, the evolution of level set function can only be impacted at the location of user's label, so that the evolution process can be accurately carried out under the influence of the user's labels.

Since the level set function ϕ and u have very different scales, it probably leads to failure if we get their direct multiplication. Here, we make a test as follows. We use the same initial method of level function in [8] that the value of the location inside the subject object is -4 and the outside is 4. So ϕ has different scale to u that led to a bad result of segmentation shown in Fig. 3.

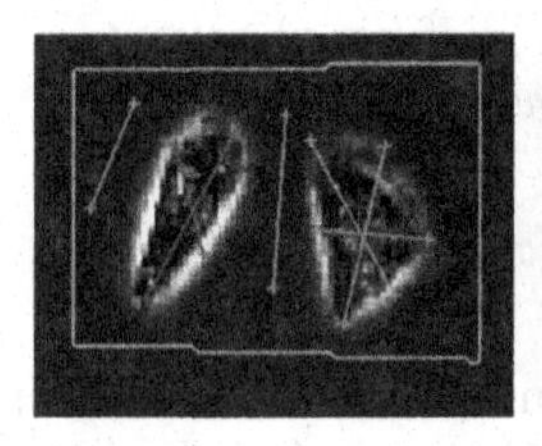
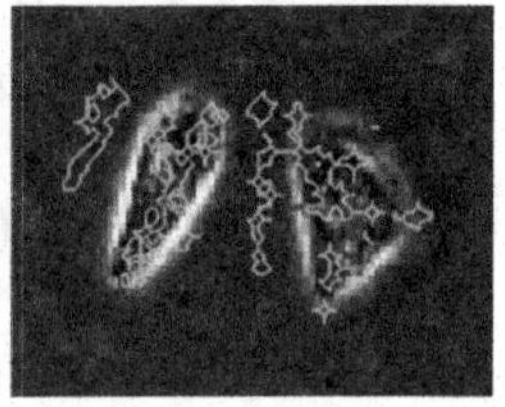

Fig. 3. The segmentation result with different scale. Different scale between level set function and user's label matrix will led to a bad result.

In order to make ϕ and u remain at the same scale, we first use a sigmoid function (17) (shown in Fig. 4) to normalized the ϕ. And then we shift it to keep the same scale with $u(x, y)$ in $[-1, 1]$, by using the expression $2\sigma(\phi) - 1$. In our proposed method, we construct the E_u as follow:

$$E_u(\phi, u) = \int -u(2\sigma(\phi) - 1)dxdy \tag{16}$$

$$\sigma(\phi) = \frac{1}{e^{-\phi} + 1} \tag{17}$$

According to Euler Lagrange equation (6), we get the new $L(\phi)$ as follow:

$$\begin{aligned} L(\phi) = &\mu[\Delta\phi - div(\frac{\nabla\phi}{|\nabla\phi|})] + \lambda\delta(\phi)div(g\frac{\nabla\phi}{|\nabla\phi|}) + vg\delta(\phi) \\ &+ 2k(u\sigma(\phi)(1 - \sigma(\phi))) \end{aligned} \tag{18}$$

The level set function process the evolution by (8). As shown in Fig. 2(c), using our method can get satisfactory result for users.

The method is user-friendly and can be cooperative by the users, it can make the evolution be processed repeatedly under the feedback of user until we get an ideal result, the segmentation process with user's feedback is expressed in Fig. 5.

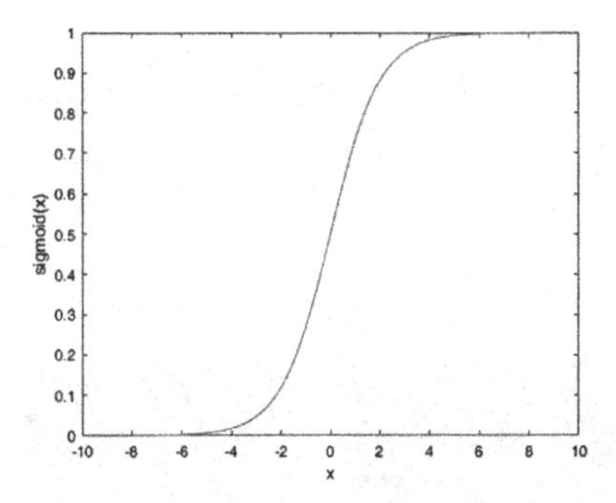

Fig. 4. The sigmoid function.

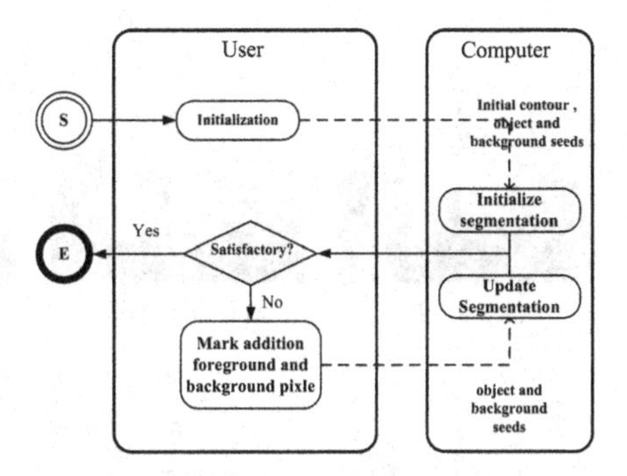

Fig. 5. Real-time feedback process of interactive level set segmentation.

4 Experiment Results

Our experiments are implemented from two aspects. First, we test on effect of the weight for our regularization term. And then, we gives experiments on interactive segmentation.

4.1 The Weight of the Regularization Term

In energy function (13), k indicates the extent to which user's mark plays a role in the segmentation process. We fix the weight of other part as mentioned in [8], and make experiments use different value of k. Figure 6 shows that we use different value of k to segment images with the same user's mark. From the result, we can see that at the beginning, with the increase of k, the segmentation results is significantly improved, and when k increases over a number range, the improvement effect is obviously reduced, even decreases. Experiment shows that when the value of k is selected to 100, we can achieve the best performance.

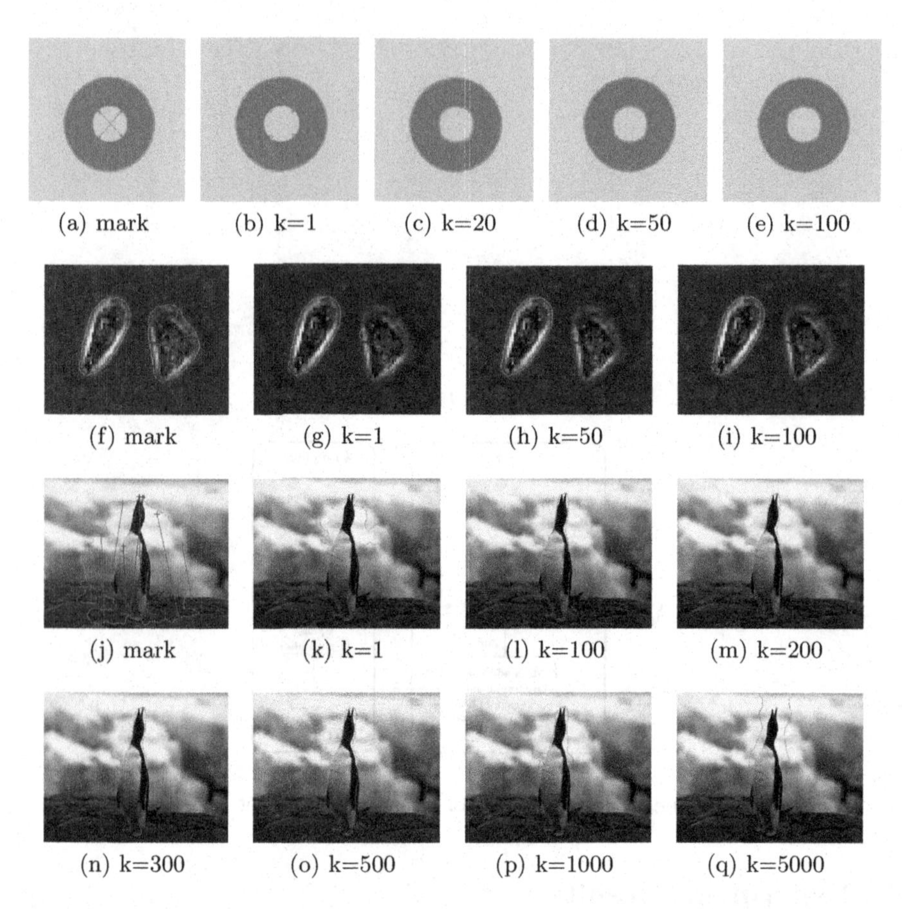

Fig. 6. The result of different value of k. For the first row and the second row, when $k = 100$ makes best performance. For the third and the forth rows, when $k \geq 100$ and $k \leq 1000$ the results have no significant changes, when $k = 5000$ the effect is obviously decreases.

4.2 Interactive Segmentation Process

Figure 8 shows the interactive segmentation process of objects in the 5 images. User draw a initial contour in the image first, then make some marks for foreground and background. After the first evolution, user can mark the pixels in the image that is error segmented and the algorithm can continue evolving on the basis of the segmentation result last time. By the feedback process several times repeatedly, the algorithm can give user their satisfied result. Obviously, if we make some mark as seeds before the first segmentation, the number of interactive would be decreased.

4.3 Behavioral Predictability

[11] given some evaluation indices for interactive segmentation algorithm, and shown some corresponding evaluation process. During the process of evaluation, many users were invited to do some interactive segmentation. According to the feedback of user, they are very sensitive to the operate experience. They liked that small localized marks only have a local effect. Conversely, users disliked algorithms in which small additions to the markup could cause large differences to the segmentation. Figure 7 shows the behavioral predictability of our method.

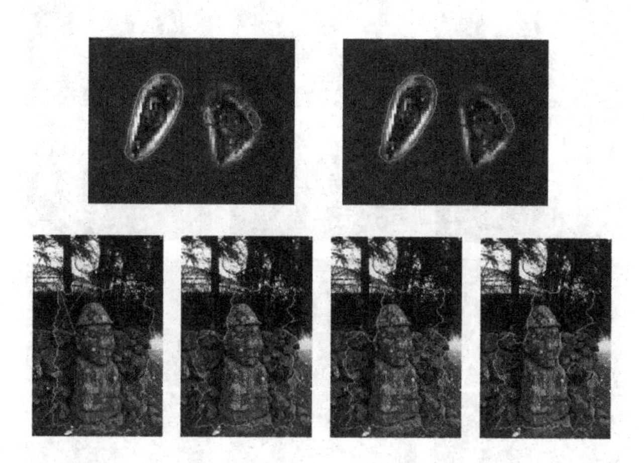

Fig. 7. Behavioral predictability. In the first row, only the part we marked as background was changed from foreground to background after segmentation process. From the pictures in the second row, we can see that the segmentation result only changed in the place we marked.

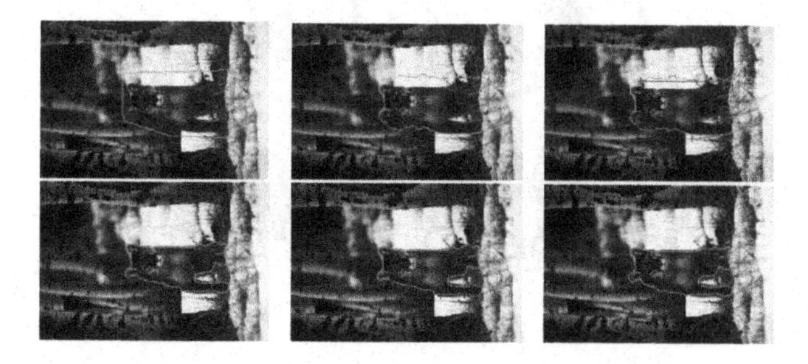

Fig. 8. Interactive segmentation process. Here we show five interactive segmentation processes respectively, the user can get their satisfied results by the feedback process. The object in the second segmentation process is a bird which is very similar with background, there is no clear boundaries, we can extract the bird after some the process of mark-evolution.

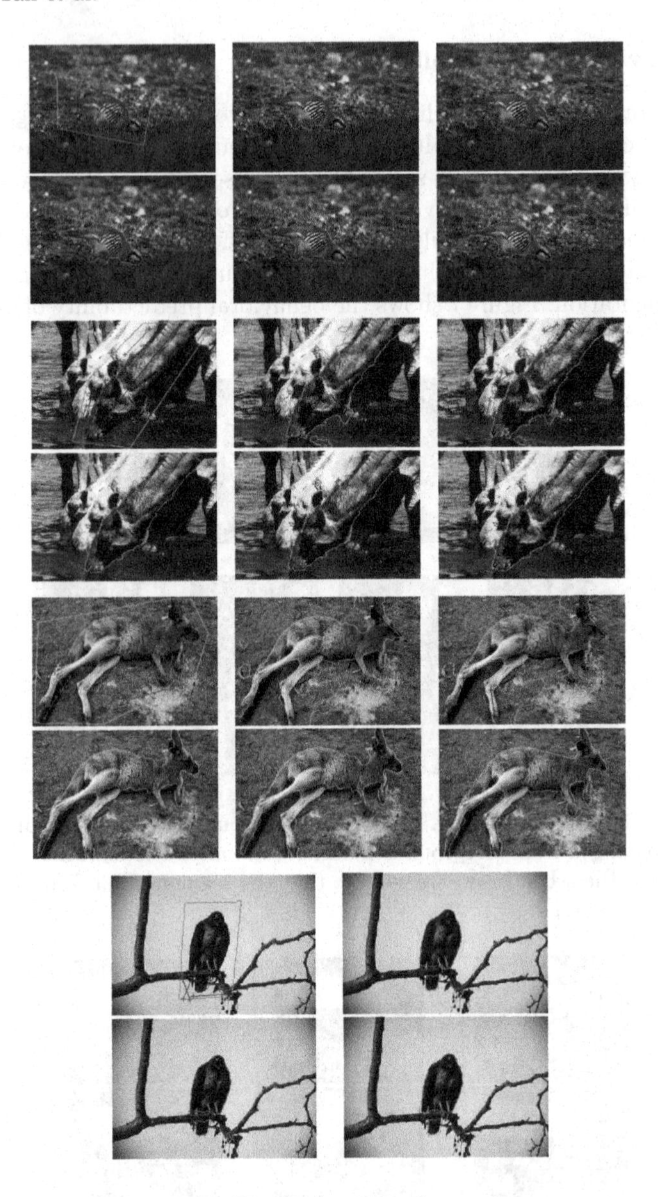

Fig. 8. (*continued*)

5 Conclusion

In this paper, we construct a novel regularization term with user's mark. Our method Solves the problem that user can't extract some specified object in the image due to the local optimum. Using our proposed method, the user can control the process of segmentation accurately. Our method also support the do the mark-evolution process repeatedly until the user get his satisfied result. In future, we will consider how to deal with region-based model by using our new regularization term.

References

1. Boykov, Y.Y., Jolly, M.P.: Interactive graph cuts for optimal boundary amp; region segmentation of objects in N-D images. In: Proceedings of Eighth IEEE International Conference on Computer Vision. ICCV 2001, vol. 1, pp. 105–112 (2001)
2. Caselles, V., Catté, F., Coll, T., Dibos, F.: A geometric model for active contours in image processing. Numer. Math. **66**(1), 1–31 (1993)
3. Caselles, V., Kimmel, R., Sapiro, G.: Geodesic active contours. Int. J. Comput. Vis. **22**(1), 61–79 (1997)
4. Chan, T.F., Vese, L.A.: Active contours without edges. IEEE Trans. Image Process. **10**(2), 266–277 (2001)
5. He, J., Kim, C.-S., Jay Kuo, C.-C.: Interactive image segmentation techniques. SpringerBriefs in Electrical and Computer Engineering. Springer Nature, Heidelberg (2013). pp. 17–62
6. Kass, M., Witkin, A.P., Terzopoulos, D.: Snakes: active contour models. Int. J. Comput. Vis. **1**(4), 321–331 (1988)
7. Li, C., Kao, C.Y., Gore, J.C., Ding, Z.: Implicit active contours driven by local binary fitting energy. In: Proceedings of IEEE Conference on Computer Vision and Pattern Recognition, pp. 1–7, June 2007
8. Li, C., Xu, C., Gui, C., Fox, M.D.: Level set evolution without re-initialization: a new variational formulation. In: Proceedings of IEEE Computer Society Conference on Computer Vision and Pattern Recognition (CVPR 2005), vol. 1, pp. 430–436, June 2005
9. Li, C., Xu, C., Gui, C., Fox, M.D.: Distance regularized level set evolution and its application to image segmentation. IEEE Trans. Image Process. **19**(12), 3243–3254 (2010)
10. Liu, Y., Yizhou, Y.: Interactive image segmentation based on level sets of probabilities. IEEE Trans. Vis. Comput. Graph. **18**(2), 202–213 (2012)
11. McGuinness, K., O'Connor, N.E.: Toward automated evaluation of interactive segmentation. Comput. Vis. Image Underst. **115**(6), 868–884 (2011)
12. Mumford, D., Shah, J.: Optimal approximations by piecewise smooth functions and associated variational problems. Commun. Pure Appl. Math. **42**(5), 577–685 (1989)
13. Nguyen, T.N.A., Cai, J., Zhang, J., Zheng, J.: Robust interactive image segmentation using convex active contours. IEEE Trans. Image Process. **21**(8), 3734–3743 (2012)
14. Niessen, W.J.: Geometric partial differential equations and image analysis. IEEE Trans. Med. Imaging **20**(12), 1426–1427 (2001)
15. Osher, S., Sethian, J.A.: Fronts propagating with curvature-dependent speed: algorithms based on Hamilton-Iacobi formulations. J. Comput. Phys. **79**(1), 12–49 (1988)
16. Sethian, J.A.: Curvature and the evolution of fronts. Commun. Math. Phys. **101**(4), 487–499 (1985)
17. Shen, J., Du, Y., Li, X.: Interactive segmentation using constrained Laplacian optimization. IEEE Trans. Circ. Syst. Video Technol. **24**(7), 1088–1100 (2014)

18. Vese, L.A., Chan, T.F.: A multiphase level set framework for image segmentation using the Mumford and Shah model. Int. J. Comput. Vis. **50**(3), 271 (2002)
19. Wang, T., Wang, H., Fan, L.: A weakly supervised geodesic level set framework for interactive image segmentation. Neurocomputing **168**, 55–64 (2015)
20. Zhu, Y., Cheng, S., Goel, A.: Interactive segmentation of medical images using belief propagation with level sets. In: 2010 IEEE International Conference on Image Processing. Institute of Electrical and Electronics Engineers (IEEE), September 2010

Modified Object Detection Method Based on YOLO

Xia Zhao(✉), Yingting Ni, and Haihang Jia

School of Electronics and Information Engineering, Tongji University, Shanghai, China
zhaoxia@tongji.edu.cn

Abstract. YOLO (You Only Look Once), the 2D object detection method, is extremely fast since a single neural network predicts bounding boxes and class probabilities directly from full images in one evaluation. However, it makes more localization errors and its training velocity is relatively slow. Benefiting from the thoughts of cluster center in super-pixel segmentation and anchor box in Faster R-CNN, in this paper, we propose a modified method based on YOLO (shorted for M-YOLO). First, we substituted YOLOs last fully connected layer for a convolutional layer, on which the cluster boxes (some anchor boxes centered on cluster center) can completely cover the whole image at the beginning of training. As a result, the new structure can speed up the training process. Second, we increase the number of divided grids i.e. cluster centers, from 7×7 to the maximum 17×17, as well as the number of predicted bounding boxes, i.e. anchor boxes, from 2 to the maximum 9 for each grid cell. The measure can improve the IOU performance. Simultaneously, we also put forward a new kind of NMS (non-max suppression) to solve the problem aroused by M-YOLO. The experimental results show that M-YOLO improves the localization accuracy by about 10%, the convergence speed of the training process is also improved.

Keywords: Deep learning · Object detection · Cluster center
Anchor box

1 Introduction

Convolution Neural Networks (CNNs) had been widely used in 1990s, but then fell out of fashion with the rise of Support Vector Machines etc. [1]. In 2012, Hinton et al. won the first place on the ImageNet Large Scale Visual Recognition Challenge (ILSVRC) by CNN. Their success led to a new wave of research, and more and more researchers use CNNs to improve the performance of image recognition and object detection.

In [1], Girshick et al. propose R-CNN (Region proposal Convolution Neural Networks). The method uses Selective Search to first generate around 2000 region proposals for the input image and computes features for each proposal using a CNN. Then the category-specific linear SVMs are employed to classify each

J. Yang et al. (Eds.): CCCV 2017, Part III, CCIS 773, pp. 233–244, 2017.
https://doi.org/10.1007/978-981-10-7305-2_21

region. After classification, post-processing is used to refine the bounding boxes and eliminate redundant detections. Compared with the traditional detection methods R-CNN achieves excellent object detection accuracy on PASCAL VOC 2007 - about 58.5%. But it has a notable drawback that detection is slow C needing 47s for each image, because R-CNN performs a ConvNet forward pass for each region proposal, without sharing computation.

For fixing the disadvantages of R-CNN, [2] put forward the Fast R-CNN. The net-work takes a whole image and multiple regions of interest (RoIs) as input. The method first computes a convolutional feature map for the entire input image using the CNN. Then, each object proposal, the RoI pooling layer extracts a fixed-length feature vector from the feature map. Finally, each feature vector is fed into fully connected layers (FCs), which has two output vectors: softmax probabilities and per-class bounding-box regression offsets. Compared with R-CNN, the algorithm only takes once feature extraction for an input image, thus increasing the detection speed. At runtime, the network processes images in 0.3s (excluding object proposal time) while achieving higher detection quality with mAP 66%.

Though Fast R-CNN has reduced the detection time, it still depends on region proposal algorithms to predict object bounds, which is the computational bottleneck in detection systems. In [3], Girshick et al. introduce a Region Proposal Network (RPN) that is used to predict object bounds. The RPN is a fully convolutional network sharing convolutional features with the detection network, so that the marginal cost for generating proposals is small. By merging RPN and Fast R-CNN into a single network, Ross proposes an end-to-end object detection framework, named Faster R-CNN. The approach comparatively improves detection speed, up to 7 frames per second (FPS), while achieving object detection accuracy with mAP 73.2%. However, in real-time detection tasks, the speed is still unable to meet requirements.

In [4], Redmon presents YOLO (You Only Look Once), a new method to object detection. The system models object detection as a regression problem. It divides the image into 7×7 grids and for each grid cell predicts 2 bounding boxes, confidence for those boxes and 20 class probabilities. Since the method unifies the separate components of object detection into a single neural network, the YOLO model processes images in real-time at 45 FPS with mAP 63.4%. However, instead of generating region proposals, YOLO directly predicts multiple bounding boxes from the input image, which makes more localization errors and slow convergence speed.

SSD (Single Shot MultiBox Detector) is the current state-of-the-art object detection system [5]. The approach evaluates a set of default boxes of different aspect ratios at each location in several feature maps with different scales. At prediction time, the network generates category scores and box offsets for each default box by using small convolutional filters. Since it eliminates bounding box proposals and the subsequent pixel or feature resampling stage, SSD is much faster than methods that utilize proposal step and has comparable accuracy (58 FPS with mAP 72.1% on VOC2007 test, vs Faster R-CNN 7 FPS with mAP 73.2%). But SSD still performs poorly, when detecting small objects.

In the practice application, many scenes contain small objects to be detected and the intersection over union (IOU) between the predicted box and the ground truth is very important for some capture operations. Considering these factors and the characteristics of YOLO, we will have some modifications on YOLO and the new approach is named M-YOLO. First, combining the thought of the cluster center in super-pixel image segmentation with the thought of the anchor box in Faster R-CNN, M-YOLO generates the cluster boxes which can completely cover the whole image. Therefore, there is a smaller gap between predicted box and ground truth box than that of YOLO at the beginning of training. Second, the method substitutes YOLOs last fully connected layer for a convolutional layer, which generates category scores and box offsets for each cluster box. The new structure can speed up the training process. At the same time, the C-NMS (class-based non-max suppression) is designed, in order to solve the problem that the same object is identified as different categories. Compared with YOLO, M-YOLO improves the location accuracy and convergence speed, while keeping the detection accuracy.

The main contents of this paper are as follows: In Sect. 2, we briefly introduce YOLO as well as its problems. In Sect. 3, the new object detection approach M-YOLO is described in detailed. The performances of YOLO and M-YOLO are compared in Sect. 4.

2 YOLO

The detection process of YOLO is shown in Fig. 1. First of all, the input image is resized to 448×448. Then runs a single convolutional network on the image, bounding boxes as well as its confidence scores will be obtained. Finally, the NMS (non-max suppression) is used to threshold the resulting detections by the models confidence, the final class probabilities and bounding box coordinates are obtained as shown in the most right of Fig. 1.

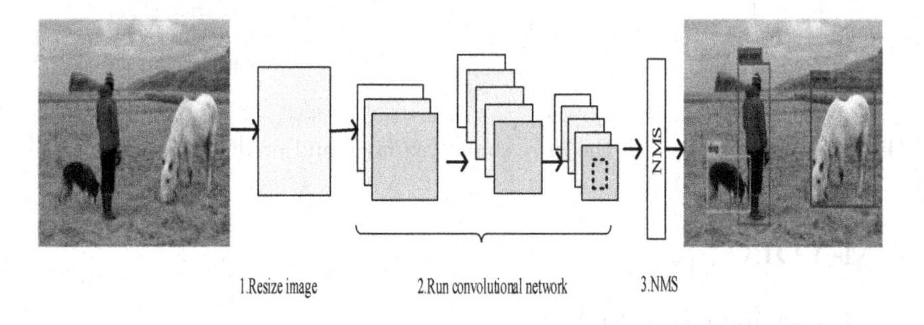

Fig. 1. The detection process of YOLO

YOLO divides the input image into 7×7 grids. Each grid cell corresponds to 2 bounding boxes. Each grid cell predicts one set of class probabilities (including

20 classes) regardless of the number of boxes. And each bounding box consists of 5 predictions: x, y, w, h and confidence. The (x, y) coordinates represent the center of the box relative to the bounds of the grid cell; the (w, h) represent the width and height of the box relative to the whole image. Thus, each grid has $(4 + 1) \times 2 + 20 = 30$ predictions and whole image has $7 \times 7 \times 30 = 1470$ predictions.

YOLOs CNN includes 25 convolutional layers, 4 pooling layers, 1 dropout layer, 1 fully connected layer and 1 detection layer. The dimension of the fully connected layer is 1470, denoting the position and confidence of the predicted bounding boxes and its class probabilities. For randomly initialized CNN, the output of the fully connected layer is also random at the beginning of the training process, i.e. there is a large gap between predicted box and ground truth box in the beginning. The sketch shown in Fig. 2, reveals the mapping between the fully connected layer and predicted box on the image. It needs a lot of iterations to get close to the ground truth box, resulting in slow convergence speed.

On the other hand, YOLO proposes far fewer bounding boxes, only $7 \times 7 \times 2 = 98$ per image, so its localization error is big.

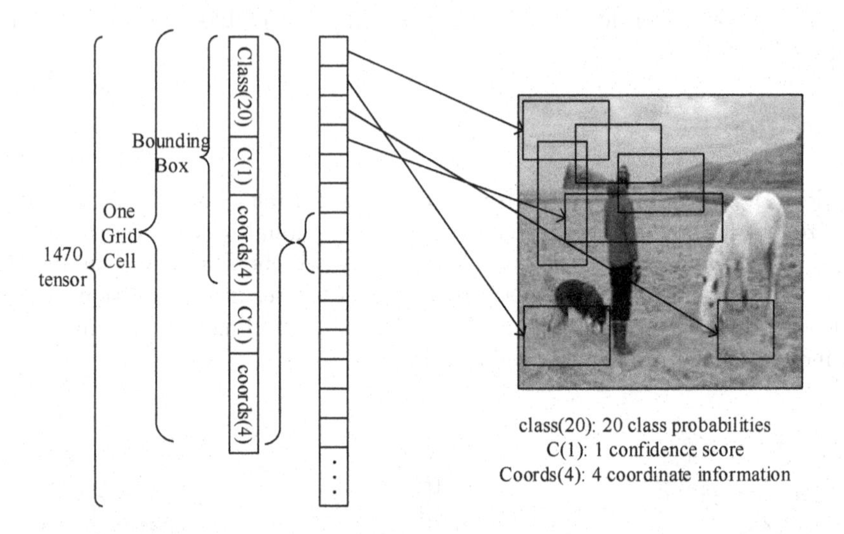

Fig. 2. Mapping between the fully connected layer and predicted box of YOLO

3 M-YOLO

3.1 The Principle of M-YOLO

In this paper, we propose a modified object detection method based on YOLO, i.e. M-YOLO, to improve the location accuracy and convergence speed. Benefiting from the thoughts of cluster center in the super-pixel segmentation and anchor box in Faster R-CNN, we introduce novel bounding boxes, called cluster boxes, that serve as references at multiple scales and aspect ratios. Simultaneously, the new approach uses a convolutional layer to replace YOLOs fully connected layer.

The Anchor Box is first introduced in the Faster R-CNN [3]. Each feature point in the last convolution layer of RPN is as an anchor. And for each anchor, 9 kinds of anchor boxes can be pre-extracted by using 3 different scales and 3 different aspect ratios (Fig. 3). Compared with YOLO's 2 predicted bounding boxes, the anchor boxes of Faster R-CNN take into account the objects with different scales and aspect ratios. Therefore, on the basis of YOLO, we prepared to increase the number of predicted borders at multiple scales and aspect ratios, in order to improve the location accuracy.

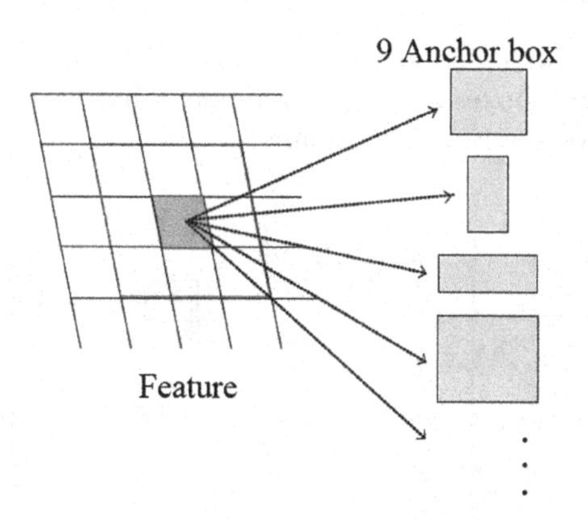

Fig. 3. Anchor boxes of the Faster R-CNN

Additionally, if the anchors in Faster R-CNN are mapped to the original image, they are some evenly distributed feature points in the image. Inspired by this, we managed to directly select some evenly distributed points from the input image, and take them as the center of predicted borders.

Super-pixel refers to an irregular pixel block with a certain visual representation, which consists of adjacent pixels with similar characteristics such as texture, color, brightness and so on [6]. In SLIC-based super-pixel segmentation, the first step is to select orderly and evenly distributed points in the input image as the cluster centers. The cluster centers can be fine-tuned to avoid appearing on the boundary of different color blocks. Using cluster centers as the initial centers of the iterative algorithm, the super-pixel with cluster center can be obtained after several iterations (Fig. 4). Due to the character of cluster centers, our method directly uses them as the center of predicted borders.

In this paper, our approach selects $n \times n$ cluster centers on the original image and predicts m bounding boxes with multiple scales and aspect ratios at each clustering center. The scale of the bounding boxes can be 1×1, 2×2, 4×4, the aspect ratios can be 1:1, 1:2, 2:1. The sketch map of cluster boxes is shown in Fig. 5, where $n = 7$, $m = 9$. The number of cluster centers and predicted boxes is important parameters affecting the detection results. We increase the number

Fig. 4. Cluster centers and Super-pixel segmentation (Color figure online)

of cluster centers, from 7×7 to the maximum 17×17, as well as the number of predicted bounding boxes, from 2 to the maximum 9 for each grid cell. The measure can improve the IOU performance.

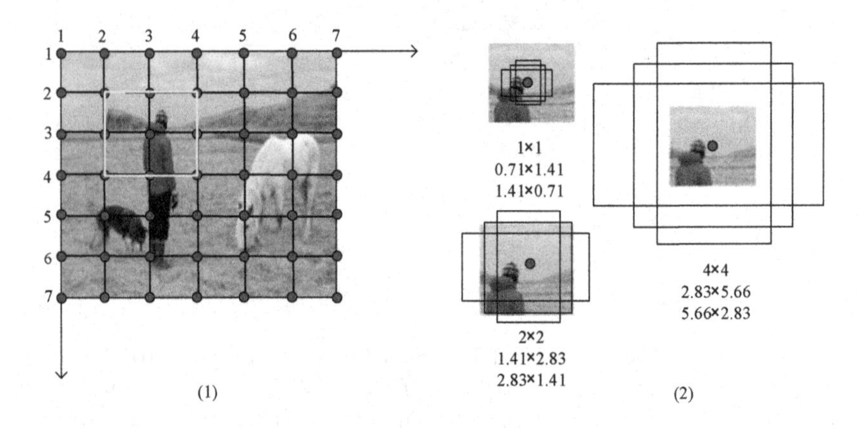

Fig. 5. The sketch map of cluster boxes

On bounding box predictions, each grid cell of YOLO shares 20 classification information since it only predicts two boxes and can only have one class. This constraint limits the number of nearby objects that the model can predict. In our method, all the boxes of each cluster center do not share 20 classification information, in order to detect nearby different objects.

M-YOLO uses the cluster boxes as the references and substitutes YOLOs last fully connected layer for a convolutional layer, on which the cluster boxes have stable sequences. The network generates category scores and box offsets for each cluster box. Since cluster boxes are orderly and evenly distributed in the input image, the gap between predicted box and ground truth box is as small as possible at the beginning of the training process. Thus this model performs well when trained and benefits convergence speed. The sketch mapping between the new conventional layer and predicted box on the image is shown in Fig. 6.

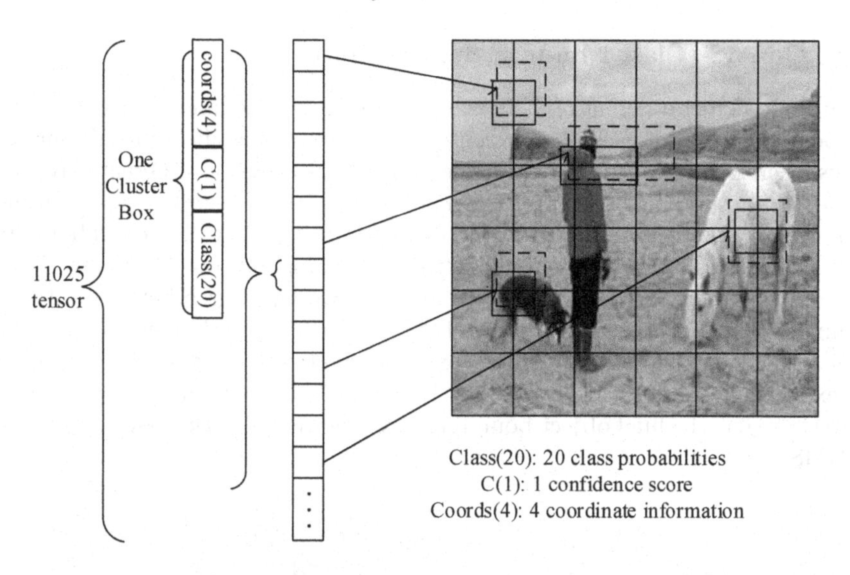

Fig. 6. Mapping between the new conventional layer and predicted box of M-YOLO

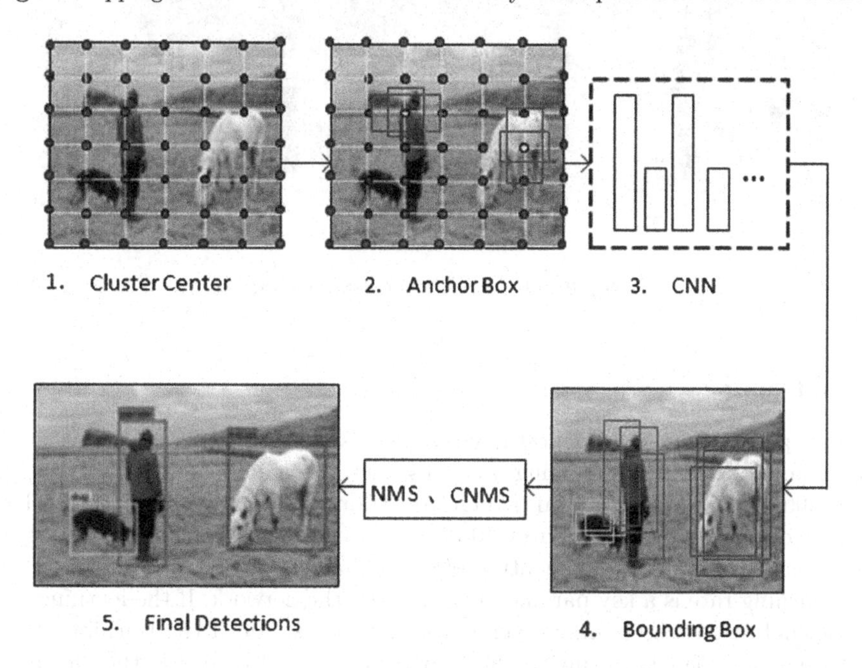

Fig. 7. The object detection process of M-YOLO

The object detection process of M-YOLO is shown in Fig. 7. It selects $n \times n$ clustering centers on the original image and predicts m bounding boxes with multiple scales and aspect ratios at each clustering center. Then, the image is input to the CNN to predict confidence, class probabilities and box offsets for each cluster box. Finally, use NMS to eliminate redundant detections.

3.2 C-NMS (Class Non-max Suppression)

In order to detect different objects in a wide range of scales and aspect ratios, the 9 bounding boxes of each cluster center do not share classification information. Thus, there is a problem that the same object is identified as different categories by adjacent bounding boxes. As shown in Fig. 8(1), the object 'horse' is identified as horse or cow. To solve this problem, we introduce the C-NMS. For all bounding boxes of adjacent cluster centers, if the class probabilities is greater than a certain probability score (we use 0.2) and the overlap degree between adjacent bounding boxes is greater than a threshold (we use 0.5), we think the same object is identified as different categories by redundant bounding boxes. According to experimental experience, we directly choose the bounding box having the largest area as the final object bounding box. Figure 8(2) is the result after using C-NMS.

Fig. 8. The result after using C-NMS

3.3 Training

In this paper, the experimental environment is Ubuntu and we use the Darknet framework for all training and testing. We train the network on the VOC2007train + VOC2007val + VOC2012train + VOC2012val dataset, which consist of about 15k images over 20 object categories. Throughout training we use a batch size of 64, a momentum of 0.9 and a decay of 0.0005.

Learning rate is a key parameter in training the network. If the learning rate is too high, the weights may diverge optimal values; and if the learning rate is too small, it takes long time to find optimal values. Figure 9 is the change of learning rate for M-YOLO and the LR is 0.0005. According to the increase in iteration times, learning rate is constantly changing. To prevent model divergence caused by unstable gradients, we start at a relatively small learning rate. Then we slowly raise the rate from LR to 10LR. We continue training with 10LR for 14000 iterations, then LR for 10000 iterations, and finally 0.1LR for 10000 iterations.

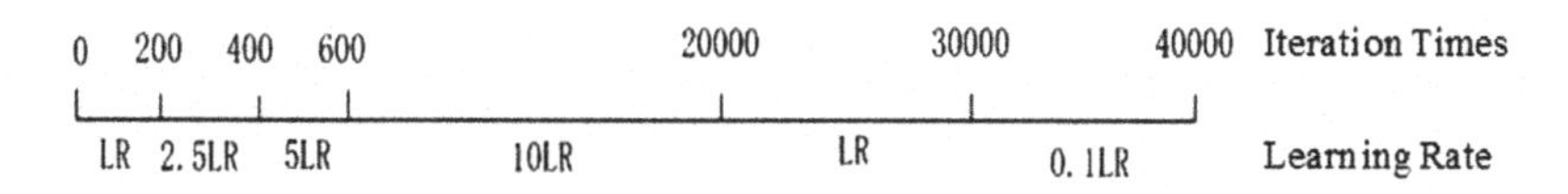

Fig. 9. The change of learning rate

4 Experimental Results

In our model, we need to determine two parameters: the number of cluster centers (n) and that of bounding boxes for each cluster center (m). The number of cluster centers changes from 7×7 to 17×17. The bounding boxes are set according to the setting rules of anchor boxes in Faster R-CNN. Simultaneously, in view of YOLO's 2 boxes and Faster R-CNNs 9 boxes, we also set the number of boxes to 5 for comparison. We comprehensively evaluate it on the VOC2007 detection dataset, consisting of about 5k test images. Performance metrics include the mean average precision (mAP), recall rate (R), the intersection over union (IOU). We also provide results on the VOC 2007 for YOLO.

As we can see from Table 1, the model M-YOLO-15-5 ($m = 15$, $n = 5$) scores 61.07% mPA, which is the best of all models and slightly higher than YOLO. The model M-YOLO-17-9 ($m = 17$, $n = 9$) has highest IOU (79.99%). Compared with YOLO, it improves IOU about 10% and has more accurate object bounding boxes. Additionally, M-YOLO-17-9 also pushes recall rate to 89.95%, which is more 13% than that of YOLO.

Figure 10(1) is the curve of mAP with the number of cluster centers n and the bounding box number m. It can be seen that mAP reaches the peak, when $n = 15$, $m = 5$. Figure 10(2) shows the curve of IOU with the number of cluster

Table 1. PASCAL VOC2007 test detection results

Method	Boxes	Cluster center	mAP	IOU	Recall
YOLO	2	7	60.36	69.66	76.55
M-YOLO	5	7	46.86	63.59	68.85
		9	54.15	72.20	80.47
		11	57.18	74.46	82.52
		13	60.40	76.78	85.58
		15	61.07	77.88	87.56
		17	60.74	78.71	87.93
	9	7	47.45	66.77	69.48
		9	55.31	74.87	81.72
		11	58.43	77.04	84.50
		13	60.54	78.65	87.11
		15	60.77	79.65	88.88
		17	60.25	79.99	89.53

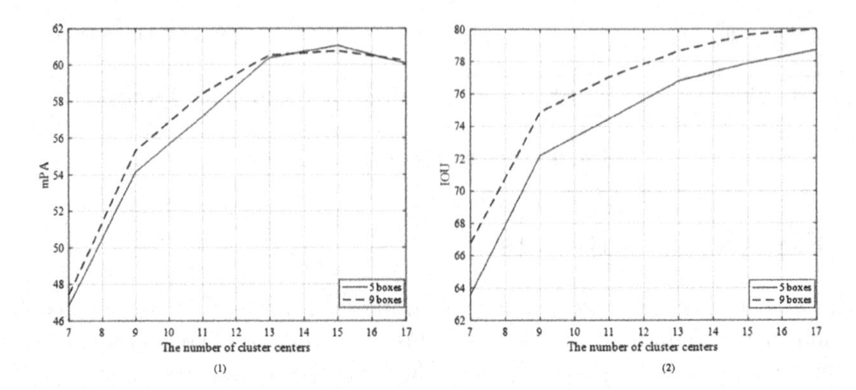

Fig. 10. The curve of performance metrics with the number of cluster centers (n) and the border number (m)

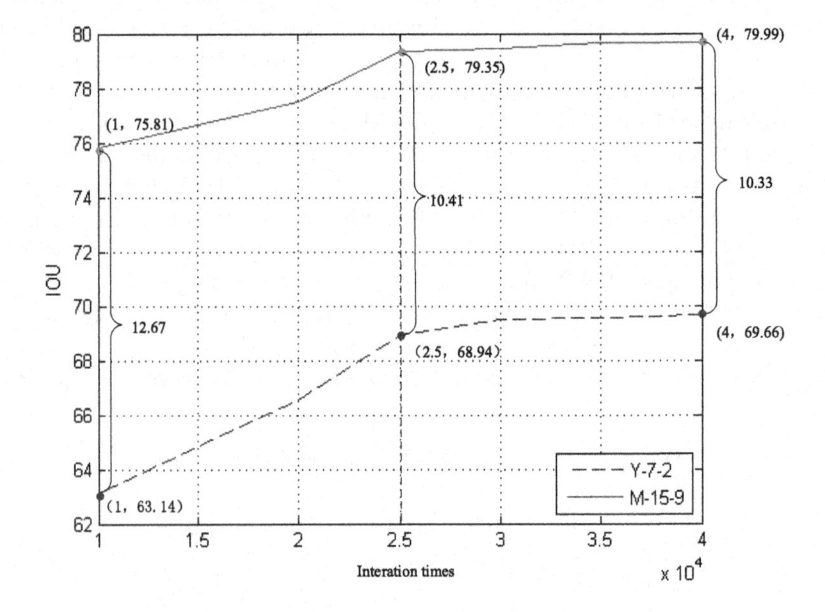

Fig. 11. The IOU values of different iteration times

centers n and the border number m. We can see that the IOU tends to be stable as the bounding box number increasing, and the IOU value of 9 bounding boxes is higher than that of 5 bounding boxes as a whole. Figure 11 shows the IOU values of different iteration times. When the iteration times are 10,000, the IOU of M-YOLO-15-5 is 12.67% higher than that of the YOLO. However, when the iteration times are 40,000, the IOU of M-YOLO-15-5 is 10.33% higher than that of the YOLO. It can be seen that the model M-YOLO-15-9 accelerates the convergence speed. Figure 12 shows selected examples of object detection results on the VOC 2007 test set using the M-YOLO system.

Fig. 12. Examples of object detection results of M-YOLO

5 Conclusion

In this paper, based on the YOLO, we propose an improved object detection method M-YOLO, which utilizes the cluster center in the super-pixel segmentation and the Anchor Box of Faster RCNN. The experimental result confirms that M-YOLO improves the accuracy of object bounding boxes by about 10% and the recall rate, while keeping the detection accuracy.

References

1. Girshick, R., Donahue, J., Darrell, T., Malik, J.: Rich feature hierarchies for accurate object detection and semantic segmentation. In: Computer Vision and Pattern Recognition, pp. 580–587 (2014)
2. Girshick, R.: Fast R-CNN. In: International Conference on Computer Vision, pp,. 1440–1448 (2015)
3. Ren, S., He, K., Girshick, R., Sun, J.: Faster R-CNN: towards real-time object detection with region proposal networks. IEEE Trans. Pattern Anal. Mach. Intell. **39**(6), 1137–1149 (2017)

4. Redmon, J., Divvala, S., Girshick, R., Farhadi, A.: You only look once: unified, real-time object detection. In: Computer Vision and Pattern Recognition, pp. 779–788 (2016)
5. Liu, W., Anguelov, D., Erhan, D., Szegedy, C., Reed, S., Fu, C.Y., Berg, A.C.: SSD: single shot multibox detector. ECCV **1**(2016), 21–37 (2016)
6. Achanta, R., Shaji, A., Smith, K., Lucchi, A., Fua, P., Süsstrunk, S.: SLIC superpixels compared to state-of-the-art superpixel methods. IEEE Trans. Pattern Anal. Mach. Intell. **34**(11), 2274–2282 (2012)

Scene Text Detection with Text Statistical Characteristics and Deep Neural Network

Yanyun Qu, Xiaodong Yang(✉), and Li Lin

Xiamen University, Xiamen, China
yyqu@xmu.edu.cn, 4490736262@qq.com, 379484437@qq.com

Abstract. Scene text recognition is a hot topic in the field of computer vision. Inspired by the success of the Single Shot Multibox Detection (SSD) on generic object detection, the architecture of SSD is implemented on scene text detection. SSD does not do well on text detection, because scene text as an object is usually smaller than a generic object and SSD cannot detect small objects well. Thus, the statistic analysis for scene text is made. Based on statistic characteristics of scene text, we propose a method named Text-SSD to detect scene text. Moreover, in order to boost the detection accuracy, multi-scale image are used to learn the multi-scale models. The voting based non-maximum suppressing is made for a candidate text region. The experimental results show that our method achieved the state-of-the-art performance on the benchmark dataset ICDAR2013 in the detection accuracy. Moreover, when using a single model, our method achieves the fastest speed compared with several latest text detection method based on deep neural network. Thus, experimental results demonstrate our method is efficient on scene text detection.

Keywords: Scene text detection · Text statistical characteristic · SSD

1 Introduction

Due to its widely applications in image retrieval, visual navigation, and scene understanding, etc., scene text detection has attracted more and more attentions [10–12]. Though tremendous efforts have been devoted to text detection, scene text detection in a wild is still a challenging task because of unconstrained environments. Moreover, scene text is flexible distributed with the changes of fonts, style and layout, and so on. Until now, the text detection accuracy is not high, which will greatly influence the successive text recognition.

Great progresses have been witnessed in object detection in recent years. Inspired by the successes of deep learning on image classification and speech recognition, object detection is solved in an end-to-end way based on a convolutional neural network, which is different from the pipeline scheme of traditional object detection methods. Especially, Single Shot Multibox Detector (SSD) has

J. Yang et al. (Eds.): CCCV 2017, Part III, CCIS 773, pp. 245–254, 2017.
https://doi.org/10.1007/978-981-10-7305-2_22

achieved the breakthrough performance in detection accuracy. And SSD is suitable for generic object detection such as dogs, bikes, persons, etc. However, it cannot do well on text detection if it is implemented directly on scene text detection. The failure is due to the following two reasons: (1) Scene text is usually smaller than generic objects. In ILSVRC and VOC, an object often makes up no smaller than 10% of the whole image. However, scene text often makes up much smaller than a generic object. (2) The aspect ratio of scene text is different from a generic object. Scene text is usually like a horizontal thin bar, while a generic object can be bounded by a rectangle. In order to make SSD correctly detect text, we treat scene text as an object, and we mine the text statistical characteristics and design a text-specified default box according to text statistical characteristics in the SSD framework, which is named Text-SSD. The text-specified default box can efficiently get the discriminative features of scene text, and make significant improvements of scene text detection in accuracy and speed.

The remainder of this paper is organized as follows. We introduce related work in Sect. 2. In Sect. 3, we detail how to mine the text statistical characteristics and how to design the default boxes. Experimental results are presented in Sect. 4, and we conclude in Sect. 5.

2 Related Work

In recent years, efforts have been devoted to scene text detection [1–8]. There are two typical classes of text detection methods, one of which are traditional methods and the other of which are deep learning based methods. Most of the traditional methods solve the text detection problem in a pipeline way. Yao et al. [11] used SWT to extract the connected components and then filtered out non-character regions by using a Random Forest classifier combined with character-level features. After that, they connected the candidate character regions into strings according to the similarity of geometric structures, spatial layouts, etc. Finally, they filtered out non-text regions by using a string-level classier. Neumann and Matas [10] extracted External Regions (ER) of an image as candidate regions. After that, the incremental features of an ER are extracted, and then, a two-stage classifier is learned from the training data to filter out the non-text candidate ER. In the first stage, a real AdaBoost classifier formed by decision trees, is implemented. In the second stage, a SVM classifier with RBF kernel is implemented. Finally, the exhaustive search [9] is used to find out the scene text regions. This method is more efficient and robust compared with other text detection methods before it.

With the upsurge of deep learning, deep learning based text detection methods have surged. They can be divided into two classes according to their framework: pipeline methods and end-to-end methods. The former uses the traditional bottom-up text detection framework, in which CNN is used to extract the text features instead of hand-crafted features [1,4,16]. The latter adopts the end-to-end framework which are used for generic object detection or segmentation

[14,15,17,18] for scene text detection. Compared to pipeline methods, they have significantly improved scene text detection in both accuracy and speed. Huang et al. [1] adopted CNN combined with MSER for text detection. In this method, MSER-tree was firstly constructed according to which candidate regions were extracted. After that, CNN was implemented to filter out the non-text regions. Furtherly, text lines were constructed based on simple features such as intensity, color, height and width etc. This method improves the robustness of MSER-based text detection methods. Jaderberg et al. [16] used CNN together with sliding window scheme for scene text detection. Each sliding window was scored by the CNN classifier and a saliency map is computed. According to the saliency map, text lines are formed. In addition, CNN can be extended to solve the 62-class classification problem. In other word, this method can not only be used for text detection, but also for text recognition in an end-to-end way.

He et al. [17] firstly transforms the text detection problem to a segmentation problem. Fully Connected Network [24] is used for text detection. The method used two CNNs, the first CNN was used to detect a text block, and the second CNN was used to extract the text lines in each text block.

Yao et al. [18] also regarded the text detection as a semantic segmentation problem which was solved by a deep neural network named HED [19]. And it can be used for multi-task text recognition, such as text proposal detection, single character recognition, and multiple character detection.

Tian et al. [15] proposed the Connectionist Text Proposal Network (CTPN) combined with the Recurrent Neural Network (RNN) for text detection. CTPN firstly used improved RPN (region proposal network) network to select the character candidate, and then used anchors with fixed width to detect the candidate area of text. After that, the feature vectors corresponding to the anchors in the same row are concatenated into a sequence, which is then put into RNN for text recognition. At last, the fully connection layer was used for classifying and regression, and the correct candidate regions are merged into a text line.

Liao et al. [14] modified the Single Shot MultiBox Detector (SSD) [13] for text detection. The convolutional filters and the aspect ratio of the sliding window are adjusted. Furthermore, a multi-scale inputs were used for improving the detection effect, but the speed is greatly reduced. Liao's method is the closest related work to ours. However, their method does not design the default box depending on the text statistical characteristics but depending on the empirical results. Thus, the model can be improved for text detection.

3 Multi-scale Text-SSD

In this section, we introduce how to modify the original SSD for scene text detection. As we know, the text scale is a very important factor for text detection, and any single model of text detection cannot do well on all text scales. Thus, in order to boost the detection accuracy, a multi-scale SSD model is designed for scene text detection, which is named muti-scale Text-SSD. As shown in Fig. 1, the proposed approach has three steps: (1) An input image is rescaled into three scales: $512 * 512$, $700 * 512$, $700 * 300$. (2) For each scale, a SSD model is designed for text detection. (3) A voting fusion strategy is used to obtain the text candidate regions.

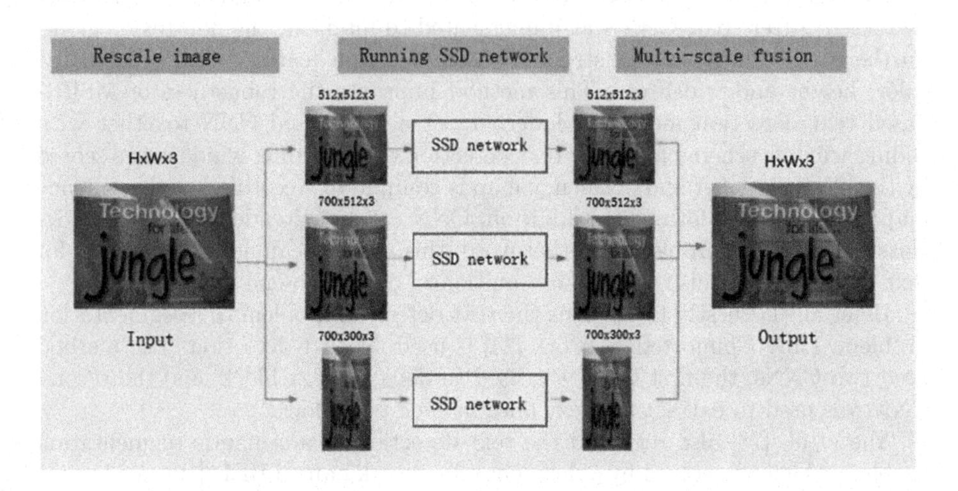

Fig. 1. The framework of multi-scale Text-SSD

3.1 Brief Review of SSD

SSD includes five parts: (1) The first part includes the first five convolution modules (conv1 to conv5_3) of VGG-16. (2) In the second part, the two fully connected layers fc_6, fc_7 of VGG-16 are modified as full convolution layers. (3) In the third part, four convolution modules from conv8 to conv11 are added, each of which contains 1×1 convolutions and 3×3 convolutions. (4) In the fourth part, the classification decision and its responding regression boxes of the last four convolution layers from conv8 to conv11 are output, in which 3×3 convolution modules are used to predict the classification label and the position of candidate regions. (5) Finally, the non-maximal suppression is performed to obtain the final result. SSD uses the sliding window scheme on a feature map instead of the original image. In addition, SSD uses the idea of Anchor in Faster R-CNN, so that the location of each sliding window corresponds to an anchor with different scales and aspect ratios, which is called default box. Default boxes are used in the six convolution layers in SSD: conv4_3, conv7, conv8_2, conv9_2, conv10_2, and conv11_2, and in each of the last four convolution layers, the default boxes are only used in the $3 * 3$ convolution for class label prediction. There are five initial aspect ratios in SSD: $a_r = \{1, 2, 3, \frac{1}{2}, \frac{1}{3}\}$. The parameter s_k is denoted the scale of the kth convolution layer. In the shallowest convolution layer, s_{min} is set to be 20%, and in the deepest convolution layer, s_{max} is set to be 90%. In the middle convolution layers, the scale is formulated as,

$$s_k = s_{min} + \frac{s_{max} - s_{min}}{m - 1}(k - 1), k = 1, 2, ..., m \tag{1}$$

where k is the order of the convolution layer in which default boxes are implemented. There are six default boxes in each position of a feature map, and the width and height are formulated as, $w_k = s_k\sqrt{a_r}$ and $h_k = s_k/\sqrt{a}$.

3.2 Limit of SSD on Text Detection

Here, we firstly find out what result in the failure of SSD on text detection. We implement SSD with its default settings which are shown in Table 1 on text detection. Some results are shown in Fig. 2. The bounding boxes signed in green are results of the original SSD detection, and the ground truth boxes are signed in red. It shows that SSD do not work well on text detection because SSD cannot deal with small objects.

This empirical results demonstrates that the inappropriate setting of the default box lead to a high missed rate and error rate. Thus, the original SSD can not be implemented directly on scene text detection, and default box setting should be modified to adapt to scene text detection.

Table 1. Default box settings in scales for each layer on ICDAR2013

Name	Conv4_3	Fc7	Conv6_2	Conv7_2	Conv8_2	Conv9_2
SSD	10–20	20–37	37–54	54–71	71–88	88–100
Text-SSD	5–10	10–25	25–40	40–55	55–70	70–85

Fig. 2. Results of the original SSD on text detection

3.3 Text Statistical Characteristics

Default boxes in SSD have two parameters: scales and aspect ratios. In order to be able to set reasonable parameters of default boxes for text detection, the distribution of text sizes and aspect ratios are investigated, which guide us to design the default box.

We make statistics for scene text on training data of ICDAR2013: the distributions of text width, the distribution of text height, the distribution of text area and the distribution of the text aspect ratio. Figure 3(a) and (b) shows the histograms of text widths and heights, respectively, Fig. 3(c) shows the histogram of text area, and Fig. 3(d) shows the histogram of text aspect ratios. Observing the statistical analysis of scene text, we can see that most text regions is relatively small, and the text shape is prone to long strip. The text areas averagely make up 4% of an image. The average of the width ratio between the text and the image is 26.9% and the height of the text is smaller than 10% of the height of an image. Thus, the initial default boxes settings exist three problems: (1) In the shallow layer (conv4_3) of neural network, the scale is too small to cover small object. (2) In the deep layer of neural network (conv8_2, conv9_2), the scale is too large, and too many backgrounds fall within the scale range. Thus, it introduces noises and results in the low overall accuracy. (3) Most text regions are detected successfully owing to the default box setting in the middle layers (fc7, conv6_2, conv7_2).

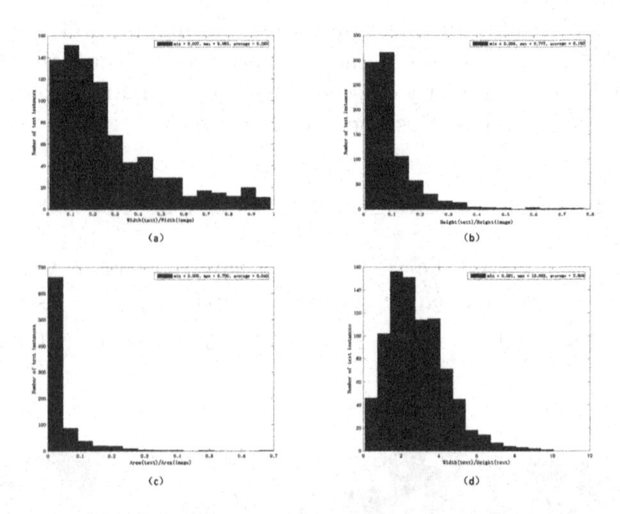

Fig. 3. Text statistical characteristics. (a) Text width ratio histogram. (b) Text height ratio histogram. (c) Text area ratio histogram. (d) Text aspect ratio histogram

We design the default box with the text-driven setting for text detection. In the text-driven setting, we modify the initial default box setting and reduce the scale of the default box in the shallow layer and make the width and height of the default boxes in the middle layers change linearly with the indicator of their layers, in order that the candidate text regions can be detected uniformly in all the layers. In Table 1, we give the scale of the default box. The first row shows the original default box setting, and the second row shows the scale setting according to the text-driven setting. Compared with the original SSD setting, there are fewer default boxes in the text-driven setting. We also set the aspect ratio of the default box as $a_r = \{1, 4, \frac{1}{4}\}$ according to the text statistical characteristics. In our experiment, we use nine default boxes with the three scales and three aspect ratios at each position.

3.4 Multi-scale Detection Fusion

When a query image is tested, it will be input in the three different Text-SSD models with three different scales. We will fuse the three results from three Text-SSDs based on non-maximum suppression. The algorithm is divided into the four steps: (1) We sort the detected bounding boxes in a descending order according to their confidence values. (2) We calculate the Jaccard overlap for each pairwise bounding boxes. (3) We remove the bounding box with a low confidence score and accumulate the voting scores for all overlapping bounding boxes. (4) After updating the score, the bounding box with the score of confidence lower than threshold is eliminated. Compared with the original non-maximal suppression, we can see that the voting based multi-scale fusion algorithm is more robust to noise.

4 Experimental Results

In this section, we implement the proposed multi-scale Text-SSD on ICDAR2013. ICDAR2013 is the competing text detection dataset and contains 229 training images and 233 testing images. We evaluate the text localization performance of Text-SSD by the standard criteria used in the competitions of ICDAR2013: the recall, the precision, and the F-score and we use the F-score as the final evaluation score. There are two standard evaluation methods used in ICDAR2013: IC13 and ICDet. We compare Text-SSD with 12 state-of-the-art methods, the six methods were published in 2015, and the other six methods are published in 2016. We also compare the single scale Text-SSD and multi-scale Text-SSD. As shown in Table 2, multi-scale Text-SSD ranks the first in the criteria of IC13 and ranks the second in the criteria of ICDet in the compared text detection methods. Multi-scale SSD is higher by about 2% than Gupta's method [5] which ranks the second in the criteria of IC13 and lower by about 1.2% than CPTN [15] which ranks the second in the criteria of ICDet. However, multi-scale Text-SSD is faster than Gupta's method. Single scale Text-SSD ranks the fifth in both the criteria of IC13 and ICDet, but it is the fastest in all the compared methods. Single scale Text-SSD achieve 11.6 Fps.

We also compare Textbox [14] which is the closest work to Text-SSD. The difference between ours and Textbox is that Text-SSD use text-specified default box setttings according to text statistical characteristics. We use fewer default boxes in SSD than Textbox. As shown in Table 3, Text-SSD is superior to TextBox in both single scale and multiple scales. Single scale Text-SSD is higher by 1.29% and 2.18% than single scale Textbox in terms of F-score in IC13 and ICDet, respectively. Multi-scale Text-SSD is higher by 1.24% and 0.83% than multi-scale Text-SSD in terms of F-score in IC13 and ICDet, respectively. Moreover, the speed of Text-SSD is faster than Textbox in both single scale and multiple scales. To sum up, Text-SSD is effective and robust in scene text detection.

Table 2. Comparison of state-of-the-art text detection methods on ICDAR2013

Name	Year	IC13			ICDet			Speed (FPS)
		Recall	Precision	F-score	Recall	Precision	F-score	
Text_Flow[3]	2015	75.89	85.15	80.25				<1
Neumann [21]	2015	72.4	81.8	77.1				3
Neumann [22]	2015	71.3	82.1	76.3				3
Busta [2]	2015	69.3	84	76.8				6
Zhang [20]	2015				74	88	80	<0.1
Yin [4]	2015	65.11	83.98	73.35				3
Zhang [8]	2016				78	88	83	<1
Gupta [5]	2016	76.4	**93.8**	84.2	75.5	92	83	<1
CTPN [15]	2016	74	93	82	83	**93**	**88**	7.1
SSD [13,14]	2016	60	80	69	60	80	69	10
Cho [23]	2016	78.45	86.26	82.17				7.69
He [6]	2016	73	93	82				<1
Single Text-SSD		76.62	86.57	81.29	77.1	87.98	82.18	**11.6**
Multi-scale Text-SSD		**82.83**	89.95	**86.24**	**83.18**	90.82	86.83	3.75

Table 3. Comparison between Text-SSD and TextBoxes on ICDAR2013

Method	IC13			ICDet			Speed (FPS)
	Recall	Precision	F-score	Recall	Precision	F-score	
Fast Textboxes [14]	74	86	80	74	88	80	11.1
TextBoxes [14]	83	88	85	83	89	86	1.37
Single Text-SSD	76.62	86.57	81.29	77.1	87.98	82.18	**11.63**
Multi-scale Text-SSD	**82.83**	**89.95**	**86.24**	**83.18**	**90.82**	**86.83**	3.75

5 Conclusion

In this paper, Text-SSD is proposed to detect scene text in which the default box is designed for scene text according to the text statistical characteristics. Moreover, in order to boost the performance of text detection, multi-scale Text-SSDs is used and the output are fused based on voting. Text-SSD is implemented on ICDAR2013, the experimental results demonstrate that the proposed method is superior to the state-of-the-art methods not only in the detection accuracy but also in the running time.

Acknowledgements. This work was supported by the National Natural Science Foundation of China under Grant 61373077.

References

1. Huang, W., Qiao, Y., Tang, X.: Robust scene text detection with convolution neural network induced MSER trees. In: Fleet, D., Pajdla, T., Schiele, B., Tuytelaars, T. (eds.) ECCV 2014. LNCS, vol. 8692, pp. 497–511. Springer, Cham (2014). https://doi.org/10.1007/978-3-319-10593-2_33
2. Busta, M., Neumann, L., Matas, J.: FASText: efficient unconstrained scene text detector. In: Proceedings of the IEEE International Conference on Computer Vision, pp. 1206–1214 (2015)
3. Tian, S., Pan, Y., Huang, C., et al.: Text flow: a unified text detection system in natural scene images. In: Proceedings of the IEEE International Conference on Computer Vision, pp. 4651–4659 (2015)
4. Yin, X.-C., Pei, W.-Y., Zhang, J., et al.: Multi-orientation scene text detection with adaptive clustering. IEEE Trans. Tattern Anal. Mach. Intell. **37**(9), 1930–1937 (2015)
5. Gupta, A., Vedaldi, A., Zissermana.: Synthetic data for text localisation in natural images. In: Proceedings of the IEEE Conference on Computer Vision and Pattern Recognition, pp. 2315–2324 (2016)
6. He, T., Huang, W., Qiao, Y., et al.: Text-attentional convolutional neural network for scene text detection. IEEE Trans. Image Process. **25**(6), 2529–2541 (2016)
7. Jaderberg, M., Simonyan, K., Vedaldi, A., et al.: Reading text in the wild with convolutional neural networks. Int. J. Comput. Vis. **116**(1), 1–20 (2016)
8. Zhang, Z., Zhang, C., Shen, W., et al.: Multi-oriented text detection with fully convolutional networks. In: Proceedings of the IEEE Conference on Computer Vision and Pattern Recognition, pp. 4159–4167 (2016)
9. Neumann, L., Matas, J.: Text localization in real-world images using efficiently pruned exhaustive search. In: International Conference on Document Analysis and Recognition (ICDAR), pp. 687–691. IEEE (2011)
10. Neumann, L., Matas, J.: Real-time scene text localization and recognition. In: IEEE Conference on Computer Vision and Pattern Recognition (CVPR), pp. 3538–3545. IEEE (2012)
11. Yao, C., Bai, X., Liu, W., et al.: Detecting texts of arbitrary orientations in natural images. In: IEEE Conference on Computer Vision and Pattern Recognition (CVPR), pp. 1083–1090. IEEE (2012)
12. Huang, W., Lin, Z., Yang, J., et al.: Text localization in natural images using stroke feature transform and text covariance descriptors. In: Proceedings of the IEEE International Conference on Computer Vision, pp. 1241–1248 (2014)
13. Liu, W., Anguelov, D., Erhan, D., Szegedy, C., Reed, S., Fu, C.-Y., Berg, A.C.: SSD: single shot multibox detector. In: Leibe, B., Matas, J., Sebe, N., Welling, M. (eds.) ECCV 2016. LNCS, vol. 9905, pp. 21–37. Springer, Cham (2016). https://doi.org/10.1007/978-3-319-46448-0_2
14. Liao, M., Shi, B., Bai, X., et al.: TextBoxes: a fast text detector with a single deep neural network. arXiv preprint arXiv:161106779 (2016)
15. Tian, Z., Huang, W., He, T., He, P., Qiao, Y.: Detecting text in natural image with connectionist text proposal network. In: Leibe, B., Matas, J., Sebe, N., Welling, M. (eds.) ECCV 2016. LNCS, vol. 9912, pp. 56–72. Springer, Cham (2016). https://doi.org/10.1007/978-3-319-46484-8_4
16. Jaderberg, M., Vedaldi, A., Zisserman, A.: Deep features for text spotting. In: Fleet, D., Pajdla, T., Schiele, B., Tuytelaars, T. (eds.) ECCV 2014. LNCS, vol. 8692, pp. 512–528. Springer, Cham (2014). https://doi.org/10.1007/978-3-319-10593-2_34

17. He, T., Huang, W., Qiao, Y., et al.: Accurate text localization in natural image with cascaded convolutional text network. arXiv preprint arXiv:160309423 (2016)
18. Yao, C., Bai, X., Sang, N., et al.: Scene text detection via holistic, multi-channel prediction. arXiv preprint arXiv:160609002 (2016)
19. Xie, S., Tu, Z.: Holistically-nested edge detection. In: Proceedings of the IEEE International Conference on Computer Vision, pp. 1395–1403 (2015)
20. Zhang, Z., Shen, W., Yao, C., et al.: Symmetry-based text line detection in natural scenes. In: Proceedings of the IEEE Conference on Computer Vision and Pattern Recognition, pp. 2558–2567 (2015)
21. Neumann, L., Matas, J.: Efficient scene text localization and recognition with local character refinement. In: 13th International Conference on Document Analysis and Recognition (ICDAR), pp. 746–750. IEEE (2015)
22. Neumann, L., Matas, J.: Real-time lexicon-free scene text localization and recognition. IEEE Trans. Pattern Anal. Mach. Intell. **38**(9), 1872–1885 (2016)
23. Cho, H., Sung, M., Jun, B.: Canny text detector: fast and robust scene text localization algorithm. In: Proceedings of the IEEE Conference on Computer Vision and Pattern Recognition, pp. 3566–3573 (2016)
24. Long, J., Shelhamer, E., Darrell, T.: Fully convolutional networks for semantic segmentation. In: Proceedings of the IEEE Conference on Computer Vision and Pattern Recognition, pp. 3431–3440 (2015)

Object Identification

Palmprint Recognition Using Sparse Representation of Variable Window-Width Real-Valued Gabor Feature

Mengwen Li, Huabin Wang(✉), Jian Zhou, and Liang Tao

School of Computer Science and Technology, Anhui University, Hefei 230601, China
limengwen628@163.com, {wanghuabin,taoliang}@ahu.edu.cn,
swjtuzhoujian@163.com

Abstract. This paper proposed a simple but effective palmprint recognition algorithm using improved Real-valued Discrete Gabor Transform (RDGT) and Sparse Representation based Classification (SRC) method. Compared to the existing palmprint recognition methods based on the spatial texture feature of palmprint, the proposed variable window-width real-valued Gabor transform extract the palmprint feature by space-frequency analysis. Given Gauss window as the analysis window, in addition, the window-width is dynamically adjusted according to the local variance of the palmprint image when solving the coefficients of RDGT. Then test sample can be sparsely represented in an overcomplete dictionary composed by training samples. Experimental results on PolyU Palmprint Database and PolyU M_B Database demonstrate the effectiveness of our proposed method.

Keywords: Palmprint recognition · Space-frequency analysis
Real-valued discrete Gabor transform
Sparse representation-based classification

1 Introduction

Facing the increasing information security problems, biometric technology attracts more and more attention for its unique advantages. Among all, palmprint recognition has been studied deeply in the past ten years. Compared with other biometric technologies, palmprint recognition has the advantages of high recognition rate, simple equipment and easy acceptance by users.

At present, many popular palmprint recognition methods extract spatial texture feature using local feature descriptors. LBP, HOG and WLD are typical local descriptors, for example. Li and Kim [1] improved the Local Tetra Pattern (LTrP), and presents Local Micro-structure Tetra Pattern (LMTrP) by comparing the relationship between the reference pixels and their surrounding pixels with a certain thickness along the horizontal and vertical directions. Hong et al. [2] extract HOG histogram from eight different local coordinate systems. It can overcome

J. Yang et al. (Eds.): CCCV 2017, Part III, CCIS 773, pp. 257–267, 2017.
https://doi.org/10.1007/978-981-10-7305-2_23

the bad effects of image blur, translation and rotation. Jia et al. [3] proposed Histogram of Oriented Lines (HOL), which is a modification of HOG, extracting the line features and direction of palmprint by linear filtering. It robust to small range of translation and rotation. Zhang et al. [4] use block-wise statistics of CompCode as features, and then CRC_RLS (collaborative representation based classification with regularized least square) method for classification. Luo et al. [5] proposed Local Line Directional Patterns (LLDP) to encode the orientation information generated by liner filtering. In [6], WLD method [7] is applied to palmprint recognition for the first time, and proposed Line feature Weber Local Descriptor (LWLD) combining the characteristics of palmprint image. Bounneche et al. [8] used multi-resolution log-Gabor filter to filter palmprint images, which made up for some shortcomings of Gabor filter. Fei et al. [9] proposed DOC (Double-orientation code) method to represent the orientation feature of palmprint and designed an effective nonlinear angular matching scheme.

As you can see, the palmprint recognition methods mentioned above pay more attention to the spatial texture characteristics of palmprint image. However, these methods have a large amount of computation, long recognition time and high feature dimension. Thus, in this paper, the frequency characteristics of palmprint are considered. Observing palmprint images, we find that palmprint has rich palm lines. Therefore, its frequency is spatial-varying. Gabor transform is one of important methods of time (space) frequency analysis. The coefficients reveal the local frequency distribution of a signal or an image. This advantage of Gabor transform has been widely used in various aspects of signal and image processing, such as speech recognition, signal detection, image compression, texture analysis, image segmentation and recognition. The traditional complex-valued discrete Gabor transform (CDGT) algorithm is complex and difficult to implement in hardware and software for complex operations. In this paper, real-valued discrete Gabor transform (RDGT) [10] proposed by Tao et al. is used to solve the transform coefficients.

But the traditional real-valued discrete Gabor transform uses a single window with a fixed width, and has fixed space-frequency resolution. Restricted by the relation between time-width and bandwidth, the spatial resolution and frequency resolution of discrete Gabor transform can not achieve best at the same time. In order to improve the recognition accuracy, in this paper, we adjusted the width of window function adaptively according to the local characteristics of palmprint images and proposed variable window-width RDGT.

The remainder of this paper is organized as follows. Section 2 presents feature extraction based on variable window-width RDGT. Section 3 describes the sparse representation of variable window-width real-valued Gabor feature for palmprint recognition. Section 4 shows the experimental results. Section 5 concludes the paper.

2 Variable Window-Width RDGT

2.1 Real-Valued Discrete Gabor Transform

When image represented by 2-D Gabor transform, the characteristics of human eye and visual system can be effectively combined with image compression coding, and the transform coefficients have less redundancy compared with the original image data. Given an image $\boldsymbol{I}(x,y)$, $x=0,1,\ldots,X-1$, $y=0,1,\ldots,Y-1$, and dividing it into $K\times L$ non overlapping lattices of dimensions $M\times N$ such that $X=KM$ and $Y=LN$. Then the image $\boldsymbol{I}(x,y)$ can be expanded as follow

$$\boldsymbol{I}(x,y)=\sum_{k=0}^{K-1}\sum_{l=0}^{L-1}\sum_{m=0}^{M-1}\sum_{n=0}^{N-1}a(k,l,m,n)g_{klmn}(x,y) \tag{1}$$

The real-valued Gabor basis function is defined as

$$g_{klmn}(x,y)=\tilde{g}(x-kM,y-kN)\cdot cas\left\{2\pi\left[\frac{mx}{M}+\frac{ny}{N}\right]\right\} \tag{2}$$

where $cas(x)=cos(x)+sin(x)$ is Hartley's cas function. The Gabor transform coefficients $a(k,l,m,n)$ can obtained by

$$a(k,l,m,n)=\sum_{x=0}^{X-1}\sum_{y=0}^{Y-1}\boldsymbol{I}(x,y)h_{klmn}(x,y) \tag{3}$$

where the real-valued auxiliary biorthogonal function is given by

$$h_{klmn}(x,y)=\tilde{h}(x-kM,y-lN)\cdot cas\left\{2\pi\left[\frac{mx}{M}+\frac{ny}{N}\right]\right\} \tag{4}$$

Once given a synthesis window as shown in Fig. 1, its biorthogonal analysis window can be obtained by numerical method [11,12], shown in Fig. 2. With the analysis window, we can calculate coefficients $a(k,l,m,n)$ using the fast 2D-DHT

$$\begin{aligned}a(k,l,m,n)&=\sum_{x=0}^{X-1}\sum_{y=0}^{Y-1}\boldsymbol{I}(x,y)\tilde{h}(x-kM,y-lN)\cdot\left\{2\pi\left[\frac{mx}{M}+\frac{ny}{N}\right]\right\}\\&=\sum_{j_1=0}^{M-1}\sum_{j_2=0}^{N-1}\left\{\sum_{i_1=0}^{K-1}\sum_{i_2=0}^{L-1}R_{mn}\left(i_1M+j_1,i_2N+j_2\right)\right\}\\&\quad\cdot cas\left\{2\pi\left[\frac{mj_1}{M}+\frac{nj_2}{N}\right]\right\}\end{aligned} \tag{5}$$

where $R_{mn}(x,y)=\boldsymbol{I}(x,y)\tilde{h}(x-kM,y-lN)$, $x=i_1M+j_1$, $y=i_2N+j_2$.

2.2 Variable Window-Width RDGT

From the Fig. 2, the analysis window obtained by numerical method is disperse, which is not suitable for extracting palmprint feature. Fortunately, the synthesis

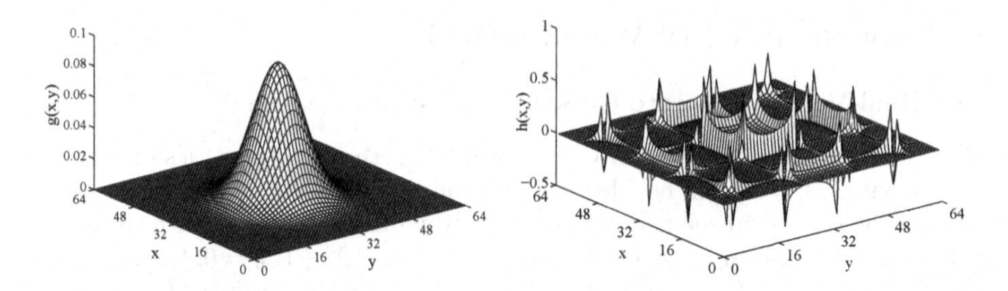

Fig. 1. Synthesis window

Fig. 2. Analysis window

window and analysis window can exchange, and we only need the RDGT coefficients. So we can define a more centralized analysis window directly. Since the Gauss function has the smallest product of the effective time-width and bandwidth in the Heisenberg Uncertainty Principle, its distribution is most concentrated in the time-frequency plane, so the Gauss window is chosen in this paper.

The window-width q in Gauss window function is used to adjust the spatial resolution and frequency resolution. When q is fixed, the space-frequency resolution is fixed too. For palmprint image, some areas have abundant lines and complex frequency components; while some regions are relatively smooth, and the frequency characteristics are not obvious. Therefore, the q should be dynamically adjusted in order to get more discriminative features. In this paper, we adopt variance to measure the local variations of palmprint image, as shown in Fig. 3. The larger variance is, the more dramatic palmprint image changes, and the smaller variance is, the more slowly palmprint image changes. In this way, the q can be changed according to the variance of local image. Now, we redefine the window function as follow

$$h_{klmn} = \tilde{h}_{kl}(x - kM, y - lN) \cdot cas\left\{2\pi\left[\frac{mx}{M} + \frac{ny}{N}\right]\right\} \tag{6}$$

$$\tilde{h}_{kl}(x, y) = \sum_i \sum_j h_{kl}(x + iX, y + jY) = \tilde{h}_{kl}(x + X, y + Y) \tag{7}$$

$$h_{kl}(x, y) = \frac{\sqrt{2}}{q} \cdot exp\left\{-\pi\left[\left(x - \frac{X}{2}\right)^2 + \left(y - \frac{Y}{2}\right)^2\right] / q^2\right\} \tag{8}$$

The window-width q is defined as follow

$$q = F\left(\boldsymbol{I}_m(u, v)\right) = \sum_u \sum_v \left|\boldsymbol{I}_m(u, v) - \lambda\right| / MN \tag{9}$$

$$\lambda = \sum_u \sum_v \boldsymbol{I}_m(u, v) / MN, \quad (u = 0, 1, \ldots, M - 1, v = 0, 1, \ldots, N - 1) \tag{10}$$

$$\boldsymbol{I}_m(u, v) = \boldsymbol{I}(s, t) \tag{11}$$

where

$$s = (kM + X/2)\%X, \ldots, (kM + X/2 + M - 1)\%X$$
$$t = (lN + Y/2)\%Y, \ldots, (lN + Y/2 + N - 1)\%Y \tag{12}$$

Once the width of window function adjusted adaptively according to the local characteristics of palmprint, more palmprint feature data can be extracted. In this way, we make improve on the traditional RDGT. In this way, we make improve on the traditional RDGT.

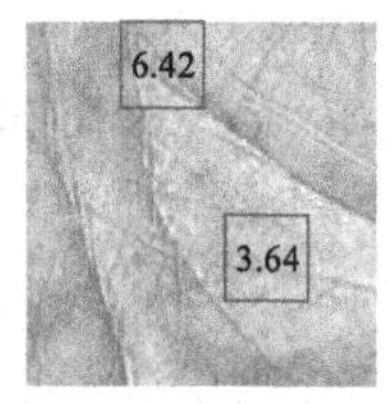

Fig. 3. Local variance

Fig. 4. The energy distribution

2.3 Variable Window-Width Real-Valued Gabor Feature of Palmprint

Dividing the palmprint image into $K \times L$ non overlapping lattice of dimension $M \times N$ (critical sampling), obtaining four-dimensional coefficients through the variable window-width RDGT. Let k and l stay the same, m takes from 1 to M, n takes from 1 to N. Then we will obtain a coefficient matrix of dimensions $M \times N$, representing the energy distribution of the (k, l) lattice. For example, Fig. 4 shows the energy distribution of the RDGT coefficients, when $K = L = 8, M = N = 8$. Let

$$e(k,l) = \sum_{m=1}^{M} \sum_{n=1}^{N} |a(k,l,m,n)|^2, (k = 1, \ldots, K, l = 1, \ldots, L) \tag{13}$$

Calculate $e(k,l)$ and form a $h-$dimensional vector $d = (e_1, e_2, \ldots, e_h)(h = K \times L)$. In order to eliminate the influence of image resolution, the vector d is normalized by the l_2 norm $d_N = d/\|d\|_2$. In summary,the steps of extracting real-valued Gabor features for palmprint image are as follow

1. Dividing the $\boldsymbol{I}(x, y)$ into $K \times L$ non overlaping lattice of dimensions $M \times N$, then the variance $q_i (i = 1, 2, \ldots, K \times L)$ of each lattice is calculated according to Eqs. (9)–(12).
2. The window function $h_{kl}(x, y)$ is generated with the window-width q_i according to Eq. (8).
3. Calculating the coefficients according to Eqs. (5)–(7).
4. Each e_i is computed to form a feature vector d.
5. Normalizing the vector d by the l_2 norm.

3 Sparse Representation of Variable Window-Width Real-Valued Gabor Feature

Sparse Representation based Classification (SRC) method has a series of excellent advantages such as high recognition rate and robustness [13–15]. In this work, we present a new method for palmprint recognition. Figure 5 shows the process of the proposed method. The steps are as follow

1. Given a training sample set (including k class), extract the palmprint features according to Sect. 2.3 and form the matrix $\boldsymbol{A}$

$$\boldsymbol{A} = [\boldsymbol{A}_1, \boldsymbol{A}_2, \ldots, \boldsymbol{A}_k] \in \mathbb{R}^{h\times(n_i\times k)} \tag{14}$$

where, h denotes the dimension of the feature vector, n_i denotes the number of training samples for each class.
2. Given the test sample, extract the real-valued feature $\boldsymbol{y}$.
3. The following l^1 minimization problem is solved to obtain the sparse representation coefficients

$$\hat{\boldsymbol{x}} = argmin\,\|\boldsymbol{x}\| \quad \boldsymbol{Ax} = \boldsymbol{y} \tag{15}$$

4. Obtain the reconstruction residual

$$r_i(\boldsymbol{y}) = \|\boldsymbol{y} - \boldsymbol{A}_i\boldsymbol{x}_i\|_2 \quad (i = 1, .., k) \tag{16}$$

5. Obtain the label of the test sample

$$\hat{i} = \arg\min_i r_i(\boldsymbol{y}) \tag{17}$$

In this paper, DALM (Dual Augmented Lagrangian Method) is used to solve the sparse representation coefficients.

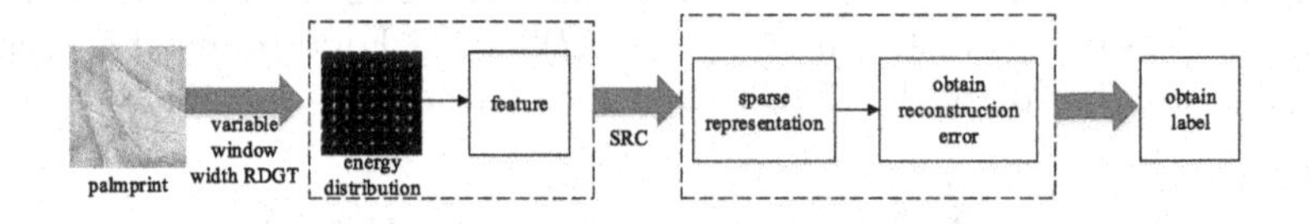

Fig. 5. The proposed algorithm processing

4 Experiments

Experiments are conducted on PolyU Palmprint Database and PolyU M_B Database. PolyU Palmprint Database contains 386 different palmprint.The palmprint images were collected in two sessions, and in each session, about 10 palmprint images were captured from each palm. PolyU M_B consists of 6000 images of 500 palms. The palmprint images were also collected in two sessions, and in each session, about 6 palmprint images were captured from each palm. In the experiments, the palmprint collected in the first stage is used as the training set, and the palmprint collected in the second stage is used as the test set.

4.1 Improvement Analysis

In this section, we compare the traditional RDGT and variable window-width RDGT on palmprint recognition. The result is shown in Table 1. The parameters are set as $K = L = 8$, $q = 16$. Table 1 indicates that the recognition rate has been greatly improved by using the variable window-width RDGT. The results show that the method proposed in this paper can better represent space-frequency characteristic of palmprint image compared with the traditional RDGT. In addition, the reconstruction residuals in the SRC method actually measure the similarity between the given test samples and the training samples. We analyze the distribution of the within-class reconstruction residuals and between-class reconstruction residuals. Figures 6 and 7 demonstrate that within-class reconstruction residuals and between-class reconstruction residuals are separated. The within-class reconstruction residuals is concentrated in the left half, the between-class reconstruction residuals is concentrated in the right half. It indicates that, the proposed method can effectively classify different palmprint.

Table 1. The effect of proposed method on PolyU

Algorithm	Recognition rate
RDGT	96.79%
Proposed	99.87%

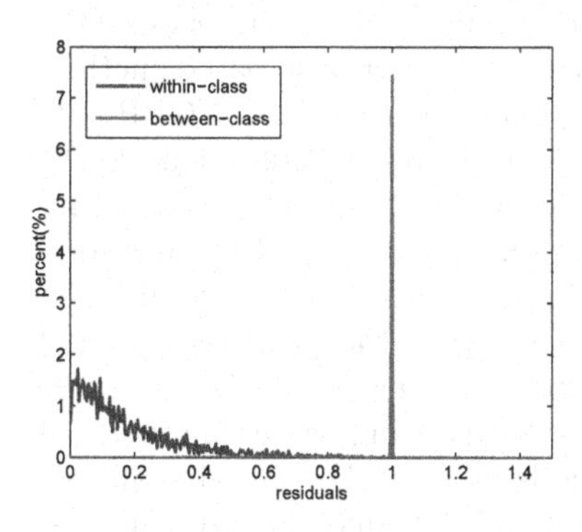

Fig. 6. PolyU

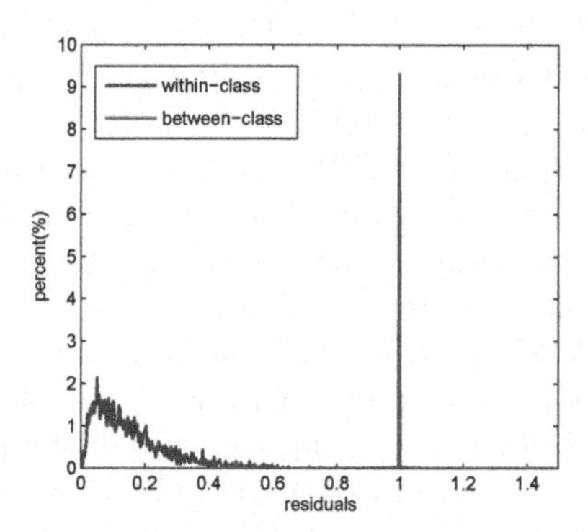

Fig. 7. PolyU M_B

4.2 Parameter Setting

When extracting variable window-width real-valued Gabor features, K and L are important parameters. In order to achieve the best recognition rate, different parameters are selected for experiments to find the optimal parameter settings.

When $K = L = 8$ or $K = L = 16$, the dictionary $\boldsymbol{A}$ formed by the feature vectors of training samples meet the requirement of sparsity. While $K = L = 32$, $\boldsymbol{A}$ does not satisfy the sparsity requirement, dimensionality reduction must be carried out. PCA method is used to reduce the dimension to 600, 500, 400 and 300 respectively, and make recognition under different dimension. Table 2 shows, when $K = L = 16$, the recognition rate will up to 100% on PolyU palmprint database; when $K = L = 8$, the recognition rate will up to 99.97% on PolyU M_B database. The proposed method in this paper can meet the demand of palmprint recognition at present.

Table 2. The test results under different parameters

$K \times L$	Dimension	PolyU	PolyU M_B
32×32	600	99.95%	99.90%
32×32	500	99.92%	99.90%
32×32	400	99.92%	99.93%
32×32	300	99.84%	99.90%
16×16	256	100%	99.93%
8×8	64	99.87%	99.97%

4.3 Performance Comparison of Palmprint Recognition Methods Based on SRC

In this section, we will extract palmprint features using some popular methods. The subspace based method such as Eigenpalms [16], Fisherpalms [17]. The statistically based method DCT [18]. The local descriptor based method such as LBP [19], HOG [3], HOL [3], LLDP [5], Gabor Wavelet [20], LGBP [19], WLD [7]. Then the SRC method is used to identify the test samples. Due to the high dimensionality of the feature vectors, the dictionary $\boldsymbol{A}$ formed by the feature vectors of training samples does no meet the requirement of sparsity. Therefore, dimensionality reduction is also needed. The experimental results are given in Table 3 and Fig. 8. From these results we can see that the proposed method usually achieves a higher recognition rate against the other methods under the same dimension. Eigenpalms and Fisherpalms are two important methods of palmprint feature extraction. These two methods analyze the spatial structure of palmprint images and map the high-dimensional data into low dimensional vectors. However, this method is easily affected by noise, resulting in low recognition rate. DCT method extracts the frequency characteristics of palmprint images, but the spatial information is omitted. HOL, LLDP and WLD methods use the directional information, and the feature dimension is higher. During the feature dimension reduction, a lot of information is lost. The proposed method takes the spatial and frequency characteristics into account. This method has obtain the higher recognition rate and the better noise robustness compared with traditional methods.

Table 3. The recognition rate of different methods on PolyU

Method	Dimension	600	500	400	300
Eigenpalms	-	95.28%	95.26%	95.18%	95.03%
Fisherpalms	-	94.84%	95.26%	95.85%	96.40%
DCT	1024	93.50%	92.93%	91.81%	89.56%
LBP	3776	95.21%	95.00%	94.79%	94.43%
HOG	2352	98.34%	98.32%	98.29%	98.34%
HOL	2352	99.53%	99.53%	99.53%	99.56%
LLDP	9216	99.69%	99.72%	99.69%	99.69%
Gabor wavelet	8029	97.23%	97.23%	97.20%	97.20%
LGBP	9440	83.96%	83.91%	84.35%	84.61%
WLD	4608	92.18%	92.20%	91.89%	91.68%
Proposed	1024	99.95%	99.92%	99.92%	99.84%

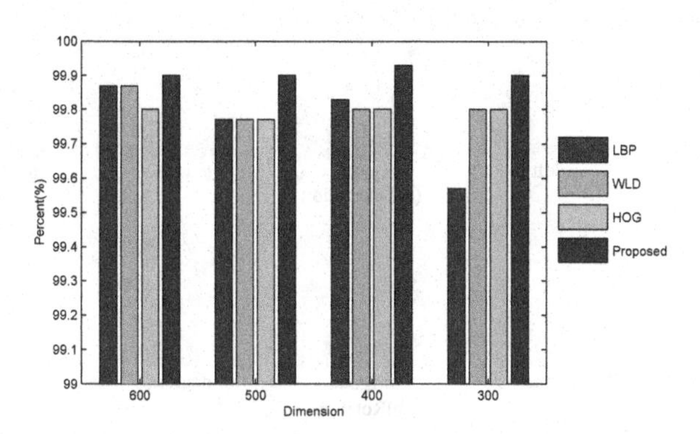

Fig. 8. The recognition rate of different methods on PolyU M_B

4.4 Computational Complexity

In this paper, all experiments are carried on MATLAB 2010 in PC with CPU 3.20 GHz, RAM 4 GB. In order to analyze the computational complexity, we compare the computational cost of the proposed method with state-of-the-art palmprint recognition methods. In Table 4, the computational time of the feature extraction and matching for different palmprint recognition methods are listed. Due to the simple feature extraction scheme, the feature extraction speed of the proposed method is faster than that of other methods.

4.5 Robustness Experiment

In this section, we will design some translation and rotation experiments to test the robustness of the proposed method. First, we carry out a simple translation test. The test palmprint is moved with left-ward shift 5 pixels, right-ward shift

Table 4. Comparison of recognition time of different palmprint recognition methods

Methods	Extraction (ms)	Matching (ms)	Total (ms)
HOL	176.81	79.40	256.21
LLDP	204.04	297.60	501.64
LGBP	407.56	213.84	621.40
Proposed	30.20	134.98	165.18

5 pixels, up-ward shift 5 pixels, down-ward shift 5 pixels, shown in Fig. 9(a). Then rotation test is carried out. The test palmprint is rotated with 3° and 5°, as shown in Fig. 9(b). Only 4 palmprint images are not correctly identified in these experiments. The experimental results show that the proposed method is robust to small range translation and rotation variations.

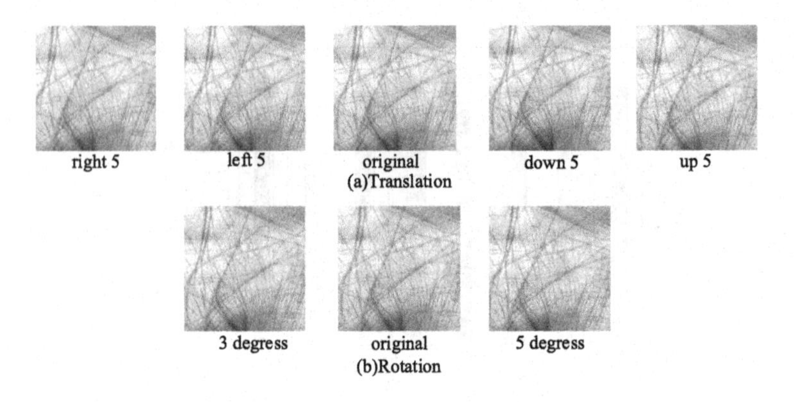

Fig. 9. The translation and rotation changes

5 Conclusion

In this paper, unlike many palmprint recognition methods based on spatial texture features, proposed a simple and effective palmprint recognition method using sparse representation of variable window-width real-valued Gabor feature. In order to analyze palmprint image in space-frequency domain, the traditional RDGT has been improved according to the characteristic of palmprint. Extensive experimental results demonstrate that the proposed method can achieve a competitive performance comparing with the state-of-art palmprint recognition methods. But in this paper, the processing of the coefficients of RDGT is relatively simple. In the next step, we will study how to make better use of the information to further improve the performance of the algorithm.

References

1. Li, G., Kim, J.: Palmprint recognition with local micro-structure tetra pattern. Pattern Recogn. **61**, 29–46 (2017)
2. Hong, D., Liu, W., Wu, X.: Robust palmprint recognition based on the fast variation Vese-Osher model. Neurocomputing **174**, 999–1012 (2016)
3. Jia, W., Hu, R.X., Lei, Y.K.: Histogram of oriented lines for palmprint recognition. IEEE Trans. Syst. Man Cybern. Syst. **44**(3), 385–395 (2014)
4. Zhang, L., Li, L., Yang, A.: Towards contactless palmprint recognition: a novel device, a new benchmark, and a collaborative representation based identification approach. Pattern Recogn. **69**, 199–212 (2017)
5. Luo, Y.T., Zhao, L.Y., Zhang, B.: Local line directional pattern for palmprint recognition. Pattern Recogn. **50**(C), 26–44 (2016)
6. Luo, Y.T., Zhao, L.Y., Jia, W.: Palmprint recognition method based on line feature Weber local descriptor. J. Image Graph. **21**(2), 0235–0244 (2016)
7. Chen, J., Shan, S., He, C.: WLD: a robust local image descriptor. IEEE Trans. Pattern Anal. Mach. Intell. **32**(9), 1705–1720 (2010)
8. Bounneche, M.D., Boubchir, L., Bouridane, A.: Multi-spectral palmprint recognition based on oriented multiscale log-Gabor filters. Neurocomputing. **205**(C), 274–286 (2016)
9. Fei, L., Xu, Y., Tang, W.: Double-orientation code and nonlinear matching scheme for palmprint recognition. Pattern Recogn. **49**(C), 89–101 (2016)
10. Tan, M., Gu, J.J., Hu, X.Y.: 2-D DHT-based fast Gabor transform for image processing. In: Second IITA International Conference on Geoscience and Remote Sensing, vol. 1, pp. 372–375. IEEE (2010)
11. Hu, X.Y., Tao, L., Wang, H.B.: An efficient image watermarking scheme based on 2-D real valued discrete Gabor transform. In: International Conference on Computer Application and System Modeling, vol. 2, pp. 277–281. IEEE (2010)
12. Tao, L., Kwan, H.K., Gu, J.J.: Filterbank-based fast parallel algorithms for real-valued discrete Gabor expansion and transform. In: International Symposium on Circuits and Systems DBLP, pp. 2674–2677 (2010)
13. Wright, J., Yang, A.Y., Ganesh, A.: Robust face recognition vial sparse representation. IEEE Trans. Pattern Anal. Mach. Intell. **31**(2), 210–227 (2008)
14. Lu, C.Y., Min, H., Gui, J.: Face recognition via weighted sparse representation. J. Vis. Commun. Image Represent. **24**(2), 111–116 (2013)
15. Lu, C.Y., Huang, D.S.: A new decision rule for sparse representation based classification for face recognition. Neurocomputing **116**(10), 265–271 (2013)
16. Lu, G., Zhang, D., Wang, K.: Palmprint recognition using eigenpalms features. Pattern Recogn. Lett. **24**(9–20), 1436–1467 (2003)
17. Wu, X., Zhang, D., Wang, K.: Fisherpalms based palmprint recognition. Pattern Recogn. Lett. **24**(15), 2829–2838 (2003)
18. Laadjel, M., AI-Maadeed, S., Bouridane, A.: Combining fisher locality preserving projections and passband DCT for efficient palmprint recognition. Neurocomputing **152**, 179–189 (2015)
19. Zhang, W., Shan, S., Gao, W.: Local Gabor binary pattern histogram sequence (LGBPHS): a novel non-statistical model for face representation and recognition. In: Tenth IEEE International Conference on Computer Vision, vol. 1, pp. 786–791. IEEE (2005)
20. Jaswal, G., Nath, R., Kaul, A.: Textrure based palm Print recognition using 2-D Gabor filter and sub space approaches. In: International Conference on Signal Processing, Computing and Control, pp. 344–349 (2016)

Robust Multi-view Common Component Learning

Jiamiao Xu, Xinge You(✉), Shi Yin, Peng Zhang, and Wei Yuan

School of Electronics and Information Engineering,
Huazhong University of Science and Technology, Wuhan 430074, China
youxg@mail.hust.edu.cn

Abstract. In many computer vision applications, one object usually exists more than one data representation. This paper focuses on the specific problem of cross-view recognition, which aims to recognize objects from different views. A majority of representative works mainly attempt to seek a common subspace, in which the Euclidean distance of within-class data is short. Intuitively, the recognition performance will be better if the various data from the same object have completely same representation in the common space. Therefore, this paper proposes robust multi-view common component learning (RMCCL) algorithm, which learns multiple linear transforms to extract the common component of multi-view data from the same instance. To enhance the discriminant ability and robustness of subspace, we introduce binary label matrix technology and serve Cauchy loss as our error measurement. RMCCL can be decomposed into two subproblems by Alternating optimization method, and each subproblem can be optimized by Iteratively Reweight Residuals (IRR) technique. Extensive experiments in both two-view and multi-view datasets demonstrate that the our method outperforms other state-of-the-art approaches.

Keywords: Cross-view recognition · Multi-view learning
Common space

1 Introduction

In many computer vision applications, one object can be captured by different sensors or observed at diverse viewpoints, e.g., [15]. Thus, one object often exists multiple data representation, usually denoted as multi-view data. Recently, cross-view recognition, which aims to recognize the samples from completely heterogeneous views, becomes more and more significant. However, due to big gap possibly existing between views, the recognition performance will be poor by directly recognizing samples from diverse views.

In order to better handle the cross-view recognition problems, many excellent and effective works have been done. Most of approaches can be attributed to metric learning methods or subspace learning approaches. Owing to the large

J. Yang et al. (Eds.): CCCV 2017, Part III, CCIS 773, pp. 268–279, 2017.
https://doi.org/10.1007/978-981-10-7305-2_24

difference between views, Euclidean distance metric are not applicable. To handle this problem, metric learning methods aim to learn a kind of new similarity measurement manners. Nevertheless, subspace learning try to find a common space shared by multiple views, in which Euclidean distance metric can work. Certainly, there are also many works can be utilized to handle the cross-view recognition problems, such as transfer learning [6,30]. Because our proposed method belongs to subspace learning, we will introduce this kind of methods in detail. Regarding subspace learning, according to the number of views, this kind of approaches can be grouped into two-view and multi-view approaches.

1.1 Two-view Subspace Learning Approaches

The canonical correlation analysis (CCA) [9] is proposed in 1936, and is the earliest method of cross-view subspace learning. CCA attempts to find two linear transformations, one for each view, such that the correlation coefficient between two views is maximized. To handle information dissipation caused by CCA, Chen et al. [2] propose continuum regression (CR) model. Experiments on cross-modal retrieval demonstrates that CR model can better address the cross-view problems. Heterogeneous face recognition is a specific but important subproblem in the cross-view recognition problems. To better address this subproblem, in [19,21], PLS is employed to conduct effective feature selection. CCA is only a unsupervised subspace learning approach. To take full advantage of label information, multi-view fisher discriminant analysis (MFDA) [3,4] effectively utilizes labels information to learn informative projections. Experimental results show that MFDA works better than previous unsupervised methods. For heterogeneous face recognition, Conventional approaches, which bring in a middle conversion stage, are easy to cause performance degradation. In order to better handle this problem, Lin and Tang [17] propose common discriminant feature extraction (CDFE) to simultaneously learn two transformation, which transform two-view samples to common subspace. Certainly, in addition to previously mentioned approaches, there are some excellent works [1,16,23,25].

1.2 Multi-view Subspace Learning Approaches

No doubt that we can use two-view subspace learning approaches to solve multi-view recognition problem by one-versus-one strategy. To efficiently handle this problem, we tend to design multi-view subspace learning methods, which can transform multiple view samples into a latent subspace, simultaneously. In [18,20], Multi-view Canonical Correlation Analysis (MCCA) is designed as a multiple versions of the CCA. In addition, Sim et al. [22] present Multimodal Discriminant Analysis (MMDA) to decompose samples into independent modes. Recently, In [11], kan et al. propose a novel method named multi-view discriminant analysis (MvDA), which tries to project the multi-view data into a common space, in which within-class samples are close to each other and between-class samples are far away. Kan et al. believe that there exists some relations between different transformations. To exploit these relations, kan et al. [12] develop view-consistency version of MvDA in 2016, which is called MvDA-VC. Besides, low-rank representation (LRR) is good at digging relations between views. Therefore, recently, LRR is also be utilized to handle the cross-view problems, such as SRRS [14], LRCS [5] and LRDE [13].

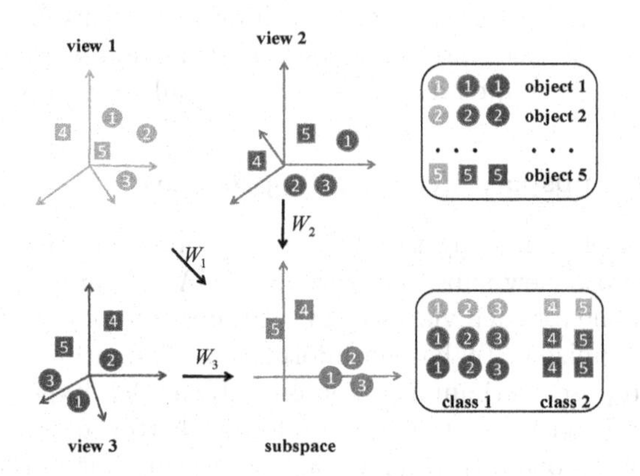

Fig. 1. Illustration of our idea. There are five objects in the Figure, and each object has three views. Each number denotes one object, and each color stands for one view. The first three objects from circle class and the last two from rectangle class. As we have seen, three view data are mapped to a unified space by W_1, W_2, W_3 projection matrices, in which various data from one object have unique representation. Furthermore, circle class is far from rectangle class and within-class objects are close to each other. (Color figure online)

As mentioned above, one object can generate multiple view data. However, existing subspace learning approaches can only minimize the difference of within-class samples, i.e., the distance of data from the same instance is close in the unified subspace. Intuitively, we can get a better performance for cross-view recognition problem if the heterogeneous data from one object have completely same representation in the latent common subspace. For this reason, this paper learns multiple linear projection matrices, one for each view, to extract the common component of the heterogeneous data. In this paper, to strengthen the robustness to noise, we employ Cauchy loss [27] as the error measurement in our algorithm. Through the above two thoughts, we still can't learn a discriminative common space. Therefore, we introduce a non-negative label matrix [28] to enlarge the margins between different classes and diminish the discrepancy of samples in the same class. Finally, the proposed method is applied to several cross-view recognition tasks. Experimental results on two heterogeneous face databases demonstrate that the proposed method outperforms previous cross-view recognitions approaches. The fundamental idea of our algorithm is illustrated in Fig. 1.

2 Robust Multi-view Common Component Learning

In this section, we first present the problem formulation, and then bring in label matrix and Cauchy loss to enhance discriminative ability and robustness of algorithm.

2.1 Problem Formulation

Let $\mathcal{X} = \{x_i^1, x_i^2, \ldots, x_i^n\}_{i=1}^m$ stands for a dataset containing m objects, and each object exists n samples, usually denoted as n views in this area. $x_i^v \in \mathcal{R}^{d_v}$ denotes the vth view of the ith object embedded in d_v dimension space. Generally, there are a large gap between different views, i.e., the between-class samples in a view might have higher similarity than within-class samples from different views. That will lead to poor performance. To better solve this problem, we expect that multiple samples from one object have completely uniform feature representation. Certainly, we can also consider that we extract component features in the process. We denote the linear transformations as $W = \{W_v\}_{v=1}^n$, where $W_v \in \mathcal{R}^{d_i \times d}$ is the transformation matrix of ith view, and projection results are $\mathcal{Z} = \{z_i\}_{i=1}^m$, where $z_i \in \mathcal{R}^d$ is the mapping result of ith object embedded in d dimension common space.

It is obvious that our objective is to learn a series of view projection matrices $\{W_v\}_{v=1}^n$ to extract common component z. A straightforward method is minimizing the empirical risk $z_i - W_v^{\mathrm{T}} x_i^v$ in the whole dataset:

$$\min_{W} \frac{1}{mn} \sum_{i=1}^{m} \sum_{v=1}^{n} \left|\left|z_i - W_v^{\mathrm{T}} x_i^v\right|\right|_2^2 + c_1 \sum_{v=1}^{n} ||W_v||_F^2 + c_2 \sum_{i=1}^{m} ||z_i||_2^2 \tag{1}$$

As can be seen in Eq. (1), least square error are used to extract common features of multiple view data, and regularization items are employed to penalize view projection matrices W and generating samples z_i, where $c_1 > 0$, $c_2 > 0$ are the balance parameters, which are presented to prevent overfitting.

No doubt that Eq. (1) is able to extract the common component of multiple views from one object. However, the constraint of Eq. (1) is too weak, so that it is not enough to obtain a discriminant subspace. To improve the performance of algorithm, like most previous works, we expect to unite within-class samples and separate between-class samples in the common space. Certainly, it is usual that manifold learning methods [7,8] can be use to improve the discriminant ability of algorithms. However, we adopt a simple but effective methods in this paper. We introduce binary label matrix to enhance the discriminant ability of our algorithm.

2.2 Least Square Regression for Multi-class Classification

Given a dataset $\mathcal{X} = \{(x_i, y_i)\}_{i=1}^m$, where x_i is the ith sample of dataset $\mathcal{X}$, and each input feature x_i corresponds to a target vector y_i. As usual, the regression problem can be expressed as

$$\min_{W,b} \sum_{i=1}^{m} ||W^{\mathrm{T}} x_i + b - y_i||_2^2 + \frac{\lambda}{2} ||W||_F^2 \tag{2}$$

where $W \in \mathcal{R}^{d \times p}$ is weight matrix, $b \in \mathcal{R}^p$ is the bias vector and λ is the regularized parameter, which is used to prevent overfitting. In the regression problem, the target y_i in Eq. (2) is continuous. Certainly, by setting target to 0 and 1, Eq. (2) can be extended to two-class classification. Regarding multi-class classification, we can adopt one-versus-one policy to generalize Eq. (2). However, we will have to address multiple subproblems generated by this policy.

In order to use regression model to better handle multi-class classification problem, we introduce label matrix in this paper. Supposing $\mathcal{X}$ is from c classes, and x_i is from class j. Then, for sample x_i, its target vector is $y_i = [0, \ldots, 0, 1, 0, \ldots, 0]^{\mathrm{T}}$ with only the jth element equal to one, where $y_i \in \mathcal{R}^c$. The target vector constructed by this rule is called as binary label matrix. Then, Eq. (2) can be generalized to multi-class classification easily.

As a matter of fact, numerous works utilize linear regression model to address multi-class classification problem by bringing in binary label matrix [28]. In this paper, we utilize it to restraint common space z learning in Eq. (1). Thus, Eq. (1) can be rewritten as

$$\min_{W} \frac{1}{mn} \sum_{i=1}^{m} \sum_{v=1}^{n} ||z_i - W_v^{\mathrm{T}} x_i^v||_2^2 + c_1 \sum_{v=1}^{n} ||W_v||_F^2 + c_2 \sum_{i=1}^{m} ||z_i - y_i||_2^2 \tag{3}$$

where y_i is the binary label vector of x_i. Supposed that the training set is from c classes, we can regard y_i as a point embedded in c dimension space. In this space, within-class samples correspond to the same point and between-class samples are forced to be separated. Therefore, we can obtain a discriminant space by using label matrix. We might as well call Eq. (3) as MCCL-L2 to distinguish what we will propose below.

2.3 Cauchy Loss

Least square loss is generally used as error measurement, as we have done in Eq. (3). However, thoroughly theory studies [10] show that least square estimator is not robust to noise, thus the performance of cross-view recognition will be seriously degraded when the data is not clean. Xu et al. [27] indicates that Cauchy loss is more robust than L_2 loss. In order to strength the robustness of algorithm, we employ Cauchy loss to instead of L_2 loss, then the objective function can be rewritten as

$$\min_{W} \frac{1}{mn} \sum_{i=1}^{m} \sum_{v=1}^{n} \log \left(1 + \frac{||z_i - W_v^{\mathrm{T}} x_i^v||_2^2}{c^2} \right) + c_1 \sum_{v=1}^{n} ||W_v||_F^2 + c_2 \sum_{i=1}^{m} ||z_i - y_i||_2^2 \tag{4}$$

where c is a constant. The Eq. (4) jointly models the relationships between each view space $\mathcal{X}^v$ and the common subspace Z by a robust algorithm. It can be expected that we can acquire multiple linear projections by solving this problem with an input of multiple view data. Equation (4) is our method in this paper, named Robust Multi-view Common Component Learning.

3 Optimization

Through alternating optimization method, problem (4) can be decomposed into two subproblems. Inspired by generalized Weiszfeld's method [24], Xu et al. [27] develop Iteratively Reweight Residuals (IRR) technology. In this paper, we utilize IRR to efficiently optimize the subproblems, respectively.

Firstly, fix view projection matrices $\{W_v\}_{v=1}^{n}$ and data representation in common subspace $\{z_j\}_{j=1}^{m}, j \neq i$, then Eq. (4) is decomposed into subproblem over z_i, which can be written as

$$\min_{z_i} \mathcal{J} = \frac{1}{n}\sum_{v=1}^{n} \log\left(1 + \frac{||z_i - W_v^{\mathrm{T}} x_i^v||_2^2}{c^2}\right) + c_2\,||z_i - y_i||_2^2 \tag{5}$$

setting the gradient of $\mathcal{J}$ with respect to z_i to zero, we have

$$\sum_{v=1}^{n} \frac{z_i - W_v^{\mathrm{T}} x_i^v}{c^2 + ||z_i - W_v^{\mathrm{T}}||_2^2} + nc_2\,(z_i - y_i) = 0 \tag{6}$$

which can also be rewritten as

$$\left(\sum_{v=1}^{n} \frac{1}{c^2 + ||z_i - W_v^{\mathrm{T}} x_i^v||_2^2} + nc_2\right) z_i = \left(\sum_{v=1}^{n} \frac{W_v^{\mathrm{T}} x_i^v}{c^2 + ||z_i - W_v^{\mathrm{T}} x_i^v||_2^2} + nc_2 y_i\right) \tag{7}$$

where $r^v = z_i - W_v^{\mathrm{T}} x_i^v$ is referred to as the residual. A weight function is then defined as

$$Q = \left[\frac{1}{c^2 + ||r^1||_2^2}, \ldots, \frac{1}{c^2 + ||r^n||_2^2}\right] \tag{8}$$

Then, put together Eqs. (7) and (8)

$$z_i = \left(\left(\sum_{v=1}^{n} Q_v\right) + nc_2\right)^{-1} \left(\left(\sum_{v=1}^{n} W_v^{\mathrm{T}} x_i^v\right) + nc_2 y_i\right) \tag{9}$$

It is obvious that Q is the function of z_i, we thus utilize IRR technology to iteratively update z_i using Eq. (9) with an initial value until convergence. The procedure is described in Algorithm 1.

By fixing all common subspace data points $\{z_i\}_{i=1}^{m}$ and view projection matrix $\{W_k\}_{k=1}^{n}, k \neq v$, Eq. (4) is reduced to the following subproblem over projection matrix W_v:

$$\min_{W_v} \mathcal{J} = \frac{1}{m}\sum_{i=1}^{m} \log\left(1 + \frac{||z_i - W_v^{\mathrm{T}} x_i^v||_2^2}{c^2}\right) + c_1\,||W_v||_F^2 \tag{10}$$

Setting the derivation of $\mathcal{J}$ with respect to W_v to zero. Finally, the projection matrix W_v can be updated by

$$W_v = \left(\left(\sum_{i=1}^{m} x_i^v Q_i\,(x_i^v)^{\mathrm{T}}\right) + mc_1\right)^{-1} \sum_{i=1}^{m} x_i^v Q_i z_i^{\mathrm{T}} \tag{11}$$

Algorithm 1. Solving Eq. (9) by IRR Method

Input: The view data of object i: $x_i^1, x_i^2, \ldots, x_i^n$; projection matrices: $\{W_v\}_{v=1}^n$; the binary label vector of object i: y_i; and the initial value: z_i^0
Output: z_i
1: **for** $k = 1, \ldots, itermax$ **do**
2: 1. calculate residuals $\{r^v\}_{v=1}^n$ using z_i^{k-1};
3: 2. update the weight function Q through Eq. (11);
4: 3. using Eq. (12) to update z_i^k;
5: **if** the estimates of z_i converge **then**
6: break;
7: **end if**;
8: **end for**;
9: **Return:** z_i

where Q_i is the ith element of Q, and Q can be written as

$$Q = \left[\frac{1}{c^2 + ||r_1||_2^2}, \ldots, \frac{1}{c^2 + ||r_m||_2^2} \right] \tag{12}$$

where $r_i = z_i - W_v^{\mathrm{T}} x_i^v$. Considering Q_i depends on W_v, we thus iteratively update W_v using Eq. (12). The iterative procedure is similar to Algorithm 1.

4 Experiments

In this section, the proposed method RMCCL is evaluated on two heterogeneous face databases, including the Heterogeneous Face Biometrics (HFB) database and the CMU Pose, Illumination, and Expression (PIE) database (CMU PIE). The representative methods CCA [9], MCCA [20], LRCS [5], MvDA [11] and MvDA-VC [12] and MCCL-L2 are brought in for comparison and the traditional single-view methods PCA and LDA serve as baselines. It is worth noting that MCCL-L2 algorithm is proposed by replacing Cauchy loss in Eq. (4) with L_2 loss. It can be regard as a middle algorithm of RMCCL. We expect to prove Cauchy loss works better than L_2 loss.

4.1 Experimental Settings

One object can be captured by different sensors, such as visual light camera and near-infrared camera. There are 100 individuals on the HFB dataset, in which each subject has four visual light photos and four near infrared photos. We use 70 individuals to train the RMCCL model, and the rest are utilized to evaluate the performance of our algorithm. Some selected faces from the HFB database can be seen in Fig. 2. Regarding the CMU PIE database, five poses, i.e., C11, C29, C27, C05 and C37, are chosen as the multiple view data, and each subject has four face images under every pose. Some selected faces under five poses can be

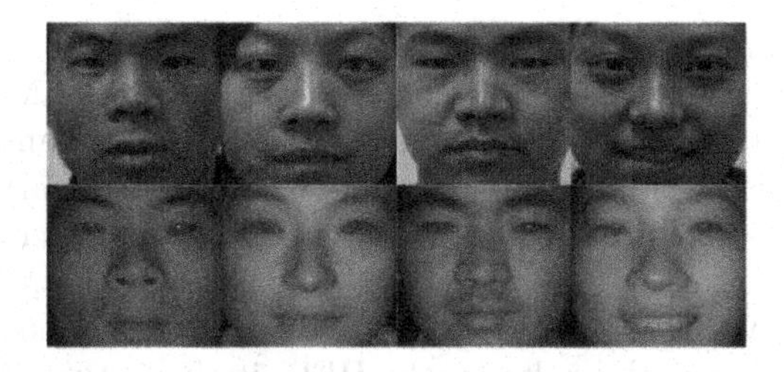

Fig. 2. Selected subjects from the HFB dataset. The first row is visual light images, and the second row is near infrared images

seen in Fig. 3. 45 people are chosen to train the model, and the rest are employed to evaluate algorithm.

The HFB and the CMU PIE heterogeneous face datasets are cropped according standard protocol and are resized to 32×32 and 64×64, respectively. It is worth noting that, in the HFB or the CMU PIE dataset, there are multiple schemes to divide datasets into training set and testing set. To make the results more convincing, all experiments are repeated ten times by randomly dividing these into two parts. The average results are regarded as the final accuracies (mACC). Furthermore, all methods employ principle component analysis to realize dimensionality reduction, and each approach sets dimension to get the best accuracy.

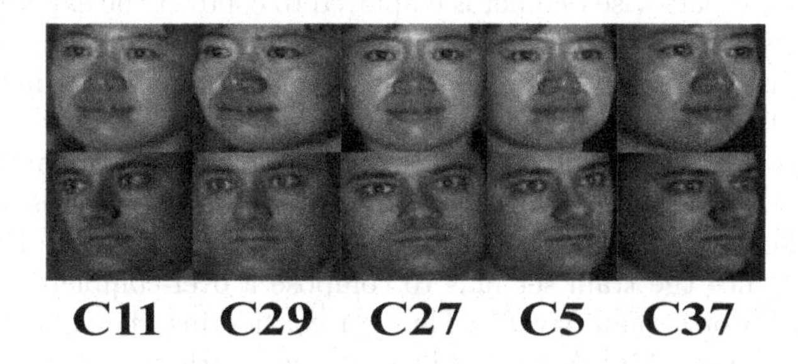

Fig. 3. Selected subjects from the CMU PIE dataset. C11, C29, C27, C05 and C37 poses are selected as the views.

4.2 Face Recognition Between Visual Light and Near Infrared Images

In this experiment, we evaluate the performance of RMCCL on the HFB, in which objects are captured by different sensors. The proposed methods RMCCL and MCCL-L2 are compared to classic works including PCA, LDA, CCA [9], MCCA [20], LRCS [5], MvDA [11], MvDA-VC [12]. The experimental results are shown in Table 1.

As can be seen, MvDA-VC is the best competitive method with an high recognition accuracy of 55.88%. However, our proposed method RMCCL achieves a highest recognition rate of 61.46%. Note that, because low rank representation methods usually select train set as the over-complete dictionary, LRCS fails to work when dataset is small. Therefore, it obtains a recognition rate of 33.21%. It is just a bit better than MCCA. What's more, our method RMCCL has better performance than MCCL-L2. Obviously, it accords with the theoretical analysis. Through the experimental results on the HFB, it shows that RMCCL outperforms other subspace learning methods in two-view case.

Table 1. The average recognition results (%) on the HFB dataset

	PCA	LDA	CCA	LRCS	MvDA	MvDA-VC	MCCL-L2	RMCCL
NIR-VIS	9.17	13.00	30.83	31.42	41.75	56.42	59.42	61.67
VIS-NIR	8.17	16.33	31.42	35.00	45.00	55.33	60.50	61.25
Average	8.67	14.67	31.13	33.21	43.38	55.88	59.96	**61.46**
Std	2.10	**1.75**	4.41	7.00	8.87	5.34	3.80	3.36

4.3 Face Recognition Across Poses

In the experiments, the CMU PIE dataset is utilized to evaluate face recognition across poses, and pair-wise manner is employed to conduct the experiment. Since each object owns five view data, it generates $5 \times 4 = 20$ recognition accuracies. Then, we utilize the mACC as final evaluation index. The Experimental results on the CMU PIE dataset can be seen in Table 2.

As can be seen, due to unite the within-class samples and separate between-class samples in the latent unified space, the MvDA outperforms PCA, LDA, MCCA on the CMU PIE database. As we have explained in the HFB dataset experiments, since the train set fails to compose a over-complete basic, LRCS performs even worse than MvDA. Through considering the relation of multiple view projections, MvDA-VC further improves performance by 11.7%. It is worth noting that, proposed method RMCCL outperforms MvDA-VC with an absolute improvement by 6.1%. In addition, RMCCL also performs better than MCCL-L2 with an improvement by 2.8%. It demonstrates that Cauchy loss has better robustness than L_2 loss. Experimental results show that our method RMCCL performs better than several state-of-the-art algorithms on the CMU PIE dataset. Furthermore, it demonstrates that our approach is also suitable for multi-view data.

Table 2. The average recognition results (%) on the CMU PIE dataset

Gallery	Probe	PCA	LDA	MCCA	LRCS	MvDA	MvDA-VC	MCCL-L2	RMCCL
C11	C29	67.1	20.2	48.0	72.3	69.3	79.7	79.1	81.0
	C27	29.6	8.3	26.5	33.2	46.7	58.3	61.3	64.2
	C05	13.9	6.1	22.3	19.5	37.0	48.2	51.2	58.0
	C37	12.7	5.1	20.5	14.7	37.4	49.6	46.3	51.6
C29	C11	69.2	19.8	44.2	69.8	67.7	80.4	77.4	79.5
	C27	42.3	14.6	43.2	51.6	57.6	77.0	78.5	80.0
	C05	21.5	5.8	33.6	29.0	47.8	62.7	69.9	73.4
	C37	17.6	6.0	24.9	22.0	42.9	55.8	54.6	59.1
C27	C11	26.5	9.2	29.6	33.0	47.7	57.0	62.2	63.2
	C29	40.1	15.7	46.2	51.2	59.2	75.0	81.6	82.3
	C05	39.9	11.5	41.6	48.7	58.8	69.5	77.2	82.2
	C37	26.6	8.8	27.4	35.1	44.0	57.8	58.2	63.7
C05	C11	11.2	5.9	23.2	19.8	37.4	48.2	52.4	54.6
	C29	20.4	6.1	28.8	25.4	47.1	61.7	69.6	71.5
	C27	34.8	13.4	35.0	46.3	60.4	70.0	75.1	78.2
	C37	65.7	29.0	48.8	73.7	65.7	74.8	78.2	81.5
C37	C11	12.2	5.4	17.2	15.1	37.8	47.9	48.3	51.1
	C29	20.0	5.9	19.6	22.1	40.1	52.4	53.0	59.0
	C27	24.1	8.2	22.9	35.1	47.8	56.2	57.9	61.8
	C37	72.5	26.3	45.3	76.7	68.5	74.1	83.0	82.6
Average		33.4	11.6	32.4	39.7	51.1	62.8	66.0	**68.9**
Std.		2.7	1.7	2.3	3.0	2.1	2.7	1.7	**1.5**

5 Conclusion

In this paper, we propose RMCCL algorithm, which learns multiple linear transforms to extract the common component of multi-view data from the same instance. To enhance the discriminant ability and robustness of subspace, we introduce binary label matrix technology and serve Cauchy loss as our error measurement. Experimental results show that our approach outperforms several state-of-the-art algorithms. In future, we'll further analyze the robust and convergence of algorithm. Certainly, in order to better address various data, we can also bring kernel technology [26,29] to develop the kernel version of RMCCL.

Acknowledgment. This work was supported partially by National Key Technology Research and Development Program of the Ministry of Science and Technology of China (No. 2015BAK36B00), in part by the Key Science and Technology of Shen zhen (No. CXZZ20150814155434903), in part by the Key Program for International S&T Cooperation Projects of China (No. 2016YFE0121200), in part by the National Natural Science Foundation of China (No. 61571205), in part by the National Natural Science Foundation of China (No. 61772220).

References

1. Chen, N., Zhu, J., Xing, E.P.: Predictive subspace learning for multi-view data: a large margin approach. In: Advances in Neural Information Processing Systems, pp. 361–369 (2010)
2. Chen, Y., Wang, L., Wang, W., Zhang, Z.: Continuum regression for cross-modal multimedia retrieval. In: 2012 19th IEEE International Conference on Image Processing (ICIP), pp. 1949–1952. IEEE (2012)
3. Diethe, T., Hardoon, D.R., Shawe-Taylor, J.: Constructing nonlinear discriminants from multiple data views. In: Balcázar, J.L., Bonchi, F., Gionis, A., Sebag, M. (eds.) ECML PKDD 2010. LNCS (LNAI), vol. 6321, pp. 328–343. Springer, Heidelberg (2010). https://doi.org/10.1007/978-3-642-15880-3_27
4. Diethe, T., Hardoon, D.R., Shawe-Taylor, J.: Multiview fisher discriminant analysis. In: NIPS Workshop on Learning from Multiple Sources (2008)
5. Ding, Z., Fu, Y.: Low-rank common subspace for multi-view learning. In: 2014 IEEE International Conference on Data Mining (ICDM), pp. 110–119. IEEE (2014)
6. Gopalan, R., Li, R., Chellappa, R.: Domain adaptation for object recognition: an unsupervised approach. In: 2011 IEEE International Conference on Computer Vision (ICCV), pp. 999–1006. IEEE (2011)
7. Gui, J., Jia, W., Zhu, L., Wang, S.L., Huang, D.S.: Locality preserving discriminant projections for face and palmprint recognition. Neurocomputing **73**(13), 2696–2707 (2010)
8. Gui, J., Sun, Z., Jia, W., Hu, R., Lei, Y., Ji, S.: Discriminant sparse neighborhood preserving embedding for face recognition. Pattern Recogn. **45**(8), 2884–2893 (2012)
9. Hotelling, H.: Relations between two sets of variates. Biometrika **28**(3/4), 321–377 (1936)
10. Huber, P.J.: Robust Statistics. Springer, Heidelberg (2011). https://doi.org/10.1007/978-3-642-04898-2_594
11. Kan, M., Shan, S., Zhang, H., Lao, S., Chen, X.: Multi-view discriminant analysis. In: Fitzgibbon, A., Lazebnik, S., Perona, P., Sato, Y., Schmid, C. (eds.) ECCV 2012. LNCS, vol. 7572, pp. 808–821. Springer, Heidelberg (2012). https://doi.org/10.1007/978-3-642-33718-5_58
12. Kan, M., Shan, S., Zhang, H., Lao, S., Chen, X.: Multi-view discriminant analysis. IEEE Trans. Pattern Anal. Mach. Intell. **38**(1), 188–194 (2016)
13. Li, J., Wu, Y., Zhao, J., Lu, K.: Low-rank discriminant embedding for multiview learning. IEEE Trans. Cybern. **47**(11), 3516–3529 (2017)
14. Li, S., Fu, Y.: Robust subspace discovery through supervised low-rank constraints. In: Proceedings of the 2014 SIAM International Conference on Data Mining, pp. 163–171. SIAM (2014)
15. Li, S.Z., Lei, Z., Ao, M.: The HFB face database for heterogeneous face biometrics research. In: IEEE Computer Society Conference on Computer Vision and Pattern Recognition Workshops, CVPR Workshops 2009, pp. 1–8. IEEE (2009)
16. Li, W., Wang, X.: Locally aligned feature transforms across views. In: Proceedings of the IEEE Conference on Computer Vision and Pattern Recognition, pp. 3594–3601 (2013)
17. Lin, D., Tang, X.: Inter-modality face recognition. In: Leonardis, A., Bischof, H., Pinz, A. (eds.) ECCV 2006. LNCS, vol. 3954, pp. 13–26. Springer, Heidelberg (2006). https://doi.org/10.1007/11744085_2

18. Nielsen, A.A.: Multiset canonical correlations analysis and multispectral, truly multitemporal remote sensing data. IEEE Trans. Image Process. **11**(3), 293–305 (2002)
19. Rosipal, R., Krämer, N.: Overview and recent advances in partial least squares. In: Saunders, C., Grobelnik, M., Gunn, S., Shawe-Taylor, J. (eds.) SLSFS 2005. LNCS, vol. 3940, pp. 34–51. Springer, Heidelberg (2006). https://doi.org/10.1007/11752790_2
20. Rupnik, J., Shawe-Taylor, J.: Multi-view canonical correlation analysis. In: Conference on Data Mining and Data Warehouses (SiKDD 2010), pp. 1–4 (2010)
21. Sharma, A., Jacobs, D.W.: Bypassing synthesis: PLS for face recognition with pose, low-resolution and sketch. In: 2011 IEEE Conference on Computer Vision and Pattern Recognition (CVPR), pp. 593–600. IEEE (2011)
22. Sim, T., Zhang, S., Li, J., Chen, Y.: Simultaneous and orthogonal decomposition of data using multimodal discriminant analysis. In: 2009 IEEE 12th International Conference on Computer Vision, pp. 452–459. IEEE (2009)
23. Tzimiropoulos, G., Zafeiriou, S., Pantic, M.: Subspace learning from image gradient orientations. IEEE Trans. Pattern Anal. Mach. Intell. **34**(12), 2454–2466 (2012)
24. Voß, H., Eckhardt, U.: Linear convergence of generalized Weiszfeld's method. Computing **25**(3), 243–251 (1980)
25. Wang, K., He, R., Wang, W., Wang, L., Tan, T.: Learning coupled feature spaces for cross-modal matching. In: Proceedings of the IEEE International Conference on Computer Vision, pp. 2088–2095 (2013)
26. Xiao, M., Guo, Y.: Feature space independent semi-supervised domain adaptation via kernel matching. IEEE Trans. Pattern Anal. Mach. Intell. **37**(1), 54–66 (2015)
27. Xu, C., Tao, D., Xu, C.: Multi-view intact space learning. IEEE Trans. Pattern Anal. Mach. Intell. **37**(12), 2531–2544 (2015)
28. Xu, Y., Fang, X., Wu, J., Li, X., Zhang, D.: Discriminative transfer subspace learning via low-rank and sparse representation. IEEE Trans. Image Process. **25**(2), 850–863 (2016)
29. You, X., Guo, W., Yu, S., Li, K., Príncipe, J.C., Tao, D.: Kernel learning for dynamic texture synthesis. IEEE Trans. Image Process. **25**(10), 4782–4795 (2016)
30. Yu, S., Abraham, Z.: Concept drift detection with hierarchical hypothesis testing. In: Proceedings of the 2017 SIAM International Conference on Data Mining, pp. 768–776. SIAM (2017)

Person Re-identification on Heterogeneous Camera Network

Jiaxuan Zhuo[1,3,4], Junyong Zhu[2,3,4], Jianhuang Lai[2,3,4(✉)], and Xiaohua Xie[2,3,4]

[1] School of Electronics and Information Technology, Sun Yat-sen University, Guangzhou, China
[2] School of Data and Computer Science, Sun Yat-sen University, Guangzhou, China
stsljh@mail.sysu.edu.cn
[3] Guangdong Key Laboratory of Information Security Technology, Guangzhou, China
[4] Key Laboratory of Machine Intelligence and Advanced Computing, Ministry of Education, Sun Yat-sen University, Guangzhou, China
zhuojx5@mail2.sysu.edu.cn, {zhujuny5,xiexiaoh6}@mail.sysu.edu.cn

Abstract. Person re-identification (re-id) aims at matching person images across multiple surveillance cameras. Currently, most re-id systems highly rely on color cues, which are only effective in good illumination conditions, but fail in low lighting conditions. However, for security issues, it is very important to conduct surveillance in low lighting conditions. To remedy this problem, we propose using depth cameras to perform surveillance in dark places, while using traditional RGB cameras in bright places. Such a heterogeneous camera network brings a challenge to match images across depth and RGB cameras. In this paper, we mine the correlation between two heterogeneous cues (depth and RGB) on both feature-level and transformation-level. As such, depth-based features and RGB-based features are transformed to the same space, which alleviates the problem of cross-modality matching between depth and RGB cameras. Experimental results on two benchmark heterogeneous person re-id datasets show the effectiveness of our method.

Keywords: Person re-identification · Multi-modality matching
Heterogeneous camera network

1 Introduction

Nowadays, the Closed Circuit Television (CCTV) has been widely deployed in many security-sensitive places such as individual house, museum, bank, etc. Because of the economical or privacy issues, there are always non-overlapping regions between different camera views. Therefore, re-identifying pedestrians across different camera views is a critical and fundamental problem for intelligent video surveillance such as cross-camera person searching and tracking. Such a problem is called person re-identification (re-id).

J. Yang et al. (Eds.): CCCV 2017, Part III, CCIS 773, pp. 280–291, 2017.
https://doi.org/10.1007/978-981-10-7305-2_25

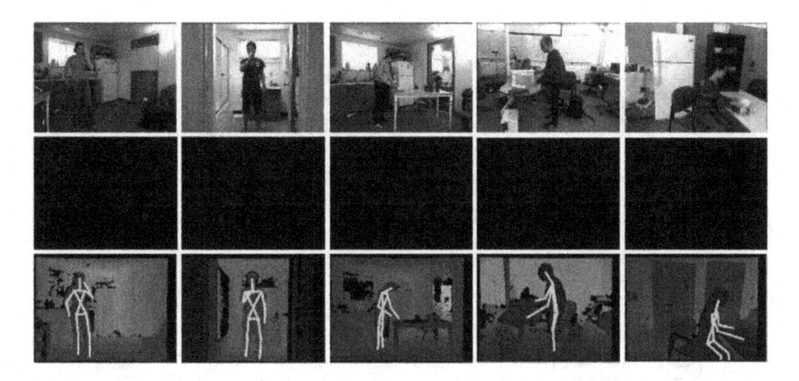

Fig. 1. Illustration of RGB and depth images. Row 1 shows the RGB images in the bright environment. Row 2 shows the RGB images in the dark environment. Row 3 shows the depth images in the dark environment.

Currently, most CCTV systems are based on RGB cameras, and thus the corresponding re-id approaches mainly rely on appearance features. However, in dark environment, appearance features may be unreliable since limited information perceived by RGB cameras. Hence, it is necessary to apply new devices to the dark environment. One alternative solution is to use depth cameras, such as Kinect [1,8,9]. Depth cameras provide depth information and body skeleton joints which are invariant to illumination changes (see Fig. 1). That is, depth cameras remain valid in the dark environment. Depth information is more related to human body shape, and thus beneficial to the re-id in the dark environment [9]. RGB cameras and depth cameras, therefore, form a heterogeneous surveillance network. Previous works in person re-id focus on either RGB camera network [2–7] or depth camera network [8,9], yet none of them addresses the heterogeneous camera network that contains both RGB and depth cameras. In this paper, we focus on how to match pedestrians across depth and RGB cameras in the heterogeneous surveillance network, which has not been studied before.

Along with the line of the traditional person re-id framework [3–6,9], our cross-modality re-id system contains two phases: feature extraction and similarity measurement. Different from [3–6,9], the key idea behind our approach is to mine the correlation between different modalities. In the feature extraction phase, considering that color and texture features [3–7] are not available in depth cameras, we propose to extract body shape information which intrinsically exists in both RGB and depth images. Specifically, for RGB images, we respectively extract two kinds of typical edge gradient features, the classic Histogram of Oriented Gradient (HOG) [7] and the recent proposed Scale Invariant Local Ternary Patterns (SILTP) [5]. Both of them are widely used for shape descriptors. For depth images, we extract the Eigen-depth feature that is recently proposed for Kinect-based person re-id [9]. Note that, both edge gradient feature and Eigen-depth feature describe human body shape and thus can reduce the discrepancy between two kinds of different-modality features. Unfortunately, this correlation is far from a good solution to the cross-modality matching. To address this

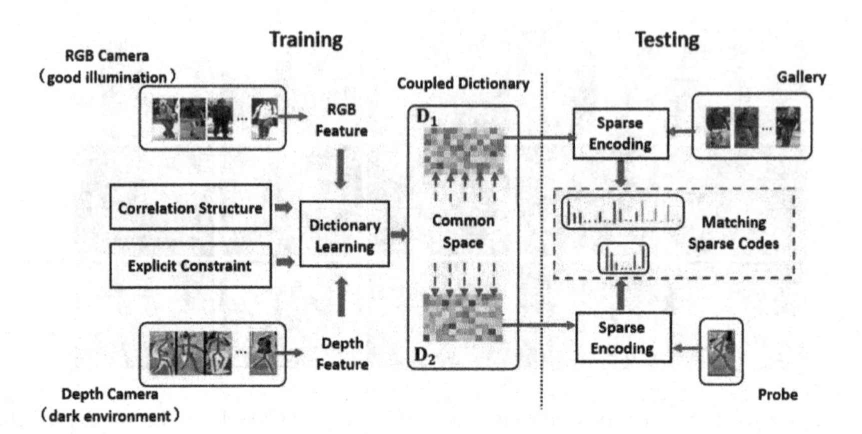

Fig. 2. Overview of our proposed approach. In the training phase, labeled image pairs from RGB and depth cameras are used to jointly learn the discriminative coupled dictionaries optimized by correlation structure and explicit constraint term. In the testing phase, we encode the features of different modalities through the coupled dictionaries as new representations for matching.

problem, we need to mine more correlation between the two modalities. Therefore, we propose a dictionary learning based algorithm that transforms edge gradient feature and Eigen-depth feature into sparse codes that share a common space. In this way, the similarity of edge gradient feature and Eigen-depth feature can be measured with the learned sparse codes. Figure 2 shows the overview of our approach.

In this paper, we identify the dark environment problem in person re-id suffered by unreliable and limited information from RGB cameras, and address this key problem through a novel cross-modality matching approach. To summarize, our contributions include:

- It is a new attempt to the re-id task across depth and RGB modality that we propose a dictionary learning based method to encode different-modality body shape features (edge gradient feature and Eigen-depth feature) into a common space.
- To enforce the discriminability of the learned dictionary pair, we design an explicit constraint term for dictionary learning so that our approach is more discriminative than several contemporary dictionary learning methods.
- Experiments on two heterogeneous person re-id benchmark datasets show the effectiveness of our approach.

2 Proposed Method

2.1 Problem Specification

For the training phase, $F_1 = [f_{11}, f_{21}, \ldots, f_{i1}]$ and $F_2 = [f_{12}, f_{22}, \ldots, f_{i2}]$ denote the gallery and the probe descriptor matrices, respectively, where f_{ij} is the feature set of the i_{th} training sample. F_1 and F_2 are from two heterogeneous cameras (depth camera and RGB camera) belonging to two different modalities with different dimensions, d_1 and d_2. The goal is to learn the dictionaries $D_1 \in \mathbb{R}^{d_1 \times k}$ and $D_2 \in \mathbb{R}^{d_2 \times k}$ jointly, where k is the size of sparse code. Let $C_1 = [c_{11}, c_{21}, \ldots, c_{i1}]$ and $C_2 = [c_{12}, c_{22}, \ldots, c_{i2}]$ denote the sparse codes of F_1 and F_2, each column of which $c_{ij} \in \mathbb{R}^k$ is the sparse code representing the i_{th} sample.

For the testing phase, feature matrices, $F_G = [f_1^G, f_2^G, \ldots, f_i^G]$ and $F_P = [f_1^P, f_2^P, \ldots, f_i^P]$, extracted from the gallery and the probe, and correspondent sparse codes are $C_G = [c_1^G, c_2^G, \ldots, c_i^G]$and $C_P = [c_1^P, c_2^P, \ldots, c_i^P]$, respectively.

2.2 Correlative Dictionary Learning

In traditional dictionary learning problem [10], the smaller reconstruction error contributes to a superior dictionary. Hence, we learn a dictionary pair by minimizing two sets of reconstruction errors. Besides, we constrain sparse codes C_1 and C_2 by L_1 regularization similar to sparse representation [10]. To prevent overfitting, we additionally incorporate L_2 regularization for dictionaries and formulate the optimization problem as:

$$\begin{aligned} \underset{D_1, D_2, C_1, C_2}{\arg\min} \ \{ &\|F_1 - D_1 C_1\|_F^2 + \|F_2 - D_2 C_2\|_F^2 + \\ &\lambda_C \|C_1\|_1 + \lambda_C \|C_2\|_1 + \lambda_D \|D_1\|_F^2 + \lambda_D \|D_2\|_F^2 \} \end{aligned} \tag{1}$$

where λ_C and λ_D are regularization parameters to balance the terms.

According to Least Square Semi-Coupled Dictionary Learning (LSSCDL) [11], L_1 regularization on sparse codes is more likely to destroy correlation structure of features, which is suggested to be replaced by L_2 regularization. Many researches [11–14] have proved that L_2 regularization can also play the effect of sparse representation. Therefore, in this paper, we also use L_2 regularization on sparse codes to improve Eq. (1).

Because of the difference between RGB-based and depth-based features, the direct matching results are always unsatisfied. So we capture the correlation between those same persons of cross-modality and penalize those largely different-class scatter. We consider minimizing the Euclidean distance between two sparse codes, namely $\|C_2 - C_1\|_F^2$, to develop the correlation between two dictionaries. Therefore, the objective function is given by:

$$\begin{aligned} \underset{D_1, D_2, C_1, C_2}{\arg\min} \ \{ &\|F_1 - D_1 C_1\|_F^2 + \|F_2 - D_2 C_2\|_F^2 + \lambda \|C_2 - C_1\|_F^2 + \\ &\lambda_C \|C_1\|_F^2 + \lambda_C \|C_2\|_F^2 + \lambda_D \|D_1\|_F^2 + \lambda_D \|D_2\|_F^2 \} \end{aligned} \tag{2}$$

where λ is a positive value which controls the tradeoff between the reconstruction errors and the distance between sparse coding matrices.

In our model, we seek for a discriminative dictionary pair that is able to be discriminative between the same pair and different pairs. We perform this discriminability by enforcing the constraint on the sparse coefficients corresponding to the learning dictionaries. Let $d_{ii} = c_{i1} - c_{i2}$ denote the Euclidean distance between sparse coefficients corresponding to the gallery and the probe of the same person i, and $d_{ij} = c_{i1} - c_{j2}$ denote the same form of different persons i and j. Specifically, we optimize our model such that the distance between the same person is much smaller than different persons, namely,

$$d_{ii} < d_{ij}, \forall j \neq i, \forall i \tag{3}$$

Thus we optimize the objective function by imposing explicit constraint term:

$$\begin{aligned} \text{s.t.} \quad & \|c_{i1} - c_{i2}\|_2^2 < \|c_{i1} - c_{j2}\|_2^2 \\ & \forall j \neq i, \forall\, i \end{aligned} \tag{4}$$

To simplify optimization, we build the objective function as convex function. Therefore, the constraint term could be modified as:

$$\begin{aligned} \text{s.t.} \quad & \|c_{i1} - c_{i2}\|_2^2 < s_1, \forall\, i \\ & \|c_{i1} - c_{j2}\|_2^2 < s_2, \forall j \neq i, \forall\, i \end{aligned} \tag{5}$$

where s_1 and s_2 are two constants, and $s_1 \ll s_2$. s_1 and s_2 are used to limit the distance between the samples.

In summary, the optimization problem of dictionary learning is described as:

$$\begin{aligned} \underset{D_1,D_2,C_1,C_2}{\arg\min} \; & \sum_{i=1}^{n} \{\|f_{i1} - D_1 c_{i1}\|_2^2 + \|f_{i2} - D_2 c_{i2}\|_2^2 + \lambda \|c_{i2} - c_{i1}\|_2^2 \\ & + \lambda_C \|c_{i1}\|_2^2 + \lambda_C \|c_{i2}\|_2^2 + \lambda_D \|D_1\|_F^2 + \lambda_D \|D_2\|_F^2\} \\ \text{s.t.} \quad & \|c_{i1} - c_{i2}\|_2^2 < s_1, \forall\, i \\ & \|c_{i1} - c_{j2}\|_2^2 < s_2, \forall j \neq i, \forall\, i \end{aligned} \tag{6}$$

We employ the alternating optimization algorithm to solve Eq. (6). Specically, we alternatively optimize over D_1, D_2, C_1 and C_2 one at a time, while fixing the other three. Firstly fix D_1, D_2, C_2 and use CVX [20] to optimize a column c_{i1} of C_1. The way to optimize C_2 is similar to C_1. And then we get D_1 and D_2 at gradient algorithm, which are given by

$$D_1 = (F_1 C_1^T)(C_1 C_1^T + \lambda_D I)^{-1} \tag{7}$$

$$D_2 = (F_2 C_2^T)(C_2 C_2^T + \lambda_D I)^{-1} \tag{8}$$

where I is a $k \times k$ identity matrix.

In this way, we alternatively optimize over D_1, D_2, C_1 and C_2 until convergency. The algorithm for training is summarized in Algorithm 1.

Algorithm 1. Correlative dictionary learning

Input: feature matrices F_1 and F_2, size of sparse code k, parameters λ, λ_C, λ_D, $s1$, $s2$
Output: coupled dictionaries D_1 and D_2
Initialize: D_1, D_2, C_1, C_2
while not converge **do**
Step 1: fix D_1, D_2, C_2, update each column of C_1 using CVX by Eq. (6)
Step 2: fix D_1, D_2, C_1, update each column of C_2 using CVX by Eq. (6)
Step 3: fix D_2, C_1, C_1, update D_1 by Eq. (7)
Step 4: fix D_1, C_1, C_2, update D_2 by Eq. (8)
end while

2.3 Person Re-identification by Our Framework

Using the correlative dictionary pair D_1 and D_2, we can obtain the sparse representations of the gallery and the probe. According to Eq. (6), the sparse code coefficients $C_G = [c_1^G, c_2^G, \ldots, c_i^G]$ and $C_P = [c_1^P, c_2^P, \ldots, c_i^P]$ can be respectively obtained by

$$\arg\min_{c_i^G} \left\| f_i^G - D_1 c_i^G \right\|_2^2 + \lambda_G \left\| c_i^G \right\|_2^2, \forall\, i \tag{9}$$

$$\arg\min_{c_i^P} \left\| f_i^P - D_2 c_i^P \right\|_2^2 + \lambda_P \left\| c_i^P \right\|_2^2, \forall\, i \tag{10}$$

where λ_G and λ_P are regularization parameters to balance the terms for the gallery and the probe, respectively.

We use CVX to solve the problems in Eqs. (9) and (10). The algorithm for testing is summarized in Algorithm 2. Finally, the learned sparse codes C_G and C_P are taken as correlative reconstructive features to identity matching by computing the similarity according to the Euclidean distance. In this way, the computational efficiency of identity matching is the same as the standard sparse representation in person re-id [21,22].

Algorithm 2. Sparse representation

Input: feature matrices F_G and F_P, coupled dictionaries D_1 and D_2, parameters λ_G, λ_P
Output: sparse code coefficients C_G and C_P
Initialize: C_G, C_P
Step 1: obtain each column of C_G using CVX by Eq. (9)
Step 2: obtain each column of C_P using CVX by Eq. (10)
Step 3: normalize C_G and C_P

3 Experiment

3.1 Datasets and Features

Datasets. We evaluate our approach on two RGB-D person re-id datasets RGBD-ID [19] and BIWI RGBD-ID [15] collected by Kinect cameras.

BIWI RGBD-ID [15] has three groups, namely "*Training*", "*Still*" and "*Walking*", which respectively contains 50, 28 and 28 humans with different clothing. Each person has 300 frames of RGB images, depth images and skeletons. We use the complete "*Training*" and "*Still*" set and hence there are 78 samples in total. And then we select one frame including RGB and depth for each sample. By convention, we randomly choose about half of the samples, 40 pedestrians for training and the other for testing.

RGBD-ID [19] contains 79 identities with five RGB images, five point clouds and skeletons. We randomly sample approximately half of people (41 identities) in "*Walking1*" for training and the rest for testing because groups "*Walking1*" and "*Walking2*" contain the same person with different frontal views. Only one frame with all information for each person is randomly selected to experiment.

Features. Torso and head are segmented from each image and divided into 6 $\times$ 2 rectangular patches by image preprocessing. We obtain integral features of each image by combining local features of each patch. To test the ability of our model adapted to different representations, we consider two kinds of representative edge features, HOG [7] and SILTP [5] as RGB-based features in our experiment. For each depth image, we combine Eigen-depth feature with skeleton information for complete representation as depth-based features [9]. Both HOG and SILTP features capture local region human body shape, as well as depth-based features. Note that RGB-based features and depth-based features belong to different modalities.

3.2 Experiment Settings

Methods for Comparison. To evaluate the effectiveness of our approach, we compare our method with Least Square Semi-Coupled Dictionary Learning (LSSCDL) [11], Canonical Correlation Analysis (CCA) [16]. We also set a baseline which is the multi-modality matching result without any connection between RGB-based and depth-based features for comparison. CCA is a coherent subspace learning algorithm which projects two sets of random variables to the correlated space so as to maximize the correlation between the projected variables in correlated space. LSSCDL is a similar dictionary learning based algorithm, which learns a pair of dictionaries and a mapping function efficiently to investigate the intrinsic relationship between feature patterns. Recently, CCA and LSSCDL have been applied to re-id problem of matching people across disjoint camera views, involving multi-view or multi-modality tasks. CCA and LSSCDL can be used to address the multi-modality matching problem because they can provide connection between uncorrelated variables.

Evaluation Metrics. Recognition rates at selected ranks and the histograms are used to evaluate the performance. The rank-n rate represents the expectation of finding the correct match in the top n matches [17] and rank-1 rate plays an important role to determine the performance of re-id. To ensure fair comparison, the same training and testing samples are used in all methods and the experiments are conducted 10 times to gain the average results.

Parameter Settings. In the following experiments, we set $k = 100$, $\lambda = 0.1$, $\lambda_C = \lambda_D = 0.001$, $\lambda_G = \lambda_P = 0.01$, $s_1 = 0.1$, $s_2 = 100$ for our method. All parameters of other methods are set as suggested in their papers [11,16].

3.3 Experiment Results

Result on BIWI. To prove the universal applicability of our approach, we extract two kinds of typical RGB-based features, HOG and SILTP, to match depth-based features, respectively. Each experiment is carried out in two cases. One is depth-based features for the gallery and RGB-based features for the probe. The other is the reverse. The results are shown in Table 1 and Fig. 3. It can be seen that our method largely outperforms the baseline, which shows the effectiveness of our method to address the multi-modality matching problem. Compared with CCA, our method can establish closer connection than CCA. The main reason is that our method allows screening the vital information and reduces the influence of invalid elements by sparse representation while CCA can not. Our method also outperforms LSSCDL generally, which demonstrates that the explicit constraint term enforces the discriminability of the learning dictionary pair.

Result on RGBD-ID. In RGBD-ID, people's head has be blurred in each RGB images, so the problem becomes more challenging. Following the protocol of experiment on BIWI, we compare the methods in [11,16] using the same feature as that on BIWI in two cases. Table 1 and Fig. 4 show that our method presents the best performance in rank-1 rate. Note that, the margins between the proposed model, CCA, and LSSCDL are small, because the blurred images

Table 1. Recognition rates (%) of cross HOG/SILTP and depth feature on BIWI and RGBD-ID dataset. F_g: Gallery; F_p: Probe; D: Depth; H: HOG; S: SILTP.

	Feature	F_g-D,F_p-H				F_g-H,F_p-D				F_g-D,F_p-S				F_g-S, F_p-D			
	Rank	1	5	10	20	1	5	10	20	1	5	10	20	1	5	10	20
BIWI	RANDOM	2.63	12.11	23.16	50.53	3.15	15.00	26.32	56.58	2.37	12.63	25.53	52.89	2.63	15.53	28.42	55.00
	CCA[13]	6.31	24.21	40.79	64.47	6.31	27.63	40.79	63.68	8.42	26.32	41.58	65.26	6.58	27.37	45.00	72.90
	LSSCDL[11]	8.42	27.11	46.05	70.79	7.11	28.42	41.32	61.32	9.47	24.21	40.26	64.74	7.37	29.47	50.26	74.74
	Ours	11.84	28.42	44.47	70.00	11.32	30.26	48.16	70.26	12.11	26.32	41.58	68.16	9.21	26.32	46.05	71.05
RGBD-ID	RANDOM	2.11	12.63	27.11	55.26	2.63	10.26	23.16	51.05	2.89	13.42	26.32	53.16	1.32	10.26	24.47	51.58
	CCA[13]	7.11	19.74	32.11	56.58	4.74	17.89	30.53	55.53	7.37	17.11	31.05	55.79	6.84	20.79	33.16	58.42
	LSSCDL[11]	7.63	10.47	31.32	57.11	8.16	20.00	34.74	57.37	8.16	18.42	31.32	56.05	7.89	17.11	28.95	53.68
	Ours	11.05	20.00	35.00	57.37	8.94	21.31	36.05	60.79	8.95	19.47	31.32	55.79	8.68	20.79	33.42	60.79

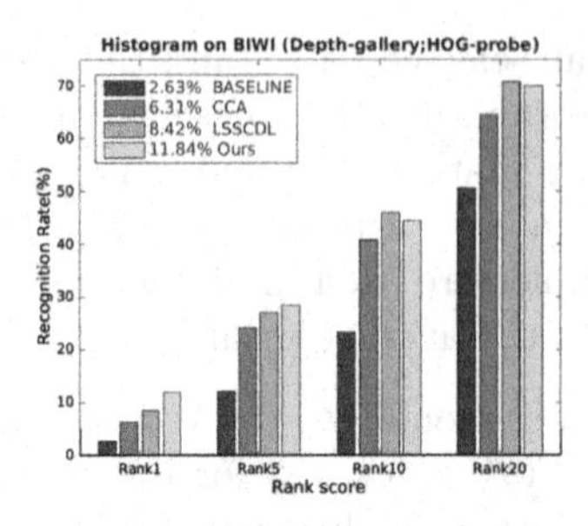

(a) Depth-gallery, HOG-probe

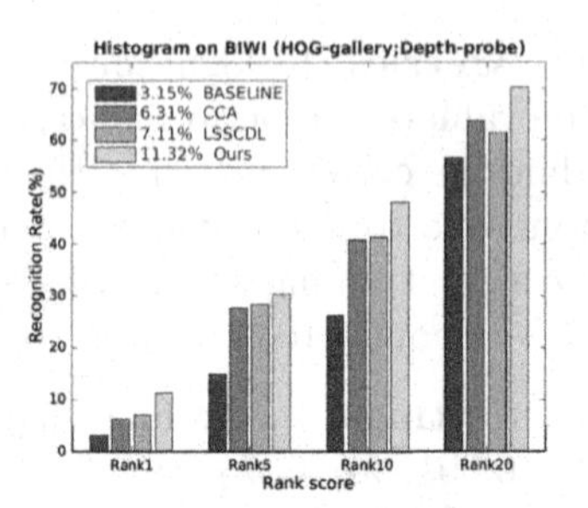

(b) HOG-gallery, Depth-probe

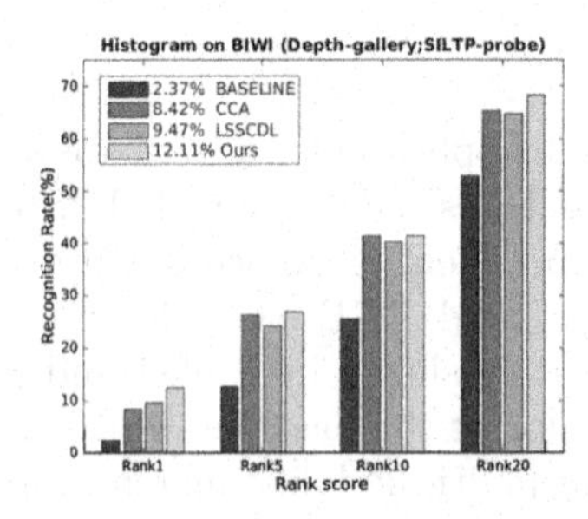

(c) Depth-gallery, SILTP-probe

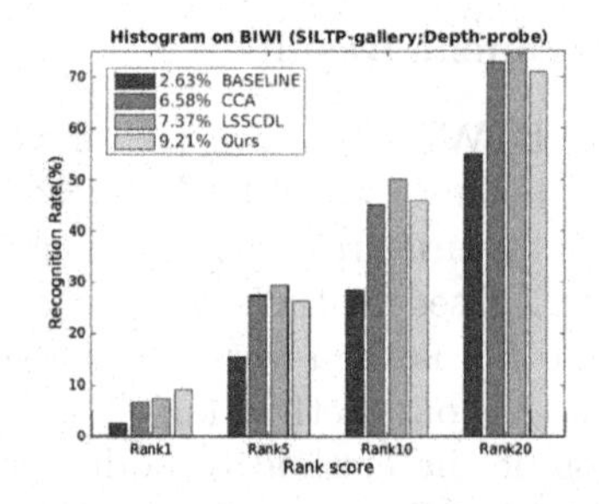

(d) SILTP-gallery, Depth-probe

Fig. 3. Histogram and rank-1 rate on BIWI in two cases.

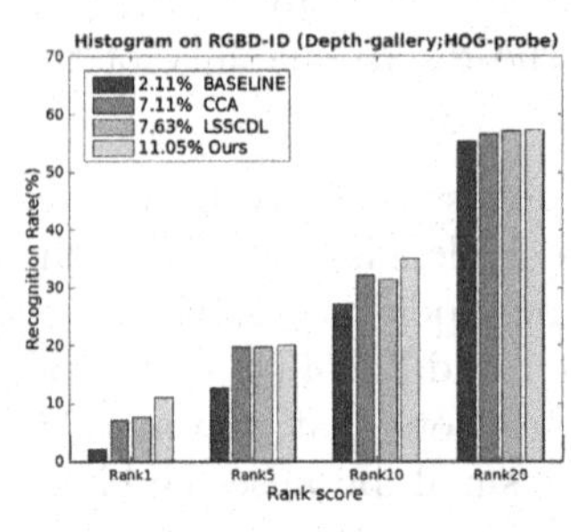

(a) Depth-gallery, HOG-probe

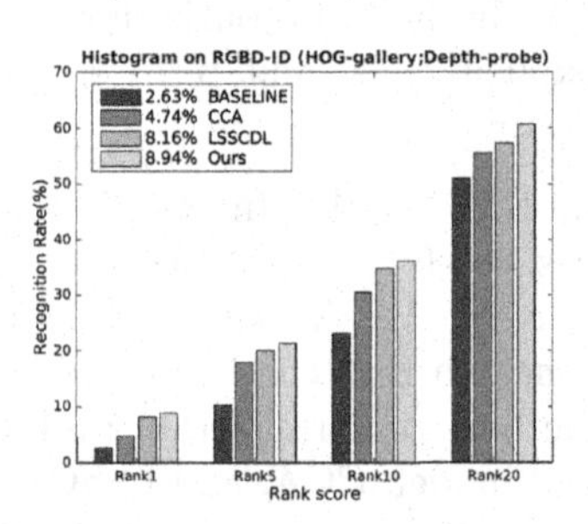

(b) HOG-gallery, Depth-probe

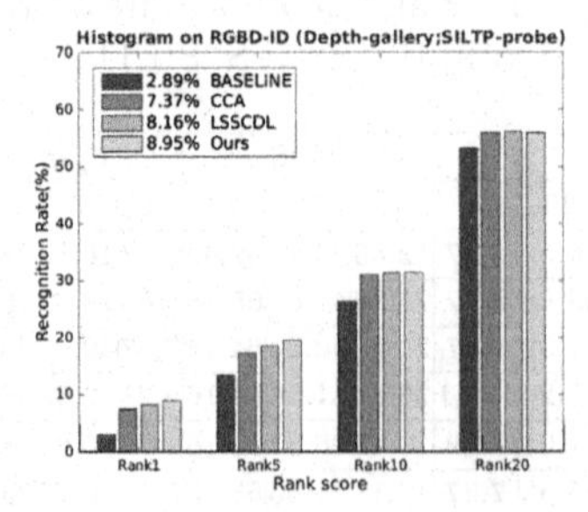

(c) Depth-gallery, SILTP-probe

(d) SILTP-gallery, Depth-probe

Fig. 4. Histogram and rank-1 rate on RGBD-ID in two cases.

may significantly degrade the discriminability of edge gradient features and thus reduce the correlation between two heterogeneous modalities. With such a weak correlation, the margins between these models cannot be large.

3.4 Effect of Feature Dimensions

We further evaluate the effect of the dimensions of reconstructive features by adjusting the number of the dimensions on BIWI dataset. In particular, we change the dimensions of reconstructive features from 50 to 500 and observe the performance of CCA, LSSCDL and our method. The results in Fig. 5 reflect that (1) reconstructive features with low dimensions outperform those with high dimensions on the whole. The reason may be attributed to the fact that high dimensions are more likely to cause overfitting when the number of training samples is small [18]. (2) The explicit constraint term in Eq. (5) allows to mine more discriminant features, making our method more stable and effective than CCA and LSSCDL in different dimensions.

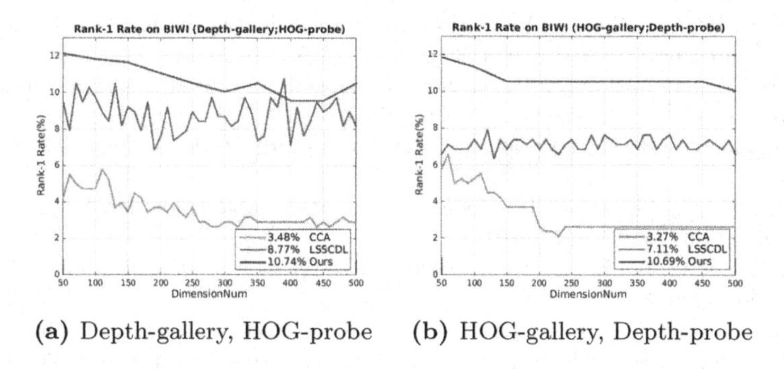

(a) Depth-gallery, HOG-probe (b) HOG-gallery, Depth-probe

Fig. 5. Rank-1 rate on BIWI with variable dimensions of reconstructive feature in two cases and mean average of rank-1 rate.

4 Conclusion

In this paper, we have extended the traditional RGB camera based person re-identification problem to a RGB and depth based cross-modality matching problem. Such a problem is critical when video analysis is needed in heterogeneous camera network. To the best of our knowledge, it is the first attempt for the person re-id to deal with the situation across RGB and depth modality. We have also proposed an effective approach to solve this cross-modality matching problem. It jointly learns coupled dictionaries for RGB and depth camera views. The two views are linked by imposing the two dictionaries to be representative and discriminative. In the testing phase, sparse codes are used for the matching person images across RGB and depth modality. Experimental results on two benchmark heterogeneous person re-id datasets show the effectiveness and superiority of the proposed approach for multi-modality re-id problem.

In the future, we will carefully integrate correlative dictionary learning into a deep convolutional neural network to jointly learn more robust feature representations and a cross-modality distance metric in an end-to-end way.

Acknowledgements. This work was is partially supported by Guangzhou Project (201604046018).

References

1. Microsoft Kinect. http://www.xbox.com/en-us/kinect/
2. Chen, S., Guo, C., Lai, J.: Deep ranking for person re-identification via joint representation learning. TIP **25**(5), 2353–2367 (2016)
3. Kviatkovsky, I., Adam, A., Rivlin, E.: Color invariants for person reidentification. PAMI **35**(7), 1622–1634 (2013)
4. Chen, Y., Zheng, W., Lai, J.: Mirror representation for modeling view-specific transform in person re-identification. In: IJCAI (2015)
5. Liao, S., Hu, Y., Zhu, X., Li, S.: Person re-identification by local maximal occurrence representation and metric learning. In: CVPR (2015)
6. He, W., Chen, Y., Lai, J.: Cross-view transformation based sparse reconstruction for person re-identification. In: ICPR (2016)
7. Dalal, N., Triggs, B.: Histograms of oriented gradients for human detection. In: CVPR (2005)
8. Albert, H., Alahi, A., Li, F.: Recurrent attention models for depth-based person identification. In: CVPR (2016)
9. Wu, A., Zheng, W., Lai, J.: Robust depth-based person re-identification. TIP **26**(6), 2588–2603 (2017)
10. Lisanti, G., Masi, I., Bagdanov, A.D., et al.: Person re-identification by iterative re-weighted sparse ranking. IEEE Trans. Pattern Anal. Mach. Intell. **37**(8), 1629–1642 (2015)
11. Zhang, Y., Li, B., Lu, H., Irie, A., Ruan, X.: Sample-specific SVM learning for person re-identification. In: CVPR (2016)
12. Shi, Z., Wang, S.: Robust and sparse canonical correlation analysis based L2; p-norm. J. Eng. **1**(1) (2017)
13. Yuan, X., Li, P., Zhang, T.: Gradient hard thresholding pursuit for sparsity-constrained optimization. In: ICML (2014)
14. Shi, X., Guo, Z., Lai, Z., et al.: A framework of joint graph embedding and sparse regression for dimensionality reduction. IEEE Trans. Image Process. **24**(4), 1341–1355 (2015)
15. Munaro, M., Fossati, A., Basso, A., Menegatti, E., Van Gool, L.: One-shot person re-identification with a consumer depth camera. In: Gong, S., Cristani, M., Yan, S., Loy, C.C. (eds.) Person Re-Identification. ACVPR, pp. 161–181. Springer, London (2014). https://doi.org/10.1007/978-1-4471-6296-4_8
16. An, L., Kafai, M., Yang, S., Bhanu, B.: Reference-based person reidentification. In: AVSS (2013)
17. Chen, Y., Zheng, W., Lai, J., Yuen, P.: An asymmetric distance model for cross-view feature mapping in person re-identification. IEEE Trans. Circuits Syst. Video Technol. **27**, 1661–1675 (2016)

18. Zhang, L., Xiang, T., Gong, S.: Learning a discriminative null space for person re-identification. In: CVPR (2016)
19. Barbosa, I.B., Cristani, M., Del Bue, A., Bazzani, L., Murino, V.: Re-identification with RGB-D sensors. In: Fusiello, A., Murino, V., Cucchiara, R. (eds.) ECCV 2012. LNCS, vol. 7583, pp. 433–442. Springer, Heidelberg (2012). https://doi.org/10.1007/978-3-642-33863-2_43
20. Grant, M., Boyd, S.: Graph implementations for nonsmooth convex programs. In: Blondel, V.D., Boyd, S.P., Kimura, H. (eds.) Recent Advances in Learning and Control, pp. 95–110. Springer, London (2008). https://doi.org/10.1007/978-1-84800-155-8_7
21. Jing, X., Zhu, X., Wu, F., et al.: Super-resolution person re-identification with semi-coupled low-rank discriminant dictionary learning. In: CVPR (2015)
22. An, L., Chen, X., Yang, S., et al.: Sparse representation matching for person re-identification. Inf. Sci. **355**, 74–89 (2016)

Multiple Metric Learning in Kernel Space for Person Re-identification

Tongwen Xu, Yonghong Song(✉), and Yuanlin Zhang

Institute of Artificial Intelligence and Robotics,
Xi'an Jiaotong University, Xi'an 710049, China
xutongwen2014@stu.xjtu.edu.cn, {songyh,ylzhangxian}@mail.xjtu.edu.cn

Abstract. In this paper, we present multiple metric learning in kernel space to preferably get more discriminative metrics. Usually, the kernel-based approaches exploit kernel trick to map vectors that usually thousands of dimensions into high dimensional space to enhance their linear capacity. But it could loss the discriminative information in the process of projecting. To address this problem, we propose to map these feature vectors into kernel space respectively and the metrics are learned in their corresponding space. The Relief algorithm is modified here to get the weights and we can get the final result by weighting multiple metrics based on features. Experiments on the public datasets demonstrate the performance of our proposed method outperforms some state of the art methods.

Keywords: Person re-identification · Multiple metric learning
Kernel space

1 Introduction

Person re-identification is the task of finding out the same identity in the field of non-overlapping cameras [1,2,15]. Due to illumination variations, human pose changes and different background of multiple cameras, its hard to match people from one camera to another. Although lots of approaches have been proposed from various standpoints, it is still an open problem.

There are two common ways to solve problems above for person re-identification, feature representation [3–5,27,28] and metric learning [4,6,7,16, 17]. The feature representation mainly focused on how to describe the discriminative information of the object. A color invariant model was proposed in [3] which designed a variable of the change is smaller than the original change to robust illumination variations. Lisanti [4] exploited five color and texture low level features to complementarily represent individuals. In [5], a novel salient color name was presented to describe colors which projected the pixels to probability distribution over sixteen color names. Besides, it is important to learn an efficient metric to measure the distance between two vectors [6,29,30]. The metric learned from the algorithm [6] which can minimize the distance

J. Yang et al. (Eds.): CCCV 2017, Part III, CCIS 773, pp. 292–302, 2017.
https://doi.org/10.1007/978-981-10-7305-2_26

between features of pairs of true matches and maximizes the distance the same between pairs of wrong matches greatly improves the performance of the metric learning in person re-identification. The KISS (Keep It Simple and Straightforward) algorithm [6] was conducted the deeply research and was found that the commonness of two vectors could also describe the similarity of them [7].

Generally, the metric is learned from the feature vectors that exacted from images. However, due to the non-linear capability of these features, the metric learned is not discriminative. To solve this problem, the kernel-based approaches [4,8,9,18] exploited kernel trick to map these features into kernel space to enhance the linear capability and then in kernel space, the metric learning is carried on. In [8], the feature vector was projected into kernel space and the metric was learned by KISS algorithm. To get an efficient kernel space, in [9], multiple kernels were used to map features into more than one subspace. These approaches prove the effectiveness of kernel trick mapping and achieve good performances. However, if the dimension of features mapped is high, the kernel representation gotten by kernel trick is not suitable and it may loss the discriminative information in the process of calculating the kernel matrix. Besides, different features may not be modeled by the same transformation function [10]. The jointly feature space may also be too complex to be robustly handled by a single metric. Therefore we propose to map these features respectively and the metric is learned in their corresponding kernel space. Finally, multiple metric measures can be weighted to get the final result.

The overview in Fig. 1 depicts the architecture of our proposed work. Firstly, the color, shape and texture features are extracted from images. And then they are mapped into kernel space respectively. In each subspace, the similarity measure is done using LSSL (Large Scale Similarity Learning) algorithm [7] to get distance based on features. Finally, the distances are weighted which the weights were already learned using modified Relief algorithm to get the result.

The main contributions in this paper are threefold: (1) multiple metric learning in kernel space is proposed to solve the problem of losing discriminative information in the process of producing kernel matrix. And the Relief algorithm [11] is modified here to get the weights of metrics. (2) Grey-world normalization is introduced to enhance discrimination of the color descriptors. (3) The experiments are deeply carried on to analyze aspects of our approach and the final results outperform the state-of-the-arts on two benchmarks.

2 Our Approach

2.1 Person Representation

Color histograms [19] are very important features and are also common used in person re-identification. But, they are sensitive to changing illuminations and the color responses of cameras. The Grey-world normalization [12] is a common and effective technique to address this problem which assumes that the average color of a picture caught by the camera is grey. It divides each channel of every

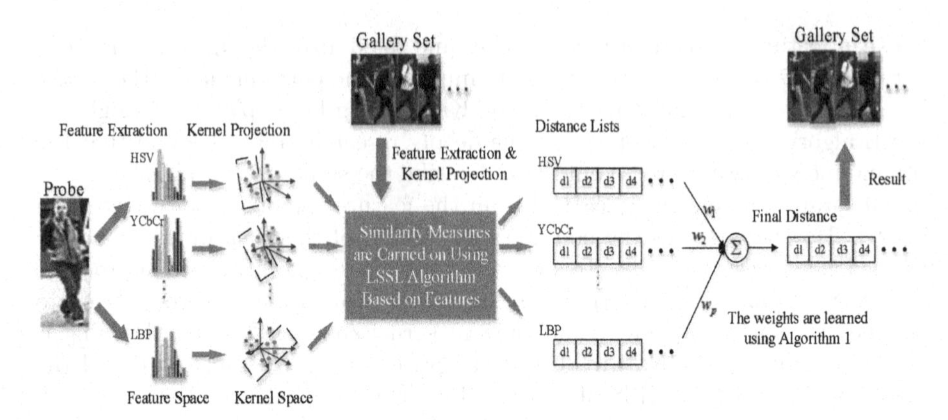

Fig. 1. The overview of the proposed approach.

pixel by the average of this channel in RGB color space, i.e. $R_1 = R/mean(R_1)$. If the illumination and color responses change, the RGB values of pixels may change a lot, but $R_1G_1B_1$ could change smaller than RGB. So the Grey-world normalization is robust to the changes. Here, we introduce this technique in RGB color space and then transformed into nRGB color space to enhance its robustness. The experiments following show its effectiveness.

Besides, the other three color spaces: HSV, RGS, YCbCr are also used to complementarily describe the color information. For each channel, 16-bin histogram is extracted from one of six non-overlapping strips in each image [6] i.e. $16 \times 6 \times 3 = 288$ dimensions for three channels color space. For shape and texture features, HOG and LBP are used [4], but we only remove 8 pixels up and down of images in order to decrease background pixel influence.

After extracted features from images, these features are projected into kernel spaces respectively using Chi-Square exponential kernel to get their kernel representation [4].

2.2 Weighted Multiple Metrics in Kernel Space

In this section, multiple metric learning in kernel space based on features is proposed. The different distances gotten by LSSL algorithm for features are weighted to be the final distance:

$$d_f = \sum_{n=1}^{p} w_n d_n \tag{1}$$

where p is the number of features and d_n is the corresponding distance with features. w_n denotes the weight and $\sum_{n=1}^{p} w_n = 1$.

To make sure the weights of distances, the importance of features should be de-pendent on. The more important the feature is, the larger weight should be given. In order to get w_n , the classic algorithm Relief [11] can be used to calculate the importance of each feature. But in the re-id problem, the Relief algorithm needs to be modified because it usually can deal with two classification problem. To solve this problem we can put the images belonging to the same person into one category and the other images into another. Besides, the sampling number in the Relief is not needed here, because the number of images of the same individual is few, e.g. two. We put all the training samples into the process of calculating the weights and the weights are cumulative sum. Finally, the values of the importance of each feature are normalized to get the weights. The specific algorithm is showed in following.

Algorithm 1. Modified Relief Algorithm

Input: Training Samples S; Training Samples Pair Number num; Features Number p;
Output: Importance Value Vector $w = [w_1, w_2, ...w_p]$;
1: **for** $i = 1, 2, ...p$ **do**
2: $w_i = 0$;
3: **for** $j = 1, 2, ...num$ **do**
4: Choose the j_{th} training sample from S, labeled as A;
5: Choose the sample belonging to the same category closest to A, labeled as B;
6: Choose the sample belonging to the different category closest to A, labeled as C;
7: Choose the sample farthest to A, labeled as D;
8: Calculate $md = (d_{AC} - d_{AB})/(d_{AD} - d_{AB})$;
9: Calculate $w_i = w_i + md$;
10: **end for**
11: **end for**

2.3 Metric Learning Algorithm

In our approach, LSSL [7] is used to learn the metrics in kernel spaces. Given two samples x and y , the similarity distance is defined by LSSL as:

$$d = (x + y)^T M(x + y) - \lambda(x - y)^T W(x - y) \tag{2}$$

λ is a constant. M and W are two metrics learned as following.

$$M = \sum\nolimits_{mD}^{-1} - \sum\nolimits_{mS}^{-1} \tag{3}$$

$$W = \sum\nolimits_{eS}^{-1} - \sum\nolimits_{eD}^{-1} \tag{4}$$

where

$$\sum\nolimits_{mS} = 1/N \sum_{i=1}^{N} (x_i + y_i)(x_i + y_i)^T \tag{5}$$

$$\sum\nolimits_{eS} = 1/N \sum_{i=1}^{N} (x_i - y_i)(x_i - y_i)^T \tag{6}$$

$$\sum\nolimits_{mD} = \sum\nolimits_{eD} = 1/(2N) \sum_{i=1}^{N} [(x_i + y_i)(x_i + y_i)^T + (x_i - y_i)(x_i - y_i)^T] \tag{7}$$

In this paper, Algorithm 1 is used to get the weights of features. But in the experiments, it is noticed that there were different maximum distances with different features. So the multiple distances are not simply weighted to get the final distance. The maximum distances based on features should be normalized to the same and then were weighted.

3 Experiments and Discussions

Our proposed approach is tested on two public datasets: VIPeR [13] and CUHK01 [14]. Images have been normalized to pixels, is set to 1.5 on two datasets. The result of the average of cumulative match characteristic (CMC) curve for 10 trials of each experiment is reported in this work.

3.1 Datasets

VIPeR dataset[1] is now considered the most challenging dataset which is also widely used in the field of person re-identification. It contains 1264 images from 632 pedestrians in two disjoint cameras with different illumination conditions and viewpoints. 632 pedestrians are randomly split into two groups, 316 pedestrians for training, the others for testing.

CUHK01 dataset[2] contains 971 individuals from two non-overlapping cameras. Each person has two images with different angles per camera. The dataset is also randomly split into two parts, 485 persons for training, the others for testing.

3.2 Comparison with State-of-the-Art Methods

In this section, the performance of our method compared with other some state-of-the-art algorithms on two datasets is shown. In Fig. 2, Tables 1 and 2 as we can see, the performance of our proposed algorithm on VIPeR and CUHK01 outperforms other methods. Accurately, the rank-1 rate is increased from 40.7% to 42.69% after we use the multiple metric learning in kernel space. In Table 2, our method greatly improves the accuracy on CUHK01 dataset. The reasonable explanation is that the form of projecting features expresses more information than the traditional kernel mapping. Besides, the Grey-world normalization and the weights learned from nor-malized modified Relief also prove their effectiveness.

[1] http://vision.soe.ucsc.edu/?q=node!178.

[2] http://www.ee.cuhk.edu.hklxgwang/CUHKjdentification.html.

3.3 Evaluation and Analysis

Effect of Grey-world normalization. To illustrate the effectiveness of Grey-world normalization fairly, we also use the KISS metric learning which was used in [6]. The four color feature vectors are concatenated to be a single vector. As shown in Fig. 3, the rank-1 recognition rate of color features extracted from VIPeR without normalization is 29.56%. After Grey-world normalization on images, the rate ups to 34.37%. In Fig. 4, the rank-1 rate of color features increases from 36.54% to 39.13%. The experiments show that the results with normalization are better than that without. This indicates that the normalization is robust to illuminations changing and color responses of cameras and it can effectively improve the accuracy of color features.

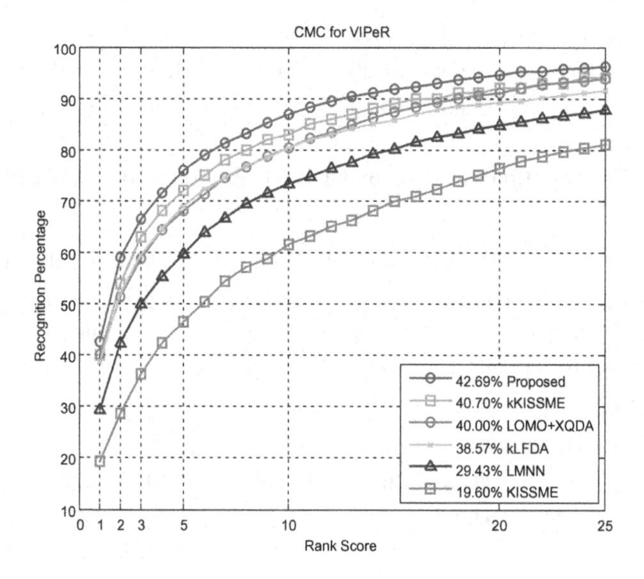

Fig. 2. CMC curves of some state-of-the-art methods on VIPeR dataset. The rank-1 identification rates are marked before the name of each curve.

Effect of multiple metric learning. Different from that multiple vectors are concatenated to be a single vector, the features were projected respectively and the corresponding metric is learned for each feature. For example, in Fig. 3, the four color feature vectors are concatenated to be a 1152 dimension vector and its rank-1 matching rate is 34.37%. Now, we learn four metrics for four color feature and their distances are weighted to be the final distance. The rate ups to 38.07%, improving 3.7% comparing with 34.37%. The HOG and LBP features without multiple metric achieve 5.16%, while the rate improves 2.28% to 7.44% using two metrics learning. Similarly, in Fig. 4 the experiments on CUHK01 dataset, the result of multiple metrics in color, shape and texture features are better than that without. This is because the features are mapped respectively instead

Table 1. The identification rates (%) on VIPeR at rank 1, 10, 20, 50, 100 are listed.

Method	Rank-1	Rank-10	Rank-20	Rank-50	Rank-100
Ours	42.69	86.99	94.65	98.80	99.65
KEPLER [10]	42.41	82.37	90.70	97.06	98.89
kKISSME [8]	40.07	83.95	92.08	98.02	-
XQDA [16]	40.00	80.51	91.08	98.54	99.72
kLFDA [18]	38.58	69.15	80.44	89.15	96.33
KCCA [4]	37.25	84.59	92.78	98.48	99.72
BSFR [20]	34.20	79.70	90.20	98.10	-
SalMatch [21]	30.16	65.54	79.15	91.49	98.10
RCCA [22]	30 75	87	96	99	
MtMCML [23]	28.83	75.82	88.51	-	-
KISSME [6]	19.60	62.20	74.92	91.80	98.00
PRDC [24]	15.66	53.86	70.09	87.79	92.84

Table 2. The identification rates (%) on CUHK01 at rank 1, 10, 20, 50, 100 are listed.

Method	Rank-1	Rank-10	Rank-20	Rank-50	Rank-100
Ours	49.75	83.77	91.15	96.16	98.37
KEPLER [10]	38.85	76.12	84.72	84.72	96.58
kKISSME [8]	36.10	72.61	81.90	-	-
SalMatch [22]	28.45	55.67	67.95	92.85	96.58
PatMatch [22]	20.53	41.09	51.56	72.46	87.91
TML [25]	20.53	56.61	69.62	85.74	93.75
SDALF [26]	9.90	30.33	41.03	55.99	67.39

of projecting a high dimensional vector. The multiple metrics can express more discri-minative information to avoid loss information in the process of mapping.

Effect of Relief algorithm. In this small section, the four common algorithms used to make sure the weights are introduced to compare their performances. Adopt Weight refers to adopt weight learning and we get the rank-1 39.78%. It is because adopt weight algorithm only consider the rank-1 to get the importance of the feature and it do not consider the distinguished ability of features. Then the Most Probability is carried on. The result is much worse than our method. In Fig. 5, the performance of normalized modified relief is better than the modified relief. This is because we consider there are different maximum distances with different features and we normalize them to solve it. So the weights by normalized modified relief can get better complementarities.

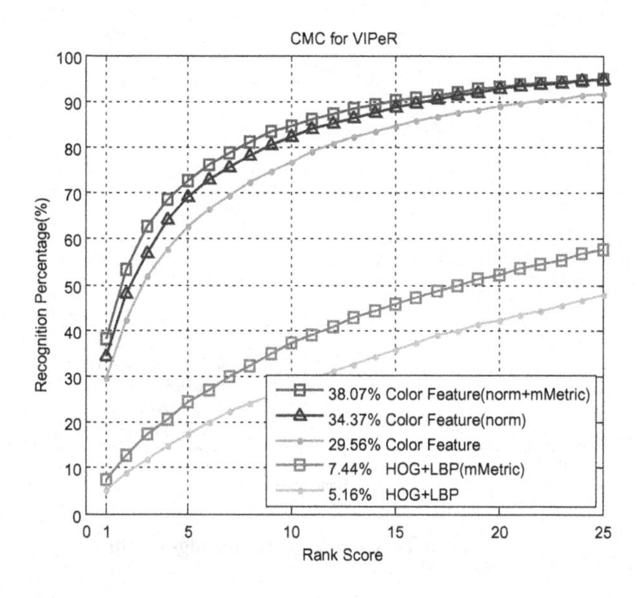

Fig. 3. The Effectiveness of Grey-world normalization and multiple metric learning in color, HOG and LBP features on VIPeR.

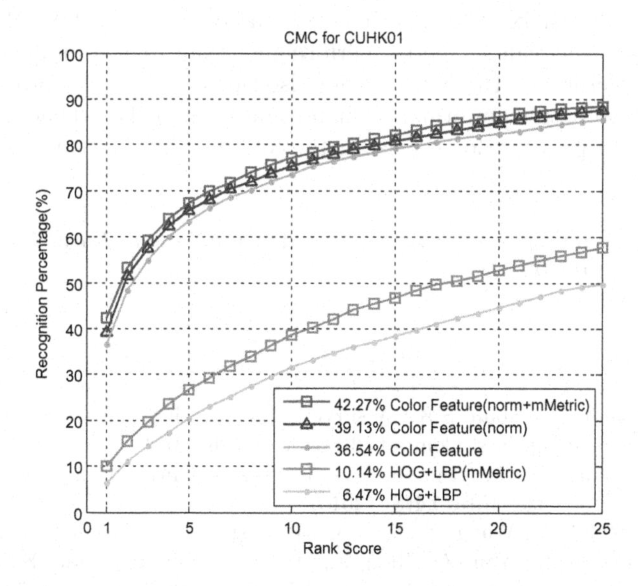

Fig. 4. The effectiveness of Grey-world normalization and multiple metric learning in color, HOG and LBP features on CUHK01.

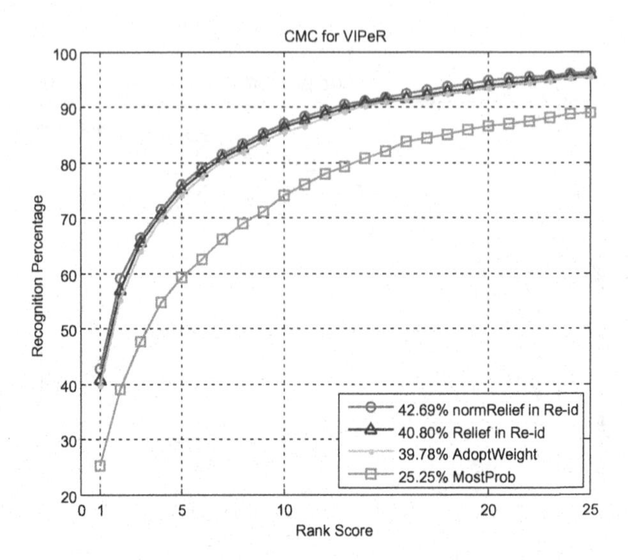

Fig. 5. The effectiveness of Relief algorithm.

4 Conclusion

In this paper, we have proposed multiple metric learning in kernel space to address the problem of losing discriminative information in the process of calculating the kernel matrix. The experiments demonstrate the results of projecting features respectively can get better performance than that of mapping one high dimensional vector and the experiments also prove that our approach is effective and achieve better performance than some state-of-the-art methods. In the future, feature representation especially the color invariant feature will be deeply investigated.

Acknowledgement. This work was supported by the National Natural Science Foundation of China (91520301).

References

1. Bedagkar-Gala, A., Shah, S.K.: A survey of approaches and trends in person re-identification. Image Vis. Comput. **32**(4), 270–286 (2014)
2. Zheng, L., Yang, Y., Hauptmann, A.G.: Person re-identification: past, present and future. J. Latex Class Files **14**(8) (2015)
3. Chen, Y., Zhao, C., Wang, X., Gao, C.: Robust color invariant model for person re-identification. In: You, Z., Zhou, J., Wang, Y., Sun, Z., Shan, S., Zheng, W., Feng, J., Zhao, Q. (eds.) CCBR 2016. LNCS, vol. 9967, pp. 695–702. Springer, Cham (2016). https://doi.org/10.1007/978-3-319-46654-5_76
4. Lisanti, G., Masi, I., Del Bimbo, A.: Matching people across camera views using kernel canonical correlation analysis. In: ICDSC, pp. 1–6 (2014)

5. Yang, Y., Yang, J., Yan, J., Liao, S., Yi, D., Li, S.Z.: Salient color name for person re-identification. In: ECCV, pp. 536–551 (2014)
6. Kostinger, M., Hirzer, M., Wohlhart, P., Roth, P.M., Bischof, H.: Large scale metric learning from equivalence constraints. In: CVPR, pp. 2288–2295 (2012)
7. Yang, Y., Liao, S., Lei, Z., Li, S.Z.: Large scale similarity learning using similar pairs for person verification. In: AAAI, pp. 3655–3661 (2016)
8. Qi, M., Tan, S., Wang, Y., Liu, H., Jiang, J.: Multi-feature subspace and kernel learning for person re-identification. Acta Automatica Sinica **42**(2), 299–308 (2015)
9. Syed, M.A., Jiao, J.: Multi-kernel metric learning for person re-identification. In: IEEE International Conference on Image Processing, pp. 784–788 (2016)
10. Martinel, N., Micheloni, C., Foresti, G.L.: Kernelized saliency-based person re-identification through multiple metric learning. IEEE Trans. Image Process. **24**(12), 5645–5658 (2015)
11. Kira, K., Rendell, L.A.: The feature selection problem: traditional methods and a new algorithm. In: AAAI, pp. 129–134 (1992)
12. Satta, R.: Appearance descriptors for person re-identification: a comprehensive review. Eprint Arxiv (2013)
13. Gray, D., Brennan, S., Tao, H.: Evaluating appearance models for recognition, reacquisition, and tracking. In: Proceedings of PETS Workshops (2007)
14. Li, W., Zhao, R., Wang, X.: Human reidentification with transferred metric learning. In: Proceedings of ACCV, pp. 31–44 (2012)
15. Chen, X., Huang, K., Tan, T.: Object tracking across non-overlapping views by learning inter-camera transfer models. Pattern Recogn. **47**(3), 1126–1137 (2014)
16. Liao, S., Hu, Y., Zhu, X., et al.: Person re-identification by local maximal occurrence re-presentation and metric Learning. Comput. Vis. Pattern Recognit. **8**(4), 2197–2206 (2015)
17. Tao, D., Guo, Y., Song, M., et al.: Person re-identification by dual-regularized KISS metric learning. IEEE Trans. Image Process. **25**(6), 1–1 (2016)
18. Xiong, F., Gou, M., Camps, O., et al.: Person re-identification using kernel-based metric learning methods. ECCV **2014**, 1–16 (2014)
19. Zeng, M., Wu, Z., Tian, C., et al.: Efficient person re-identification by hybrid spatiogram and covariance descriptor. In: CVPR 2015, pp. 48–56 (2015)
20. Liu, H., Ma, L., Wang, C.: Body-structure based feature representation for person re-identification. In: IEEE International Conference on Acoustics, Speech and Signal Processing, IEEE, pp. 1389–1393 (2015)
21. Zhao, R., Ouyang, W., Wang, X.: Person re-identification by salience matching. In: IEEE International Conference on Computer Vision, IEEE, pp. 2528–2535 (2013)
22. An, L., Kafai, M., Yang, S., et al.: Reference-based person re-identification. In: IEEE International Conference on Advanced Video and Signal Based Surveillance, IEEE, pp. 244–249 (2013)
23. Ma, L., Yang, X., Tao, D.: Person re-identification over camera networks using multi-task dis-tance metric learning. IEEE Trans. Image Process. **23**(8), 3656–70 (2014). A Publication of the IEEE Signal Processing Society
24. Zheng, W.S., Gong, S., Xiang, T.: Reidentification by relative distance comparison. IEEE Trans. Pattern Anal. Mach. Intell. **35**(3), 653 (2013)
25. Li, W., Zhao, R., Wang, X.: Human reidentification with transferred metric learning. In: Lee, K.M., Matsushita, Y., Rehg, J.M., Hu, Z. (eds.) ACCV 2012. LNCS, vol. 7724, pp. 31–44. Springer, Heidelberg (2013). https://doi.org/10.1007/978-3-642-37331-2_3

26. Bazzani, L., Cristani, M., Murino, V.: Symmetry-driven accumulation of local features for human characterization and re-identification. Comput. Vis. Image Underst. **117**(2), 130–144 (2013)
27. Gray, D., Tao, H.: Viewpoint invariant pedestrian recognition with an ensemble of localized features. In: Forsyth, D., Torr, P., Zisserman, A. (eds.) ECCV 2008. LNCS, vol. 5302, pp. 262–275. Springer, Heidelberg (2008). https://doi.org/10.1007/978-3-540-88682-2_21
28. Ma, B., Su, Y., Jurie, F.: Covariance descriptor based on bio-inspired features for person re-identification and face verification. Image Vis. Comput. **32**(6), 379–390 (2014)
29. Prosser, B., Zheng, W.-S., Gong, S., Xiang, T., Mary, Q.: Person re-identification by support vector ranking. In: BMVC (2010)
30. Li, W., Wang, X.: Locally aligned feature transforms across views. In: IEEE Conference on Computer Vision and Pattern Recognition (2013)

Multi-angle Insulator Recognition Method in Infrared Image Based on Parallel Deep Convolutional Neural Networks

Zhenbing Zhao[1(✉)], Xiaoqing Fan[1], Yincheng Qi[1], and Yongjie Zhai[2]

[1] School of Electrical and Electronic Engineering, North China Electric Power University, Baoding 071003, China
zhaozhenbing@ncepu.edu.cn

[2] Department of Automation, North China Electric Power University, Baoding 071003, China

Abstract. Insulator is the most common equipment in power system, and the recognition of insulator in infrared image is affected seriously under unconstrained condition with angle change. Due to the insulators' angle diversification and the manually-designed features' limitation, the existing algorithms have a problem of low accuracy in classification; therefore how to obtain the rotation invariant representations to cope with these adverse conditions is a very important problem. Deep Convolutional Neural Networks (DCNNs) have established a remarkable performance in image classification. The deep features obtained at the top fully-connected layer of the DCNNs (FC-features) exhibit rich global semantic information and are extremely effective in image classification. In this paper, we present a rotation invariant representation generation method named PFE-FDS (Parallel Feature Extraction and Feature Dimension Selection) for infrared insulator recognition which is based on parallel DCNNs FC-features extraction as well as feature sorting and dimension selection based on mutual information to eliminate redundancy. Then the SVM (Support Vector Machine) classifier is trained on our standard infrared insulator dataset for classification. The experimental accuracy shows that the PFE-FDS can improve the accuracy of multi-angle object recognition.

Keywords: Infrared insulators
Parallel deep convolutional neural networks
Rotation invariance representations · Feature dimension selection

1 Introduction

Insulator is the most common equipment in power system which is made of non-conducting material. It is used to support the electrical conductors and shield them from the ground or other conductors. The failure of insulators would be the direct threat to the stability and safety of the system [1]. With the advantages of

J. Yang et al. (Eds.): CCCV 2017, Part III, CCIS 773, pp. 303–314, 2017.
https://doi.org/10.1007/978-981-10-7305-2_27

being non-contact and non-destructive, infrared imaging technology is efficient for monitoring and evaluating the thermal condition of insulators. According to statistics, tripping accidents caused by insulator fault accounted for the 81.3% of transmission line accident [2]. Therefore, monitoring insulator status regularly and detecting insulator fault timely is crucial. Accurate and efficient recognition of insulators is the premise of realizing the intelligent detection. Generally, insulators are diversified with different orientations, and using common feature representations for recognition may not be accurate, so obtaining robust rotation invariant representations is necessary.

The pictures in Fig. 1 show the positive and negative training samples as well as the test samples after multi-angle rotating.

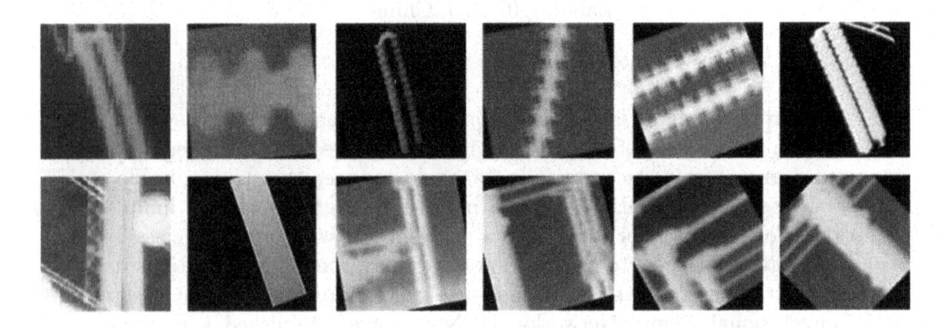

Fig. 1. Positive and negative samples. The two rows show the positive and negative images for training and testing with different orientations, respectively

Over the last few years, there has been some progress in insulators recognition. Yao et al. [3] proposed a zero value insulators recognition method under different pollution levels and different humidity conditions by combining the feature of insulator strings' relative temperature distribution characteristics and extracted from the artificial neural networks. Zhao et al. [4] extracted insulator outline from aerial insulator image based on non sampling contourlet transform. Jin et al. [5] extracted surface area and detected by using the optimal entropy threshold segmentation method. Ye [6] achieved object localization by using feature points matching between object image and template image based on SIFT (Scale Invariant Feature Transform) feature. Yen [7] studied the recognition and localization of insulators based on HOG (Histogram of Oriented Gradient) feature and SVM.

The above methods adopted the traditional hand-crafted feature and had a common problem of low accuracy, large calculation, being sensitive to the rotation of the angle. With the development of deep learning technology, more and more attention has been focused on the recognition of insulators based on the DCNNs. In this paper, we present a rotation invariant representation generation method for infrared insulator image named PFE-FDS which is based on parallel DCNNs FC-features extraction as well as feature sorting and dimension selection based on mutual information to eliminate redundancy.

2 Related Work

Recent researches show that DCNNs models can not only characterize large data variations but also learn a compact and discriminative feature representations when the size of the training data is sufficiently large, and it has good performance for object recognition and localization tasks [8]. Feature representations play an important role in computer vision, and have been widely used in many computer vision tasks [9]. An ideal feature representation should meet two basic characteristics, high quality representation and low computational cost, and it needs to capture important and unique information in images and be robust to various transformations.

After the AlexNet was proposed, more and more representative works have emerged in the structural optimization of the deep learning model. Zeiler and Fergus [10] get the new networks structure named ZF-net by reducing the size of the first convolution kernel in AlexNet from $11 * 11$ to $7 * 7$. Deep Convolutional Neural Networks have established an overwhelming presence in image classification starting with the 2012 ImageNet Large Scale Visual Recognition Challenge. In the VGG [11] model, $3 * 3$ convolution kernels are used in convolution layers, which can significantly reduce the number of parameters and improve the discrimination. Residual Net (ResNet) [12] is a 152 layer networks invented by He et al., which was ten times deeper than others. Following the path VGG introduces, ResNet explores deeper structure with simple layer. Lin et al. [13] proposed a DeepBit32 model by introducing a new layer named fc8_kevin which encodes different representations of certain angle and obtains binary representations with certain rotation invariance.

DCNNs impose high computational burdens both at training and testing time, and training requires collecting and annotating large amounts of data. Recently, the second method has drawn much attention for image representation. Deep feature activations extracted from a pre-trained CNN model have been successfully utilized as general feature extractor for image representation. To get a generic representation, after a series of convolutional filtering and pooling, the neural activations from first or second fully connected layers (FC-layers) are extracted from a pre-trained CNN model. Gong et al. [14] proposed a certain scale pooling method to improve the rotation invariance of features. Yoo et al. [15] utilize Fisher Vector encode method for polymerizing multi-scale deep feature. The research of [16] shows that the ability of expressing features can be enhanced by integrating the representation of multiple layers. Tan et al. [17] proposed a Feature Generating Machine, which learns a binary indicator vector to show whether one dimension is useful or not. Deep representations generated from CNN model have achieved great success in object recognition. More feature selection methods [18–20] are proposed for different types of applications.

Current deep representations with small angle range rotation invariance can not meet the requirement of multi-angle insulators recognition. So this paper presents a rotation invariant representation generation method named PFE-FDS for infrared insulator image which is based on parallel DCNNs FC-features extraction as well as feature sorting and dimension selection based on mutual information to eliminate redundancy. The details are described in Sect. 3.

3 Proposed Method

In this section, we propose a novel method named PFE-FDS to recognize insulators. Firstly, enter the input image into parallel DCNNs made up of pre-trained VGG16 model and DeepBit32 model, and extract different FC-layer feature representations. Then combine the representations directly and sort feature representations based on mutual information. After that select the dimension in line with the above sorting results. Finally the SVM classifier is trained on our standard infrared insulator dataset for classification. The overall framework is shown in Fig. 2.

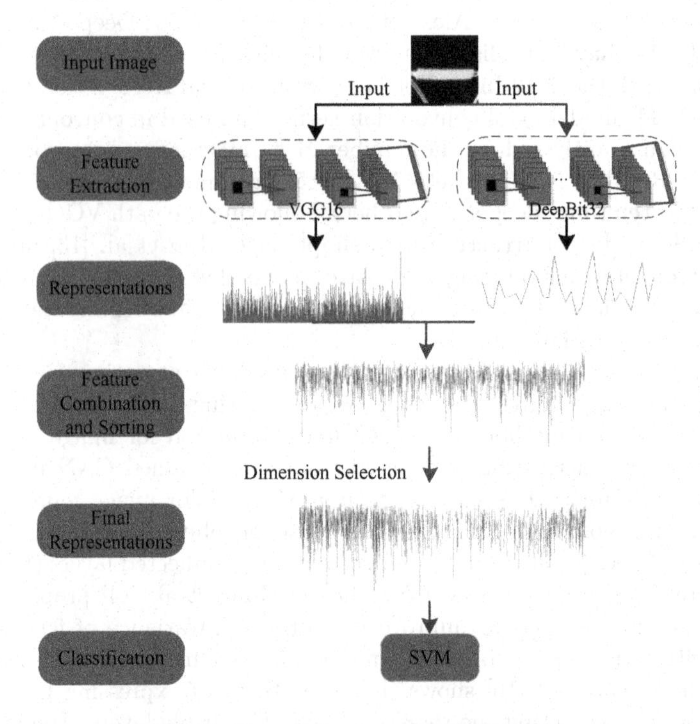

Fig. 2. Overall framework of PFE-FDS

3.1 Parallel Deep Feature Extraction

The research of [16] shows that the ability of expressing features can be enhanced by the integration of multiple layers. Inspired by this, our method extracts FC-features from parallel DCNNs. A typical DCNNs is made up of several convolutional layers, followed by pooling layers, fully-connected layers and a softmax decision layer.

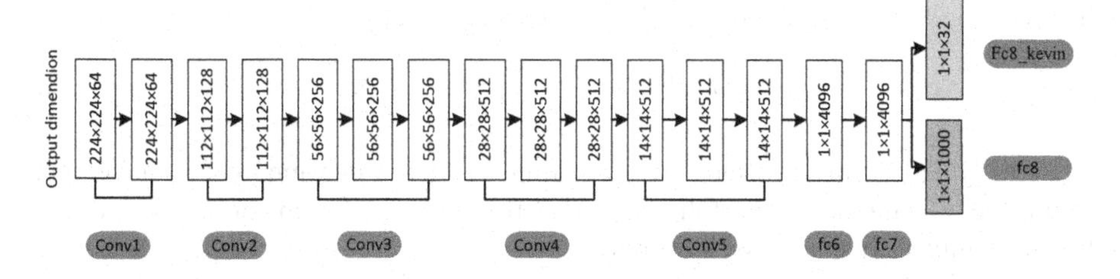

Fig. 3. Architecture of the DeepBit32 and VGG16 model trained on ImageNet 2012 classification dataset used in this paper. Each layer, represented by a box, is labeled with the size $R_l * C_l * K_l$ of its output in (1).

In Fig. 3 we illustrate the DCNN model. It consists of convolutional layers, max-pooling layers, fully connected layers and a softmax decision layer (It can also be called fc8 layer). At any given layer l, the layer's output data is an $R_l * C_l * K_l$ array

$$\left[x_{ij}^{l} \in R^{k_l}\right]_{i=1,\ldots,R_l, j=1,\ldots,C_l} \tag{1}$$

that is the input to the next layer, with the input to the first layer being an RGB image of size $R_0 * C_0$ and $K_0 = 3$ color channels. The fully connected layers can be seen as convolutional layers with kernels having the same size as the layer's input data.

While these methods adopted only the deep aspect of DCNNs, our goal is to combine the advantages of both approaches. The feature representations we utilize are extracted from fc6 layer in VGG16 model and fc8 layer in DeepBit32 model. The feature representations are complementary in discrimination and rotation.

3.2 Feature Combination and Sorting

In this section, we combine the features representations directly and form a representation of 4128 dimension. Because Principal Component Analysis (PCA) can not handle high-order correlation data, and can not be personalized optimization, so we utilize a feature selection algorithm based on mutual information for sorting. Although feature selection algorithm based on mutual information is not a new algorithm, it is the first application in the feature sorting of parallel DCNNs and power equipment recognition. Mutual Information is taken as the basic criterion to find Max-Relevance and Min-Redundancy between features [21]. The mutual information I of two variables x and y is defined based on their joint probabilistic distribution $p(x,y)$ and the respective marginal probabilities $p(x)$, $p(y)$:

$$I(x, y) = \sum_{i,j} p(x_i, y_j) \log \frac{p(x_i, y_j)}{p(x_i)\, p(y_j)} \tag{2}$$

Max-Relevance is to search features satisfying (3), **S** denote the subset of features we are seeking, which approximates $D(S,c)$ with the mean value of all mutual information values between feature x_i and class label c:

$$\max D(S,c), D = \frac{1}{S}\sum_{x_i \in S} I(x_i, c) \tag{3}$$

It is likely that features selected according to Max-Relevance could have rich redundancy. Therefore, the following Min-Redundancy condition can be added to select mutually exclusive features.

$$\min R(S), R = \frac{1}{S^2}\sum_{x_i, x_j \in S} I(x_i, x_j) \tag{4}$$

We define a new equation (5) for sorting. Then we use V_i denote the value calculated by (5) of the ith dimension feature, the values are arranged in descending order, and the ranking results are stored in the matrix **A** and the features are stored in **S** corresponds to the matrix **A**.

$$V_i = \max(D(S,c)) - R(S) + \frac{D(S,c)}{R(S)}, i = 1, 2...4128 \tag{5}$$

3.3 Feature Dimension Selection

In image classification, the generated feature representations dimension is high and has rich redundancy. So we select the dimension of features **S** according to the classification results of SVM. The recognition result of VGG16_fc6 combine DeepBit32_fc8 is considered as a baseline. We select feature representations from matrix A in the top n for testing, the accuracy of recognition is the highest when n equals 3994.

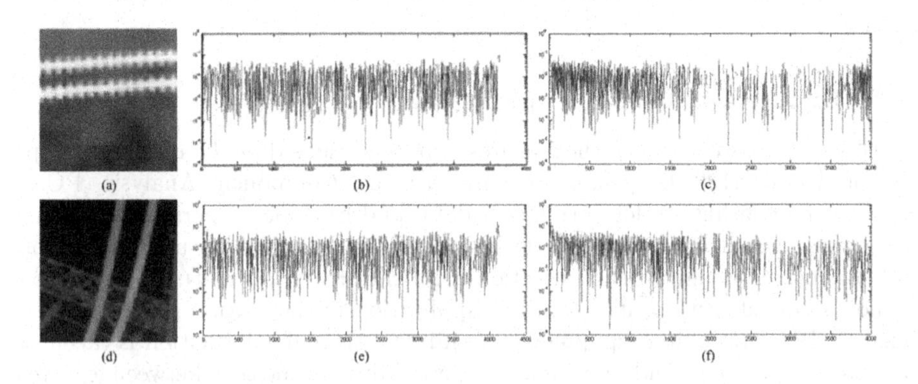

Fig. 4. Visualization of insulator and its feature representations

As shown in Fig. 4, the picture (a) is the original insulator image, the picture (b) represents its feature representations of two model combined directly, the picture (c) represents its feature representations with dimension is 3994 after feature sorting and dimension selecting. The picture (d), (e), (f) is the feature representations of the corresponding negative sample.

4 Experiments Results and Analysis

In this section, we begin by introducing our infrared insulator datasets. Then, we evaluate our rotation invariant feature representations generation method on our standard infrared insulator datasets. In order to verify the practicability of this method, we selected two kinds of indoor scene for testing, and the effect is excellent.

4.1 Datasets

Due to the difficulty of obtaining insulators infrared image, and there is no public infrared image datasets, we use a large number of infrared images collected from insulator inspection system to build the insulator infrared image datasets. In the task of recognizing insulator, the infrared image datasets consists of 672 insulator samples and 1012 background samples. These original images are getting from the power substations varying from 110 kV to 500 kV level. Due to the limited samples of the insulator, we rotate the images manually for testing during the experiments.

We divide the dataset into two parts: 70% of this dataset for training and the remaining 30% for testing. All the training samples are labeled with "positive"and "negative", respectively.

4.2 Multi-angle Infrared Insulators Recognition

We visualize the feature maps of each convolutional layers of VGG16 model in Fig. 5. From the Fig. 5, we found that the neuron will response to the edge when the insulator image rotated and it will influence the recognition accuracy, but this influence is caused by human, can be ignored.

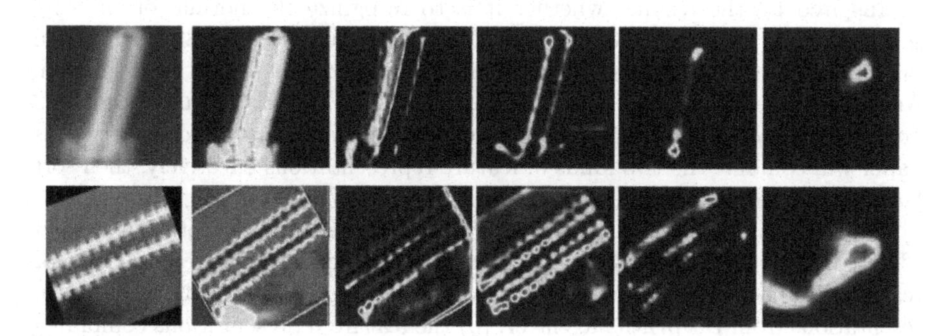

Fig. 5. Neural activation feature maps of each convolutional layers. The top line is normal insulator, the bottom line is the insulator rotated 30°

We extract feature representations from different FC-layers such as fc6 layer and fc8 layer in a DCNNs. Then we carry out the same operation on the different deep models such as VGG16, AlexNet, and we also experiment on some traditional feature descriptors like Speeded Up Robust Features (SURF), Oriented FAST and Rotated BRIEF (ORB) and Binary Robust Invariant Scalable Keypoints (BRISK). Classification tasks are implemented on the normal samples and the samples rotating 30° respectively. The experimental results are shown in Fig. 6.

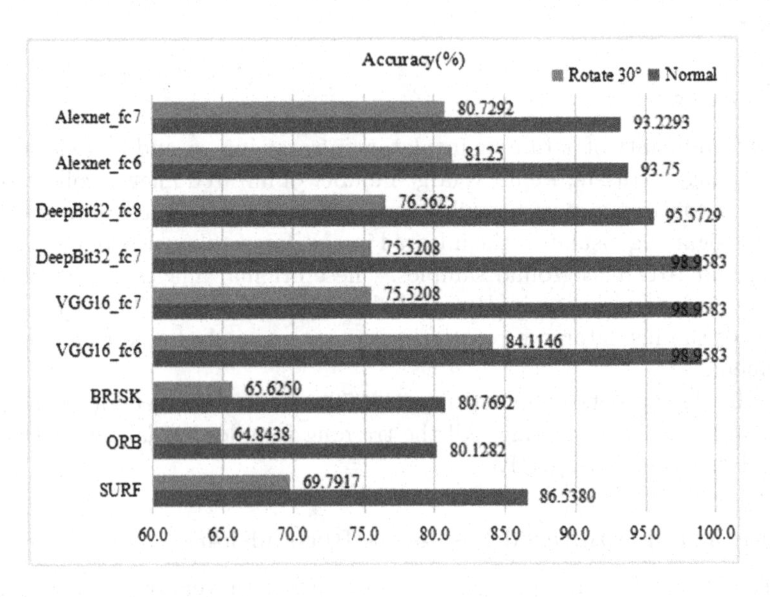

Fig. 6. Recognition results of different layers in different deep models (Color figure online)

Inspired by the results, whether it is to recognize the normal or rotating samples, the accuracy of VGG16 is the highest. Although DeepBit32 has rotation invariance in certain degree, its recognition performance is not good enough. And we discover the feature representations extract from VGG16 and DeepBit32 are complementary due to its introduced process. Based on this discovery, this paper will combine the two kinds of feature representations effectively, then sort features and select dimensions.

The test samples of datasets are rotated respectively and the rotation angle is 5°, 10°, 15°, 20°, 25°, 30°, 45°, 60°, then we utilize following representations for testing: (1) the feature representations from VGG16 fc6 layer; (2) the feature representations from DeepBit32 fc8 layer; (3) the feature representations combining two model representations directly, named P-DCNNs (parallel DCNNs); (4) the representations consist of two model representations, then conduct feature selection and no dimension selection, named PFE-FS (Parallel Feature Extraction and Feature Sorting); (5) the representations consist of two feature representations, then conduct feature selection and dimension selection which dimension is 3994, named PFE-FDS. The classification and intuitive results are shown in Table 1 and Fig. 7.

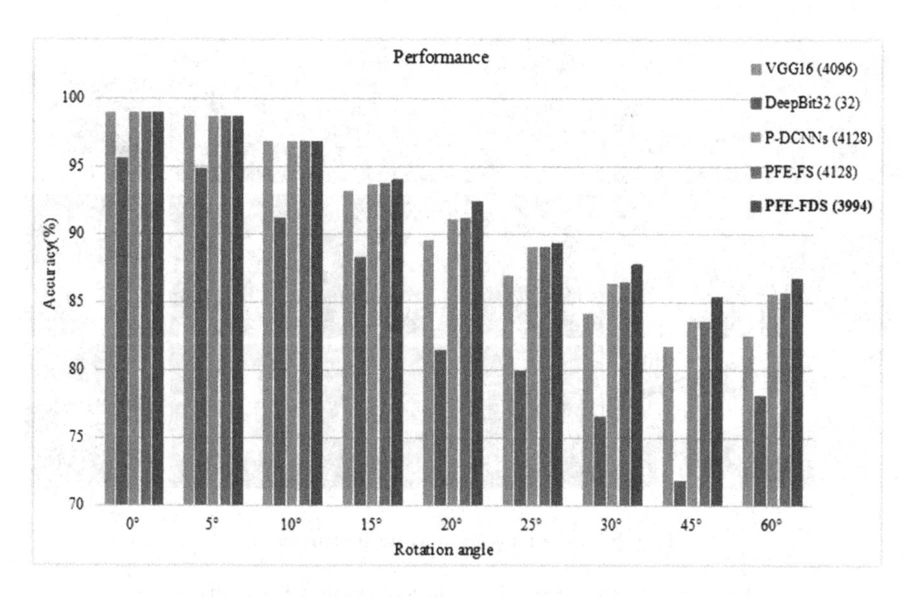

Fig. 7. Classification results of our method on multi-angle (Color figure online)

Table 1. Recognition results of insulators with multi-angle

Angles (°)	Accuracy (%)				
	VGG16 (4096)	DeepBit32 (32)	P-DCNNs (4128)	PFE-FS (4128)	PFE-FDS (3994)
0	98.9593	95.5729	98.9593	98.9593	98.9593
5	98.6979	94.7917	98.6979	98.6979	98.6979
10	96.8750	91.1667	96.8750	96.8750	96.8750
15	93.2292	88.2812	93.7500	93.7500	94.0104
20	89.5833	81.5104	91.1458	91.1458	92.4429
25	86.9792	79.9479	89.0625	89.0625	89.3229
30	84.1146	76.5625	86.4583	86.4583	87.7604
45	81.7708	71.8750	83.5938	83.5938	85.4167
60	82.5521	78.1250	85.6771	85.6771	86.7188

For the normal infrared insulators, the recognition accuracy of a single deep model has been very high, so the use of PFE-FDS is not much improvement, but with the angle increases, the recognition accuracy increases faster.

4.3 Multi-angle Scene Recognition

In order to verify the validity of the method we proposed, we select two kinds of similar indoor scene named airport inside and bar from the datasets published for the task of indoor scene recognition. The number of samples is 608 and 603 respectively, we divide the dataset into two parts: 70% of this dataset for training and the remaining 30% for testing, and we performed the same operation on the test set like the above. The scene dataset and its feature map of each convolutional layer are shown in Fig. 8, and experimental results are shown in Table 2.

Fig. 8. Indoor image and its feature map

Table 2. Recognition results of indoor scene with multi-angle

Angles (°)	Accuracy (%)		
	P-DCNNs (4128)	PFE-FS (4128)	PFE-FDS (4004)
0	93.3649	93.3649	96.6825
5	92.8910	92.8910	95.2607
10	91.9431	91.9431	96.2085
15	94.3128	94.3128	96.6825
20	93.3649	93.3649	93.3649
25	92.4171	92.4171	92.4171
30	90.0474	90.0474	95.2607
45	83.4123	83.4123	85.7820
60	85.7820	85.7820	88.6256

For indoor scene recognition task, the recognition accuracy is highest when the dimension is 4004. The recognition results indicate that our proposed method outperforms the other two methods in precision.

From the results of the above experiments, we can see that the accuracy of our proposed method is not only higher than the two separate models, but higher than the accuracy of combining representations directly, and the contrast results of last two columns reflect the necessity select the feature and dimension. The experimental results of indoor scene recognition show that our proposed method is not only effective for multi-angle infrared insulator recognition, but also can be applied to multi-angle visible image recognition.

5 Conclusion

Infrared imaging has advantages in inspecting abnormal heating in electrical equipment, and it is efficient, reliable and non-destructive. However, most of the recognition and detection are conducted manually. With a great quantity of insulators to be inspected, the automatic recognition and localization method is needed.

In the light of the problem that the insulator recognition method is sensitive to the change of angles, and the recognition accuracy is low, a feature representation method for infrared insulators is proposed. Because of the few samples of infrared insulators, the data requirements of training model can not be met. We introduce the feature representations generation method PFE-FDS based on parallel DCNNs. The method doesn't need to do any direct finetune which needs a lot of time, realize the leap from feature designing to feature learning, then sort the feature and select its dimension to obtain the feature representations with robust rotation invariance and no redundancy. The high accuracy shows the efficiency of the proposed method. Then the method will be applied to practice.

Acknowledgments. This work was supported in part by the National Natural Science Foundation of China under grant number 61401154, by the Natural Science Foundation of Hebei Province under grant number F2016502101, and by the Fundamental Research Funds for the Central Universities under grant number 2015ZD20.

References

1. Li, P., Huang, X., Zhao, L., Zhu, Y.: Research and design of transmission line condition monitoring agent. Proc. Csee **33**(16), 153–161 (2013)
2. Tong, W., Yuan, J., Li, B.: Application of image processing in patrol inspection of overhead transmission line by helicopter. Power Syst. Technol. **34**(12), 204–208 (2010)
3. Yao, J., Guan, S., Lu, J., Jiang, Z., Zhao, C., Xia, D., Qian, Y.: Identification of zero resistance insulators by combining relative temperature distribution characteristics with artificial neural network. Power Syst. Technol. **36**(2), 170–174 (2012)
4. Zhao, Z., Jin, S., Liu, Y.: Aerial insulator image edge extraction method based on NSCT. Chin. J. Sci. Instrum. **33**(9), 2045–2052 (2012)
5. Jin, L., Da, Z., Duan, S., Yao, S.: Contamination grades measurement of insulators based on image color feature fusion. J. Tongji Univ. **42**(10), 1611–1617 (2014)
6. Ye, H., Liu, Z.-G., Han, Z.-W., Yang, H.-M.: Fracture detection of ear pieces of catenary support devices of high-speed railway based on sift feature matching. J. Chin. Railw. Soc. **36**(2), 31–36 (2014)
7. Yan, L.: Insulator location and recognition algorithm based on hog characteristics and SVM. J. Transp. Eng. Inf. **13**(4), 53–60 (2015)
8. Zhang, T., Dong, Q., Hu, Z.: Pursuing face identity from view-specific representation to view-invariant representation. In: IEEE International Conference on Image Processing, pp. 3244–3248 (2016)
9. Krizhevsky, A., Sutskever, I., Hinton, G.E.: Imagenet classification with deep convolutional neural networks. In: International Conference on Neural Information Processing Systems, pp. 1097–1105 (2012)

10. Zeiler, M.D., Fergus, R.: Visualizing and understanding convolutional networks. In: Fleet, D., Pajdla, T., Schiele, B., Tuytelaars, T. (eds.) ECCV 2014. LNCS, vol. 8689, pp. 818–833. Springer, Cham (2014). https://doi.org/10.1007/978-3-319-10590-1_53
11. Simonyan, K., Zisserman, A.: Very deep convolutional networks for large-scale image recognition. Computer Science (2014)
12. He, K., Zhang, X., Ren, S., Sun, J.: Deep residual learning for image recognition, pp. 770–778 (2015)
13. Lin, K., Lu, J., Chen, C.-S., Zhou, J.: Learning compact binary descriptors with unsupervised deep neural networks. In: IEEE Conference on Computer Vision and Pattern Recognition, pp. 1183–1192 (2016)
14. Gong, Y., Wang, L., Guo, R., Lazebnik, S.: Multi-scale orderless pooling of deep convolutional activation features. In: Fleet, D., Pajdla, T., Schiele, B., Tuytelaars, T. (eds.) ECCV 2014. LNCS, vol. 8695, pp. 392–407. Springer, Cham (2014). https://doi.org/10.1007/978-3-319-10584-0_26
15. Yoo, D., Park, S., Lee, J.-Y., Kweon, I.S.: Multi-scale pyramid pooling for deep convolutional representation. In: Computer Vision and Pattern Recognition Workshops, pp. 71–80 (2015)
16. Kulkarni, P., Zepeda, J., Jurie, F., Perez, P.: Hybrid multi-layer deep CNN/aggregator feature for image classification. In: IEEE International Conference on Acoustics, Speech and Signal Processing, pp. 1379–1383 (2015)
17. Tan, M., Wang, L., Tsang, I.W.: Learning sparse SVM for feature selection on very high dimensional datasets. In: International Conference on Machine Learning, pp. 1047–1054 (2010)
18. Huang, K., Aviyente, S.: Wavelet feature selection for image classification. IEEE Trans. Image Process. Publ. IEEE Sign. Process. Soc. **17**(9), 1709 (2008)
19. Sun, Z., Wang, L., Tan, T.: Ordinal feature selection for iris and palmprint recognition. IEEE Trans. Image Process. Publ. IEEE Sign. Process. Soc. **23**(9), 3922–3934 (2014)
20. Zhang, Y., Wu, J., Cai, J.: Compact representation of high-dimensional feature vectors for large-scale image recognition and retrieval. IEEE Trans. Image Process. Publ. IEEE Sign. Process. Soc. **25**(5), 2407 (2016)
21. Mandal, M., Mukhopadhyay, A.: An improved minimum redundancy maximum relevance approach for feature selection in gene expression data. Procedia Technol. **10**(1), 20–27 (2013)

Enhanced Deep Feature Representation for Person Search

Jinfu Yang[1,2], Meijie Wang[1,2(✉)], Mingai Li[1,2], and Jingling Zhang[1,2]

[1] Faculty of Information Technology, Beijing University of Technology, Beijing, China
wangmeijie1992@163.com

[2] Beijing Key Laboratory of Computational Intelligence and Intelligent System, Beijing, China

Abstract. Person re-identification (re-id) has attracted widespread attention due to its application and research significance. However, since the person re-id puts the cropped images as input, it is far from the real-world scenarios just like person search which aims at matching a target person from a gallery of the whole scene images. Person search is more difficult but more practical and meaningful. In this paper, we propose a new person search network with an enhanced feature representation. Our network mainly consists two parts, a pedestrian proposal net and an identification net. In the identification net, we utilize hand-crafted features and Convolutional Neural Network (CNN) features to get more discriminative and compact features. Experiments on a large-scale benchmark dataset demonstrate our network gets better performance than others counterparts.

Keywords: Person search · Hand-crafted features · CNN features Feature representation

1 Introduction

Person re-identification (re-id) aims at matching the target person from different non-overlapping camera views which has been studied extensively in recent years [3,13,14,17,21,26]. It is widely used in video surveillance, robot and image retrieval. The most challenging problem is how to correctly match the target person from a gallery images under the complex variations of human poses, significant changes in view angle, lighting, occlusion, resolution, background clutter and etc.

There are plentiful existing datasets and methods on the person re-id, but person re-id has a large difference with the real-world application. In most benchmarks [5,10,11,13,15], there are only pedestrian images which made by manually cropped, as shown in Fig. 1(a), however, finding the target person in a gallery of whole scene images is the practical goal (Fig. 1(b)). That is, most of the person re-identification methods suppose that the pedestrian detectors can perfectly

J. Yang et al. (Eds.): CCCV 2017, Part III, CCIS 773, pp. 315–327, 2017.
https://doi.org/10.1007/978-981-10-7305-2_28

(a) Person re-identification: match person with perfected cropped pedestrians

(b) Person search: find the target person in the whole scene images

Fig. 1. Comparison of person re-identification with person search. According to real-world applications, the person search is more different but more practical and meaningful.

crop the pedestrian images, but it is unavailable. Existing pedestrian detectors would inevitably produce detection errors, which significantly harm the final performance of person search.

The conventional approaches of person search break down it into two separate tasks–pedestrian detection and person re-id. In 2014, Xu *et al.* [23] first proposed a unified framework for person search. They jointly modeled the commonness of people and uniqueness of the queried person. But a searching strategy based on sliding window brings high computational cost and blocks the scalability. And only using the hand-crafted feature maps limits the performance. In 2017, Xiao *et al.* [22] made a big step towards to the real-world application. They use a single CNN for person search. An Online Instance Matching (OIM) loss function is proposed to effectively train the network and be scalable to a mass of identities. However, the features extracted in [22] do not contain enough color features and texture features to distinguish different people. As shown in Fig. 2, the first line of Fig. 2 is the ground truth, the blue bounding box (the left one) is the target person and the green bounding boxes are the matched images with the target person. The second line is the predicted results, the green one is the correct result, and the orange ones are the false results. As we known, concatenating hand-crafted appearance features, e.g. RGB, HSV colorspaces and LBP descriptor designed to overcome appearance variations in different views for person re-identification is more distinctive and reliable [20].

In this paper, we propose a novel person search network which could effectively fuse the hand-crafted feature maps and deep feature maps. Hand-crafted features are the Ensemble of Local Features, named ELF16 and deep features

Fig. 2. Some example results of person search. It is clear that the features [22] do not contain enough color features and texture features to distinguish between different people. (Color figure online)

are CNN features. The proposed network consists of two parts: a pedestrian proposal net and an identification net. The identification net can extract more complementary features for person search. Experimental results on a large-scale benchmark dataset demonstrate the effectiveness of our enhanced deep feature representation.

2 Method

Our network architecture is shown in Fig. 3. A given whole scene image is fed into a basic CNN stage as input. And the raw pixels are transformed to CNN features through the basic CNN stage. Next these features are fed into a pedestrian proposal net to predict bounding boxes of candidate pedestrians. Then they are fed into a RoI-Pooling layer [6], followed by an identification net to extract L2-normalized features. The identification net includes two parts. One is the rest

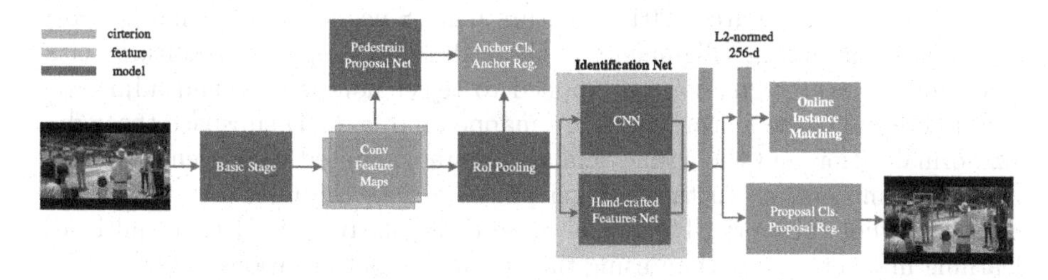

Fig. 3. The architecture of person search network. An input image is fed into the basic stage to extract feature maps. Then the pedestrian proposal net is used to generate bounding boxes of candidate person, followed by an identification net that contains a CNN and a hand-crafted feature net. Finally, the extracted features are projected to a L2-normalized 256-d subspace, and Online Instance Matching loss function is employed to train the network.

part of ResNet-50, the other is the hand-crafted features net. We combine these two parts into a full-fledged feature map via Buffer Layer and Fusion Layer. At training stage, in order to supervise the identification net, the OIM loss function is employed to train the pedestrian proposal net via a multi-task manner. At testing stage, we compute the distances between the target person with each of the gallery person. In the next sections, we introduce the details of the CNN model structure and the hand-crafted features as well as the details of the fusion process.

2.1 Model Structure

Our model is built upon the ResNet-50 [9]. The first layer of the ResNet-50 has a 7×7 convolution filter, named conv1. The following layers consist of four blocks. Each of layers has 3, 4, 6, 3 residual units, named conv2_x to conv5_x. We regard the layers from conv1 to conv4_3 as the basic stage. The input image produces 1024 feature channels through the basic stage, and the resolutions of the feature maps is 1/16 of the original image.

In order to detect pedestrian bounding boxes, we build a pedestrian proposal network upon these features. In order to make the feature maps more suitable for pedestrian detection task, we use a $3 \times 3 \times 512$ filter to transform the feature maps. Following [18], at each feature map location we adopt 9 anchors, and use a softmax classifier and a linear regression. The softmax classifier is to predict whether each anchor is a pedestrian or not and the linear regression is to adjust their bounding boxes. In order to reduce redundancy, we utilize a non-maximum suppression (NMS) on these bounding boxes and then reserve the top 128 bounding boxes as our pedestrian proposals.

In the cause of finding the target person in these proposals, we build an identification net upon these proposals to further extract the feature maps that contain CNN feature maps and hand-crafted feature maps, and compare with the features of the target person. First, we use an RoI-Pooling layer [6] to extract a $14 \times 14 \times 1024$ feature map from the basic stage for each proposal. Then the extracted feature maps are fed into the identification net and a global average pooling layer to generate a 2048D feature map. Since there are some malposition, distortion and misalignments in the pedestrian proposals, a softmax classifier and a linear regression are applied to reject non-persons and adjust the location. Finally, the feature maps are mapped into a 256D subspace through a L2-normalization and the cosine similarities are computed to inference whether it is the same person. In the training phase, we use the OIM loss function to train the whole network. The whole network is jointly trained in a multi-task learning manner, rather than using the alternative optimizations in [18].

2.2 The Detail of the Identification Net Structure

The identification net contains two parts as shown in Fig. 4. The first part is the rest of the ResNet-50, i.e., conv4_4 to conv5_3, following as [22]. The convolution in pedestrian proposal network allows part displacement and visual changes to

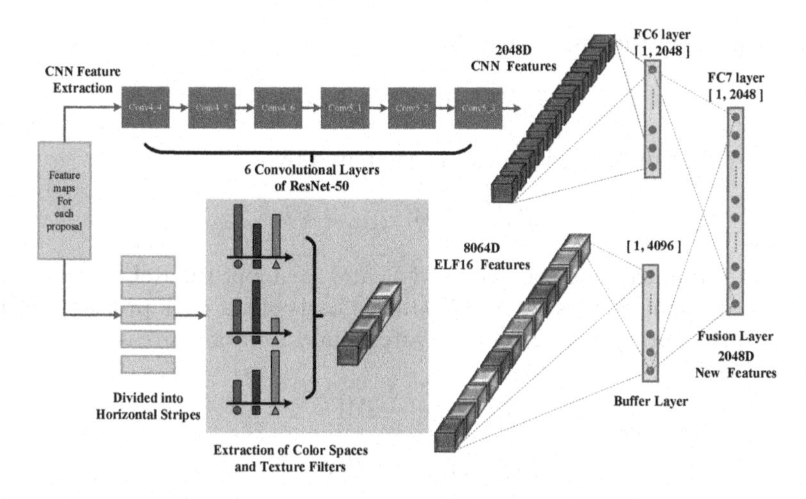

Fig. 4. ELF16 feature maps and CNN feature maps of the identification net. (Color figure online)

be alleviated in this part. At the same time, convolution kernels provide different descriptions for pedestrian images. In addition, pooling and LRN layers provide nonlinear expression of corresponding description, which significantly reduces the overfitting problem. So we build the identification net upon pedestrian proposal network and RoI-Pooling layer to extract more discriminatory. The second part extracts additional hand-crafted features from the same feature map, followed by a Buffer Layer which is a 4096D-output fully connection layer. A fusion layer is also used to combine feature maps of the two parts into a full-fledged feature map.

In the lower part of Fig. 4. ELF [8] is extracted. In a general way, the input image of the pedestrian is partitioned into several horizontal stripes. Each horizontal stripe extracts different color and texture information. Specifically, we partition the feature maps of each pedestrian proposal into 16 horizontal stripes. And the color feature maps and the texture feature maps are extracted respectively from each horizontal stripe. The color features contain RGB, HSV, LAB, XYZ, YCbCr and NTSC and the texture features contain Gabor, Schmid and LBP. Each channel of the above features extracts a L1-normalized 16D histogram. All histograms are concatenate together to form a single feature vector, named ELF16.

In order to effectively fuse CNN feature maps with ELF16, we use a buffer layer and a fusion layer. In our network, the ELF16 could affect the parameters of the whole network. Generally speaking, the fusion feature maps should be more discriminative than both CNN feature maps and ELF16. It is important for the buffer layer in our network, since it bridges the gap between the ELF16 and CNN feature maps with enormous difference, moreover, it can ensure the convergence.

If the input of the fusion layer is

$$x = [ELF16, CNN\_Features] \tag{1}$$

Then the output of the fusion layer is computed by:

$$Z_{Fusion}(x) = h(W_{Fusion}^{T} x + b_{Fusion}) \tag{2}$$

The activation function is denoted as $h(\cdot)$. And we adopt the ReLU and dropout layers, the dropout ratio is set as 0.5. On the basis of the back propagation(BP) algorithm, the parameters of lth layer after a new iteration are as following:

$$\begin{aligned} W_{new}^{(l)} &= W^{(l)} - \alpha[(\frac{1}{m}\Delta W^{(l)}) + \lambda W^{(l)}] \\ b_{new}^{(l)} &= b^{(l)} - \alpha[(\frac{1}{m}\Delta b^{(l)})] \end{aligned} \tag{3}$$

where parameters α, m and λ are set as [1].

2.3 How Do ELF16 and CNN Feature Maps Make More Complementary?

We effectively fuse CNN feature maps which do not contain enough color features and texture features to distinguish different people and ELF16 to get better performance. If the parameters of the whole network are affected by the ELF16, i.e., the gradient of the network parameters are adjusted according to ELF16, then CNN feature maps and ELF16 could be more complementary with each other, as the buffer layer and fusion layer are used to make our features more discriminative in different images.

Denote ELF16 as $\tilde{x}$ and conv5_3 feature map as x, and the weight connecting the jth node in n^{th} node in $(n+1)^{th}$ layer as W_{ij}^{n}. Denote $Z_j^n = \sum_j W_{ji}^{n-1} a_i^{n-1}$ where $a_i^{n-1} = h\left(Z_i^{n-1}\right)$. Then

$$\delta_i^n = \frac{\partial J}{\partial Z_i^n} \tag{4}$$

Note that $Z_j^6 = \sum_j W^{n-1} x_i^{n-1}$. According to back propagation, $\tilde{x}$ influence $\frac{\partial J}{\partial W_{ij}^5}$. According to this, ELF16 $\tilde{x}$ could affect the parameters of CNN feature maps to make more complementary. And

$$\frac{\partial J}{\partial W_{ij}^5} = x_j \delta_j^6 \tag{5}$$

where

$$\delta_i^6 = \left(\sum_j W_{ji}^6 \delta_j^7\right) h'\left(Z_i^6\right) \delta_j^7 = \left(\sum_k W_{kj}^7 \delta_k^8\right) h'\left(Z_j^7\right) \tag{6}$$

At the same time, $\tilde{x}$ could affect δ_j^7 by,

$$Z_k^7 = \sum_j W_{kj}^6 a_j^6 + \sum_j \tilde{W}_{kj}^6 \tilde{a}_j^6 \tag{7}$$

where

$$\tilde{a}_j^6 = h\left(\sum \tilde{W}_{ji}^5 \tilde{x}_i\right) \tag{8}$$

3 Experiments

3.1 Dataset

We use a large-scale person search dataset [22] to evaluate our approach. The dataset comes from hundreds of scenes of street in an urban city and movie snapshots with the variations of lighting, viewpoints, and background conditions. The dataset contains 18,184 images, 8,432 identities, and 96,143 pedestrian bounding boxes. The statistics of the dataset are shown in Table 1. However, the images which the person who appear with half bodies or abnormal poses are not be annotated.

Table 1. Statistics of the dataset with respect to data sources and training/test splits.

Source/split	# image	# pedestrians	# identities
StreetSnap	12,490	75,845	6,057
Movie & TV	5,694	20,298	2,375
Training	11,206	55,272	5,532
Test	6,978	40,871	2,900
Overall	18,184	96,143	8,432

3.2 Evaluation Protocols and Metrics

The dataset is split into a training set and a test set, and there are no overlapped images or labeled identifies between the two sets. Further, the test identity instances are split into queries and galleries. We randomly choose one of these identity instances as the query for each of the 2900 test identifies. At the same time, all the images that contain the other instances and some images which do not contain this person with randomly sample constitute the homologous gallery set. Different queries correspond different galleries and jointly covered all the 6,978 test images.

We use cumulative matching characteristic (CMC top-K) and mean averaged precision (mAP) to evaluate the proposed method. The CMC top-K is inspired from the person re-id task, when one can match with anyone of the top-K predicted bounding boxes which overlaps with intersection-over-union (IoU) greater or equal to 0.5 of the ground truths. The mAP is inherited from the object detection task to judge the validity of predicted bounding boxes.

3.3 Experiment Settings

We use a pre-trained caffemodel that implemented the person search network in [22]. We regard the first convolution layer and the batch normalization layers as constant affine transformations in the basic stage. All the loss weights are the same. There are two scene images in each mini-batch. The initial learning rate is set as 0.00001, decreased by one-tenth of the initiative after 40K iterations. And the network took 50K iterations to converge.

We compare our method with traditional methods which regard the person search as two separate task, that is, pedestrian detection task and person re-id task. In our experiments, there are 3 pedestrian detection and 5 person re-id methods, besides the person search network [22]. For pedestrian detection, we employ the off-the-shelf detectors, including CCF [24], ACF [4] and Faster R-CNN [18] with ResNet-50.

For person re-id, we employ popular feature representations of re-id which are DenseSIFT-ColorHist (DSIFT) [25], Bag of Words (BoW) [26] and LOMO [14]. And every feature representation and a specific distance metric are jointly used. The distance metrics contain Euclidean, Cosine similarity, KISSME [10] and XQDA [14]. We use the remaining net without the pedestrian proposal network in [22], and give the cropped pedestrian images as input images to train a baseline re-id method that named IDNet.

3.4 Comparison with Detection and Re-ID

We compare our method with the above mentioned methods which include a person search framework that was proposed by Xiao *et al.* in [22] and 15 baseline combinations that separate the person search problem into pedestrian detection and person re-id tasks. The results are shown in Tables 2 and 3. It is shown that our network outperforms the others. Thanks to the joint optimization of the pedestrian detection and person re-id, we can see that both the person search method [22] and ours outperforms the others by a large margin.

Table 2. Comparison of our network with separate pedestrian detection + person re-identification method and the person search network that was proposed by Xiao *et al.* [22] in terms of CMC top-1.

CMC top-1 (%)	CCF	ACF	CNN	GT
DSIFT + Eudlidean	11.7	25.9	39.4	45.9
DSIFT + KISSME	13.9	38.1	53.6	61.9
BoW + Cosine	29.3	48.4	62.3	67.2
LOMO + XQDA	46.4	63.1	74.1	76.7
IDNet	57.1	63.0	74.8	78.3
Xiao *et al.* [1]	—	—	78.7	80.5
Ours	—	—	**80.6**	**82.8**

Table 3. Comparison of our network with separate pedestrian detection + person re-identification method and the person search network that was proposed by Xiao *et al.* [22] in terms of mAP.

mAP (%)	CCF	ACF	CNN	GT
DSIFT + Eudlidean	11.3	21.7	34.5	41.1
DSIFT + KISSME	13.4	32.3	47.8	56.2
BoW + Cosine	29.9	42.4	56.9	62.5
LOMO + XQDA	41.2	55.5	68.9	72.4
IDNet	50.9	56.5	68.6	73.1
PSNet	—	—	75.5	77.9
Ours	—	—	**77.8**	**80.2**

From Tables 2 and 3 for each person re-id method, we can see that the person search performance are significantly affected by the detectors. It may not a good idea to use an off-the-shelf detector to employ the existing person re-id methods for the real-world person search. Otherwise, the pedestrian detector will impede the recall of the better person re-id methods.

On the other hand, the performance of different person re-id methods is consistent with all the detectors. We can see that IDNet has similar performance with LOMO+XQDA when employing ACF and CNN detectors or GT. However, it is significantly better than LOMO+XQDA method when using the CCF detector. We can also see that our person search has better performance than Xiao *et al.* [22], since our network can extract more effective features to distinguish different people.

The performance of person search is directly affected by the detector in the separate network. Poor performance of detector could not get a good performance of person search task. The two parts could not affect each other, i.e. the pedestrian detection network only affect the person re-id network, and the person re-id network could affect the pedestrian detection network. However, in our work, these two parts could affect each other to make two parts more complementary.

3.5 Experiments on Effectiveness of Fusion Features

We evaluated our features on three benchmark datasets of person re-identification to demonstrate the effectiveness of our fusion features.

Datasets and Experimental Settings. We use three publicly available datasets: VIPeR [7], CUHK01 [12] and PRID450s [19]. Each dataset was captured from two disjoint camera views under significant misalignment, light change and body part distortion. There is a brief introduction for the datasets in Table 4.

Table 4. Statistics of the dataset with respect to data sources and training/test splits.

Dataset	# identities	# images	# images in training sets	# camera views
VIPeR	632	1,264	316	2
CUHK01	971	3,884	485	2
PRID450s	450	900	225	2

In each experiment, the training set was half of the identities via randomly selected and the rest as testing set. During training, we learn the projection matrix W using the training set. During testing, we get the final features by $x' = W^T x$ then measure the distance between two inputs. In order to make the results reliable and stable, we repeated 10 times for each experiment and then computed the average Rank-i accuracy rate. CMC was present in Fig. 5.

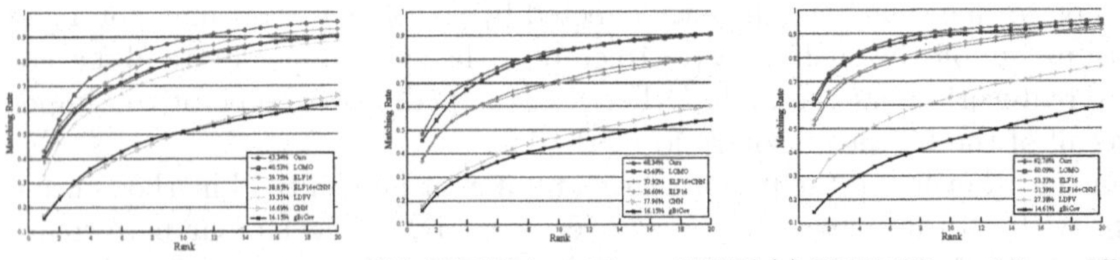

(a) VIPeR on Mirror KMFA (b) CUHK01 on Mirror KMFA (c) PRID450s on Mirror KMFA

Fig. 5. CMC Curves of our experiments on three datasets with Mirror KMFA metric learning algorithm.

During testing, we regard one image which was chosen from View 2 as a probe and we also regard all images which in View 1 as the gallery. However, in the CUHK01 dataset, there are two images of the same person in one view, one image of each identity was randomly chosen and regard it as the gallery.

In [2], Chen *et al.* proposed a high-performance metric learning algorithm named Mirror Kernel Marginal Analysis (KMFA) for person re-id. Our experiments employ this metric learning algorithm, and the setting as the [2].

Features. In our experiments, we use six methods to extract feature maps, which contain LDFV [15], gBiCov [16], CNN features, LOMO features [14], ELF16 features and our features. Specifically, the CNN features are the same as the upper network of the Fig. 4. And our features are the fusion features of the Fig. 4. Because of the copyrights of the LDFV features code, we only experiment the LDFV features on VIPeR dataset. In order to extract features efficiently, we resized all images to 224 × 224.

In addition, we enhanced the features representation (CNN+ELF16) to prove the effectiveness of our approach. The features of CNN+ELF16 are the concatenation of CNN features to ELF16.

Experiments on the Features. The Fig. 5(a)–(c) are the results of our experiments on VIPeR, CUHK01 and PRID450s with Mirror KMFA, respectively. We can see that our features outperform other standalone features. And our features significantly outperformed than CNN+ELF16 features, which maybe according to this two reasons:

1. The CNN features of the experiment were trained to make them complementary with the ELF16 features, which it is not optimal to simply cascade CNN features with ELF16.

2. We use the buffer layer and fusion layer to automatically tune the parameters of the network, and the performance of the fused features is better than others.

4 Conclusion

In this paper, we propose an enhanced feature representation for person search. It jointly handles the pedestrian detection and person re-id tasks into a single CNN. And in person re-id part we jointly use CNN feature maps and ELF16 which include color features and texture features. Then the buffer layer and fusion layer could regularize the network to make the identification net more focus on extracting complementary features maps. Experiments on a large-scale person search dataset show the effectiveness of our person search network.

Acknowledgments. This work is partly supported by the National Natural Science Foundation of China under Grant nos. 61573351, and 81471770.

References

1. Bottou, L.: Stochastic gradient descent tricks. In: Montavon, G., Orr, G.B., Müller, K.-R. (eds.) Neural Networks: Tricks of the Trade. LNCS, vol. 7700, pp. 421–436. Springer, Heidelberg (2012). https://doi.org/10.1007/978-3-642-35289-8_25
2. Chen, Y.C., Zheng, W.S., Lai, J.: Mirror representation for modeling view-specific transform in person re-identification. In: IJCAI, pp. 3402–3408. Citeseer (2015)
3. Chu, X., Ouyang, W., Li, H., Wang, X.: Structured feature learning for pose estimation. In: Proceedings of the IEEE Conference on Computer Vision and Pattern Recognition, pp. 4715–4723 (2016)
4. Dollár, P., Appel, R., Belongie, S., Perona, P.: Fast feature pyramids for object detection. IEEE Trans. Pattern Anal. Mach. Intell. **36**(8), 1532–1545 (2014)
5. Felzenszwalb, P.F., Girshick, R.B., McAllester, D., Ramanan, D.: Object detection with discriminatively trained part-based models. IEEE Trans. Pattern Anal. Mach. Intell. **32**(9), 1627–1645 (2010)
6. Girshick, R.: Fast R-CNN. In: Proceedings of the IEEE International Conference on Computer Vision, pp. 1440–1448 (2015)
7. Gray, D., Brennan, S., Tao, H.: Evaluating appearance models for recognition, reacquisition, and tracking. In: Proceedings of the IEEE International Workshop on Performance Evaluation for Tracking and Surveillance (PETS), vol. 3 (2007)

8. Gray, D., Tao, H.: Viewpoint invariant pedestrian recognition with an ensemble of localized features. In: Forsyth, D., Torr, P., Zisserman, A. (eds.) ECCV 2008. LNCS, vol. 5302, pp. 262–275. Springer, Heidelberg (2008). https://doi.org/10.1007/978-3-540-88682-2_21
9. He, K., Zhang, X., Ren, S., Sun, J.: Deep residual learning for image recognition. In: Proceedings of the IEEE Conference on Computer Vision and Pattern Recognition, pp. 770–778 (2016)
10. Koestinger, M., Hirzer, M., Wohlhart, P., Roth, P.M., Bischof, H.: Large scale metric learning from equivalence constraints. In: 2012 IEEE Conference on Computer Vision and Pattern Recognition (CVPR), pp. 2288–2295. IEEE (2012)
11. Kviatkovsky, I., Adam, A., Rivlin, E.: Color invariants for person reidentification. IEEE Trans. Pattern Anal. Mach. Intell. **35**(7), 1622–1634 (2013)
12. Li, W., Zhao, R., Wang, X.: Human reidentification with transferred metric learning. In: Lee, K.M., Matsushita, Y., Rehg, J.M., Hu, Z. (eds.) ACCV 2012. LNCS, vol. 7724, pp. 31–44. Springer, Heidelberg (2013). https://doi.org/10.1007/978-3-642-37331-2_3
13. Li, W., Zhao, R., Xiao, T., Wang, X.: DeepReID: deep filter pairing neural network for person re-identification. In: Proceedings of the IEEE Conference on Computer Vision and Pattern Recognition, pp. 152–159 (2014)
14. Liao, S., Hu, Y., Zhu, X., Li, S.Z.: Person re-identification by local maximal occurrence representation and metric learning. In: Proceedings of the IEEE Conference on Computer Vision and Pattern Recognition, pp. 2197–2206 (2015)
15. Ma, B., Su, Y., Jurie, F.: Local descriptors encoded by fisher vectors for person re-identification. In: Fusiello, A., Murino, V., Cucchiara, R. (eds.) ECCV 2012. LNCS, vol. 7583, pp. 413–422. Springer, Heidelberg (2012). https://doi.org/10.1007/978-3-642-33863-2_41
16. Ma, B., Su, Y., Jurie, F.: Covariance descriptor based on bio-inspired features for person re-identification and face verification. Image Vis. Comput. **32**(6), 379–390 (2014)
17. Paisitkriangkrai, S., Shen, C., van den Hengel, A.: Learning to rank in person re-identification with metric ensembles. In: Proceedings of the IEEE Conference on Computer Vision and Pattern Recognition, pp. 1846–1855 (2015)
18. Ren, S., He, K., Girshick, R., Sun, J.: Faster R-CNN: towards real-time object detection with region proposal networks. In: Advances in Neural Information Processing Systems, pp. 91–99 (2015)
19. Roth, P.M., Hirzer, M., Köstinger, M., Beleznai, C., Bischof, H.: Mahalanobis distance learning for person re-identification. In: Gong, S., Cristani, M., Yan, S., Loy, C.C. (eds.) Person Re-Identification. ACVPR, pp. 247–267. Springer, London (2014). https://doi.org/10.1007/978-1-4471-6296-4_12
20. Wu, S., Chen, Y.C., Li, X., Wu, A.C., You, J.J., Zheng, W.S.: An enhanced deep feature representation for person re-identification. In: 2016 IEEE Winter Conference on Applications of Computer Vision (WACV), pp. 1–8. IEEE (2016)
21. Xiao, T., Li, H., Ouyang, W., Wang, X.: Learning deep feature representations with domain guided dropout for person re-identification. In: Proceedings of the IEEE Conference on Computer Vision and Pattern Recognition, pp. 1249–1258 (2016)
22. Xiao, T., Li, S., Wang, B., Lin, L., Wang, X.: Joint detection and identification feature learning for person search. In: Proceedings of the CVPR (2017)
23. Xu, Y., Ma, B., Huang, R., Lin, L.: Person search in a scene by jointly modeling people commonness and person uniqueness. In: Proceedings of the 22nd ACM international conference on Multimedia, pp. 937–940. ACM (2014)

24. Yang, B., Yan, J., Lei, Z., Li, S.Z.: Convolutional channel features. In: Proceedings of the IEEE International Conference on Computer Vision, pp. 82–90 (2015)
25. Zhao, R., Ouyang, W., Wang, X.: Unsupervised salience learning for person re-identification. In: Proceedings of the IEEE Conference on Computer Vision and Pattern Recognition, pp. 3586–3593 (2013)
26. Zheng, L., Shen, L., Tian, L., Wang, S., Wang, J., Tian, Q.: Scalable person re-identification: a benchmark. In: Proceedings of the IEEE International Conference on Computer Vision, pp. 1116–1124 (2015)

Joint Temporal-Spatial Information and Common Network Consistency Constraint for Person Re-identification

Zhaoxi Cheng, Hua Yang(✉), and Lin Chen

Institute of Image Communication and Network Engineering,
Department of Electronic Engineering, Shanghai Jiao Tong University,
Shanghai 200240, China
{czx536,hyang,SJChenLin}@sjtu.edu.cn

Abstract. Recent research of person Re-identification most focuses on exploring person appearance feature and distance measure between specific cameras pair. In this paper, a joint temporal-spatial information and common network consistency constraint framework is proposed to improve the re-identification performance in all pairwise cameras. First, a correction function is introduced for describing the influence factor of temporal-spatial information on similarity scores. Then the amended similarity score strategy is provided to tradeoff between the person appearance and temporal-spatial information. Finally the whole global optimization problem of the jointing temporal-spatial and common consistence constraints is solved by integer programming method. Using the multi-cameras RAiD dataset, the experiment results validate that the proposed framework obtains significantly better performance compared to the state of the art camera network person re-identification methods.

Keywords: Person re-identification · Temporal-spatial information
Network consistency

1 Introduction

Recently, person re-identification (re-ID) has become increasingly popular in the community. Its aim is seeking a person of interest in previous time and other cameras. There are many general challenges such as scale, altitude, illumination and view angle changes because the person images are obtained from different cameras deployed in non-overlapping scenes.

Facing these problems, most existing traditional person re-identification approaches focus on constructing distinctive visual features and learning an optimal metric. The feature-based methods measure the similarity of person in images with features [1–4]. These features mostly are color feature, shape feature, gradient feature, texture feature, or special features obtained by learning, such as BIF [5], covariance descriptors [6] and LOMO [7]. It is difficult to achieve better effect by using standard distance measurement, treating each of

J. Yang et al. (Eds.): CCCV 2017, Part III, CCIS 773, pp. 328–339, 2017.
https://doi.org/10.1007/978-981-10-7305-2_29

the apparent features equally. To solve this problem, some researchers have proposed methodologies based on metric learning [8–10]. In a framework based on metric learning, a series of training data is used to learn a non-Euclidean measure so that the distance between the features of the correct match is the smallest and the distance between the mismatched pairs is greatest. Lately, a cross-view quadratic discriminant analysis (XQDA) method is proposed [7], which achieves good performance.

Both of the above two approach only depend on visual information and the effect is not very satisfactory. Further, deep learning method is proposed to learn better feature automatically, leading to some better results. CNN (Convolutional Neural Networks) has been widely utilized to extract spatial information of the single-shot pedestrians and achieved success [11–13], since it establish multidimensional model of pedestrians and can provide a better and robust feature representation. However in practical surveillance scenes, there are many cameras distributing in a large area of region, each of these cameras cover its own region which is non-overlapping with other camera's region. In the situation where a person is captured by difference camera, the time span may be very large, leading to a problem that the appearance change of a individual person may be very significant. At this time, it is difficult to provide sufficient identity discrimination using visual information purely. Among the images of persons, there are temporal-spatial information, and among the cameras, there are network consistent information. All the method mentioned above is mainly regard person re-identification problem as a retrieval task based on camera pair, with no consideration of camera network information. It remains a problem of great value to propose a re-identification framework considering the camera networks to improve the accuracy.

To address the problem, in 2014, Abir et al. first consider camera network for person re-identification [14], introducing the concept of consistent in network. First make a hypothesis: each pedestrian in each camera just appeared once, then define global similarity as the objective function and take network consistency as constraint. Finally the problem is transformed into integer linear programming problem. For each camera pairs in this camera network, the result has improve compared with traditional method. But there are also some limitation that its constraints is relatively strong, and it did not use the spatial-temporal information. Huang et.al. proposed a method of converting the spatial-temporal information and the similarity score obtained by the person re-identification of camera pair into probability [15]. They use probability to represent similarity and temporal-spatial information, then multiple the two probability and use the result as modified similarity. They obtain some improvement in person re-identification but the method doesn't consider the consistent problem in camera network.

Motivated by above mentioned works, we propose a camera network person re-identification framework, which combine the visual information, temporal-spatial information and consistent information and lead to an improvement of the camera pairwise re-identification performance between all the cameras. Our contribution is summarized by the following three aspects. (1) A novel framework of camera network person re-identification system with a unified symbol architecture is introduced. The framework we proposed joint the temporal-spatial information, network consistency and pairwise person re-identification method. (2) A correction function in introduced describing the influence factor of temporal-spatial information on similarity scores. And an amended similarity score strategy is provided to tradeoff between the person appearance and temporal-spatial information. Specially addition of the correction function and the original pairwise similarity score is proposed, with a coefficient controlling the ratio between original score and influence factor. (3) Jointing temporal-spatial information, common consistence constraints and pairwise person re-identification method, the whole global optimization problem is deviate which is solved by integer programming method, lead to the final result of camera network person re-identification problem.

2 Our Method

2.1 Overall Framework

In this section we describe the camera network person re-identification system. The proposed method can be represented using a framework diagram shown in Fig. 1. For each camera pair, original similarity score between every two person in different camera are computed by using existing methods of feature representation and metric learning. Temporal-spatial information is represented by correction function defined in (3). The consistency is represented as a constraint of an optimization problem which joint original score and temporal-spatial information. Finally we transform the whole problem into an integer programming problem.

2.2 Proposed Method

Symbol definition. First of all, we should define a set of symbol that unify all discussion below. Suppose there are M Cameras $\{C_1, C_2, ..., C_M\}$ in a camera network. Each camera control its own region $\{region1, region2, regionM\}$ and has a corresponding scenes position coordinates $\{(x_1, y_1), (x_2, y_2), ..., (x_M, y_M)\}$, which is the central location of its view region.

Then there are $C_M^2 = \frac{m(m-1)}{2}$ camera pairs theoretically, For each pair, there is a corresponding distance between the camera pair $d_{ij} = ((x_i - x_j)^2 + (y_i - y_j)^2)^{\frac{1}{2}}$ $(i < j)$. The intuitive diagram of this network consist of M cameras is shown in Fig. 2(a).

For arbitrary camera C_i, there are n_i $(i = 1, 2, ..., M)$ person having ever appeared. We denote P_j^i as the jth person captured by the ith camera, then the n_i person captured by the ith camera is specifically $\{P_1^i, P_2^i, ...P_j^i, ..., P_{n_i}^i\}$, each of which has a corresponding capture time. The capture time of person P_j^i is t_j^i,

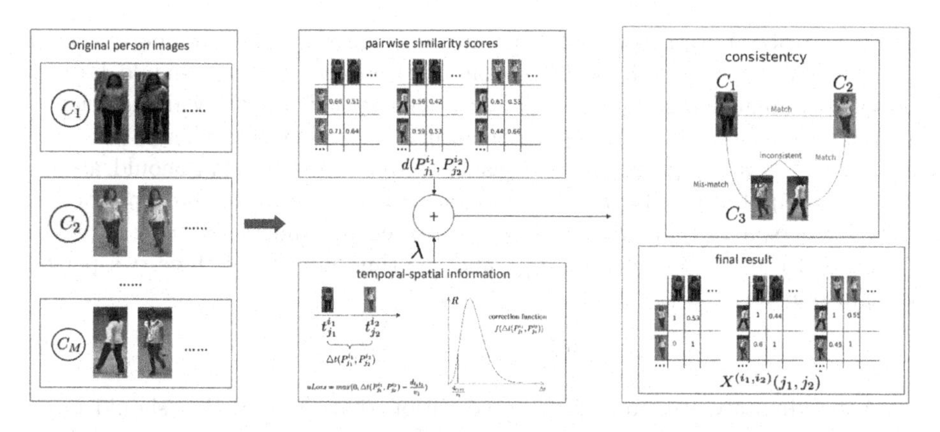

Fig. 1. Framework diagram: pairwise similarity scores is calculated by traditional method. Temporal-spatial information is modelled as correction function. Then pairwise similarity scores and correction function are added, with a coefficient λ to tradeoff the two addends. Finally the sum and network consistency are jointed, leading to the whole global optimization problem.

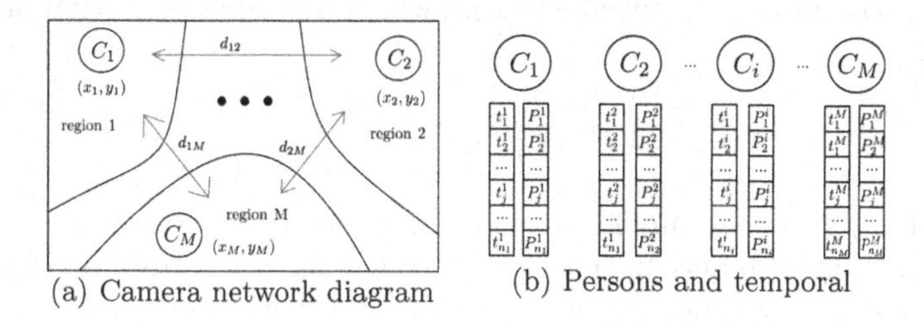

Fig. 2. Symbol architecture of the camera network person re-identification framework

where $i = 1, 2, ..., M$, $j = 1, 2, ..., n_i$. These information can be described by a graph shown as Fig. 2(b).

For simply discussing, we first assume that the same N person are present in each of the M cameras. Now for each camera pair, for example (i_1, i_2), $(i_1, i_2 = 1, 2, ..., M)$. We can obtain the similarity score of each person pair in different camera by performing feature representation and metric learning. For person $P_{j_1}^{i_1}$ and $P_{j_2}^{i_2}$, its similarity score is denoted as $d(P_{j_1}^{i_1}, P_{j_2}^{i_2})$, $(j_1, j_2 = 1, 2, ..., N)$. We can put all these similarity score of persons between camera (i_1, i_2) into an $N \times N$ matrix, called similarity matrix, denoted as $D^{(i_1,i_2)}$. The entry $D^{(i_1,i_2)}(a, b)$ represent $d(P_a^{i_1}, P_b^{i_2})$. For each camera pair we can obtain a similarity matrix, finally all the $\frac{M(M-1)}{2}$ similarity matrices can be obtained. We define a overall similarity matrix $\boldsymbol{D}$.

$$\boldsymbol{D} = [D^{(1,2)}, D^{(1,3)}, ..., D^{(1,M)}, D^{(2,1)}, D^{(2,2)}, ..., D^{(2,M)}, ..., D^{(M-1,M)}] \quad (1)$$

Joint temporal-spatial constraint and consistency. This temporal-spatial constraint is Built on a hypothesis that the velocity of person should be in a reasonable range depend on practical occasions, specifically the walking velocity should have an upper bound v_1 under normal circumstances. Then if person A and person B are captured by two camera, their time interval should also in a range computed by distance of two camera and velocity bound. Specifically, consider person $P_{j_1}^{i_1}$ and $P_{j_2}^{i_2}$, they are captured by camera Ci_1 and C_{i_2}. The distance of these two camera is $d_{i_1 i_2}$ and the time interval of these two person is defined as formula (2).

$$\triangle t(P_{j_1}^{i_1}, P_{j_2}^{i_2}) = t_{j_1}^{i_1} - t_{j_2}^{i_2} \tag{2}$$

Approximately, we can assume a condition that $\triangle t(P_{j_1}^{i_1}, P_{j_2}^{i_2})$ should be in the range of interval $(\frac{d_{i_1 i_2}}{v_1}, \infty)$. If the condition is not satisfied, these two person is not the same person very likely, then the similarity score $d(P_{j_1}^{i_1}, P_{j_2}^{i_2})$ is changed to zero, else similarity score $d(P_{j_1}^{i_1}, P_{j_2}^{i_2})$ is amended by adding a correction term $R(P_{j_1}^{i_1}, P_{j_2}^{i_2})$. The correction term is a function (Eq. 3) of time interval $\triangle t(P_{j_1}^{i_1}, P_{j_2}^{i_2})$ defined in interval $(\frac{d_{i_1 i_2}}{v_1}, \infty)$ which we called correction function in this paper.

$$R(P_{j_1}^{i_1}, P_{j_2}^{i_2}) = f(\triangle t(P_{j_1}^{i_1}, P_{j_2}^{i_2})), \qquad \triangle t(P_{j_1}^{i_1}, P_{j_2}^{i_2}) \in (\frac{d_{i_1 i_2}}{v_1}, \infty) \tag{3}$$

The function f can be in any form as long as it satisfy the following condition: f is first an increment function, then f reach the peak at a certain point, and then f become a decrease function, finally when $\triangle t(P_{j_1}^{i_1}, P_{j_2}^{i_2}) \to \infty$, $f \to 0$. We can consider it as a factor reflecting a kind of probability, such that $\triangle t(P_{j_1}^{i_1}, P_{j_2}^{i_2})$ is more likely at the middle of interval $(\frac{d_{i_1 i_2}}{v_1}, \infty)$ than at the two ends. In this paper we use Chi-square distribution probability density function as the correction function, as shown in Fig. 3(a), which satisfy the condition discussed above. Inspired by the velocity approximation of person mentioned above, we can define a function called upper loss($uLoss$) as formula (4).

$$uLoss = max(0, \triangle t(P_{j_1}^{i_1}, P_{j_2}^{i_2}) - \frac{d_{i_1 i_2}}{v_1}) \tag{4}$$

Then the similarity $d(P_{j_1}^{i_1}, P_{j_2}^{i_2})$ is corrected to $d^{'}(P_{j_1}^{i_1}, P_{j_2}^{i_2})$ according to the addition of original score and the correction term as shown in Eq. (5). The thought of the addition relation is similar as the rate distortion theory in information theory. Its main advantage is that we can use an factor to control the contribution ratio between original score and correction term I

$$d^{'}(P_{j_1}^{i_1}, P_{j_2}^{i_2}) = (d(P_{j_1}^{i_1}, P_{j_2}^{i_2}) + \lambda R(P_{j_1}^{i_1}, P_{j_2}^{i_2}))I(uLoss \neq 0) \tag{5}$$

where $I()$ is indicative function, λ is a coefficient that control the proportion between similarity and correction term. So the similarity is added by a correction term, but if the time interval is beyond the pre-specified range, the similarity

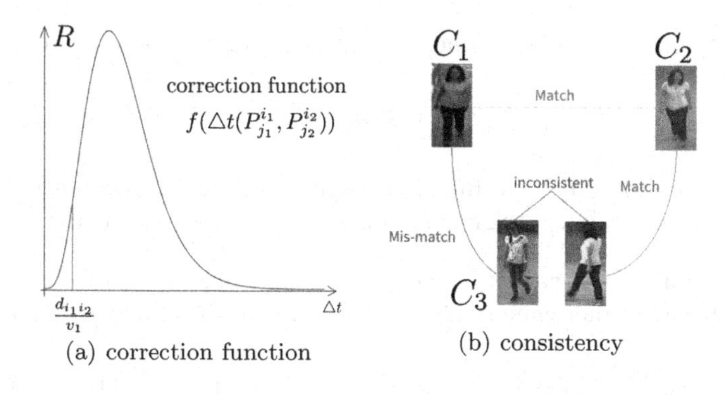

(a) correction function (b) consistency

Fig. 3. Joint temporal-spatial information and consistency

is changed to zero. After traversing all the entry of $D^{(i_1,i_2)}$, we can get similarity matrix $D^{'(i_1,i_2)}$ as median result. Repeating the above process we can get all the $\frac{M(M-1)}{2}$ median similarity matrices and the overall corrected similarity matrix $\boldsymbol{D}^{'}$.

$$\boldsymbol{D}^{'} = [D^{'(1,2)}, D^{'(1,3)}, ..., D^{'(1,M)}, D^{'(2,1)}, D^{'(2,2)}, ..., D^{'(2,M)}, ..., D^{'(M-1,M)}] \quad (6)$$

After getting the overall corrected similarity matrix $\boldsymbol{D}^{'}$, we use consistent constraint to further optimize the similarity matrix. The thought of this part is almost from the previous work [14]. Here we just give the train of thought shortly.

For each camera pair C_{i_1}, C_{i_2}, we define an assignment variable $x(P_{j_1}^{i_1}, P_{j_2}^{i_2})$ for all person pair $P_{j_1}^{i_1}$ and $P_{j_2}^{i_2}$, $(j_1, j_2 = 1, 2, ..., N)$.

$$x(P_{j_1}^{i_1}, P_{j_2}^{i_2}) = \begin{cases} 1 & \text{if } P_{j_1}^{i_1} \text{ and } P_{j_2}^{i_2} \text{ are the same person} \\ 0 & \text{otherwise} \end{cases} \quad (7)$$

We can put all these assignment variable of persons between camera (i_1, i_2) into an $N \times N$ matrix, called assignment matrix, denoted as $X^{(i_1,i_2)}$. Its entry $X^{(i_1,i_2)}(j_1, j_2)$ equal to 1 if $P_{j_1}^{i_1}$ and $P_{j_2}^{i_2}$ are the same person, equal to 0 otherwise.

For each camera pair we can obtain a assignment matrix, finally all the $\frac{M(M-1)}{2}$ assignment matrices can be obtained. We define a overall assignment matrix $\boldsymbol{X}$.

$$\boldsymbol{X} = [X^{(1,2)}, X^{(1,3)}, ..., X^{(1,M)}, X^{(2,1)}, X^{(2,2)}, ..., X^{(2,M)}, ..., X^{(M-1,M)}] \quad (8)$$

For the camera pair (i_1, i_2), the sum of the similarity scores of assignment is given by the equation below.

$$\sum_{j_1,j_2} d^{'}(P_{j_1}^{i_1}, P_{j_2}^{i_2}) x(P_{j_1}^{i_1}, P_{j_2}^{i_2})$$

Summing over all possible camera pairs the global similarity score can be written as

$$C = \sum_{i_1<i_2} \sum_{j_1,j_2} d^{'}(P^{i_1}_{j_1}, P^{i_2}_{j_2})x(P^{i_1}_{j_1}, P^{i_2}_{j_2}) = D^{'} X^{T} \tag{9}$$

Equation (9) is the objective function whom we want to maximize. There are two kinds of constraints: assignment constraint and loop constraint.

1. **Assignment constraint:** A person from any camera C_{i_1} can have only one match from another camera C_{i_2}. mathematically, $\forall x(P^{i_1}_{j_1}, P^{i_2}_{j_2}) \in \{0,1\}$

$$\begin{cases} \sum_{j_1} x(P^{i_1}_{j_1}, P^{i_2}_{j_2}) = 1 & \forall j_2 = 1,2,...,N \quad \forall i_1 < i_2 \in \{1,2,...,M\} \\ \sum_{j_2} x(P^{i_1}_{j_1}, P^{i_2}_{j_2}) = 1 & \forall j_1 = 1,2,...,N \quad \forall i_1 < i_2 \in \{1,2,...,M\} \end{cases} \tag{10}$$

   As a result only one element per row and per column is 1 in each assignment matrix $X^{(i_1,i_2)}$.

2. **Loop constraint:** Figure 3(b) illustrate the inconsistency phenomenon: For a triplet of cameras $\{C_1, C_2, C_3\}$, suppose $(P^1_{j_1}, P^2_{j_2})$, $(P^2_{j_2}, P^3_{j_3})$ and $(P^1_{j_1}, P^3_{j_4})$ is matched by similarity score independently, when these assignments are combined together, it leads to an infeasible scenario: $P^3_{j_3}$ and $P^3_{j_4}$ are the same person.
   This infeasibility can be corrected by formula (11), called loop constraint [14].

$$\begin{aligned} x(P^{i_1}_{j_1}, P^{i_2}_{j_2}) \geq x(P^{i_2}_{j_2}, P^{i_3}_{j_3}) + x(P^{i_3}_{j_3}, P^{i_1}_{j_1}) - 1 \\ \forall j_1, j_2 = 1,2,...,N \quad i_1 < i_2 < i_3 \in \{1,2,...,M\} \end{aligned} \tag{11}$$

The overall optimization problem is shown below. The constraints of this integer programming problem reflect the consistent information.

$$\arg\min_{x(P^{i_1}_{j_1}, P^{i_2}_{j_2})} \sum_{i_1<i_2<M} \sum_{i,j=1}^{N} x(P^{i_1}_{j_1}, P^{i_2}_{j_2})(d(P^{i_1}_{j_1}, P^{i_2}_{j_2}) + \lambda R(P^{i_1}_{j_1}, P^{i_2}_{j_2}))I(uLoss \neq 0) \tag{12}$$

$$\begin{aligned} s.t. \quad & \sum_{j_1} x(P^{i_1}_{j_1}, P^{i_2}_{j_2}) = 1 \qquad \forall j_2 = 1,2,...,N \quad \forall i_1 < i_2 \in \{1,2,...,M\} \\ & \sum_{j_2} x(P^{i_1}_{j_1}, P^{i_2}_{j_2}) = 1 \qquad \forall j_1 = 1,2,...,N \quad \forall i_1 < i_2 \in \{1,2,...,M\} \\ & x(P^{i_1}_{j_1}, P^{i_2}_{j_2}) \geq x(P^{i_2}_{j_2}, P^{i_3}_{j_3}) + x(P^{i_3}_{j_3}, P^{i_1}_{j_1}) - 1 \\ & \forall j_1, j_2 = 1,2,...,N \quad i_1 < i_2 < i_3 \in \{1,2,...,M\} \end{aligned} \tag{13}$$

The objective function (12) and constraints (13) form a binary integer program (BIP). Mathematically, this BIP problem has a mature solution approach to solve X [16].

After getting X, we can further modify similarity $d^{'}(P^{i_1}_{j_1}, P^{i_2}_{j_2})$ to the final version $d^{''}(P^{i_1}_{j_1}, P^{i_2}_{j_2})$ according to the Eq. (14). Finally, we obtain the final overall similarity matrix $D^{''}$.

$$d''(P_{j_1}^{i_1}, P_{j_2}^{i_2}) = \begin{cases} d'(P_{j_1}^{i_1}, P_{j_2}^{i_2}) & \text{if} x(P_{j_1}^{i_1}, P_{j_2}^{i_2}) = 0 \\ 1 & \text{if} x(P_{j_1}^{i_1}, P_{j_2}^{i_2}) = 1 \end{cases} \tag{14}$$

3 Experiment and Discussion

3.1 Experiment Setup

Data Set and Evaluation. Common dataset of state of the art person re-identification, such as VIPeR [17], CUHK01 [18], is obtained from two camera that is not satisfied our demand. In paper [14], whose issue is about camera network person re-identification, the author create the RAiD [14] dataset and use RAiD to perform experiment. The RAiD dataset is collected from four cameras covering large areas, camera 1 and 2 is indoor, camera 3 and 4 is outdoor, so it has large illumination variation. There are 43 person walking through this camera network, 41 of which appear in all the 4 cameras. We can just use these 41 subjects to unfold our experiment.

The final results are shown in terms of Cumulative Matching Characteristic (CMC) curves reflecting recognition rate.

Pairwise Similarity Score Generation. There are a lot of method to generate camera pairwise similarity score. In this paper we aim at discussing the role of the combination of temporal-spatial information and consistent constraint in camera network, so the selection is not very important, as long as stay constant through the whole experiment. In this paper we starts with extracting appearance features in the form of HSV color histogram from the images of the targets. Then we generate the similarity scores by using ICT [19], a recent work where pairwise re-identification was posed as a classification problem in the feature space formed of concatenated features of persons viewed in two different cameras and use RBF kernel SVM as classifier. For RAiD data set, we use 21 persons for training RBF kernel SVM while the rest 20 were used in testing. We take 10 images each person for both training and testing. The SVM parameter were selected using 4-fold cross validation. For each test we ran 5 independent trials and report the average results.

Temporal-spatial Information Generation. In order to verify the role of temporal-spatial information, dataset should provide relative labels for each person, specifically time when the corresponding person is captured by a camera, and the location of each camera. Without losing rationality, we can take some hypothesis for these temporal-spatial information. Firstly, because every person appears in all these 4 cameras, for each person we can generate a random permutation of $\{1, 2, 3, 4\}$ representing the path of the corresponding person. For example, suppose the generating permutation is $\{2, 4, 1, 3\}$, the meaning is that the person first captured in the camera network by camera C_2 when he is walking in this camera network, then was captured by the sequence of C_4, and then

C_1, finally C_3. Secondly, we give each person a random velocity in the range of (0.8 m/s, 1.2 m/s) by experience. Thirdly, four camera is given an random coordinate (x, y) in the range that $x \in (-500\,\mathrm{m}, 500\,\mathrm{m})$, $y \in (-500\,\mathrm{m}, 500\,\mathrm{m})$, and the distance between each camera pair can be calculated. Finally, For each person we assume a time, denoted as t_0, when the person was first captured in this camera network. t_0 is random variable and follow normal distribution, actually all sample is in the range (100 s, 300 s). Given the generated information mentioned above, we can calculate the precision time when each person was captured by each camera. Approximately, we get the temporal-spatial information of each person in RAiD dataset.

3.2 Experiment Results and Analysis

We compare the performance for camera pairs 1-2, 1-3, 1-4, 2-3, 2-4 and 3-4 respectively, as shown in Figs. 4(a)–(f). Here we denoted temporal-spatial as "T-S". The parameters of experiment are v_1, λ and degrees of freedom n, Our choice is that $v_1 = 2.4\,\mathrm{m/s}$, $lambda = 10$ and $n = 700$. There are two cases, one is for camera pair 1-2 and for camera pair 3-4 where appearance variation is not so much, the other is for camera pair 1-3, 1-4, 2-3, 2-4 where there is significant lighting variation. As can be seen, the proposed method greatly improves CMC compared with conventional methods (ICT), and the method of network consistent re-identification (NCR) for both the case. The improvements are obvious after introducing spatial-temporal and consistent information. For the indoor camera pair 1-2 and the outdoor camera pair 3-4, where lighting variation is not significant, the rank1, rank2 and rank5 performance is shown in Table 1.

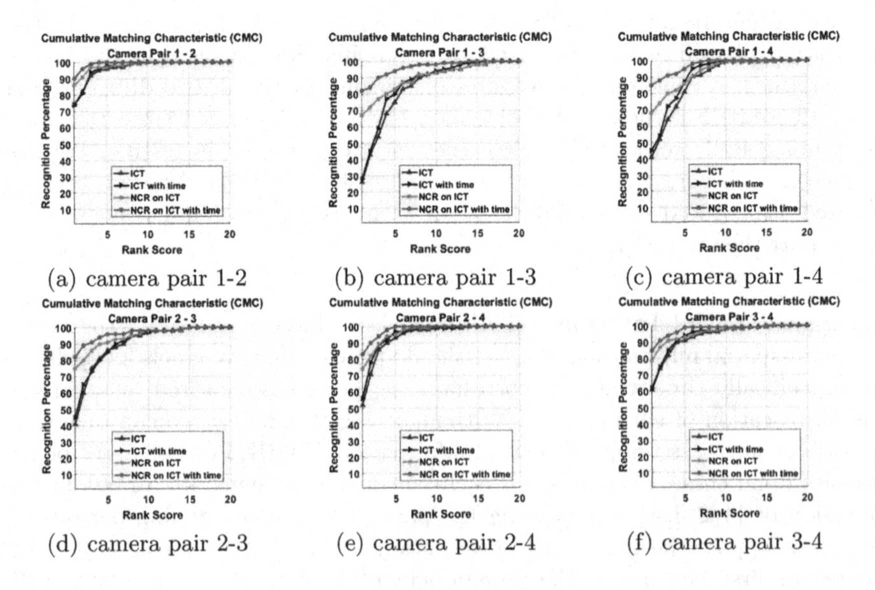

(a) camera pair 1-2 (b) camera pair 1-3 (c) camera pair 1-4

(d) camera pair 2-3 (e) camera pair 2-4 (f) camera pair 3-4

Fig. 4. CMC curves for each camera pairs

Table 1. Comparison with state-of-the-art NCR methods on RAiD datasets (%).

camera piars	1-2			1-3			1-4			2-3			2-4			3-4		
CMC Rank R	R=1	R=2	R=5	R=1	R=2	R=5	R=1	R=2	R=5	R=1	R=2	R=5	R=1	R=2	R=5	R=1	R=2	R=5
ICT [19]	74	82	96	26	43	75	41	52	81	41	60	86	52	71	93	61	74	91
ICT with T-S	74	81	98	28	45	80	45	54	87	45	65	85	56	78	97	62	75	94
NCR [14]	86	91	97	67	72	83	68	74	86	75	80	91	74	82	95	79	87	93
NCR with T-S	90	96	100	82	84	94	85	88	95	82	89	96	83	90	100	85	91	99

As can be seen, the proposed method applied on similarity scores generated by ICT (NCR on ICT with T-S) achieve 90% and 85% rank 1 performance respectively.

For the camera pair 1-3, 1-4, 2-3 and 2-4, all of which is indoor-outdoor pair where lighting variation is significant relatively, the rank1, rank2 and rank5 performance is shown in Table 1. The proposed method applied on similarity scores generated by ICT (NCR on ICT with T-S) achieve 82%, 85%, 82% and 83% rank 1 performance respectively.

It can further be seen that when we apply only the T-S information to ICT, normally there is a slightly improvement of the rank 1 performance compared to their original rank 1 performance. This phenomenon reflect the effect of temporal-spatial information in the auxiliary aspect. Additionally, it can be seen that for camera pairs with large illumination variation (i.e. 1-3, 1-4, 2-3 and 2-4) the performance improvement is significantly large relatively. For example, for camera pair 1-3, the rank 1 performance boosts up to 82% on application of NCR with T-S to ICT compared to their original rank 1 performance of 26% respectively. And with the help of T-S information, the rank 1 performance bring about a 15% increment compared with the NCR rank 1 performance of 67%. So it is meaningful to introduce the combination of temporal-spatial information and consistent information in the issue of person re-identification in camera network.

4 Conclusion

This paper puts forward a new framework that add temporal-spatial information in the system of camera network person re-identification. The proposed method boosts camera pairwise re-identification performance. The future directions of our research is the occasion where the hypotheses that same N person are present in each of the M cameras is removed. In additional, we will try to apply our approach to bigger networks.

Acknowledgments. This work was supported in Science and Technology Commission of Shanghai Municipality (STCSM, Grant Nos. 15DZ1207403, 17DZ1205602, NSF61771303).

References

1. Martinel, N., Micheloni, C.: Re-identify people in wide area camera network. In: 2012 IEEE Computer Society Conference on Computer Vision and Pattern Recognition Workshops, pp. 31–36. IEEE (2012)
2. Bazzani, L., Cristani, M., Murino, V.: Symmetry-driven accumulation of local features for human characterization and re-identification. Comput. Vis. Image Underst. **117**(2), 130–144 (2013)
3. Liu, C., Gong, S., Loy, C.C., Lin, X.: Person re-identification: what features are important? In: Fusiello, A., Murino, V., Cucchiara, R. (eds.) ECCV 2012. LNCS, vol. 7583, pp. 391–401. Springer, Heidelberg (2012). https://doi.org/10.1007/978-3-642-33863-2_39
4. Cheng, D.S., Cristani, M., Stoppa, M., Bazzani, L., Murino, V.: Custom pictorial structures for re-identification. In: BMVC, vol. 1, p. 6 (2011)
5. Ma, B., Su, Y., Jurie, F.: BiCov: a novel image representation for person re-identification and face verification. In: British Machive Vision Conference, 11 p. (2012)
6. Bak, S., Corvee, E., Bremond, F., Thonnat, M.: Boosted human re-identification using Riemannian manifolds. Image Vis. Comput. **30**(6), 443–452 (2012)
7. Liao, S., Hu, Y., Zhu, X., Li, S.Z.: Person re-identification by local maximal occurrence representation and metric learning. In: Proceedings of the IEEE Conference on Computer Vision and Pattern Recognition, pp. 2197–2206 (2015)
8. Yang, L., Jin, R.: Distance metric learning: a comprehensive survey. Mich. State Univ. **2**, 78 (2006)
9. Alavi, A., Yang, Y., Harandi, M., Sanderson, C.: Multi-shot person re-identification via relational stein divergence. In: 2013 IEEE International Conference on Image Processing, pp. 3542–3546. IEEE (2013)
10. Farenzena, M., Bazzani, L., Perina, A., Murino, V., Cristani, M.: Person re-identification by symmetry-driven accumulation of local features. In: 2010 IEEE Conference on Computer Vision and Pattern Recognition (CVPR), pp. 2360–2367. IEEE (2010)
11. Xiao, T., Li, H., Ouyang, W., Wang, X.: Learning deep feature representations with domain guided dropout for person re-identification. In: Proceedings of the IEEE Conference on Computer Vision and Pattern Recognition, pp. 1249–1258 (2016)
12. Ahmed, E., Jones, M., Marks, T.K.: An improved deep learning architecture for person re-identification. In: Proceedings of the IEEE Conference on Computer Vision and Pattern Recognition, pp. 3908–3916 (2015)
13. Shi, H., Yang, Y., Zhu, X., Liao, S., Lei, Z., Zheng, W., Li, S.Z.: Embedding deep metric for person re-identification: a study against large variations. In: Leibe, B., Matas, J., Sebe, N., Welling, M. (eds.) ECCV 2016. LNCS, vol. 9905, pp. 732–748. Springer, Cham (2016). https://doi.org/10.1007/978-3-319-46448-0_44
14. Das, A., Chakraborty, A., Roy-Chowdhury, A.K.: Consistent re-identification in a camera network. In: Fleet, D., Pajdla, T., Schiele, B., Tuytelaars, T. (eds.) ECCV 2014. LNCS, vol. 8690, pp. 330–345. Springer, Cham (2014). https://doi.org/10.1007/978-3-319-10605-2_22
15. Huang, W., Hu, R., Liang, C., Yu, Y., Wang, Z., Zhong, X., Zhang, C.: Camera network based person re-identification by leveraging spatial-temporal constraint and multiple cameras relations. In: Tian, Q., Sebe, N., Qi, G.-J., Huet, B., Hong, R., Liu, X. (eds.) MMM 2016. LNCS, vol. 9516, pp. 174–186. Springer, Cham (2016). https://doi.org/10.1007/978-3-319-27671-7_15

16. Schrijver, A.: Theory of Linear and Integer Programming. Wiley, Hoboken (1998)
17. Gray, D., Brennan, S., Tao, H.: Evaluating appearance models for recognition, reacquisition, and tracking. In: Proceedings of IEEE International Workshop on Performance Evaluation for Tracking and Surveillance (PETS), vol. 3 (2007)
18. Li, W., Zhao, R., Wang, X.: Human reidentification with transferred metric learning. In: Lee, K.M., Matsushita, Y., Rehg, J.M., Hu, Z. (eds.) ACCV 2012. LNCS, vol. 7724, pp. 31–44. Springer, Heidelberg (2013). https://doi.org/10.1007/978-3-642-37331-2_3
19. Avraham, T., Gurvich, I., Lindenbaum, M., Markovitch, S.: Learning implicit transfer for person re-identification. In: Fusiello, A., Murino, V., Cucchiara, R. (eds.) ECCV 2012. LNCS, vol. 7583, pp. 381–390. Springer, Heidelberg (2012). https://doi.org/10.1007/978-3-642-33863-2_38

Building Extraction by Stroke Width Transform from Satellite Imagery

Li Xu[1,2](✉), Mingming Kong[3], and Bing Pan[3]

[1] The Postdoctoral Station, Southwest Jiaotong University, Chengdu 610039, China
xuli198607153764@163.com
[2] The Postdoctoral Station, Xihua University Based on Collaboration Innovation Center of Sichuan Automotive Key Parts, Chengdu 610039, China
[3] The Center for Radio Administration Technology Development, Xihua University, Chengdu 610039, China
Kongming000@126.com, 350618364@qq.com

Abstract. This paper proposes a novel building extraction method from satellite imagery. Intuitively, the symmetry and regularity of architecture could be used to detect as building area, which benefited from the stroke width transform (SWT) algorithm. Meanwhile, the roof of building are very different with rural area (such as vegetation, wild, etc.) in color space, which can be partitioned by K-means clustering method. The area clustering can obtain the consistency region to complement the discontinuity from SWT algorithm detection. Then, A superpixel generation algorithm is adapted to yield the color distribution of building region, and final building area is able to extract accurately. Different with existing methods, the proposed method performs on a single source satellite image without any other supplement information. Experiment on a large number of satellite imagery demonstrates the efficiency of the proposed method for building extraction.

Keywords: Building extraction · Satellite imagery
Stroke width transform · Kmeans algorithm

1 Introduction

Wireless communication system is explosively growth application in urban environment. An appropriate radio propagation model is significant to predict the signal communication which is influenced by building's shadowing, reflection, absorption and fading. Building distribution is an important factor to modify the radio propagation model in simulation environment. Building density analysis is valuable for predicting the signal propagation and visualization, and simulation parameters of radio propagation model can be modified [1–5]. Therefore, accuracy automation building extraction from satellite imagery is helpful to design the wireless propagation model.

J. Yang et al. (Eds.): CCCV 2017, Part III, CCIS 773, pp. 340–351, 2017.
https://doi.org/10.1007/978-981-10-7305-2_30

The satellite imagery has many advantages, such as the macroscopic imaging, low cost of approximating in horizontal projection objects and easily to be digitized, which became more and more popular to research in computer vision [6,7]. In recent years, many researches have been focusing on building extraction from image. These methods can be coarsely divided into three classes. The first one is extracting building by some human interaction (semi-automatic) [8–11]. Based segmentation method requires much more user workload, which is impractical for the large segmentation task. Tian [11] adopts optimal parameters (shape, spectral bands, from-related) from the given type of objects (road, building, sports filed) to process the multi-resolution segmentation on high-resolution satellite image. The experiment proves that this algorithm is capable of producing with the comparable result. However, Suggested optimal parameters are usefulness and should be preliminarily demonstrated.

In order to decrease the user workload, some methods are proposed to extract building from different types imagery information (single-view or multi-views), such as optical image and radar data [10,12]. Rottensteiner [13] creates a 3D building model form the high-resolution lidar data. The skew error distribution function is introduced to compute the digital terrain mode by separating points of building from other object classes. After analyzing the height difference with digital surface model, building is derived by curvature-based segmentation techniques. Guo [14] locates the boundary of building by a snake-based approach from the high-resolution IKONOS satellite imagery and captures the height data by airborne laser scanning system. Teimouri [15] identifies an optimal fusion of radar and optical image. The decision-level fusion is used to assess the classification of panchromatic, multi-spectral, and radar images. Then, author applies an artificial neural network to classify those feature, and the decision level of integration for the building detection purpose.

Since multi-source image is expensive to obtain, which limited to be widely applied for ordinary users. Then, the third method according to the principle of elements acquirement (region-based or edge-based) are proposed, that could be reduced the cost of equipment. There are many methods to automatically extract the building area by this way, such as morphological-based methods [16–18], region and edge detection-based methods [19–22]. Huang and zhang [16,17] rely on the morphological shadow index (MSI), and the jointly use morphological building index (MBI) for the spatial constraint of building. The dual-threshold approach of their research improves for building extraction form background without collection any training samples and process of supervised leaning. Their method was proved more accurately than the support vector machine interpretation. Jin [18] presented an automated building-extraction strategy via mathematical morphology informations (the structural, contextual, and spectral) from the high-resolution satellite imagery. This system need not provide any initial algorithm parameters, training set and pre-classification. Jiang and Zhang [19] propose a semi-automatic building extraction method to get building by edge detection and region-driven. This method result heavily relies on the image's spectral information. Lin [20] describes a system to detect the building in aerial scenes. After Low-level segmentation, a perceptual grouping approach is used to collect fragment segments from shape properties, and the shadow is used to form and verify the hypotheses region. Gamba [21] classifies the urban area mapping separately by a neural network and an adaptive Markov random field model. Decision fusion process combines with the classification mapping to exploit objects boundaries. Lee [22] uses the classification result of Ikonos multi-spectral image and Hough transformation to detect and form the building boundary. Extraction error is mainly caused by the road class as well as the occlusion and shadow. These methods usually utilize color, space, textual, and spectral information to extract building area. In other works, the successful building extraction of these existing region-based or edge-based methods is usually based on the assumption that the source image is in high-resolution and easily distinguished form the target area. However, when the image is in low resolution and with some noise information, such as the available google or network satellite mapping with slight fog and intense illumination, these methods will be not suitable to segment building region from around the environment.

Stroke width is related to the linear feature which has applied into extracting road's networks. Doucette [23] presents a self-supervised road classification method to detect road centerline from high resolution multi-spectral imagery. Candidate road centerline is detected by the anti-parallel-edge. It is crucial for this method that needing the sufficient number of correct roads as training samples. The author reviews a variety of techniques which identified the linear, planimetric feature form imagery based on spectral, spatial and radiometric rules. Epshtein [24] detected the text in many fonts and languages use stroke width transform (SWT). This method suggests that the SWT algorithm combined with pixel dense estimation is efficient to extract the object with same stroke (symmetry region) in natural images. Quackenbush [25] also presents a review of the system for the linear feature extraction from imagery, that regarded as significantly requirement to automation detecting interest objects ongoing research from remotely sensed imagery.

In this paper, we propose a novel automatic method to extract building region from the single-view and in low-resolution satellite imagery, such as Fig. 1. The figure contains a complex environment, such as old building, playground, green landscape and waste land. Some of building is low and covered with vegetation. Our method is motivated by segmenting the symmetry and regularity architecture from this satellite mapping, and analyzed this region's building density to modify the wireless propagation model. This study introduces SWT algorithm [24] to detect the hypothetical region, and fuses with k-means algorithm to achieve the presence of building from satellite imagery. The framework of building segmentation contains several steps. The first step uses the morphology transformation to add redundant boundary points to Canny operator [26], which benefited for calculating image pixels' stroke width. The second step calculates images' SWT spectral to detect the symmetry region from satellite imagery. K-means algorithm is used to obtain building region by color information. Then, the possibility building region is generated. Integrated SWT spectral with the

Fig. 1. The satellite imagery.

classification result, building boundary can be totally acquired. The third step fusing second result could wipe out some interference region. Then, combined with the originate imagery mapping, Building region is grown into a fine segmentation result by a simple linear iterative clustering algorithm. For speciality of this satellite imagery, we compare our method result with K-means and Isodata clustering algorithm. The experiment results prove that our method is efficiently improvement building exaction.

The paper is organized as follows. In Sect. 2, our proposed method and the detailed of stroke width generation are introduced. In Sect. 3, experiment result and performance evaluation are presented. Finally, conclusions are given in Sect. 4.

2 Proposed Method

Our goal is to obtain the building region and evaluation its density. Therefore, this method could be performed as jointly estimating the symmetry region and color clustering problem. Intuitively, the symmetry region is applied to distinguish the character of building, and it can be detected by SWT algorithm. Besides, building always has different features with rural region in color space. We perform k-means algorithm in color clustering to removing interference. After intergrading those detected result, the simple linear iterative clustering algorithm [27] is used to maintain the building region consistency. Based on the consideration above, an unsupervised jointly framework is proposed as shown in Fig. 2. From the flowchart, we known that this method is designed simply and could be automation segmentation building region from satellite imagery.

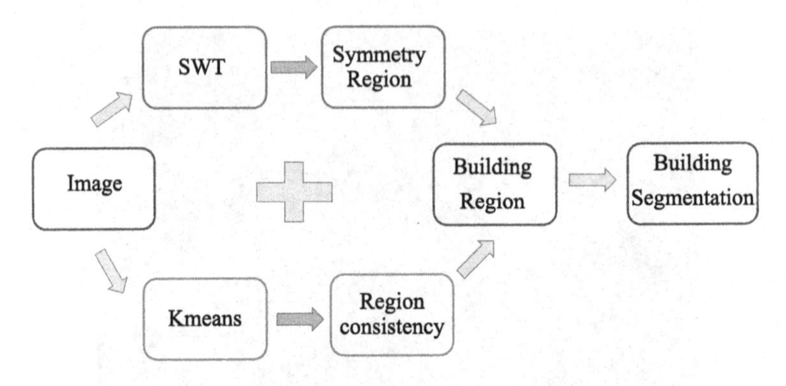

Fig. 2. Flowchart of the proposed method.

2.1 Stroke Width Generation

The stroke width of image pixel is a locate description operator based on Canny edge detector. It has been extensively applied to extract the symmetry region (letter or road) in nature scenes. Neighboring pixels with the same stroke width may be grouped together. Relied on those similar stroke width of pixels, approximately symmetry region would be clustering from image. Building's roof has a good symmetry in geometry morphology. Therefore, focusing on building's roof is helpful to reduce a number of interference region.

Stroke width is defined as the length of line and rough perpendicular to the orientation of stroke. Its calculation should be firstly detected image edge by Canny operator. Those symmetry pixels located in edge are found. Then, the length of pixels in line is assigned as the same stroke width. Image pixels with same width would be in high spectral distribution. Building region is mainly identified by those image pixels' stroke width.

Intuitively, stroke width is calculated from the image edge boundary. A gradient direction d_p belongs to the pixel p on edge. If other pixel can be found from edge and signed as q, and the orientation of d_q follow the ray $r = p+n \cdot d_p$, where $n > 0$. The gradient direction d_q is roughly opposite to $d_q = -d_p \pm \pi/6$. Then, the SWT spectral of output image along with the segment $[p, q]$ is assigned the width with $\|\overrightarrow{p-q}\|$. If a lower value already is appeared, pixel's width would be chose the lower one as shown in Fig. 3.

If the pixel q is not existed, or d_p does not opposites with d_q, the ray would be discarded. If pixels are in corner, and the average width of pixels is m, the length of pixels longer than m would be replaced by m as shown in Fig. 3.

The output of the SWT is a image, which contained the width of the stroke with each pixel as the same size of input image. Based on those theories, a detailed description of each step of the method is provided in the following section.

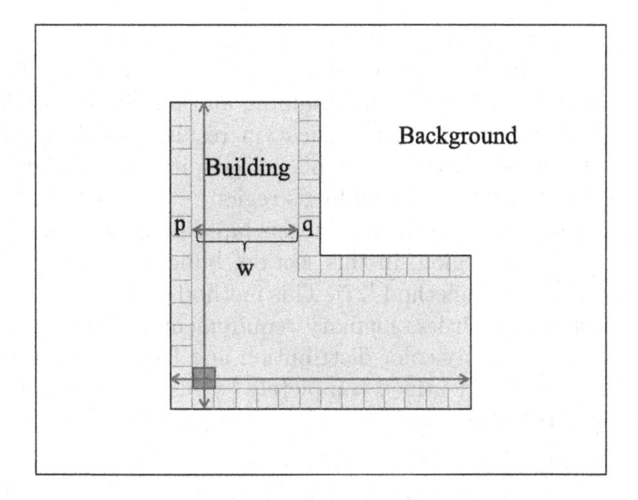

Fig. 3. Implementation pixels with stroke width using SWT algorithm.

2.2 Building Boundary Extraction and Fusion

After calculating image pixels' stroke width, Those symmetry region appear more intensity than others. In this section, pixels group together into building candidates via the distribution of pixels' width. Edge detection determines the calculation scope of SWT algorithm, and is mainly influenced by the bright light and noise (shadow) in satellite imagery. Several tests prove that the more region's edge detection, the better for the later building region extraction. Therefore, satellite imagery is necessary to be preprocessed before finding the building region.

In mathematical morphology, open operator is used to expand feature and close some of gaps in image. It followed with subtract operator could remove some of small or narrow elements, without influencing the large one. After those preprocessing, the uniformity of image background is added. Edge detection would become more complicated, so that building region would be appeared more obviously. Road and lawn region would be lower density in the distribution of SWT spectrum. The aim of this step could be to reduce several dissymmetry region.

Based on the edge detection, image's SWT spectral is calculated. Neighbor pixels would be grouped together with the similar stroke width. For allowing more elaborate and perspective distortions of the symmetry region. If two neighboring pixels ratio does not exceed 3.0, the stroke with smoothly varying width will be grouped together. Those component region may contain building. A small set of rule is also employed by extracting the geometric character of building. The threshold of pixels' stroke width (b) is set less than 30. After this threshold, road and lawn region (with less stroke wind) would be rejected from image.

Image's SWT spectral obtains a lot of scattered region by Canny operator. To reduce the hypothesis building region, we classify image into two groups by

K-means algorithm based on image's color mapping. Building's roof and road are homogenous in color and with much higher light than others. Therefore, the classification result is easily obtaining building and road from satellite imagery. We combine SWT spectral with classification result, and then the outline of building can be detected clearly. After this merged, irregular objects and mostly of roads are eliminated from the building's region.

After above preprocessing, the hypothesis building region is brought up and scattered distribution in region. In this section, building boundary is extracted by a superpixel generation method [27]. This method generates superpixel based on K-means clustering with less memory requirement and strong performance. Superpixels contain similarly color distribution and labeled, respectively. Then, building region would be identified completely by fusing the building boundary with superpixel distribution.

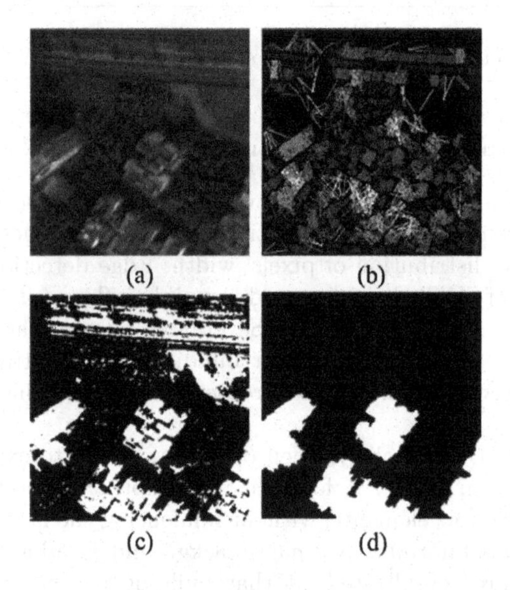

Fig. 4. An example of this method: (a) The original satellite image. (b) The SWT mapping. (c) The result of k-means classification. (d) The result of proposed method.

In order to analysis the advantage of proposed method, an example of subjective result is shown in Fig. 4. We can see that this area contains many complex object region, such as building, road, tree and others shadow in Fig. 4(a), it can be seen that the probability map by SWT obtains the building symmetric region, but introduces some disconnected region in Fig. 4(b). Meanwhile, the result obtained by kmeans extracts building region as consistency in color. However, it introduces many noise region such as the flat region in (Fig. 4(c)). Combined the SWT with k-means result, most of dissymmetry region is wiped out. Building region would be recognized. While, the result in (Fig. 4(d)) successfully segments building region and reduces noise region after integrating building boundary and superpixel distribution.

3 Experiment and Analysis

The experiment is tested on Google earth satellite map with 1.07 metric/pixel spatial resolution. The dataset contains 224 images are special representative to predict the signal communication as shown in Fig. 4. The building region contains many different shapes, such as square and strip. Moreover, there are many tree and grass surrounding the old building, which seriously interfered building segmentation.

Iterative self-organizing data analysis technique algorithm (Isodata) is clustering algorithm based on the k-means, which added the merged and divided of clustering result. Isodata is an adaptive method to adjust the clustering number. Isodata and K-means are basic clustering methods. For segmenting the building density of satellite image with low resolution and single information, most of building excitation methods are failure. Therefore, we use the Isodata and k-means clustering algorithm comparative with our purposed method.

Performance Evaluation. In order to demonstrate the effective of our proposed method, we use the intersection-over-union (IOU) to evaluate the segmentation performance, which is widely applied for image segmentation. The groundtruth of building region is manual annotation. IOU is defined as follows:

$$precision = \frac{A \bigcap B}{A \bigcup B} \quad (1)$$

where A is the binary mask which denoted the segmentation region by our method. B denotes the groundtruth. The higher the iou is, the better the performance is.

Some subjective result and objective result are shown in the following parts. Figure 5(a) is the groundtruth of satellite imagery. Figure 5(b)–(d) show the different segmentation results of Isodata, K-means, and our method, respectively. From this figure, we can see that the result of Isodata algorithm produces much noise due to the light point in original image. Meanwhile, the result of K-means algorithm obtain much more compact building than the result of Isodata. However, it produce some false positive in the middle bottom of the image, i.e., extracted the grass region as building. The result of proposed method is better than K-means and Isodata. It could not only remove the noise point but also avoid to segment the grass region as building. Meanwhile, the objective results are shown in Table 1. It can be seen that our method performs better than Isodata and K-means algorithm. The result demonstrates the effectiveness of the proposed method. The K-means result contain lots of building region, which contained some Non-architectural area. Therefor, it has much larger rise of IOU than Isodata. The value IOU of our method is improved limitedly. However,

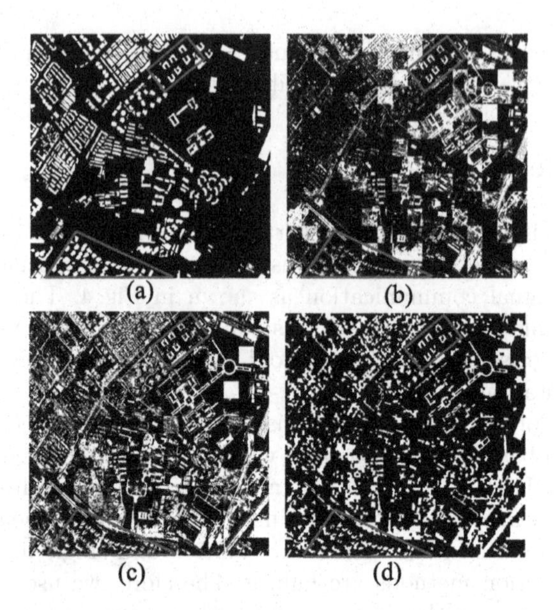

Fig. 5. Result of building extraction for our method and compared methods. (a) The groundtruth. (b) Isodata algorithm results. (c) K-means algorithm result. (d) Our method result.

Table 1. IOU result.

Method	Building segmentation result (%)
Isodata algorithm	31.30
K-means algorithm	35.19
Our method	36.37

the segmentation result show that our method proves more similarly building region. The main reason is that building region loses some part of true territory caused by in the last step fused with superpixels region. As shown in Fig. 5(d), the new building region could be partitioned completely with much higher floor.

In this paper, stroke width of image pixel is used to provide the gradient information of building boundary. K-means algorithm detects more compactness region than Isodata. So we combined image pixels' width with the K-means classification result to discard some noise. The building region could be detected very well by our algorithm. Comparing the extraction performance with K-means algorithm, Most of roads, grassland and bare ground could be deserted. The shadow of building could add the contrast with surrounding areas. Test result shows that image with much higher building could be enhanced the performance, which influenced by the shadow. Nevertheless, some of artificial road with good symmetry is hardly eliminated completely by this method. The experiment result also shows that our method is poor performance for the image only containing road or lawn region. Form the original satellite imagery, A lots of old building with a dark brown color roof was also made to be labeled mistakenly in test.

4 Conclusions

For modifying the wire radio propagation model parameters, loaded those satellite images is used to extracted the building density from google earth with low resolution and single spectre. In this paper, we proposed a simple automation method to segment building regions epically for an arbitrarily satellite mapping. We introduce SWT algorithm to detect the symmetry region, distinguished the character of building from around region, and cluster the similar color region via K-means algorithm. Finally, fused those result with the originate imagery mapping, building region is grown into a fine segmentation result. The proposed method could extract building region automatically and doesn't need any high-level information or others prior knowledge. Compared with the object-oriented remotely sensed building extraction, it has a more application area. Based on the value of building density, wireless transform models will be modified in further researcher.

Acknowledgments. This work is supported by the Typical Spectrum of Radio Signal (szjj2016-092). The Postdoctoral Station at Xihua University Based on Collaboration Innovation Center of Sichuan Automotive Key Parts. The National Natural Science Foundation (61372187, 61473239), and Laboratory of Intelligent Network Information Processing, (No. szjj2015-061).

References

1. Erceg, V., Rustako, A.J., Roman, R.S.: Diffraction around corners and its effects on the microcell coverage area in urban and suburban environments at 900 MHz, 2 GHz, and 4 GHz. IEEE Trans. Veh. Technol. **43**(3), 762–766 (1994)
2. Kim, S.C., Guarino, B.J., Willis, T.M., et al.: Radio propagation measurements and prediction using three-dimensional ray tracing in urban environments at 908 MHz and 1.9 GHz. IEEE Trans. Veh. Technol. **48**(3), 931–946 (1999)
3. Souley, A.K.H., Cherkaoui, S.: Realistic urban scenarios simulation for ad hoc networks. In: 2nd International Conference on Innovations in Information Technology (IIT05) (2005)
4. Sommer, C., Eckhoff, D., German, R., et al.: A computationally inexpensive empirical model of IEEE 802.11 p radio shadowing in urban environments. In: 2011 Eighth International Conference on Wireless On-Demand Network Systems and Services (WONS), pp. 84–90. IEEE (2011)
5. Letourneux, F., et al.: Dual-polarized channel measurement and modeling in urban macro- and small-cells at 2 GHz. In: 2017 IEEE Wireless Communications and Networking Conference (WCNC), San Francisco, CA, pp. 1–6 (2017)
6. Albert, A., Kaur, J., Gonzalez, M.: Using convolutional networks and satellite imagery to identify patterns in urban environments at a large scale. arXiv preprint arXiv:1704.02965 (2017)

7. Pelta, R., Chudnovsky, A.A.: Spatiotemporal estimation of air temperature patterns at the street level using high resolution satellite imagery. Sci. Total Environ. **579**, 675–684 (2017)
8. Mayunga, S.D., Zhang, Y., Coleman, D.J.: Semi-automatic building extraction utilizing Quickbird imagery. In: Proceedings of the ISPRS Workshop CMRT, vol. 13, pp. 1–136 (2005)
9. Bypina, S.K., Rajan, K.S.: Semi-automatic extraction of large and moderate buildings from very high-resolution satellite imagery using active contour model. In: IEEE International on Geoscience and Remote Sensing Symposium (IGARSS), pp. 1885–1888. IEEE (2015)
10. Khesali, E., Zoej, M.J.V., Mokhtarzade, M., et al.: Semi automatic road extraction by fusion of high resolution optical and radar images. J. Indian Soc. Remote Sens. **44**(1), 21–29 (2016)
11. Tian, J., Chen, D.M.: Optimization in multi-scale segmentation of high-resolution satellite images for artificial feature recognition. Int. J. Remote Sens. **28**(20), 4625–4644 (2007)
12. Hedman, K., Wessel, B., Stilla, U.: A fusion strategy for extracted road networks from multiaspect SAR images. U. Stilla, F. Rottensteiner und S. Hinz (Herausgeber), Int. Arch. Photogramm. Remote Sens. CMRT05 **36**(Part 3), W24 (2005)
13. Rottensteiner, F., Briese, C.: A new method for building extraction in urban areas from high-resolution LIDAR data. Int. Arch. Photogramm. Remote Sens. Spat. Inf. Sci. **34**(3/A), 295–301 (2002)
14. Guo, T., Yasuoka, Y.: Snake-based approach for building extraction from high-resolution satellite images and height data in urban areas. In: Proceedings of the 23rd Asian Conference on Remote Sensing, pp. 25–29 (2002)
15. Teimouri, M., Mokhtarzade, M., Valadan Zoej, M.J.: Optimal fusion of optical and SAR high–resolution images for semiautomatic building detection. GISci. Remote Sens. **53**(1), 45–62 (2016)
16. Huang, X., Zhang, L.: A multidirectional and multiscale morphological index for automatic building extraction from multispectral GeoEye-1 imagery. Photogramm. Eng. Remote Sens. **77**(7), 721–732 (2011)
17. Huang, X., Zhang, L.: Morphological building/shadow index for building extraction from high–resolution imagery over urban areas. IEEE J. Sel. Top. Appl. Earth Obs. Remote Sens. **5**(1), 161–172 (2012)
18. Jin, X., Davis, C.H.: Automated building extraction from high-resolution satellite imagery in urban areas using structural, contextual, and spectral information. EURASIP J. Adv. Sig. Process. **2005**(14), 745309 (2005)
19. Jiang, N., Zhang, J.X., Li, H.T., et al.: Semi-automatic building extraction from high resolution imagery based on segmentation. In: International Workshop on Earth Observation and Remote Sensing Applications, EORSA 2008, pp. 1–5. IEEE (2008)
20. Lin, C., Huertas, A., Nevatia, R.: Detection of buildings using perceptual grouping and shadows. In: Proceedings of the 1994 IEEE Computer Society Conference on Computer Vision and Pattern Recognition, CVPR 1994, pp. 62–69. IEEE (1994)
21. Gamba, P., Dell'Acqua, F., Lisini, G., et al.: Improved VHR urban area mapping exploiting object boundaries. IEEE Trans. Geosci. Remote Sens. **45**(8), 2676–2682 (2007)
22. Lee, D.S., Shan, J., Bethel, J.S.: Class-guided building extraction from Ikonos imagery. Photogramm. Eng. Remote Sens. **69**(2), 143–150 (2003)

23. Doucette, P., Agouris, P., Stefanidis, A.: Automated road extraction from high resolution multispectral imagery. Photogramm. Eng. Remote Sens. **70**(12), 1405–1416 (2004)
24. Epshtein, B., Ofek, E., Wexler, Y.: Detecting text in natural scenes with stroke width transform. In: 2010 IEEE Conference on Computer Vision and Pattern Recognition (CVPR), pp. 2963–2970. IEEE (2010)
25. Quackenbush, L.J.: A review of techniques for extracting linear features from imagery. Photogramm. Eng. Remote Sens. **70**(12), 1383–1392 (2004)
26. Canny, J.: A computational approach to edge detection. IEEE Trans. Pattern Anal. Mach. Intell. **6**, 679–698 (1986)
27. Achanta, R., Shaji, A., Smith, K., et al.: SLIC superpixels compared to state-of-the-art superpixel methods. IEEE Trans. Pattern Anal. Mach. Intell. **34**(11), 2274–2282 (2012)

Siamese Cosine Network Embedding for Person Re-identification

Jiabao Wang(✉), Yang Li, and Zhuang Miao

College of Command Information Systems,
PLA Army Engineering University, Nanjing 210007, China
jiabao_1108@163.com, solarleeon@outlook.com, emiao_beyond@163.com

Abstract. In person re-identification, feature embedding is the key point for new coming identities. Most state-of-the-art models adopt the features learned by convolutional neural networks (CNNs) to do similarity comparison. However, the learned features are not good enough for new identities because CNNs are designed for classification of class-known objects, not for similarity comparison of any two identities. To improve feature embedding, we propose a pairwise cosine loss based on cosine similarity measurement. Subsequently, we design a Siamese cosine network embedding (SCNE) to learn deep features for person re-identification. It is based on the Siamese architecture, with intra-class input pairs and joint supervision by the softmax loss and the pairwise cosine loss. Experimental results show that our SCNE achieves the state-of-the-art performance on the public Market1501 and CUHK03 person re-ID benchmarks.

Keywords: Person re-identification · Feature embedding
Pairwise cosine loss · Siamese network · Convolutional neural network

1 Introduction

Given a person of interest as query, person re-identification (re-ID) is aim to determine whether the person has been observed by another camera [9,16,17,20,25,26]. It is a completely different problem from classification, which can be considered as a close-set problem [6,8,11]. However, person re-ID needs to search a new person which has to be treated as a new class because it has never appeared in the training dataset. So person re-ID requires good features to represent the new identities. It is an unclose-set challenging problem.

Recently, the features learned by convolutional neural networks (CNNs) have been widely used for person re-ID [25]. However, these features are not good enough for person re-ID because CNNs is designed for classification of class-known objects, not for similarity comparison of any two identities. As shown in Fig. 1(a), the 2D features are learned by CNNs with only softmax loss on MNIST dataset by LeNets++ [19], where we find that the features fill the whole feature space and have an uniform and flat distribution for each class samples.

J. Yang et al. (Eds.): CCCV 2017, Part III, CCIS 773, pp. 352–362, 2017.
https://doi.org/10.1007/978-981-10-7305-2_31

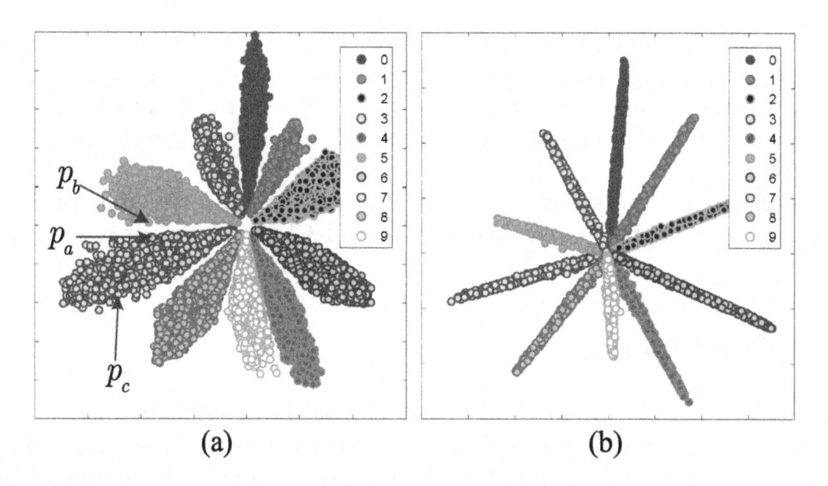

Fig. 1. The distributions of the features learned by the CNN models with (a) the softmax loss, (b) the softmax loss and our proposed pairwise cosine loss, on the MNIST dataset. In (a), p_a and p_c are two intra-class features while p_b has a different label with them. For classification, all of them can be classified correctly. But the cosine similarity between p_a and p_c is lower than that between p_a and p_b, which is bad for person re-ID. In (b), the compact intra-class distribution is fit for cosine similarity comparison.

This is inappropriate for person re-ID, because two intra-class features may have a relatively low similarity even if they are classified correctly. More importantly, there is no extra space for new identities. If we want the CNNs to learn more discriminative features for new identities, we need to compact the intra-class distribution of the learned features for the existing classes.

To compact the intra-class distribution, we propose a new pairwise cosine loss to measure the similarity between two intra-class features. As the features learned by the existing CNNs have an angle distribution as illustrated in Fig. 1(a), so it is desired to use a cosine loss to learn features and also utilize the cosine similarity for feature comparison in the evaluation stage. As shown in Fig. 1(b), by using our proposed pairwise cosine loss, the angle distribution of the features from the known classes are indeed compact. Hence, a lot of room is spared for describing new incoming identities. Another contribution of this paper is to design a novel network based on the Siamese network, which inputs only positive pair of images and pulls their features closer as possible. It is different from the exist methods, which inputs both positive and negative pairs [9,20].

In this paper, we design a Siamese cosine network embedding (SCNE), to learn the discriminative features for person re-ID. Compared to previous networks, we make the learned features not only separable but also compact. Our contributions are:

- A pairwise cosine loss is proposed to compact the distribution of the intra-class features. It is appropriate for cosine similarity comparison in person re-ID application.

- We design the SCNE to learn discriminative features by the joint supervision of the softmax loss and the pairwise cosine loss. The input pairs of our proposed network only have the positive pairs, without the negative pairs. This is because the inter-class separation can be achieved by the softmax loss in CNNs.
- Experimental results show that our approach achieves the state-of-the-art performance on the public Market1501 and CUHK03 person re-ID benchmarks.

2 Related Works

Our SCNE is inspired by the work of [26], where the identification loss and the verification loss are used for training. The former is the same as the softmax loss, and the latter is a variant of the center loss, where the added Square layer is an Euclidean distance for each dimension of the features. In evaluation, the similarity is computed by the cosine distance, so it is not good enough when the network is supervised by the center loss. However, the pairwise cosine loss we proposed in this paper is consistent with the similarity comparison. So it could achieve better performance than the work of [26].

Several works solved the person re-ID problem based on Siamese network, such as [5,16,17,22]. The work of [17] adopted the Long Short-Term Memory (LSTM) for memorizing the spatial dependencies of the divided regions in a person image. The Siamese network architecture is used for comparing the input pair images by a contrastive loss function. The contractive loss is to repel dissimilar inputs and attract similar inputs. The work of [16] also used the Siamese network for comparing features across pairs of images. It adopted a gating function to selectively emphasize the fine common local patterns in a person image. The work of [5] is also very similar to our work, but it used the GoogLeNet [14] as the base network. And more, a loss specific dropout unit is proposed to have a pairwise-consistent dropout for the verification subnet. This special designed network has achieved great performance. All above works used the negative input pair and the positive input pair to learn the network, which is different from our only positive input pair.

Besides, the work of [22] also used a cosine distance for Siamese network, but they adopted it as a connection function for the cost function. They treated the output of the network as a binary-class classification problem just for similar measurement. It is naturally a verification network, it has been proved that it is not good enough for person re-ID, without the identification network [26]. In this paper, we propose to combine two identification networks by the pairwise cosine loss, which can separate inter-class features and effectively compact intra-class features.

3 Siamese Cosine Network Embedding (SCNE)

3.1 The Proposed Pairwise Cosine Loss

Suppose the input image of CNNs is $\mathbf{x}_i$ and its label is y_i. The input $\mathbf{f}_i$ of the last fully-connection (FC) layer is always used as feature to represent $\mathbf{x}_i$ for similarity comparison. In the last FC layer, suppose the parameters is $\mathbf{W}^j, j = 1, \ldots, C$, where C is the number of the output, and then the output is $o_i^j = (\mathbf{W}^j)^T \mathbf{f}_i$. If we want the jth output to be maximum, we need to maximize the value o_i^j. For the widely used softmax log-loss, we have

$$L_s = -\sum_{i=1}^{N} \log \frac{\exp(o_i^{y_i})}{\sum_{k=1}^{C} \exp(o_i^k)} \tag{1}$$

where $o_i^{y_i}$ is the output value at the label y_i position, and N is the number of the samples.

Obviously, the softmax log-loss just separates the features into different class without compacting the intra-class features effectively. The problem boils down to develop an efficient loss function to compact the feature distribution of each class. Intuitively, based on the angle distribution of the features learned by the ImageNet pre-trained CNNs, the model is going to minimize the cosine loss of the two intra-class features produced by the input pair in Siamese network, to pull the intra-class features close to each other. The cosine similarity measurement could be adopted to achieve better performance for person re-ID.

To this end, we propose the pairwise cosine loss function, as formulated in (2):

$$L_c = \sum_{i=1}^{N} (1 - \cos(\mathbf{f}_i^a, \mathbf{f}_i^b)) \tag{2}$$

where $\cos(\mathbf{f}_i^a, \mathbf{f}_i^b) = \frac{(\mathbf{f}_i^a)^T \mathbf{f}_i^b}{||\mathbf{f}_i^a|| ||\mathbf{f}_i^b||} = \left(\frac{\mathbf{f}_i^a}{||\mathbf{f}_i^a||}\right)^T \left(\frac{\mathbf{f}_i^b}{||\mathbf{f}_i^b||}\right)$, $\mathbf{f}_i^a$ and $\mathbf{f}_i^b$ are the deep learned features of the input pair, and $\frac{\mathbf{f}_i^a}{||\mathbf{f}_i^a||}$ and $\frac{\mathbf{f}_i^b}{||\mathbf{f}_i^b||}$ are the l_2 normalized features. This loss function has a cosine part, which is the cosine value between $\mathbf{f}_i^a$ and $\mathbf{f}_i^b$. It effectively characterizes the intra-class cosine variation if the pair images have the same label. So it requires that the input of our Siamese network must have only the positive pair.

To learn and update the parameters of our network, we need to compute the gradient of L_c with respect to $\mathbf{f}_i^a$ and $\mathbf{f}_i^b$ to conduct the back propagation algorithm. The gradients are given as follows,

$$\frac{\partial L_c}{\partial \mathbf{f}_i^a} = \frac{1}{||\mathbf{f}_i^a||} \left(\cos(\mathbf{f}_i^a, \mathbf{f}_i^b) \frac{\mathbf{f}_i^a}{||\mathbf{f}_i^a||} - \frac{\mathbf{f}_i^b}{||\mathbf{f}_i^b||} \right) \tag{3}$$

$$\frac{\partial L_c}{\partial \mathbf{f}_i^b} = \frac{1}{||\mathbf{f}_i^b||} \left(\cos(\mathbf{f}_i^a, \mathbf{f}_i^b) \frac{\mathbf{f}_i^b}{||\mathbf{f}_i^b||} - \frac{\mathbf{f}_i^a}{||\mathbf{f}_i^a||} \right) \tag{4}$$

In (3) and (4), $\cos(\mathbf{f}_i^a, \mathbf{f}_i^b)$, $\frac{\mathbf{f}_i^a}{||\mathbf{f}_i^a||}$ and $\frac{\mathbf{f}_i^b}{||\mathbf{f}_i^b||}$ can be pre-computed in the forward pass, and they will be re-used in back propagation for efficient computation purpose. It has to be noted that there is no parameters in our added pairwise cosine loss layer.

3.2 Joint Optimization

If one wants to compact the intra-class features while keeping them separated, the softmax log-loss in (1) and the pairwise cosine loss in (2) should be combined. The joint objective function of the two losses is given as follows:

$$L = L_s + \lambda L_c \tag{5}$$

where the parameter λ is used for balancing the two losses. The softmax log-loss can be considered as a special case when $\lambda = 0$.

If we only use the softmax log-loss as supervision, the learned features would contain large intra-class variations. On the other hand, if we only supervise CNNs by the pairwise cosine loss, the learned features will be degraded to zeros or lines (At this point, the cosine loss is very small). Simply using either of them could not achieve discriminative feature learning. So it is necessary to combine them.

3.3 Architecture of the Designed SCNE

Our network is based on the Siamese network. Figure 2 briefly illustrates the architecture of the proposed network, where the parameter shared layers can be replaced by ImageNet pre-trained CNN layers. The network consists of two parameter shared CNN streams, two modified FC layers and three losses. The features extracted by the network are used as the descriptors, which directly supervised by two softmax losses and one pairwise cosine loss. The softmax loss is used for class prediction and the pairwise cosine loss is used for compacting the intra-class variation. The high level feature $\mathbf{f}_i^a$, $\mathbf{f}_i^b$ are merged in our added pairwise cosine loss layer, which has no parameters. The ImageNet pre-trained CNN model can be taken as AlexNet [8], VGGNet [11] or ResNet [6]. In this paper, we take Res50Net as the baseline for comparing with the-state-of-arts.

In order to finetune the network on different person re-ID datasets, we replace the final FC layer of the pre-trained Res50Net model with a $1 \times 1 \times 2048 \times n$ dimensional FC layer, where n is the number of training identities in the training dataset. Given an input pair of intra-class images resized to 224×224, the network predicts the identities of the two images and computes the pairwise cosine loss for them. The pairwise cosine loss layer is coupled with the last FC layer and affects the distribution of the learned features.

4 Experiments

4.1 Datasets and Preparation

The proposed model is tested on two large-scale person re-ID benchmarks, Market1501 [24] and CUHK03 [9].

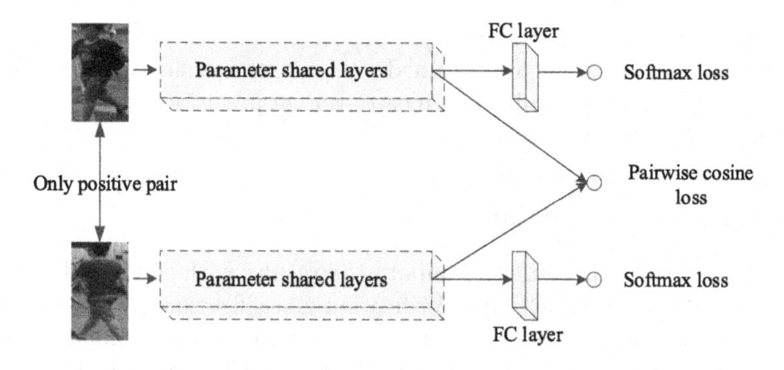

Fig. 2. The architecture of our proposed SCNE.

Market1501 dataset has 32668 images of 1501 identities. According to the dataset setting, 12936 images of 751 identities are for training and 19732 images of 750 identities and distractors are for testing. The images are cropped by the deformable part model (DPM) [4] detector automatically and are closer to the realistic setting. The evaluation is followed by the dataset baseline setting.

CUHK03 dataset consists of 13164 cropped images of 1467 identities collected in the CUHK campus. The bounding boxes detected by DPM detector are closer to realistic setting and are used in experiments. Following the given setting, the dataset is partitioned into a training set of 1367 identities and a testing set of 100 identities. The experiments are repeated with 20 random splits. In evaluation, we randomly select 100 images from 100 identities under another camera as galley.

The training images are resized to 256×256 uniformly, and subtract the mean image computed from all the training images. For adapting to the input of the Res50Net network, we cropped the images at 224×224. The training images are randomly mirrored horizontally. We get the batch of the training images randomly and online sample another same label images to compose an intra-class input pair.

4.2 Implementation Setting

The MatconvNet package [18] is used for training and testing. The epoch is set to 30 epochs. We adopt the mini-batch stochastic gradient descent to update the parameters of our network. The batch size is set 64 pairs. The learning rate is initialzed as 0.01 and set to 0.001 after 15 epochs, and 0.0001 for the final 5 epochs. There are three objectives in our network. All the gradients produced by every objectives respectively and added together by different weights. We assign 0.5 for the two gradients produced by two softmax log-losses and 1 for the gradient produced by the pairwise cosine loss.

For testing, we extract features by only activating one stream at the output before the FC layer in our fine-tuned model. Given an input image with size 224×224, we feed forward the image to the network and get the corresponding descriptor at the output of the 'pool5' layer for Res50Net. Once the descriptors for query and gallery sets are obtained, we sort the cosine distance between two sets to get the final result. The mean average precision (mAP) and rank-1 accuracy are used for evaluation.

4.3 Results on Market1501

On the Market-1501 dataset, we compare the results with state-of-the-art algorithms, in which PersonNet [20], Verification-Classification [26], DeepTransfer [5], Gated Reid [16] and S-LSTM [17] are all based on the Siamese network and have achieved the state-of-the art performance. SMOAnet [2] uses synthetic data to train a Inception network, while GAN ResNet [27] use the generative adversarial networks (GAN) to generate unlabeled samples for learning better models. Both of them can be thought as a variant of data augmentation.

Table 1. Comparison with the state-of-the-art methods on Market1501.

Method	SQ		MQ	
	mAP	rank-1	mAP	rank-1
PersonNet [20]	18.57	37.21	–	–
DADM [12]	19.6	39.4	25.8	49.0
CAN [10]	24.43	48.24	–	–
MultiRegion [15]	26.11	45.58	32.26	56.59
SLSC [3]	26.35	51.90	–	–
FisherNet [21]	29.94	48.15	–	–
S-LSTM [17]	–	–	35.3	61.6
Gated Reid [16]	39.55	65.88	48.45	76.04
SOMAnet [2]	47.89	73.87	56.98	81.29
GAN ResNet [27]	56.23	78.06	68.52	85.12
Verif.-Classif. [26]	59.87	79.51	70.33	85.47
DeepTransfer [5]	**65.5**	**83.7**	**73.8**	**89.6**
ResNet Basel. [25]	51.48	73.69	63.95	81.47
Ours SCNE (ResNet)	63.50	83.25	71.27	88.42

The single query (SQ) and multiple query (MQ) results are reported in Table 1. Our SCNE achieves 83.25% rank-1 accuracy and 63.50% mAP under the single query mode and 88.42% rank-1 accuracy and 71.27% mAP under the multiple query mode, which is the second among all the above results. It greatly outperforms Gated Reid [16] and S-LSTM [17] methods, which used the Siamese network without combining classification loss and verification loss. Our method also outperforms Verification-Classification [26], which used a Euclidean loss for verification. It's not good enough for similarity comparison by using cosine similarity measurement. The best method is the DeepTransfer [5], which adopted a different designed dropout strategy to combine classification loss and verification loss, based on the GoogLeNet base network.

4.4 Results on CUHK03

On the CUHK03 dataset, there are two types of evaluations, single shot (SS) and multiple shots (MS).

In single shot setting, we compare with ImprovedDeep [1], PersonNet [20], Verification-Classification [26], Pose Invariant [23], DNN-IM [13], SOMAnet [2], GAN ResNet [27], CNN-FRW-IC [7], DeepTransfer [5] and ResNet baseline [25]. We randomly select 100 images from 100 identities under another camera as gallery and report the mAP and rank-1 accuracy in Table 2. We achieve rank-1 accuracy = 85.1%, mAP = 83.3%, which is the excellent result compared with above methods.

Table 2. Comparison with the state-of-the-art methods on CUHK03.

Method	SS		MS	
	mAP	rank-1	mAP	rank-1
ImprovedDeep [1]	–	45.0	–	–
S-LSTM [17]	–	–	46.3	57.3
Gated Reid [16]	–	–	58.8	68.1
PersonNet [20]	–	64.8	–	–
Verif.-Classif. [26]	71.2	66.1	68.2	73.1
Pose Invariant [23]	71.3	67.1	–	–
DNN-IM [13]	–	72.0	–	–
SOMAnet [2]	–	72.4	–	**85.9**
GAN ResNet [27]	77.4	<u>73.1</u>	<u>77.4</u>	73.1
CNN-FRW-IC [7]	–	82.1	–	–
DeepTransfer [5]	**84.1**	–	–	–
ResNet Basel. [25]	75.8	71.5	–	–
Ours SCNE (Res50Net)	<u>83.3</u>	**85.1**	**88.1**	<u>82.0</u>

In multiple shot setting, all the images from another camera are used as gallery and the number of the candidate images is about 500. This evaluation is much closer to image retrieval and alleviate the unstable effect caused by random gallery selection. We compare with S-LSTM [17], Gated Reid [16], Verification-Classification [26], SOMAnet [2] and GAN ResNet [27] on the mAP and rank-1 accuracy. Our SCNE achieves rank-1 accuracy = 82.0%, mAP = 88.1%, which is also very competitive.

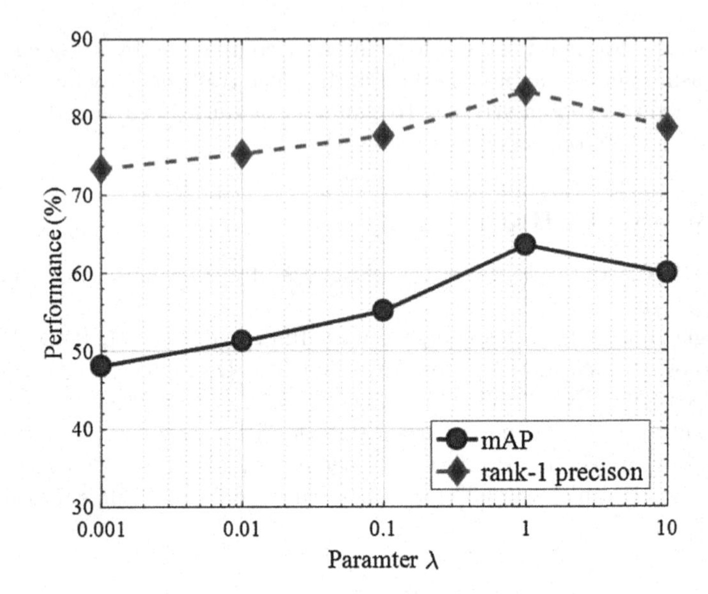

Fig. 3. The performance of our SCNE as different parameter λ.

4.5 Parameter Sensitivity Analysis

As the parameter λ dominates the balance of the pairwise cosine loss and the softmax loss, it is essential to our SCNE. So we conduct experiments to investigate the influence of the parameter λ on the Market1501 dataset. The results are reported in Fig. 3. From Fig. 3, we find that a proper λ can achieve the best mAP and rank-1 accuracy. A good performance is achieved when $\lambda = 1$.

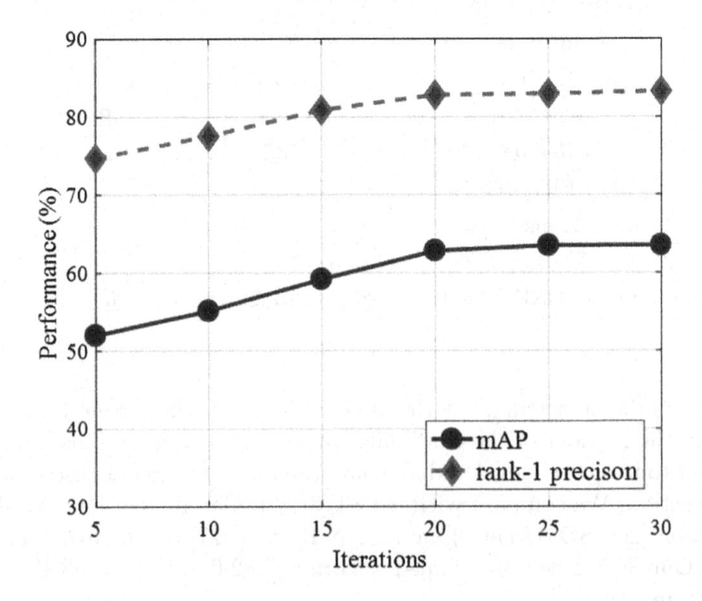

Fig. 4. The performance of our SCNE as the training iterations.

Besides, we also report the performance change of our SCNE as the iteration increases in training in Fig. 4. From Fig. 4, we can find that the performance rise slowly after 20 epoches.

5 Conclusion

In this paper, we propose a pairwise cosine loss to compact the distribution of the intra-class features and design the SCNE to learn the discriminative features for person re-ID. Our SCNE is trained by the joint supervision of the softmax loss and the pairwise cosine loss. Compared to previous networks, we make the learned features not only separable but also compact. Experimental results show that our approach achieves the state-of-the-art performance on the public Market1501 and CUHK03 person re-ID benchmarks. Since our SCNE is apt for similarity comparison, so we will apply it to identity retrieval in the further.

Acknowledgment. This work has been supported by the National Natural Science Foundation of China (61402519), and partially supported by the Natural Science Foundation of Jiangsu Province (BK20150721).

References

1. Ahmed, E., Jones, M., Marks, T.K.: An improved deep learning architecture for person re-identification. In: IEEE International Conference on Computer Vision and Pattern Recognition, pp. 3908–3916 (2015)
2. Barbosa, I.B., Cristani, M., Caputo, B., Rognhaugen, A., Theoharis, T.: Looking beyond appearances: synthetic training data for deep CNNs in re-identification. CoRR abs:1701.03151 (2017)
3. Chen, D., Yuan, Z., Chen, B., Zheng, N.: Similarity learning with spatial constraints for person re-identification. In: IEEE International Conference on Computer Vision and Pattern Recognition, pp. 1268–1277 (2016)
4. Felzenszwalb, P.F., Girshick, R.B., Mcallester, D., Ramanan, D.: Object detection with discriminatively trained part-based models. IEEE Trans. Pattern Anal. Mach. Intell. **32**(9), 1627 (2010)
5. Geng, M., Wang, Y., Xiang, T., Tian, Y.: Deep transfer learning for person re-identification. CoRR abs:1611.05244 (2016)
6. He, K., Zhang, X., Ren, S., Sun, J.: Deep residual learning for image recognition. In: IEEE International Conference on Computer Vision and Pattern Recognition, pp. 770–778 (2015)
7. Jin, H., Wang, X., Liao, S., Li, S.Z.: Deep person re-identification with improved embedding and efficient training. CoRR abs:1705.03332 (2017)
8. Krizhevsky, A., Sutskever, I., Hinton, G.E.: Imagenet classification with deep convolutional neural networks. In: Advances in Neural Information Processing Systems, pp. 1097–1105 (2012)
9. Li, W., Zhao, R., Xiao, T., Wang, X.: Deepreid: deep filter pairing neural network for person re-identification. In: IEEE International Conference on Computer Vision and Pattern Recognition, pp. 152–159 (2014)

10. Liu, H., Feng, J., Qi, M., Jiang, J., Yan, S.: End-to-end comparative attention networks for person re-identification. IEEE Trans. Image Process. **26**(7), 3492–3506 (2016)
11. Simonyan, K., Zisserman, A.: Very deep convolutional networks for large-scale image recognition. CoRR abs:1409.1556 (2014)
12. Su, C., Zhang, S., Xing, J., Gao, W., Tian, Q.: Deep attributes driven multi-camera person re-identification. In: Leibe, B., Matas, J., Sebe, N., Welling, M. (eds.) ECCV 2016. LNCS, vol. 9906, pp. 475–491. Springer, Cham (2016). https://doi.org/10.1007/978-3-319-46475-6_30
13. Subramaniam, A., Chatterjee, M., Mittal, A.: Deep neural networks with inexact matching for person re-identification. In: Advances in Neural Information Processing Systems, pp. 2667–2675 (2016)
14. Szegedy, C., Liu, W., Jia, Y., Sermanet, P., Reed, S., Anguelov, D., Erhan, D., Vanhoucke, V., Rabinovich, A.: Going deeper with convolutions. In: IEEE International Conference on Computer Vision and Pattern Recognition, pp. 1–9 (2015)
15. Ustinova, E., Ganin, Y., Lempitsky, V.: Multiregion bilinear convolutional neural networks for person re-identification. CoRR abs:1512.05300 (2015)
16. Varior, R.R., Haloi, M., Wang, G.: Gated siamese convolutional neural network architecture for human re-identification. In: Leibe, B., Matas, J., Sebe, N., Welling, M. (eds.) ECCV 2016. LNCS, vol. 9912, pp. 791–808. Springer, Cham (2016). https://doi.org/10.1007/978-3-319-46484-8_48
17. Varior, R.R., Shuai, B., Lu, J., Xu, D., Wang, G.: A siamese long short-term memory architecture for human re-identification. In: Leibe, B., Matas, J., Sebe, N., Welling, M. (eds.) ECCV 2016. LNCS, vol. 9911, pp. 135–153. Springer, Cham (2016). https://doi.org/10.1007/978-3-319-46478-7_9
18. Vedaldi, A., Lenc, K.: Matconvnet:convolutional neural networks for matlab. In: ACM International Conference on Multimedia, pp. 689–692 (2015)
19. Wen, Y., Zhang, K., Li, Z., Qiao, Y.: A discriminative feature learning approach for deep face recognition. In: Leibe, B., Matas, J., Sebe, N., Welling, M. (eds.) ECCV 2016. LNCS, vol. 9911, pp. 499–515. Springer, Cham (2016). https://doi.org/10.1007/978-3-319-46478-7_31
20. Wu, L., Shen, C., Hengel, A.V.D.: Personnet: person re-identification with deep convolutional neural networks. CoRR abs:1601.07255 (2016)
21. Wu, L., Shen, C., Hengel, A.V.D.: Deep linear discriminant analysis on fisher networks: a hybrid architecture for person re-identification. Pattern Recogn. **65**, 238–250 (2017)
22. Yi, D., Lei, Z., Li, S.Z.: Deep metric learning for practical person re-identification. In: International Conference on Pattern Recognition, pp. 34–39 (2014)
23. Zheng, L., Huang, Y., Lu, H., Yang, Y.: Pose invariant embedding for deep person re-identification. CoRR abs:1701.07732 (2017)
24. Zheng, L., Shen, L., Tian, L., Wang, S., Wang, J., Tian, Q.: Scalable person re-identification: A benchmark. In: IEEE International Conference on Computer Vision. pp. 1116–1124 (2015)
25. Zheng, L., Yang, Y., Hauptmann, A.G.: Person re-identification: Past, present and future. CoRR abs:1610.02984 (2016)
26. Zheng, Z., Zheng, L., Yang, Y.: A discriminatively learned cnn embedding for person re-identification. CoRR abs:1611.05666 (2016)
27. Zheng, Z., Zheng, L., Yang, Y.: Unlabeled samples generated by gan improve the person re-identification baseline in vitro. CoRR abs:1701.07717 (2017)

Person Re-identification with End-to-End Scene Text Recognition

Kamlesh, Pei Xu(✉), Yang Yang, and Yongchao Xu

The School of Electronic Information and Communications,
Huazhong University of Science and Technology (HUST), Wuhan 430074, China
kdnarwani@hotmail.com, xupei_chn@163.com, {yangzw,yongchaoxu}@hust.edu.cn

Abstract. Person re-identification (Re-ID) has become increasingly popular in vision community. Many previous works rely on either feature representation learning and/or metric learning. Different from classical methods, we find that some text in images could be considered as a key cue for differentiating persons under some circumstances (e.g., racing bib number of marathon participants). Based on this observation, we propose to simplify the person Re-ID problem into an end-to-end text recognition problem. Thanks to many powerful state-of-the-art text recognition systems, we can largely improve the efficiency and accuracy of person re-identification in such circumstances. Moreover, we collect a dataset consisting of 9706 marathon images and propose an appropriate measurement to benchmark person identification. Our work provides a promising perspective to person Re-ID and end-to-end text recognition fields, showing also high potentials for video surveillance and image retrieval.

Keywords: Person re-identification · Text detection
End-to-end text recognition · Image retrieval · Video surveillance

1 Introduction

Recently, person re-identification (Re-ID) has received considerable attention in computer vision research community. The mainstream state-of-the-art methods for re-identifying person can be categorized into three classes: (1) Feature representation based methods, which extract some handcraft visual features (e.g., color, texture, shape) to define a distinctive descriptor to recognition. Some representative works are [7,20]. (2) Metric learning based methods, which shift the focus from feature selection based efforts to improve Re-ID by learning appropriate distance metrics in the sense of maximizing matching accuracy. These methods are also considered as relative ranking and manifold-based affinity learning problems in many works [1,14]. (3) Deep learning based methods. With the success of deep convolutional neural networks (CNN) in many vision problems,

Kamlesh and P. Xu have contributed equally to this work.

J. Yang et al. (Eds.): CCCV 2017, Part III, CCIS 773, pp. 363–374, 2017.
https://doi.org/10.1007/978-981-10-7305-2_32

several tentative works using CNN for person Re-ID have been proposed recently onto the task to learn to tell different persons since video-based re-id has a large data volume, e.g.,[3,17].

Different from those mainstream Re-ID methods, we notice that under some circumstances every person is associated with a unique text label (e.g., racing bib number). This text label can be considered as an important and effective cue to identify a person. Thus the person Re-ID problem can be turned into a text detection and recognition problem, which is much easier thanks to a great performance of current end-to-end text recognition in the wild.

In this paper, we propose to identify a person by an end-to-end recognition on text that could differentiate one person from others. The main difference between traditional person Re-ID task and our person identification task is depicted in Fig. 1.

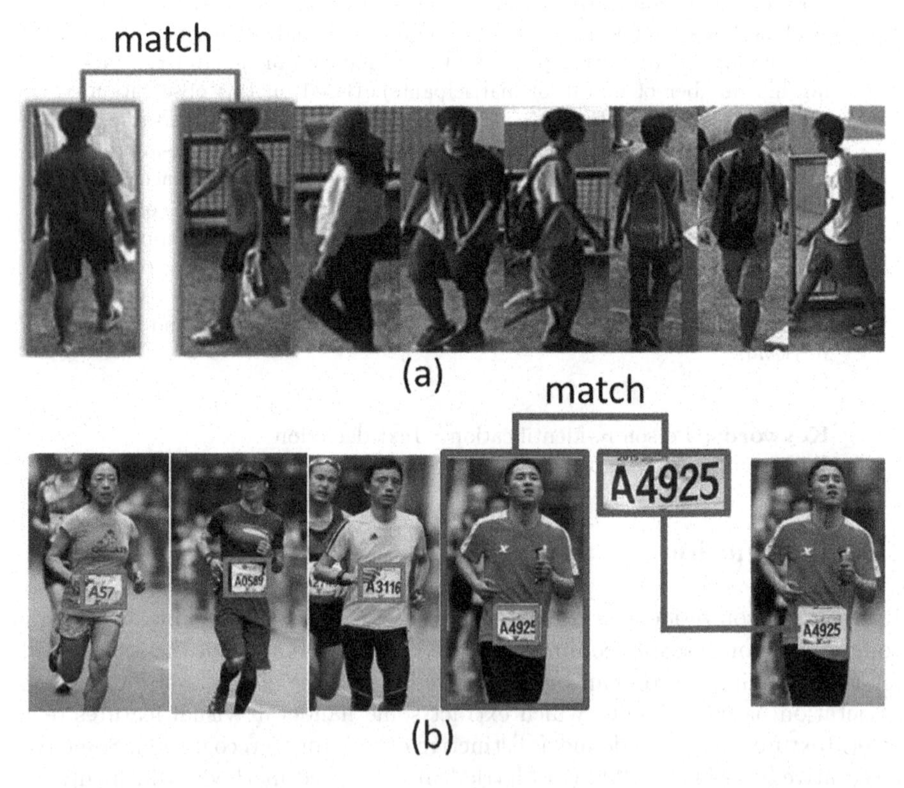

Fig. 1. Comparison of classical person re-identification in (a) and our task in (b), where we aim to identify specific person using end-to-end scene text recognition method.

The proposed idea of using text recognition system to achieve person Re-ID is extremely useful in races such as marathons. Recently, marathons are organized throughout the world and attract many broadcasters and individuals, which results in a large volume of photograph and video data. It is difficult or even impossible to manually identify any particular participant based only on the visual appearance. So each participant is given a unique Racing Bib Number (RBN) designating that contestant. Bib numbers are often printed on a piece of cloth or paper and attached to the body part (e.g., chest) of contestants. We propose to identify a specific person via the text information in terms of RBN recognized using an end-to-end scene text recognition system. This would make it much easier to identify a specific person, search the photo shot of a queried RBN from a large dataset of photos or video recording competition results.

More specifically, since RBN is usually presented as horizontal text attached to the chest of participants, we propose to adopt a state-of-the-art horizontal scene text detector TextBoxes [11] to spot RBN's location. Contrary to most conventional text detectors requiring multiple post-processing steps, TextBoxes is an end-to-end trainable and detects efficiently and accurately horizontal text from scene images. The detection step is followed by an efficient text recognition algorithm called CRNN [15] to transcript detected text. CRNN is also end-to-end trainable and performs well in recognizing scene text. Since RBN usually has a clean background, hopefully, the pipeline of combining TextBoxes and CRNN would recognize correctly RBN from the input image.

The main contributions of this paper are three folds:

(1) We first proposed an alternative and novel idea for person Re-ID via a sophisticated horizontal end-to-end scene text recognition. We proposed an appropriate matching criterion to adapt the text recognition task to person identification.
(2) We collected 9706 marathon images to establish a large and reliable benchmark dataset for person Re-ID task using text cues. To the best of our knowledge, there is no public dataset dedicated for this topic. This dataset will be soon made publicly available.
(3) We conducted a number of experiments confirming the usefulness of the proposed idea which identifies person with end-to-end scene text recognition. The experimental results show that the proposed idea has a considerable potential in person identification, pedestrian search, and video surveillance.

The rest of the paper is organized as follows. Some related work is provided in Sect. 2, followed by our method in Sect. 3. We present in Sect. 4 some experimental results. Finally, we conclude in Sect. 5.

2 Related Work

Person Re-ID is a fundamental task and also a difficult problem in computer vision. This is because the same person may look quite different in different images and two persons may look very similar when exposed to complex backgrounds, viewpoint variations, and so on. Many works about this topic focus on extracting representative visual features for images and/or designing a robust distance metric to compare those features.

In this paper, we focus on identifying participants from Marathon race images. Since each person is associated with a unique RBN, in this case, the person Re-ID can be turned into a text detection and recognition problem. We shortly review some text detection and recognition works in the following. More details can be found in these two comprehensive review papers.

2.1 Text Detection

Text detection aims to localize text regions in images, often in the form of word bounding boxes. Many works have been proposed to solve this issue. They can be divided into three categories: texture based methods [6,10,21], connected-component based methods [4,5,13], and deep learning methods [9,11,19]. Recently, deep learning based methods have been the mainstream and achieved great improvements. In this paper, we adopt TextBoxes [11], and end-to-end trainable method [11] which achieves high performance by applying SSD [12] to text detection. Compared to other deep learning based methods, TextBoxes based on SSD framework uses a fully convolutional pipeline and non-maximum suppression as post processing without extracting regions of interest (ROI). TextBoxes (see [11] for more details) has advantages in the following aspects:

(1) *Hierarchical feature maps for detection*: TextBoxes uses feature map outputs of 6 convolutional layers to detect text of multiple scales. The input image of TextBoxes is of size $300 * 300$, and the sizes of extracted feature maps are $39 * 39$, $20 * 20$, $10 * 10$, $5 * 5$, $3 * 3$, $1 * 1$ respectively.
(2) *Fully convolutional kernel for prediction*: TextBoxes substitutes fully-connected layers with fully-convolutional layers as predictor, which largely reduces computation burden. Different from the $3 * 3$ convolutional predictor used in SSD, TextBoxes uses $1 * 5$ convolution kernels as the predictor, which better fits long text lines.
(3) *Multi-scale aspect ratios for default boxes*: TextBoxes adapts SSD framework by defining 6 aspect ratios for default boxes, including 1, 2, 3, 5, 7, and 10, which better caters to long text lines in natural images.

2.2 Text Recognition

Text recognition aims to translate text regions into human or machine readable character sequences. The text recognition methods can be roughly grouped into two categories: character-based [2,16,18] and word-based recognition [9,15]. In this paper, we adapt a state-of-the-art word-based method called CRNN [15]. CRNN is an end-to-end trainable neural network for image-based sequence recognition, which takes an image as input and outputs recognized string. Its architecture consists of three parts: (1) Convolutional layers which extract a feature sequence from the input image; (2) Recurrent layers which predict a label distribution for each frame; (3) Transcription layer which translates the per-frame predictions into the final label sequence. CRNN (see [15] for more details) shows its superiority in sequence recognition as follows:

(1) It can be directly learned from sequence labels (e.g., words), requiring no detailed annotations (e.g., characters);
(2) It requires neither handcraft features nor preprocessing steps, including binarization or segmentation, component localization, etc.
(3) It is unconstrained to the lengths of sequence-like objects, requiring only height normalization in both training and testing phases.

3 Method

3.1 Text Detection of RBN

Compared to scene text, text of racing bib numbers has the following differences:

(1) Scene text in the wild may appear in the form of a single word, a long text line, or curved lines, which shows large variations both in size and aspect ratio. However, although text of RBN may vary in size, its aspect ratio does not change significantly. This is because RBN is custom made to fixed width and height for a competition.
(2) Scene text in the wild often undergoes complex background and noises. Unlikely, RBN has distinguished appearance between black text and pure color background. It is easier to detect RBN text than scene text in the wild.

Based on these two differences, we propose to adapt TextBoxes [11] and make the following changes to detect text of RBN:

(1) Original TextBoxes uses 6 aspect ratios for default boxes (see Sect. 2.1). In the case of RBN text detection, almost all the RBN texts have an aspect ratio lower than 5 even for twisted and deformed cases. In fact, based on our observation, the aspect ratio for RBN text is near 3. So we only reserve aspect ratios 1, 2, and 3 of TextBoxes shortly reviewed in Sect. 2.1. In this way, we can achieve higher accuracy and meanwhile remove unnecessary computations.
(2) Since the size of images in our dataset is 1200×800, resizing them to 300*300 may fail to capture RBN of small size due to low resolution. So we alternately resize input images to 300×300 and 700×700 every 200 batches. Hopefully, higher resolution of inputs helps to detect small texts explicitly.
(3) Finally, original TextBoxes uses 1×5 convolutional kernels as predictor of score and offsets for scene text detection in the wild. In the case of RBN text, we do not need such a long kernel. Instead, we use 1×3 convolutional kernel.

The architecture of our adapted TextBoxes is illustrated in Fig. 2.

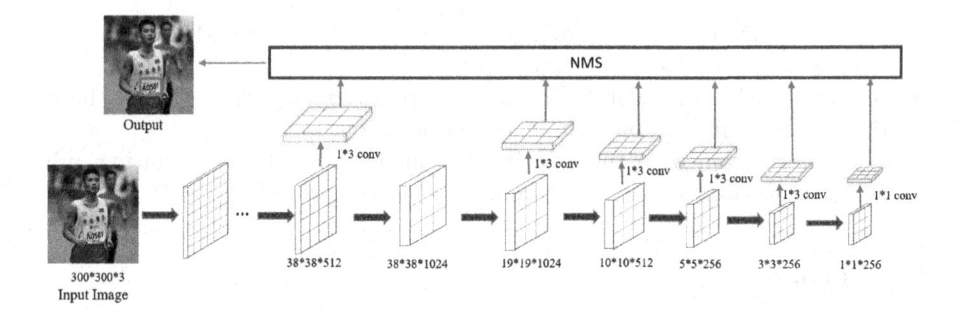

Fig. 2. Architecture of adapted TextBoxes [11] for RBN text detection.

3.2 Text Recognition of RBN

We adopt CRNN [15] shortly reviewed in Sect. 2.2 to recognize text bounding boxes produced by modified TextBoxes detailed in Sect. 3.1. Since RBN text is mainly composed of pure numbers or an English character followed by several numbers, which is a special case of general scene text concerned in original CRNN. For scene text recognition, original CRNN uses the synthetic dataset (Synth, 800K) released by Jaderberg *et al.* [8] as the training data and tests on all other real-world test datasets without any fine-tuning on their training data. The images in Synth dataset have a distinguished dictionary, which mainly consists of English characters, thus very likely to work poorly on our task if using directly the released trained model. Consequently, we propose to use an alternative synthetic dataset to retrain the model of CRNN. With the dictionary of our marathon dataset, we employ the method in [8] to synthesize 100K images looking authentic compared to marathon dataset. We retrain the pretrained model released by CRNN with such generated 100K synthetic images and then fine-tune the model with our train data.

4 Experiments

4.1 Dataset Creation

We have collected by ourselves 9706 images from different events of marathon under varying imaging environments. We divide this dataset into two subsets: training set containing 8706 images and testing set with the other 1000 images. The collected dataset exhibits following multiple challenges: multiple candidates of different scales, deformed text with low illumination, and text in blur or in low resolution. For all images, text regions are manually labeled with their bounding boxes $(x1, y1, x2, y2, x3, y3, x4, y4)$, corresponding strings (s), and an indicator (dif) denotes whether the underlying text is difficult to recognize. This amounts up to a vector of 10 entries. The dictionary of the annotations consists mainly of numbers and a few English characters.

We believe our collected dataset can serve as a standard benchmark for end-to-end text recognition as well as for person re-identification where a unique bib number identifies each individual runner.

4.2 Evaluation Protocols

Detection Measurement. We apply classical and popular precision, recall, and F-measure metrics to assess text detection performance. A detection box dt is considered to match a ground truth box gt if the intersection-over-union IoU exceeds a given threshold (set to 0.5 in all our experiments). The IoU is defined as $IoU = \frac{|gt \bigcap dt|}{|gt \bigcup dt|}$, where $|\cdot|$ stands for the cardinality. When more than one detection is matched to the same ground truth, only the detection with the maximum IoU is kept, the rest are matched to null (i.e., unmatched). This means we stick to one-to-one match strategy. Then the precision P, recall R, and $F_{measure}$ are computed by

$$P = \frac{TP}{TP + FP} \tag{1}$$

$$R = \frac{TP}{TP + FN} \tag{2}$$

$$F_{measure} = \frac{2 * P * R}{P + R} \tag{3}$$

where TP, FP, and FN are the number of hit detection boxes, incorrectly identified detection boxes (unmatched detections), and missing ground truth boxes(unmatched ground truth) respectively.

End-to-End Recognition Measurement. We use edit distance ed as evaluation metric for recognition. Following the same matching strategy as detection process, we calculate the edit distances between all matching pairs. If a detection is matched to null, then an empty string will be taken as the ground truth text. Similarly, if a ground truth is matched to null, we also calculate its edit distance to empty string. The edit distances are summarized and divided by the number of test images. The resulting average edit distance AED is given by:

$$AED = \frac{\sum_{i=1}^{N_t} \sum_{j=1}^{N_m} ed(dt, gt))}{N_t}, \tag{4}$$

where N_t is the number of test images, N_m is the number of matched pairs. This average edit distance AED is considered as the metric. A smaller AED means a better performance.

Person Identification Measurement. AED is an appropriate measurement for overall end-to-end recognition, but not adapted to measure the accuracy of person identification. Because AED calculates the edit distance between the recognized sequence and the corresponding ground truth sequence. Whereas, in the case of person Re-ID via RBN recognition, only when all characters of the recognized text are completely the same as the RBN, the identification can be deemed as finding correctly the queried identity. So we introduce a more precise measurement of matching criterion for person identification. Let *hit* denote whether we identify the right person corresponding to the queried RBN or not. If the *ed* of a recognized string is 0, *hit* is equal to 1, otherwise 0. We use ID_Fm, ID_P, ID_R defined in the following to measure the identification performance on the overall test dataset:

$$ID\_P = \frac{\sum_{i=1}^{N_t}\sum_{j=1}^{N_r} hit}{\sum_{i=1}^{N_t} N_r} \tag{5}$$

$$ID\_R = \frac{\sum_{i=1}^{N_t}\sum_{j=1}^{N_r} hit}{\sum_{i=1}^{N_t} N_g} \tag{6}$$

$$ID\_Fm = 2 * \frac{ID\_P * ID\_R}{ID\_P + ID\_R} \tag{7}$$

where N_r is the number of extracted boxes to be recognized on a single test image, N_g denotes the number of ground truth boxes correspondingly, ID_P, ID_R, ID_Fm for precision, recall, and F_measure of identification respectively. A larger ID_Fm means a better performance on person identification through end-to-end text recognition.

4.3 Implementation Details

RBN Detection. We conduct two comparison experiments in detection experiment depending on whether to use synthetic detection data to pretrain or not:

(1) For experiment using synthetic data, the model is trained for 500k iterations using a learning rate 0.001 for the first 40k iterations, and 0.0001 for the rest iterations. The batch size for SGD is 32 for $300 * 300$ input. Then the model is fine-tuned on real marathon data for 14.1k iterations with the learning rate set to 0.0001. The batch size is set to 8 for resized images into $300 * 300$ or $700 * 700$ alternatively.
(2) For experiment without using synthetic data, we consider it as a raw training on pretrained vgg-16 model. The model is trained for 23.5k iterations with the learning rate set to 0.001 for the first 3.5k iterations, and 0.0001 for the rest 20k iterations. The batch size for SGD is set to 8.

RBN End-to-End Recognition. We first use 100k synthetic images specifically synthesized for marathon dataset to train on the released pretrained model by CRNN for 50k iterations. Then we use real data to train for 330k iterations. The two trainings are both optimized by ADADELTA to automatically calculate per-dimension learning rates.

Experimental Environment. The text detection method is implemented using Caffe, and the recognition method is implemented using Torch. All the experiments are conducted on a workstation with a 1.9 GHz Intel(R) Xeon(R) E5-2609 CPU, 64 GB RAM and a NVIDIA GTX TitanX GPU.

Fig. 3. Performance of our proposed method to identify specific person using an end-to-end scene text recognition method.

4.4 Results

Since there is no previous work on the person Re-ID task using an end-to-end scene text recognition method, we evaluate our proposed Textboxes + CRNN pipeline on self-collected marathon dataset, and compare it to another component based detection pipeline FCN + CRNN. Since they both use the same recognition model CRNN, we compare both the detection and recognition performance.

Some quantitative results are depicted in Table 1. It can be seen that our proposed method achieves remarkable performance, which is better than the other baseline in both detection and recognition performance. This improvement is consistent with the detection performance gap between TextBoxes and FCN. And the ID_Fm of the best model TCRM (TextBoxes + CRNN 0+ Raw + Multi) can reach above 0.7, nearly 1% superior to FCN + CRNN pipeline. Furthermore, our end-to-end text recognition method is very fast, the detection phase takes 0.076 s for testing in single scale (0.19 s for testing in multi scale), and the recognition phase takes 0.95 s for testing with a lexicon.

As demonstrated by these results, an end-to-end scene text recognition can also be applied to person identification and achieve remarkable performance.

Table 1. The performance on the marathon dataset. The sign in method name TC for TextBoxes + CRNN, R/S for raw finetuning on vgg16 model or train first with synthetic data on vgg16 and then finetune, M/S for testing in multi/single scale. The compare method in the table is tested by FCN + CRNN. AED (resp. AED_LEX) for average edit distance without (resp. with) a lexicon, and ID_Fm (resp. ID_Fm_LEX) for Fmeasure without (resp. with) lexicon.

Method	P	R	F_measure	AED	AED_LEX	ID_fm	ID_fm_LEX
TCRM	**0.9294**	0.8931	**0.9109**	**1.529**	**3.389**	**0.6167**	**0.7054**
TCSM	0.9156	**0.8993**	0.9074	1.638	3.463	0.6038	0.6989
TCRS	0.9197	0.8785	0.8986	1.662	3.526	0.5978	0.6929
TCSS	0.9226	0.8764	0.8989	1.718	3.646	0.5957	0.6909
FCN + CRNN	0.9026	0.8767	0.8895	1.691	3.577	0.5954	0.6957

Some qualitative illustrations of the best identification performances are shown in Fig. 3. We can see that the proposed end-to-end scene text recognition handles well many different circumstances, such as text regions of different scales, unbalanced illumination, low resolution, deformed text, block or blur, etc. Furthermore, we can see that some recognition result can surpass human recognition. For many text regions, the recognition can even learn more than human can annotate. For those person whose text regions are blocked or with incomplete annotation, we can rank the recognition results according to edit distance and give top 10 results. Towards these candidates of most possibilities, we can further involve standard person re-identification methods to identify persons precisely.

Fig. 4. Some failure cases of our proposed method. Red boxes and black annotations are correct end-to-end detection. Green Boxes represent missing boxes. Red annotations stands for wrong recognized characters. (Color figure online)

Some failure cases are also shown in Fig. 4, where we miss a part of boxes or falsely recognize texts. The proposed method does not handle well small text in blur and low resolution, unbalanced illumination. Thus the proposed person identification is still far from ideal and has lots of potentials to improve. One perspective is to integrate feature representation of image or metric learning method in the proposed method to improve the performance for person identification.

5 Conclusion

We have presented in this paper a novel alternative method for person identification based on current sophisticated end-to-end scene text recognition system. We adopted TextBoxes and CRNN to accomplish this task. Moreover, we have collected a large dataset of marathon images for benchmarking the proposed method. This dataset will be soon made publicly available for combination of text recognition and person identification. Furthermore, we proposed an appropriate measurement for evaluation identification performance and compared with another pipeline to confirm our proposal. Experimental results demonstrate promising results of the proposed method, and a high potential to explore in person Re-ID, video surveillance, and image retrieval. In the future, we plan to integrate some traditional person Re-ID methods into our proposed pipeline to further improve identification performance on our benchmark and make more comprehensive comparisons.

Acknowledgements. This work was supported by National Natural Science Foundation of China 61703171 and 61222308.

References

1. Bai, S., Bai, X., Tian, Q.: Scalable person re-identification on supervised smoothed manifold. CoRR abs/1703.08359 (2017)
2. Bissacco, A., Cummins, M., Netzer, Y., Neven, H.: PhotoOCR: Reading text in uncontrolled conditions. In: Proceedings of ICCV (2013)
3. Bromley, J., Bentz, J.W., Bottou, L., Guyon, I., LeCun, Y., Moore, C., Säckinger, E., Shah, R.: Signature verification using a "siamese" time delay neural network. IJPRAI **7**(4), 669–688 (1993)
4. Busta, M., Neumann, L., Matas, J.: Fastext: efficient unconstrained scene text detector. In: Proceedings of ICCV (2015)
5. Epshtein, B., Ofek, E., Wexler, Y.: Detecting text in natural scenes with stroke width transform. In: Proceedings of CVPR (2010)
6. Goto, H., Tanaka, M.: Text-tracking wearable camera system for the blind. In: 10th International Conference on Document Analysis and Recognition (2009)
7. Gray, D., Tao, H.: Viewpoint invariant pedestrian recognition with an ensemble of localized features. In: Forsyth, D., Torr, P., Zisserman, A. (eds.) ECCV 2008. LNCS, vol. 5302, pp. 262–275. Springer, Heidelberg (2008). https://doi.org/10.1007/978-3-540-88682-2_21
8. Jaderberg, M., Simonyan, K., Vedaldi, A., Zisserman, A.: Synthetic data and artificial neural networks for natural scene text recognition. CoRR abs/1406.2227 (2014)
9. Jaderberg, M., Simonyan, K., Vedaldi, A., Zisserman, A.: Reading text in the wild with convolutional neural networks. IJCV **116**(1), 1–20 (2016)
10. Li, H., Doermann, D.S., Kia, O.E.: Automatic text detection and tracking in digital video. IEEE Trans. Image Process. **9**(1), 147–156 (2000)
11. Liao, M., Shi, B., Bai, X., Wang, X., Liu, W.: TextBoxes: a fast text detector with a single deep neural network. In: Proceedings of AAAI (2017)

12. Liu, W., Anguelov, D., Erhan, D., Szegedy, C., Reed, S., Fu, C.-Y., Berg, A.C.: SSD: single shot multibox detector. In: Leibe, B., Matas, J., Sebe, N., Welling, M. (eds.) ECCV 2016. LNCS, vol. 9905, pp. 21–37. Springer, Cham (2016). https://doi.org/10.1007/978-3-319-46448-0_2
13. Neumann, L., Matas, J.: A method for text localization and recognition in real-world images. In: Kimmel, R., Klette, R., Sugimoto, A. (eds.) ACCV 2010. LNCS, vol. 6494, pp. 770–783. Springer, Heidelberg (2011). https://doi.org/10.1007/978-3-642-19318-7_60
14. Prosser, B.J., Zheng, W., Gong, S., Xiang, T.: Person re-identification by support vector ranking. In: BMVC 2010 (2010)
15. Shi, B., Bai, X., Yao, C.: An end-to-end trainable neural network for image-based sequence recognition and its application to scene text recognition. TPAMI **39**(11), 2298–2304 (2016)
16. Wang, T., Wu, D.J., Coates, A., Ng, A.Y.: End-to-end text recognition with convolutional neural networks. In: Proceedings of ICPR (2012)
17. Wu, L., Shen, C., van den Hengel, A.: PersonNet: person re-identification with deep convolutional neural networks. CoRR abs/1601.07255 (2016)
18. Yao, C., Bai, X., Shi, B., Liu, W.: Strokelets: A learned multi-scale representation for scene text recognition. In: Proceedings of CVPR (2014)
19. Zhang, Z., Zhang, C., Shen, W., Yao, C., Liu, W., Bai, X.: Multi-oriented text detection with fully convolutional networks. In: Proceedings of CVPR (2016)
20. Zhao, R., Ouyang, W., Wang, X.: Unsupervised salience learning for person re-identification. In: Proceedings of CVPR (2013)
21. Zhong, Y., Karu, K., Jain, A.K.: Locating text in complex color images. Pattern Recogn. **28**(10), 1523–1535 (1995)

Hierarchical Structure Construction Based on Hyper-sphere Granulation for Finger-Vein Recognition

Jinfeng Yang[1(✉)], Yuqing Yang[2], Zhiyuan Liu[1], and Yihua Shi[1]

[1] Tianjin Key Lab for Advanced Signal Processing, Civil Aviation University of China, Tianjin, China
jfyang@cauc.edu.cn

[2] Tianjin University of Commerce, Tianjin, China

Abstract. Recently, the finger-vein (FV) trait has attracted substantial attentions for personal recognition in biometric community, and some FV-based biometric systems have been well developed in real applications. However, the recognition efficiency improvement over a large-scale database remains a big practical problem. In this paper, we propose an efficient and powerful hierarchical model based on hyper-sphere granular computing (HsGrC) for saving recognition cost. For a given FV database, samples are first viewed as atomic granules for building a basic hyper-sphere granule set. Using HsGrC, several different granule sets with multi-granularities are then generated by hyper-sphere granulation. To build a hierarchical structure of granule sets with granularity variation, a new quotient space relationship is established considering recognition efficiency improvement. Experimental results over a large finger-vein image database demonstrate that the proposed hierarchical model performs very well in computing cost reduction as well as recognition accuracy improvement.

Keywords: Finger-vein recognition · Granular computing · Biometrics

1 Introduction

Biometric identification technology has progressed highly over the past three decades. In recent years, the finger-vein trait, as an emerging biometric pattern, has increasingly attracted popular attention [3,14,15]. Compared with some conventional biometric traits, e.g., fingerprint [5,16], face [20], palmprint [4,6,23], iris [2], the finger-vein pattern itself is forgery-proof, biologically active and highly acceptable for users [7,22]. Hence, finger-vein recognition are very suitable for many critical security applications, such as ATM, access control, border crossing, etc.

Great progress has been made on FV recognition in practice, but recognition efficiency always was an important problem to be addressed besides accuracy,

J. Yang et al. (Eds.): CCCV 2017, Part III, CCIS 773, pp. 375–386, 2017.
https://doi.org/10.1007/978-981-10-7305-2_33

especially under the circumstance of great user population. Hierarchy is a significant strategy that has long been used for computing efficiency improvement in many fields. Ross and Sunder [17] used statistical features of iris textures to generate the hierarchical partition structure. Wang and Xu [19] introduced a hierarchy scheme for brain big data processing. In finger-vein recognition, Tan et al. [18] presented a hierarchy model with two layers based on appearance feature and content feature of finger-vein images. In this model, feature extraction operation makes the computational cost increasing greatly over a large image database. Recently, quotient space theory (QST) has drawn substantial attention and been applied for reducing time cost in some applications [24–26]. For example, in handwritten numeral recognition, a common hierarchy construction approach has been established based on QST [1]. In this paper, a fuzzy similarity matrix was first generated, and then by measuring the similarity with different thresholds, several quotient spaces can be obtained for building a hierarchical structure suitable for recognition efficiency improvement. Unfortunately, similarity-matrix based methods for hierarchical structure construction is very time consuming over a large database so that recognition operation always cost greatly. To address this problem, exploring some fast approaches for quotient space generation is necessary in practice.

In this paper, we propose a novel hierarchical model construction strategy based on hyper-sphere granular computing. Granular computing (GrC) is a methodology that has been rapidly developed and widely used in information analysis [11,12,21]. The basic idea of GrC is to use information granulation for complex problem solving. Hyper-sphere granular computing (HsGrC) is a kind of GrC which treats each finger-vein image as a hyper-sphere granule. Based on HsGrC, the center of a granule can be positioned by a feature vector, and its initial granularity can be set to zero. However, if the dimension of a feature vector is too high, it is also difficult to readily describe a granule and make granulation directly in high dimension space. Thus, in order to make HsGrC more suitable for modeling image granules, dimension-reduction methods are usually necessary for vectoring images. Based on the above consideration, for a given finger-vein image database, the image samples are first vectorized using PCA coefficients, and then represented by a hyper-sphere granule with zero granulation. By making multiple granulations, we can obtain several granule sets based on HsGrC. Each granule set with a certain granularity is factually corresponding to a quotient space, which is the foundation of hierarchical structure in quotient space theory. Thus, by establishing a full coverage criterion, we can build a topological relations of these quotient spaces as a hierarchical model for finger-vein recognition. Hence, this paper tries to use HsGrc to handle FV recognition problem with the help of simple feature extraction.

2 Hierarchical Model Construction

In this section, we first review hyper-sphere granulation and detail the proposed construction method of hierarchical model. Subsequently, PCA is introduced for image feature dimension reduction. In addition, we compare the proposed model with the existing hierarchical structure. This will be introduced in quotient space theory in Sect. 4.

2.1 Granule Set Generation Based on Hyper-sphere Granulation

HsGrC is an new method which transforms all the data to be processed into different hyper-sphere granules and make a problem solution in granule space. So, based on HsGrC, we can regard each image sample as an atomic hyper-sphere granule with zero granularity, and its extracted feature vector as the atomic granule center. Thus, an initial granule set with the finest granularity that contains all the image samples can be obtained. Next, by measuring the similarities between the atomic granules and using union operation, a coarser granular set composed of the atomic granule can be generated. In this way, the obtained new granules are with larger granularity than the atomic granules. For union operation, it is defined as [8–10]

$$G = G_1 \cup G_2 = [C, R] = [\frac{1}{2}(P+Q), \frac{1}{2} \parallel P - Q \parallel] \tag{1}$$

$$P = C_1 - r_1(C_{12}/\|C_{12}\|) \tag{2}$$

$$Q = C_2 + r_2(C_{12}/\|C_{12}\|) \tag{3}$$

where $G_1 = (C_1, r_1)$ and $G_2 = (C_2, r_2)$ are two hyper-sphere granules, $C_1 = (x_1, x_2, \ldots, x_n)$ and $C_2 = (y_1, y_2, \ldots, y_n)$ are the centers of G_1 and G_2, r_1 and r_2 are granularities of G_1 and G_2. G is an union granule, C and R respectively represent its center and granularity. C_{12} denotes the vector from C_1 to C_2. A 2-dimensional union operation between two granules is schematically illustrated in Fig. 1. For example, $G_1 = (2, 6, 2)$ and $G_2 = (5, 7, 3)$ represent two hyper-spherical granules in 2-dimensional space, the union hyper-spherical granule is $G(3.97, 6.66, 4.08)$ denoted by the red circle in Fig. 1.

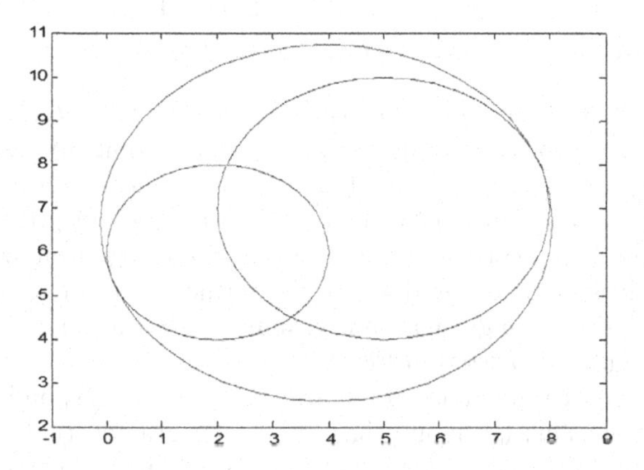

Fig. 1. Union operation between two granules [9] (Color figure online)

Here, the hyper-sphere granulation process is briefly introduced as follows. First, a sample set $S = \{x_i|i = 1, 2, \ldots, n\}$ in n-dimensional space is used to build an atomic granule set $AGS = \{G_i = (x_i, 0)|i = 1, 2, \ldots, n\}$. This means x_i in S can be represent as a hyper-sphere $G_i = (x_i, 0)$ centered at x_i with zero radius. Second, let GS be a new granule set, we transfer an atomic granule G_1 from AGS into GS. Third, for the rest atomic granule G_i, let d_{ij} represent a distance measure between G_i and CG_j belonging to GS, if d_{ij} is minimal and the granularity of the union of G_i and CG_j is less than or equal to a given granularity threshold, CG in GS is replaced by the union granule of Gi and CG_j, otherwise the atomic granule G_i is moved to GS from AGS until AGS is emptied. By these operations, a new granule set GS with larger granularity can be re-generated accordingly. Thus, making multi-granulations to an atomic granule set can generate a cluster of hyper-sphere granule sets. For a recognition task, its solution should be hidden among the obtained cluster of hyper-sphere granule sets.

2.2 Hierarchical Model Construction Based on Full Coverage Criterion

In order to implement a recognition task effectively and efficiently, according to QST, a hierarchical model should be constructed by establishing a rule of connecting these multiple hyper-sphere granule sets together. Here, we design a full coverage criterion to hierarchically model the obtained granule sets. For convenience, a hierarchical structure with two layers is used as a example. Assume that $GS_1 = \{CG_{1i}|i = 1, 2, \ldots, x\}$ denotes the first layer with coarse granularity, $GS_2 = \{CG_{2j}|j = 1, 2, \ldots, y\}(x \leq y)$ denotes the second layer with fine granularity, and RCG_{1ij} represent an another fine granule set describing the relationship with the coarse granule CG_{1i}, RCG_{1ij} must be satisfying the following conditions,

$$RCG_{1ij} = \{CG_{1i} \subseteq (CG_{2j} \cup CG_{2k} \cup \ldots \cup CG_{2w})\} \quad (4)$$
$$(j \neq k \neq \ldots \neq w; j, k, w = 1, 2, \ldots, y),$$
$$RCG_{1ij} = \{CG_{2j} \cap CG_{1i} \neq \emptyset | j = 1, 2, \ldots, y\}. \quad (5)$$

The two conditions mean that a relationship of CG_{1i} and CG_{2j} can be built if CG_{1i} share any atomic granule with CG_{2j}. According to this rule, all the granules $RCG_{1ij}(j \in 1, 2, \ldots, y)$ related with CG_{1i} must contain more atomic hyper-sphere granules than CG_{1i}. Thus, this strategy can greatly reduce the uncertainty of finding a correct solution so that the recognition accuracy can be improved effectively. Hence, based on the above rules, we can readily make a hierarchical construction for a given atomic granule set. For intuitive understanding, a schematic diagram is plotted in Fig. 2.

The first layer with maximal granularity ($\rho = \infty$) corresponds to the coarsest granular level. This layer only has one union granule that contains all of the atomic granules, which means all the atomic granules belongs to the same class. The last layer with minimal granularity ($\rho = 0$) corresponds to the finest

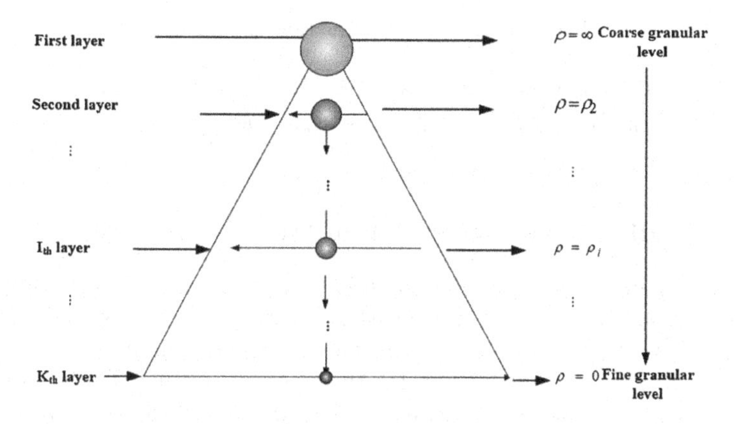

Fig. 2. Hierarchical model based on HsGrC

granular level which considers all the atomic granules inhomogeneous. Using full coverage criterion, the relationships between two adjacent layers can be built, and a hierarchical model can be established level-wisely.

2.3 Feature Dimension Reduction

Directly treating a FV image as a hyper-sphere granules can make the feature dimension too high. This is not good for hyper-sphere granulation since both memory consumption and computational burden can increase greatly. Hence, for improving the recognition efficiency effectively, the dimension of features should be reduced properly.

Here, the principal component analysis method is used for dimensionality reduction of FV images. PCA is a very effective feature dimension reduction method to obtain some informative features of high-dimensional datasets [13]. In addition, PCA-based FV feature extraction is also more favorable to verify the performance of the hierarchical model, since PCA cannot behave effectively on discrimination information extraction.

3 Experiments in FV Recognition

3.1 Dimension Reduction

As motioned above, it is necessary to reduce the dimension of the FV samples. Moreover, due to the dimension after PCA will have a great influence on the accuracy and efficiency of hyper-sphere granular computing, thus it is also important to choose a suitable dimensionality. In this paper, six thresholds of the cumulative energy are tested, as shown in Table 1. Considering both feature discrimination and calculation efficiency, 85% cumulative energy is chosen. This means the dimensionality of all the FV images is reduced from 18,000 to 40. Thus, each FV sample can be represented as an atomic hyper-sphere granule with zero granularity, and positioned by a 40 dimension vector.

Table 1. Dimension reduction using PCA

Cumulative energy	99%	95%	90%	85%	80%	75%
Dimensionality	761	146	68	40	25	16

3.2 Recognition Using the Euclidean Distance Measure

For a given FV image database, based on HsGrC, a part of its image samples are used to establish a hierarchical model with L layers, and the rest samples are used for testing. Let $LG_\ell(\ell = 1, 2, L)$ be a granule set corresponding to the ℓth layer, all the atomic granules are labeled for training and testing. By hierarchically measuring the similarities between a testing granule and the training granules, we can predict the class of the testing sample efficiently. For convenience, the Euclidean distance is used for the similarity measure since this measure can effectively reveal the potentiality of the proposed model in recognition accuracy. Here, the Euclidean Distance is defined as

$$d_{ij}(G_i, G_j) = \|C_i - C_j\|_2 - R_i - R_j \tag{6}$$

where G_i and G_j are two granules, C_i and C_j are two vectors respectively positioning G_i and G_j, and R_i and R_j are two granularities respectively corresponding to G_i and G_j.

Assume that $SG_\ell(\ell = 1, 2, L)$ is an empty granule set with labels, the recognition processing is detailed as follows.

Step1. For a testing granule G_i, the Euclidean Distance Measure d_{ij} is computed between G_i and any G_j^1 of LG_1. If $d_{ij} \leq 0$ is satisfying, the granule G_j^1 is moved from the first layer LG_1 into SG_1.
Step2. Then, in the ℓth layer ($\ell \geq 2$), by computing d_{in} between G_i and G_n^ℓ of LG_ℓ and using the full coverage criterion defined by Eqs. (4) and (5), the granule G_n^ℓ is moved from LG_ℓ into SG_ℓ if $d_{in} \leq 0$ is satisfying.
Step3. Repeat **Step2.** for $\ell < L$ till $\ell = L$. Thus, every granule in SG_L is labeled. Minimizing the d_{im} between G_i and G_m^L can find the class label corresponding to G_i. The obtained label is the final recognition result of granule G_i.

To illustrate the proposed recognition scheme intuitively, a schematic diagram is plotted in Fig. 3. In fact, in a layer of the established hierarchical model, the granularities of granules are not identical except for the atomic granules in the last layer (with the finest granularity), since the granularity thresholds only are the upper bounds. The green balls of each layer, as shown in Fig. 3, imply that the Euclidean distance between the granules and a test granule is not bigger than zero. The arrowed solid or dot lines indicate some indispensable steps must be taken for finding a solution. But an effective way of finding the final answer should be the path denoted by the red and solid arrowed lines. Moreover, it is usually redundant to find an answer in the last layer with the finest granularity.

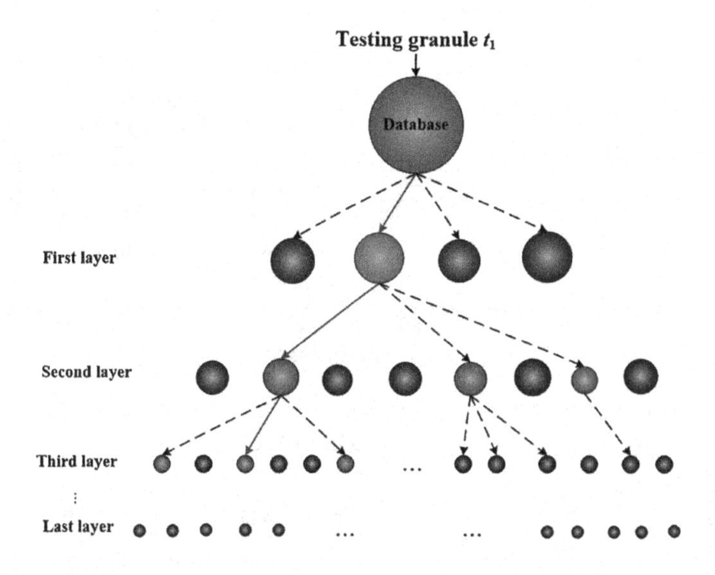

Fig. 3. The recognition procedure (Color figure online)

Matching operation between the coarse granules and a testing granules can also find the correct answer. This is therefore helpful to greatly reduce the matching cost in a recognition task.

3.3 Some Experimental Results

Here, a homemade database that contains 5000 finger-vein images from 500 individual fingers is used for performance test. In order to verify the performance of hierarchical structure thoroughly, the FV samples are divided into two sets: training set and testing set. The training set is used to establish the hierarchical model and the testing set is used to test the efficiency and accuracy of the hierarchical model. For a finger contributing 10 images, we select 7 FV images of it as training data and the rest as probe data. Hence, the training set contains 3500 images, and the testing set has 1500 images. We mainly analyze and discuss the performance of hierarchical model based on HsGrC from training accuracy (Tr), testing accuracy (Ts), training time (tr), testing time (ts).

In order to validate the hierarchical model adequately, we first establish a single layer structure. By observing the performance showed in Fig. 4, we find the single layer model already has a good performance in both accuracy and efficiency when $\rho = 25$. And in a large threshold range, the performance is relatively reliable. Then, using $\rho = 25$ as the first granularity threshold, we build a two layers hierarchical model by adding another bigger granularity threshold than 25. Some results are shown in Fig. 5. Compared Fig. 5 with Fig. 4, we can clearly find the recognition efficiency list in Fig. 5 is almost double of that list in Fig. 4. Hence, building a hierarchical model based on HsGrC is beneficial for saving time cost.

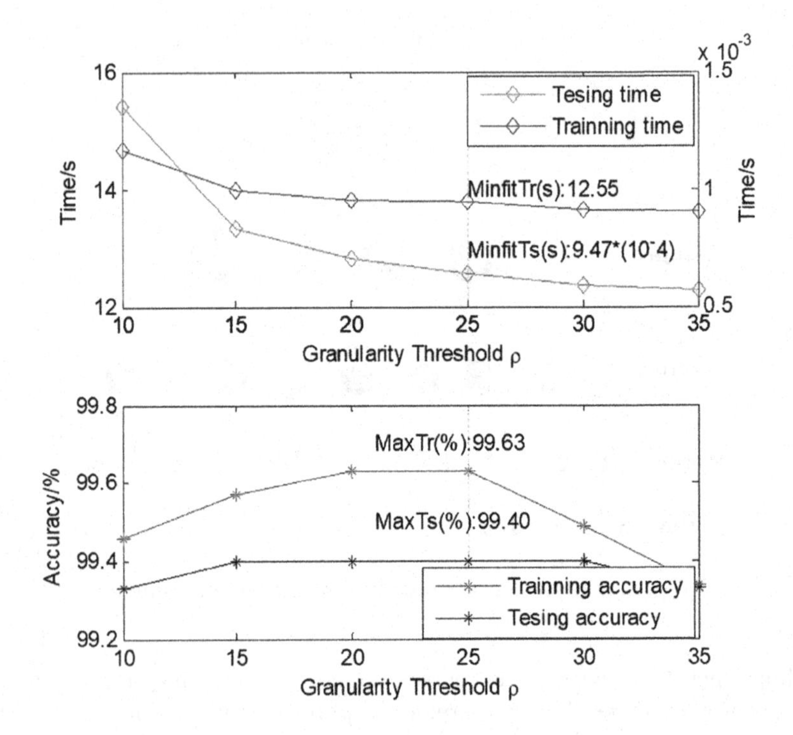

Fig. 4. Performance of single layer model

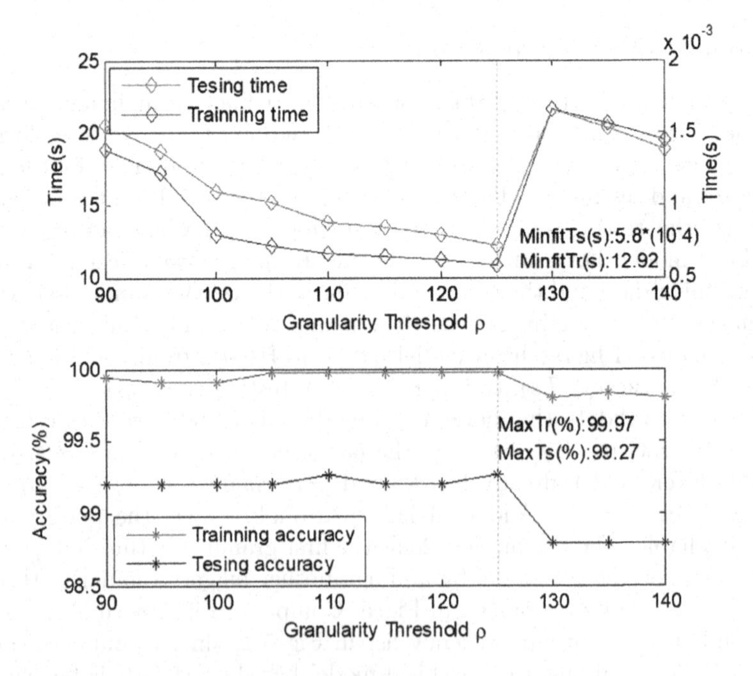

Fig. 5. Performance of two layers model

Therefore, we find the two layers hierarchical model is more suitable for FV recognition. But what about three layers? And whether this model can be used for another recognition problem? We will discuss these problems in Sect. 4.

4 More Experimental Validations and Discussions

In this section, we first discuss the performance of three layers model for FV recognition. In addition, we compare the existing construction method of hierarchy that proposed in quotient space theory with our construction model. Furthermore, another pattern, finger-knuckle-print (FKP) is adopted here to verify the recognition performance of the proposed model.

4.1 Three Layers Model for FV Recognition

On the basis of the threshold $\rho = 125/25$, we establish a three layers model by setting a larger threshold than the second layer $\rho = 125$. Some results are list in Table 2.

Table 2. Performance of three layers hierarchical model

Threshold	Tr (%)	Ts (%)	tr (s)	ts (s)
150/125/25	86.46	85.93	80.29	7.2e−3
200/125/25	73.97	73.53	244.36	0.0287
250/125/25	74.00	73.40	241.88	0.0288

From Table 2, we can see that the three layers model with the threshold $\rho = 150/125/25$ can achieve the optimal performance. However, compared with the two layers hierarchical model, the performance of the three layers model is really not satisfying. For the poor accuracy both on the training set and the testing set, we consider that the granules with large granularity $\rho = 150$ may greatly suffer discrimination. To some extent, we can observe this phenomenon in Fig. 5. When the threshold is larger than $\rho = 125$, both training accuracy and testing accuracy have a down tendency, which means improper granularity may lead to a wrong classification. As for computation cost increase, it may cause by the full coverage criterion. If we set a large granularity threshold, lower layer granules will build relationships with the larger granule, and all of these lower layer granules need to be matched for find an answer. Thus, the cost of testing and training may increase accordingly.

4.2 Three Layers Model for FV Recognition

As motioned in Sect. 1, the granule set formation method that using fuzzy similarity relation matrix (FSRM) has been successfully used in handwritten numeral

Table 3. Performance of different hierarchy construction methods

Method	Tr (%)	Ts (%)	tr (s)	ts (s)
FSRM	100	99.73	287.1	4.3e−3
SHsGrC	99.63	99.40	12.55	9.47e−4
DHsGrC	99.97	99.27	12.92	5.8e−4

recognition. We have evaluated this method on our finger-vein database and compared with the proposed HsGrC method, the experimental results are list in Table 3.

In the experiment of FSRM, we have chosen Euclidean distance strategy to measure the similarity of FV images after PCA and generate the fuzzy similarity relation matrix. The similarity threshold ρ_1 is from 0.980 to 0.995 with step 0.005. If a threshold value is given, a corresponding granule set can be formed accordingly. This method achieves the optimization performance if $\rho_1 = 0.990$, the results are list in the first line in Table 3. However, with the similar accuracy, the single model and the two layers model based on HsGrC have a much better performance both on testing time and training time.

4.3 Hierarchical Model Based on HsGrC for FKP Recognition

Here, the used finger-knuckle-print images are also captured by our homemade imaging device. We preprocess the FKP images and the FV image using a same method, the filtered image is shown in Fig. 6(b). A homemade database also containing 5000 FKP images from 500 individual fingers is established. For a finger contributing 10 images, we select 7 FKP images of it as training data and the rest as probe data. Hence, the training set contains 3500 images, and the testing set has 1500 images. In fact, the FKP images and FV images were captured simultaneously using a homemade imaging device.

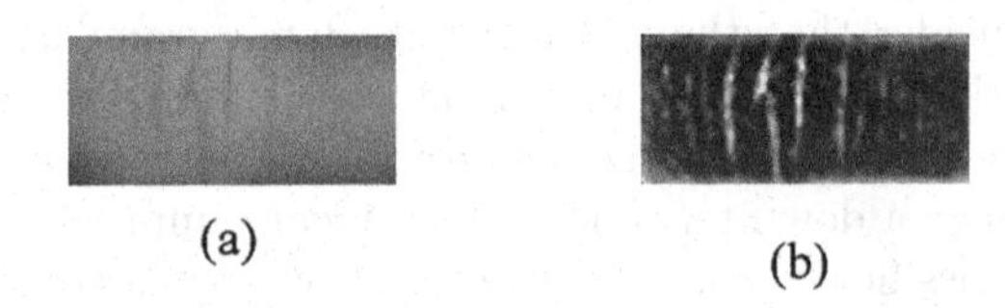

Fig. 6. Image acquisition. (a) A normalized FKP image. (b) The enhanced result.

Here we also adopt PCA for feature dimension reduction, and choose the cumulative energy of 90% which reduces the dimensionality to 73. Then using the proposed hierarchical model, the performances of FKP recognition are listed in Table 4. Compared with the single layer model ($\rho = 0.1$), the two layers structure shows a better performance with threshold $\rho = 2/0.1$. Especially,

Table 4. Recognition performance on FKP images

Threshold	Tr (%)	Ts (%)	tr (s)	ts (s)
0.1	**96.17**	**99.47**	**16.46**	**1.2e−3**
0.5/0.1	99.20	97.33	27.88	1.5e−3
1/0.1	99.60	98.07	21.07	9.4e−4
2/0.1	**99.94**	**99.13**	**15.51**	**8.0e−4**
2.5/0.1	99.86	99.13	18.21	1.2e−3

the training accuracy has improved 3.77%. Table 4 has evaluated the capability of the proposed model in FKP recognition.

5 Conclusion and Future Work

In this paper, we have proposed a novel hierarchical model based on HsGrC. The experimental results have shown that the proposed method performed well on both accuracy and efficiency in both FV and FKP recognition. In order to simplify the construction method of the model, PCA method was adopted for feature vector generation. To improve the performance of the proposed method effectively, some suitable feature extraction methods should be used for different patterns recognition in future.

References

1. Chen, J.: Research on granulation technologies and problem solving methods based on quotient space theory (2014)
2. Daugman, J.: How iris recognition works. IEEE Trans. Circ. Syst. Video Technol. **14**(1), 21–30 (2004)
3. Ding, Y., Zhuang, D., Wang, K.: A study of hand vein recognition method. In: Proceedings of the IEEE International Conference on Mechatronics and Automation, vol. 4, pp. 2106–2110 (2005)
4. Huang, D., Jia, W., Zhang, D.: Palmprint verification based on principal lines. Pattern Recogn. **41**(4), 1316–1328 (2008)
5. Jain, A., Flynn, P., Ross, A.: Handbook of Biometrics. Springer, New York (2007). https://doi.org/10.1007/978-0-387-71041-9
6. Kong, A., Zhang, D., Kamel, M.: Palmprint identification using feature-level fusion. Pattern Recogn. **39**(3), 478–487 (2006)
7. Kono, M., Ueki, H., Umemura, S.: Near-infrared finger vein patterns for personal identification. Appl. Opt. **41**(35), 7429–7436 (2002)
8. Liu, H., Li, L., Wu, C.: Color image segmentation algorithms based on granular computing clustering. Int. J. Sig. Process. Image Process. Pattern Recogn. **7**(1), 155–168 (2014)
9. Liu, H., Liu, C., Wu, C.: Granular computing classification algorithms based on distance measures between granules from the view of set. Comput. Intell. Neurosci. (2014)

10. Liu, H., Zhang, F., Wu, C., Huang, J.: Image superresolution reconstruction via granular computing clustering. Comput. Intell. Neurosci. **1**(50) (2014)
11. Liu, Q.: Granular language and its deductive reasoning. Commun. Inst. Inf. Comput. Mach. **5**(2), 63–66 (2002)
12. Liu, Q., Liu, Q.: Approximate reasoning based on granular computing in granular logic. In: International Conference on Machine Learning and Cybernetics, Hoboken, USA, vol. 3, pp. 1258–1262 (2002)
13. Maadooliat, M., Huang, J., Hu, J.: Integrating data transformation in principal components analysis. J. Comput. Graph. Stat. **24**(1), 84–103 (2015)
14. Miura, N., Nagasaka, A.: Feature extraction of finger-vein pattern based on repeated line tracking and its application to personal identification. Mach. Vis. Appl. **15**(4), 194–203 (2004)
15. Miura, N., Nagasaka, A.: Extraction of finger-vein patterns using maximum curvature points in image profiles. IEICE - Trans. Inf. Syst. **90**, 185–1194 (2007)
16. Ratha, N., Bolle, R.: Automatic Fingerprint Recognition Systems. Springer-Verlag New York, Inc., New York (2004). https://doi.org/10.1007/b97425
17. Ross, A., Sunder, M.: Block based texture analysis for iris classification and matching. In: Computer Vision and Pattern Recognition Conference, pp. 30–37 (2010)
18. Tan, D., Yang, J., Shi, Y., Xu, C.: A hierarchal framework for finger-vein image classification. In: Asian Conference on Pattern Recognition, pp. 833–837 (2013)
19. Wang, G., Xu, J.: Granular computing with multiple granular layers for brain big data processing. Brain Info. **1**, 1–10 (2014)
20. Wechsler, H.: Reliable Face Recognition Methods - System Design, Implementation and Evaluation. Springer, Boston (2006). https://doi.org/10.1007/978-0-387-38464-1
21. Xie, G., Liu, J.: A review of the present studying state and prospect of granular computing. Software **32**(3), 5–10 (2011)
22. Yang, J., Shi, Y., Yang, J.: Personal identification based on finger-vein features. Comput. Hum. Behav. **27**(5), 1565–1570 (2010)
23. Zhang, D., Kong, W.K., You, J., Wong, M.: Online palmprint identification. IEEE Trans. Pattern Anal. Mach. Intell. **25**(9), 1041–1050 (2003)
24. Zhang, L., Zhang, B.: The theory and application of problem solving (1990)
25. Zhang, L., Zhang, B.: Theory of fuzzy quotient space. J. Softw. **14**, 770–776 (2003)
26. Zhang, L., Zhang, B.: Fuzzy reasoning model under quotient space structure. Inf. Sci. **173**(4), 353–364 (2005)

Photography and Video

Imaging Model and Calibration of Lytro Light Field Camera

Huixian Duan[1,2], Jun Wang[1], Lei Song[1,3(✉)], Na Liu[1], and Lin Mei[1]

[1] Cyber Physical System R&D Center,
The Third Research Institute of Ministry of Public Security, Shanghai, China
songlei9312@126.com
[2] Shanghai International Technology and Trade United Co., Ltd., Shanghai, China
[3] Shenzhen Key Laboratory of Media Security,
Shenzhen University, Shenzhen, China

Abstract. Light field cameras with microlens array are being paid more and more attention, due to their light gathering and post-capture processing. Calibrating light field cameras is still an open and challenging problem. In this paper, we present a imaging model and a calibration method for Lytro light field camera. Firstly, based on the image formation for pinhole camera and the theory of light propagation, we model the imaging process of Lytro light field, and drive a homogenous intrinsic matrix, which represents the relationship between each pixel on the image plane and its corresponding light field. Then, based on the derived intrinsic matrix, a calibration algorithm using the checkerboard grid for Lytro light field camera is proposed. At last, extensive experimental results have demonstrated the correctness of the imaging model of Lytro light field camera and the effectiveness of our calibration algorithm.

Keywords: Lytro light field camera · Imaging model · Intrinsic matrix Calibration

1 Introduction

Due to the increased depth of field and light gathering relative to traditional cameras, plenoptic camera [3] are paid more and more attention. What's more, plenoptic camera will play an important role in computer vision applications, such as refocus, occlusion removal, closed-form visual odometry and so on [4,6,7,13,22]. For the plenoptic camera, little attention has been paid on the calibration of light field, which is the most basic and essential part.

Plenoptic cameras includes several types: microlens array, mask-based cameras, freeform collections of cameras [9,16,18–21]. Ng [18] proposes a hand-held light field camera by placing a micro-lens array in front of the camera sensor. For each micro-lens, a tiny sharp image of the len aperture can be obtained and the directional distribution of incoming rays through it can be estimated. Based on Ng's model, Georgiev and Lumsdaine [9] make some modifications, that is,

J. Yang et al. (Eds.): CCCV 2017, Part III, CCIS 773, pp. 389–400, 2017.
https://doi.org/10.1007/978-981-10-7305-2_34

the micro-lens array is interpreted as an imaging system lying in the focal plane of the main camera lens. Compared with Ng's model, this system can capture a light field image with higher spatial resolution and lower angular resolution. Based on these methods by a micro-lens array, commercial light-field cameras such as Lytro [1] and Raytrix [2] have been released.

There are a lot of work about the calibration of grids or freeform collections of multiple cameras [19]. However, these methods introduce more degrees of freedom, which are not necessary to describe the microlens-based plenoptic camera. Based on the ray transfer matrix analysis, Georgiev et al. [10] derive a plenoptic camera model. For the plenoptic camera model, Johannsen et al. [15] propose a metric calibration method through a dot pattern with a known grid size, and a depth distortion correction for the focused light-field cameras. In more details, Dansereau et al. [8] describe a decoding, calibration and rectification procedure for lenselet-based plenoptic cameras. They derive a homogeneous 5D intrinsic matrix relating each recorded pixel to its corresponding ray in 3D space, which can be applied to the commercial Lytro light field. Bok et al. [5] present a novel method of geometric calibration of micro-lens-based light-field cameras using line features. Instead of using sub-aperture images, they utilize raw images directly for calibration. Inspired by this work [8,22], based on the image formation for pinhole camera and the theory of light propagation, we model the imaging process of Lytro light field, and drive a novel intrinsic matrix.

In this paper, we derive a imaging model of Lytro light field camera, and present a calibration method using the checkboard grid. In theory, the microlens array can be modeled as an array of pinholes and the main len can be modeled as a thin len. According to these assumptions, the imaging model and the intrinsic matrix of Lytro light field camera are derived, which represent the relationship between each pixel and its corresponding light field. Then, based on the derived intrinsic matrix, we propose a calibration algorithm using the checkerboard grid for Lytro light field camera. At last, experiments have validated the correctness of our imaging model of Lytro light field camera and the effectiveness of the proposed calibration algorithm.

This paper is organized as follows: Sect. 2 reviews the image formation for pinhole camera and the representation of 4D light field. Section 3 presents a imaging model for Lytro light field camera and derive a intrinsic matrix. In Sect. 4, the calibration method based on the checkerboard grid for Lytro light field camera is described in detail. Experimental results are shown in Sect. 5. Finally, Sect. 6 presents some conclusion remarks.

2 Preliminaries

A bold letter denotes a vector or a matrix. Without special explanation, a vector is homogenous coordinates. In the following, we briefly review the image formation for pinhole camera in [14], and the representation of 4D light field in [11,17].

2.1 Pinhole Camera

Let the intrinsic parameter matrix of the pinhole camera be

$$\mathbf{K}_c = \begin{pmatrix} r_c f_c & s & u_0 \\ 0 & f_c & v_0 \\ 0 & 0 & 1 \end{pmatrix},$$

where r_c is the aspect ratio, f_c is the focal length, $(u_0, v_0, 1)^T$ denoted as $\mathbf{p}$ is the principal point, and s is the skew factor. Under the pinhole camera model, a space point $\mathbf{M}$ is projected to its image point $\mathbf{m}$ by [14]

$$\mathbf{m} = \lambda \mathbf{K}_c [\mathbf{R}, \mathbf{t}] \mathbf{M}, \tag{1}$$

where λ is a scalar, $[\mathbf{R}, \mathbf{t}]$ includes a rotation matrix and a translation, $\mathbf{K}_c$ is the intrinsic matrix.

2.2 The Representation of 4D Light Field

The light field is a vector function that describes the amount of light flowing in every direction through every in space. The direction of each ray is given by the 5D plenoptic function, and the magnitude of each ray is given by the radiance. Instead of 5D plenoptic function, Gortler et al. [11] sample and reconstruct a 4D function, which is called a Lumigraph. The Lumigraph is a subset of the complete plenoptic function.

Levoy and Hanrahan [17] propose a new representation of the light field, the radiance as a function of position and direction, in regions of space free of occluders (free space). In free space, the light field is a 4D, not a 5D function. As shown in Fig. 1, the coordinate system on the first plane is (u, v) and on the second plane is (s, t). An oriented line is defined by connecting a point on the uv plane to a point on the st plane. Thus, the light field can be represented through two parallel planes or a 4D vector $(u, v, s, t)^T$. That is, when the light passes through the first plane, record its position information $(u, v, 1)^T$; when the light passes through the second plane, record its direction information $(s, t)^T$.

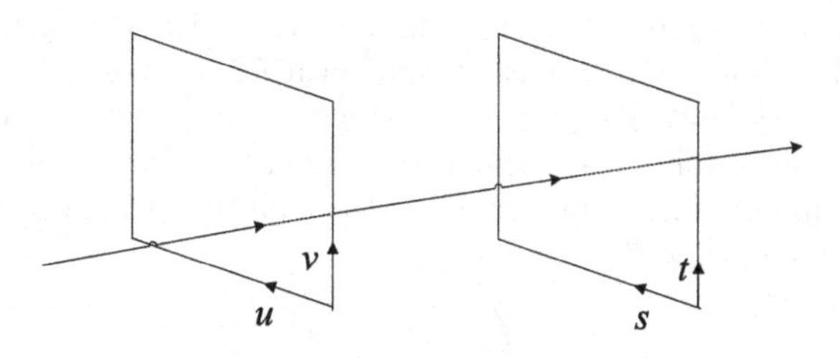

Fig. 1. The representation of 4D light field.

3 Imaging Model of Lytro Light Field Camera

In theory, though each pixel of a plenoptic camera integrates light from a volume, each pixel is approximated as integrating along a single ray [12]. In additional, the microlens array is modeled as an array of pinholes. Based on these assumptions, in this section, according to the theory of light propagation, we present the imaging model of Lytro light field camera, and derive its intrinsic matrix, which represents the relationship between each pixel and its corresponding light field.

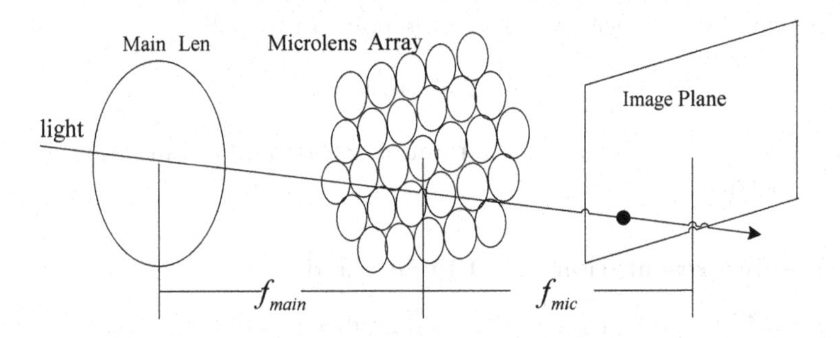

Fig. 2. The image formation for Lytro light field.

For Lytro light field camera, the microlens array is hexagonally packed. As shown in Fig. 2, microlens array is placed on the imaging plane of the conventional camera main lens. Then, optically sensitive device (CCD or CMOS) is parallel to the microlens array, and the distance between them is the focal length of the microlens array. At last, by the microlens array, the light is mapped to the corresponding area on the optically sensitive device.

Assume that the microlens array includes $W \times W'$ micronlens. Denote the image plane of one microlen (k, l) be $\mathbf{U}_{mic(k,l)}$. Set up the coordinate systems on the microlen (k, l) and its image plane $\mathbf{U}_{mic(k,l)}$ respectively, as shown in Fig. 3. In the image coordinate system, the image center $\mathbf{o}_{mic(k,l)}$ as the origin;the line through the origin and orthogonal to the image plane as $z_{mic(k,l)}$, which points to the optically sensitive device. The microlen coordinate system is parallel to the image coordinate system and the origin is the optical center $\mathbf{O}_{mic(k,l)}$.

With the development of camera manufacturing technology, there are same parameters for each microlen in the microlens array. What's more, the aspect ratio is 1, the skew factor is 0 and the principal point is close to the image center. Thus, the principal point can be estimated through the image center. Therefore, let the intrinsic matrix of the microlen be

$$\mathbf{K}_{mic} = \begin{pmatrix} f_{mic} & 0 & u^0_{mic} \\ 0 & f_{mic} & v^0_{mic} \\ 0 & 0 & 1 \end{pmatrix}.$$

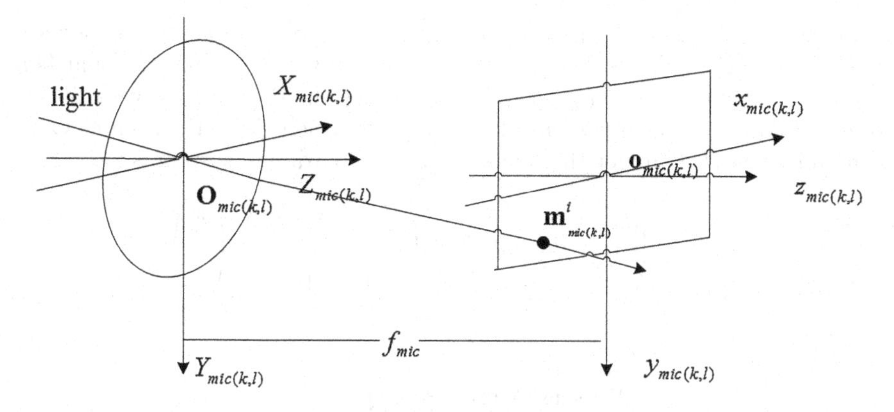

Fig. 3. The image format for the microlens (k, l).

According to the pinhole camera imaging model (Eq. (1)), under the microlen coordinate system, the coordinate of the image point $\mathbf{m}^i_{mic(k,l)}$ is

$$\mathbf{M}^i_{mic(k,l)} = \mathbf{K}_c^{-1}\mathbf{m}^i_{mic(k,l)}$$

$$= \begin{pmatrix} \frac{1}{f_{mic}} & 0 & -\frac{u^0_{mic}}{f_{mic}} \\ 0 & \frac{1}{f_{mic}} & -\frac{v^0_{mic}}{f_{mic}} \\ 0 & 0 & 1 \end{pmatrix} \begin{pmatrix} u_i \\ v_i \\ 1 \end{pmatrix}.$$

where $(u_i, v_i, 1)^T$ is the coordinate of the image point $\mathbf{m}^i_{mic(k,l)}$ in the image coordinate system $\{\mathbf{o}_{mic(k,l)} - x_{mic(k,l)}, y_{mic(k,l)}, z_{mic(k,l)}\}$.

Because the microlen (k, l) is parallel to its image plane $\mathbf{U}_{mic(k,l)}$, by the representation of 4D light field, the homogeneous representation of the corresponding light field of the image point $\mathbf{m}^i_{mic(k,l)}$ is

$$L_{m^i_{mic(k,l)}} = (0_{mic(k,l)}, 0_{mic(k,l)}, \frac{u_i}{f_{mic}}, \frac{v_i}{f_{mic}}, 1)^T, \tag{2}$$

where $\mathbf{O}_{mic(k,l)} = (0_{mic(k,l)}, 0_{mic(k,l)}, 1)^T$ records the location of the light, which is the optical center of the microlen (k, l), and $\mathbf{M}^i_{mic(k,l)}$ records the direction of the light.

The light field $L_{m^i_{mic(k,l)}}$ (Eq. (2)) is equivalent to

$$L_{m^i_{mic(k,l)}} = \begin{pmatrix} 0 & 0 & 0 & 0 & 0 \\ 0 & 0 & 0 & 0 & 0 \\ 0 & 0 & \frac{1}{f_{mic}} & 0 & -\frac{u^0_{mic}}{f_{mic}} \\ 0 & 0 & 0 & \frac{1}{f_{mic}} & -\frac{v^0_{mic}}{f_{mic}} \\ 0 & 0 & 0 & 0 & 1 \end{pmatrix} \begin{pmatrix} 0_{mic(k,l)} \\ 0_{mic(k,l)} \\ u_i \\ v_i \\ 1 \end{pmatrix}$$

$$= \mathbf{L}(\mathbf{K}_{mic}) \times \mathbf{n}^i_{mic(k,l)}. \tag{3}$$

In the following, we set up a microlens array coordinate system, as shown in Fig. 4. Denote the light field on the micronlens array as L_{mic} and the light field on the micronlen (k, l) as $L_{mic(k,l)}$. Assume that the distance between optical centers of two adjacent microlens along X_{mic} is d, then the optical center $\mathbf{O}_{(k,l)}$ of the microlen (k, l) under the microlens array coordinate system is

$$\mathbf{O}_{(k,l)} = \begin{pmatrix} (k - \frac{W'+1}{2})d \\ (l - \frac{W+1}{2})d \\ 1 \end{pmatrix} = \begin{pmatrix} d & 0 & -\frac{W'+1}{2}d \\ 0 & d & -\frac{W+1}{2}d \\ 0 & 0 & 1 \end{pmatrix} \begin{pmatrix} k \\ l \\ 1 \end{pmatrix}. \tag{4}$$

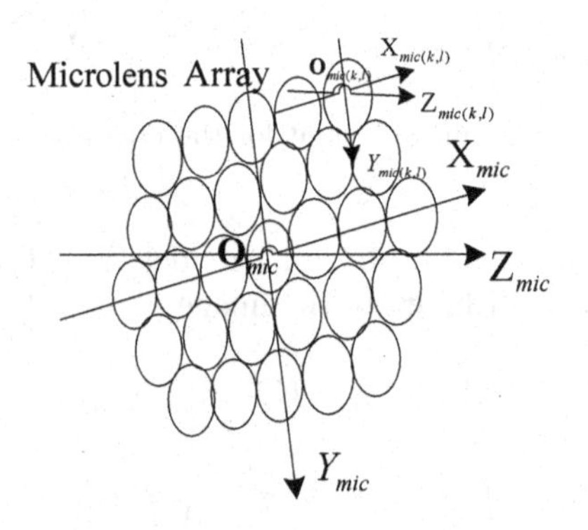

Fig. 4. The coordinate system on the microlens array.

Therefore, in the light field L_{mic}, the corresponding light field of the image point $\mathbf{m}^i_{mic(k,l)}$ changes to

$$L_{m^i_{mic}} = V_{mic}(\mathbf{O}_{mic(k,l)}) \times \mathbf{L}(\mathbf{K}_{mic}) \times \mathbf{n}^i_{mic(k,l)}, \tag{5}$$

where

$$\mathbf{V}_{mic}(\mathbf{O}_{mic(k,l)}) = \begin{pmatrix} 1 & 0 & 0 & 0 & (k - \frac{W'+1}{2})d \\ 0 & 1 & 0 & 0 & (l - \frac{W+1}{2})d \\ 0 & 0 & 1 & 0 & 0 \\ 0 & 0 & 0 & 1 & 0 \\ 0 & 0 & 0 & 0 & 1 \end{pmatrix}.$$

Denote $\hat{\mathbf{n}}^i_{mic(k,l)} = (k, l, u_i, v_i, 1)^T$, by Eqs. (3) and (4), Eq. (5) is equivalent to

$$L_{m^i_{mic}} = \begin{pmatrix} d & 0 & 0 & 0 & -\frac{W'+1}{2}d \\ 0 & d & 0 & 0 & -\frac{W+1}{2}d \\ 0 & 0 & \frac{1}{f_{mic}} & 0 & -\frac{u_{mic}}{f_{mic}} \\ 0 & 0 & 0 & \frac{1}{f_{mic}} & -\frac{v_{mic}}{f_{mic}} \\ 0 & 0 & 0 & 0 & 1 \end{pmatrix} \begin{pmatrix} k \\ l \\ u_i \\ v_i \\ 1 \end{pmatrix} = \mathbf{K}_{light} \times \hat{\mathbf{n}}^i_{mic(k,l)}, \tag{6}$$

where k, l are the indices of the microlens through which a light passes, and $\mathbf{K}_{light}$ is called the intrinsic matrix of Lytro light field camera.

4 Calibration of Lytro Light Field

In this section, based on the intrinsic matrix $\mathbf{K}_{light}$ of Lytro light field camera, we propose a calibration method using the checkerboard pattern with known dimensions.

For the microlen (k, l), let $\mathbf{m}^{ij}_{mic(k,l)}$ be the extracted corner points on the checkerboard image plane $\mathbf{U}_{mic(k,l)}$, $i = 1, 2, \ldots, N_{mic(k,l)}$, $j = 1, 2, \ldots, \hat{N}$, $k = 1, 2, \ldots, W$ and $l = 1, 2, \ldots, W'$. $N_{mic(k,l)}$ refers to the number of extracted corner points on the checkerboard image plane $\mathbf{U}_{mic(k,l)}$, and $\hat{N}$ refers to the number of the checkerboard images taken by Lytro light field. According to the camera calibration method proposed in [23], we can obtain the $W \times W' \times \hat{N}$ homography matrix $\mathbf{H}^{j}_{mic(k,l)}$, and the $W \times W' \times \hat{N}$ equations about the intrinsic parameters of the microlen $\mathbf{K}_{mic}$. Then, through SVD decomposition, the intrinsic matrix $\mathbf{K}_{mic}$ can be estimated.

What's more, based on the estimated intrinsic matrix $\mathbf{K}_{mic}$, we can compute the rotation matrix $\mathbf{R}^{j}_{mic(k,l)}$ and the translation vector $\mathbf{t}^{j}_{mic(k,l)}$. Therefore, the distance d between optical centers of two adjacent microlens along X_{mic} can be estimated as follows:

$$d = \frac{1}{\hat{N} \times W \times W'} \sum_{j=1}^{\hat{N}} \sum_{l=1}^{W'} \sum_{k=1}^{W-1} \|\mathbf{t}^{j}_{mic(k,l)} - \mathbf{t}^{j}_{mic(k+1,l)}\|. \tag{7}$$

The method to calibrate the Lytro light field is outlined as follows:

Step 1: Take a few images $\hat{N}$ of the checkerboard under different orientations by moving either the plane or Lytro light field camera;
Step 2: For each checkerboard image, extract the corner points $\mathbf{m}^{ij}_{mic(k,l)}$, where $i = 1, 2, \ldots, N_{mic(k,l)}$, $j = 1, 2, \ldots, \hat{N}$, $k = 1, 2, \ldots, W$ and $l = 1, 2, \ldots, W'$;
Step 3: Estimate the intrinsic matrix $\mathbf{K}_{mic}$, the rotation matrix $\mathbf{R}^{j}_{mic(k,l)}$ and the translation vector $\mathbf{t}^{j}_{mic(k,l)}$ by the method in [23];
Step 4: Decode the image points $\mathbf{m}^{ij}_{mic(k,l)}$ into 4D light field $\hat{\mathbf{n}}^{i}_{mic(k,l)}$ by the method in [8];
Step 5: Obtain the distance d between optical centers of two adjacent microlens along X_{mic} by Eq. (7);
Step 6: Calibrate the intrinsic matrix of Lytro light field $\mathbf{K}_{light}$ by Eq. (6);
Step 7: Initial the light field camera pose $\mathbf{T}^{j} = (\mathbf{R}^{j}, \mathbf{t}^{j})$ as follows:

$$\mathbf{R}^{j} = \frac{1}{W \times W'} \sum_{l=1}^{W'} \sum_{k=1}^{W} \mathbf{R}^{j}_{mic(k,l)}, \mathbf{t}^{j} = \frac{1}{W \times W'} \sum_{l=1}^{W'} \sum_{k=1}^{W} \mathbf{t}^{j}_{mic(k,l)}; \tag{8}$$

Step 8: Optimize the intrinsic matrix $\mathbf{K}_{light}$ of Lytro light field and the camera pose $\mathbf{T}^{j}$, $j = 1, 2, \ldots, \hat{N}$, through minimizing the following object function:

$$argmin_{\mathbf{K}_{light}, \mathbf{T}} = \sum_{j=1}^{\hat{N}} \sum_{i=1}^{N_{mic(k,l)}} \|\phi^{i}_{c}(\mathbf{K}_{light}, \mathbf{T}^{j}), \mathbf{P}^{ij}\|^{ray},$$

where $\mathbf{P}^{ij}$ refers to the known 3D feature location on the checkerboard grid, the light field camera pose $\mathbf{T} = (\mathbf{T}^1, \mathbf{T}^2, \ldots, \mathbf{T}^{\hat{N}})^T$, and $\|\|^{ray}$ is the ray reprojection error described in [8].

5 Experiments

In this section, we perform a number of experiments with real images to evaluate the correctness of the imaging model of Lytro light field camera and the performance of our calibration algorithm.

Firstly, make the checkerboard grid, which the grid space is 7.22 mm × 7.22 mm and the board size is 19 × 19. As shown in Fig. 5, the checkerboard grid images are captured under different orientations by moving the plane or Lytro light field camera, which are downloaded from http://www-personal.acfr.usyd.edu.au/ddan1654/PlenCalCVPR2013DatasetC.zip. There are 12 checkerboard grid images are used to calibrate the intrinsic matrix of Lytro light field camera.

Fig. 5. Twelve real images of the checkerboard captured by Lytro light field.

Next, using the calibration algorithm proposed in Sect. 4, the intrinsic matrix $\mathbf{K}_{light}$ of Lytro light field and the camera pose $\mathbf{T}^j$, $j = 1, 2, \ldots, \hat{N}$ are estimated. Figure 6 shows the estimated light field camera pose. In Fig. 7, the left image refers to the initial estimation of light field camera pose by Eq. (8), and the right image refers to the optimized estimation of Lytro camera pose through minimizing the ray reprojection error.

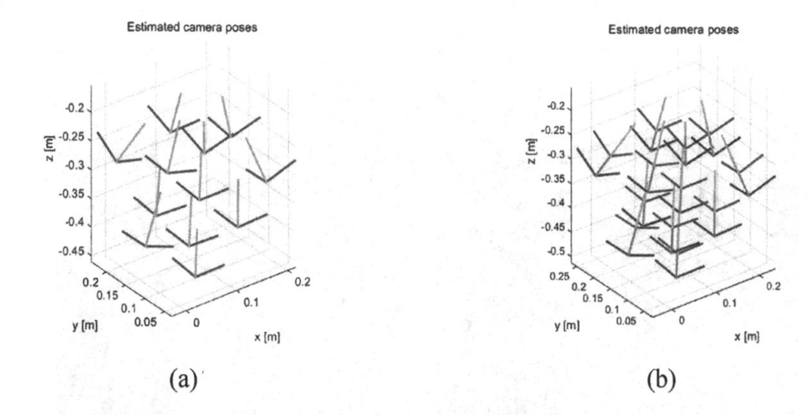

(a) (b)

Fig. 6. (a): The gray refers to the initial estimation of Lytro camera pose; (b): the green refers to the optimized estimation of Lytro camera pose. (Color figure online)

At last, integrating the lens distortion model proposed in [8], the estimated intrinsic matrix $\mathbf{K}_{light}$ is used to rectify the light field images, as shown in Figs. 7 and 8. Figure 7(a) shows the initial image of the checkerboard grid captured by Lytro light field camera, and Fig. 7(b) shows the rectified image of the checkerboard grid based on the estimated intrinsic matrix $\mathbf{K}_{light}$. Figure 8(a) is the initial light field image captured in the natural environment, and Fig. 8(b) is the rectified real image through the estimated intrinsic matrix $\mathbf{K}_{light}$. From Figs. 7 and 8, especially from the enlarged part in Fig. 8, we can be seen that the initial light field image is well rectified. Therefore, our imaging model of Lytro light field is correct and the proposed calibration algorithm is effective.

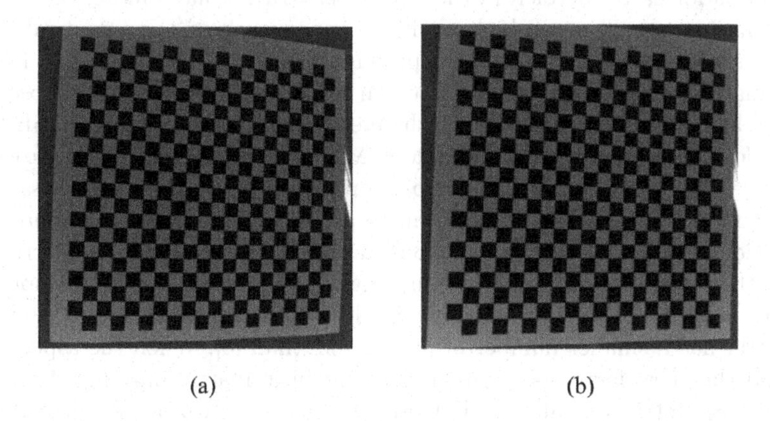

(a) (b)

Fig. 7. (a) The initial image of the checkerboard grid under Lytro light field camera, (b) the rectified image of the checkerboard grid using the estimated intrinsic matrix.

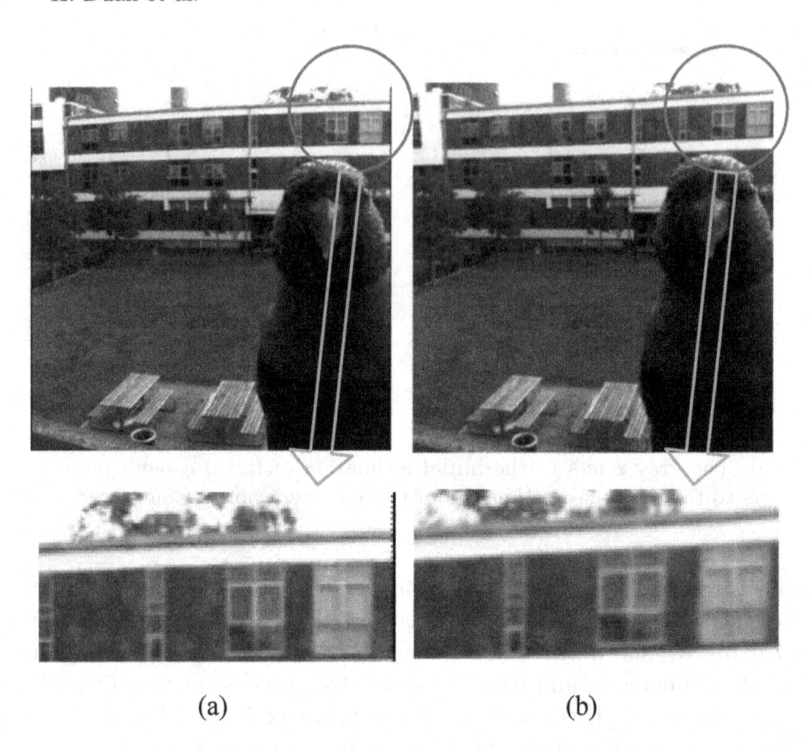

Fig. 8. (a) The initial light field image captured in the natural environment, (b) the rectified real image through the estimated intrinsic matrix.

With the development of camera manufacturing technology, there are same parameters for each microlen in the microlens array. What's more, the some of the intrinsic parameters can be known in advance, such as: the aspect ratio r_{mic} is 1; the skew factor s_{mic} is 0 and the principal point $\mathbf{p}_{mic} = (u^0_{mic}, v^0_{mic}, 1)^T$ can be estimated by the image center. Based on these assumptions, our proposed calibration algorithm is compared with the method proposed in [8]. The SSE (Sum of the Squares Error) and RMSE (Root Mean Square Error) ray reprojection errors are shown in Table 1. The comparison results verify the correctness of our imaging model for Lytro light field camera and the effectiveness of the proposed calibration algorithm. Although our SSE and RMSE ray reprojection errors are almost the same as those of [8], our imaging model and calibration method have two advantages. That is, our imaging model is easy to understand and the intrinsic matrix has a smaller number of parameters than [8]. When the aspect ratio is 1 and the skew factor is 0, our derived intrinsic matrix only has 4 parameters. But the intrinsic matrix of Lytro light field camera proposed in [8] has 12 parameters.

Table 1. The comparison results.

Our		Dansereau
Initial (SSE)	$2.43944\,\text{mm}^2$	$2.47627\,\text{mm}^2$
Rectified (SSE)	$2.42449\,\text{mm}^2$	$2.4349\,\text{mm}^2$
Initial (RMSE)	0.107202 mm	0.108008 mm
Rectified (RMSE)	0.106873 mm	0.107102 mm

6 Conclusions

In this paper, we model the image processing of Lytro light field and propose a calibration algorithm based on the checkerboard grid. Firstly, according to the imaging process of Lytro light field, the relationship between each pixel and its corresponding light field are derived. Secondly, the homogenous intrinsic matrix of Lytro light field is established. Secondly, based on the intrinsic matrix, a calibration algorithm by the checkerboard grid for Lytro light field camera is proposed. Then, there are 12 checkerboard grid images under different orientations are used to estimate the intrinsic matrix. At last, the camera pose and the rectified images are obtained based on the estimated intrinsic matrix to evaluate the correctness of our imaging model and the effectiveness of our calibration method. The experimental results have shown that our imaging model of Lytro light field camera is correct and the proposed calibration algorithm is effective.

Acknowledgments. The authors of this paper are members of Shanghai Engineering Research Center of Intelligent Video Surveillance. This work is sponsored by the National Natural Science Foundation of China (61403084, 61402116); by the Project of the Key Laboratory of Embedded System and Service Computing, Ministry of Education, Tongji University (ESSCKF 2015-03); by the Shanghai Rising-Star Program (17QB1401000); and by the special funds for the basic R&D business expenses of the central level public welfare scientific research institutions (C17384) "The Construction of Standard Video Dataset and Intelligent Video Evaluation Platform".

References

1. The lytro camera. https://www.lytro.com
2. The raytrix camera. https://www.rautrox.de
3. Adelson, E., Wang, J.: Single lens stereo with a plenoptic camera. IEEE Pattern Anal. Mach. Intell. **14**(2), 99–106 (1992)
4. Bishop, T., Favaro, P.: The light field camera: extended depth of field, aliasing and superresolution. IEEE Pattern Anal. Mach. Intell. **34**, 972–986 (2012)
5. Bok, Y., Jeon, H., Kweon, I.: Geometric calibration of micro-lens-based light-field cameras using line features. In: IEEE European Conference on Computer Vision, pp. 47–61 (2014)
6. Dansereau, D., Bongiorno, D., Pizarro, O., Williams, S.: Light field image denoising using a linear 4D frequency-hyperfan all-in-focus filter. In: IEEE International Society for Optics and Photonics (2013)

7. Dansereau, D., Mahon, I., Pizarro, O., Williams, S.: Plenoptic flow: closed-form visual odometry for light field cameras. In: IEEE Intelligent Robots and Systems, pp. 4455–4462 (2011)
8. Dansereau, D., Pizarro, O., Williams, S.: Decoding, calibration, and rectification for lenselet-based plenoptic cameras. In: IEEE Computer Vision and Pattern Recognition, pp. 1027–1034 (2013)
9. Georgiev, T., Lumsdaine, A.: Reducing plenoptic camera artifacts. Comput. Graph. Forum **29**(6), 1955–1968 (2010)
10. Georgiev, T., Lumsdaine, A., Goma, S.: Plenoptic principal planes. In: Computational Optical Sensing and Imaging (2011)
11. Gortler, J., Grzeszczuk, R., Szeliski, R., et al.: The lumigraph. In: Annual Conference on Computer Graphics and Interactive Techniques, pp. 43–54 (1996)
12. Grossberg, M., Nayar, S.: The raxel imaging model and ray-based calibration. Int. J. Comput. Vis. **61**, 119–137 (2005)
13. Harris, M.: Focusing on everything - light field cameras promise an imaging revolution. Spectrum **4**, 44–50 (2012)
14. Hartley, R., Zisseman, A.: Multiple view geometry in computer vision. Cambridge University Press, UK (2003)
15. Johannsen, O., Heinze, C., Goldluecke, B., Perwaß, C.: On the calibration of focused plenoptic cameras. In: Grzegorzek, M., Theobalt, C., Koch, R., Kolb, A. (eds.) Time-of-Flight and Depth Imaging. Sensors, Algorithms, and Applications. LNCS, vol. 8200, pp. 302–317. Springer, Heidelberg (2013). https://doi.org/10.1007/978-3-642-44964-2_15
16. Lanman, D.: Mask-based light field captuer and display. Ph.D. thesis, Brown University (2012)
17. Levoy, M., Hanraahan, P.: Light field rendering. In: Annual Conference on Computer Graphics and Interactive Techniques, pp. 31–42 (1996)
18. Ng, R., Levoy, M., Bredof, M., Horowitz, M., Hanrahan, P.: Light field photography with a hand-held plenoptic camera. Computer Science Technical repot (2005)
19. Svoboda, T., Martinec, D., Pajdla, T.: A convenient multicamera self-calibration for virtual environments. Teleoperators Virtual Environ. **14**, 407–442 (2005)
20. Wilburn, B., Joshi, N., Vaish, V., Talvala, E., Antunez, E., Barth, A., Adams, A., Horowitz, M., Levoy, M.: High performance imaging using large camera arrays. Trans. Graph. **24**, 765–776 (2005)
21. Xu, Z., Ke, J., Lam, E.: High-resolution lightfield photography using two masks. Opt. Express **20**, 10971–10982 (2012)
22. Zhang, X., Li, C.: Caliration and imaging model of light field camera with microlens array. Acta Opt. Sin. **34**, 95–107 (2014)
23. Zhang, Z.: A flexible new technique for camera calibration. Trans. Pattern Anal. Mach. Intell. **22**, 1330–1334 (2000)

Bayesian Inference Based Choquet Integral Method

Chunxiao Ren[1,2(✉)]

[1] Shandong Science and Technology Exchange Center, No. 607 Shunshua Rd, High-Tech Development Zone, Jinan 250101, China
alanren@163.com

[2] Shandong Engineering Research Center of Special Optical Fiber and Cable, No. 1 Optical Fiber and Cable Industrial Park, Yanggu, China

Abstract. The non-additive measure provides a useful tool for many problems in different communities. The Choquet integral has been successfully used for many applications. However, their applicability is restricted due to the exponential computational complexity. In this paper, a novel polynomial method is proposed to solve the parameter estimation problem for Choquet integral. The basic idea of our method is to regard the problem as a sequential one at first; and then we use Bayesian inference method to solve the problem. Using our method, the computational complexity for the non-additive measure is reduced from $O((n+K)*2^{2n})$ to $K*O(n^2logn)$. Specifically, the semantic information of the 2^n variables is not lost. This method can be utilized to real-time applications. We provide statistical performance guarantees for the proposed method. As a real-world application, cross-layer design of wireless multimedia networks is optimized by the proposed method. The experiments show that the performance achieved by using this method is better than that of traditional methods.

Keywords: Non-additive measures · Choquet integral
Bayesian inference · Maximum likelihood estimation · Cross-layer

1 Introduction

Within the fields of economics, finance, computer science and decision theory there is an increasing interest in the problem how to replace the additivity property of probability measures by that of monotonicity or, more generally, a non-additive measure. Several types of integrals with respect to non-additive

C. Ren—The author would like to thank Dr. Song Ci, Dr. Haifeng Guo, Dr. Wei An, Dr. Dalei Wu, Dr. Haiyan Luo and Ms. Yuxiao Wu for their helpful comments and valuable suggestions which improved the quality of this paper. This paper is supported by Shandong Outstanding Young Scientists Foundation under Grant No. BS2013DX047 and Shandong Key Research and Development Project under Grant Nos. 2016GGX101042, 2016JMRH0331.

J. Yang et al. (Eds.): CCCV 2017, Part III, CCIS 773, pp. 401–415, 2017.
https://doi.org/10.1007/978-981-10-7305-2_35

measures, also known as fuzzy measures [30], were developed for different purposes in various works [2,14,21,32,37]. The Choquet integral, as a popular representation of the non-additive measure, has been successfully used for many applications such as information fusion [7], multiple regressions [33], classification [15,38], multicriteria decision making [9], image and pattern recognition [20,39], and data modeling [12].

In general, the basic idea to solve the non-additive model based on Choquet integral is a two-step procedure. The first step is to reduce the non-linear multi-regression model to the traditional linear multi-regression model by converting each n-dimensional vector to a 2^n-dimensional one, which is defined over the powerset of attributes; and thus, the second step is to solve the linear model by using various numerical indices and optimization method [27]. So far, numerous related works have been developed [16,19,22,23,26].

However, the numbers of variables involved in solving the non-additive model increases exponentially with n and so will the computational time. Thus, the use of non-additive measure in practical applications is clearly curbed by this exponential complexity [24]. Several approaches to deal with this problem are known. The notion of k-additivity proposed by Grabisch [6,10,11,13] enables to find a trade-off between the complexity of representation and the richness of the modeling. In [40], Yan developed a hierarchical Choquet integral to model the data. These techniques, however, are all working on solving the complexity problem through discarding or integrating the 2^n variables to smaller ones. The results obtained through these techniques are all approximate solutions. On the other hand, these results cannot be mapping back to the original 2^n-dimensional space.

To solve the computational complexity problem, several approaches are known as follows. The notion of k-additivity proposed by Grabisch [6,10,11,13] enables to find a trade-off between the complexity of representation and the richness of the modeling. In [40], Yan developed a hierarchical Choquet integral to model the data. These techniques, however, are all working on solving the complexity problem through converting the 2^n-dimensional space to a smaller but less precise one. In the original 2^n-dimensional space, the practical Choquet integral problem is still not solved.

In this paper, we propose a novel polynomial method to solve the parameter estimation problem for Choquet integral. The basic idea of our method is to regard the problem as a sequential one at first; and then we use Bayesian inference method to solve the problem. The parameters of Choquet integral are updated after each data sample presentation. This update procedure is finished in a low-dimensional space mapping from original one. We provide statistical performance guarantees for the proposed method. The experiments show that the performance achieved by using this method is better than that of traditional methods. A special benefit of our method is that the 2^n variables are retained. So the semantic information of the 2^n variables is not lost. Based on our previous works [19,25,35,36], this sequential method is very suitable to real-world applications.

The main contributions of the paper can be summarized as follows. (1) A novel polynomial method is proposed to solve the parameter estimation problem for Choquet integral. Using our method, the computational complexity for the non-additive measure is reduced from $O((n+K)*2^{2n})$ to $K*O(n^2 logn)$. (2) As a case of combining sequential concept and Choquet integral, this method can be utilized to real-time applications. (3) As a real-world application, cross-layer design of wireless multimedia networks is optimized by the proposed method.

The remainder of the paper is organized as follows. In Sect. 2, we introduce the preliminaries which provides a mathematical setting for fuzzy measures. Section 3 gives the Bayesian inference based algorithm to estimate the parameters of large-scale Choquet integral. A benchmark dataset test and a real-world application are provided to illustrate the proposed algorithms in Sect. 4. Finally, summary and future works are drawn in Sect. 5.

2 Choquet Integral

2.1 Basic Definitions

Let us consider a multi-feature problem described by a finite set of alternatives $A = \{a_1, a_2, \ldots\}$ and a finite set of features $F = \{f_1, f_2, \ldots\}$. Each alternative $a_i \in A$ can be associated with a profile $(\mu_1, \ldots, \mu_n) \in \mathcal{R}^n$, where, for any $f_j \in F, \mu_j \in \mathcal{R}$ represents the utility of a_i related to feature j.

In general, one can compute an aggregation score $H(f, w)$ by a linear basis function model which takes into account the weights of importance of the feature.

$$H(f, w) = \sum_{j=1}^{n} w_j \phi_j(f) = w^T \phi(f) \tag{1}$$

where $w = (w_1, \ldots, w_n)^T$ and $\phi = (\phi_1, \ldots, \phi_n)^T$. The weight w is also regarded as a Lebesgue measure on a singleton f. However, the assumption of independence among features is rarely verified. In order to take into account a flexible representation of complex interaction phenomena among features, it is useful to substitute to the weight vector w a non-additive set function μ on $N = 1, \ldots, n$, called Choquet capacity [34], allowing defining a weight not only on the importance of each feature, but also on the importance of each subset of features. It is clearly more powerful than the Lebesgue integral model since Lebesgue integral thus becomes a special case of the Choquet integral model. The Choquet integral is defined as follows [4,7].

Definition 1. For every space Ω and algebra $\mathcal{A}$ of subset of Ω, a set-function $\mu : \mathcal{A} \to \mathcal{R}$ is called a capacity if it satisfies the following:

(i) $\mu(\emptyset) = 0, \mu(\Omega) = 1,$
(ii) $\forall A, B \in \mathcal{A} : A \subseteq B \Rightarrow \mu(A) \leq \mu(B)$
Furthermore, a capacity μ on Ω is said to be additive if
(iii) $\forall A, B \in \mathcal{A} : \mu(A \cup B) \geq \mu(A) + \mu(B) - \mu(A \cap B)$

For each subset of feature $A \subseteq \Omega$, the number $\mu(A)$ can be regarded as the weight or the importance of A.

Definition 2. If $\phi : \Omega \to \mathcal{R}$ is a bounded A-measurable function and μ is any capacity on Ω, we define the Choquet integral of ϕ with respect to μ to be the number

$$\int_{\Omega} \phi(f)\, d\mu(f) = \int_{0}^{\infty} \mu(\{f \in \Omega : \phi(f) \geq \alpha\})\, d\alpha \\ + \int_{-\infty}^{0} [\mu(\{f \in \Omega : \phi(f) \geq \alpha\}) - 1]\, d\alpha \tag{2}$$

where the integrals are taken in the sense of Riemann. In particular, if Ω is finite and $\phi(f_1) \geq \phi(f_2) \geq \ldots \geq \phi(f_n)$, then

$$\int_{\Omega} \phi(f)\, d\mu(f) = \sum_{i=1}^{n-1} (\phi(f_i) - \phi(f_{i+1}))\, \mu(\{f_1, \ldots, f_i\}) \\ + \phi(f_n) \tag{3}$$

Definition 3. Let ϕ be a function from $X = \{x_1, \ldots x_n\}$ to $[0, 1]$. Let $\{x_{\sigma(1)}, \ldots, x_{\sigma(n)}\}$ denote a reordering of the set X such that $0 \leq \phi(x_{\sigma(1)}) \leq \ldots \leq \phi(x_{\sigma(n)})$, and let $A_{(i)}$ be a collection of subsets defined by $A_{(i)} = \{x_{\sigma(i)}, \ldots, x_{\sigma(n)}\}$. Then, the discrete Choquet integral of ϕ with respect to a fuzzy measure μ on X is defined as

$$\begin{aligned} C_{\mu}(\phi) &= \sum_{i=1}^{n} \mu(A_{(i)}) \left(\phi(x_{(i)}) - \phi(x_{(i-1)})\right) \\ &= \sum_{i=1}^{n} \phi(x_{(i)}) \left(\mu(A_{(i)}) - \mu(A_{(i+1)})\right) \end{aligned} \tag{4}$$

where we take $\phi(x_{(0)}) = 0$, $A_{(n+1)} = 0$, and $x_{(i)} = x_{\sigma(i)}$.

2.2 Existing Method

Based on the Definitions 1–3, the relationship between the Choquet integral of ϕ and the capacity μ can also be described by a new nonlinear multi-feature regression model [5]:

$$C_{\mu}(\phi) = e + \int_{\Omega} \phi(f)\, d\mu(f) + \mathcal{N}(0, \delta^2) \tag{5}$$

where e is a regression constant, ϕ is an observation of $X = \{x_1, x_2, \ldots, x_n\}$, $N(0, \delta^2)$ is a normally distributed random perturbation with expectation 0 and variance δ^2, and δ^2 is the regression residual error. The Choquet integral problem is reduced to a traditional linear multi-regression model. At the same time, specifying a general fuzzy measure requires specification of $2^n - 1$ parameters, which is clearly exponential.

For solving this linear multi-regression model, usually we need to define a loss function. Through minimizing this loss function, the parameter of the linear model can be derived. Given classes $C_1, \ldots, C_n$, Grabisch [17,18] proposed a mean squared error (MSE) criterion, where the difference between desired outputs t_i for $i = 1, \ldots, n$ and the actual outputs $C_\mu(\phi)$ is minimized under constraints [1]. The loss function L is

$$L^2 = \sum_{\phi \in C_1} \left(C_\mu(\phi) - t_1\right)^2 + \cdots + \sum_{\phi \in C_n} \left(C_\mu(\phi) - t_n\right)^2 \tag{6}$$

Given the observation data, the optimal regression coefficients μ can be determined by using regression methods. For example, in order to make the loss function minimal, the least square method, as a regression method, is used for determine $\mu_k (k = 1, 2, \ldots, 2^n - 1)$ [28].

We use a maximum likelihood estimation (MLE) [8] method to fulfill the least square method. In fact, we define the likelihood function by using Gaussian distribution:

$$p(t \,|\phi, \mu, \beta) = \mathcal{N}\left(t \,\middle|C_\mu(\phi), \beta^{-1}\right) \tag{7}$$

where t is desired outputs given by a deterministic function $C_\mu(\phi)$, and β is the precision of Gaussian random variable. Consider an observation set of inputs ϕ with corresponding desired outputs $t_1, \ldots, t_n$,

$$p(t \,|\phi, \mu, \beta) = \prod_{n=1}^{N} \mathcal{N}\left(t_n \,\middle|\mu^T \phi_n, \beta^{-1}\right) \tag{8}$$

where t and μ are column vectors

Solving for μ using MLE method, we obtain

$$\mu_{ML} = \left(\Phi^T \Phi\right)^{-1} \Phi^T \mathrm{t} \tag{9}$$

Here Φ is an $N \times M$ matrix whose elements are given by $\Phi_{nj} = \phi_j(f_n)$ [31], so that

$$\Phi = \begin{pmatrix} \phi_0(f_1) & \phi_1(f_1) & \cdots & \phi_{M-1}(f_1) \\ \phi_0(f_2) & \phi_1(f_2) & \cdots & \phi_{M-1}(f_2) \\ \vdots & \vdots & \ddots & \vdots \\ \phi_0(f_N) & \phi_1(f_N) & \cdots & \phi_{M-1}(f_N) \end{pmatrix} \tag{10}$$

To summarize, the algorithm to be employed by MLE method is as follows:

For time-complexity analysis, we assume that there are n feature and K samples. The Choquet integral space is 2^n-dimensional. In addition, we assume that the complexity of the inversion of a matrix $n \times n$ is $\Omega(n^2 logn)$ [29]. We present the time complexity using pseudo-code analysis.

Algorithm 1. MLE-based Choquet integral

Input: matrix Φ, target variable t
Output: μ

- Step 1: Compute the $\Phi^T\Phi$
- Step 2: Compute the Φ^Tt
- Step 3: Compute the $(\Phi^T\Phi)^{-1}$
- Step 4: Compute the $(\Phi^T\Phi)^{-1}\Phi^T$t

$$\begin{array}{ll} O(K(2^n)^2) & \text{for } \Phi^T\Phi \\ O(K*2^n) & \text{for } \Phi^T\text{t} \\ \Omega((2^n)^2 log 2^n) & \text{for } (\Phi^T\Phi)^{-1} \\ O((2^n)^2) & \text{for } (\Phi^T\Phi)^{-1}\Phi^T t \end{array}$$

We can rewrite this time complexity as

$$O(K*2^{2n} + K*2^n + log2*n*2^{2n} + 2^{2n})$$

Thus, we have that the time complexity for MLE-based method is $O((n+K)*2^{2n})$. Obviously, the numbers of variables involved in solving the non-additive model increases exponentially with n and so will the computational time. In fact, the use of non-additive measure in practical applications is clearly curbed by this exponential complexity [24]. As an important challenge, the practical application of the non-additive measure has been plagued by real-time computational problems for a long time.

On the other hand, the traditional method, such as the MLE-based method, which involves processing the entire samples in one go, can be inappropriate for some real-time applications in which the data observations are arriving in a continuous stream, and predictions must be made before all of the samples are acquired. For a large scale dataset, it may be suitable to use sequential algorithms in which the data samples are considered one by one, and the model parameters updated after each such presentation.

3 Our Method

A general Choquet integral is defined by $2^n - 1$ coefficients. In order to reduce the computational complexity, we developed a Bayesian inference based Choquet integral (BIBCI) method.

Now we introduce some theorems needed in this work, and these theorems also can be found in [3].

Given a marginal Gaussian distribution for ϕ and a conditional Gaussian distribution for $C_\mu(\phi)$ given ϕ in the form

$$p(\phi) = \mathcal{N}(\phi|\varphi, P^{-1}) \tag{11}$$

$$p(C_\mu(\phi)|\phi) = \mathcal{N}(C_\mu(\phi)|\mu\phi + b, Q^{-1}) \tag{12}$$

where φ, μ, and b are parameters governing the means, and P and Q are precision matrices. So the marginal distribution of $C_\mu(\phi)$ and the conditional distribution of ϕ given $C_\mu(\phi)$ are given by

$$p(C_\mu(\phi)) = \mathcal{N}\left(C_\mu(\phi)\left|\mu\varphi + b, Q^{-1} + \mu P^{-1}\mu^T\right.\right) \tag{13}$$

$$p(\phi|\,C_\mu(\phi)) = \mathcal{N}\left(\phi\left|\sum\left\{\mu^T Q\left(C_\mu(\phi) - b\right) + P\varphi\right\}, \Sigma\right.\right) \tag{14}$$

where

$$\Sigma = \left(P + \mu^T Q \mu\right)^{-1} \tag{15}$$

Based on (3)–(7), we can derive the posterior distribution of μ as

$$p(\mu\,|t) = \mathcal{N}(\mu\,|\varphi_N, S_N) \tag{16}$$

where

$$\varphi_N = S_N\left(S_0^{-1}\varphi_0 + \beta\Phi^T t\right) \tag{17}$$

$$S_N^{-1} = S_0^{-1} + \beta\Phi^T\Phi \tag{18}$$

where φ_N is prior means, S_N is prior precision matrices, β is prior uncertainty. Because the posterior distribution is Gaussian, its mode coincides with its mean. Thus the maximum posterior weight vector is simply given by $\mu_{MAP} = \varphi_N$.

Suppose that we have already observed N data points; thereby the posterior distribution over μ is given. This posterior can be regarded as the prior for the next observation. By considering an additional data point (ϕ_{N+1}, t_{N+1}) the resulting posterior distribution can be derived based on (8)–(10)

$$p(\mu\,|t_{N+1}, \phi_{N+1}, \varphi_N, S_N) = \mathcal{N}(\mu\,|\varphi_{N+1}, S_{N+1}) \tag{19}$$

with

$$S_{N+1}^{-1} = S_N^{-1} + \beta\phi_{N+1}\phi_{N+1}^T \tag{20}$$

$$\varphi_{N+1} = S_{N+1}\left(S_N^{-1}\varphi_N + \beta\phi_{N+1}t_{N+1}\right) \tag{21}$$

Since the non-additive measure in the Choquet model is defined over the powerset Ω, the reduction step basically aggregates the observed data of individual features to the observation on sets. It is clear that there are only $n (n \ll 2^n)$ non-zero elements in each 2^n-dimensional vector data. On the other hand, it is not all necessary to use the 2^n coefficients for building a Choquet integral model. In most cases, we only need a very small part of these coefficients to compute the Choquet integral. More importantly, some coefficients may not be used at all in many applications. However, for traditional Choquet integral method, the model cannot be built if we do not figure out all the coefficients.

We now develop a new approximation algorithm to reduce the computational complexity. Because most information about the relationship among features is contained in only a small fraction of all the coefficients, we can use the n non-zero element in each sample to adjust the related parameter. For each sample, the parameters of Choquet integral related with non-zero element are updated through Bayesian inference method. The basic idea behind the proposed method is that it is far easier to consider a sequence of conditional distributions than it is to obtain the marginal by integration of the joint density. Based on (19)–(21), we can update the parameters of Choquet integral model by Bayesian inference. Though only n parameters are modified in each training process, the 2^n-dimensional parameter vector will tend to be reasonable after certain rounds of training. Meanwhile, this algorithm only updates the coefficients that the training data support. It is possible to acquire a reasonable model which maybe has a few coefficients much less than 2^n. It then executes the following algorithm.

We present the time complexity using pseudo-code analysis. We can rewrite this time complexity as

$$K * O(n + n^2 + n^2 + n^2 logn + n^2 + n + n + n + n)$$

Thus, the time complexity for our method is $K * O(n^2 logn)$.

Our method uses the idea of mapping high-dimensional space distributions to lower one to update the utility parameters.

Given an n-dimensional vector in original non-additive space, it can be mapped to a 2^n-dimensional linear space. If the our method use K samples to achieve convergence, and we define a learning rate as λ_p under certain precision p, the frequency of training of each element in an n-dimensional vector is generally in a direct ratio to n:

$$\frac{nK}{2^n} \approx \lambda_p n \tag{22}$$

Algorithm 2. Bayesian Inference Based Choquet Integral Method

Input: matrix Φ, target variable t

Output: μ

- Initialize model parameters $\{\beta, \mu_0\}$
- For k=1,...,N
  1. Sample the data ϕ_k and target variable t_k.
  2. Obtain the n-dimensional non-zero vector $\phi_k^{(n)}$, $\mu_{k-1}^{(n)}$and $t_k^{(n)}$
  3. Initialize n-dimensional prior precision matrices $\{S^{(n)}\}$
  4. Compute the posterior precision matrices $S_k^{(n)}$:

$$\left(S_k^{(n)}\right)^{-1} = \left(S^{(n)}\right)^{-1} + \beta\phi_k^{(n)}\left(\phi_k^{(n)}\right)^T$$

  5. Compute the posterior $\mu_k^{(n)}$:

$$\mu_k^{(n)} = S_k^{(n)}\left(\left(S^{(n)}\right)^{-1}\mu_{k-1}^{(n)} + \beta\phi_k^{(n)}t_k^{(n)}\right)$$

  6. Update the n related parameters back to μ_k

Usually, our method needs K samples to achieve the precision p, and K can be estimated by:

$$K \approx 2^n \lambda_p \tag{23}$$

4 An Application Example

For wireless multimedia networks, cross-layer design has been regarded as one of the most effective and efficient ways to provide quality of service (QoS). At the application layer, prediction mode and quantization parameter (QP) in video encoding are two critical design variables [H.264 2005]. At the physical layer, modulation and coding schemes (MCS) have been adopted to achieve a good tradeoff between transmission rate and transmission reliability. PSNR indicates the performance of wireless multimedia networks. The channel signal-to-interference-noise ratio (SINR) reflects the channel conditions

Using non-additive measure, we can estimate the parameters $\mu_k(k = 1, 2, \ldots, n)$. In fact, μ_k indicates the importance of coefficients k. In previous works [19,25,35,36], we have illustrated that the parameters of model is stable under certain scenario, and the optimization using non-additive measure for cross-layer design is effective when channel condition is known.

Let us further consider this issue for real-world applications. In fact, it is very hard to acquire the real-time channel condition information in wireless multimedia networks community. In other words, the system is usually unaware of the state transitions. For example, we can only obtain the coefficients of QP and MCS for a wireless multimedia networks system. The optimal configuration of QP and MCS is largely determined by SINR under the current scenario. However, the system is not able to obtain the current SINR.

For cross-layer design optimization of wireless multimedia networks, the three problems mentioned above all exist. The traditional non-additive measure method is not able to handle this multi-modal, real-time and high complexity problem.

4.1 The Performance of the Application

The MLE and our methods were applied to the real-time transmission of an individual video bitstream across a multi-hop 802.11a/e wireless network. We first discuss the regression experiments and then the real-time application experiments.

Regression. In this part, the data set contained 8064 3-D samples, each containing one PSNR value from each of the three variables (Mac_length, QP and AMC) used in the cross-layer optimization problem. Two algorithms, MLE and Bayesian inference based Choquet integral (BIBCI) methods, were considered for non-additive measure.

To evaluate the methods as regression algorithms, three runs of a 10-fold cross-validation (in a total of 30 simulations) were applied for the MLE method. As real-time methods, our methods were applied directly to real-time prediction. The overall performance is evaluated by MAD and RMSE. We also conducted the paired t-tests against true PSNR value confirmed the statistical significance of each method.

The results are shown in Table 1. Under the MAD and RMSE criterion, the BIBCI method outperforms the MLE method. For the time consumption, the MLE method is better than BIBCI method. This is an interesting outcome, since the matrix of variables can be calculated directly under low-dimensional situation, with no need for accumulated calculations. However, from the MAD and RMSE points of view, the best option is still our method.

Table 1. The experiment results in terms of the MAD errors and t-test

	MAD	RMSE	Mean	t	p_{one}	p_{two}	Time
MLE(1)	4.935	6.775	29.35	−0.008	0.497	0.994	0.00042
MLE(2)	4.936	6.776	29.35	0.0004	0.5	1.0	0.00023
MLE(3)	4.931	6.769	29.44	0.00004	0.5	1.0	0.00039
BIBCI	1.788	2.694	29.34	0.09	0.497	0.93	6.71

Real-Time Application. In this part, we simulate the real time wireless multimedia dataset. In this dataset, channel condition is unstable and constantly changing. We draw samples to simulate the real-life situation where the channel condition changes from bad to good or vice versa. There are 1300 samples in this dataset, and it includes 2 period of channel condition changing.

For better illustration, we process the dataset using kernel smoothing as shown in Figs. 1 and 2. The channel condition is just utilized to acquire the optimal PSNR. In real-life situation, the channel condition is hidden and unknown.

For the MLE method, the utility of variable can be calculated by traversing all the options. For our dataset, the utility of Mac_length equals to 28.56, the utility of QP equals to 35.73 and the utility of AMC equals to 38.40. So the distortion performance is more sensitive to the AMC than to other variables. In this test, system optimizes the AMC configuration all the time for the MLE method.

For our method, we calculate the utilities for each slice, and consider the variable with the largest utility value as the most important variable. For each slice, system always optimizes the most important variable. In order to avoid local optimum, we update the utilities by 8 random samples before dealing with each slice.

The results are shown in Fig. 3. We can see clearly that the performance of our method is closer to the optimal than traditional Choquet integral. We also can find that the accuracy of the first 200 samples is not very satisfactory,

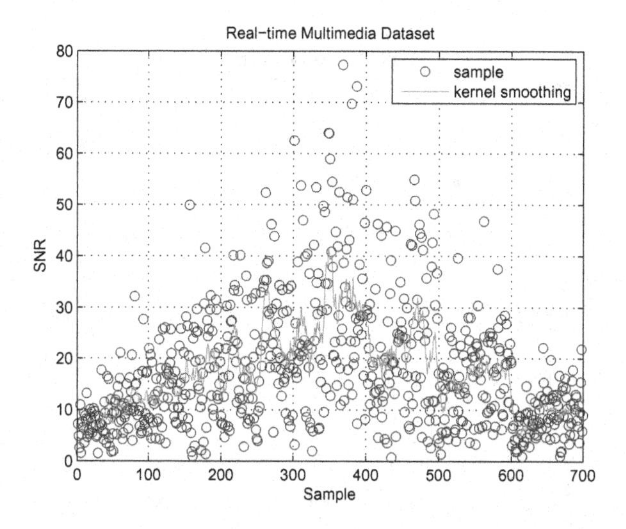

Fig. 1. The kernel smoothing of the real-time wireless multimedia dataset

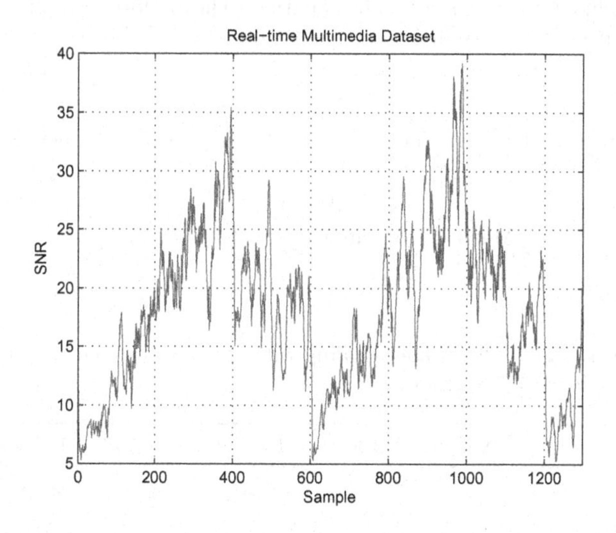

Fig. 2. The SNR of the real-time wireless multimedia dataset

and this phenomenon does not happen in second period, which means the our algorithm turns to stable within $9 * 200$ updates under this dataset.

The performances of MLE and our methods are shown in Table 2. Larger values result in better multimedia system. To illustrate the quality of the reconstructed videos, we show some sample video frames dealing with the optimal configuration, the MLE-based method and our method, respectively. As shown in Table 3, we can see that the performance achieved by using our method is better than that of MLE-based method.

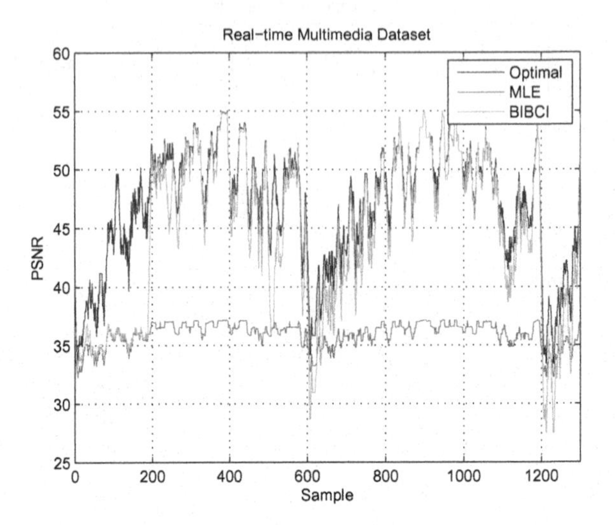

Fig. 3. The simulation on the real-life situation. The performance of our method is closer to the optimal than traditional Choquet integral.

Table 2. The mean performance of MLE and BIBCI methods

	Optimum	MLE	BIBCI
Mean PSNR	46.05	35.77	42.71

Table 3. Some sample video frames dealing with the optimal configuration, the MLE-based method and BIBCI method.

Optimum	Non-optimized	MLE-based method	BIBCI method

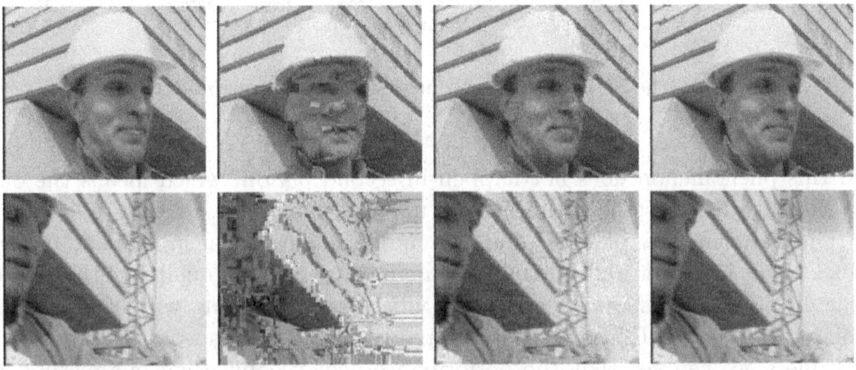

5 Conclusion

In this paper, we propose a novel polynomial method to solve the parameter estimation problem for Choquet integral. The proposed approach allows training parameters of Choquet integrals without requiring entire input samples. The computational complexity of the proposed algorithm is low and the number of parameters needed to compute is only linear rather than exponential with the number of inputs. Using our method, the computational complexity for the non-additive measure is reduced from $O((n+K)*2^{2n})$ to $K*O(n^2logn)$. Specifically, the semantic information of the 2^n variables is not lost. This method can be utilized to real-time applications. As a real-world application, cross-layer design of wireless multimedia networks is optimized by the proposed method. The experiments show that the performance achieved by using this method is better than that of traditional methods.

References

1. Andres, M., Paul, G.: Minimum classification error training for choquet integrals with applications to landmine detection. IEEE Trans. Fuzzy Syst. **16**(1), 225–238 (2008)
2. Aumann, R., Shapley, L., Aumann, R.: Values of Non-atomic Games. Princeton University Press, Princeton (1974)
3. Bishop, C., et al.: Pattern Recognition and Machine Learning. Springer, New York (2006)
4. Choquet, G.: Theory of capacities. Ann. Inst. Fourier **5**(131–295), 54 (1953)
5. Ci, S., Guo, H.: Quantitative dynamic interdependency measure and significance analysis for cross-layer design under uncertainty. In: Proceedings of 16th International Conference on Computer Communications and Networks, ICCCN 2007, pp. 900–904 (2007)
6. Fujimoto, K.: New characterizations of k-additivity and k-monotonicity of bi-capacities. In: 2nd International Conference on Soft Computing and Intelligent Systems and 5th International Symposium on Advanced Intelligent Systems, SCIS-ISIS 2004 (2004)
7. Gilboa, I., Schmeidler, D.: Additive representations of non-additive measures and the Choquet integral. Ann. Oper. Res. **52**(1), 43–65 (1994)
8. Golub, G., Van Loan, C.: Matrix Computations. Johns Hopkins University Press, Baltimore (1996)
9. Grabisch, M.: Fuzzy integral in multicriteria decision making. Fuzzy Sets Syst. **69**(3), 279–298 (1995)
10. Grabisch, M.: k-order additive discrete fuzzy measures and their representation. Fuzzy Sets Syst. **92**(2), 167–189 (1997)
11. Grabisch, M.: The interaction and Mobius representations of fuzzy measures on finite spaces, k-additive measures: a survey. In: Fuzzy Measures and Integrals Theory and Applications, pp. 70–93 (2000)
12. Grabisch, M.: Modelling data by the Choquet integral. Stud. Fuzziness Soft Comput. **123**, 135–148 (2003)
13. Grabisch, M., Labreuche, C.: Bi-capacities-I: definition, Mobius transform and interaction. Fuzzy Sets Syst. **151**(2), 211–236 (2005)

14. Grabisch, M., Nguyen, H., Walker, E.: Fundamentals of Uncertainty Calculi with Applications to Fuzzy Inference. Springer, Dordrecht (1995). https://doi.org/10.1007/978-94-015-8449-4
15. Grabisch, M., Nicolas, J.: Classification by fuzzy integral: performance and tests. Fuzzy Sets Syst. **65**(2–3), 255–271 (1994)
16. Grabisch, M., Raufaste, E.: An empirical study of statistical properties of the Choquet and Sugeno integrals. IEEE Trans. Fuzzy Syst. **16**(4), 839 (2008)
17. Grabisch, M., Sugeno, M.: Multi-attribute classification using fuzzy integral. In: IEEE International Conference on Fuzzy Systems 1992, pp. 47–54 (1992)
18. Grabisch, M., Sugeno, M., Murofushi, T.: Fuzzy Measures and Integrals: Theory and Applications. Springer-Verlag, New York Inc., Secaucus (2000)
19. Guo, H., Zheng, W., Ci, S.: Precise determination of non-additive measures. In: 2008 IEEE Conference on Cybernetics and Intelligent Systems, pp. 911–916 (2008)
20. Keller, J., Gader, P., Hocaoglu, A.: Fuzzy integrals in image processing and recognition. In: Fuzzy Measures and Integrals: Theory and Applications, pp. 435–466 (2000)
21. Klement, E., Mesiar, R., Pap, E.: A universal integral as common frame for Choquet and Sugeno integral. IEEE Trans. Fuzzy Syst. **18**(1), 178–187 (2010)
22. Kojadinovic, I.: Minimum variance capacity identification. Eur. J. Oper. Res. **177**(1), 498–514 (2007)
23. Kojadinovic, I.: Quadratic distances for capacity and bi-capacity approximation and identification. 4OR: Q. J. Oper. Res. **5**(2), 117–142 (2007)
24. Kojadinovic, I., Labreuche, C.: Partially bipolar Choquet integrals. IEEE Trans. Fuzzy Syst. **17**(4), 839–850 (2009)
25. Luo, H., Argyriou, A., Wu, D., Ci, S.: Joint source coding and network-supported distributed error control for video streaming in wireless multihop networks. IEEE Trans. Multimed. **11**(7), 1362–1372 (2009)
26. Marichal, J.: An axiomatic approach of the discrete Choquet integral as a tool toaggregate interacting criteria. IEEE Trans. Fuzzy Syst. **8**(6), 800–807 (2000)
27. Miranda, P., Grabisch, M.: Optimization issues for fuzzy measures. Int. J. Uncertain. Fuzziness Knowl. Based Syst. **7**(6), 545–560 (1999)
28. Rao, C., Mitra, S.: Generalized Inverse of the Matrix and Its Applications. Wiley, New York (1971)
29. Rota, G.: On the foundations of combinatorial theory I. Theory of Möbius functions. Probab. Theory Relat. Fields **2**(4), 340–368 (1964)
30. Sugeno, M.: Theory of fuzzy integrals and its applications. Tokyo Institute of Technology (1974)
31. Tveit, A.: On the complexity of matrix inversion. Mathematical Note, IDI, NTNU, Trondheim, Norway (2003)
32. Verma, N., Hanmandlu, M.: Additive and nonadditive fuzzy hidden Markov models. IEEE Trans. Fuzzy Syst. **18**(1), 40–56 (2010)
33. Wang, Z., Guo, H.: A new genetic algorithm for nonlinear multiregressions based on generalized Choquet integrals. In: Proceedings of the FUZZIEEE 2003, pp. 819–821 (2003)
34. Wang, Z., Leung, K., Klir, G.: Applying fuzzy measures and nonlinear integrals in data mining. Fuzzy Sets Syst. **156**(3), 371–380 (2005)
35. Wu, D., Ci, S., Wang, H.: Cross-layer optimization for video summary transmission over wireless networks. IEEE J. Sel. Areas Commun. **25**(4), 841–850 (2007)
36. Wu, D., Ci, S., Luo, H., Guo, H.: A theoretical framework for interaction measure and sensitivity analysis in cross-layer design. ACM Trans. Model. Comput. Simul. (2010)

37. Wu, W., Leung, Y., Mi, J.: On generalized fuzzy belief functions in infinite spaces. IEEE Trans. Fuzzy Syst. **17**(2), 385–397 (2009)
38. Xu, K., Wang, Z., Heng, P., Leung, K.: Classification by nonlinear integral projections. IEEE Trans. Fuzzy Syst. **11**(2), 187–201 (2003)
39. Xu, K., Wang, Z., Wong, M., Leung, K.: Discover dependency pattern among attributes by using a new type of nonlinear multiregression. Int. J. Intell. Syst. **16**(8), 949–962 (2001)
40. Yan, N., Wang, Z., Chen, Z.: Classification with Choquet integral with respect to signed non-additive measure. In: ICDMW, pp. 283–288 (2007)

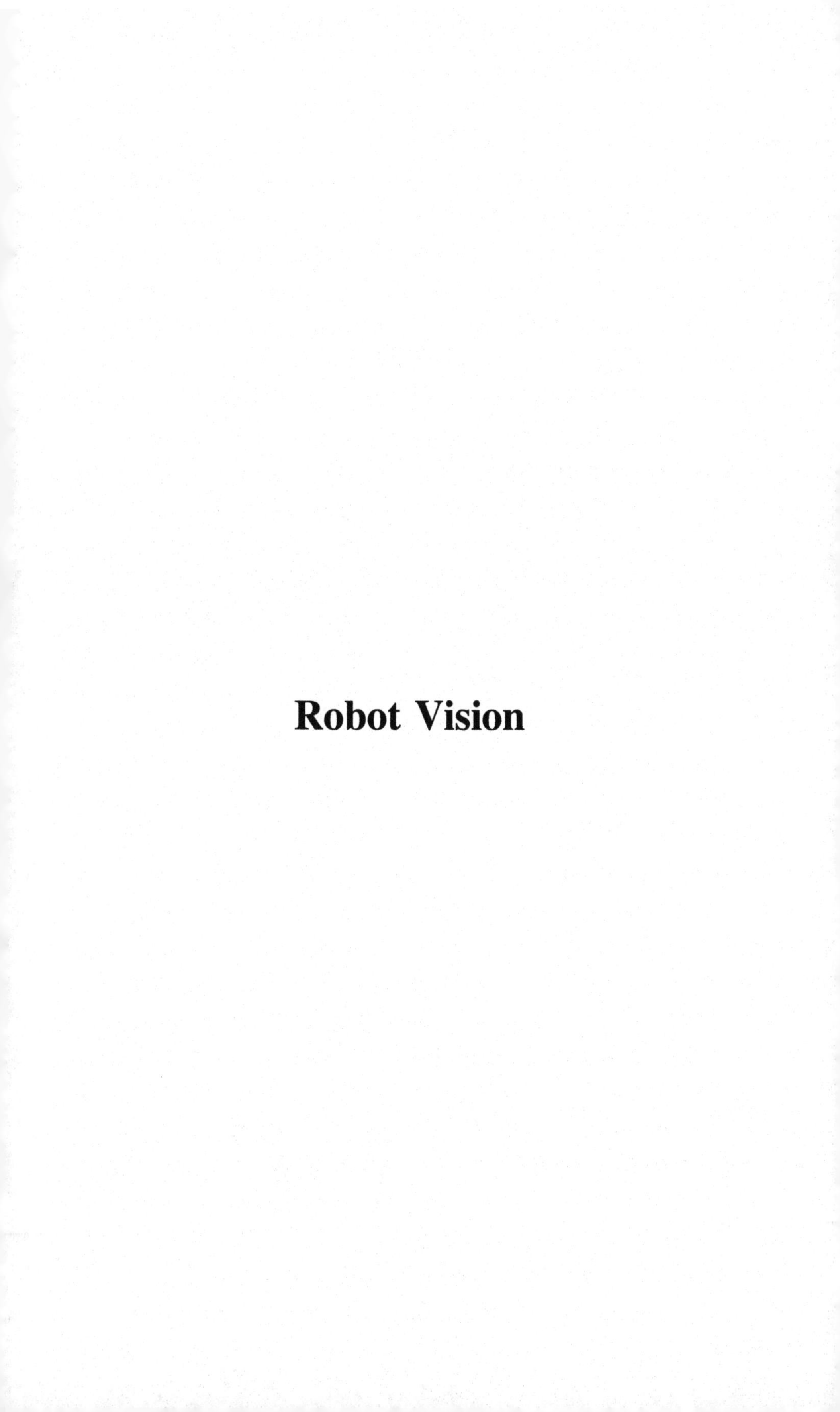

Robot Vision

Statistical Degradation Analysis for Real-Time Tracking in Severely Degraded Videos

Yuan Feng, Sheng Liu[(✉)], Chao Wang, Gaoxuan Ying, Kejie Yin, and ShengYong Chen

College of Computer Science and Technology, Zhejiang University of Technology, Hangzhou 310023, Zhejiang, People's Republic of China
edliu@zjut.edu.cn

Abstract. Recently, Correlation Filter have been widely applied in object tracking, by combining different features, correlation filter based methods can robustly track the target in various situations like illumination change, target deformation, motion blur and camera defocus. However, those methods can hardly deal with non-uniform degradations. In this paper, we proposed a Statistical Degradation Model to handle those issues, in our model, we apply colour-based statistical feature for deformation and defocus, gradient distributions for illumination change and colour degradation, degradation distribution for fast motion and motion blur. By combining those features our method could achieve obvious improvement in severely degraded videos, and outperform some state-of-the-art methods in popular benchmark OTB2013.

Keywords: Statistical analysis · Color distribution
Gradient distribution · Degradation distribution · Real-time tracking

1 Introduction

Object tracking is a significant part of computer vision, it has a lots of applications in robotics applications, human behaviour analysis, traffic surveillance system, military object detection, augmented reality, etc. However, object tracking has become a very challenging task. To solve this problem, people recently proposed correlation filter to track the target, which brings outstanding improvement. However, when it encounter with severe degradations, those state-of-the-art algorithms can not achieve satisfying results. Actually, few people take the degradation information itself into the tracking procedure.

In this paper, we proposed a new Statistical Degradation Model. Three different features are introduced in our model to cope with the challenging tracking task, colour distribution (histogram of RGB) for the deformation of the target, gradient distribution (HOG) for the illumination change, degradation distribution (histogram of Motion Blur direction) for fast motion and motion blur. For each feature, we learn a feature score of the target. Then, those three feature

J. Yang et al. (Eds.): CCCV 2017, Part III, CCIS 773, pp. 419–429, 2017.
https://doi.org/10.1007/978-981-10-7305-2_36

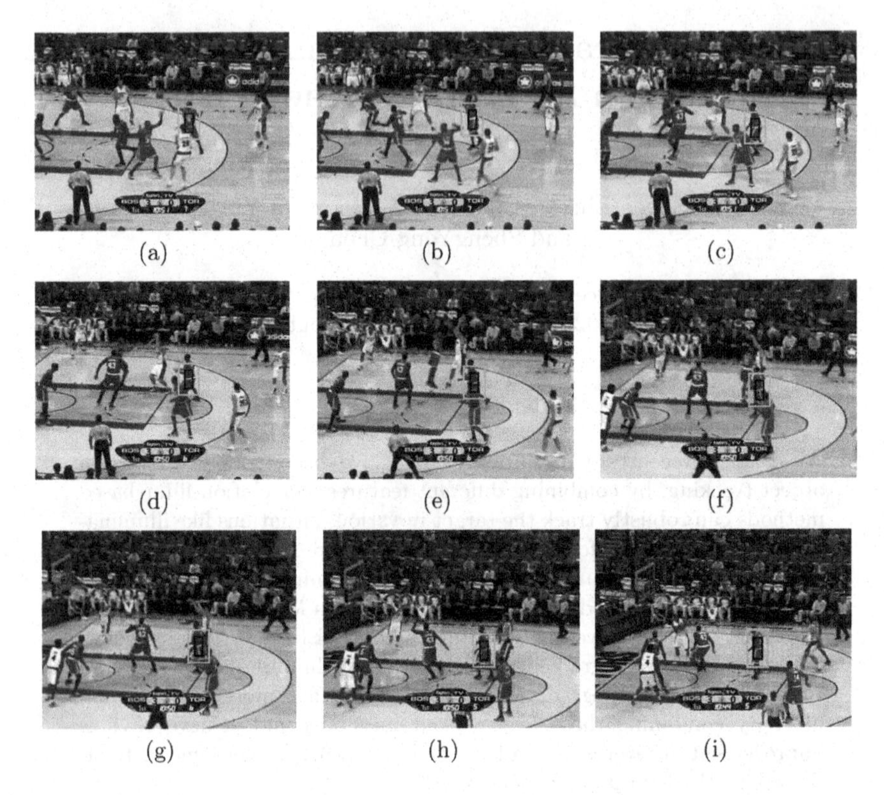

Fig. 1. (a)–(h) are the tracking result of proposed method in severely degraded situation. (The yellow rectangles annotates our result) (Color figure online)

scores are combined into our model, and the final score which indicates the target location is obtained. By using the state-of-the-art framework, our method can easily achieve real-time tracking at 80 FPS.

Further more, the application of the degradation distribution makes our method outperforms other methods, especially in severe degraded situations, like fast motion and motion blur sequences.

2 Related Works

Nowadays, when tracking object people often use an on-line object detection algorithm. Algorithm for robustness improvement of the results [21], Struck [12] has concise formulations, with the purpose to find the minimum of the localized structured object output [3]. However, the various features and amount of the training samples make those algorithms lack of efficient.

Correlation filter [2] is to find the minimum in all the cyclic shifts of the positive example from the least squares loss. This doesn't seem to be a proper approximation to the current real world, by using the Fourier domain and dense

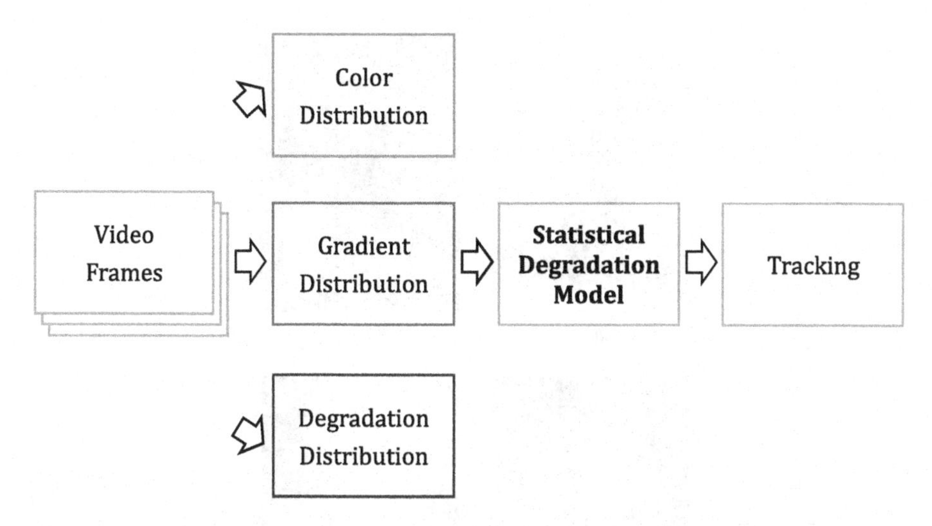

Fig. 2. Overview of our algorithm (Firstly, we calculate the colour, gradient and degradation distribution of the frame sequence. Secondly, we combine those distribution into our Statistical Degradation Model. Finally, we use correlation filter to track the target.)

sampling examples and high-dimensional feature images, it can easily achieve real-time. The disadvantage of correlation filters is that they are limited to learning from all cyclic shifts. Several recent works on [4,11,13] have attempted to solve this problem, especially spatial regularization, and some [9] formulation has shown good tracking results.

However, this is at the expense of real-time operations. [5], they extend to multiple feature channels, so the HOG feature [7] enables the technology to implement the most advanced performance [16] in VOT14. DSST [8] Challenger winners incorporate multi scale templates using 1D correlation filters to distinguish scale space tracking. Current correlation filters based method [17] are inherently limited in learning rigid template problems. This is a problem when a target undergoes shape deformation in sequence. Perhaps the simplest way to achieve robustness to deformation is to use a representation [19] insensitive to shape changes. The image histogram has this property because they discard the position of each pixel.

In fact, the histogram can be considered orthogonal to the correlation filter because the correlation filter is learned from the cyclic shift, while the histogram is invariant for cyclic shifts [18,20]. However, separate histograms are usually insufficient to distinguish objects from backgrounds.

The primary alternative to achieving deformation robustness is to learn a deformable models. We think that learning deformable models from a single video is very meaningful, and the only monitoring is the position in the first frame, so a simple bounding box is adopted. Although our approach is superior to the recent sophisticated parts based model [6,22] in benchmarking, deformable models have richer representations that can do tracking better. The degradation information is used in [10], which bring improvement of the parts based method [6]. Thus, the degradation extracting method is applied in our model.

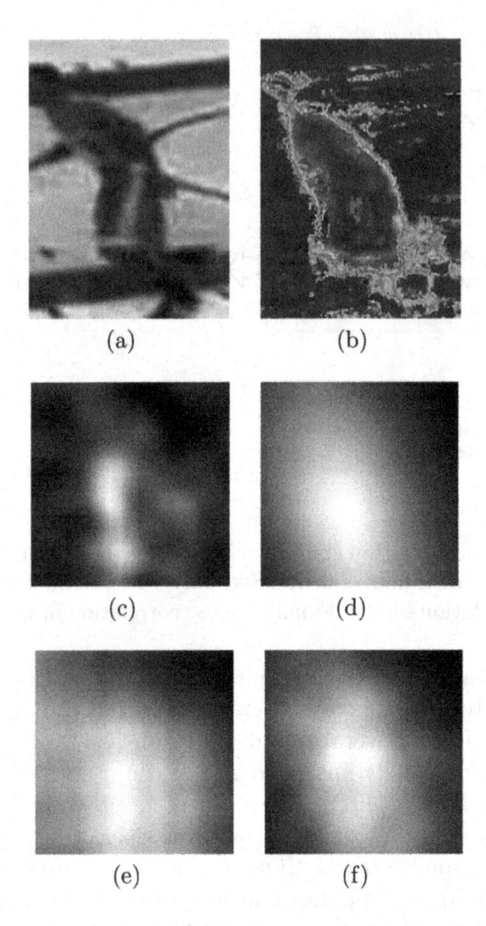

Fig. 3. (a) is the input frame patch, (b) is the per-pixel score of the colour testing, (c) is the gradient score, (d) is the colour score, (e) is the degradation score and (f) is the final score. (Color figure online)

3 Statistical Degradation Analysis

3.1 Statistical Degradation Model

In our Statistical Degradation Model, we use the i to be the number of frame, we choose l_i to denote the bounding-box, so that it can determines the target position in the frame x_i. The bounding-box l_i is chosen from the set S_i to find the maximum score:

$$l_i = \arg\max_{l \in S_i} f(T(x_i, l); \rho_{i-1}). \tag{1}$$

T is a transformation function of the frame, $f(T(x, l); \rho_{i-1})$ can assign the score to the bounding-box l in frame x by the designed parameters ρ. The designed parameters should be given to find the minimum of the loss function

$L(\rho; X_i)$ that object depends in those on the former frames and the position of the object in those frames $X_i = \{(x_j, l_j)\}_{j=1}^{i}$:

$$\rho_i = \arg\min_{\rho \in Q} \{L(\rho; X_i) + \lambda G(\rho)\}. \tag{2}$$

The parameters space is Q. In this section, we use regularization term $G(\rho)$ with a weight λ to keep the model complexity and eliminate over-fitting. The position l_1 is the object position in the first frame. To get a real-time speed, functions f and function L is given to find the position of the object efficiently, accurately and reliably.

The score function is proposed to combine template, histogram and degradation scores:

$$f(x) = \gamma_{grad} f_{grad}(x) + \gamma_{hist} f_{hist}(x) + \gamma_{degr} f_{degr}(x). \tag{3}$$

The whole model parameter $\rho = (\alpha, \beta, \omega)$, because λ_{grad}, λ_{hist} and λ_{degr} is the implicit in α, β and ω. Loss of the training that can be optimized to find the best parameters, which is considered to be the weighted linear combination of the losses in each image:

$$L(\rho, x_t) = \sum_{i=1}^{T} w_i \ell(x_i, l_i, \rho). \tag{4}$$

And the image loss function in each image could be :

$$U(x, l, \rho) = d(l, \arg\max_{m \in S} f(T(x, m); \rho)), \tag{5}$$

in which $d(l, m)$ denotes the cost of the dedicated rectangle m while the right rectangle is l.

Those three feature scores is learnt in our model:

$$\begin{aligned}
\alpha_i &= \arg\min_{\alpha} \{L_{grad}(h; X_t) + \frac{1}{2}\lambda_{grad}\|\alpha\|^2\}, \\
\beta_i &= \arg\min_{\beta} \{L_{hist}(\beta; X_t) + \frac{1}{2}\lambda_{hist}\|\beta\|^2\}, \\
\omega_i &= \arg\min_{\omega} \{L_{degr}(\omega; X_t) + \frac{1}{2}\lambda_{degr}\|\omega\|^2\},
\end{aligned} \tag{6}$$

where α denotes the colour distribution scores, β the gradient distribution scores and ω the degradation distribution scores.

Finally, we give out a overall combination of the three scores, using $\lambda_{grad} = 1 - \eta_1 - \eta_2$, $\lambda_{hist} = \eta_1$ and $\lambda_{degr} = \eta_2$. The maximum final score is considered as the centre of the target in current frame. The Fig. 3 is a visual view of the overall parameters and the final score.

Fig. 4. (a) is the origin image input, (b) is the degradation map of the input image, the gray level indicates the directions of the patches, (c) is the sketch of the normalized directions, (d) is the histogram of directions of the input image.

3.2 Statistical Degradation Features

In this paper, three different statistical features are applied to describe the target in object tracking progress. Colour-based statistical feature is used to cope with deformation and defocus, gradient distributions is combined to eliminate the influence of illumination change, and degradation distribution is used to deal with fast motion and motion blur that.

Colour distribution. The RGB histogram score is calculated from the samples in each image, with using the correct location as a positive sample. We use W to denote the set pairs (m, y) of rectangular box m and the corresponding regression result $y \in R$, including the positive sample $(p, 1)$. And the image loss in each frame is then

$$\ell_{hist}(x, p, \beta) = \sum_{(m,y)\in W} \left(\beta^T \left[\sum_{u\in H} \psi_{T(x,m)}[u] \right] - y \right)^2 . \tag{7}$$

Gradient distribution. Obtained by HOG feature, with a correlation filter formulation using least-squares, the image loss in each frame is

$$\ell_{grad}(x, l, \alpha) = \left\| \sum_{n=1}^{N} \alpha^n \star \phi^n - y \right\|^2, \tag{8}$$

where α^n is the channel n of multi-channel frame image α, ϕ is the short form of $\phi_{T(x,l)}$, y is the expected score (usually we use a maximum value 1 Gaussian function at the first time), and $\star$ denotes the periodic cross-correlation.

Degradation distribution. The Degradation information D is calculated by a in-plant function of autocorrelation [10] for position (x, y) in a local patch P that centred the location:

$$g(x, y) = \sum_{(x_i, y_i) \in P} \left[I(x_i, y_i) - I(x_i + \Delta x, y_i + \Delta y) \right]^2, \tag{9}$$

where $I(x_i, y_i)$ means the gradient of $I(x_i, y_i)$ in part p, Δx and Δy is the deviation in the direction of x and y. The image of motion direction D is generated as Fig. 4(b).

Then we normalize the matrix value to 8 directions {0, 22.5, 45, 67.5, 90, 112.5, 135, and 157.5}, we take the most amount of the direction as the direction of the patch, then the degradation distribution of the image part is obtained by calculate the histogram of the degradation image D.

The degradation distribution is calculated as D, and the degradation score is calculated as the same way as the RGB histogram score. And the image loss in each frame is

$$\ell_{degr}(x, l, \omega) = \sum_{(m,y) \in W} \left(\omega^T \left[\sum_{u \in D} \psi_{T(x,m)}[u] \right] - y \right)^2. \tag{10}$$

4 Experiments

We used dual Intel(R) Xeon(R) CPU E5-2670 Server with graphic card GTX 1080Ti to execute ours and others' methods. In order to confirm the efficiency and accuracy of our algorithm, we run the methods on the sequences OTB2013. Most of the challenges of Object Tracking are contained in the OTB2013, such as: Illumination Variation, Scale Variation, Occlusion, Deformation, Motion Blur, Fast Motion, In-Plane Rotation, Out-of-Plane Rotation, Out-of-View, Background Clutters, Low Resolution.

In Table 1, we report the values of the most important parameters we use. And in the Table 2, we give out the experiment results of our method with different value of the parameter.

The **bold** number in the Table 2 indicate the best result of our method in the OTB2013, with $\eta_1 = 0.5$ and $\eta_2 = 0.6$.

Table 1. Parameters in our method

Colour features	RGB
Colour histogram bins	$32 \times 32 \times 32$
Degradation histogram bins	$8 \times 8 \times 8$
Merge factor η_1	0.5
Merge factor η_2	0.6

Table 2. Experiment results of different value of the parameters.

	No.	1	2	3	4	5	6	7
Parameters	η_2	0.5	0.5	0.5	0.5	0.5	0.5	0.5
	η_1	0.3	0.4	0.5	0.6	0.7	0.8	0.9
Fast motion	OPE	0.506	0.52	0.56	0.54	0.521	0.536	0.517
	SRE	0.484	0.493	0.498	0.503	0.5	0.483	0.478
	TRE	0.51	0.521	0.527	0.532	0.54	0.533	0.525
Motion blur	OPE	0.523	0.522	0.574	0.545	0.529	0.542	0.503
	SRE	0.496	0.509	0.51	0.513	0.509	0.486	0.478
	TRE	0.529	0.534	0.544	0.547	0.555	0.542	0.527
Deformation	OPE	0.57	0.594	0.58	0.622	0.616	0.606	0.601
	SRE	0.549	0.559	0.562	0.579	0.575	0.561	0.565
	TRE	0.623	0.634	0.629	0.638	0.644	0.638	0.64
Overall	OPE	0.577	0.596	0.607	**0.611**	0.594	0.578	0.58
	SRE	0.558	0.557	0.555	**0.555**	0.554	0.55	0.545
	TRE	0.607	0.613	0.61	**0.618**	0.613	0.597	0.594

4.1 Experiment Analysis

Our algorithms performs better when obvious variation of appearance and structure occurs, as shown in Fig. 5.

We mainly contrast our tracking results to Staple [2] and other state-of-the-art algorithms in OTB2013, and we will give detailed analysis in the following to explain the advantage of our method.

Severe movements, such as fast motion, motion blur and non-uniform degradation, are usually grim challenge for object tracking, these abnormal movements usually do not follow the movement hypothesis.

Most of the algorithms fail tracking the target when the target accelerate or change the direction of motion in non-uniform degradation datasets. Also, Staple [2] suddenly fail to track the target when the target abruptly turn to another direction which makes the target very fuzzy and take a lot of useless information.

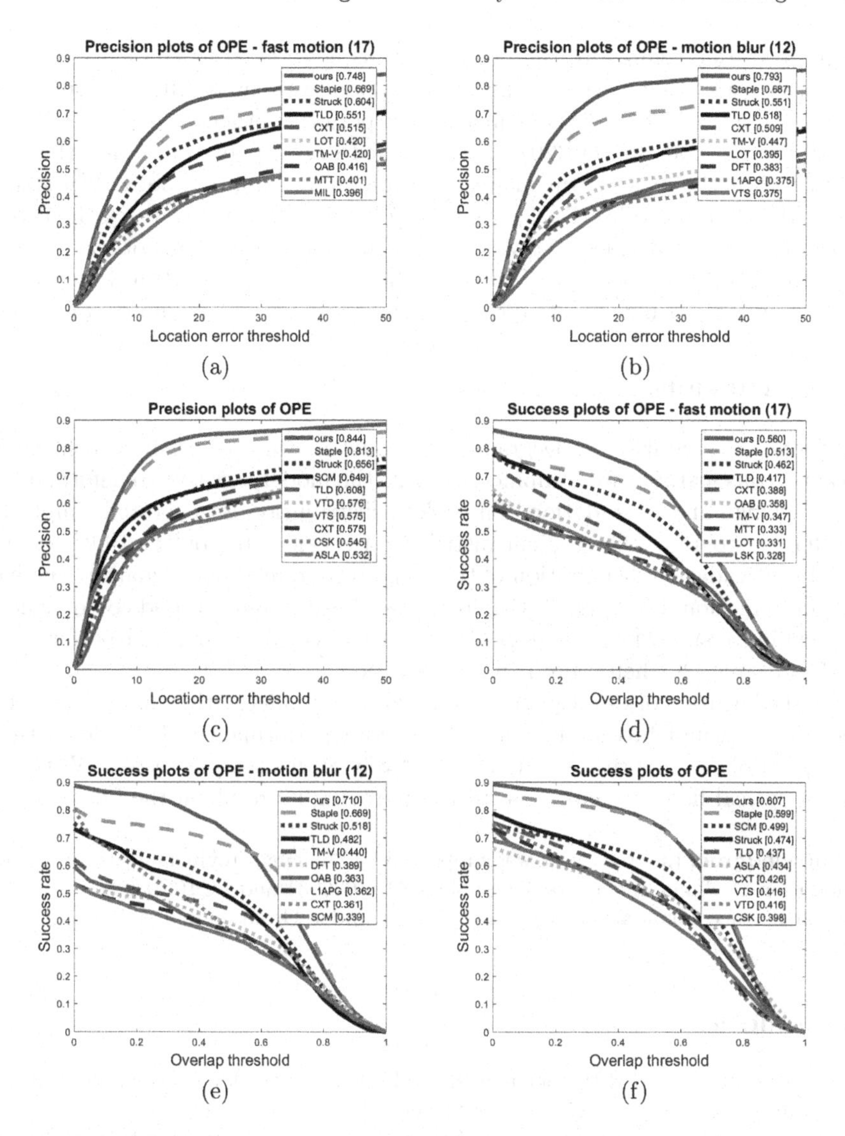

Fig. 5. These figures are the result of some popular methods in OTB2013, (a)–(c) are the precision plot of fast motion datasets, motion blur datasets and all datasets, (d)–(f) are the Success rate plot of fast motion datasets, motion blur datasets and all datasets (The red curve is the result of our method which has the best performance, the green curve is the state-of -the-art method Staple [2] and the score in the legend is the average score of the OTB2013). (Color figure online)

As what is shown in Fig. 5, our method performs much better than other methods in the severely degraded videos, that is because we have apply degradation information into the tracking procedure. The degradation information is a misplaced resource, and it can be useful by combining it in to the tracking progress. In our algorithm, the degradation is considered as the motion direction, dislike the HOG feature, motion direction gives more information of the moving target, which makes our method outperforms others.

All the methods in OTB2013 is contrasted in our paper, and only the best ten of the methods is shown in Fig. 5, common tracking methods (like struck [12], MIL [1], TLD [15], CT [23], KCF [14]) inevitably failed to track the target in severely degraded videos. Obviously, our method outperforms other methods very much. However, because of the limitation of HOG and RGB histogram, our method performs similar to Staple [2] when the sequence have no degradation.

5 Conclusion

We proposed a Statistical Degradation Model in this paper, in which, three advantageous features are combined to make the model sensitive to deformation, colour change and degradation. The colour distribution is generated simply by the RGB histogram, the gradient distribution is calculated by the HOG feature and the degradation distribution of the target is obtained by calculating the histogram of motion direction. With those three features our method could achieve outstanding result when the degradation of the video occurs, and performs as good as Staple [2] when there is no degradation.

In the future work, we plan to improve our model with real-time Optical Flow. That would further increase the overall performance of our algorithm. The speed of our algorithm is approximatively 80 frames per second. With the help of optical flow, we look forward to improving its result in the future.

Acknowledgments. This work was supported by Zhejiang Provincial Natural Science Foundation of China under Grant number LY15F020031 and LQ16F030007, National Natural Science Foundation of China (NSFC) under Grant numbers 11302195 and 61401397.

References

1. Babenko, B., Yang, M.H., Belongie, S.: Robust object tracking with online multiple instance learning. IEEE Computer Society (2011)
2. Bertinetto, L., Valmadre, J., Golodetz, S., Miksik, O., Torr, P.: Staple: Complementary Learners for Real-Time Tracking, pp. 1401–1409 (2015). arXiv http://arxiv.org/abs/1512.01355
3. Blaschko, M.B., Lampert, C.H.: Learning to localize objects with structured output regression. In: Forsyth, D., Torr, P., Zisserman, A. (eds.) ECCV 2008. LNCS, vol. 5302, pp. 2–15. Springer, Heidelberg (2008). https://doi.org/10.1007/978-3-540-88682-2_2
4. Boddeti, V.N., Kanade, T., Kumar, B.V.K.V.: Correlation filters for object alignment. In: IEEE Conference on Computer Vision and Pattern Recognition, pp. 2291–2298 (2013)
5. Bolme, D.S., Beveridge, J.R., Draper, B.A., Lui, Y.M.: Visual object tracking using adaptive correlation filters. In: Computer Vision and Pattern Recognition, pp. 2544–2550 (2010)

6. Cai, Z., Wen, L., Lei, Z., Vasconcelos, N., Li, S.Z.: Robust deformable and occluded object tracking with dynamic graph. IEEETrans. Image Process. **23**(12), 5497 (2014). A Publication of the IEEE Signal Processing Society
7. Dalal, N., Triggs, B.: Triggs, b.: Histograms of oriented gradients for human detection. In: CVPR, vol. 1, no. 12, pp. 886–893 (2005)
8. Danelljan, M., Hger, G., Khan, F.S., Felsberg, M.: Accurate scale estimation for robust visual tracking. In: British Machine Vision Conference, pp. 65.1–65.11 (2014)
9. Danelljan, M., Khan, F.S., Felsberg, M., Weijer, J.V.D.: Adaptive color attributes for real-time visual tracking. In: IEEE Conference on Computer Vision and Pattern Recognition, pp. 1090–1097 (2014)
10. Feng, Y., Liu, S., Zhang, S.B.: Structured degradation model for object tracking in non-uniform degraded videos. In: Tan, T., Li, X., Chen, X., Zhou, J., Yang, J., Cheng, H. (eds.) CCPR 2016. CCIS, vol. 662, pp. 345–355. Springer, Singapore (2016). https://doi.org/10.1007/978-981-10-3002-4_29
11. Galoogahi, H.K., Sim, T., Lucey, S.: Multi-channel correlation filters. In: IEEE International Conference on Computer Vision, pp. 3072–3079 (2013)
12. Hare, S., Saffari, A., Torr, P.H.S.: Struck: structured output tracking with kernels. In: IEEE International Conference on Computer Vision, pp. 263–270 (2012)
13. Henriques, J.F., Carreira, J., Rui, C., Batista, J.: Beyond hard negative mining: efficient detector learning via block-circulant decomposition. In: IEEE International Conference on Computer Vision, pp. 2760–2767 (2013)
14. Henriques, J.F., Rui, C., Martins, P., Batista, J.: High-speed tracking with kernelized correlation filters. IEEE Trans. Pattern Anal. Mach. Intell. **37**(3), 583–596 (2014)
15. Kalal, Z., Mikolajczyk, K., Matas, J.: Tracking-learning-detection. IEEE Computer Society (2012)
16. Kristan, M., et al.: The visual object tracking VOT2014 challenge results. In: Agapito, L., Bronstein, M.M., Rother, C. (eds.) ECCV 2014. LNCS, vol. 8926, pp. 191–217. Springer, Cham (2015). https://doi.org/10.1007/978-3-319-16181-5_14
17. Ma, C., Yang, X., Zhang, C., Yang, M.H.: Long-term correlation tracking. In: IEEE Conference on Computer Vision and Pattern Recognition, pp. 5388–5396 (2015)
18. Nummiaro, K., Koller-Meier, E., Van Gool, L.: Object tracking with an adaptive color-based particle filter. In: Van Gool, L. (ed.) DAGM 2002. LNCS, vol. 2449, pp. 353–360. Springer, Heidelberg (2002). https://doi.org/10.1007/3-540-45783-6_43
19. Possegger, H., Mauthner, T., Bischof, H.: In defense of color-based model-free tracking. In: Computer Vision and Pattern Recognition, pp. 2113–2120 (2015)
20. Pérez, P., Hue, C., Vermaak, J., Gangnet, M.: Color-based probabilistic tracking. In: Heyden, A., Sparr, G., Nielsen, M., Johansen, P. (eds.) ECCV 2002. LNCS, vol. 2350, pp. 661–675. Springer, Heidelberg (2002). https://doi.org/10.1007/3-540-47969-4_44
21. Wu, Y., Lim, J., Yang, M.H.: Online object tracking: a benchmark. In: IEEE Conference on Computer Vision and Pattern Recognition, pp. 2411–2418 (2013)
22. Xiao, J., Stolkin, R., Leonardis, A.: Single target tracking using adaptive clustered decision trees and dynamic multi-level appearance models. In: Computer Vision and Pattern Recognition, pp. 4978–4987 (2015)
23. Zhang, K., Zhang, L., Yang, M.H.: Real-time compressive tracking. In: European Conference on Computer Vision, pp. 864–877 (2012)

A Two-Layered Pipeline for Topological Place Recognition Based on Topic Model

Xingliang Dong[1], Jing Yuan[1,2](✉), Fengchi Sun[3], and Yalou Huang[3]

[1] College of Computer and Control Engineering, Nankai University, Tianjin, China
dongxingliang_1@163.com, yuanj@nankai.edu.cn
[2] Information Technology Research Base of Civil Aviation Administration of China, Civil Aviation University of China, Tianjin, China
[3] College of Software, Nankai University, Tianjin, China
{fengchisun,huangyl}@nankai.edu.cn

Abstract. In this paper, a two-layered pipeline for topological place recognition is proposed. Based on the Bag-of-Words (BoW) algorithm, Latent Dirichlet Allocation (LDA) is applied to extract the topic distribution of single images. Then, the environment is partitioned into different topological places by clustering the images in the topic space. By selecting the relevant topological place for the newly acquired image according to its topic distribution in the first level, we can reduce the number of images to perform the image-to-image matching in the second step while preserving the accuracy. We present evaluation results for City Centre data set using Precision-Recall curves, which reveal that our algorithm outperforms other state-of-the-art algorithms.

Keywords: Place recognition · Topological place discovery
Latent Dirichlet Allocation

1 Introduction

Visual place recognition [1] with mobile robots is an important problem for vision-based navigation and simultaneous localization and mapping (SLAM) [2, 3]. It is a technique that the mobile robot uses the images collected by the visual sensor to determine whether the current place is a new place or the one that has been visited. During the last decade, many novel methodologies have been presented in the literature aiming to address the place recognition task. A well known method is the Bag-of-Words (BoW) algorithm [4–7], which is borrowed from the realm of information retrieval. The framework of BoW is described below. Firstly, all of the local descriptors are collected, and then a clustering

J. Yuan—This work was sponsored in part by Natural Science Foundation of China under grant 61573196, in part by the Natural Science Foundation of Tianjin under grant 15JCYBJC18800 and 16JCYBJC18300 and in part by the Open Project Foundation of Information Technology Research Base of Civil Aviation Administration of China under grant CAAC-ITPB-201503.

J. Yang et al. (Eds.): CCCV 2017, Part III, CCIS 773, pp. 430–441, 2017.
https://doi.org/10.1007/978-981-10-7305-2_37

algorithm is run on this data set. After clustering, each cluster center is selected as a visual word. All cluster centers are selected to form a visual dictionary. A histogram of the term frequency is extracted as image features. This approach is initially inspired for solving image retrieval problems, which represents an image as a vector by defining the term frequency of a visual word.

However, although an image is represented as a collection of visual words, there are still some noisy visual words, which are useless for place recognition. Aiming at this problem, we introduces LDA [8] into place recognition task. The procedure of LDA applied in our work is shown in Fig. 1. Each visual topic is viewed as a mixture of various visual words. Via LDA, the visual topic distribution of each image is assigned according to the visual words extracted from it. The effect of noisy visual words on image features can be reduced. Afterwards, Affinity Propagation (AP) [9] algorithm is performed in visual topic space to partition the environment into different topological places. Then, each topological place is represented by the cluster center, which is the most representative image in this cluster. When a new image is acquired by the mobile robot, Nearest Neighbor (NN) algorithm is applied to find the best matching topological place. Finally, an image-to-image matching is performed to find the final loop closure. The contributions of this work are as follows.

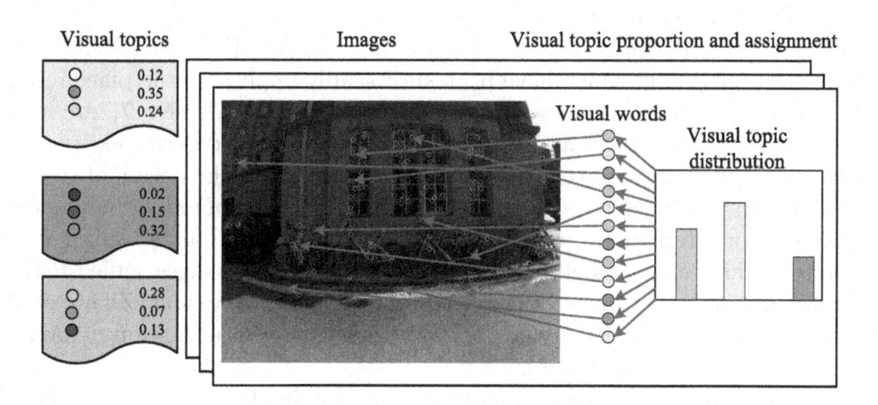

Fig. 1. The procedure of LDA in our work.

(1) The novelty of the presented method lies upon the partition of the matching procedure into two stages: initially between the image and the topological places and later between single images. In the first stage, the images are automatically divided into different topological places by clustering. Rather than matching all of the images in the image database, the time cost of image matching in the first stage is greatly reduced by matching all of the topological places. And then, the images belong to the relevant topological place are chosen as matching candidates. In the second stage, geometrical validation is performed to find the final loop closure between the newly acquired image and matching candidates.

(2) To the best of our knowledge, this is the first attempt at applying LDA in the place recognition task. The visual topic distribution of the image in visual topic space is not only used to discover topological places in the environment, but also to recognize topological places at the first step. Using the visual topic distribution of the image to represent image features, rather than the term frequency, the effect of noisy visual words can be reduced.

The rest of the paper is organized with the following structure: Sect. 2 briefly discusses the related work in the area of appearance-based place recognition. In Sect. 3 the proposed algorithm is described in detail, while Sect. 4 provides experimental results on City Centre data set. Finally, Sect. 5 serves as an epilogue for this paper where our final conclusions are drawn.

2 Related Work

BoW, as a well established technique, has been widely used in the past decade to address the place recognition task [3–7,10–15]. The conventional BoW algorithms are composed of three major steps, selection of local descriptors, generation and organization of the visual dictionary and the measure of similarity between two different images.

There are many kinds of local descriptors, such as SIFT [16], SURF [17], BRIEF [18], ORB [19], which have been successfully applied to the place recognition task. The real-valued descriptors, such as SIFT [16], SURF [17], are more robust against illumination and rotation, while the binary descriptors, such as BRIEF [18], ORB [19], are very fast to compute. Recently, the notion of the use of local descriptors by matching consecutive images, instead of individual images, has been reported in the literatures. Kawewong et al. [5,11] aims to find some descriptors which appear together in a couple of successive images, called PIRF, which are invariant to change of the environment to a large extent. Zhang et al. [14] catches the change of two matched local descriptors from motion dynamics to generate the visual dictionary.

The visual dictionary generation process can be divided into two categories: one is offline and the other is online and incremental. The work presented in [3,4,6,10,12,13,15] can be placed in the first category. This kind of algorithms is easy to generate a visual dictionary by using a clustering algorithm, such as K-Means [4,10], hierarchical K-Median [3,6,12,13,15]. However, it is sensitive to the appearance of the environments. When facing an new environment, the visual dictionary constructed by a known data set may be invalid. The online and incremental algorithms [5,7,11,14] can solve this problem. Accordingly, additional computational cost of update and maintenance of the visual dictionary has to be introduced.

The measure of similarity between two different images is usually calculated with L1-score [3,5,6,11,12,15]. As well, probabilistic models can be applied to calculate the similarity between two different images, such as FAB-MAP [4,10]. Both kinds of similarity can be utilized to find matching candidates.

However, because of the exchangeability of visual words, the geometrical consistency among the local descriptors is ignored, which produces a lot of false matches. Therefore, geometrical validation [2,3,6,12,13,15] is always implemented between the image and matching candidates to find the final loop closure.

Recently, some of two-layered place recognition algorithms [13,15,20], have been proposed. The first level is to find correlations between the image and different image sets, and the images belonging to the relevant image set are chosen as matching candidates. The second level performs geometrical validation between the image and these matching candidates to find the final loop closure. Mohan et al. [13] decomposed the set of images into independent environments and selected the best matching environment before the place recognition. Bampis et al. [15] segmented the image data set into different image sets based on their spatiotemporal proximity. Korrapati et al. [20] adopted image sequence partitioning techniques to group panoramic images together as different image sets.

Despite their success achieved in term of place recognition, there are still lots of noisy visual words in a single image, which are affect the performances of place recognition. Therefore, in this paper, in order to further reveal the nature of the image, LDA [8] is introduced into the place recognition task to mining the visual topic distribution of the image in the visual topic space. Then, through clustering the topic distribution of the image in the topic space, we can find all topological places in the environment. At last, a two-layered pipeline for topological place recognition is proposed, one of which is to find correlations between the image and topological places and the other is geometrical validation on these individual images.

3 Approach

The preparation before our proposed algorithm includes the offline creation of a visual dictionary. In this paper, PIRF [5,11] descriptors are extracted to yield a generic training set. Then, the open source implementation called OpenFABMAP [21] is utilized to create a visual dictionary with default parameters. Finally, we create 3982 words in the final dictionary. In the following, we first provide a brief description of LDA and its usage in extracting visual topics from images. Subsequently, the process of discovering topological places in the environment is discussed in detail. Finally, we elaborate the two levels of matching: the first level is to find correlations between the image and topological places and the second one is to match these individual images.

3.1 Latent Dirichlet Allocation

In the following, we briefly introduce the application of LDA in our work. Suppose we are given a collection D of images $\{I_1, I_2, \cdots, I_M\}$. Each image I is a collection of PIRF descriptors $I = \{\omega_1, \omega_2, \cdots, \omega_N\}$, where ω_i is the visual word index representing the i-th PIRF descriptor. A visual word is the basic item from a visual dictionary indexed by $\{1, 2, \cdots, V\}$.

The LDA model assumes there are K underlying latent visual topics. Each visual topic is represented by a multinomial distribution over the V visual words. An image is generated by sampling a mixture of these visual topics, then sampling visual words conditioning on a particular visual topic. The generative process of LDA for an image I in the collection can be formalized as follows:

I. Choose $\theta \sim \text{Dir}(\alpha)$;
II. For each of the N visual words ω_n:
   (a) Choose a visual topic $z_n \sim \text{Mult}(\theta)$;
   (b) Choose a visual word ω_n from $\omega_n \sim \text{p}(\omega_n|z_n, \beta)$, a multinomial probability conditioned on z_n.

$\text{Dir}(\alpha)$ represents a Dirichlet distribution with the parameter α, and $\text{Mult}(\theta)$ represents a Multinomial distribution with the parameter θ. The parameter θ indicates the mixing proportion of different visual topics in a particular image. α is the parameter of a Dirichlet distribution that controls how the mixing proportions θ vary among different images. β is the parameter of a set of Dirichlet distributions, each of them indicates the distribution of visual words within a particular visual topic. Learning a LDA model from a collection of images $D = \{I_1, I_2, \cdots, I_M\}$ involves finding α and β that maximize the log likelihood of the data $l(\alpha, \beta) = \sum_{d=1}^{M} \log P(I_d|\alpha, \beta)$. This parameter estimation problem can be solved by Gibbs sampling [22]. Then, each image I is represented by a distribution of different visual topics $T_i = \{t_1, t_2, \cdots, t_K\}$, where t_j is the probability that the i-th image contains the j-th visual topic.

3.2 Clustering and Topological Place Discovery

In practical situations, a topological place can be defined as an image set, which contains a group of images which are similar to each other in appearance and continuous in actual positions. Thus, by clustering images in the visual topic space, the environment can be partitioned into different image sets, each of which represents a topological place. However, the number of topological places of the environment is usually hard to specify in advance. Therefore, we choose Affinity Propagation (AP) [9] as the clustering algorithm, which do not need to specify the cluster number in advance and can work for any meaningful measure of similarity between data samples, to discover different topological places in the environment.

In this paper, the similarity between two images contains two parts: one is cosine similarity and the other is the spatial potential function. The definition is shown in (1).

$$sim(i, j) = \frac{T_i \cdot T_j}{||T_i|| \cdot ||T_j||} + \frac{1}{e^{|i-j|}} \tag{1}$$

where i, j are indices of images, T_i is the visual topic distribution of image I_i, $T_i \cdot T_j / ||T_i|| \cdot ||T_j||$ is the cosine similarity between I_i and I_j and $e^{-|i-j|}$ is the spatial potential function between I_i and I_j. The cosine similarity part represents that images with similar visual topic distributions can be clustered together. The space potential function guarantees that images that are contiguous in the environment will be clustered together.

After clustering, each cluster represents a topological place. Each cluster center, which is defined as an *exemplar* in AP, is chosen as the representative image of the topological place, which is used in the first step of place recognition.

3.3 Two-Layered Topological Place Recognition

After discovering all topological places in the environment, each topological place is represented by the *exemplar* of each cluster, which is the most representative image in each cluster. When a new image I_t is acquired, we can use the well-trained LDA model to infer its visual topic distribution. Then, we match I_t with all topological places to determine which topological place it belongs to as our first level of recognition. The similarity between the image and the topological place is calculated by cosine similarity. The best matching topological place $P_q = \{I_i | i = k, \cdots, k+m\}$ is found by Nearest Neighbor (NN) algorithm. All images belonging to P_q are chosen as matching candidates.

In the final step, geometrical validation is implemented between the related images using PIRF descriptors. For each $I_i \in P_q$, we adopt RANSAC to compute the fundamental matrix between I_t and I_i. If it fails to provide a fundamental matrix, we reject the loop closure between I_t and I_i. At the same time, if the number of inliers of PIRF descriptors of I_i is less than a constant f, the loop closure between I_t and I_i is also rejected. If more than one I_i meets the above criteria, the one with the most inliers will be chosen as the final loop closure.

4 Experimental Results

In this section, we present the experimental results of the proposed method and compare them with other state-of-the-art methods, including FAB-MAP 2.0 and DBoW2.

4.1 Experimental Setup

We implement the proposed algorithm on City Centre data set collected by Cummins and Newman [4]. The robot traveled twice around a loop with total path length of 2 km along public roads near the city center. The data set includes 1237 image pairs, which were collected by two cameras facing the left and right of the robot, respectively. The data set contains many dynamic objects, such as traffic and pedestrians. The ground truth of the trajectory is shown in Fig. 2. The yellow curve is the robot travelling trajectory, which is obtained by GPS. The loop part of this data set is from the 677th image pairs to the 1237th image pairs.

In this paper, we put SIFT [16] descriptors into the PIRF 2.0 [11] framework. The size of the sliding window in the PIRF 2.0 framework is set to 2.

Fig. 2. Motion trajectory of the robot in the City Centre data set. (Color figure online)

As mentioned above, the open source implementation called OpenFABMAP [21] is utilized to create the visual dictionary with default parameters. Finally, we are able to create 3982 words in the final visual dictionary. The number of visual topics is set to 300 in this paper. α is chosen small enough to ensure that each image contains one visual topic as far as possible. β is chosen small enough to make sure that each visual word belongs to only one visual topic. α and β are both set to 0.01.

4.2 Validity of Topological Places

Experimental results of topological place discovery for the first loop of the City Centre data set are shown in Fig. 3, which is drawn by GPS. In the first loop, 143 different topological places are discovered in total. Continuous marks with a same color represent a topological place. We use different colors to make a distinction between two consecutive topological places. It can be seen that images belonging to the same topological place are continuous in actual positions.

Figure 4 shows the left side images of the 71st–73rd topological places of the City Centre data set. It is easy to find the differences among the three topological places. The *exemplar* of each topological place is marked by a red box. It can be observed that the appearance of each *exemplar* is significantly different with others. Thus, the experimental results demonstrate that the proposed topological place discovery algorithm can automatically discover different topological places in the environment and ensure that the images, which belong to a same topological place, are similar with each other in appearance and continuous in terms of physical positions.

Fig. 3. Topological places discovered from the first loop of the City Centre data set.

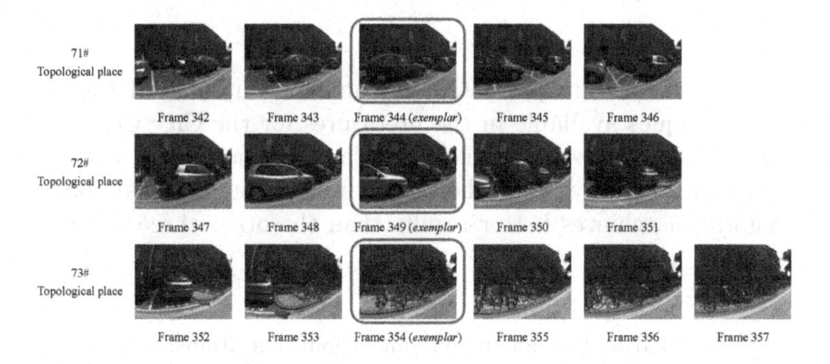

Fig. 4. Three obtained topological places on City Centre data set (left side of image pairs only). (Color figure online)

4.3 Overall Performance

With a view to measuring the performance of our system, we have chosen to use Precision-Recall curves. Precision is defined as the ratio between the correct and the total number of detected loop closures. On the other hand, recall is equal to the ratio of the correctly detected loop closures, to the total number of loops the data set contains.

$$Precision = \frac{TP}{TP + FP} \tag{2}$$

$$Recall = \frac{TP}{TP + FN} \tag{3}$$

where the TP is true positive, i.e., the correctly detected loop closures, FP is false positive, i.e., the incorrectly detected loop closures, and FN is false negative, i.e., the loop closures that the system erroneously discards. Figure 5 shows the Precision-Recall curve for the City Centre data set. We obtained this curve by varying the constant f. It can be seen that our proposed algorithm retains 71.83% Recall with 100% Precision.

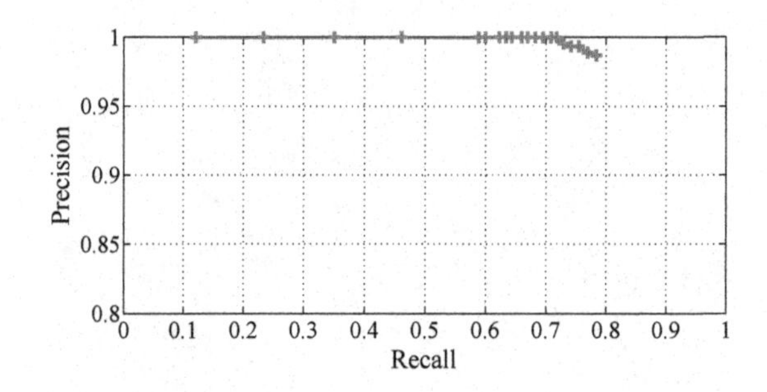

Fig. 5. Precision-Recall curve, for City Centre data set, illustrating the recognition performance of the proposed method.

Comparative results are presented in Table 1 with some of the most well established techniques available in the literatures for the City Centre data set, viz. [6,10,12]. The achieved performance of the aforementioned algorithms was obtained straightforwardly from their respective papers. It can be observed that our algorithm achieves better results than the other three state-of-the-art algorithms.

Table 1. Comparison among four algorithms.

Data set	Approaches	Precision (%)	Recall (%)
City Centre	DBoW2 [6]	100	30.61
	FAB-MAP 2.0 [10]	100	38.77
	Mur-Artal [12]	100	43.03
	Proposed method	**100**	**71.83**

Finally in Fig. 6, example frames on City Centre data set are presented and lines depict final corresponding PIRF descriptors. Figure 6(a) shows an correct loop closure that our algorithm is able to accept. Although the visual angle of the robot has changed, our algorithm is still able to detect the loop closure between the 697th frame and the 173rd frame. An incorrect loop closure that our algorithm is able to reject is shown in Fig. 6(b). It can be observed that the 951st frame and the 353rd frame are quite similar with each other in some local

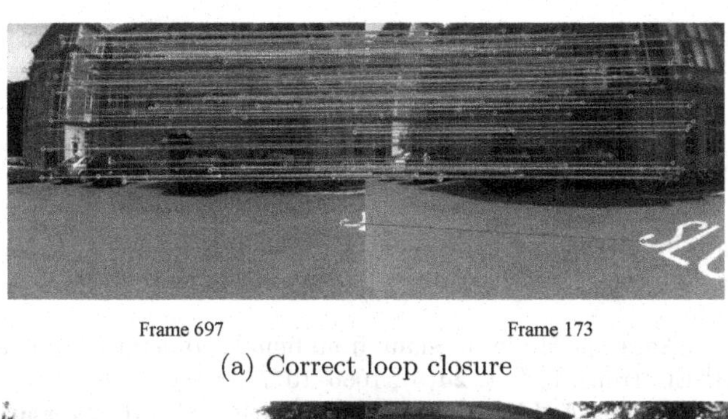

Frame 697 Frame 173

(a) Correct loop closure

Frame 951 Frame 393

(b) Incorrect loop closure

Fig. 6. Example frames on City Centre data set.

regions. That means there are a lot of common visual words between the 951st frame and the 353rd frame. Because of the exchangeability of visual words, the visual topic distributions of these two frames assigned by LDA are similar with each other. Nevertheless, our algorithm is still able to reject the loop closure between the 951st frame and the 353rd frame according to the number of inliers in PIRF descriptors. Therefore, the experimental results demonstrate that the proposed algorithm can achieve a high performance in place recognition task, even facing dynamic environments with large variability.

5 Conclusions

In this paper, LDA is introduced to assign the visual topic distributions of each image. Image features are represented by the visual topic distributions in the visual topic space. Then, by clustering images in the visual topic space, the environment can be automatically partitioned into different topological places. When a new image is acquired, a two-layered pipeline for topological place recognition is proposed to improve the performance of place appearance. Compared with the state-of-the-art place recognition algorithms, our algorithm is more robust against variation of the place appearance. Furthermore, a complete representation for the visual topics can help the robot to better understand and interpret the environments, even to extract the semantic cues, which can bridge the gaps of understanding and expression for the environmental information between the human and the robot.

References

1. Lowry, S.M., Sünderhauf, N., Newman, P., Leonard, J.J., Cox, D.D., Corke, P.I., Milford, M.J.: Visual place recognition: a survey. IEEE Trans. Robot. **32**(1), 1–19 (2016)
2. Konolige, K., Agrawal, M.: Frameslam: from bundle adjustment to real-time visual mapping. IEEE Trans. Robot. **24**(5), 1066–1077 (2008)
3. Mur-Artal, R., Montiel, J.M.M., Tardós, J.D.: ORB-SLAM: a versatile and accurate monocular SLAM system. IEEE Trans. Robot. **31**(5), 1147–1163 (2015)
4. Cummins, M., Newman, P.M.: FAB-MAP: probabilistic localization and mapping in the space of appearance. Int. J. Robot. Res. **27**(6), 647–665 (2008)
5. Kawewong, A., Tongprasit, N., Tangruamsub, S., Hasegawa, O.: Online and incremental appearance-based SLAM in highly dynamic environments. Int. J. Robot. Res. **30**(1), 33–55 (2011)
6. Gálvez-López, D., Tardós, J.D.: Bags of binary words for fast place recognition in image sequences. IEEE Trans. Robot. **28**(5), 1188–1197 (2012)
7. Khan, S., Wollherr, D.: IBuiLD: incremental bag of binary words for appearance based loop closure detection. In: IEEE International Conference on Robotics and Automation, ICRA 2015, Seattle, WA, USA, 26–30 May 2015, pp. 5441–5447 (2015)
8. Blei, D.M., Ng, A.Y., Jordan, M.I.: Latent Dirichlet allocation. J. Mach. Learn. Res. **3**, 993–1022 (2003)
9. Frey, B.J., Dueck, D.: Clustering by passing messages between data points. Science **315**(5814), 972–986 (2007)
10. Cummins, M., Newman, P.M.: Appearance-only SLAM at large scale with FAB-MAP 2.0. Int. J. Robot. Res. **30**(9), 1100–1123 (2011)
11. Kawewong, A., Tongprasit, N., Hasegawa, O.: Pirf-Nav 2.0: fast and online incremental appearance-based loop-closure detection in an indoor environment. Robot. Auton. Syst. **59**(10), 727–739 (2011)
12. Mur-Artal, R., Tardós, J.D.: Fast relocalisation and loop closing in keyframe-based SLAM. In: 2014 IEEE International Conference on Robotics and Automation, ICRA 2014, Hong Kong, China, 31 May–7 June 2014, pp. 846–853 (2014)
13. Mohan, M., Gálvez-López, D., Monteleoni, C., Sibley, G.: Environment selection and hierarchical place recognition. In: IEEE International Conference on Robotics and Automation, ICRA 2015, Seattle, WA, USA, 26–30 May 2015, pp. 5487–5494 (2015)
14. Zhang, G., Lilly, M.J., Vela, P.A.: Learning binary features online from motion dynamics for incremental loop-closure detection and place recognition. In: 2016 IEEE International Conference on Robotics and Automation, ICRA 2016, Stockholm, Sweden, 16–21 May 2016, pp. 765–772 (2016)
15. Bampis, L., Amanatiadis, A., Gasteratos, A.: Encoding the description of image sequences: a two-layered pipeline for loop closure detection. In: 2016 IEEE/RSJ International Conference on Intelligent Robots and Systems, IROS 2016, Daejeon, South Korea, 9–14 October 2016, pp. 4530–4536 (2016)
16. Lowe, D.G.: Distinctive image features from scale-invariant keypoints. Int. J. Comput. Vis. **60**(2), 91–110 (2004)

17. Bay, H., Ess, A., Tuytelaars, T., Gool, L.J.V.: Speeded-up robust features (SURF). Comput. Vis. Image Underst. **110**(3), 346–359 (2008)
18. Calonder, M., Lepetit, V., Strecha, C., Fua, P.: BRIEF: binary robust independent elementary features. In: Daniilidis, K., Maragos, P., Paragios, N. (eds.) ECCV 2010. LNCS, vol. 6314, pp. 778–792. Springer, Heidelberg (2010). https://doi.org/10.1007/978-3-642-15561-1_56
19. Rublee, E., Rabaud, V., Konolige, K., Bradski, G.R.: ORB: an efficient alternative to SIFT or SURF. In: IEEE International Conference on Computer Vision, ICCV 2011, Barcelona, Spain, 6–13 November 2011, pp. 2564–2571 (2011)
20. Korrapati, H., Courbon, J., Mezouar, Y., Martinet, P.: Image sequence partitioning for outdoor mapping. In: IEEE International Conference on Robotics and Automation, ICRA 2012, 14–18 May 2012, St. Paul, Minnesota, pp. 1650–1655 (2012)
21. Glover, A.J., Maddern, W.P., Warren, M., Reid, S., Milford, M., Wyeth, G.: OpenFABMAP: an open source toolbox for appearance-based loop closure detection. In: IEEE International Conference on Robotics and Automation, ICRA 2012, 14–18 May 2012, St. Paul, Minnesota, pp. 4730–4735 (2012)
22. Griffiths, T.L., Steyvers, M.: Finding scientific topics. Proc. Nat. Acad. Sci. U.S.A. **101**(Suppl 1), 5228–5235 (2004)

Monocular Dense Reconstruction Based on Direct Sparse Odometry

Libing Mao, Jiaxin Wu, Jianhua Zhang(✉), and Shengyong Chen

College of Computer Science and Technology, Zhejiang University of Technology, Hangzhou 310023, ZheJiang, People's Republic of China
zjh@zjut.edu.cn

Abstract. Monocular dense reconstruction plays more and more important role in AR application. In this paper, we present a new reconstruction system, which combines the Direct Sparse Odometry (DSO) and dense reconstruction into a uniform framework. The DSO can successfully track and build a semi-dense map even in low texture environment. The dense reconstruction is built on the fast superpixel segmentation and location consistency. However, a big gap between the semi-dense map and the dense reconstruction still needs to be bridged. To this end, we develop several elaborate methods including map points selection strategy, container for data sharing, and coordinate system transforming. We compare our system with a state-of-the-art monocular dense reconstruction system DPPTAM. The comparison experiments run on the public monocular visual odometry dataset. The experimental results show that our system has better performance and can run robustly, effectively in indoor and outdoor scenarios.

Keywords: Monocular SLAM · Dense reconstruction
Map points selection · Shared container · Coordinate system transforming

1 Introduction

More and more AR applications introduce SLAM for camera tracking. However, only tracking camera position is still not sufficient for a satisfactory AR application, because the virtual model cannot have real interactions with real scene. For example, current AR applications always use the scene as 2D background, and the virtual model is levitated above the scene. This extremely influences the user experience.

For more realistic user experience, the 3D model of scene should be reconstructed. Thus, both scene and virtual model live in a 3D space, and they can interact smoothly.

Some dense reconstruction methods based on SLAM are proposed. For example, KinectFusion [7] and ElasticFusion [12] use the depth data from Kinect to track the 3D posture of the sensor and reconstruct the geometric 3D model

J. Yang et al. (Eds.): CCCV 2017, Part III, CCIS 773, pp. 442–452, 2017.
https://doi.org/10.1007/978-981-10-7305-2_38

of the physical scene in real time. However, the extra sensors need more cost, computation resource, and more power. Some monocular camera methods are also proposed. For example, DTAM [8] proposes a dense model to track and achieves real-time performance by using current commodity GPU. However, it is not suitable for AR application, because AR application usually runs on mobile device. DPPTAM [1] is a direct monocular dense reconstruction system based on single CPU, which has impressive results in some indoor scenarios. But as for some outdoor scenarios, especially when the direction of the camera movement is relatively large, it will lose camera tracking.

Therefore, we propose a new dense reconstruction system, by combing the Direct Sparse Odometry (DSO) [3] and dense reconstruction pipeline. DSO has robust performance for camera tracking, low computation cost and very precise optimization mechanism. For example, it can successfully track camera even in low texture or dark scene. Furthermore, it can also generate a semi-dense map and provide sufficient map points for further dense reconstruction. However, the DSO is not developed for dense reconstruction, there are still some big gaps between semi-map and dense reconstruction. First, we cannot distinguish which point to use because DSO generates four types of points in each keyframe. Second, it costs much memory to save all information of each keyframe. Simultaneously, we should keep important content after the shared data exchange. Third, the transformation between the image coordinate system and the camera coordinate system may have difference. Therefore, we design some elaborate strategies to bridge these gaps. We first design a map point selection in DSO. At the same time a container is proposed to exchange the shared data between DSO and dense mapping thread. We also unify the same coordinate system. We compare our algorithm with DPPTAM in a public monocular visual dataset. The results show that our algorithm is more robust both in outdoor and indoor environments.

2 Related Work

Several dense reconstruction methods have been proposed, which can be roughly divided into three categories. The first one is based on depth camera. KinectFusion [7] fuses the depth data streamed from a Kinect camera into a single global implicit surface model. Kitinuous [11] is a complete three-dimensional reconstruction system, which combines the loop closure detection and the loop optimization. ElasticFusion [12] can reconstruct surfel-based maps of room scale environments with a RGB-D camera. Besides, they both use the iterative closest point algorithm to calculate the camera pose. These algorithms need much computation resource. GPU support is always used in general case to make system real-time.

The second category is based on GPU. Methods in this category calculate every pixel to recover the structure and consequently have to use GPU to accelerate the computation. DTAM [8] refers to dense tracking and mapping in real-time, which based on direct tracking. It is very robust to feature deletion and image blurring. However, since DTAM recovers dense depth maps for each pixel and adopts global optimization, the computational complexity is very huge. Even with GPU acceleration, the efficiency of the expansion is still low. REMODE [9] refers to probabilistic, monocular dense reconstruction in real time which combines Bayesian estimation and recent development on convex optimization to get a more accurate depth map. The approach runs on a CUDA-based laptop computer.

The third category is based on CPU. But the dense reconstruction algorithm based on CPU is a big challenge. Fastfusion [10] is a pure reconstruction algorithm which can fully generates mesh map and runs in a CPU. INTEL instruction set is used to accelerate calculation. However, the algorithm must rely on rotation and translation matrix which are generated by a SLAM system or visual odometry. At the same time the accurate depth value is needed first by a RGB-D camera before reconstruction. In Greene et al. [6] proposed a multi-resolution depth estimation and spatial smoothing process to determine the scale of the different texture. This approach increases the reconstruction density and quality, and saves more computational resource simultaneously. DPPTAM [1] refers to dense piecewise planar tracking and mapping from a monocular sequence. Alejo Concha et al. made the assumption that homogeneous-color regions belong to approximately planar areas. And then they combined low texture planar areas of the superpixel segmentation with semi-dense depth map. The highly textured image areas in DPPTAM are mapped using standard direct mapping techniques in [4].

3 DSO-Dense Reconstruction System

3.1 System Review

Our system consists of three parts. The first part is the DSO system, we can call it as the front end. It provides camera positions, key frames and semi-dense maps to the system. The second part is the dense reconstruction pipeline, which consists of super-pixel consistency, robust plane estimation, dense mapping and dense map. This part is learned from DPPTAM and is called as the back end. However, because the DSO is not developed for dense reconstruction, the first two parts cannot directly be combined. Therefore, a bridge part is necessary, which is the third part of our system. In this part, we design a map point selection strategy to choose those map points that are suitable for determining the positional relationship between the points and the contour. We also design a container that extracts data from the front end, and then transfers these data to the backend. Furthermore, due to the difference of coordinate system between front end and back end, our bridge part also transforms the coordinate system for them (Fig. 1).

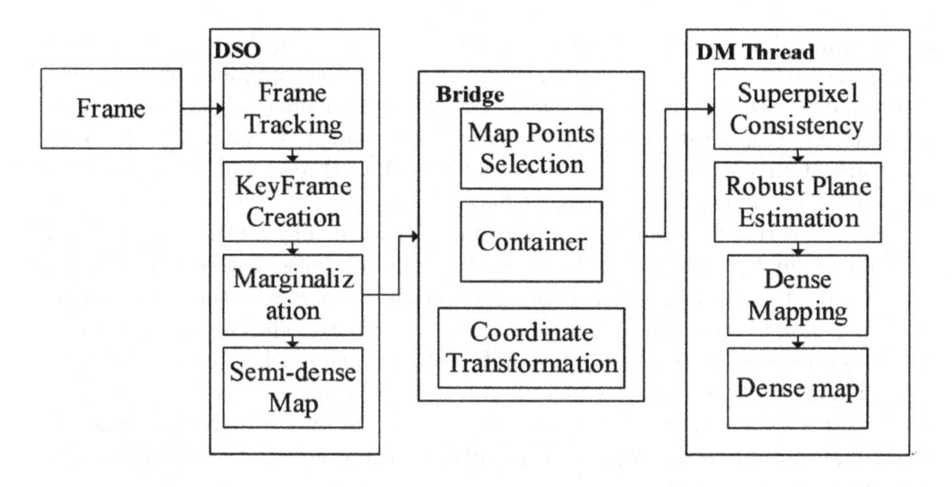

Fig. 1. Our system overview.

3.2 The Front End

When a new frame is coming, high-gradient points are extracted and the photometric error between a reference frame and a target frame is calculated. The weighted sum of squared differences (SSD) is calculated over a small neighborhood of pixels. Then the full photometric error over all frames and points is counted [3]. In order to obtain a more accurate camera position, a sliding window with 7 active keyframe is kept and the total photometric error is optimized in this sliding window using the Gauss-Newton algorithm. The keyframe creation strategy of [3] is used to keep important messages through all over the frames. Three criteria are considered to create keyframe:

$$f := \left(\frac{1}{n} \sum_{i=1}^{n} \|\boldsymbol{p} - \boldsymbol{p}'\|^2 \right)^{\frac{1}{2}} \quad (1)$$

$$f_t := \left(\frac{1}{n} \sum_{i=1}^{n} \|\boldsymbol{p} - \boldsymbol{p}_t'\|^2 \right)^{\frac{1}{2}} \quad (2)$$

$$a := \left| log \left(e^{(a_j - a_i)} t_j {t_i}^{-1} \right) \right| \quad (3)$$

Equation 1 is measured by the mean square optical flow when the field of view changes. $\boldsymbol{p}, \boldsymbol{p}'$ represents for extracted points from the last keyframe to the latest frame, respectively. Equation 2 is measured by the mean flow without rotation when camera translation causes occlusions and disocclusions. In this situation, f may be small. $\boldsymbol{p_t}'$ represents for the warped point position with rotation matrix. Equation 3 is measured by the relative brightness factor between two frames when the camera exposure time changes significantly, t_i, t_j represent for the exposure times of the images and a_i, a_j represent for the brightness transfer function parameters. A new keyframe is extracted when $\omega_f f + \omega_{f_t} f_t + \omega_a a > 1$ and f, f_t, a represent the relative weight between three indicators.

Keeping all keyframe data in the system seems unreality because of the limit of the memory. So the marginalization using the schur complement is needed. After marginalization we prepare the keyframe image, camera pose, camera parameters and semi-dense points into a data buffer which is constituted by a list container. Our container makes sure that the DSO thread and dense mapping thread can exchange data freely (see Sect. 3.5).

After selecting keyframes, the map points are formed by estimating the depth of high-gradient points in keyframes. A point in semi-dense map is defined as the point where the source pixels ray impinges on the surface. And the inverse depth of a map point is estimated by a Gaussian framework. The original DSO only generates a sparse map, which is not sufficient for the following dense reconstruction. Therefore, two strategies including 8-pixel residual pattern [3] and keeping more active points in the sliding window are used to make our result denser. In particular, marginalization points and optimization points in the sliding window are taken into this strategy.

3.3 The Back End

In this section, the exchanged data from the container is already gotten, we use a graphics-based segmentation method in [5]. As for the pixel gradient in an image, its always existing where color suddenly changed on the edge of the object. So we can determine the position relation between semi-point clouds and superpixel contours as follow:

$$\boldsymbol{P_i} = \lambda_i K^{-1} \boldsymbol{p_i} \tag{4}$$

$$distance(\boldsymbol{P_i}, \boldsymbol{S_i}) < \varepsilon \tag{5}$$

where Eq. 4 denotes the standard pinhole model [2]. $\boldsymbol{P_i}$ denotes the 3D location. λ_i denotes the depth constant. K denotes the camera parameters. $\boldsymbol{p_i}$ denotes the 2d position in the image. $\boldsymbol{S_i}$ in Eq. 5 denotes the superpixel in an image. Threshold ε denotes the neighbor eight pixels around the position of the projection point. If the distance is less than ε, then the projection point is belong to the superpixel contour. The plane Π is estimated by the points and contours which are defined in the previous paragraph. Every contour is calculated in the image. Singular value decomposition (SVD) and random sample consensus (RANSAC) are used to fit a robust plane similar to [1]. Three evaluation criteria are proposed to judge the quality of the estimated plane. Normalized residuals test uses the distance consistency ratio.

$$\frac{dis\left(\boldsymbol{P_u}, \boldsymbol{\Pi_i}\right)}{dis\left(\boldsymbol{P_i}, \boldsymbol{P_j}\right)} < \xi \tag{6}$$

where the molecular stands for distance between the 3D points to the plane. The denominator stands for the distance between the 3D points to themselves, and the threshold ξ is -0.05. Degenerated cases exclude error contours by solving the degenerate rank in the SVD. Active Search is a very important part to reduce the error between 3D contours and 2D superpixels. After superpixel segmentation, we cannot distinguish which superpixel corresponds to the contour in a single view because every contour may have at least two neighboring superpixels. So the re-projection error between the re-projected contour and the contours of the potential matches in neighbor frames is calculated for accurate search. But low overlapping in the re-projection is rejected. Then a global energy function which contains three terms is put forward to dense mapping [1].

$$E_p = \int \omega_1 C\left(v, \rho\left(\boldsymbol{v}\right)\right) + G\left(v, \rho\left(v\right)\right) + \frac{\omega_2}{2} M\left(v, \rho\left(\boldsymbol{v}\right), \rho_\tau\left(v\right)\right) \partial v \tag{7}$$

ω_1, ω_2 represent for relative importance between the photometric, Manhattan/piecewise-planar and smoothness costs.

$$C\left(\boldsymbol{v}, \boldsymbol{\rho}\left(\boldsymbol{v}\right)\right) = \frac{1}{|I_s|} \sum_{j=1, j \neq r}^{n} \left\| \boldsymbol{I}_r\left(v\right) - \boldsymbol{I}_j\left(\boldsymbol{T}_{rj}\left(v, \rho\right)\right) \right\|_1 \tag{8}$$

Color difference between the reference image and the set of short-baseline images is calculated for photometric error. As for each pixel $\boldsymbol{v}$ in image $\boldsymbol{I_r}$, it is then backprojected at an inverse distance ρ and projected in other close image $\boldsymbol{I_j}$.

$$G\left(v, \rho\left(v\right)\right) = e^{-\beta \|\nabla \boldsymbol{I}_r(u)\|_2} \|\nabla\| \rho\left(v\right) \|_\delta \tag{9}$$

The preceding index item is the Huber norm of the gradient of the inverse depth map where β is a constant. The latter part decreases the regularization strength across superpixel contours.

$$M\left(v, \boldsymbol{\rho}\left(\boldsymbol{v}\right), \rho_\tau\left(v\right)\right) \partial v = \omega \|\rho\left(v\right) - \rho_\tau\left(v\right)\|_2^2 \tag{10}$$

ρ_τ represents for the inverse depth prior of superpixels. The third constraint is measured through distance between each point and the estimated planar prior ρ_τ. In our strategy there is a situation where the keyframe is extracted with a small parallax (See the second criterion in Sect. 3.2). So the approach in [1] is used to reject the large error areas. The informative index of an area decides whether we should discard our candidate areas or not. After then, our superpixel contours become the robust, accurate dense maps. For convenience, we display our dense map in world coordinate and then combine with semi-dense map in Sect. 3.2 directly.

3.4 Bridge Part

Map Points Selection. DSO has candidate point activation to stabilize the points number across all active frames combined. It then generates four types of points: marginalization points, hessian points, immature points and hessian-outlier points. Marginalization points are generated by schur complement. Some candidate points need to be activated and replace the marginalized points at the same time. Hessian points are active points in DSO, which are used in windowed optimization [3]. Immature points are the original candidate points for

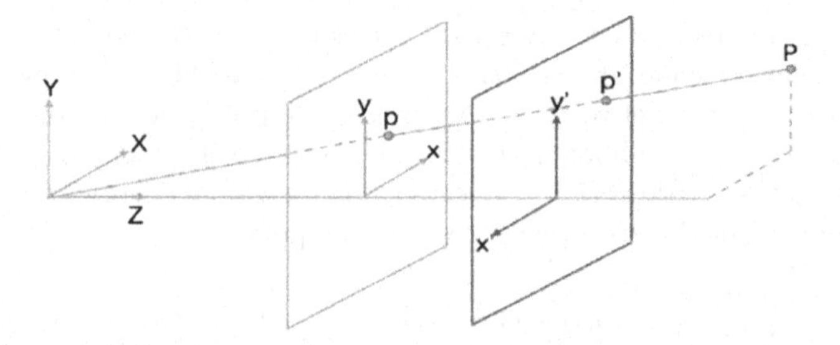

Fig. 2. The blue coordinate represents for standard pinhole model used in DSO and the red coordinate represents for pinhole model changed in DPPTAM. The red coordinate is supposed to be the same location as blue one. We take them apart for illustrating clearly. (Color figure online)

roughly camera tracking. They initialize the depth of points once the points are activated. Hessian-outlier points are potential outliers which are removed by searching along the epipolar line during camera tracking. If we put all four types of points to dense mapping thread, we may get a big reconstruction error in our system. Taking into account the different properties of the four kinds of points, we propose a map points selection strategy (see Sect. 3.5).

Container. When DSO generates a keyframe we cannot pass data to dense mapping thread im-mediately, because dense mapping thread may be running. So using container to keep keyframe data is very necessary. But if we keep all keyframe information, it will progressive increase the cost in our system. After then the system memory will be crashed. So the deleting and updating mechanisms of container must be applied to reduce the burden of the system. Simultaneously, when our dense mapping system thread runs end in one time, we should exchange the contain data to this thread.

Coordinate Systems. We observed that the coordinate systems in DSO and the dense reconstruction are different. As shown in Fig. 2, where the X axis position of the point $\boldsymbol{p}^{'}$ become the opposite number of the point $\boldsymbol{p}$. So if we combine them directly, the dense reconstruction results will separate from other DSO semi-dense results. Thus we cannot get a perfect final reconstruction result. Therefore, in the bridge part, the camera position and map points position are also transformed from the DSO coordinate system to the dense reconstruction coordinate system.

3.5 Implementation Details

According to Sect. 3.4, DSO has four types of points in each keyframe, and among them the first two types of points are suitable for generating the denser scene. Because they are more stable. Although increasing the density has little benefit in terms of accuracy or robustness, denser point clouds have a certain impact on position determination between 2D points and contours as well as plane estimation. For trade-off speed and efficiency, we finally choose 8000 active points across all active frames in this paper. Our list container includes 5 list structures. An index id, a gray image, a rotation matrix, a translation matrix and accurate semi-dense point clouds of the keyframe are contained in each structure. A three-channel image from keyframe in DSO is prepared first; then, the rotation, translation matrix are transformed from world to camera; finally, semi-dense points are converted from camera coordinate to world coordinate. And then we put them together as one piece of the list container. Once dense mapping thread is over, we exchange data right away. Besides, when the coming data exceed the limit of a container, the oldest data is removed immediately according to the index id. Owning to this container, the memory consumption of the system is in a certain range and we also maintain the key information in recent keyframes.

4 Experimental Results

Our experiments is evaluated on the Monocular Visual Odometry Dataset published by TUM [3]. We first show the intermediate results about the location consistency between semi-dense points and superpixel contours (see Fig. 3).

Fig. 3. One scene from sequence30 in dataset (left to right: keyframe, superpixel segmentation, semi-dense point cloud, dense reconstruction result).

The left image in Fig. 3 is the keyframe extracted in DSO (see Sect. 3.2). The third image in Fig. 3 is the semi-dense point cloud corresponding to the first one (see Sect. 3.3), the picture shows that those points almost corresponding to the border of the object in the image. The second image in Fig. 3 is the superpixel segmentation result in Sect. 3.3. In this picture, the planes in the real world are in the same color. The rightmost image in Fig. 3 is the dense reconstruction result after Sect. 3.5.

The white highlighted area is a part of our final dense result. These dense planes almost fill the black area in the third image in Fig. 3. Our results show

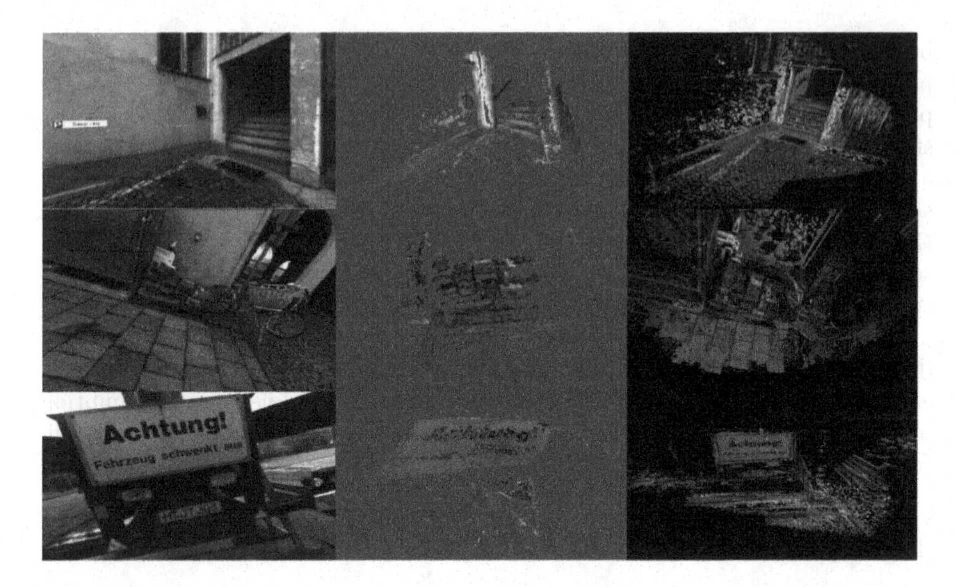

Fig. 4. Compare results (from left to right: keyframe of the scene, DPPTAM result, our result (red); from top to bottom: sequence33, sequence30, sequence20). (Color figure online)

Fig. 5. Some of our dense reconstruction results (white highlight). (from top to bottom: sequence49, sequence20, sequence30).

that our algorithm can solve the gap between the high gradient points and the low texture areas. Second, we show our results which is compared with DPPTAM. In all our experiments we have used the same parameters which have been given in the DPPTAM (re-projection error threshold, overlapping threshold etc.). Simultaneously, our system runs for a camera resolution of 640 × 480 pixels. It must be mentioned that DSO relies on a global camera. So we just use the undistortion image which is produced by DSO. And then put them into a ROS bag for DPPTAM. In most of datasets (sequence20, 26, 30 etc.), the camera position only changes a little after frame tracking initialization. But when the camera turns to another perspective and the rotation of camera is a little big, DPPTAM is weak to track the subsequent frame. And then the reconstruction pipeline will be broken. So some results of DPPTAM only have the reconstruction result at the beginning. Thus, we just show the compare results in the first few keyframes in Fig. 4. Our results in red color rebuild the low texture area successfully (see Fig. 4 right column). Figure 5 show our reconstruction of entire scene and the white high-light is our results. Our algorithm can reconstruct the streets, walls, and the surface of the building. Our algorithm runs on a 2.3 GHz Intel Core i5-6300HQ processor and 8.0 GB of RAM memory of a Lenovo Y700-15ISK laptop. Superpixel extraction using the algorithm in [1] takes around 0.18 s per image.

5 Conclusion

In this paper, we proposed a monocular dense reconstruction system that is effective and faster than the current state-of-the-art. We combined the state-of-the-art visual odometry and dense reconstruction pipeline by an elaborated bridge part. By comparison with state-of-the-art monocular method, our system can successfully run on the indoor, outdoor environments, even in textureless scenes. The experimental results indicated that the direct sparse odometry is well combined with the superpixel-based plane detection. Furthermore, our system is running on CPU and is realtime, therefore, it is possible to be applied in many fields.

References

1. Concha, A., Civera, J.: Dpptam: dense piecewise planar tracking and mapping from a monocular sequence. In: 2015 IEEE/RSJ International Conference on Intelligent Robots and Systems (IROS), pp. 5686–5693. IEEE (2015)
2. Delage, E., Lee, H., Ng, A.: Automatic single-image 3D reconstructions of indoor Manhattan world scenes. Robot. Res. 305–321 (2007)
3. Engel, J., Koltun, V., Cremers, D.: Direct sparse odometry. IEEE Trans. Pattern Anal. Mach. Intell. (2017)
4. Engel, J., Schöps, T., Cremers, D.: LSD-SLAM: large-scale direct monocular SLAM. In: Fleet, D., Pajdla, T., Schiele, B., Tuytelaars, T. (eds.) ECCV 2014. LNCS, vol. 8690, pp. 834–849. Springer, Cham (2014). https://doi.org/10.1007/978-3-319-10605-2_54
5. Felzenszwalb, P.F., Huttenlocher, D.P.: Efficient graph-based image segmentation. Int. J. Comput. Vis. **59**(2), 167–181 (2004)
6. Greene, W.N., Ok, K., Lommel, P., Roy, N.: Multi-level mapping: real-time dense monocular slam. In: 2016 IEEE International Conference on Robotics and Automation (ICRA), pp. 833–840. IEEE (2016)

7. Newcombe, R.A., Izadi, S., Hilliges, O., Molyneaux, D., Kim, D., Davison, A.J., Kohi, P., Shotton, J., Hodges, S., Fitzgibbon, A.: Kinectfusion: real-time dense surface mapping and tracking. In: 2011 10th IEEE international symposium on Mixed and augmented reality (ISMAR), pp. 127–136. IEEE (2011)
8. Newcombe, R.A., Lovegrove, S.J., Davison, A.J.: Dtam: dense tracking and mapping in real-time. In: 2011 IEEE International Conference on Computer Vision (ICCV), pp. 2320–2327. IEEE (2011)
9. Pizzoli, M., Forster, C., Scaramuzza, D.: Remode: probabilistic, monocular dense reconstruction in real time. In: 2014 IEEE International Conference on Robotics and Automation (ICRA), pp. 2609–2616. IEEE (2014)
10. Steinbrücker, F., Sturm, J., Cremers, D.: Volumetric 3D mapping in real-time on a CPU. In: 2014 IEEE International Conference on Robotics and Automation (ICRA), pp. 2021–2028. IEEE (2014)
11. Whelan, T., Kaess, M., Fallon, M., Johannsson, H., Leonard, J., McDonald, J.: Kintinuous: Spatially Extended Kinectfusion (2012)
12. Whelan, T., Leutenegger, S., Salas-Moreno, R., Glocker, B., Davison, A.: Elasticfusion: dense slam without a pose graph. In: Robotics: Science and Systems (2015)

Three Dimensional Object Segmentation Based on Spatial Adaptive Projection for Solid Waste

Yeqiang Qiu(✉), Jiawei Chen, Jianshuang Guo, Jianhua Zhang(✉), Sheng Liu, and Shengyong Chen

College of Computer Science and Technology, Zhejiang University of Technology, Hangzhou 310023, Zhejiang, People's Republic of China
qiuyeqiang@hotmail.com, zjh@zjut.edu.cn

Abstract. Automatically waste sorting is usually resorted to robot arm to grasp solid waste. However, the solid waste is prone to being stacked up, and they are difficult to segment and consequently to be grasped. To resolve this problem, we present a novel spatial adaptive projection method based on only one RGB-D sensor to accurately segment the piled solid waste. Through the RGB-D sensor, the point cloud of waste is obtained firstly. Then it will be projected into different planes to get an optimized image which is easy to be segmented. The projection parameters are automatically computed based on a mathematical optimization framework. After segmenting stacked waste into several isolated objects, we re-project them into the point cloud. We evaluate the effectiveness and accuracy of our method in a large dataset of stacked solid waste, the results shows the performance of the proposed method is satisfactory.

Keywords: Automatically waste sorting · Spatial adaptive projection
Stacked waste segmentation · Robot arm grasping

1 Introduction

As long as the growing influence to environment by more and more city solid waste, how to effectively deal with these solid waste has received increasing attention from the government and society. Recycling solid waste is a good choice, but the efficiency of classification and recycling by human is too low, so a common solution is resorted to the automatically grasping by robot arm.

However, before the robot arm can grasp an object, it should have the object information. In some traditional, planning-based robot grasping methods, they typically use a priori about object models to acquire the necessary shape and pose information [5,6,10,13]. These approaches can only handle the limited objects with knowing models and geometric shape in advance. They are not suitable for grasping solid waste, because models and geometric shapes of solid waste are diverse. In addition, a mass of solid waste with similar color and texture is usually stacked together when using robot arm to grasp. Therefore, traditional object detection and segmentation based on 2D images is also not suitable for them.

J. Yang et al. (Eds.): CCCV 2017, Part III, CCIS 773, pp. 453–464, 2017.
https://doi.org/10.1007/978-981-10-7305-2_39

Thus, the point cloud of stacked solid waste should be obtained through Lidar or depth camera for segmentation. However, because the solid waste is stacked, it is still difficult to segment the point cloud obtained from a fixed point of view. To solve this problem, an approach is to increase the number of sensors which are fixedly installed in some different point of views. Nevertheless, this solution has two shortages. At first, it increases the cost. Second, fixed installation cannot satisfy all different condition of stacked solid waste.

To solve this problem, we present a novel smart strategy that only uses one cheap depth camera with fixed installation. This strategy is inspired by our careful observation. We observe the point cloud, if it is difficult to segment from the fixed point of view, may be segmented easily from other point of view. Therefore, we just need to rotate the point cloud to find a good point of view, from which the point cloud of the connecting regions between stacked objects is relatively sparse, as shown in Fig. 1. Figure 1(b) is the point cloud observed from the depth camera which is fixed installation. It is obvious that this point cloud is difficult to segment. But if we observe the point cloud from another point of view, as shown in Fig. 1(c), it is easy to segment. Therefore, the core idea of the presented strategy is to find the optimized point of view.

To this end, we develop a novel adaptive projection method. It uses the point cloud obtained by the fixed installed depth camera as input, and projects the point cloud into different planes. This procedure simulates a camera to move on a sphere and take pictures for the stacked solid waste. But our method does not need to move the camera to cover the whole sphere, only move along a special path which is determined by our optimization algorithm. After obtaining a projected image which is the easiest one to segment objects, this procedure finishes. We then segment objects in this image, re-project into the point cloud, and finally obtain the point cloud for different objects. The proposed method is evaluated on a large dataset which we collected from a real production line. The experimental results demonstrate that our method is satisfactory.

This paper is organized as follows. In the next section we give a brief review about point cloud segmentation. In Sect. 3, the space adaptive projection point cloud segmentation method is presented and visualized for each individual steps. Section 4 provides the experimental results of the proposed method. Finally, we summarize the main achievements of this work and draw some conclusion.

2 Related Work

There are many existing point cloud segmentation methods provided by the Point Cloud Library [11], which aims to fit specific object, like plane model, cylinders model and spheres model. Region growing segmentation algorithm is to merge the points that are close enough in terms of the smoothness constraint. Euclidean Cluster Extraction algorithm make use of nearest neighbors and implement a clustering to separate objects. In [14], a progressive morphological filter is implemented for segmenting ground points. The algorithm in [3] is for scale-based segmentation of unorganized point clouds, it performs a scale based segmentation of a given input point cloud to find the points that belong to the given scale parameter. While in the actual process, the solid waste is not individually lying on the conveyor belt. In some complicated cases, such as multiple objects stacked, the algorithm mentioned above can not handle these situations well. In this paper, we focus on the problem of segmenting stacked objects, and therefore these methods are not suitable.

The algorithm in [2,12] are state-of-the-art methods, known as LCCP and CPC. They both use Voxel Cloud Connectivity Segmentation (VCCS) to over-segments point cloud into patches [9], a recent method for generating super-voxels (a 3D analog of super-pixels). LCCP then segment the adjacent patches by classifying whether the connection between two patches is convex or concave based on Extended Convexity Criterion and Sanity criterion, finally employ Region growing algorithm to cluster small regions into larger objects. CPC introduced Locally Constrained Directionally Weighted RANSAC algorithm and applied it on the edge cloud extracted from the adjacent patches to segment the point cloud. These two methods are able to segment stacked objects. However they are not adaptive segmentation methods, and different parameters need to be set for different data. Therefore, these two methods are not suitable for us and we will show it in detail in the experiment.

The modern 3D semantic segmentation methods like in [4,8] need to train 3D models using a large amount of 3D ground truth annotated point cloud, which is usually computationally and time-consuming. Moreover, the model of waste object are too diverse, so it is also not suitable for us.

3 Method

As observed from raw data shown in Fig. 1(a), the solid waste is on the belt where the belt is a planar surface. It is necessary to segment the belt and extract the objects point cloud above the belt before separating them. The algorithm in [6,7,9] can fit a plane into a set of neighboring points. However, the belt is not all flat due to the mechanical alignment error, where the four corners is a little higher than the plane. If segmenting the 4 corners, some thin objects would be segmented as plane. So we use the 3D background-difference method instead. This method is similar to the image background-difference method. A frame with no object on the belt is used as the background frame, and we perform differential operation between the current frame and background frame to obtain the object cloud. Then Euclidean Cluster Extraction algorithm is used to segment the objects into clusters, which is our target point cloud to be segmented.

Our segmentation algorithm includes four parts: (a) Point cloud projection, where the point cloud is projected into a plane and converting it to an image. (b) Objective function construction, where a function that represents the quality of projected images is constructed. The quality measures how easily to segment the object and how accurate the segmented result. For the same segmentation object, the lower objective function value, the higher quality of projected image, and the easier to segment. (c) Optimization, where the mathematical optimization is used to find the optimal projected image, which has the minimum objective

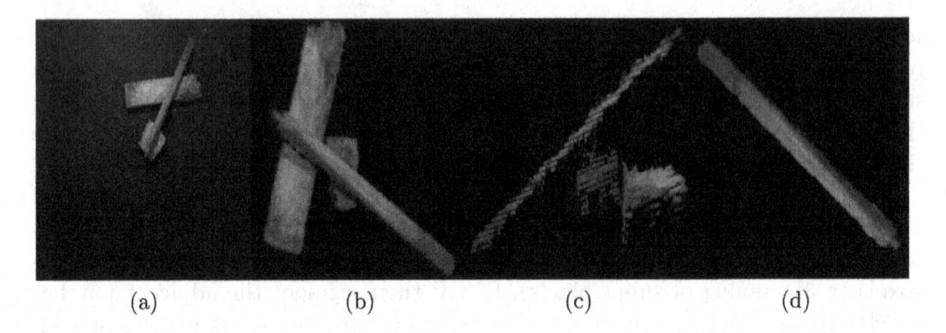

(a) (b) (c) (d)

Fig. 1. (a) Is the raw point cloud, (b) is our segmentation object extract from (a), (c) is the optimized view of (b) and can segments three objects easily, (d) is the segmentation result for robot grasping

function value. (d) Segmenting object, where the optimal projected image is segmented, and re-projected into the original point cloud to obtain objects.

3.1 Point Cloud Projection

The first step is projecting point cloud into images, which is the basic of the whole algorithm. Before projecting the point cloud, it needs to be filtered. The raw point cloud data generated by Lidar or depth camera has noise and outliers. We use statistical technique to eliminate outliers and reduce noise [15], which is based on the estimation of a set of points in a given neighborhood. In Eq. 1, P^f represent the point cloud after filtering, P^{raw} is the raw point cloud, the point outside $d_{thresh}.\sigma$ distance from the mean distance μ is outliers. σ is the standard deviation. The neighborhood is set to 50 and d_{thresh} is set to 1.0. The neighborhood and d_{thresh} are fixed for different types of solid waste.

$$P^f = \left\{p_i^r \in P^{raw} | \overline{dist}\left(p_i^r, p_j^r\right) < \mu + d_{thresh}.\sigma\right\} \tag{1}$$

After filtering outliers, we also need to smooth the point cloud, for the reason of the raw point cloud structure. As shown in Fig. 2(a), it is stratified and may cause an exception value when building the objective function, because it may form the bad projected cloud and undesired projected image. We employ the Moving Least Squares algorithm to smooth the point cloud [1], which can also correct the small errors that statistical filter cannot remove. As shown in Fig. 2(d), the point cloud is not stratified after smoothing, and both the projected point cloud and projected image are look well.

Through the pre-processing, we obtain the point cloud P^f. Now the point cloud can be projected into a plane. The projecting procedure is as follows. P^f is first translated to the coordinate system where the origin is the center of P^f. Then a sphere with the radius r is constructed and the tangent plane of each point in the upper hemisphere is used as projecting plane. Thus all projected points coordinates in each tangent plane can be calculated by two angle parameters θ and ϕ, and the radius r.

First, the maximum and minimum values of X, Y and Z axes are calculated and denoted as a set $3dminmax = \{x_{min}, x_{max}, y_{min}, y_{max}, z_{min}, z_{max}\}$. Thus, as shown in Eq. 2, the center of P^f and the translation vector T can be computed. Through T, the transformed point cloud P^t is obtained.

$$\begin{cases} p_{center} = \left(\frac{(max.x+min.x)}{2}, \frac{(max.y+min.y)}{2}, \frac{(max.z+min.z)}{2}\right) \\ T = \left[-p_{center}.x \quad -p_{center}.y \quad -p_{center}.z\right]^{\mathrm{T}} \\ P^t = \left\{p_i^f \in P^f | p_i^f = p_i^f + T\right\} \end{cases} \tag{2}$$

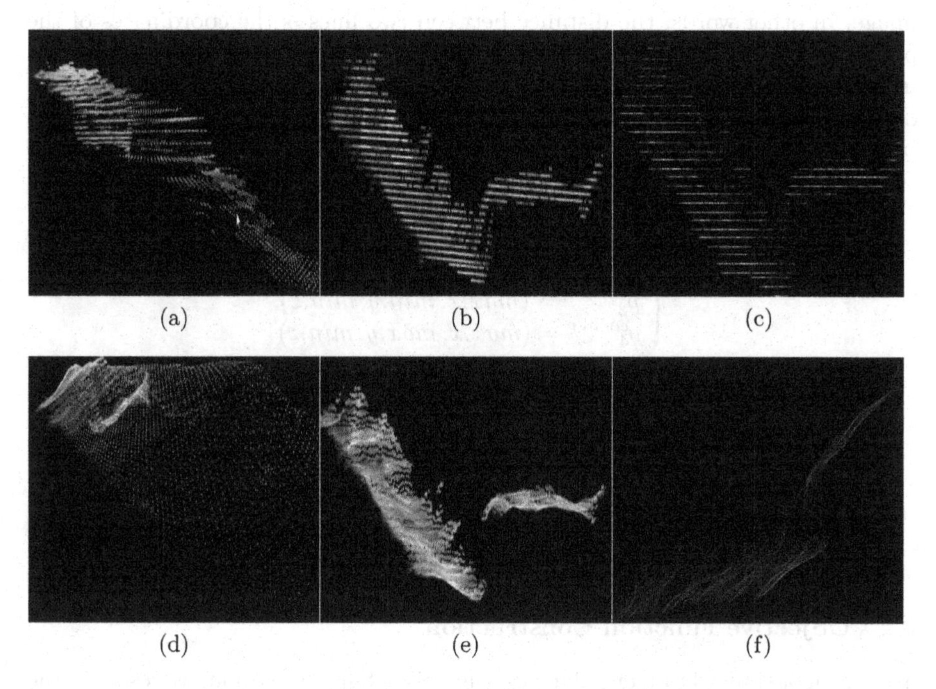

Fig. 2. (a) Is the raw point cloud, (b) is our segmentation object extract from (a), (c) is the optimized view of (b) and can segments three objects easily, (d) is the segmentation result for robot grasping

Then we build a sphere with a radius of r, the center of the sphere is at the origin of the coordinate system. The projecting plane is all the plane tangent to the sphere. The spherical coordinate system is employed to represent the planes. Denoting PP as the projecting plane set, θ as the angle between positive Z axis and the normal vector of the projecting planes, ϕ as the angle between positive X axis and the projecting line of the normal vector projecting in the XOY, PP can be computed by Eq. 3:

$$PP = \left\{ pp_i = \begin{matrix} r * \sin\theta * \cos\phi * x \\ +r * \sin\theta * \cos\phi * y \\ +r * \cos\theta * z \end{matrix} \middle| \begin{matrix} \theta \in [0, 360] \\ \phi \in [0, 360] \end{matrix} \right\} \tag{3}$$

Next, the point cloud P^t can be projected accordingly. As shown in Fig. 2(b) and (e), they are the projected point cloud from one angle. After rotating the point cloud parallel to XOY, we converting it to an image. First get the *3dmin-max* of the point cloud, through which to determine the 4 corner points and the size of the image. As formulated in Eqs. 4 and 5, the ratio of the length and width of the image is determined by the difference between the X and the Y. Assuming the difference in X is greater than the difference between Y, so the line formed by p_3^{corner} and p_1^{corner} is the width edge l_{width} of the image, and the line formed by p_2^{corner} and p_1^{corner} is the height edge l_{height} of the image. The distance of one 3D point to the two lines determines their position in the image. In other words, the distance between two lines is the coordinates of the pixel point to the image. If a point is in the same pixel point, the pixel value will increase 40. Thus, the more point in the same position, the brighter the position of the projected image. As shown in Fig. 2(c) and (f), some of the image are bright, while others are relatively darker. The place where is black means no point in this position.

$$\begin{cases} p_1^{corner} = (min.x, min.y, min.z) \\ p_2^{corner} = (min.x, max.y, min.z) \\ p_3^{corner} = (max.x, min.y, min.z) \\ p_4^{corner} = (max.x, max.y, min.z) \end{cases} \tag{4}$$

$$\begin{cases} image.width = \overline{max}\,(max.x - min.x, max.y - min.y) \\ image.height = \overline{min}\,(max.x - min.x, max.y - min.y) \end{cases} \tag{5}$$

$$Pixel\,(x, y) = 40 * \overline{num}\left(\overline{comp}\left(\left(\overline{dist}\left(p_i^f, l_{height}\right), \overline{dist}\left(p_i^f, l_{width}\right)\right), (x, y)\right)\right) \tag{6}$$

3.2 Objective Function Construction

To find an optimized tangent plane for the point cloud projection, we resort to the optimization technique to obtain the plane parameters. Therefore, an objective function should be constructed in advance. This objective function consists of the plane parameters which determining the view angle, and the result of this objective function should be small if the projection result is segmented easily and accurately given current plane parameters.

As shown in Fig. 2(c) and (f), points in the projected image is relatively sparse and difficult to segment. Therefore, a 5×5 Gaussian kernel is applied to inflate the image. Then a threshold operation is performed in the projected image and a binary image can be obtained, as shown in Fig. 3. If objects in one image is easy to segment, the objects must scatter and consequently the bounding box around all objects will be large. However, for segmenting accurately, the separation between objects must be big. Therefore, the ratio between objects area and the length of bounding box can be used as our objective function. It can be seen from this figure, the good view angles are in Fig. 3(c) and (d), where the ratio between the contour area (white pixels) and length of contour bounding box (red line) is small. Thus, we firstly have the following objective function as defined in Eq. 7, where $carea\,(\theta,\phi)$ is the contour area of the projected image, and $ilength\,(\theta,\phi)$ is the perimeter of the projected image.

$$\begin{gathered}\arg\min_{\theta,\phi} f\,(\theta,\phi)\\ where,\, f\,(\theta,\phi) = carea\,(\theta,\phi)\,/ilength\,(\theta,\phi)\end{gathered} \tag{7}$$

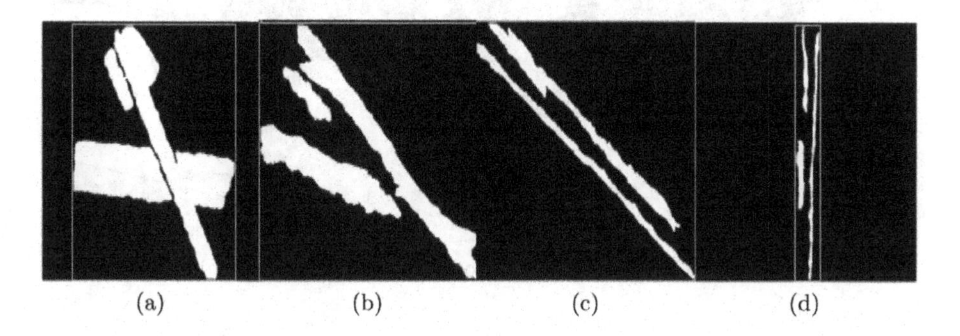

Fig. 3. Four projected image from four angles (Color figure online)

However, by comparison the last two projected image, i.e. Fig. 3(c) and (d), the better one is Fig. 3(d), but the ratio computed by Eq. (7) is higher. This means just using Eq. (7) is not sufficient. The reason behind this is that the number of isolated objects will also influence the results. Therefore, the number of objects also needs to be taken into consideration. Finally, the objective function is updated as shown in Eq. 8, where $ccoe\,(\theta,\phi)$ is the contours coefficient. The contour coefficient is related to the number of contours, but we constrain it not bigger than 3 for avoiding the over-segmentation.

$$\begin{gathered}\arg\min_{\theta,\phi} f\,(\theta,\phi)\\ where,\, f\,(\theta,\phi) = carea\,(\theta,\phi)/\,(ilength\,(\theta,\phi) * ccoe\,(\theta,\phi))\end{gathered} \tag{8}$$

3.3 Optimization

In order to obtain the best projecting angle quickly, the steepest descent method is employed to optimize the objective function. In Fig. 4, we show the objective function value from all projecting angles. The point within the red regions are the optimal projecting value, where their gradient values $g\,(\theta,\phi)$ approach zero. The steepest descent method takes the negative direction of the gradient as the search direction and searches continuously in that direction until the gradient value tends to zero. In Eq. 10, X is the angle (θ,ϕ). λ is the search step length

Fig. 4. Object function prototype (Color figure online)

and it is the key of the steepest descent method to find the optimized result rapidly. If λ is too large, it may be divergent. Otherwise, the convergence speed will be too slow. Therefore, the golden section method is employed to solve the problem as to find the value make $\omega(\theta, \phi) = 0$. $\lambda = 0$. It is a one-dimensional optimization method, and searches for the optimal value by narrowing the search interval. When finding out the value, return to Eq. 10, until $f(X_{k+1}) = 0$. Thus, we obtain the best projecting angle X_{k+1}.

$$g(\theta, \phi) = \begin{bmatrix} \lim\limits_{\Delta\theta \to 0} \frac{f(\theta+\Delta\theta,\phi)-f(\theta,\phi)}{\Delta\theta} \\ \lim\limits_{\Delta\phi \to 0} \frac{f(\theta,\phi+\Delta\phi)-f(\theta,\phi)}{\Delta\phi} \end{bmatrix} \tag{9}$$

$$f(X_{k+1}) = f(X_K + \lambda * (-g(X_k))) \tag{10}$$

$$\omega(\lambda) = f(X_k + \lambda * (-g(X_k))) \tag{11}$$

There are some points as shown in the yellow region of Fig. 4. Their gradient values approach zero, while their values is not the minimum or maximum. They are called Saddle point. In order to avoid these saddle points, we select 8 angles for projection before choosing the initial projection angle. And the minimum point of the objective function value will be set as the initial point for optimization. The eight angles are $(0, 45), (90, 45), (180, 45)$, $(270, 45), (45, 90), (135, 90), (225, 90), (315, 90)$. Finally, we start from the initial point to find the optimal projecting angle.

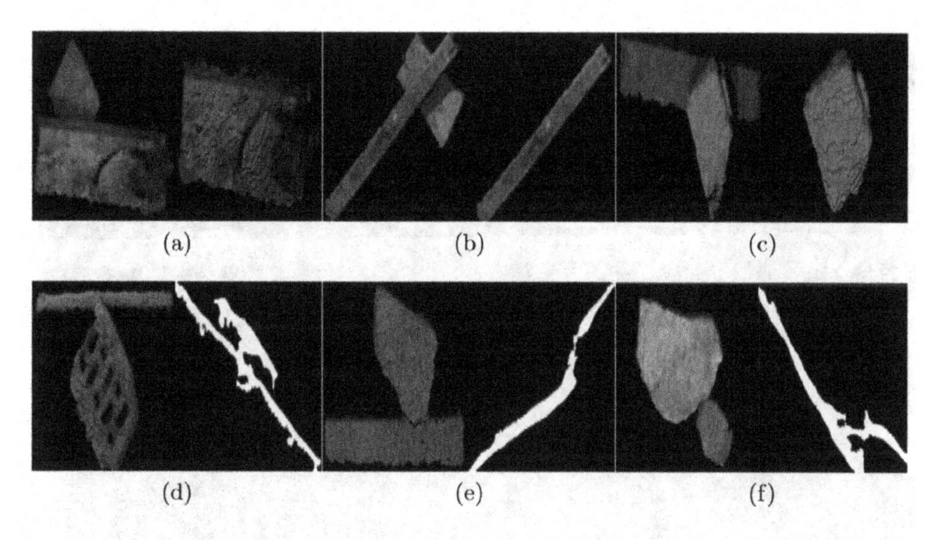

Fig. 5. Segmentation result

3.4 Segment

After we obtain the optimal projecting angle and projected image. From the optimized projected image, it is easy to segment objects, because most of these objects are separated, as shown in the first row of Fig. 5. Even in the case that objects are adjacent, it still can be segmented simply. As shown in the second row of Fig. 5, we apply the morphological method to segment objects. Next we re-project these segmented results into the point cloud, and obtain the final results (Fig. 6).

$$P^{obj} = \left\{ p^f \in P^f | \overline{project} \left(p_i^r \right) \in C_j \right\} \tag{12}$$

4 Experiment

Since there is no solid waste dataset, we use our own dataset to experiment. In practice, we use an ASUS Xtion as the depth camera to obtain the point cloud. We use the conveyor belt to transport solid waste. It first arrives at the visual area where the depth camera is fixed above the belt, and then reaches the grasp area where the robot is located. In the visual area we obtain the RGB image and point cloud data of solid waste. We obtained 189 sets of data as experimental subjects.

As shown in Fig. 7, they are the three representative experiment results. Data1 contains two objects, one is brick, and the other is wood. The LCCP and CPC segment the bricks into two objects, while our approach can segment it well. Data2 contains two stacked tricks, these three methods can segment it into two objects. But the segments result of LCCP and CPC are not accurate enough,

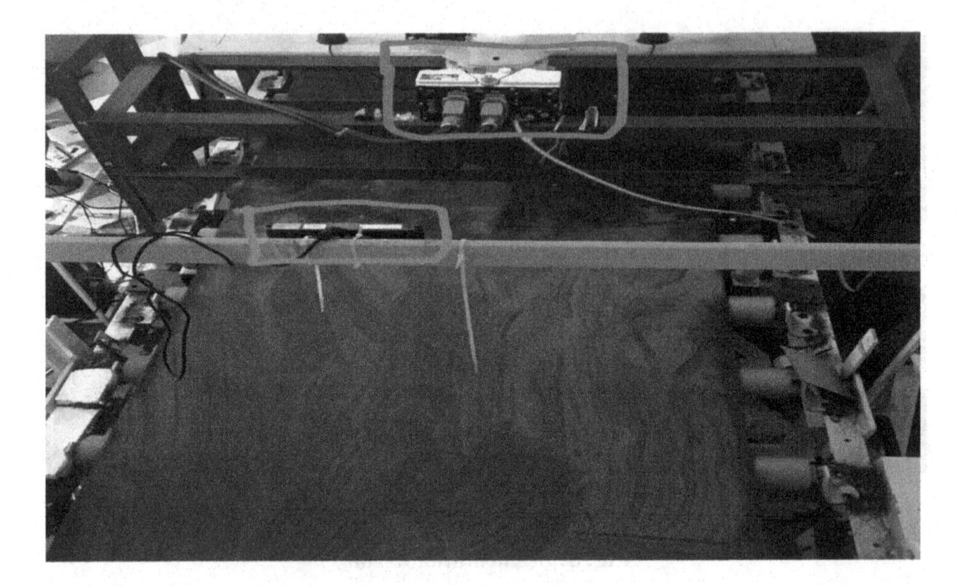

Fig. 6. Experimental scene, the red area is the depth camera ASUS Xtion, the blue area is the robot arm Kawasaki RS010L (Color figure online)

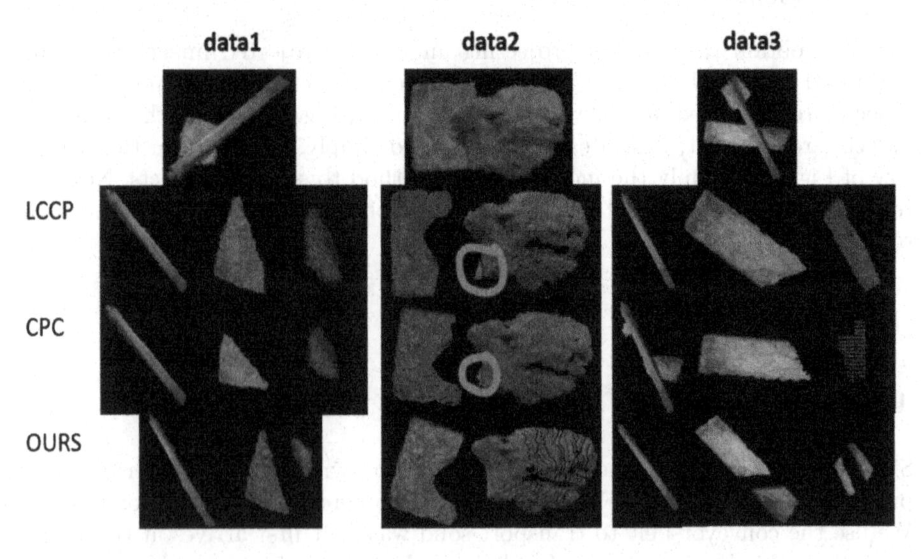

Fig. 7. Three representative experiment results (Color figure online)

the red region does not belong to the brick. While our result does not have this part. Data3 contains three objects, both LCCP and CPC method segment it into more than three objects under the same set thresholds with the data1 and data2. It indicates that they are not self-adaptive, which is the main reason why they are not suitable for our solid waste automatic sorting.

Table 1. Segmentation results comparison

Method	Over-segmentation rate	Under-segmentation rate	Segmentation accuracy
LCCP	31.7%	6.3%	91.9%
CPC	27.5%	4.2%	93.5%
OURS	5.3%	7.9%	98.7%

Table 1 shows the results of the experiment under 189 set of data. Over-segmentation means the method will segment one object into multiple objects, while under-segmentation represents the method can not separate multiple objects. Segmentation accuracy means the proportion of the part belong to the object when segment it into correct objects. We compared the three segment results with the manual segment results, and got the above data. It shows our method get better result in the waste solid segmentation.

5 Conclusion

In this paper, we proposed a segmentation algorithm based on adaptive spatial projection for the solid objects in the belt, which is a model-free segmentation algorithm and not limited by the given object model, the world knowledge, or the ability to extract the supporting planes. The algorithm can segment the stacked objects adaptively, which is achieved by projecting the point cloud into images in an optimal angle. The optimal angle is obtained by constructing object function to represent the segmentation result. And steepest descent method and golden section method are used to get the optimal angle rapidly. Finally we segment the object from the optimal projecting image. More importantly, our approach is adaptive and can handle the stacked objects that transported by conveyor belt.

References

1. Alexa, M., Behr, J., Cohen-Or, D., Fleishman, S., Levin, D., Silva, C.T.: Computing and rendering point set surfaces. IEEE Trans. Vis. Comput. Graph. **9**(1), 3–15 (2003)
2. Christoph Stein, S., Schoeler, M., Papon, J., Worgotter, F.: Object partitioning using local convexity. In: Proceedings of the IEEE Conference on Computer Vision and Pattern Recognition (CVPR), pp. 304–311 (2014)
3. Ioannou, Y., Taati, B., Harrap, R., Greenspan, M.: Difference of normals as a multi-scale operator in unorganized point clouds. In: Proceedings of the International Conference on 3D Imaging, Modeling, Processing, Visualization and Transmission (3DIMPVT), pp. 501–508 (2012)
4. Kahler, O., Reid, I.: Efficient 3D scene labeling using fields of trees. In: Proceedings of the IEEE Conference International Conference on Computer Vision (ICCV), pp. 3064–3071 (2013)

5. Kuehnle, J., Verl, A., Xue, Z., Ruehl, S., Zoellner, J., Dillmann, R., Zoellner, R.: 6D object localization and obstacle detection for collision-free manipulation with a mobile service robot. In: IEEE Conference on Robotics and Automation (ICRA), pp. 1–6 (2009)
6. Narayanan, V., Likhachev, M.: Perch: perception via search for multi-object recognition and localization. In: IEEE Conference on Robotics and Automation (ICRA), pp. 5052–5059 (2016)
7. Pham, T.T., Eich, M., Reid, I., Wyeth, G.: Geometrically consistent plane extraction for dense indoor 3D maps segmentation. In: Proceedings of the IEEE Conference on Intelligent Robots and Systems (IROS), pp. 4199–4204 (2016)
8. Pham, T.T., Reid, I., Latif, Y., Gould, S.: Hierarchical higher-order regression forest fields: An application to 3D indoor scene labelling. In: Proceedings of the IEEE Conference Computer Vision and Pattern Recognition (CVPR), pp. 2246–2254 (2015)
9. Rabbani, T., Van Den Heuvel, F., Vosselmann, G.: Segmentation of point clouds using smoothness constraint. Int. Arch. Photogramm. Remote Sens. Spat. Inf. Sci. **36**(5), 248–253 (2006)
10. Rusu, R., Bradski, G., Thibaux, R., Hsu, J.: Fast 3D recognition and pose using the viewpoint feature histogram. In: Proceedings of the IEEE Conference on Intelligent Robots and Systems (IROS), pp. 2155–2162 (2010)
11. Rusu, R., Cousins, S.: 3D is here: Point Cloud Library (PCL). In: IEEE Conference on Robotics and Automation (ICRA), pp. 1–4 (2011)
12. Schoeler, M., Papon, J., Worgotter, F.: Constrained planar cuts-object partitioning for point clouds. In: Proceedings of the IEEE Conference on Computer Vision and Pattern Recognition (CVPR), pp. 304–311 (2014)
13. Sun, M., Bradski, G., Xu, B.-X., Savarese, S.: Depth-encoded hough voting for joint object detection and shape recovery. In: Daniilidis, K., Maragos, P., Paragios, N. (eds.) ECCV 2010. LNCS, vol. 6315, pp. 658–671. Springer, Heidelberg (2010). https://doi.org/10.1007/978-3-642-15555-0_48
14. Zhang, K., Chen, S., Whitman, D., Shyu, M., Yan, J., Zhang, C.: A progressive morphological filter for removing nonground measurements from airborne lidar data. IEEE Trans. Geosci. Remote Sens. **41**(4), 872–882 (2003)
15. Zhang, Z.: Iterative point matching for registration of free-form curves and surfaces. Int. J. Comput. Vis. **13**(2), 119–152 (1994)

RGB-D Based Object Segmentation in Severe Color Degraded Environment

Chao Wang, Sheng Liu(✉), Jianhua Zhang, Yuan Feng, and Shengyong Chen

College of Computer Science and Technology, Zhejiang University of Technology, Hangzhou 310023, Zhejiang, China
edliu@zjut.edu.cn

Abstract. In the robotic waste sorting lines, most existing segmentation methods will fail due to irregular shapes and color degradation of solid waste. Especially in the cases of adhesion and occlusion, errors may frequently occur while labeling the ambiguous regions between solid waste objects. In this paper, we propose an efficient RGB-D based segmentation method for construction waste segmentation in harsh environment. First, an efficient background modeling strategy is designed to separate the solid waste regions from the cluttered background. Second, we propose an ambiguous regions extracting method to deal with the problem of adhesion and occlusion. Finally, a relabeling method is developed for ambiguous regions and a high precision segmentation will be obtained. A dataset of construction materials consists of RGB-D images is built to evaluate the proposed method. Results show that our approach outperforms other state-of-the-art methods in harsh environment.

Keywords: Solid waste recycling · Object segmentation
Ambiguous region extraction

1 Introduction

Construction waste recycling is most sustainable way to manage waste materials generated during construction and remodeling. Robotic waste sorting is one of the key technology in construction waste recycling. And, solid waste objects segmentation is the core technology in robotic waste sorting. Traditional treatment of construction waste is landfill, which will cause serious air and soil contamination. With the rapidly increasing amount, situation will be increasingly severe. So there is growing interest in studying the construction waste recycling. The appearance of robots provides a new, more efficient solution to this problem as they can grasp object quickly and work continuously. Image segmentation algorithms, especially with depth information, are indispensable in robotic waste sorting as they can offer solid waste objects' information of positions and contours [10–12,15].

Custom image segmentation algorithms are not applicable in industrial field. Most 2D image segmentation methods [2,3] use features such as color and contour. Though they achieve a good performance on some datasets, they cannot

J. Yang et al. (Eds.): CCCV 2017, Part III, CCIS 773, pp. 465–476, 2017.
https://doi.org/10.1007/978-981-10-7305-2_40

deal with complex industrial scene. An example image of construction waste scene is shown in Fig. 1 (a). The surface of the conveyor belt is covered with dust and solid waste objects are not all isolated. Because of the color degradation caused by deposition of dust particles on the surface of the solid waste, custom 2D image segmentation methods lose their effect. With the advent of depth sensors, image segmentations with the depth information became a research focus. However, recent work on RGB-D image process has been targeted to semantic segmentation or labeling [1,4,16]. These methods aimed to assign a category label to each pixel of an image and complete segmentation and classification of whole scene. As for the task of solid waste sorting, objects contours and positions are indispensable to find best grab points and angles. Most semantic segmentation methods lack the concept of object instance, they can not separate adhesive objects with same category label. Furthermore, deep learning is widely used in semantic labeling such as [5,13]. However, solid waste objects are volatile in shape and color degraded so that it is difficult to collect a representative training set. It is needed to develop a new algorithm for solid waste objects segmentation in harsh environment.

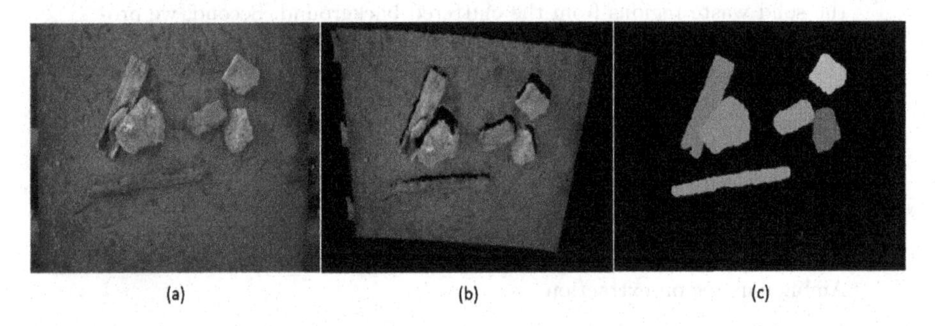

Fig. 1. A segmentation example of proposed algorithm. (a) Original image of cluttered belt scene. (b) Point cloud of cluttered belt scene. (c) Result of proposed algorithm.

In this paper, we propose an efficient RGB-D based segmentation method for construction waste segmentation in harsh environment. Our contributions in this paper are summarized as follows: (1) we propose a strategy to extract ambiguous regions to separate adhesive and occluded objects and perform pixel-level relabeling on ambiguous regions to get a high precision result. (2) We also build a unique challenging dataset of construction waste.

The paper is structured as follows. The next section discusses the presented work in RGB-D image segmentations. Sect. 3 shows the structure of the proposed method, and explains more detailed about every module. Evaluation and results are shown in Sect. 4, before the work ends with a conclusion in Sect. 5.

2 Related Work

In 3D scene, a single object consists of several planes. Segmenting an image in separate objects means labeling each plane. The approach proposed by Holz et al. [6] compute local surface normal and cluster the pixels in different planes in normal space. And the result planes are segmented and classified in different objects in both normal space and spherical coordinates. Analogously, [16] propose a multiscale-voxel algorithm to do plane extraction. Then the result planes are combined with depth data and color data to apply graph-based image segmentation. In [12], Richtsfeld et al. preprocess input RGB-D image based on surface normal, get surface patches by using a mixture of planes and NURBS (non-uniform rational B-splines) and find the best representation for the given data by model selection [8]. Then, they construct a graph from surface patches and relations between pairs of plane patches and perform graph cut to arrive at object hypotheses segmented from the scene. Irregular shapes of solid waste objects will lead to over-segmentation by plane-based methods. Moreover, plane-based methods do not deal well with touched objects caused by adhesion and occlusion.

An object is also defined as a compact region enclosed by a closed edge which is called object contour. Segmenting objects means finding effective object contours in image. In [10], Mishra et al. build a probabilistic edge map obtained by color and depth cues, and then select only closed contours that correspond to objects in the scene while reject duplicate and non-object closed contours by the fixation-based segmentation framework [9]. Toscana and Rosa [15] use a modified canny edge detector to extract robust edges by using depth information and two simple cost functions are proposed for combining color and depth cues to build an undirected graph, which is partitioned using the concept of internal and external differences between graph regions. As color degradation in construction waste images is severe, algorithms based on color cues lose their effect.

3 Our Algorithm

3.1 System Overview

Our algorithm consists of three major parts: Background Modeling, Ambiguous Regions Extraction and Ambiguous Regions Relabeling. Figure 2 shows the processing chain of those parts in detail. Background modeling builds a background model based on depth information, and foreground mask is got by comparison between object point cloud and background model. Each connected region in foreground mask is defined as a local mask. Ambiguous regions extraction is performed on local mask to extract ambiguous regions to separate adhesive and occluded objects. At last, multiple adjacent superpixel sets are selected to relabel ambiguous regions on pixel-level. In followed parts, we introduce these three modules in more details.

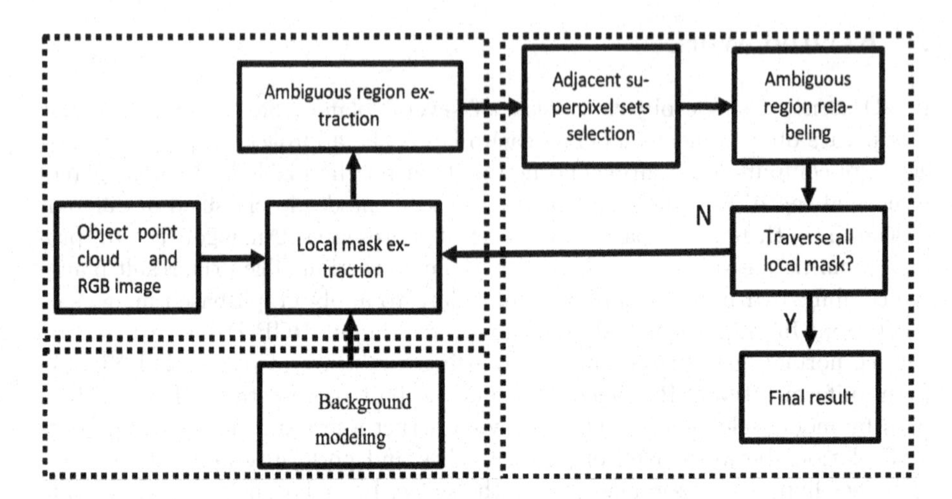

Fig. 2. System overview: background modeling, Ambiguous Regions Extraction, Ambiguous Regions Relabeling.

3.2 Background Modeling

To separate objects from the dusty conveyor belt, a depth-based background subtraction algorithm is used. Most existing methods for background subtraction build a color background model by a number of background frames [14]. However, in industrial scene, camera is fixed while the conveyor belt is running continuously. Color-based model is unreliable in industrial scene as background colors are changing extremely. Instead, depths gained by 3D sensors are more stable, so it is easy to think of using depth information as a cue to achieve background subtraction. Though depth information is more credible than color, a simple subtraction between background point cloud and object point cloud is ineffective. The reasons come that conveyor belt is deform because of the weight of objects and the degree of deformation is not coincident on different parts. Moreover, depth is also influenced by the vibration caused by the running of conveyor belt. So we build a depth-based background Gaussian Mixture Model to remove the background from the image.

Adaptive Gaussian Mixture Model proposed in [7] re-investigates the update equations and utilizes different equations at different phases. This allows algorithm learn faster and more accurately as well as adapt effectively to changing environments. We use depth information to build Adaptive Gaussian Mixture Model of background conveyor belt. Each pixel is modeled by a mixture of K (K is a small number from 3 to 5) Gaussian distributions. We find that $K = 3$ is enough in our scenario by experiment as the depth varies within a certain range. Background conveyor belt frames are inputted to build background model, and when a new object RGB-D image comes, a comparison is processed between the point cloud and the model to generate foreground mask of solid waste objects.

3.3 Ambiguous Region Extraction

In foreground mask, there are several connected regions. Each connected region is defined as a local mask, and it is difficult to judge whether a local mask is single object or touched objects caused by adhesion and occlusion. Furthermore, touched objects separating is also a difficulty. Our algorithm solves these two problems by extract ambiguous regions based on inner edges and SLIC superpixels.

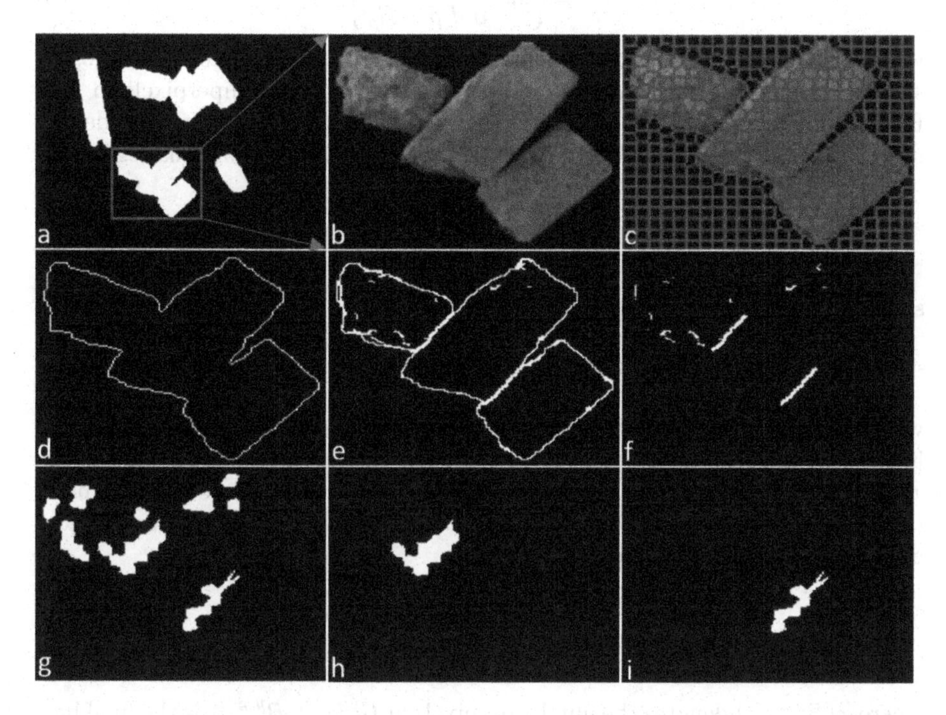

Fig. 3. The process of ambiguous region extraction. (a) Foreground mask. (b) Local image without background. (c) The result of SLIC. (d) The contour of local mask. (e) Local edge map. (f) Local inner edge map. (g) Borderline superpixels. (h) and (i) are ambiguous regions.

Firstly, for each local image, a SLIC superpixel segmentation is performed on and the superpixel set is extracted and defined as $S = \{s_1, s_2, s_3, \cdots, s_{n-1}, s_n\}$ based on the result of SLIC. s_i denotes a superpixel and it is a pixel set as it consists of multiple pixels. Local edge map is also generated based on computing depth gradient and defined as E_m. Let F_c denotes the contour of local mask, and then, inner edge map is computed by

$$E_{inner} = E_m - F_c \bigoplus C_{2*k+1}, \quad (1)$$

where E_{inner} is inner edge map and $F_c \bigoplus C_{2*k+1}$ denotes a dilation operation on F_c with kernel size of $2*k+1$. Inner edge pixel set E_p is extracted based on E_{inner} by

$$E_p = \{p(x,y)|E_{inner}(x,y) = 255\}, \tag{2}$$

where $E_{inner}(x,y)$ means the pixel value of column x and row y in E_{inner}. Then, we combine inner edge pixel set and SLIC superpixel set to extract borderline superpixels by

$$B_{sp} = \left\{ s_k \middle| \begin{array}{c} p \in s_k \\ and\ p \in E_p \end{array} \right\}, \tag{3}$$

where B_{sp} is a set of borderline superpixels and p is the pixel in image. As shown in Fig. 3 (g), there are several connected borderline superpixels in B_{sp}. Each connected borderline superpixel set is extracted and defined as a borderline region B_{region}. After that, an iteration is used to extract ambiguous region:

$$M_{obj} = M_{local} - B^{x_{th}}_{region}, \tag{4}$$

where M_{local} is local mask and $B^{x_{th}}_{region}$ is borderline region with x_{th} times expansion. $B^{x_{th}}_{region}$ is expanded to $B^{(x+1)_{th}}_{region}$ by

$$B^{(x+1)_{th}}_{region} = B^{x_{th}}_{region} \cup A^{x_{th}}_{sp}, \tag{5}$$

where $A^{x_{th}}_{sp}$ is the set of neighboring superpixels of $B^{x_{th}}_{region}$ and $B^{x_{th}}_{region}$ is borderline region with x_{th} times expansion. The iteration is processing with increasing x from 0 to 4 and stopped when $x > 4$ or M_{obj} has two or more effective parts (have 7 superpixels or more). When the iteration is over, the importance score of borderline region is computed by

$$P_{B^{y_{th}}_{region}} = \begin{cases} (1 - \frac{\varphi(B^{y_{th}}_{region})}{\varphi(M_{local})})^y & \text{if } f = 1 \\ 0 & \text{if } f = 0 \end{cases}, \tag{6}$$

where $\varphi(B^{y_{th}}_{region})$ denotes the number of pixels in $B^{y_{th}}_{region}$, $B^{y_{th}}_{region}$ is the final borderline region and y is the final times of expansion. f denotes whether extracted borderline region can separate M_{obj} apart or not. $f = 1$ means M_{obj} has two or more effective parts and $f = 0$ mean it do not. If the score is larger than threshold $C(C = 0.4$ in our algorithm), we take this borderline region as an ambiguous region. If a local mask has ambiguous regions, it consists of two or more objects. Otherwise, the local mask is a single object.

3.4 Ambiguous Region Relabeling

A local mask contains touched objects is needed to more precisely segment by assigning every pixel a label. Extracted ambiguous regions separate local mask into multiple object parts, and effective ones are selected as object bodies. As shown in Fig. 4 (a), Pixels in different object bodies are assigned with different labels la $(la = 1, 2, 3 \ldots)$ and the ambiguous regions and unlabeled parts in OBJ_m are labeled with 0.

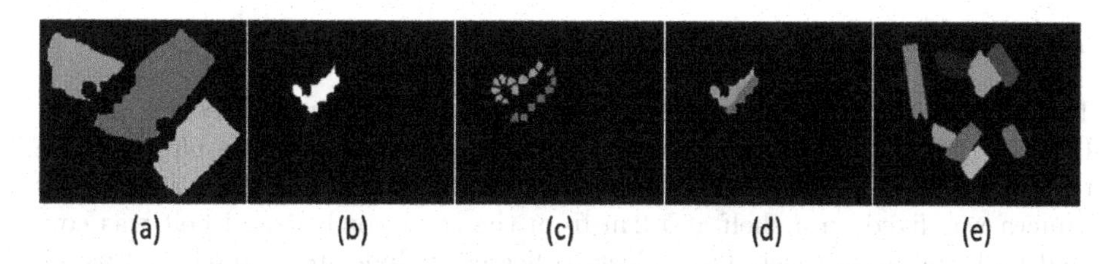

Fig. 4. The process of ambiguous regions relabeling. (a) Labeled object bodies with different labels. (b) An ambiguous region. (c) Selected adjacent superpixel sets. (d) Labeled ambiguous region. (e) Segmentation result of our method. (Color figure online)

For relabeling unclassified pixels ($la = 0$), adjacent superpixel sets are selected as representatives of objects. To an object in construction waste, different parts are not always coincident in feature space. It is not efficient to find a model to represent an object. So the superpixels adjacent to the ambiguous region are selected, and these superpixels have two or more labels. The superpixels with same label are extracted and defined as a superpixel set to represent an object. Two or more superpixel sets are extracted adjacent to an ambiguous borderline region, and then our algorithm classifies the unclassified regions on pixel-level by computing the similarity between a pixel and these sets. For every superpixel, the mean values in LAB color space, depth space and xy space are computed to represent it. So the similarity between a pixel and a superpixel is defined as:

$$d_i = \frac{w_{lab} * d_{lab} + w_{depth} * d_{depth} + w_{xy} * d_{xy}}{w_{lab} + w_{depth} + w_{xy}}, \tag{7}$$

where d_{lab} is the similarity in LAB space, d_{depth} is the similarity of depth space and d_{xy} is the similarity in xy space. These similarities are all computed based on Euclidean distance in each feature space. w_{lab}, w_{depth} and w_{xy}($w_{lab} = 4$, $w_{depth} = 3$, $w_{xy} = 3$ are chosen by experiment in our algorithm) are the weight of each similarity and i is the serial number of superpixels. Then, the similarity between a superpixel set and a pixel is defined as

$$d = \min_{0<i<=n}(d_i), \tag{8}$$

where n is the number of superpixels, d is the similarity between a superpixel set and a pixel. Each pixel in ambiguous region is reassigned with the label of most similar (the smaller the value of d, the more similar) superpixel set. After relabeling ambiguous regions, pixels in local mask are all labeled. However, there may be some areas isolated. So we refine the result by finding these independent pixels and reassigning them with the label of most neighbor pixels. After traversing all local masks, all pixels in image are labeled and solid waste objects are all separated as shown in Fig. 4 (e). For seeing single object clearly, different objects are randomly colored.

4 Experiments and Results

To evaluate the performance of algorithm, a representative dataset is needed. However, there is no available construction waste dataset. So we built a scene to simulate the working environment of the robot as shown in Fig. 5 : 3D sensor camera was fixed on a shelf at 1.2 m from the conveyor belt and belt was covered with soil and gravel. The objects collected include stone, brick and wood. These objects have irregular shapes and deposition of dusty particles on the surface of them lead to color degradation. We used ASUS Xtion PRO to capture 618 RGB-D images of construction waste. And 100 RGB-D images of dusty conveyor belt were also captured to build Adaptive Gaussian Mixture Model of background. Construction waste dataset contains two kinds of scenes. A simple scene that objects are randomly placed on the belt and every object is isolated, as is shown in Fig. 6 row one. A complex scene that objects are also randomly placed on the belt and some object are adhesive or occluded, as is shown in Fig. 6 row two to five. To evaluate our algorithm, manually labeled ground truths of objects are also provided.

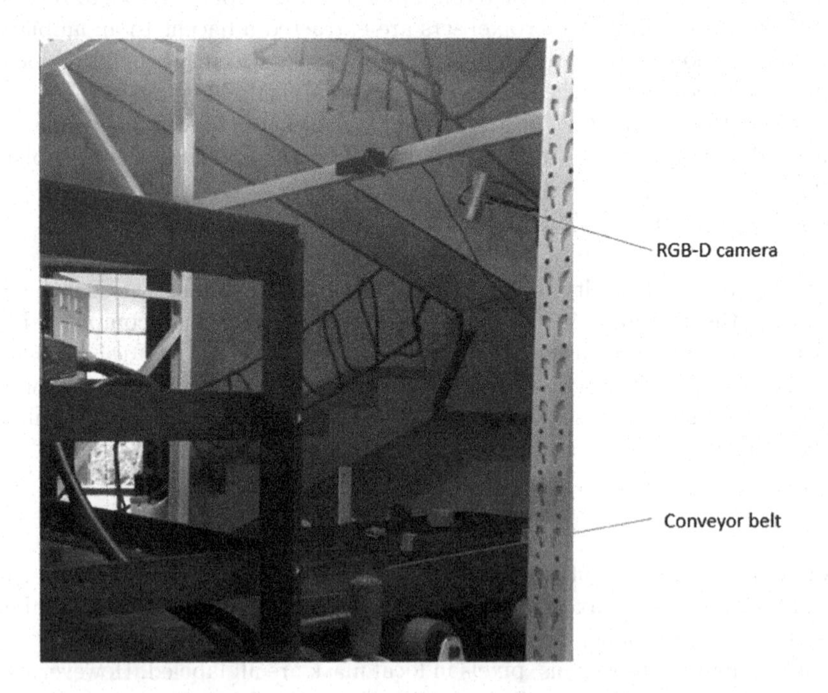

Fig. 5. The simulated scenario of robot working environment.

The performance of our segmentation method is evaluated in terms of how successful it is to segment the objects in the scene. For the task of solid waste sorting, precise object masks which contain positions and contours are important. We define R_i as the pixel set of the ith object in segmentation result,

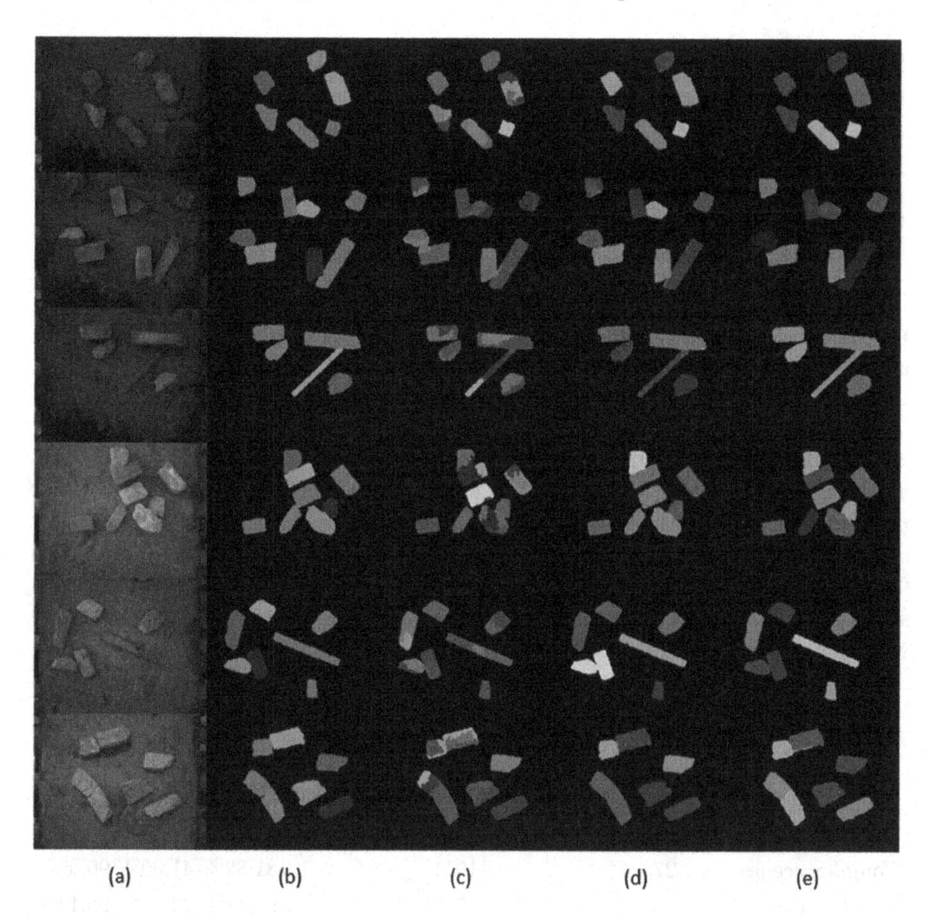

Fig. 6. Some comparison results of different methods. (a) Original images. (b) The ground truths. (c) The results of [15]. (d) The results of [12]. (e) The results of our algorithm. The input for each algorithm are RGB images and point clouds without background. The objects in results are colored randomly. (Color figure online)

G_i is the pixel set of ith object in ground truths and i is the serial number of the objects. So, we analyze the results quantitatively by

$$P_i = \frac{\varphi(I_i) - \varphi(O_i)}{\varphi(G_i)}, \tag{9}$$

where $I_i = G_i \cap R_i$, is the intersection set of G_i and R_i. $O_i = R_i - G_i$, is the pixel set when pixels are in R_i while out of G_i. $\varphi(I_i)$ means the number of pixels in I_i and P_i denotes the segmentation precision of ith object. Over-segmentation and under-segmentation both lead to a low segmentation precision (Fig. 7).

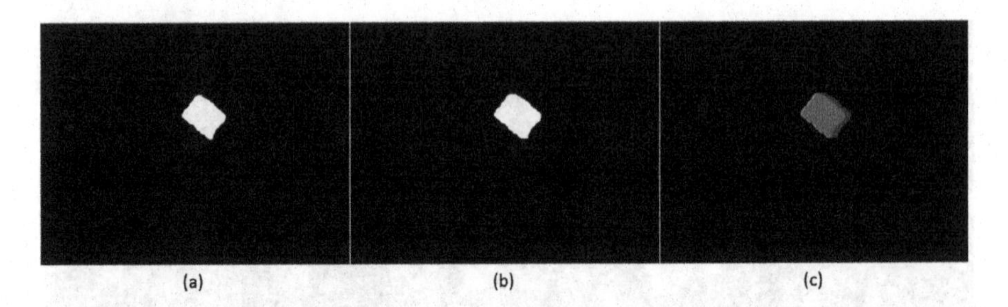

Fig. 7. (a) pixel set of G_i. (b) Pixel set of R_i. (c) Red pixel set is I_i and blue pixel set is O_i. (Color figure online)

We compare our method with other two algorithms of [12] and [15]. The average segmentation precisions of all objects are shown in Table 1. Form Table 1, our method performs better than other two algorithms over the entire dataset. Our segmentation results reach 99.14% segmentation precision in simple scenes and 90.69% segmentation precision in complex scenes. While most objects have been segmented successfully, segmentation error happens (shown in Fig. 5 last row) when some pixels in ambiguous region of two objects are not labeled correctly.

Table 1. Average precision of object segmentation. Compare with [12] and [15].

	Number of images	Number of objects	[12]	[15]	Ours
Simple scenes	345	738	93.73%	32.72%	99.14%
Complex scenes	273	1911	81.82%	41.00%	90.69%
Simple scenes without background	345	738	98.11%	72.83%	99.14%
Complex scenes without background	273	1911	87.98%	64.92%	90.69%

Because of color degradation, color-based algorithms lose their effect. Though algorithm proposed in [15] is newer than [12], it does not achieve a good performance on construction waste dataset as it depends more on color cues. Method presented in [12] computes surface patches by using a mixture of planes and NURBS and performs graph cut on patches to arrive at object hypotheses segmented, so it is plane-based and not affected by interferential color information. But it does not perform well when adhesion and occlusion happen. Our algorithm gets better segmentation results than other two algorithms as color does not play a decisive role in our algorithm and meantime a strategy is proposed to extract ambiguous regions to separate adhesive and occluded objects specially. Masks of segmented objects are provided by our algorithm as they contain necessary information which construction waste sorting needs.

5 Conclusion

In this paper, we have presented a RGB-D based segmentation method for solid waste objects segmentation in cluttered conveyor belt scene. The segmentation results provide the location and boundaries of each solid waste object for the sorting process accurately. In contrast to existing approaches, our method yield satisfactory results in color degraded environment, furthermore, it also performs quite well in the cases where adhesion and occlusion occurred between solid waste objects. Our algorithm deals with adhesion and occlusion by extracting and relabeling on the so-called ambiguous regions to generate accurate segmentation results. To evaluate the proposed algorithm, we have additionally built a dataset of construction waste. The presented results of our method show that it is promising and is looking forward to being used in robotic task of construction waste sorting.

Acknowledgments. This work was supported by Zhejiang Provincial Natural Science Foundation of China under Grant numbers LY15F020031 and LQ16F030007, National Natural Science Foundation of China (NSFC) under Grant numbers 11302195 and 61401397.

References

1. Banica, D., Sminchisescu, C.: Second-order constrained parametric proposals and sequential search-based structured prediction for semantic segmentation in RGB-D images. In: Computer Vision and Pattern Recognition, pp. 3517–3526 (2015)
2. Chen, D., Mirebeau, J.M., Cohen, L.D.: A new finsler minimal path model with curvature penalization for image segmentation and closed contour detection. In: CVPR, pp. 355–363 (2016)
3. Fu, X., Wang, C.Y., Chen, C., Wang, C., Kuo, C.C.J.: Robust image segmentation using contour-guided color palettes. In: IEEE International Conference on Computer Vision, pp. 1618–1625 (2016)
4. Gupta, S., Arbelaez, P., Malik, J.: Perceptual organization and recognition of indoor scenes from RGB-D images. In: Computer Vision and Pattern Recognition, pp. 564–571 (2013)
5. Höft, N., Schulz, H., Behnke, S.: Fast semantic segmentation of RGB-D scenes with GPU-accelerated deep neural networks. In: Lutz, C., Thielscher, M. (eds.) KI 2014. LNCS (LNAI), vol. 8736, pp. 80–85. Springer, Cham (2014). https://doi.org/10.1007/978-3-319-11206-0_9
6. Holz, D., Holzer, S., Rusu, R.B., Behnke, S.: Real-time plane segmentation using RGB-D cameras. In: Röfer, T., Mayer, N.M., Savage, J., Saranlı, U. (eds.) RoboCup 2011. LNCS (LNAI), vol. 7416, pp. 306–317. Springer, Heidelberg (2012). https://doi.org/10.1007/978-3-642-32060-6_26
7. Kaewtrakulpong, P., Bowden, R.: An improved adaptive background mixture model for real-time tracking with shadow detection. In: Remagnino, P., Jones, G.A., Paragios, N., Regazzoni, C.S. (eds.) Video-Based Surveillance Systems, pp. 135–144. Springer, Heidelberg (2002). https://doi.org/10.1007/978-1-4615-0913-4_11

8. Leonardis, A., Gupta, A., Bajcsy, R.: Segmentation of range images as the search for geometric parametric models. Int. J. Comput. Vis. **14**(3), 253–277 (1995)
9. Mishra, A., Aloimonos, Y., Fah, C.L.: Active segmentation with fixation. In: IEEE International Conference on Computer Vision, pp. 468–475 (2009)
10. Mishra, A., Shrivastava, A., Aloimonos, Y.: Segmenting simple objects using RGB-D. In: IEEE International Conference on Robotics and Automation, pp. 4406–4413 (2012)
11. Rao, D., Le, Q.V., Phoka, T., Quigley, M.: Grasping novel objects with depth segmentation. In: IEEE/RSJ International Conference on Intelligent Robots and Systems, pp. 2578–2585 (2010)
12. Richtsfeld, A., Mrwald, T., Prankl, J., Zillich, M., Vincze, M.: Segmentation of unknown objects in indoor environments. In: IEEE/RSJ International Conference on Intelligent Robots and Systems, pp. 4791–4796 (2012)
13. Silberman, N., Sontag, D., Fergus, R.: Instance segmentation of indoor scenes using a coverage loss. In: Fleet, D., Pajdla, T., Schiele, B., Tuytelaars, T. (eds.) ECCV 2014. LNCS, vol. 8689, pp. 616–631. Springer, Cham (2014). https://doi.org/10.1007/978-3-319-10590-1_40
14. Sobral, A., Vacavant, A.: A comprehensive review of background subtraction algorithms evaluated with synthetic and real videos. Comput. Vis. Image Underst. **122**, 4–21 (2014)
15. Toscana, G., Rosa, S.: Fast graph-based object segmentation for RGB-D images. arXiv preprint arXiv:1605.03746 (2016)
16. Wang, Z., Liu, H., Wang, X., Qian, Y.: Segment and Label Indoor Scene Based on RGB-D for the Visually Impaired. In: Gurrin, C., Hopfgartner, F., Hurst, W., Johansen, H., Lee, H., O'Connor, N. (eds.) MMM 2014. LNCS, vol. 8325, pp. 449–460. Springer, Cham (2014). https://doi.org/10.1007/978-3-319-04114-8_38

Structure-Aware SLAM with Planes in Man-Made Environment

Jiyuan Zhang(✉), Gang Zeng, and Hongbin Zha

Key Laboratory on Machine Perception, Peking University,
Room 2207, Science Building No. 2, Beijing 100871, China
{zhangjiyuan.eecs,zeng}@pku.edu.cn, zha@cis.pku.edu.cn

Abstract. We propose to utilize co-planar constraints in point-based simultaneous localization and mapping of man-made environment. Planes are common structures in both indoor and outdoor city scenes. They are good features in two ways: strong constraint on points, and suitable supplementary for long-distance tracking. Large structural planes in optimization-based framework can reduce drifting error by connecting many points and cameras. Our method can detect and locate planes from multiple images and sparse 3D points. No dense reconstruction or triangulation is needed. The planes are actively extended as SLAM system goes on, making our algorithm suitable for exploration-style applications. The proposed method is tested on real world indoor and outdoor long video sequences, showing the capability to significantly reduce drifting error.

Keywords: SLAM · Plane detection · Vanishing points
Color segmentation · Non-linear optimization

1 Introduction

Simultaneous Localization And Mapping (SLAM) is an important task in computer vision. It is widely used in applications such as robotics, auto driving and 3D reconstruction. Feature-based visual SLAM works [8,13,18] use feature points as input, 3D points as landmark. There are also direct methods [3,4,6] that process every pixel in image. The keypoints or pixels are always treated as independent elements, either to reduce computation cost or to be robust to errors. Only when dense reconstruction is needed [19–21] some normalization is imposed on pixels, which increases complexity greatly.

However more information can be found in images. It is necessary to utilize relation of points in visual SLAM to achieve better results. In this paper we consider general man-made environments, in which a great variety of additional structural information can be used. A SLAM system with ability to recognize and utilize these knowledge would perform better. In the proposed method we detect and locate structural 3D planes from image and sparse 3D points, and use them to help improve the SLAM result. Planes can be tracked longer than individual

J. Yang et al. (Eds.): CCCV 2017, Part III, CCIS 773, pp. 477–489, 2017.
https://doi.org/10.1007/978-981-10-7305-2_41

feature points, especially in visual exploration tasks, as shown in Fig. 1. Even if there is no co-visible feature point, the structure still provides the constraint to all frames seeing it. By employing planes, the commonly seen drifting error could be reduced, though never completely eliminated unless explicit loop closing is made.

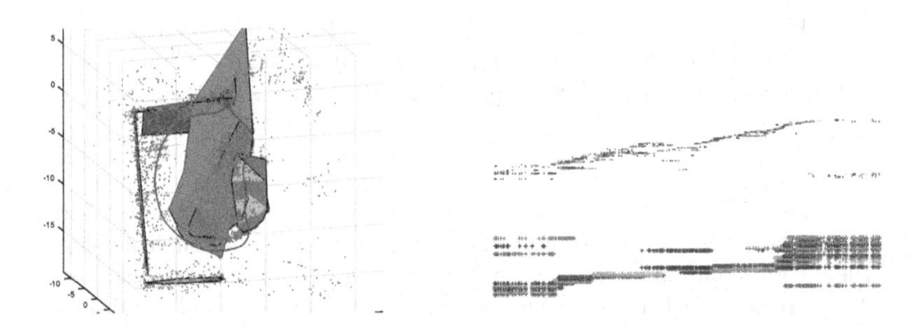

Fig. 1. Left: 3D points and planes of an outdoor scene, viewed from top. Right: the visibility of keypoints (top half, dot) and planes (bottom half, cross) through time (x-axis). The planes are much better for long-term tracking in man-made environments.

Different from points or lines, there is no "projected image" for an ideal infinite plane. It has to be reconstructed from some 3D source. A common choice is RANSAC in point cloud. Random sampling could be problematic if the cloud is sparse and not evenly distributed, which is usually the case for feature-based point landmarks. Even with a well-developed SLAM system, reconstructed sparse 3D points can only give basic impression of scene structures. Human viewers can understand the scene by adopting a lot of prior knowledge available in the input images, which is in our interest.

Our method detects potential co-planar points in a set of images, with the aid of sparse reconstruction of 3D points. First the images are over-segmented by color, as single-color regions are more likely to be planar. Such regions provide single-view co-planar information. Gathered over multiple views, the relation is more stable and ready to be clustered into co-planar groups. Then vanishing points in images are used to limit the search space of planes. The two information sources are combined to detect planes, after which structure constraints are used to associated points. The planes, points and camera poses are then jointly optimized to give consistent result.

2 Related Work

Many works have explored the idea of using structures in SLAM. Information can be as high level as objects [23], or as simple as planes or lines. It is easier to extract geometry structure from depth sensors, like RGB-D camera [10,11,24]. For monocular images, extra effort must be paid. Planar regions of texture could be tracked through homography [12,15], providing constraints for motion. However to build a meaningful and stable error function, texture must be compared, leading to a dense or semi-dense color consistency check. Texture-less planar regions have to be matched by shape. DPPTAM [3] detects and maintains such regions in 3D, but only use them for dense reconstruction. The planes do not participate in motion estimation. If indoor scenes with floor and walls are assumed, pop-up structures [28] can be detected for low-texture regions.

Pure plane-based work [29] assumes one or more large planes are present in the scene, and detects them by applying RANSAC iteratively. The planes play the major role in motion computation in the form of homography. However it is an off-line framework, batching all images together.

A recent work [22] also exploit the indoor planes in the dense SLAM system. All pixels are reconstructed as surfels, then divided to planar and non-planar types. Planes are made from combined surfels as memory-efficient representation. It is implemented in the manner of depth fusion for dense reconstruction. Our work is different in that discrete points are still kept. Planes are not employed to fill the map, but only impose constraints to points. Point-on-plane condition is represented by re-projection error, which is view-dependent. This flexibility can tolerate unstable points commonly seen in outdoor scenes, where features can be very far from camera. Also we allow planes to cross color boundary to establish long-term landmarks, helping large-scale visual odometry, while dense depth map of [22] is more suitable for small-scale indoor SLAM.

The Multilayer Feature Graph (MFG) [16] is proposed to combine multiple types of features, including point, line segment, vanishing point and plane. After point and line segment features are extracted from image, other color information is dropped. The planes are detected with RANSAC among spatial points and lines, using 3D distance measurement. All the features are detected and associated with others by various kinds of relationship. Then an overall non-linear optimization is performed to solve both structure and motion.

Previous works [2,9] using planes as features, detected directly from sparse point cloud with RANSAC. It is applicable for the small-scale AR scene, but not for larger outdoor scenes where point cloud is too sparse. Later [17] on the other hand uses point with normal to represent planes. This allows planes and points to be tracked in a unified framework of Extended Kalman Filter. The connections between co-planar points are built simply by geometry and not used in filter update.

3 Planes in Local Map

Our work is based on keypoint-based monocular SLAM framework. Sparse 3D points $\{\mathbf{X}\}$ are reconstructed first. Large-scale loop closure is not employed, so that long-term drifting will be significant. Our work focus on the newest part of the map, called the local map. It originates from the newest keyframe and all 3D points $\mathbf{X}_i$ visible in it. Then all keyframes in which any $\mathbf{X}_i$ is visible, and any other 3D points visible in these keyframes, are also included. Limited by scene co-visibility and memory issue, the number of keyframes in local map seldom reaches the scale of 100.

Planes are detected and located in the local map with points assigned. Details will be given in the rest of this section. Both points and planes are put into the structure-and-pose bundle adjustment over the local map. Let $\pi(\mathbf{X})$ be the projection from 3D to 2D, and $\mathbf{u}$ the observed feature point. In addition to conventional point re-projection error

$$e_{proj}(\mathbf{X},\mathbf{u}) = |\pi(T_{cw}\cdot\mathbf{X}) - \mathbf{u}|, \tag{1}$$

the planes introduces two new kinds of edge:

Point-on-plane projection. For a plane $\mathbf{p} = (\mathbf{n}^T, d)^T$ and a point $\mathbf{X}$ assigned with $\mathbf{p}$, the closest point

$$\mathbf{X}_\mathbf{p} = \mathbf{X} - (\mathbf{n}^T\mathbf{X} + d)\mathbf{n} \tag{2}$$

exactly on $\mathbf{p}$ is considered the additional source of projection.

$$e_{plane\_proj}(\mathbf{X},\mathbf{p},\mathbf{u}) = e_{proj}(\mathbf{X}_\mathbf{p},\mathbf{u}) \tag{3}$$

is also included for all observations $\mathbf{u}$. This error measurement imposes less weight if $\mathbf{p}$ is frontal-parallel in all views, avoiding additional view-dependent weights. Any criteria of point re-projection errors could be shared.

Spatial point to plane. Non-local 3D points not visible in local map may also be included in large planes. Those points are fixed during the optimization as anchors for plane. Point-plane distance

$$e_{space}(\mathbf{X}_{old},\mathbf{p}) = |\mathbf{n}^T\mathbf{X}_{old} + d| \tag{4}$$

is used to describe non-local points on plane relationship. This distance is included so that planes can impose constraints across longer range, both spatial and temporal.

The final objective function

$$E = \sum_{\mathbf{X}}\left[e_{proj}(\mathbf{X},\mathbf{u})^2 + \sum_{\mathbf{p}} e_{plane\_proj}(\mathbf{X},\mathbf{p},\mathbf{u})^2\right] + \lambda\sum e_{space}(\mathbf{X}_{old},\mathbf{p})^2 \tag{5}$$

is then minimized with Gauss–Newton algorithm. The parameter λ is employed to balance different units. e_{space} is measured in arbitrarily scaled spatial distance while e_{proj} in pixel. λ should be chosen according to the scene, with tolerances of spatial error σ_{space} and pixel error σ_{pixel}. σ_{space} is also used in the creation of plane candidates later in Sect. 3.3.

As planes are included, the objective function is different from conventional point-based Bundle Adjustment (BA). The error term $e_{plane\_proj}(\mathbf{X},\mathbf{p},\mathbf{u})$ modified the structure of BA by introducing terms with three parameter blocks. However by carefully sorting the blocks and using the fact that large structural planes are much fewer than points, it is still possible to solve the Jacobian equation with Schur complement in reasonable time. Timing results will be discussed in Sect. 4.1.

3.1 Single-Color Regions

It is assumed that single-color regions in image are likely to be planar, which is applicable in many man-made environments. We only need such assumption to find connections between features, which are lost from the very beginning of feature point extraction. It is possible to drop any advanced single-view plane detection algorithm here without major change of the rest of our algorithm.

Graph-based method [7] is used to create over-segmentations on input image, example shown in left of Fig. 2. Only large regions are considered. The points in the same region are more likely to be coplanar. As features are often close to color boundaries, the regions are dilated by a few pixels before deciding membership of features in this image. Then one feature can be assigned to multiple regions, which forges connection across color boundary to provide structural constraints at larger scale. It is common that a real large plane consists of several single-color regions.

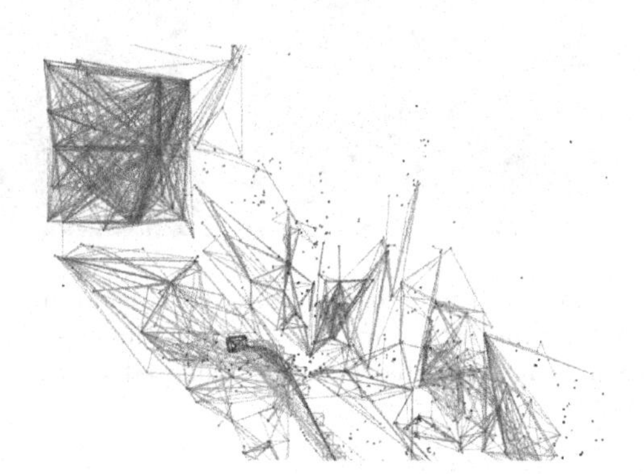

Fig. 2. Left top: input keyframe. Left bottom: color segmentation. Right: region-based similarity of 3D points, connected if seen in enough single-color regions. Best viewed in color.

The region-based similarity $\mathbf{S}(i, j)$ is defined as the number of regions in all local keyframes where 3D points $\mathbf{X}_i$ and $\mathbf{X}_j$ are both seen inside. Note this number might be larger than simply counting same-region keyframes, because one point could be covered by several *dilated* regions. The number of regions is used, instead of keyframes, to emphasis the similarity of feature pairs close to the same color boundary. Whether there is depth discontinuity or not, the color boundary itself is likely to be planar in man-made environment.

As shown in the right of Fig. 2, the accumulated $\mathbf{S}$ roughly sketches three major planes in the scene, with a few outliers.

3.2 Line-Based Structures

Straight lines are another type of common and important feature in man-made environment, being good hints for presence of planes. Parallel line groups, or Vanishing Points (VP) are used in this work to limit the normal of plane candidates.

First the lines are detected with LSD [27]. Vanishing points are clustered from each frame using J-Linkage [25]. Each VP in image represents a direction $\mathbf{v}_c$ in camera coordinates. With known camera-to-world transformation T_{wc}, the VP directions are unified in the world coordinate as $\mathbf{v}_w = R_{wc}\mathbf{v}_c$. Though not matched by image textures, observations of same VP are close in space. As frames coming in, the significant VPs are seen again and again, while secondary or erroneous directions can be simply rejected by counting.

The potential normals are computed from cross product of VPs and cleaned by merging close ones. One plane normal is supported by a VP if they are orthogonal. We use strong criteria that valid plane normals should be supported by at least two VP directions, so only major structural planes are detected. Spatial sampling of plane normal is no longer needed. Figure 3 shows the normals computed from accumulated significant VP directions.

Fig. 3. Left and middle: examples of vanishing point groups on two keyframes. Note the VP groups are not matched between images. Right: top view of potential normals in local map, computed from accumulated vanishing point directions. Best viewed in color.

3.3 Plane Candidates

We construct planes in the local map from two different sources, both using restricted plane normals from VP directions. We use MLESAC [26] to detect planes from a set of 3D points. The measurement, point-plane distance $e_{space}(\mathbf{X}, \mathbf{p})$, is easily computed given the plane normal. Only one point is sampled in each iteration. The error threshold σ_{space} is chosen according to the scale of the scene.

One way is from single-color regions. For each $\mathbf{X}_i$ one plane is detected with $\{\mathbf{X}_j | \mathbf{S}(i,j) > t_{color}\}$, with t_{color} a fixed threshold. Overlapping among clusters are allowed, so that one point may be assigned to several planes.

The other way is direct MLESAC using all 3D points in the local map. The required amount of inliers are proportional with that of local 3D points, much more than region-based one. This method is kept to recover planes with good texture, on which many keypoints but no large color regions are detected.

Many of these planes are duplicates from different sources, which is also a kind of similarity. The planes are reversely indexed by inlier points, thus each point $\mathbf{X}_i$ can have zero or more potential planes $\mathcal{P}_i$ populated from both sources.

The points are then clustered again with similarity $\mathbf{S}$, requiring not only $\mathbf{S}(i,j) > t_{local}$ but also $\mathcal{P}_i \cap \mathcal{P}_j \neq \emptyset$. We choose $t_{local} < t_{color}$ to encourage large planes. Detail of this greedy clustering is described in Algorithm 1. Note this is different from the previous clustering by allowing center of cluster changed during the expansion. The clustered planes are then compared with previous structural ones, and merged if very close.

Data: available anchor 3D points $A \subseteq \{\mathbf{X}_i\}$, potential planes for each point $\{\mathcal{P}_i\}$, region-based similarity $\mathbf{S}$, potential normals $\{\mathbf{n}\}$, threshold t_{local}
Result: clusters of points $\{C\}$ and planes $\{\mathbf{p}\}$
compute degrees of each $\mathbf{X}_i$: $D_i = \sum_{j \neq i} \mathbf{S}(i,j)$;
while $A \neq \emptyset$ **do**
  pick anchor $\mathbf{X}_a \in A$ with largest D_a;
  let $C = \{\mathbf{X}_a\} \cup \{\mathbf{X}_j | \mathbf{S}(a,j) > t_{local}\}$;
  fit best plane $\mathbf{p}$ with C and $\{\mathbf{n}\}$;
  let potential planes $\mathcal{P} = \{\mathbf{p}\}$;
  repeat
    for $\mathbf{X}_c \in C$ **do**
      let $J = \{j | \mathbf{S}(c,j) > t_{local}, \mathcal{P}_j \cap \mathcal{P} \neq \emptyset \vee \mathcal{P}_j = \emptyset\}$;
      let $C = C \cup \{\mathbf{X}_j | j \in J\}$;
      let $\mathcal{P} = \mathcal{P} \cup \bigcup_{j \in J} \mathcal{P}_j$;
    end
    fit best plane $\mathbf{p}$ with C and $\{\mathbf{n}\}$, remove outliers from C;
  until C *unchanged*;
  remove C from A;
end

Algorithm 1. Greedy clustering with color information and potential planes of each point. $|\mathcal{P}| \geqslant 1$ for each cluster.

To speed up the second clustering, the previous planes are extended first. $\mathbf{X}_i$ in local map is assigned to a previous plane $\mathbf{p}_k$ if $e_{space}(\mathbf{X}_i, \mathbf{p}_k)$ is small enough *and* there is some $\mathbf{X}_j$ assigned to $\mathbf{p}_k$ satisfies $\mathbf{S}(i,j) > t_{local}$. The occupied 3D points are not used as starting of clustering, but still can be added to other clusters (Fig. 4).

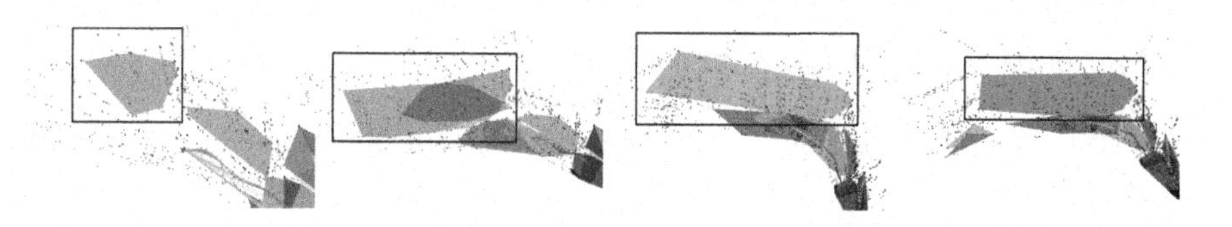

Fig. 4. The detected planes in local map near Fig. 3, timestamps from left to right. Note how the farthest plane (marked by black box) is growing in size by assigning points to it. Similar planes are actively merged.

4 Experiments

Our experimenting system is built based on ORB-SLAM [18], with explicit loop-closure disabled. Note our method does not perform loop-closure with planes. Previously tracked planes are expanded only if they share some regions in the local map, which requires continuous visibility. The process of lines and planes is inserted after local map updating, refining both 3D points and camera poses. Plane detection may fail if too few are found in single-color regions. In this case, only points are refined and ORB-SLAM goes on.

Table 1. Errors of ORB-SLAM and proposed method on sequences of dataset [5]. Bold indicates measurements *not* improved by our method.

		ORB-SLAM				Proposed			
		E_{align}	E_R	E_S	RMSE	E_{align}	E_R	E_S	RMSE
Indoor	Seq. 16	0.9709	0.6469	0.9240	0.5017	0.1792	0.3237	0.9847	0.0498
	Seq. 28	8.0150	**18.5386**	0.5773	**0.2868**	5.0024	19.5471	0.8470	0.2974
	Seq. 35	5.1530	3.4454	0.7500	0.5830	3.1686	2.8149	0.7988	0.4849
	Seq. 36	2.2379	2.9817	0.8411	0.5757	1.1187	2.1175	0.9942	0.3841
Outdoor	Seq. 23	17.6954	**1.6574**	0.3709	0.4455	3.6029	2.2916	0.7808	0.4357
	Seq. 30	2.6210	1.1645	0.8409	0.6261	1.3844	0.9717	1.0845	0.2267
	Seq. 45	2.8350	2.2240	0.8240	0.1788	1.5819	1.7018	0.9000	0.1785

Our method is tested on sequences of TUM monocular visual odometry dataset [5] with both indoor and outdoor man-made scenes. The dataset is created for evaluation of visual odometry systems. Every sequence are closed loop, with the same well-textured objects at the both ends where ground-truth of motion is provided. We follow the evaluation method of the authors by comparing the drifting of the loop. The recovered trajectory is aligned by scaled transformation $Sim(3)$ twice, with the ground-truth at the beginning and the end. Figure 5 shows some trajectories of ORB-SLAM (left) and ours (right). Two aligning transformations T_A and T_E are computed, and $T_{align} = T_A^{-1}T_E$. The alignment error measurement E_{align} is the average positional displacement of two aligned trajectories. Other useful error measurements are:

- E_R: rotation part of T_{align}, in degrees;
- E_S: scale part of T_{align}, should be close to 1;
- RMSE: average positional drifting using a single alignment, only partially available.

Comparison of error values is shown in Table 1.

Most indoor sequences are captured in narrow corridors and medium-sized offices, sometimes with going up or down stairs. The structural planes mainly comes from walls and floors, with noncontinuous texture. The major rotation

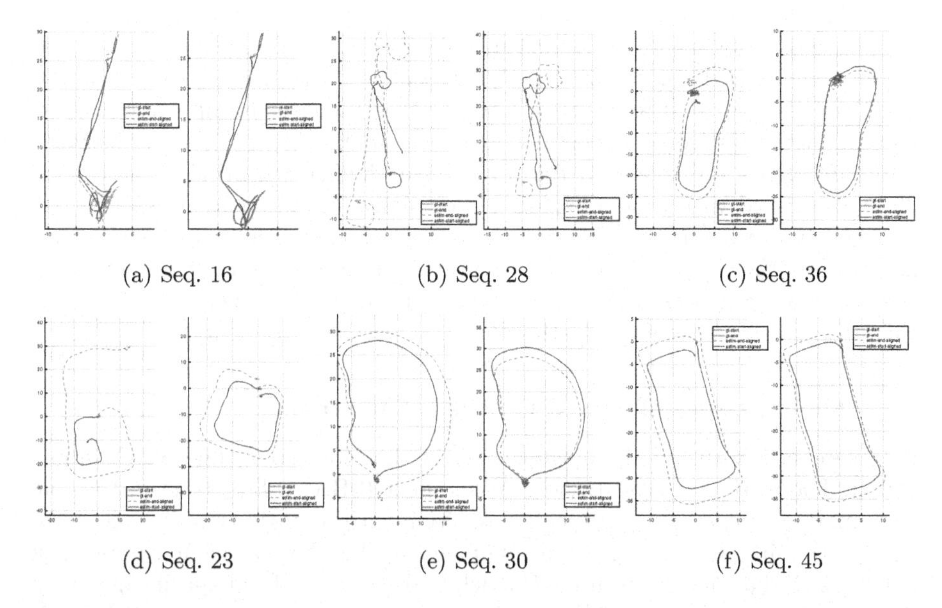

(a) Seq. 16 (b) Seq. 28 (c) Seq. 36

(d) Seq. 23 (e) Seq. 30 (f) Seq. 45

Fig. 5. Motion from indoor (top row) and outdoor (bottom row) sequences. For each sequence, ORB-SLAM at left and our result at right are aligned to ground truth at the beginning (blue, solid) and the end (red, dashed). The closer two trajectories, the better result. See experiment section for detail. (Color figure online)

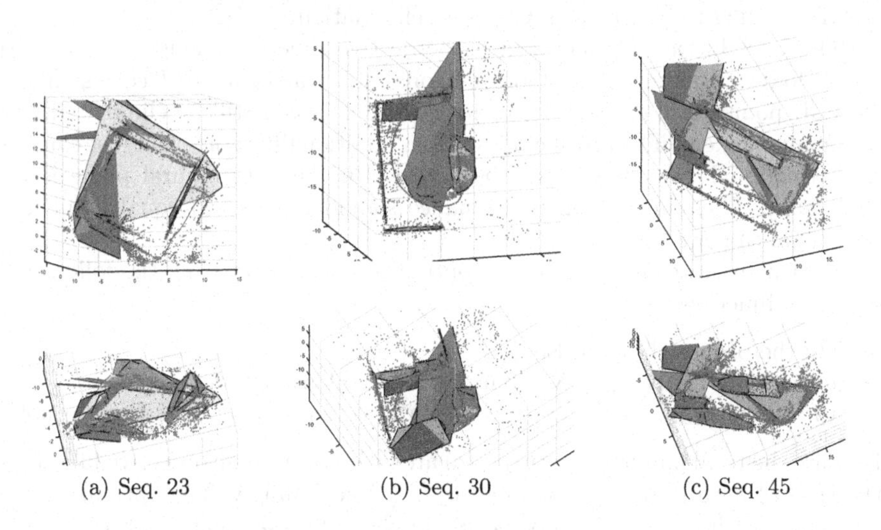

(a) Seq. 23 (b) Seq. 30 (c) Seq. 45

Fig. 6. Structural planes reconstructed for some outdoor scenes, top view and bird view.

error shown in Fig. 5(b) is introduced from climbing up spiral stairs, where walls are texture-less but circular. The significant drifting of ORB-SLAM on the stairs is not corrected with plane assumption. However the scale is correctly kept.

In the outdoor scenes the structural planes come from ground and facades. As we allow shared points between detected planes, the ground plane can grow very large even if texture changes. Long-tracked planes can improve quality of motion by forcing consistent scale, as shown in Fig. 5(d). Some of recovered structure planes are shown in Fig. 6 as convex hull of footprints of associated 3D points, filled with random color. Major planes in the scene are successfully detected and reconstructed. Most of them are real ones with texture, while some are virtual. Virtual planes do not harm SLAM results, as they provide same level of structural constraints on 3D points. It is easy to reject virtual ones by checking color consistency if textured reconstruction is necessary.

4.1 Timing Issues of Mixed Objective Function

It is necessary to finish the optimization in time for a real-time SLAM application. The work of Sparse Bundle Adjustment (SBA) [14] has shown the conventional point-and-camera objective function with re-projection error could be effectively minimized using Schur complement. The key of Schur complement is the block diagonal pattern of $J^T J$ when solving the Jacobian linear system $J\delta\mathbf{x} = \epsilon$. In typical BA problem, most degrees of freedom are occupied by 3D points. There is only one point in each error term, so that the columns of Jacobian matrix J corresponding to the points are mutually orthogonal, producing a large block-diagonal sub-matrix in $J^T J$. It is much easier to invert this sub-matrix, leading to faster block Gaussian elimination.

The introduction of planes in the objective function changes the pattern of J. However the key part, block-diagonal sub-matrix in $J^T J$ corresponding with 3D points, still exists. The variables of points can still be eliminated with Schur complement. The rest includes planes additionally, more than just cameras in conventional BA, thus it will be slower. But large structural plane is not that many. Their impact on performance should not exceed equal number of additional cameras.

To demonstrate the time cost of optimization by comparing the following 3 objective functions:

- *full*: the proposed one in Eq. 5,
- *conventional*: only $e_{proj}(\mathbf{X}, \mathbf{u})$ are included, planes ignored,
- *spatial*: use e_{space} instead of e_{proj} regardless of the point $\mathbf{X}$ is fixed or not.

Because the total number of terms $N(e)$ are different in these objective functions, the time of each iteration is divided by $N(e)$ accordingly. The optimization is performed with Google Ceres Solver [1] in C++, for maximal 5 iterations.

Figure 7 shows the number of planes and the per-term time cost in one iteration for Seq. 30. The number of active planes are kept less than 20 in this outdoor scene. The average statistics for some sequences are shown in Table 2. The major part, conventional re-projection error terms $e_{proj}(\mathbf{X}, \mathbf{u})$, are the same among three objective functions. *full* introduces extra terms with planes. There is visible performance impact that time cost per term is increased. The overall time for a local map optimization is increased but still applicable in the

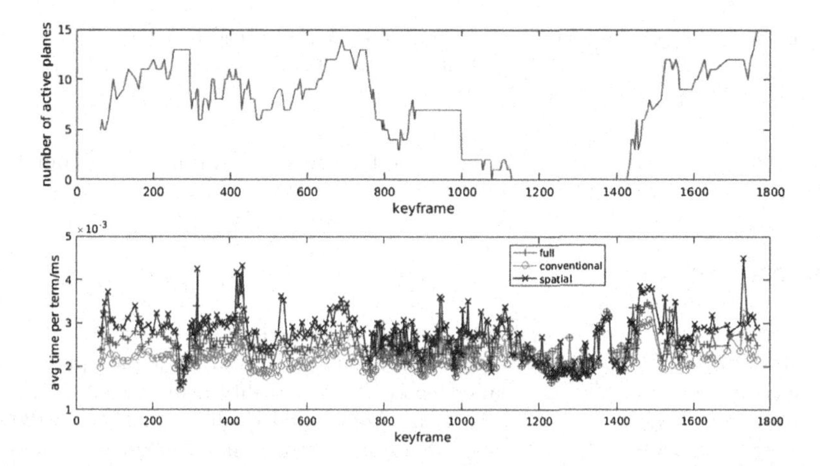

Fig. 7. Top: number of planes in the local map of Seq. 30. Bottom: time cost per term in one iteration, in milliseconds. There is no active plane around Frame 1200, where *full* falls back to *conventional.*

Table 2. Average statistics of objective function optimization. $N(e)$ is the number of that type of error term. (*conven.* short for *conventional*)

		$N(e_{proj}(\mathbf{X_p}))$	$N(e_{space})$	Frame time (ms)	Time per term (ms)
Seq. 23	*full*	7222	1393	886.5194	0.0024
$N(e_{proj}(\mathbf{X}))$	*conven.*	-	-	662.8082	0.0022
= 48587	*spatial*	-	3089	933.2013	0.0028
Seq. 30	*full*	2406	323	657.9031	0.0024
$N(e_{proj}(\mathbf{X}))$	*conven.*	-	-	548.6552	0.0022
= 40794	*spatial*	-	641	711.7544	0.0027
Seq. 45	*full*	1663	425	506.9243	0.0023
$N(e_{proj}(\mathbf{X}))$	*conven.*	-	-	420.9805	0.0021
= 32739	*spatial*	-	692	555.9996	0.0026

ORB-SLAM framework. It is possible to replace the conventional local BA with proposed objective function without major rework of a multi-threaded real-time SLAM system.

Note that the *spatial* objective function introduces less extra terms, but costs more time to optimize.

5 Conclusion

In this paper we propose a structure-aware SLAM method utilizing lines and planes in man-made environment. With clustered vanishing points and color-based image segmentation, planar structures are detected from images and sparse 3D points. Using more information from image improves visual SLAM result.

The ability to track and update planes for long range reduces the drifting error of camera trajectory. The detected planes can also be used as base of large-scale dense reconstruction.

Acknowledgments. This work is supported by National Natural Science Foundation of China (NSFC) 61375022 and 61403005.

References

1. Agarwal, S., Mierle, K., et al.: Ceres solver. http://ceres-solver.org
2. Chekhlov, D., Gee, A.P., Calway, A., Mayolcuevas, W.W.: Ninja on a plane: automatic discovery of physical planes for augmented reality using visual SLAM. In: International Symposium on Mixed and Augmented Reality, pp. 153–156 (2007)
3. Concha, A., Civera, J.: DPPTAM: dense piecewise planar tracking and mapping from a monocular sequence. Intelligent Robots and Systems, pp. 5686–5693 (2015)
4. Engel, J., Koltun, V., Cremers, D.: Direct sparse odometry. arXiv:1607.02565, July 2016
5. Engel, J., Usenko, V., Cremers, D.: A photometrically calibrated benchmark for monocular visual odometry. arXiv:1607.02555, July 2016
6. Engel, J., Schöps, T., Cremers, D.: LSD-SLAM: Large-Scale Direct Monocular SLAM, pp. 834–849. Springer International Publishing, Cham (2014)
7. Felzenszwalb, P.F., Huttenlocher, D.P.: Efficient graph-based image segmentation. Int. J. Comput. Vis. **59**(2), 167–181 (2004)
8. Forster, C., Pizzoli, M., Scaramuzza, D.: SVO: fast semi-direct monocular visual odometry. In: International Conference on Robotics and Automation, pp. 15–22 (2014)
9. Gee, A.P., Chekhlov, D., Mayolcuevas, W.W., Calway, A.: Discovering planes and collapsing the state space in visual SLAM. In: British Machine Vision Conference (2007)
10. Hsiao, M., Westman, E., Zhang, G., Kaess, M.: Keyframe-based dense planar SLAM. In: IEEE International Conference on Robotics and Automation, ICRA, Singapore, May 2017, to appear
11. Kaess, M.: Simultaneous localization and mapping with infinite planes. In: 2015 IEEE International Conference on Robotics and Automation (ICRA), pp. 4605–4611, May 2015
12. Kähler, O., Denzler, J.: Implicit feedback between reconstruction and tracking in a combined optimization approach. In: Rigoll, G. (ed.) DAGM 2008. LNCS, vol. 5096, pp. 274–283. Springer, Heidelberg (2008). https://doi.org/10.1007/978-3-540-69321-5_28
13. Klein, G., Murray, D.W.: Parallel tracking and mapping for small AR workspaces. In: International Symposium on Mixed and Augmented Reality, pp. 225–234 (2007)
14. Lourakis, M.A., Argyros, A.: SBA: a software package for generic sparse bundle adjustment. ACM Trans. Math. Softw. **36**(1), 1–30 (2009)
15. Lovegrove, S., Davison, A.J., Ibãnez-Guzmán, J.: Accurate visual odometry from a rear parking camera. In: IEEE Intelligent Vehicles Symposium, June 2011
16. Lu, Y., Song, D.: Visual navigation using heterogeneous landmarks and unsupervised geometric constraints. IEEE Trans. Rob. **31**(3), 736–749 (2015)
17. Martinezcarranza, J., Calway, A.: Unifying planar and point mapping in monocular SLAM. In: British Machine Vision Conference (2010)

18. Murartal, R., Montiel, J.M.M., Tardos, J.D.: ORB-SLAM: a versatile and accurate monocular SLAM system. IEEE Trans. Rob. **31**(5), 1147–1163 (2015)
19. Newcombe, R., Lovegrove, S., Davison, A.J.: DTAM: dense tracking and mapping in real-time. In: International Conference on Computer Vision, pp. 2320–2327 (2011)
20. Pizzoli, M., Forster, C., Scaramuzza, D.: REMODE: probabilistic, monocular dense reconstruction in real time. International Conference on Robotics and Automation, pp. 2609–2616 (2014)
21. Ranftl, R., Vineet, V., Chen, Q., Koltun, V.: Dense monocular depth estimation in complex dynamic scenes. In: Computer Vision and Pattern Recognition, pp. 4058–4066 (2016)
22. Salasmoreno, R.F., Glocken, B., Kelly, P.H.J., Davison, A.J.: Dense planar SLAM. In: International Symposium on Mixed and Augmented Reality, pp. 157–164 (2014)
23. Salasmoreno, R.F., Newcombe, R., Strasdat, H., Kelly, P.H.J., Davison, A.J.: SLAM++: simultaneous localisation and mapping at the level of objects. In: Computer Vision and Pattern Recognition, pp. 1352–1359 (2013)
24. Taguchi, Y., Jian, Y., Ramalingam, S., Feng, C.: Point-plane SLAM for hand-held 3D sensors. International Conference on Robotics and Automation, pp. 5182–5189 (2013)
25. Toldo, R., Fusiello, A.: Robust multiple structures estimation with j-linkage. In: European Conference on Computer Vision, pp. 537–547 (2008)
26. Torr, P.H.S., Zisserman, A.: MLESAC: a new robust estimator with application to estimating image geometry. Comput. Vis. Image Underst. **78**(1), 138–156 (2000)
27. Von Gioi, R.G., Jakubowicz, J., Morel, J.M., Randall, G.: LSD: a fast line segment detector with a false detection control. IEEE Trans. Pattern Anal. Mach. Intell. **32**(4), 722–732 (2010)
28. Yang, S., Song, Y., Kaess, M., Scherer, S.: Pop-up SLAM: semantic monocular plane slam for low-texture environments. In: IEEE/RSJ International Conference on Intelligent Robots and Systems (IROS). IEEE, October 2016
29. Zhou, Z., Jin, H., Ma, Y.: Robust plane-based structure from motion. In: Computer Vision and Pattern Recognition, pp. 1482–1489 (2012)

Key Words Extraction and Semantic-Based Image Retrieval on RNNs

Lifei Han and Guanghua Gu(✉)

School of Information Science and Engineering,
Yanshan University, Qinhuangdao 066004, Hebei, China
guguanghua@ysu.edu.cn

Abstract. A quick glance at an image, it is very easy for us to perceive the main semantics while it is an challenge for computers. In this paper, we proposed an interesting approach to learn the core semantics of an image and generate the key words to describe it First, we need to get the corresponding of images and text named image-sentence alignment model, the alignment model was trained by image samples and the correlative sentence descriptions to learn the correspondences between the images and the texts. Second, we picked out several key words to describe the core semantic contents of images by the core semantic extraction method. Then, the semantic-based image retrieval was performed to demonstrate the effectiveness of the correspondence between the core semantic of image and the key word. The good performance of our method is illustrated on the Microsoft COCO dataset.

Keywords: Core semantics · Key words
Semantic-based image retrieval · The core semantic extraction model

1 Introduction

Image semantic perception is a very hard task in image understanding. Even though a quick glance at an image is sufficient for a person to perceive and describe the core semantic of the image, the computer cannot have the outstanding ability. Describing the core semantic contents of images is meaningful for automated image captioning. The majority of previous work in image captioning [1,2] focus on generating the sentence descriptions to describe the image content as detailed as possible. However, the sentence descriptions generated by the model may contain some useless or error words to decrease the accuracy of the image annotations. So extracting the key words from the sentence descriptions to represent the core semantic of images becomes more and more significant. For instance, we may get "there are a group of people" when looking for the images in Fig. 1. The core semantic of the two images is "group", no matter the people are "standing" or "sitting" on the ground. The core semantic or key word is greatly useful for semantic-based image retrieval.

J. Yang et al. (Eds.): CCCV 2017, Part III, CCIS 773, pp. 490–499, 2017.
https://doi.org/10.1007/978-981-10-7305-2_42

Fig. 1. The two images have the same core semantic word "group".

Many methods have been proposed to represent the correspondences between the images and texts in high-level features. On the image side, the Convolution Neural Networks (CNNs) [3,4] recently have been presented as a powerful method for the image classification and object detection [5]. On the language side, Recurrent Neural Networks (RNNs) have been used for language model [6] and description generation [7]. Some methods have explored to describe the images by using the natural language [8–15]. Several methods for image captioning [16–23] depend on the combination of RNN language model and the image information with CNNs. Our framework is inspired by [7], which trained the model to generate the sentence descriptions of the images. Unlike [7], our work not only implements the image captioning model, but pick out the words according to the core semantic annotations of images.

2 Approach

The purpose of our model is to generate some key words that describe the core semantics of images. In the training process, the training samples are the image datasets and their corresponding sentence descriptions. First, we trained an alignment model that aligns the sentences to the visual images. Then the alignment model are treated as the training data for the Bidirectional Recurrent Neural Network (BRNN) model to generate the sentence descriptions. After that, we pick out key words of the sentence by comparing the word candidate prediction in the generation of sentence description as the core semantic annotation of image.

2.1 Image Representation

The features of images are extracted by VGG16 [23]. The neural network is pre-trained on ImageNet [24] and finetuned on the 200 classes of the ImageNet Detection Challenge [25]. The representations of images are computed by

$$\nu = W_m[CNN_{\theta_c}(I)] + b_m \quad (1)$$

where I is an image, θ_c the parameters of the CNN model and b_m the bias of our method. $CNN(I)$ returns the 4096-dimensional activations of the fully connected layer before the classifier. Each images and sentence representations will be mapped on a multimodal space. In the space, the dimensionality of the image representations is set to h-dimension.

2.2 Sentence Representation

To establish the inter-modal relationship of the images and sentence descriptions, the words in the sentence are mapped to the same h-dimensional on the embedding space. The bidirectional recurrent neural network (BRNN) is applied to learn the word representations. The BRNN takes a sentence containing N words and transforms each one into an h-dimensional vector. It is described as follows:

$$x_t = W_w \Pi_t \tag{2}$$

$$e_t = f(W_e x_t + b_e) \tag{3}$$

$$h_t^f = f(e_t + W_f h_{t-1}^f + b_f) \tag{4}$$

$$h_t^b = f(e_t + W_b h_{t+1}^f + b_b) \tag{5}$$

$$S_t = f(W_d(h_t^f + h_t^b)b_d) \tag{6}$$

The activation function f, here rectified linear unit (ReLU), is defined as $f : x \to \max(0, x)$. The index $t = (1, 2, \cdots, N)$ denotes the location information of the word in a sentence. Π_t is an indicator column vector in which the t-th word is encoded by a 1-of-k representation on the vocabulary. The weighted matrix W_w is initialized with 300-dimensional word2vec weights [26]. The BRNN consists of two processing, one moving left to right (h_t^f) and the other right to left (h_t^b). S_t is an h-dimensional representation for the t-th word. The other parameters, $W_e, W_f, W_b, W_,d$ and their corresponding bias b_e, b_f, b_b, b_d are needed to learn in the network training.

2.3 Alignment Model

For the alignment model, we use a combination of VGG16 [23] and BRNN to separately extract the features of images and their corresponding sentence descriptions. After the processing mentioned above, the images and their corresponding sentences are mapped into a h-dimensional space. In this paper, we leverage the vector $\nu_i^T S_t$ between the image i and the sentence t as similarity measure method:

$$S_{kl} = \sum_{t \in g_l} \max_{i \in g_k} \nu_i^T S_t \tag{7}$$

Here g_k is the set of images, and g_l is the set of sentences. The best alignment of image-sentence pair can be found by computing the score S_{kl}. Assuming that $k = l$ is the best corresponding pair, the final loss function $C(\theta)$ is calculated by

$$C(\theta) = \sum_k [\underbrace{\sum_l \max(0, S_{kl} - S_{kk} + 1)}_{rankimages} + \underbrace{\sum_l \max(0, S_{lk} - S_{kk} + 1)}_{ranksentences}] \quad (8)$$

Here, θ is the parameters of the alignment model containing both the CNN parameters and BRNN parameters.

Core semantic description generation. In this section, our purpose is to extract the key words relative to the core semantics of the images. First, we learn the model to generate the sentence descriptions. Set the image I and the set of sequences $(x_1, \cdots, x_T)$ as the input vectors of the mRNNs model. The hidden sequences $(h_1, \cdots, h_t)$ and the output sequence $(y_1, \cdots, y_t)$ are computed by the following iterative formula for $t = 1$ to $t = T$.

$$b_\nu = W_{hi}[CNN_{\theta_c}(I)] \quad (9)$$

$$h_t = f(W_{hx}x_t + W_{hh}h_{t-1} + b_h + 1(t = 1) \bigodot b_\nu) \quad (10)$$

$$y_t = softmax(W_{oh}h_t + b_o) \quad (11)$$

Here, y_t is the output vector which has the size of the word dictionary and one additional dimension for a special END token used to end up the generative process. Besides, each element of y_t is the probability value predicted by the model. The parameters $t, W_{hi}, W_{hx}, W_{hh}, W_{oh}$, and b_o are needed to learn.

Meanwhile, during the sentence generations, we pick out the key words to describe the core semantic contents of images. The RNNs contains the input units, output units and hidden layers. The hidden layers are charged with the most works of the RNNs model. In RNNs, there has a one-way stream of information from the input units to the hidden layer, at the same time, another one-way information stream transmits the message from hidden layer to the output units. In some cases, the RNNs may perform the message from the output units return to hidden layers, called "backward prediction"; meanwhile, the input of hidden layer also contains the state of previous hidden layer. A typical structure of RNNs simply unfold is illustrated in Fig. 2.

Note that we found an interesting thing in the process of sentence generation. During the prediction of the next word y_t, if the forward prediction message transmitted from hidden layer to the output units has a great difference compared to the backward prediction, that means the current forward prediction result can greatly describe the core semantic of images, such as the object semantic or the action semantic. The current word prediction is just the key word(s) that we want. Details as follows.

For each word of the prediction during the sentence generation, we select the top two ranking of the candidate probability values y_{ft1}, y_{ft2} when $y_{ft1} > y_{ft2}$ at the time t. For the backward propagation prediction, we also pick out the top two candidate probability y_{bt1}, y_{bt2}, when $y_{bt1} > y_{bt2}$. Comparing the two value and, if, we select the value of y_{ft1} as the output of next word. The selected word y_{ft1} can well represent the core semantic of images such as object semantic or action one. The selective word selections can be extracted based on Algorithm 1.

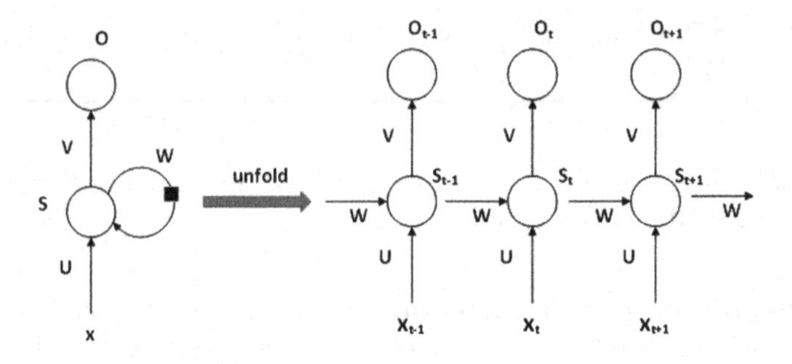

Fig. 2. A typical RNNs structure unfold on the time series of t.

Algorithm 1. Core semantic generation of images

1: **Input:** image I computed as the vector of $CNN_{\theta_c}(I)$ and the alignment model contains the corresponding of images and words.
2: **Output:** Core semantic word y of image.
3: For each prediction of next word y do
4: **Core semantic words selection**
5: If $y_{bt1} > y_{bt2}$ then
6: **Output** y_{ft1}.
7: else
8: output null
9: end if
10: **Core semantic words generation**
11: for the length of generation n do
12: if $n = 1$ then
13: output the object of image
14: else if $n = 2$ then
15: output the action of image
16: else if $n \geq 3$ then
17: output more objects of image
18: end if
19: end for
20: end for

2.4 Image Retrieval Based on Core Semantic

After the generation of the key words and the core semantic description of images, we try to apply the semantic annotation in image retrieval to find the similar images. We first generate the descriptions both the retrieval image and the images to be retrieved in database. Each image has the annotation label. Then the similar images have the same annotation label can be retrieved by compare their annotation label with that of the retrieval image.

Generally, we often set the model to generate one or two words as the key words of image, the first word explains the object of image and the second describes the action semantic of image if the behavior semantic is happened in the image. Such, when we need to find the same object of images, we just compare the first word. Similarly, comparing the second word can find the same action semantics in images.

3 Experiment

3.1 Dataset

We use the MSCOCO [24] dataset to train the models in our experiments. The dataset contains 123287 images and each of them is annotated by 5 sentences with the tool of Amazon Mechanical Turk. From the MSCOCO dataset, we choose the train2014 as the training data for our model. For the test samples, we separately choose 1000 images from Val2014 and Fliker8K datasets. All the testing images have no any annotation information.

3.2 Evaluation of Generated Core Semantic Description

The word prediction results are evaluated by 10 persons. Each person judged whether the output key words from our model are truly related to the core semantics of image. The average of the 10 statistic results is the last evaluation performance.

There may be several kinds of common annotation mistakes, such as some specific object classes or behavior semantics. For instance, "man" is probably mistaken as "woman", "bus" as "train" or "sitting" as "standing". Besides, even the word generation is correct but not describing the core semantic of images, we treat it wrong.

Considering the conditions mentioned above, the word annotation is evaluated on two datasets, Val2014 and Flickr8K. The experimental result is shown in Table 1. F-VGG is not fine-tuned on VGG16, but the fine-tuned alignment model. Seen form Table 1, the result of F-VGG is higher about 3% than that of VGG16 for the two datasets. It is demonstrated that the fine-tuned work strengthens the adjustment of the weights to get a better alignment model, certainly resulting in a better word annotation. For the intuitive view, an example of word annotations is illustrated in Fig. 3. The top two key words are selected as the annotations in the experiment. For the left image, the word "motorcycle" describes the main object, and "parked" describes the action of the main object. The right image extracts two words "cat" and "laying" to express the meaning of "a cat is laying on a sofa".

Table 1. The statistic result of the core semantic extraction model.

Dataset	VGG16 (%)	F-VGG (%)
Val2014	89.92	93.25
Flickr8K	76.5	79.5

Fig. 3. A display of the kernel image semantic description.

3.3 Semantic-Based on Image Retrieval

There has the difference in between the object-semantic retrieval and the action-semantic retrieval. If we retrieve the images based on the object-semantic, the first word annotation, that is object word, will be generated for image retrieval. If we focus on the action-semantic, the two word annotations, that are object and action word, will be generated. The second word, that is action word, is applied to retrieve the images. For the image retrieval, word annotations are first to generated for the query image and the images to be retrieved. Examples for the object-semantic and action-semantic retrieval are displayed in Figs. 4 and 5 respectively.

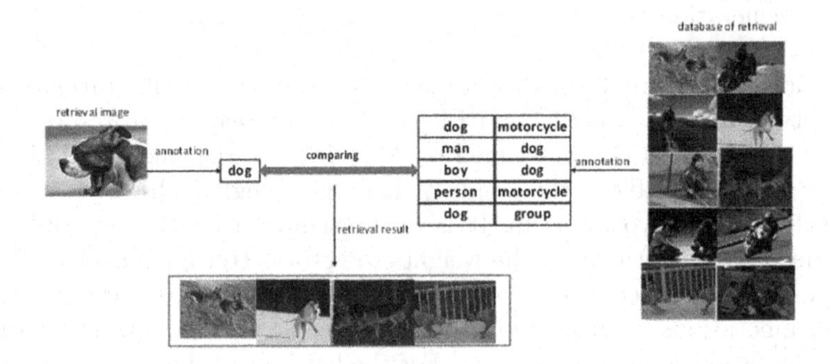

Fig. 4. Example of the object-semantic image retrieval.

From the dataset Flicker30K [25], we select 220 images as the retrieval database including 4 categories of semantic images, such as "dog", "group", "riding" and "sitting". Each category contains 55 images. The image retrieval result is shown in Table 2. We report the research result in the list with Recall@K (R@K is Recall@K) to measure the times a correct item was found among the top K results.

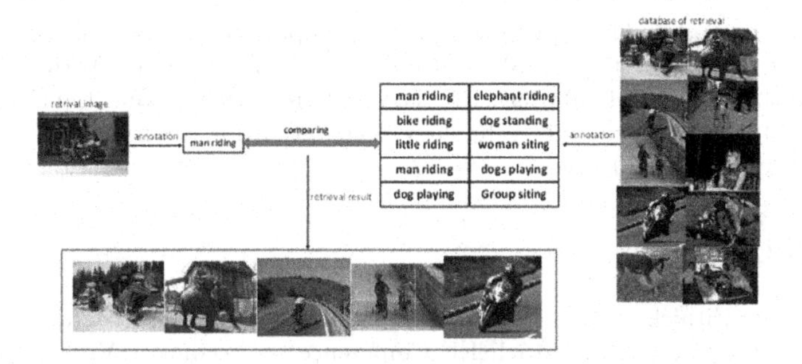

Fig. 5. Example of action-semantic image retrieval.

Table 2. Image retrieval experiment result based on core semantic.

class	R@1	R@5	R@10	R@15	R@20	R@25	R@30	R@35	R@40	
Dog	1	5	10	15	20	25	30	35	40	41
Group	1	5	10	15	20	25	30	35	40	45
Riding	1	4	8	12	19	24	28			
Sitting	1	3	6	10	12					

In Table 2, for the "dog" class, our method can annotate 41 images with the "dog" label from the 220 images, so the value of recall is about 0.75. It is easy to see, the recall is 0.82, 0.51 and 0.22 for the "group", "riding" and "sitting" class, respectively. Known from Table 2, the recall of the action class is less than that of object one. It is because the action-semantic class is more subjective than object-semantic one, which leads to a terrible retrieval result.

4 Discussions

There are two limitations for our approach. One, the size of input images in our model is required to be fixed. Some images must be resized to meet the size requirement leading to some distortions of objects. The other, in the core semantic generation processing, some words such as "a", "the", "is", "are" may be selected in the output.

5 Conclusions

We introduce a model that generates the core semantic descriptions of images based on weak labels in forms of images and sentences. During the generation of sentence description, comparing the current prediction candidates both of forward predictions and backward predictions, we can pick out the key words in the sentence generation as the core semantic of images. The semantic-based retrieval is performed on Flicker30K dataset. Experimental results show that the performance of object-semantic retrieval outperforms action-semantic retrieval.

Acknowledgments. This work was partly supported by Natural Science Foundation of China (No. 61303128), Natural Science Foundation of Hebei Province (Nos. F2013203220, F2017203169), Key Foundation of Hebei Educational Committee (ZD2017080) and Science and Technology Foundation for Returned Overseas People of Hebei Province (CL201621).

References

1. Chen, X., Zitnick, C.L.: Mind's eye: a recurrent visual representation for image caption generation. In: 2015 IEEE Conference on Computer Vision and Pattern Recognition (CVPR), pp. 2422–2431 (2014)
2. Xu, K., Ba, J., Kiros, R.: Show, attend and tell: neural image caption generation with visual attention. Computer Science, pp. 2048–2057 (2015)
3. Krizhevsky, A., Sutskever, I., Hinton, G.E.: ImageNet classification with deep convolutional neural networks. In: International Conference on Neural Information Processing Systems, vol. 25, pp. 1097–1105. Curran Associates Inc. (2012)
4. Karpathy, A., Johnson, J., Fei-Fei, L.: Visualizing and Understanding Recurrent Networks (2015)
5. Ren, S., He, K., Girshick, R.: Faster R-CNN: towards real-time object detection with region proposal networks. IEEE Trans. Pattern Anal. Mach. Intell. **39**(6), 1137–1149 (2015)
6. Mikolov, T., Karafit, M., Burget, L.: Recurrent neural network based language model. In: Conference of the International Speech Communication Association, Interspeech 2010, pp. 1045–1048, Makuhari, Chiba, Japan, DBLP, September 2010
7. Karpathy, A., Li, F.F.: Deep visual-semantic alignments for generating image descriptions. IEEE Trans. Pattern Anal. Mach. Intell. **39**(4), 664–676 (2014)
8. Barnard, K., Duygulu, P., Forsyth, D.: Matching words and pictures. J. Mach. Learn. Res. **3**(2), 1107–1135 (2003)
9. Kulkarni, G., Premraj, V., Ordonez, V.: BabyTalk: understanding and generating simple image descriptions. In: IEEE Computer Vision and Pattern Recognition, pp. 1601–1608 (2013)
10. Farhadi, A., Hejrati, M., Sadeghi, M.A., Young, P., Rashtchian, C., Hockenmaier, J., Forsyth, D.: Every picture tells a story: generating sentences from images. In: Daniilidis, K., Maragos, P., Paragios, N. (eds.) ECCV 2010. LNCS, vol. 6314, pp. 15–29. Springer, Heidelberg (2010). https://doi.org/10.1007/978-3-642-15561-1_2
11. Ordonez, V., Han, X., Kuznetsova, P.: Large scale retrieval and generation of image descriptions. Int. J. Comput. Vis. **119**(1), 46–59 (2016)
12. Socher, R., Li, F.F.: Connecting modalities: semi-supervised segmentation and annotation of images using unaligned text corpora. In: IEEE Computer Vision and Pattern Recognition, pp. 966–973 (2010)
13. Socher, R., Karpathy, A., Le, Q.V.: Grounded compositional semantics for finding and describing images with sentences. NLP Stanford Edu. (2013)
14. Kuznetsova, P., Ordonez, V., Berg, A.: Generalizing image captions for image-text parallel corpus. In: Meeting of the Association for Computational Linguistics, pp. 790–796 (2013)
15. Jia, Y., Salzmann, M., Darrell, T.: Learning cross-modality similarity for multinomial data. In: IEEE Computer Society International Conference on Computer Vision, pp. 2407–2414 (2011)
16. Mao, J.: Explain Images with Multimodal Recurrent Neural Networks. Computer Science (2014)

17. Vinyals, O., Toshev, A., Bengio, S.: Show and tell: a neural image caption generator, pp. 3156–3164 (2014)
18. Donahue, J., Hendricks, L.A., Guadarrama, S.: Long-term recurrent convolutional networks for visual recognition and description. In: IEEE Computer Vision and Pattern Recognition, vol. 39, pp. 85–91 (2015)
19. Hochreiter, S., Schmidhuber, J.: Long short-term memory. Neural Comput. **9**(8), 1735–1780 (1997)
20. Kiros, R., Salakhutdinov, R., Zemel, R.S.: Unifying Visual-Semantic Embeddings with Multimodal Neural Language Models. Computer Science (2014)
21. Fang, H., Platt, J.C., Zitnick, C.L.: From captions to visual concepts and back, pp. 1473–1482 (2014)
22. Werbos, P.J.: Generalization of backpropagation with application to a recurrent gas market model. Neural Netw. **1**(4), 339–356 (1988)
23. Simonyan, K., Zisserman, A.: Very Deep Convolutional Networks for Large-Scale Image Recognition. Computer Science (2014)
24. Lin, T.Y., Maire, M., Belongie, S.: Microsoft COCO: Common Objects in Context, pp. 740–755 (2014)
25. Young, P., Lai, A., Hodosh, M.: From image descriptions to visual denotations: new similarity metrics for semantic inference over event descriptions. NLP CS Illinois Edu. (2014)

Learning the Frame-2-Frame Ego-Motion for Visual Odometry with Convolutional Neural Network

Mingqi Qiao and Zilei Wang(✉)

Department of Automation, University of Science and Technology of China, Hefei 230027, China
qmq@mail.ustc.edu.cn, zlwang@ustc.edu.cn

Abstract. Visual odometry (VO) is one of the important components of visual SLAM systems, and some impressive works about VO have been presented recently. However, these methods mostly follow the traditional feature detection and tracking pipeline, which usually suffer from less robustness to complex scenarios. Deep learning has presented outstanding performance in various visual tasks, which has great potential to improve VO. In this paper, we discuss how to learn an appropriate estimator to predict the frame-2-frame ego-motion with convolutional neural network. Specifically, we construct a CNN model which formulates the pose regression as a supervised learning problem. Here the proposed architecture uses raw images and optical flow as input to predict the motion. As a result, the trajectories can be produced by iterative computation. We experimentally demonstrate the performance of the proposed method on public dataset, which can achieve better ego-motion estimation compared to the baselines.

Keywords: Visual odometry · Ego motion · CNNs

1 Introduction

Visual odometry (VO) as the front end of visual SLAM is a highly active area of research, which has wide applications in numerous secnarios, such as robotics, navigation, and virtual reality. In this paper, we focus on the task of monocular camera motion estimation in visual odometry. Over the past few years, some impressive works about VO have been proposed with the development of visual SLAM [9,12]. However, the results of these works mostly are far from the expected performance in real systems, especially for the complex scenarios. In recent years, deep learning has shown great success in various visual tasks, while it has not yet been well explored in visual odometry [8,25].

In general, visual odometry is implemented through computing the camera motion between consecutive frames. Specifically, the frame-2-frame ego-motion is firstly estimated by utilizing geometric theory, and then refined with other

J. Yang et al. (Eds.): CCCV 2017, Part III, CCIS 773, pp. 500–511, 2017.
https://doi.org/10.1007/978-981-10-7305-2_43

optimization strategies, such as Kalman filtering or bundle adjustment. For the geometric methods in monocular VO, some sophiscated computing frameworks have already been formed. However, these systems are not robust enough while complex scenarios are encountered, *e.g.*, the initial feature extraction process is apt to be interrupted. In addition, the problem of scale recovery is always one of main obstacles of developing monocular VO. In order to further improve the performance, more informative and robust features have always been desired, which makes the geometric algorithms fall into the bottleneck of performance now. Deep learning has achieved great success owing to the powerful ability of extracting high-level features, particularly for image understanding tasks such as classification and detection [22,24], semantic segmentation [23]. Inpired by these works, it is believed that deep neural networks can learn the representations of camera motion from large dataset. That is, the true scale and intrinsic rules of camera motion can be learned even without other information.

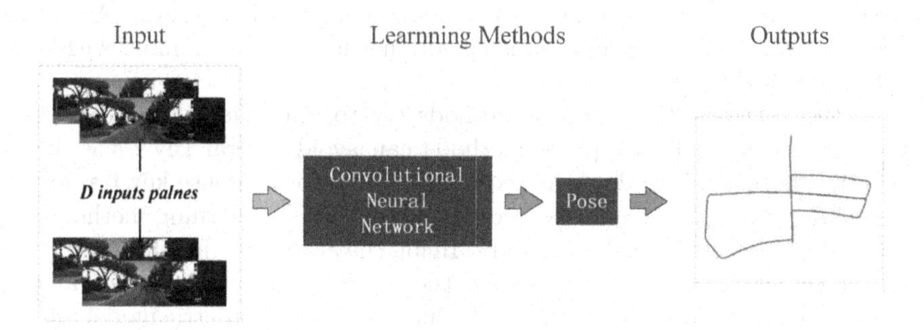

Fig. 1. Overview of the proposed method for visual odometry. A pair of consecutive images as input are fed into the trained CNN estimator to predict the frame-2-frame ego-motion, and the trajectory is producted by iterative computation finally.

Our purpose in this paper is to build a proper architecture of CNN to directly learn the frame-2-frame camera motion for VO, and consequently the trajectory can be produced via iterative computation, as shown in Fig. 1. To this end, we first construct a convolualtional neural network which uses both the image pair and optical flow as input to perform the prediction. Then to reduce the offset error, we propose to use a residual network to further boost the estimation accuracy. Finally, we experimentally verified the effectiveness of the proposed method through evaluation on public dataset.

The organization of this paper is as follows: Sect. 2 briefly explains the related works, and Sect. 3 shows the details of our proposed method. Section 4 describes the experimental results by comparing with the baseline methods. Finally, we conclude this work in Sect. 5.

2 Related Works

Here we briefly review the related works on monocular visual odometry. According to the technical routes adopted by ego-motion estimation, we can roughly divide the VO algorithms into two broad categories: *Geometric methods* and *Learning methods.*

Geometric methods can be further divided into the feature based methods and direct methods. The feature based methods rely on detecting and tracking a sparse set of salient image features [9,10], while the direct methods depend on the pixel intensity values to extract motion information [11–13]. Specifically, the feature based methods first match the feature points across the consecutive frames, and then reconstruct 3D points by triangulation. Finally, the camera pose can be estimated. Compared to the feature based methods, the direct methods can theoretically achieve better accuracy and stability because they try to use pixels of the whole image. But it is difficult for the direct methods to be used in real systems due to introducing the heavy computation [12,13]. As for the true scale, some extra information is usually needed, *e.g.*, combining with other sensor such as IMU [14].

On the contrary, the learning methods try to infer the motion estimation directly from data. This type of methods can avoid several key issues in the geometric methods, including the requirements of storing dense key frames and establishing frame-2-frame feature correspondences. The learning method early do not adopt the end-to-end framework. In [5], the authors train a KNN regressor while one image is divided into cells. Another work is to create a semi-parametric learning approach for visual odometry by incorporating geometric model into the CGP framework [6,28].

Recent years, researchers start to deal with the inter-frame problems using deep neural networks. For example, the authors propose a deep learning architecture to deal with human pose recovery problems [26,27]. Dosovitskiy *et al.* [1] design a network FlowNet to compute the optical flow between two images. DeTone *et al.* [2] propose to use deep networks to estimate the homography matrix between two images, which essentially is a regression problem. Similarly, a convolutional neural network for camera relocalization is proposed in [4], where the pose vectors are discretized and the original problem is transformed into a classification problem. In [3], the authors propose a learning method for visual odometry with CNNs, where the depth data are required but may be unavailable in real systems. A more related work to ours is P-CNN [15]. In the work, a CNN architecture is designed only using the optical flow, and the robustness of learning VO is experimentally demonstrated. Different from [15], we propose a new network using both raw images and optical flow in this paper, and consequently the better estimation performance can be achieved.

3 Methodology

In this section, we elaborate on the proposed method. Here we firstly formulate the pose regression problem, and then explain the network architecture used to estimate camera motion with the raw images and optical flow.

3.1 Problem Formulation

Given a pair of consecutive images with the resolution of $n \times m$, we want to learn a function f that is able to estimate the camera motion between them. Our network outputs a motion vector $y \in \mathcal{Y} \subset \Re^6$, given by the displacement p of the camera centre and orientation q represented by three Eular angles:

$$y = [p, q] \tag{1}$$

The input $x \in \mathcal{X} \subset \Re^{n \times m \times 3}$ is the RGB representation of raw image or dense optical flow. So, the problem is to find a function difined as:

$$f : \mathcal{X} \to \mathcal{Y} \tag{2}$$

Here the motion vector $y \in \mathcal{Y}$ is defined relative to consecutive frames. So we can select the first frame of the entire image sequence as the reference frame to create a continuous trajectory via iterative computation.

In order to regress motion, we construct and train a CNN with the Euclidean loss, *i.e.*, the following loss function is adopted:

$$\mathcal{L} = \frac{1}{N} \sum_{i=1}^{N} \| f(x_i) - y_i \|_2 \tag{3}$$

In this work, both the raw images and dense optical flow are fed into the network. In particular, the dense optical flow are extracted using the Broxs algorithm [7], which allows for the large displacements and linearization. For one time of motion estimation, a pair of consecutive frames and the corresponding dense optical flow are used, as show in Fig. 2, where the optical flow is encoded in RGB image. To more efficiently train the networks in practice, the raw images for VO are down-sampled with a resolution of 160×48, and the optical flow images are with 78×24.

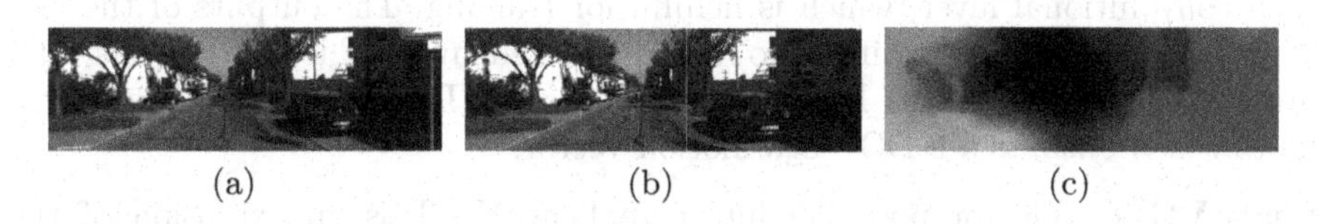

(a) (b) (c)

Fig. 2. Two subsequent frames from Seq 00 and corresponding dense optical flow. (a) and (b) show the raw images and (c) shows the dense optical flow (Color figure online).

3.2 Network Architecture for Ego-Motion Estimation

Inspired by the works for action recognition [16,17], we propose a new CNN architecture for pose estimation in this paper. Specifically, we first introduce an architecture called LearnVO-I fed by a pair of raw images, and then extend it to a new two-stream architecture called LearnVO-T with both the raw image and optical flow.

LearnVO-I. The main idea of LearnVO-I is to first process each of a pair of consecutive images by two separate and identical networks and then combine them in some middle layer. The resulting network is illustratred in Fig. 3. With this architecture, the network is constrained to produce meaningful representations for images separately and the motion estimation is performed by fusing them on a high level. Such a way is similar with the traditional matching and tracking approach.

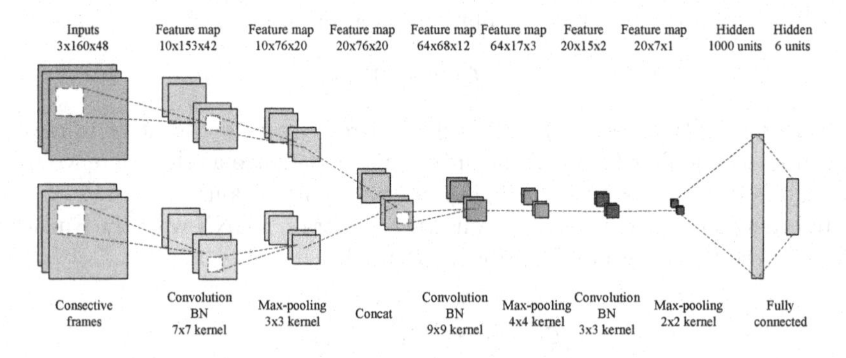

Fig. 3. Architecture of LearnVO-I. Two consective images are fed into the network and the corresponding features are extracted by CNNs. Two fully connected layers are used to produce 6-dimensional motion vectors.

Specifically, we stack convolutional layers, pooling layers, and RELU layers to construct the LearnVO-I network, and two extracted features are stacked channel-wise for the fusion. More specifically, we use several convolutional layers with max pooling to keep the salient value and meanwhile downsample the feature maps. It is worth noting that the batch normalization [18] is added after each convolutional layer, which is helpful for training. The outputs of the last convolutional layer are fed into two fully connected layers. The first one has 1000 units and is followed by a RELU activation layer. The Second one has only 6 units which outputs a 6-DOF ego-motion vector.

LearnVO-T. The proposed architecture LearnVO-T is an extension of the LearnVO-I by following the two-stream [17]. As shown in Fig. 4, the architecture uses both the two raw images and optical flow as input to predict camera motion. Overall, we can divide the architectue into two parts: optical flow stream and raw image stream, and each of them is implemented by CNNs.

Specifically, the optical flow network uses the optical flow data as input and extract features through two convolutional layers with max pooling layers. Here we combine the features from different levels by merging the outputs of the first convolutional layer and the last one. For the raw image, we use a similar network with LearnVO-I. We concatenate the flatten features from two streams to form the final representations for the followng motion estimation, and feed them into the fully connected layers. Here we similarly set the two fully connected layers with 1000 and 6 uints respectively.

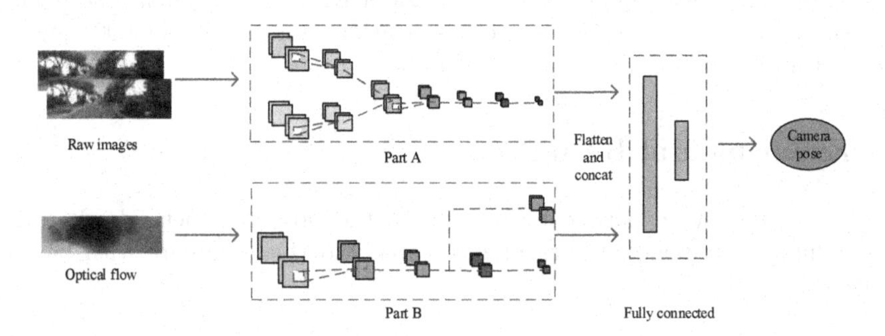

Fig. 4. Architecture of LearnVO-T network. It consists two sub parts: the raw images stream (Part A) and the optical flow stream (Part B), then they are fed into the same fully connected layers.

3.3 Boosting with Residual Network

Monocular camera VO is usually suffer from accumulate error caused by iterative computation, which is important but very difficult to solve in learning based method. Essentially, the accumulate error is derived from the inaccuracy of camera motion estimation between consecutive frames. Here we use a simple technique to make the motion estimation more accurate, and thus to some extent mitigate the interference of accumulate error for whole trajectory. We propose to train a residual network to fit the error fluctuation of the ego-motion generated by the LearnVO network, and so that we can correct the inital pose estimation. The idea is inspired by ORB-SLAM which guesses an initial value by the motion model and then optimize it. We use a similar architecture as LearnVO-T to build the residual network fed by raw images and optical flow, and the network outputs 6-dimension error compensation vector. And we combine the outputs of two network with the *add* operator in the testing phase. LearnVO-T with residual does help but the improvement in accuracy is limited as shown in Sect. 4.2 and we will explore more effective methods to deal with accumulate error in the future work.

3.4 Training Details

We have designed a network architecture which takes the paired images and optical flow as inputs to regress the ego-motion. In order to obtain a good network model, we firstly train the two networks with fully connected layers separately, and then conduct a global finetuning in which the parameters of the fully connected layers will be relearned. In practice, all weights in the convolutional layers and fully connected layers are initialized with *Xavier*. We set the parameter *use_global_stats* in BN layer as *false* at the training phase and *true* at the test phase. The networks are designed to adopt L2 loss, and *Adam* is adopted as the solver to minimize the loss. The base learning rate is set as 0.0001 and the momentum is 0.9.

4 Experimental Results

In this section, we experimentally evaluate the proposed methods. We show the results of different network architectures and then compare with baseline methods.

4.1 Dataset and Experiment Protocal

We evaluate the performance of the proposed method on the public dataset KITTI vision benchmark [19], which provide 11 sequences with the precise groundtruth trajectories in the terms of a 3×4 transformation matrix. There are all about 23000 images to be used with a resolution of 1241×376 or 1226×370. In our experiments, we transform the groudtruth to 6-DOF pose using Peter Corke's Robotic Toolbox. For the learning methods, we use the first 7 sequences to train the model and the other 3 sequences to evaluate the performence.

We choose three different methods to make comparison: a geometric monocular SLAM system (ORB-SLAM) [9], a geometric visual odometry (VISO2-M) [20], and a learning method (P-CNN VO) [15]. VISO2-M computes the trajectories through the frame-2-frame estimation without bundle adjustment, and thus it is comparable to our proposed method. Consdering the scale recovery problem, we have aligned the trajectories of ORB-SLAM, VISO2-M to the groundtruth with a similarity transformation using Horn's algorithm [21]. For fair comparison, the same optical flow data are used for P-CNN VO and our methods.

4.2 Performance Analysis

In this section, we analyze the performances of different network architectures: LearnVO-I, LearnVO-T, and LearnVO-T with residual network, and the results are shown in Fig. 5. From the results, it can be seen that LearnVO-T produces more accurate trajectories than LearnVO-I for all sequences. It is implies that raw images and optical flow represent different types of information and their combination may offer more robust features for pose estimation. The bold lines in Fig. 5 demonstrate the performance with the additional residual network. It can be seen that the results of Seqences 08 and 09 are improved and the almost same performance is achieved for Sequence 10, which show that the residual network is helpful to reduce the offset error. In the next comparison with the baselines, we will adopt LearnVO-T without residual network. On one hand, it is more fair since P-CNN VO does not have a refinement process. On the other hand, LearnVO-T can yield good enough results with a relatively simple architecture.

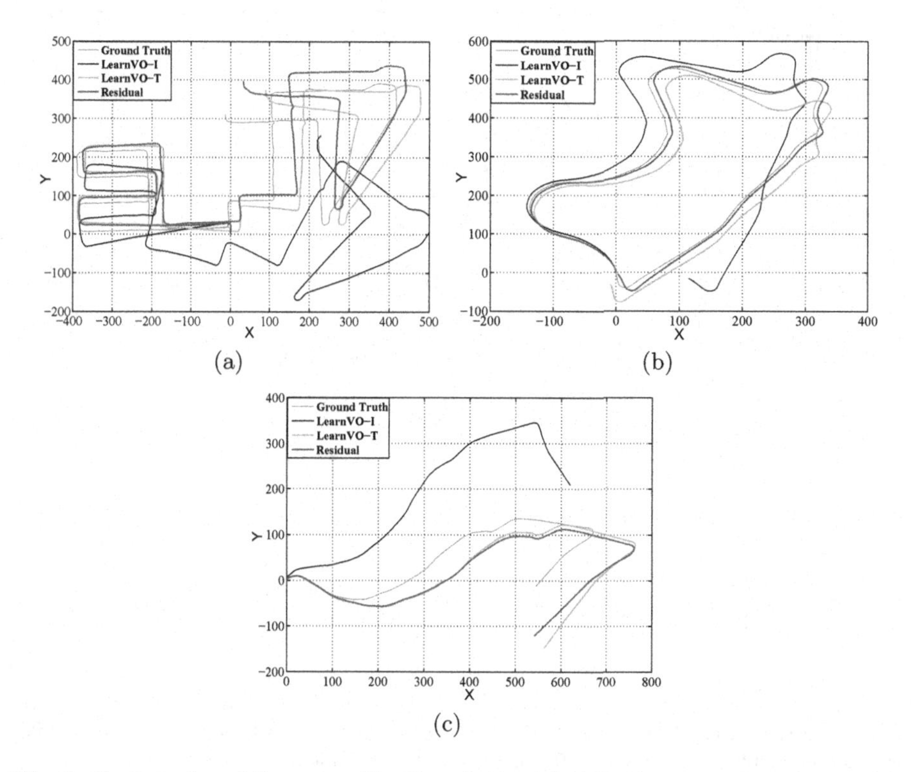

Fig. 5. Trajectories of Sequence 08, 09 and 10 with different network architectures: LearnVO-I, LearnVO-T, and LearnVO-T with residual network. (a): sequence 08, (b): sequence 09, (c): sequence 10. Best viewed electronically.

4.3 Performance Comparison

In this subsection, we evaluate the effectiveness of the proposed method by comparing with other three methods mentioned in Sect. 4.1. Here we conduct comparisons from two aspects: accuracy and computational time, which together measure the overall performance.

Accuracy. Figure 6 gives all the resulting reconstructed trajectories while Table 1 provides the translation and rotation error for different methods. The qualitative results in Fig. 6 show that our LearnVO-T can produce more accrurate trajectories for all three sequences than other methods except ORB-SLAM. Here ORB-SLAM is a complete SLAM system with the feature tracking, mapping, and loop detecting, and we are not surprised that it performes best. The result of Sequence 08 provides an interesting observation that our LearnVO-T is almost at the same accurate level with ORB-SLAM, and more robust since ORB-SLAM tracks unsuccessfully for many times in our experiments.

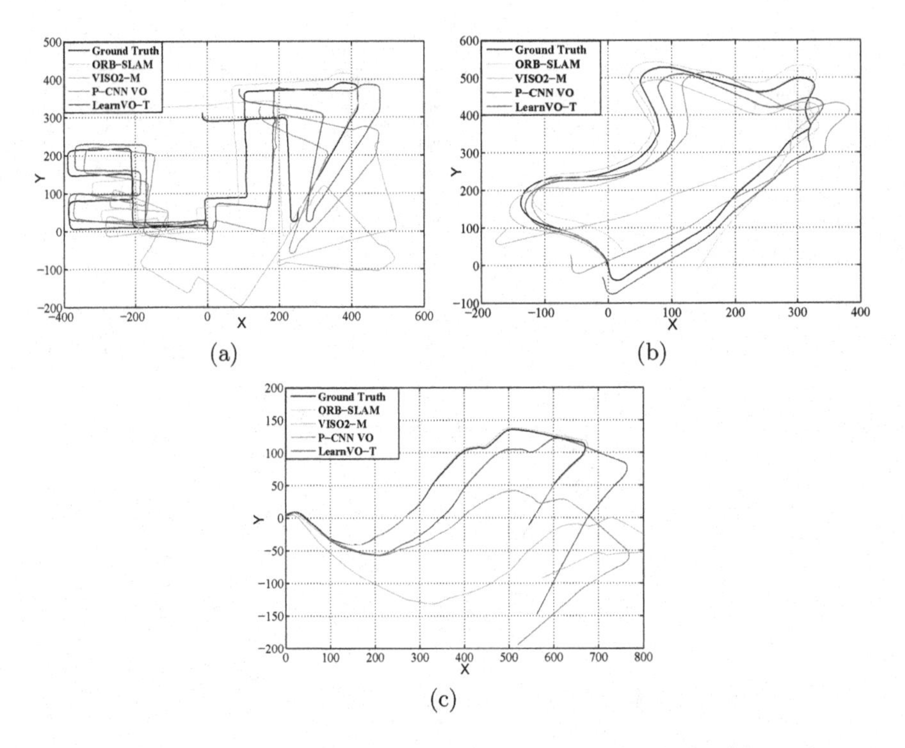

Fig. 6. Trajectories of Seqence 08, 09, and 10 for different methods. (a): sequence 08, (b): sequence 09, (c): sequence 10. Best viewed electronically.

Table 1 shows the median RMSE error [19] of the trajectories over four executions. Aside from ORB-SLAM, P-CNN VO and our LearnVO-T perform better than the geometric VISO2-M. It shows that the deep learning methods can predict the 6-DOF frame-2-frame motions as accurate as possible. In general, LearnVO-T outperforms P-CNN VO. In particular, for Sequence 10, the translation error of length reduces nearly 50% from 21.23% to 13.6%. From the results, it is convinced that the learning methods have enormous advantage for the trajectories consisting of many curves.

Computational Time. Table 2 provides the computational time for different methods, where ORB-SLAM is not included due to its incomparableness to others. Here P-CNN VO and LearnVO-T are implemented using caffe with a

Table 1. Comparison results in terms of average translation and rotation errors.

	ORB-SLAM		VISO2-M		P-CNN VO		LearnVO-T	
	Trans (%)	Rot (deg/m)	Trans (%)	Rot (deg/m)	Trans (%)	Rot (deg/m)	Trans (%)	Rot (deg/m)
08	5	0.0067	19.39	0.0393	7.6	0.0187	6.05	0.017
09	1.6	0.0024	9.26	0.0279	6.75	0.0252	6.4	0.026
10	1.2	0.0018	27.55	0.0409	21.23	0.0405	13.6	0.028

NVIDIA K40 GPU. Our LearnVO-T takes about 20 ms to estimate a camera pose, and the estimator is satisfied for real-time applications. Taking the results in Tables 1 and 2, it can be seen that compared to P-CNN VO, our method reduces the time cost using a more simple architecture without accuracy loss.

Table 2. Comparison results in terms of average computational time.

	VISO2-M	P-CNN VO	LearnVO-T
08(s)	0.187	0.021	0.017
09(s)	0.198	0.013	0.019
10(s)	0.233	0.024	0.020

5 Conclusions

In this paper, we propose a novel camera motion estimation method based on CNNs for visual odometry. Specifically, both the raw images and optical flow data are used and the corresponding CNN architectures are proposed. In addition, to reduce the offset error, we propose to use a residual network to learn the errors. We experimentally analyze and compare the performance of the proposed method on public dataset. The results verify the effectiveness of our method, which can implement the frame-2-frame ego-motion estimation well. It is believed that deep learning has a great potential to achieve accurate estimation for visual odometry or visual SLAM. In the future, we plan to explore more advanced approach for pose estimation under the deep learning framework, *e.g.*, combining with loop detection.

Acknowledgements. This work is supported partially by the National Natural Science Foundation of China under Grant 61673362 and 61233003, Youth Innovation Promotion Association CAS, and the Fundamental Research Funds for the Central Universities.

References

1. Dosovitskiy, A., Fischer, P., llg, E.: Flownet: learning optical flow with convolutional networks. In: Proceedings of the IEEE International Conference on Computer Vision, pp. 2758–2766 (2015)
2. DeTone, D., Malisiewicz, T., Rabinovich, A.: Deep image homography estimation. arXiv preprint arXiv:1606.03798 (2016)
3. Konda, K.R., Memisevic, R.: Learning visual odometry with a convolutional network. VISAPP **1**, 486–490 (2015)
4. Kendall, A., Grimes, M., Clipolla, R.: Posenet: a convolutional network for real-time 6-DOF camera relocalization. Proceedings of the IEEE international conference on computer vision, pp. 2938–2946 (2015)
5. Roberts, R., Nguyen, H., Krishnamurthi, N., Balch, T.: Memory-based learning for visual odometry. In: IEEE International Conference on Robotics and Automation, pp. 47–52 (2008)
6. Guizilini, V., Ramos, F.: Semi-parametric learning for visual odometry. Int. J. Robot. Res. **32**(5), 526–546 (2013)
7. Brox, T., Bruhn, A., Papenberg, N., Weickert, J.: High accuracy optical flow estimation based on a theory for warping. In: Pajdla, T., Matas, J. (eds.) ECCV 2004. LNCS, vol. 3024, pp. 25–36. Springer, Heidelberg (2004). https://doi.org/10.1007/978-3-540-24673-2_3
8. Cadena, C., Carlone, L., Carrillo, H., Latif, Y., Scaramuzza, D., Neira, J., Leonard, J.: Past, present, and future of simultaneous localization and mapping: toward the robust-perception age. IEEE Trans. Robot. **32**(6), 1309–1332 (2016)
9. Mur-Artal, R., Montiel, J.M.M., Tardos, J.D.: ORB-SLAM: a versatile and accurate monocular SLAM system. IEEE Trans. Robot. **31**(5), 1147–1163 (2015)
10. Klein, G., Murray, D.: Parallel tracking and mapping for small AR workspaces. In: The IEEE International Symposium on Mixed and Augmented Reality, pp. 225–234, November 2007
11. Forster, C., Pizzoli, M., Scaramuzza, D.: SVO: fast semi-direct monocular visual odometry. In: IEEE International Conference on Robotics and Automation, pp. 15–22, May 2014
12. Newcombe, R.A., Lovegrove, S.J., Davison, A.J.: DTAM: dense tracking and mapping in real-time. In: International Conference on Computer Vision, pp. 2320–2327, November 2011
13. Engel, J., Schöps, T., Cremers, D.: LSD-SLAM: large-scale direct monocular SLAM. In: Fleet, D., Pajdla, T., Schiele, B., Tuytelaars, T. (eds.) ECCV 2014. LNCS, vol. 8690, pp. 834–849. Springer, Cham (2014). https://doi.org/10.1007/978-3-319-10605-2_54
14. Leutenegger, S., Lynen, S., Bosse, M., Siegwart, R., Furgale, P.: Keyframe-based visualCinertial odometry using nonlinear optimization. Int. J. Robot. Res. **34**(3), 314–334 (2015)
15. Costante, G., Mancini, M., Valigi, P., Ciarfuglia, T.A.: Exploring representation learning with CNNs for frame-to-frame ego-motion estimation. IEEE Robot. Autom. Lett. **1**(1), 18–25 (2016)
16. Feichtenhofer, C., Pinz, A., Zisserman, A.: Convolutional two-stream network fusion for video action recognition. In: IEEE Conference on Computer Vision and Pattern Recognition, pp. 1933–1941 (2016)
17. Simonyan, K., Zisserman, A.: Two-stream convolutional networks for action recognition in videos. In: Conference on Neural Information Processing Systems, pp. 568–576 (2014)

18. Ioffe, S., Szegedy, C.: Batch normalization: accelerating deep network training by reducing internal covariate shift. In: 32nd International Conference on Machine Learning, pp. 448–456 (2015)
19. Geiger, A., Lenz, P., Urtasun, R.: Are we ready for autonomous driving? the kitti vision benchmark suite. In: IEEE Conference on Computer Vision and Pattern Recognition, pp. 3354–3361, June 2012
20. Geiger, A., Ziegler, J., Stiller, C.: Stereoscan: dense 3D reconstruction in real-time. In: Intelligent Vehicles Symposium (IV), pp. 963–968, June 2011
21. Horn, B.K.: Closed-form solution of absolute orientation using unit quaternions. JOSA A **4**(4), 629–642 (1987)
22. Ren, S., He, K., Girshick, R., Sun, J.: Faster R-CNN: towards real-time object detection with region proposal networks. In: Conference on Neural Information Processing Systems, pp. 91–99 (2015)
23. Dai, J., He, K., Sun, J.: Instance-aware semantic segmentation via multi-task network cascades. In: IEEE Conference on Computer Vision and Pattern Recognition, pp. 3150–3158 (2016)
24. Liu, W., Anguelov, D., Erhan, D., Szegedy, C., Reed, S., Fu, C.-Y., Berg, A.C.: SSD: single shot multibox detector. In: Leibe, B., Matas, J., Sebe, N., Welling, M. (eds.) ECCV 2016. LNCS, vol. 9905, pp. 21–37. Springer, Cham (2016). https://doi.org/10.1007/978-3-319-46448-0_2
25. Liu, W., Wang, Z., Liu, X., Zeng, N., Liu, Y., Alsaadi, F.E.: A survey of deep neural network architectures and their applications. Neurocomputing **234**, 11–26 (2017)
26. Hong, C., Yu, J., Wan, J., Tao, D., Wang, M.: Multimodal deep autoencoder for human pose recovery. IEEE Trans. Image Process. **24**(12), 5659–5670 (2015)
27. Hong, C., Yu, J., Tao, D., Wang, M.: Image-based three-dimensional human pose recovery by multiview locality-sensitive sparse retrieval. IEEE Trans. Industr. Electron. **62**(6), 3742–3751 (2015)
28. Guizilini, V., Ramos, F.: Semi-parametric models for visual odometry. In: IEEE International Conference on Robotics and Automation, pp. 3482–3489, May 2012

Shape Representation and Matching

A Progressive Method of Simplifying Polylines with Multi-bends Simplification Templates

Jiawei Du, Fang Wu(✉), Jinghan Li, and Haiwei He

Zhengzhou Institute of Surveying and Mapping, Zhengzhou, Henan, China
whdxdjw@126.com, wufang_630@126.com

Abstract. Polyline simplification is one of the most important and classic researches in multi-scale expression of cartography and GIS. Polyline simplification methods based on bend units get widespread attention recently. Having analyzed the bends composing characteristics and the manual simplification process of polylines, it is resonable to find that not all of invisible bend units are deleted directly. A new progressive simplification method based on multi-bends groups is proposed. Firstly, based on the threshold of the simplified scale, polylines are divided into multi-bends groups. Secondly, in order to avoid over-simplification, some reasonable bends deletion options which called multi-bends templates are proposed. However multi-bends templates is not steady and always related to the quantity of multi-bends uniquely. Considering of the minimal errors derived form polyline simplification, the best multi-bends template that result in the least displacement is selected to simplify muli-bends groups. Thirdly, polylines do not stop repeating the above two steps until all bend units of simplified polylines can be detected in the simplified scale. Experiments with real road-net data were implemented, and comparison with other algorithms is discussed. Results show advantages including: (i) The proposed algorithm is a progressive simplification process which is conform to human cognition. (ii) The proposed algorithm can preserve the main shape of the polyline with deleting invisible bend units enough. (iii) The proposed algorithm avoid unnecessary displacement by deleting bend units as little possible.

Keywords: Cartography and GIS · Multi-scale expression
Polyline simplification · Bend units deletion

1 Introduction

Polyline is one of the most basic geometric elements in electronic maps and GIS. A lot of geographic objects are expressed as polylines, such as rivers, roads, shorelines, boundaries, and so on. Automatic polylines simplification with computers can not only reduce workloads of map production greatly but also provide important technical support for GIS multi-scale expression, real-time mobile and web map expression, and spatial data integration, matching and updating.

J. Yang et al. (Eds.): CCCV 2017, Part III, CCIS 773, pp. 515–528, 2017.
https://doi.org/10.1007/978-981-10-7305-2_44

Classical polyline simplification methods focus mostly on the quantity compression of polyline nodes (e.g. [1–6]). And then, above methods are improved continuously (e.g. [7–11]). But according to [7], on the basis of visual cognition, polylines are divided to some bend units which contain structural information on geographic features of polylines. From the cognitive view, polylines are composed of a series of bend units. And through the process of manual simplification, each bend unit of polylines is deleted or preserved as a whole unit to reach the balance of overall shape similarity and all details visibility. Furthermore, from the geographical level, polyline simplification is not supposed to break the integrity of geographical features. No matter from the cognitive perspective or the functional perspective [12], polyline simplification is a gradually process of deleting subordinate bends and preserving major bends.

To simplify polylines with computers, bend units of polylines need to be recognized automatically first, and then small bend units which are not visible in simplified scale need to be deleted gradually. A lot of polyline bend units recognition methods has been proposed, such as bends recognition with inflection points (e.g. [13]), bends recognition with constrained Delaunay triangles (e.g. [14–16]), bends recognition with Oblique-Dividing [17], bends recognition base on visibility [18] and so on. Furthermore, computers need to identify each bend unit recognized before whether can be seen clearly or not at the target scale. And all invisible bend units which can not be seen clearly at the target scale are removed directly in most of current methods (e.g. [15,19]).

However, inspired by the progressive process [20] of manual simplification, not all of invisible bend units need to be deleted directly. For example, the initial polyline has six bend units, and all of the bend units are invisible at the target scale (Fig. 1(a)). If all of these six invisible bend units are deleted directly, the simplified polyline cross through the initial polyline (Fig. 1(b)). This simplified polyline is too stiff to have a little harmful effect on the geometry consistency and the cognition consistency of simplification. Probably, all bend units of simplified polylines can be seen at the target scale as soon as only parts of the invisible bend units have been deleted. As Fig. 1(c) shows, just two invisible bend units have been deleted, and all of the six invisible bend units of the initial polyline convert to two new bend units. If these two new bend units can be seen at the target scale, the simplified polyline removed these two invisible bend units can be considered as the final simplification result. And compared with Fig. 1(b), this simplification result (Fig. 1(c)) is deleted much less bends to reach the visual requirement of the target scale, and keep a better geometry morphology and more abundant geographical features that the two new bends contains. Therefore, all of invisible bend units deletion directly is prone to result in excessive polylines simplification results that contains much more uncertainties (e.g., shape dissimilarity uncertainty [21], area differences uncertainty [16], location uncertainty [22]). If these two new bend units can not be seen at the target scale, it is necessary to simplify these two invisible bend units further. Actually, bend units simplification is not a direct process that all invisible bend units must be deleted at once but rather a progressive process that all invisible bend units dont convert to new bend units until all bend units of the simplified polyline can be seen at the target scale.

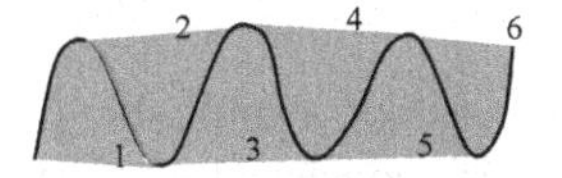

(a) Polyline and invisible bend units

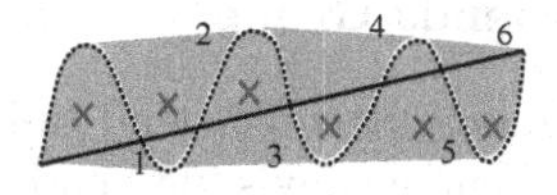

(b) All invisible bend units deletion

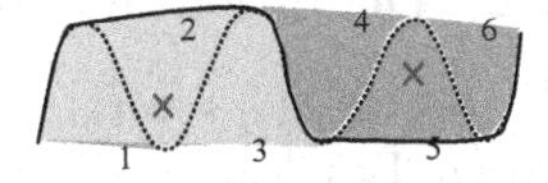

(c) Two invisible bend units deletion

Fig. 1. Problems of current invisible bend units deletion methods. (Color figure online)

The key of this article is to research how to simplify polylines progressively with bend units deletion by computer, which unnecessary bend units deletion can be avoided. The remainder of this article is organized into 4 parts: (i) Sect. 2 illustrate principles of bend units deletion during progressive polyline simplification, and introduce a new concept of multi-bends group; (ii) in Sect. 3, the strategy of the iterative polyline-simplification approach based on multi-bends groups is described in detail; (iii) Sect. 4 reports the experiments applying the proposed approach on a number of real-world data and evaluates the proposed approach in comparison to some other algorithms; and (iv) Sect. 5 discusses the overall performance and potential improvements of the proposed approach.

2 Multi-bends Group

2.1 Progressive Bend Units Deletion

Polyline is consisted of a series of bend units which are continuous and alternate along the polyline. If a bend unit, big enough, can be seen at the target scale, it will be retained. If a bend unit is too small to be seen at the target scale, it will be deleted. Polyline simplification can be considered as the process of removing small bend units and preserving characteristic bend units.

However, a shared segment exists between two adjacent bend units (e.g., the green segment of Fig. 1(a) is the shared segment between the 1st bend unit and the 2nd bend unit of the initial polyline). Due to the shared segment, if a bend unit is deleted it will be merged with its adjacent bends and a new bend unit will be created (e.g., as Fig. 1(c) shows, the 2nd bend unit is adjacent with the 1st bend unit and the 3rd bend unit. If the 2nd bend unit is deleted, a new bend unit - colored with blue- consisted with 1st, 2nd and 3rd bend unit is created).

Two principles can be concluded from the above analysis: (i) If a certain bend unit is deleted, bend units which are adjacent with the deleted bend unit will be changed but bend units which are not adjacent with the deleted bend unit will not be influenced; (ii) Bend units can not be seen at the target scale of the polyline called invisible bend units, while others can be seen at the target scale called visible bend units (i.e., visible bend units are preserved in the simplified polyline). If the target scale of the simplification is confirmed, the polyline is divided into some invisible bend units and some visible bend units uniquely.

2.2 Definition of Multi-bends Group

In this article, several contiguous invisible bend units consist of a multi-bends group. Besides, only one independent invisible bend unit also consist of a multi-bends group. A polyline contains some multi-bends units groups, and each Multi-bends group contains one invisible bend unit at least. Considering with the principle (ii) of Sect. 2.1, multi-bends groups division of a certain polyline is related to the target scale uniquely. In this paper, a single bend unit is denoted by $S(x)$, while a multi-bends group is denoted by $S[y - z]$.

Multi-bends group has some characteristics during polyline simplification as follows: (i) Every multi-bends group is between two visible bend units except multi-bends groups containing the first bend unit or the last bend unit of the polyline. It is generally assumed that before the first bend unit, following a infinite bend unit and after the last bend unit, there is also a infinite bend unit as cognition. Therefore before every multi-bends group, following a visible bend unit and after the multi-bends group, there is also a visible bend unit. (ii) Bend units deletion of polyline simplification just operates at each multi-bends group, and hierarchic bend units deletion also just operates at each multi-bends group. If the new bend still can not be seen at target scale, it also will be deleted. (iii) Both multi-bends group are isolated by a visible bend unit, so bend units deletion operates in one multi-bends group have little influence with the other multi-bends groups.

Therefore, each multi-bends group, not each bend unit, is the independent simplification unit, and polyline simplification is just related to bends deletion or preservation of all multi-bend groups.

2.3 Bend Units Recognition

Let $L = \{v_1, v_2, \ldots, v_p, \ldots, v_u\}$ $(u \geq 3)$ represent the polyline to be simplified, $\boldsymbol{v_{p-1}v_p}$ represent a vector connected from node v_{p-1} to node v_p. If F(1) is satisfied, concavity and convexity will be change at $\boldsymbol{v_pv_{p+1}}$ [13]. Algebraically, if F(2) is satisfied, segment $v_{p-1}v_pv_{p+1}$ and segment $v_pv_{p+1}v_{p+2}$ has the same concavity or convexity (i.e., v_{p-1}, v_p, v_{p+1} and v_{p+2} are in a same bend unit).

$$(\boldsymbol{v_{p-1}v_p} \times \boldsymbol{v_pv_{p+1}}) \cdot (\boldsymbol{v_pv_{p+1}} \times \boldsymbol{v_{p+1}v_{p+2}}) < 0 \tag{1}$$

$$\begin{cases} \boldsymbol{v_{p-1}v_p} \times \boldsymbol{v_pv_{p+1}} < 0 \\ \boldsymbol{v_pv_{p+1}} \times \boldsymbol{v_{p+1}v_{p+2}} < 0 \end{cases} \parallel \begin{cases} \boldsymbol{v_{p-1}v_p} \times \boldsymbol{v_pv_{p+1}} > 0 \\ \boldsymbol{v_pv_{p+1}} \times \boldsymbol{v_{p+1}v_{p+2}} > 0 \end{cases} \tag{2}$$

Several successive bend units of the polyline can be recognized by the following steps which is executed from the initial three nodes of the polyline L:

Step1: If $\boldsymbol{v_{p-1}v_p} \times \boldsymbol{v_p v_{p+1}}$ <0, v_{p-1}, v_p and v_{p+1} will be added in $S(x)$. And then set $p = p+1$; If $\boldsymbol{v_{(p+1)-1}v_{(p+1)}} \times \boldsymbol{v_{(p+1)}v_{(p+1)+1}}$>0, $v_{(p+1)+1}$ will be added in the same $S(x)$ and then step1 will be repeated; If $\boldsymbol{v_{(p+1)-1}v_{(p+1)}} \times \boldsymbol{v_{(p+1)}v_{(p+1)+1}}$<0, $S(x)$ will be recognized. And then set $x = x+1$ and jump to step2 to recognized the next bend unit; If v_{p+1} is v_u, $S(x)$, with nodes $v_{(p+1)}, \ldots, v_u$, will be recognized, and all bend units will be recognized.
Step2: If $\boldsymbol{v_{(p-1)}v_p} \times \boldsymbol{v_p v_{p+1}} < 0$, v_{p-1},v_p and v_{p+1} will be added in $S(x)$. And then set $p = p + 1$; If $\boldsymbol{v_{(p+1)-1}v_{(p+1)}} \times \boldsymbol{v_{(p+1)}v_{(p+1)+1}}$<0, $v_{(p+1)+1}$ will be added in the same $S(x)$ and then step2 will be repeated; If $\boldsymbol{v_{(p+1)-1}v_{(p+1)}} \times \boldsymbol{v_{(p+1)}v_{(p+1)+1}}$ >0, $S(x)$ will be recognized. And then set $x = x + 1$ and jump to step1 to recognized the next bend unit; If v_{p+1} is v_u, $S(x)$, with nodes $v_{(p+1)}, \ldots, v_u$, will be recognized, and all bend units will be recognized.

Using above method, polyline can be divided as $S(x)$. As Fig. 2 shows polyline can be divided as tree bend units, $S(1)$, $S(2)$, $S(3)$, using above recognition method.

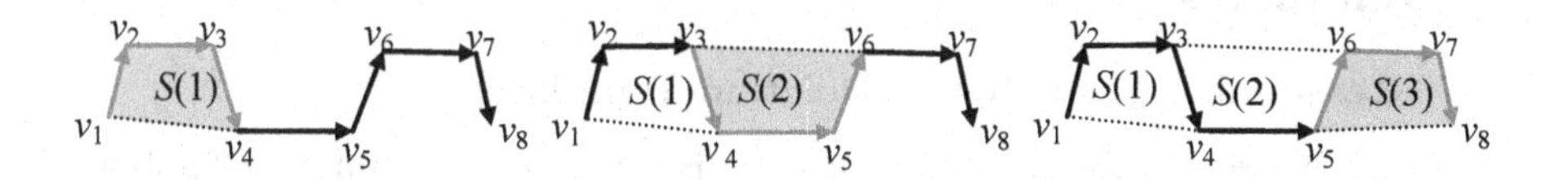

Fig. 2. Problems of current invisible bend units deletion methods.

2.4 Multi-bends Groups Division

If target scale is confirmed, multi-bends groups of polyline will be divided uniquely. Learning From experience of manual simplification, if the height and width of the bend unit are both smaller than SVS (the smallest visual size [3]), this bend unit will be not seen at the target scale. The height and the width of the bend unit can be calculated by the following:

Fig. 3. Bends geometric indexes. (Color figure online)

Bend baseline is line the from the first node to the last node of the bend unit. Bend area is the area of the polygon which is consist of the bend baseline and the bend unit segment. The height of the bend is the longest vertical distances from all nodes, from the first node to the last node, of the bend unit to the bend base line (e.g., the length of the red line in Fig. 3 is the height of each bend unit.). The width of the bend can be calculated by dividing bend area by bend height.

Calculate each bend height and bend width of $S(x)$ recognized before. If a bend height and width are both smaller than SVS at the target scale, this bend unit will be invisible at the target scale. Multi-bends groups $S[x_1 - x_n]$ can be divided as following steps:

Step1: If $S(x)$ is invisible at the target sale, $S(x)$ will be added in $S[x_1 - x_n]$. And then set $x = x + 1$, and repeat step1; If $S(x)$ is visible, $S[x_1 - x_n]$ will be added in $S[x_1 - x_n]$, and $S[x_1 - x_n]$ will be cleared; And then set $x = x + 1$, and jump to step2; If $S(x)$ is the last bend unit of the polyline, $S[x_1 - x_n]$ division are finished.
Step2: If $S(x)$ is invisible at the target sale, set $x = x + 1$ and return to step1; If $S(x)$ is visible, set $x = x + 1$, and return step2; If $S(x)$ is the last bend unit of the polyline, $S[x_1 - x_n]$ division are finished.

3 Simplfying Pollylines with Multi-bends Simplification Templates

3.1 Reasonable Multi-bends Deletion Templates

If all invisible bend units of multi-bends group are deleted, the simplified polyline will cross through the initial polyline and some problems, introduced in Sect. 1, will be conducted by this operation. Besides above attentions, the reasonable option of bend units deletion also need to keep all invisible bends in each multi-bends group changed.

Each multi-bends group is an independent simplification unit. Reasonable bend units deletion options of $S[x_1 - x_n]$ comply the following rules:

(i) Two adjacent bend units can not be deleted at the same time (i.e., because of shared segment of two adjacent bend units, reasonable bend units deletion option should to avoid deleting bend units, adjacent with each other, at the same time).
(ii) If a bend unit is deleted, the bend unit and two bend units which are adjacent with the bend unit will be merged. Therefore, two contiguous bend units which are not contained the first or the last bend unit of the multi-bends group can be preserved at the same time. While, three contiguous bend units which are not contained the first or the last bend unit of the multi-bends group can not be preserved at the same time. Neither do two contiguous bend units contained the first or the last bend unit of the multi-bends group (i.e., There are no invisible bend units not changed after reasonable bends deletion options).

The reasonable bend units deletion options of multi-bends called multi-bends deletion templates. Multi-bends deletion templates are used to simplify multi-bends groups. Furthermore, multi-bends deletion templates is related to the quantity of multi-bends uniquely.

3.2 Multi-bends Groups Simplification with the Best Template

Let $S(x) = 0$ represent $S(x)$ need to be deleted, and $S(x) = 1$ represent $S(x)$ need to be preserved. $S[x_1 - x_n]$, n indicate the bend units number of the multi-bends group, has several reasonable templates, defined as $\{Fj\}$. $\{Fj\}$ is the collection of all templates about the multi-bends groups with n bend units.

While $n = 1$, $\{Fj\} = S(1) = 0$;

While $n = 2$, $\{Fj\} = S(1) = 0, S(2) = 1$ or $S(1) = 1, S(2) = 0$;

While $n \geq 3$, If $S(x_1) = 0, S(x_2) = 1$; If $S(x_1) = 1, S(x_2) = 0$. And then, let the first bend unit need to be deleted which is $S(x_i) = 0$ as the start and execute the following steps to characterize $\{Fj\}$:

Step1: if $S(x_i) = 0, S(x_{i+1}) = 1$. While $x_{i+1} \neq x_n$, jump to step2; While $x_{i+1} = x_n$, stop here;
Step2: While $x_i + 2 \neq x_n, S(x_{i+2}) = 0$ or $S(x_{i+2}) = 1$. If $S(x_{i+2}) = 1, S(x_{i+3}) = 0$. While $x_{i+3} = x_n$, stop here and accomplished; while $x_{i+3} \neq x_n$, let $x_i = x_{i+3}$, and return to step1. If $S(x_{i+2}) = 0$, let $x_i = x_{i+2}$, and return to step1. While $x_{i+2} = x_n$, stop here and accomplished.

All Templates, $\{Fj\}$, of the muli-bends group, $S[x_1 - x_n]$, are gotten as above method (e.g. If $n = 5$, the result using above method is $\{Fj\} = \{\{1,0,1,0,1\}||\{1,0,0,1,0\}||\{0,1,0,1,0\}||\{0,1,0,0,1\}\}$, containing the whole reasonable options of bend units deletion). The best template of $\{Fj\}$ need to be selected. The best simplification option is to delete bend units as little as possible to reach the visual requirement of the target scale. (i.e., Too much errors may be caused by too many bends deletion. The best simplification option is to avoid deleting bend units which are not necessary.). Therefore, the area summation of deletion bend units in each template, Fj, need to be calculated. The smallest one is considered as the best template to simplify this multi-bends group.

3.3 Hierarchic Polylines Simplification Procedure

The initial polyline are divided several multi-bends groups. But some new bend units will merge after multi-bends groups are simplified just once. It is necessary to check these new bend units whether can be seen at the target scale (i.e., It is necessary to compare each new bend unit index, bend width and bend height, with SVS to judge whether the bend unit is invisible or not). If all of the new bend units can be seen at the target scale, the polyline simplification will be over. Therefore, polyline simplification is a repeated process of bends recognition, multi-bends groups division and multi-bends simplification. Polyline simplification process are characterized as following steps:

Step1: The initial polyline are recognized as, several continuous bend units collection, $\{S(x)\}$ according to the Sect. 2.3. Compare each bend unit index, width and height, with SVS to check whether invisible bend units are contained in $\{S(x)\}$. If some invisible bend units are contained in $\{S(x)\}$, jump to step2; Otherwise, jump to step4.

Step2: $\{S(x)\}$ are divided into $\{S[x_1 - x_n]\}$ according to the Sect. 2.4, and then jump to step3.
Step3: $\{S[x_1 - x_n]\}$ are simplified with its best deletion template chosen from all templates according to Sects. 3.1 and 3.2. And all of simplified multi-bends groups and all of visible bends will constitute a new polyline simplified once. The new polyline also can be seen as a new initial polyline and return to step1.
Step4: The polyline simplification is over, and all bend units of the simplified polyline can be seen at the target scale.

All simplification results will reach the visual requirement of the target scale according to simplifing all polylines via above steps.

4 Experiments and Evaluations

With the increasing quantity of bend units contained in multi-bends group, the time consumption of choosing the best templates will get more and more expensive. Therefore, to reach the balance between the simplification result and time consumption, learning from lots of tests, each multi-bends group contains thirty bend units at most during the following experiments (i.e. If a multi-bends group contains more than thirty bend units, the muti-bends group need to be divided into several multi-groups containing thirty bends). In other words, there are no more than thirty bend units in one multi-bend groups after division. It is through this operation that polyline simplification used the proposed method can both reach a better simplification result and a cheaper time consumption.

As a common sense, 1 mm at the map can be considered as the SVS. Furthermore, the following experiments were implemented in CSharp based on ArcGis Engine.

4.1 Progressive Simplification Procedure

The polyline, as Fig. 4 shows, at the scale of 1:250K, was simplifying to the target scale of 1:500K used the proposed method. It was necessary to repeat the hierarchic simplification procedure, according to Sect. 3.3, five times to reach the simplified result with the proposed method. Polyline was simplified gradually as the simplification procedure circled each time. The polyline did not contain invisible bend units until the process repeated five times over. To illustrate the progressive simplification clearly, as Fig. 6 shows, some parts (e.g. part i and part ii) of the polyline were zoomed in.

Therefore, polyline simplification used the proposed method is a progressive procedure which is conform to the cognitive procedure. Moreover, it is easy to ensure that the simplified polyline reaches the visual requirement, all bend units are visible, of the target scale.

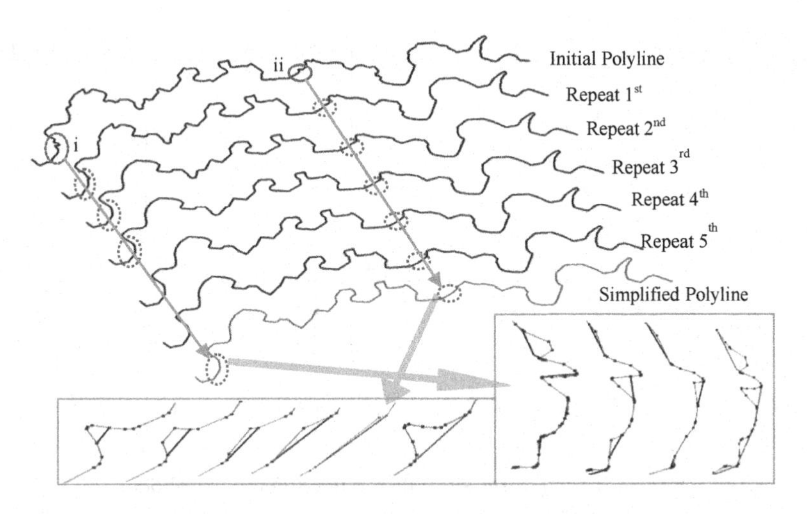

Fig. 4. Polyline progressive simplification used the proposed method.

4.2 Contrast Experiments and Qualitative Discussion

D-P algorithm is a classic simplification method, while bend units direct deletion is a common method of polyline simplification based on bend units. The method proposed in this article was compared with the D-P algorithm [1] and the direct bend units deletion method (called D-D method in this paper) that all invisible bend units were deleted via this method. To make the contrast experiments more scientific, during the following experiments SVS was considered as the threshold of both the D-P algorithm and the D-D method. In addition that, the D-D method also used the hierarchic polylines simplification procedure, according to Sect. 3.3, without multi-bends groups division.

The test dataset is the roads net at the scale of 1:250K from the region of Zhejiang province, China. The test roads, approximately 42 km * 38 km, are planarized before, and 209 polylines are contained in the test data. The test dataset was simplified by the above three methods to the target scale of 1:500K and 1:1M. Simplified results were shown as Fig. 5.

With the target scale increasing, the simplified polyline used all of the three simplification methods get simpler and simpler. Besides there was not any one invisible bend unit in all simplified results.

However, compared with all simplified results (Fig. 5) used the three methods, polylines simplified by the proposed method were more smooth and much more details which can represent more geographic features were persevered. Therefore, polylines simplified by the proposed method had a higher similarity with the initial polylines both in geometry and geography.

Furthermore, some details of Fig. 5 were zoomed in (e.g. part i simplified to 1:500K, part ii simplified to 1:1M and part iii simplified both to 1:500K and 1:1M were zoomed in) on the Fig. 6. According to Fig. 6, it was easier to find that results simplified by the proposed method was better:

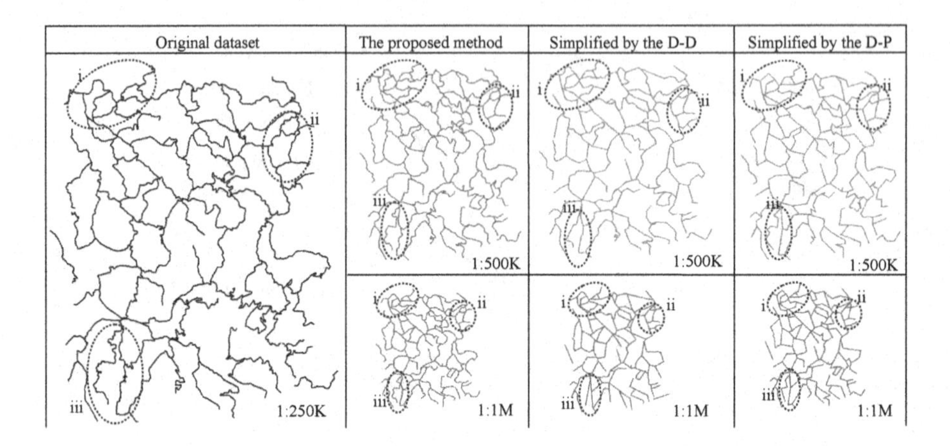

Fig. 5. Test dataset simplified with different methods.

(i) Polylines simplified by D-P focused on nodes compression were too stiff to conform the cognition of human. (ii) Polylines simplification based on bend units was conform to manual operation. But the D-D method deleted too much bend units which were not necessary to delete. The simplification result of the D-D method was smooth enough, but lots of bend units and geographical features of these bend units were omitted. (iii) Polylines simplified by the proposed method was the most closed to the results of human cognition and manual operation. Moreover, small invisible bend units were simplified enough while the obvious major bending features were preserved after simplification. Polylines simplified by the proposed method were not only more beautiful in morphology but also plentiful features in geography than other methods.

Therefore, from the qualitative aspect, the proposed simplification method is better enough.

4.3 Statistical Evaluations and Quantitative Discussion

To prove the polylines simplified by the proposed method was better than others from the quantitative and statistical view. Location errors [22] was defined as:

$$Loc_{er}(i) = \frac{Area(i)}{l(i)} \tag{3}$$

where $l(i)$ was the length of the polyline whose id is i, $Area(i)$ was the area difference between the simplified polyline and initial polyline whose id is i (i.e. $Area(i)$ represents the whole displacement between original and simplified polyline). Therefore, $Loc_{er}(i)$ is used to measure the average displacement of the polyline, $l(i)$, simplification. For the statistical assessment, $Loc_{er}(i)$ of every polyline, in test dataset, resulted from the three different method was calculated (i.e. each polyline's $Loc_{er}(i)$ of Fig. 5 was calculated). Due to the limitation of papers, some polyline $Loc_{er}(i)$ were shown as Table 1 randomly. To avoid

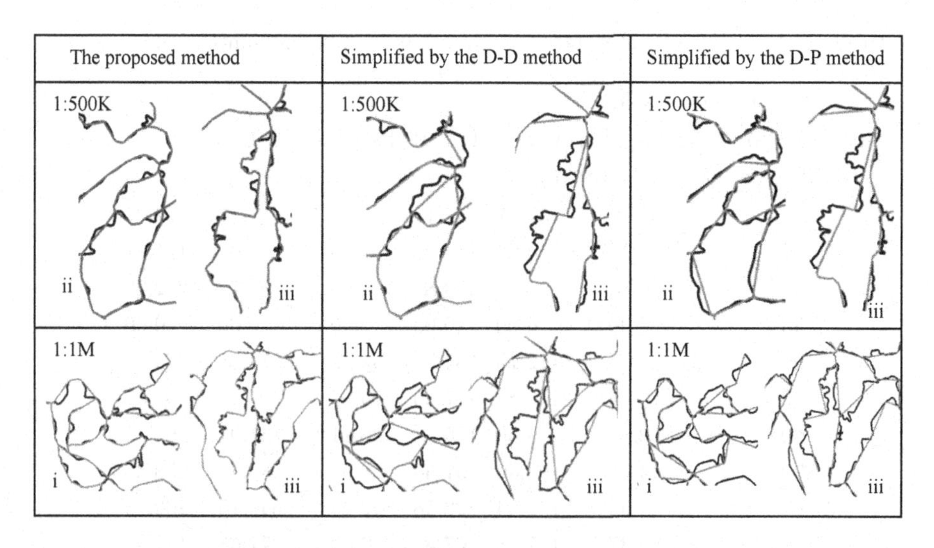

Fig. 6. Zoom in parts of simplification results.

selecting polylines intentionally, polylines whose id was several times of ten were selected (Table 1). Moreover, another two metrics, $All(Loc_{er})$ and $Eve(Loc_{er})$, originated from $Loc_{er}(i)$ were defined as

$$All(Loc_{er}) = \frac{\sum_{i=1}^{n} Area(i)}{\sum_{i=1}^{n} l(i)} \tag{4}$$

$$Eve(Loc_{er}) = \frac{\sum_{i=1}^{n} Loc_{er}(i)}{n} \tag{5}$$

$Eve(Loc_{er})$ and $All(Loc_{er})$ also can be used to measure the whole displacement resulted from the simplification of the all test dataset. Additionally, $All(Loc_{er})$ and $Eve(Loc_{er})$ of all results that the test dataset was simplified to 1:500K and 1:1M via different methods were calculated (Table 2).

Having analysed the Table 1, it was easy to found that $Loc_{er}(i)$ derived from each polyline simplified by the proposed method was smaller than other methods universally. Each polyline of test dataset simplified by the proposed method reached the visual requirement (i.e. simplified polyline did not contain any invisible bends) with less displacement derived from bends deletion. The polyline was more meandering, the polyline simplified by the proposed method was better than others (e.g. polyline $id = 40$, 50, 100, etc.). While the polyline was more simple, the polyline simplified by the proposed method had less difference with others (e.g. polyline $id = 30$, 140, 150, etc.). Besides, the gap between the original scale and the target scale was larger, the polyline simplified by the proposed method had less difference with others. Although, sometimes, when the simplified scale is big enough or the original polyline is simple eough or both, $Loc_{er}(i)$ derived from the proposed method maybe equal others, the $Loc_{er}(i)$ of the proposed method is smallest.

Table 1. $Loc_{er}(id)$ of test dataset originated from different simpification results.

Target scale	1:500K			1:1M		
Mehod	Proposed	D-D	D-P	Proposed	D-D	D-P
Polyline id	$Loc_{er}(id)$ (m)					
0	57.17	57.17	57.17	57.17	57.17	57.17
10	14.34	14.34	152.38	71.18	175.07	175.07
20	37.73	96.10	293.89	128.65	322.77	322.29
30	0.00	0.00	0.00	0.00	0.00	0.00
40	49.71	112.75	134.68	54.72	374.58	374.58
50	19.86	150.63	297.35	79.95	408.57	408.57
60	52.81	53.50	145.61	126.21	131.91	364.51
70	44.91	264.46	197.55	58.61	589.46	458.35
80	32.53	192.94	122.58	42.75	343.15	162.16
90	59.37	177.68	67.57	77.60	177.68	177.68
100	36.54	114.65	79.78	83.04	189.828	189.82
110	103.51	163.66	163.66	163.66	163.66	163.66
120	42.39	155.92	186.04	155.92	155.92	155.92
130	38.58	93.10	136.00	126.82	222.55	222.55
140	0.00	0.00	35.61	35.61	35.61	35.61
150	13.39	13.39	27.62	16.22	27.62	27.62
160	39.63	76.65	208.94	66.41	713.39	623.14
170	138.13	215.70	259.54	183.03	538.31	538.31
180	31.29	99.47	84.68	31.29	497.80	154.72
190	46.92	69.37	280.34	99.45	280.34	280.34
200	32.55	257.59	142.06	61.06	487.25	296.77

Table 2. $All(Loc_{er})$ and $Eve(Loc_{er})$ originated from different simpification results.

Target scale	1:500K			1:1M		
Mehod	Proposed	D-D	D-P	Proposed	D-D	D-P
$All(Loc_{er})$(m)	43.89	141.86	157.09	112.53	356.34	268.73
$Eve(Loc_{er})$(m)	62.15	366.60	403.12	136.41	398.77	378.65

Having analysed the Table 2, it was clear to found that the test dataset simplification results, at the scale of both 1:500K and 1:1M used the proposed method is better than others. The whole displacement of the test dataset resulted from the proposed simplification method was smaller than others.

Therefore, from the quantitative aspect, the proposed simplification method is better enough.

5 Conclusion

Not all of invisible bend units, at the target scale, need to be deleted at once. However, invisible bend units deletion is a progressive procedure. In this article, having analyzed the composing characteristics of polylines and the bend units deletion procedure of manual operation, a new simplification method based on multi-bends division is proposed. With this method the polyline to be simplified firstly divided into a chain of bend units. According to their geometric indexes and the SVS of the target scale, bend units are categorized into visible bend units and invisible bend units. Moreover, several successive invisible bend units compose of a multi-bends group. Each multi-bends group is an independent simplification units. All of reasonable bend units deletion options which is related to bend units quantity of the multi-bends group are formalized in detail, and the best one called the best multi-bends simplification template which is deleting bends as less as possible is used to simplify its corresponding multi-bends group. Supported by iteratively operated multi-bends divisions and each multi-bends deletion with the best simplification template, polylines are simplified progressively. It is not over, until all bend units of the polyline can be seen at the target scale. The proposed method is suitable for human cognition and manual simplification. According to results of our contrast experiments, the proposed polyline simplification method not only preserve the main shape accurate but also keep feature details plentiful. Furthermore, bend units are deleted as less as possible via the proposed method. It is perfect for some simplification applications with a high area-preserving constraint.

Besides the road net, the proposed simplification method is now being experimented on various types polylines of hydrographic lines, land-use boundaries, etc. On the other hand, different polyline features also have different constraints in different applications, the simplification strategy should be extended and enhanced for different domains.

Acknowledgments. Special thanks to colleagues, Haiwei He and Jinghan Li, and reviewers for their constructive comments that improved the quality of this paper substantially.

References

1. Douglas, D.P., Peucker, T.K.: Algorithm for the reduction of the number of points required to represent a digitized line or its caricature. Can. Cartogr. **10**, 112–122 (1973)
2. Jenks, G.F.: Geographic logic in line generalisiation. Cartographica **26**(1), 27–42 (1989)
3. Li, Z.L., Openshaw, S.: Algorithms for automated line generalization based on a natural principle of objective generalization. Int. J. Geogr. Inf. Syst. **6**, 373–389 (1992)
4. Visvalingam, M., Whyatt, J.D.: Line generalisation by repeated elimination of points. Cartographic **30**, 46–51 (1993)

5. Kulik, L., Duckham, M., Egenhofer, M.: Ontology-driven map generalization. J. Vis. Lang. Comput. **16**, 245–267 (2005)
6. Geoffrey, D.: Scale, sinuosity, and point selection in digital line generalization. Cartogr. Geogr. Inf. Sci. **26**, 33–54 (1999)
7. Saalfeld, A.: Topologically consistent line simplification with the Douglas-Peucker algorithm. Cartogr. Geogr. Inf. Sci. **26**, 7–18 (1999)
8. Park, W., Yu, K.: Hybrid line simplification for cartographic generalization. Pat. Recog. Let. **32**, 1267–1273 (2011)
9. Pallero, J.L.G.: Robust line simplification on the plane. Comput. Geosci.-UK **61**, 152–159 (2013)
10. Paulo, R.: Scale-specific automated line simplification by vertex clustering on a hexagonal tessellation. Cartogr. Geogr. Inf. Sci. **40**, 427–443 (2013)
11. Tienaah, T., Stefanakis, E., Coleman, D.: Contextual Douglas-Peucker simplification. GeoInformatica **69**, 327–338 (2015)
12. Chaudhry, O.Z., Mackaness, W.A., Regnauld, N.: A functional perspective on map generalisation. Comput. Environ. Urban Syst. **33**, 349–362 (2009)
13. Plazanet, C.: Measurements, characterization and classification for automated line feature generalization. In: Proceedings of the 12th International Symposium on Computer- Assisted Cartography, Charlotte, USA pp. 59–68 (1995)
14. Jones, C.B., Bundy, G.L., Ware, J.M.: Map generalization with a triangulated data structure. Cartogr. Geogr. Inf. Sci. **22**, 317–331 (1995)
15. Ai, T., Zhou, Q., Zhang, X., et al.: A simplification of ria coastline with geomorphologic characteristics preserve. Mari. Geod. **37**, 445–448 (2014)
16. Ai, T., Ke, S., Yang, M., et al.: Envelope generation and simplification of polylines using delaunay triangulation. Int. J. Geogr. Inf. Sci. **31**, 297–319 (2017)
17. Qian, H., Zhang, M., Wu, F.: A new simplification approach based on the oblique-dividing-curve method for contour lines. Int. J. Geo-Inf. **153**, 1–21 (2016)
18. Zhu, Q., Wu, F., Zhai, R.: A novel of contour line simplification algorithm base on visibility. Im. Grap. J. **14**, 359–364 (2009)
19. Cao, Z., Li, M., Cheng, L.: Multi-way trees representation for curve bends. Acta Geod. Cartogr. Sin. **42**, 602–607 (2013)
20. Guo, Q., Brandenberger, C., Hurni, L.: A progressive line simplification algorithm. Geo-Spat. Inf. Sci. **5**, 41–45 (2002)
21. Shi, W., Cheung, C.: Performance evluation of line simplification algorithms for vector generalization. Cartogr. J. **43**, 27–44 (2006)
22. Zhu, K., Wu, F., Wang, L., et al.: Improvement and assessment of Li-Openshaw algorithm. Acta Geod. Cartogr. Sin. **36**, 450–456 (2007)

Infrared and Visible Image Registration Based on Hypercolumns

Zhenbing Zhao, Lingling Zhao(✉), Yincheng Qi, Ke Zhang, and Lei Wang

School of Electrical and Electronic Engineering,
North China Electric Power University, No. 619, Yonghua North Street,
Lianchi District, Baoding 071003, Hebei, China
1521834700@qq.com

Abstract. Image registration is a challenging and critical task in computer vision and image processing. As the typical multi-modal image, infrared and visible image have greatly difference in gray scale, which makes the registration method based on hand-crafted features have a low accuracy. In this paper, we introduce a method based on hypercolumns and matching strategy. Combining different layers features in Convolutional Neural Network using hypercolumns, we can get more comprehensive and essential features to achieve a higher accuracy. Meanwhile we use coordinate difference acquired by automatic selection as the spatial distance constraint. Firstly, the key points are extracted by corner detection from the infrared and visible images. Then the features of the key points are extracted using the hypercolumns. Finally, similarity metric is performed by spatial geometric constraints. The experimental results show that the accuracy of our method is higher than that of the traditional method.

Keywords: Image registration · Hypercolumns · Multi-modal image

1 Introduction

Image registration is a challenging research field. Image registration is based on some similarity metric to determine the transformation parameters between the images, so that two or more images of the same scene obtained from different sensors, different angles, different time is transformed to the same coordinate system, and then get the best matching process on the pixel layer. Image registration has been widely used in military, remote sensing, medical, computer vision and other fields. Infrared image can locate objects with higher temperature in the scene, while the visible image provides background information, and they can be combined to realize the function of locating the high-temperature objects in the background. However, all of the existing infrared and visible image registration methods are based on hand-crafted features, it depends on good edge information and enough texture information which do not take full advantage of the essential features of image. It has a pool performance in infrared and visible image registration.

J. Yang et al. (Eds.): CCCV 2017, Part III, CCIS 773, pp. 529–539, 2017.
https://doi.org/10.1007/978-981-10-7305-2_45

In order to solve this problem, we propose a method based on hypercolumns [1]. Extracting depth features by using CNN (convolutional neural network) which can describe essential features between infrared and visible image. By selecting the feature layers of the front, middle and last, we can extract the detail information and the global information of the image, which is more conducive to identify the target in the image and obtain more detail information. The semantic information and detail information in the image can be used to describe the key points better. In the similarity measure, we add coordinate difference to constrain the search range of the matching. The method flow of this paper is using corner detection to extract key points, combining different layers features of CNN to describe key points and comparing the similarity. We will show that this method can improve the accuracy of registration.

2 Related Work

Although there are many registration methods based on hand-crafted features between infrared and visible images and depth features between the visible images, there is no one to use depth features to realize infrared and visible image registration now. We briefly review previous articles that are relevant to our work, including feature descriptors, visible images matching based on deep learning, and infrared and visible image registration.

Feature Descriptors: Feature descriptors are designed to provide discriminative expressions of salient image key points, which will be very robust to transformations such as viewpoint or illumination changes. At present, the common method is SIFT [2], SURF [3], HOG [4], BRISK [5], DAISY [6], ORB [7]. Even if these hand-crafted features are more successful, they have been defeated by learned features, such as [8–10] encoding non-linearity to map patches into descriptors. Hariharan et al. [1] proposed a hypercolumns model, which extracts the features of different layers in CNN.

Visible image registration based on deep learning: There are some trained CNN models which directly extract features from the images to match. For example, Matchnet [11] trained a Siamese CNN model to extract features, followed by a fully connected network to learn the comparison metric. Deepcompare [12] showed that training a network model focused on the image center can improve the performance of the description. Zbontar et al. [13] relied on the CNN architecture to achieve better results on narrow-baseline stereo. Altwaijry et al. [14] learned descriptor and space conversion module to match the image patches. Yi et al. [15] combined detection, orientation estimation, and feature description together by deep learning model for image matching. Altwaijry et al. [16] proposed a network model to learning feature description and key points detection to match key points. Rocco et al. [26] use CNN architecture for category-level image alignment.

Infrared and Visible image registration: The registration of infrared and visible images is a typical multi-modal image registration, which is an important step in image processing and analysis. But at present, the methods used are basically based on hand-crafted features. For example, Kristin and Anandanet [17] proposed a method based on edge subset, Coiras et al. [18] using segmentation, Kelman et al. [19] based on improved SIFT, Istenic et al. [20] relying on the Hough transform, Li and Zhou [21] using the Wavelet transform.

The registration methods based on hand-crafted features cannot express the essence of images. And the methods based on deep learning are performed under the visible images. In the infrared images, the accuracy of registration will be reduced due to the low pixel, poor contrast and bad influence of noise. These methods are based on the features from the final layer of CNN, but this layer is more sensitive to the semantic information, rough in space and insensitive to disturbance (attitude, illumination, and position), which makes performance greatly reduced. Therefore, this paper uses the hypercolumns method to obtain more comprehensive and essential feature descriptors, building the foundations for accurate registration.

3 Infrared and Visible Image Registration Based on Hypercolumns

With Infrared and visible light images from different imaging equipment, imaging mechanism is inconsistent. With access to images of time, angle and environment different, the gray-scale properties are greatly different, even the same object under the same scene, in different imaging mode, which makes the registration of infrared and visible images a challenging task. Aiming at the difficulty of infrared and visible image registration mostly based on hand-crafted features, this paper proposes a novel method of infrared and visible image registration based on deep learning. We adopt the registration method based on point features. Figure 1 provides an overview of our proposed method. This section introduces our features extraction and similarity metric algorithm in detail.

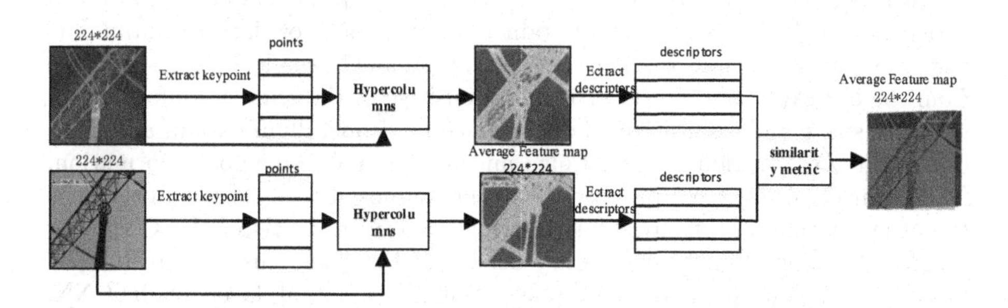

Fig. 1. Flow chart of the algorithm.

3.1 Key Points Detection

As the first step in the registration, the selection of key points largely determines the quality of the final results. As an important local image features, corner point is a feature point with two main directions in the neighborhood. Corner feature has rich information, which is easy to measure and express, and can adapt to the change of ambient light. In order to obtain the same point of infrared and visible image as much as possible, this paper uses Harris [22] to extract the key points of infrared and visible image. The results are shown in Fig. 2.

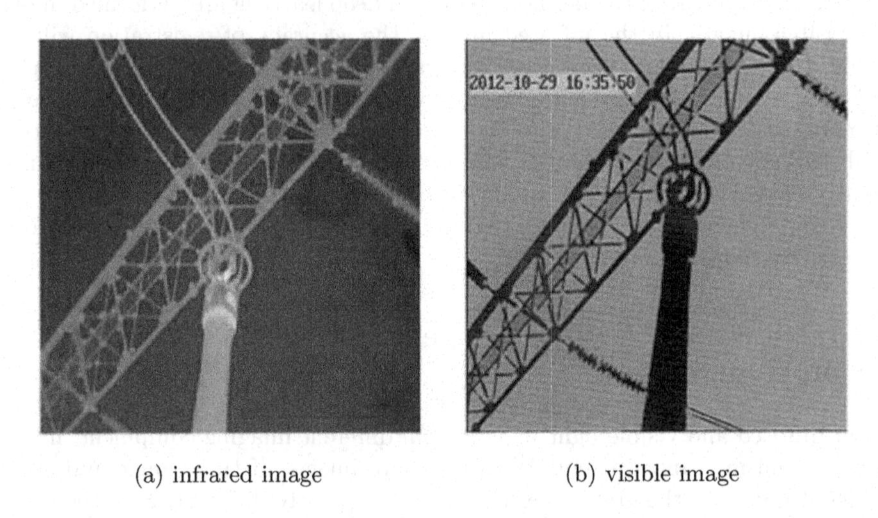

(a) infrared image (b) visible image

Fig. 2. Images after extracting points

3.2 Feature Descriptors Extraction

At present, the infrared and visible image registration methods usually use handcrafted features, having not used the deep feature. Traditional methods lack the distinction of the essential features of images. However, deep convolution neural network simulates the structure of the visual perception system to establish a rich hidden layer structure and train a large number of data to obtain the essential features of images. The deep learning model usually extracts features from the last layer, but this layer is more sensitive to the semantic information, rough in space and insensitive to disturbance (attitude, illumination, and position), while registration needs detailed information and also global information. Hariharan et al. [1] have proposed a hypercolumns model. The hypothesis is that the information of interest is distributed over all levels of the CNN and should be exploited in this way. They define the hypercolumns at a given input location as the outputs of all units above that location at all layers of the CNN, stacked into one vector. That is to say the hypercolumns at a pixel is the vector of activations of all units that lie above that pixel. It extracts the feature maps of different layers in the CNN. For achieving better features, we first fine tune the vgg16 network.

In order to fine tune feature extraction networks, we need a large number of representative infrared and visible image patches that are labeled similar and dissimilar. Because there are not so many infrared and visible image datasets, a key

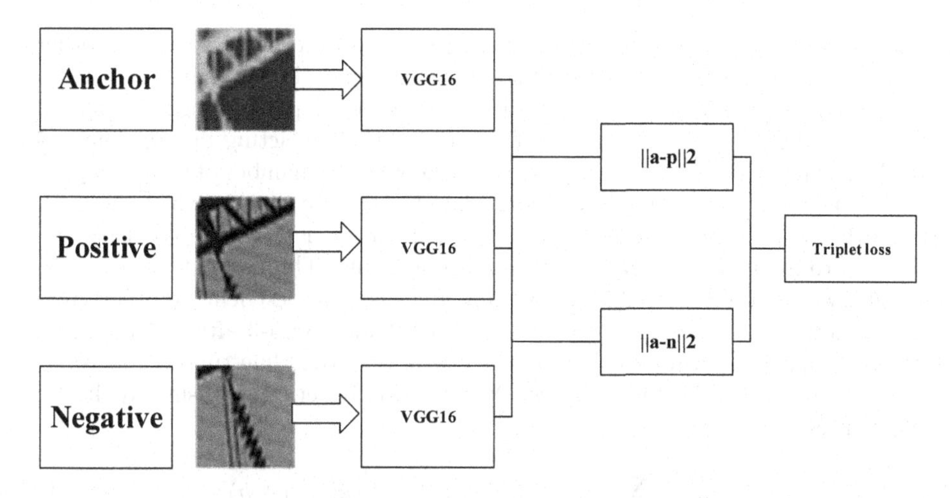

Fig. 3. Flow chart of the fine tune architecture.

step is to make datasets. We sample 15 pairs of infrared and visible images from the actual scene, using traditional methods to register one by one, and a pair of registered images were cut into many patches according to different proportions. For example, a pair of images of size $256 * 256$ can be cut into 16 patches with size of $64 * 64$ or cut into 64 patches with size of $32 * 32$. Since the size of the 15 pairs of images is different, we use different ratio segmentation to make the training data more detailed and global, and thus more representative. Finally, the image patch is resized to $64 * 64$. Since the datasets is too few, we use rotation scaling and gray value augmentation to enhance the image patches. The negative samples are derived from random selection in dissimilar images. Finally, 10 thousand pairs of positive samples and 10 thousand pairs of negative samples are generated. Each batch is chosen to have a mix of positives and negatives with a 1 : 1 ratio. The purpose of fine tune is to extract features with degree of differentiation, learn a nonlinear mapping f(x) from image patch p to feature vector R^D. For example, for a pair of image patch p1 and p2, f(p1) and f(p2) Euclidean distance between the matching patches is small and large for the not matching patches. We use the triplet loss function [23], compared with softmax, it defines the distance relationship between similar samples and heterogeneous samples in nature: the sum of all same class samples distance and a threshold is smaller than the distance between different samples. Make the similarity between patches more distinctive in this way. We named the anchor for a patch, the positive which is similar to anchor, and the negative which is not similar to anchor. In particular, fine tune should achieve that a patch p1 (anchor) is closed to the patch p2 (positive) than it is any to other patch p3 (negative). Figure 3 illustrates the fine tune architecture. The input to the network are batches of patch triplets p1, p2, p3. First, each patch uses a vgg16 network to compute its feature vectors, and three network share the same weights. Then, the feature vectors are normalized into a 1000 dimensional. Afterwards, the pairwise

Euclidean distances between the feature vectors of anchor and the two other patches are computed. The network trained with triplet loss to make the Euclidean distances of similar patches smaller and dissimilar patches larger shown in Eq. 1. We define a loss-function ℓ_T for training the extracting description network. Given triplets $T = \{t_j : (p_1^j, p_2^j, p_3^j)\}$ where j is the number of train samples. D is a Euclidean distance function defined on the feature vectors, which are computed from the image patches shown in Eq. 2. In order to ensure good results, we need to manually set triplet threshold constraints. Through extensive experiments, we set the threshold 0.2 to train. Because the best feature descriptors are obtained by using conv1-1, conv2-1, conv4-2 and conv4-3 after a large number of compare experiments. Freeze other layers, we only fine tune the conv1-1, conv2-1, conv4-2 and conv4-3 layers. Optimization is performed using stochastic gradient descent.

$$\ell_T = \frac{1}{N}\sum_j [max(0, D(p_1^j, p_2^j) - D(p_1^j, p_3^j) + h)] \tag{1}$$

$$D(p_a, p_b) = \|f(p_a) - f(p_b)\|_2 \tag{2}$$

We use the deep learning model based on the hypercolumns to combine the features of front, middle and last layers, and down-sampling to ensure the corresponding relationship between the input image and the output feature maps, which makes the feature descriptors express more comprehensive and the registration performance greatly improved. By selecting the three feature layers of the front, middle and last, we can extract the detail information and the global information of the image, which is more conducive to identify the target in the image and obtain more detail information. The semantic information and detail information in the image can be used to describe the key points better. So it is more conducive to registration of the infrared and visible images with larger spatial structure information. In this paper, we input the infrared and visible images to the fine tuned network (vgg16), extract the feature maps of conv1-1,

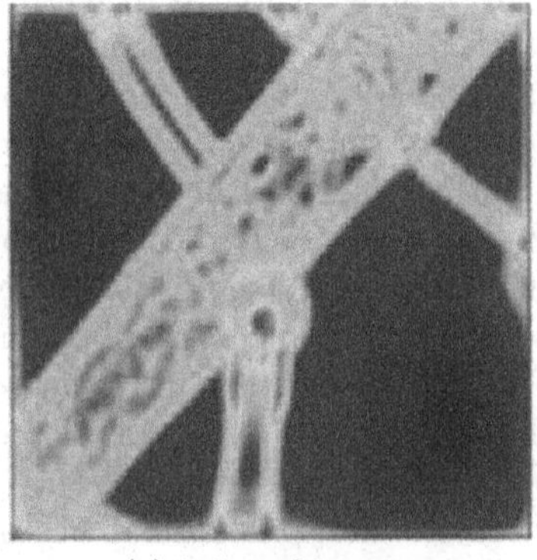

(a) infrared image

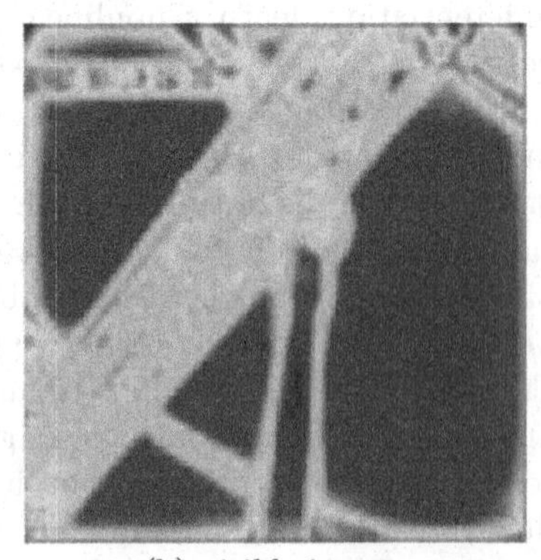

(b) visible image

Fig. 4. Images average feature map

conv2-1, conv4-2 and conv4-3 selected by extensive experiments, down-sampling all feature maps, and average them as one feature map having the same size as input image, as shown in Fig. 4. Each point on the input image is in one-to-one correspondence with each point of the average feature map. The feature value is formed by the 5 * 5 region around the key point, that is, one key point is represented by a 25-dimensional vector. Therefore, our method can obtain more comprehensive and essential feature descriptors, to build the foundations for accurate registration.

3.3 Similarity Metric

As the infrared and visible light image gray-scale information greatly different, registration is a challenging task. In order to get better results, we add the spatial distance constraint in the traditional similarity metric method. The spatial distance constraint is the coordinate difference acquired by automatic selection. After the feature descriptors are obtained from Sect. 3.2, we use the Euclidean distance as the metric and add the coordinate difference as the constraint.

Use selective search [24] to select the candidate boxes which are considered to be object in images. $(x_{1k}, y_{1k}, w_{1k}, h_{1k})$ and $(x_{2k}, y_{2k}, w_{2k}, h_{2k})$ are sequentially selected for the infrared and visible images, where x_{1k}, the abscissa of the initial point of the k-th candidate box, y_{1k} the ordinate of the initial point of the k-th candidate box,w_{1k}, the width of the k-th candidate box, h_{1k}, the height of the k-th candidate box and k,the number of candidate boxes. Next compare the size of the candidate boxes. If the difference between w_{1k} and w_{2k} and between h_{1k} and h_{2k} both in the range of (t_1, t_2), calculate the coordinate difference corresponding to x and y, such that, $a_k = x_{1k} - x_{2k}$ and $b_k = y_{1k} - y_{2k}$, where a and b are the sets of coordinate difference $a = \{a_1, a_2, a_3, \ldots, a_k, \ldots\}$ and $b = \{b_1, b_2, b_3, \ldots, b_k, \ldots\}$. Then, remove the maximum and minimum values in $\bar{a}$. Take the four values with minimum variance in a and find the average $\bar{a}$ of the four, selecting a_m as the closest value to $\bar{a}$ in a. According to a_m, get the corresponding ordinate b_m and (a_m, b_m). In the same way taking the four values

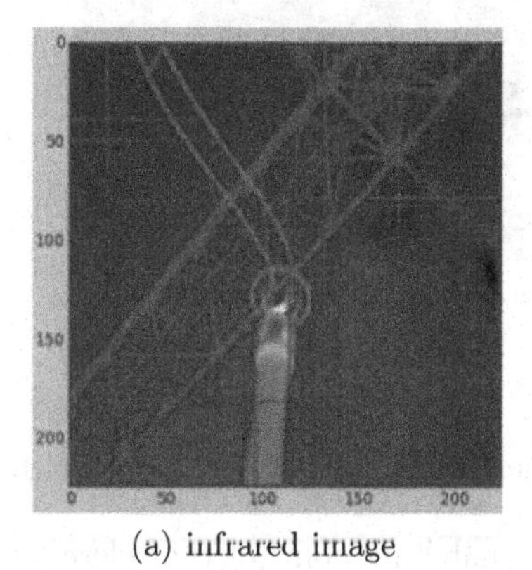

(a) infrared image

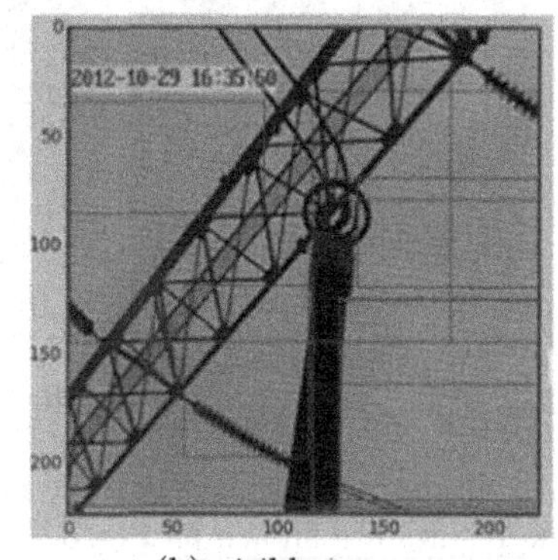

(b) visible image

Fig. 5. Results of the candidate boxes

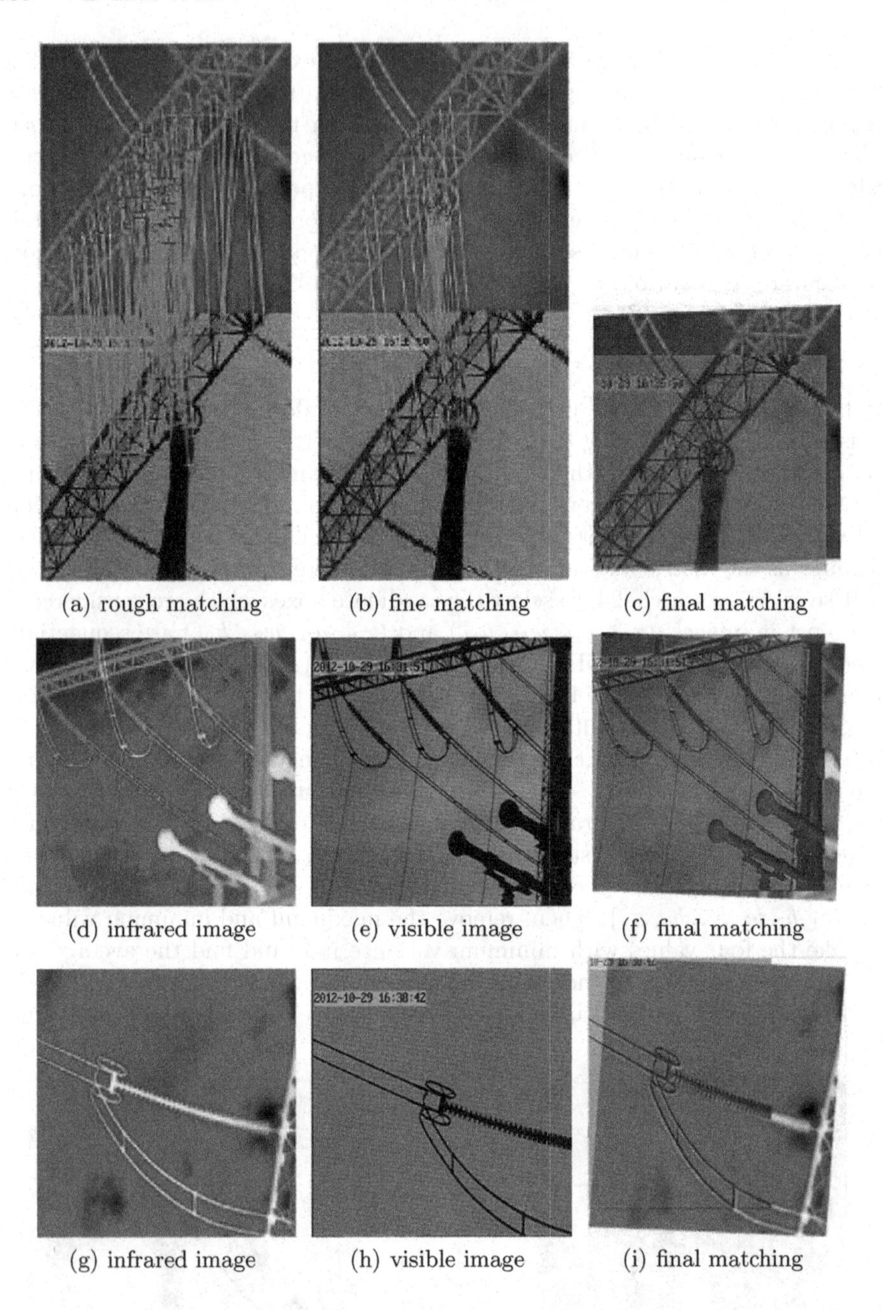

(a) rough matching (b) fine matching (c) final matching
(d) infrared image (e) visible image (f) final matching
(g) infrared image (h) visible image (i) final matching

Fig. 6. Results of three pairs

with minimum variance in b we can get another coordinate difference (a_n, b_n). The results are shown as Fig. 5.

According to the coordinate difference, the matching point of the reference image is searched within the range specified in the image to be registered. The search range in the image to be registered is $(i - a_n \pm t_3, j - b_n \pm t_3)$ and

$(i - a_m \pm t_3, j - b_m \pm t_3)$ for the point (i, j) in the reference image, where t_3 represents the size of the search range. The similarity between key points is compared according to Euclidean distance. After get the result of rough matching and use Ransac [25] for fine matching, and then transform to a coordinate system to get the results, as shown in Fig. 6. We do experiments on two pairs of images. Figure 6(a–c) are from the first pair, and Fig. 6(d–i) come from the other two.

4 Experiment

This section examines the performance of this approach. We compare our method with traditional feature based registration methods and depth features of single layer, using the number of control point pairs after Ransac, the right pairs after Ransac and correct rate as evaluation index. Meanwhile we compare against CNN geometric matching [26] which use CNN architecture for category-level image alignment. The experimental results on the three pairs shown in Fig. 7.

For verifying the effectiveness of this method, we use the SIFT, which is matched with the strategy of this paper, and the SURF of the matching strategy. In order to verify the performance of hypercolumns, we select three kinds of single-layer features as a comparison, but before and after the operation are consistent with this article, only modifying the feature extraction part.

These three layers are conv1 (representing the features of the previous layer), conv3 (representing the middle layer features), conv5 (on behalf of the latter

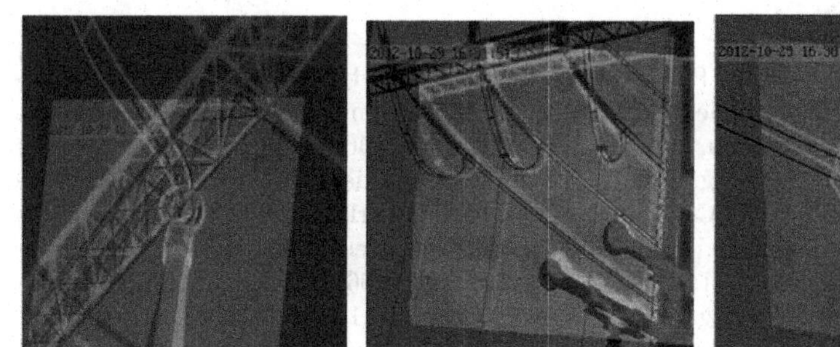
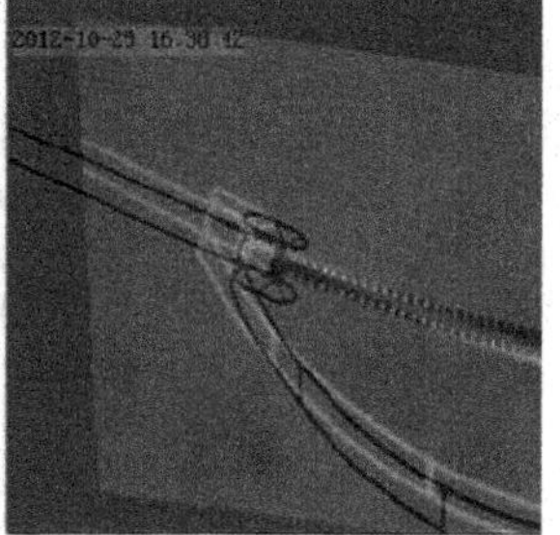

Fig. 7. Results of CNN geometric matching

Table 1. Comparative performance of our method vs. five competitors

Method	Fine match pairs	Right pairs	Accuracy
SIFT	25	4	16%
SURF	8	2	25%
Conv1	12	7	58.3%
Conv3	19	11	57.9%
Conv5	9	3	33.3%
Our method	15	13	86.7%

features). We set $t_3 = 20$. We use the number of pairs after Ransac named fine match pairs, the right pairs after Ransac and correct rate as evaluation index. The results are shown in Table 1. The experimental results show that the best registration performance can be achieved by our method.

5 Conclusion

Multi-modal image registration is a challenging problem in computer vision. In this work, we proposed a registration algorithm for infrared and visible image based on learned feature descriptors and a novel matching strategy. Combining different layers features in Convolutional Neural Network, we can get more comprehensive and essential features to achieve a higher accuracy. In our experiments on images, our algorithm outperformed state-of-the-art. Our approach serves as a step to bring Multi-modal image registration closer together with the recent progress in deep learning. We plan to further investigate the models performance on Multi-modal image.

Acknowledgments. This work was supported in part by National Natural Science Foundation of China (No. 61401154), Hebei Province Nature Science Foundation of China (No. F2016502101) and the Fundamental Research Funds for the Central Universities (No. 2015ZD20).

References

1. Hariharan, B., Arbelez, P., Girshick, R., Malik, J.: Hypercolumns for object segmentation and fine grained localization. In: IEEE Conference on Computer Vision and Pattern Recognition, CVPR, Boston, pp. 447–456 (2015)
2. Lowe, D.G.: Object recognition from local scale-invariant features. In: IEEE International Conference on Computer Vision, ICCV, Kerkyra, pp. 1999–1150 (1999)
3. Bay, H., Tuytelaars, T., Van Gool, L.: SURF: speeded up robust features. In: Leonardis, A., Bischof, H., Pinz, A. (eds.) ECCV 2006. LNCS, vol. 3951, pp. 404–417. Springer, Heidelberg (2006). https://doi.org/10.1007/11744023_32
4. Dalal, N., Triggs, B.: Histograms of oriented gradients for human detection. In: IEEE Conference on Computer Vision and Pattern Recognition, CVPR, San Diego, pp. 886–893 (2005)
5. Leutenegger, S., Chli, M., Siegwart, R.Y.: Brisk: binary robust invariant scalable keypoints. In: IEEE International Conference on Computer Vision, ICCV, Barcelona, pp. 2548–2555 (2011)
6. Tola, E., Lepetit, V., Fua, P.: A fast local descriptor for dense matching. In: IEEE Conference on Computer Vision and Pattern Recognition, CVPR, Anchorage, pp. 1–8 (2008)
7. Rublee, E., Rabaud, V., Konolidge, K., Bradski, G.: ORB: an efficient alternative to SIFT or SURF. In: IEEE International Conference on Computer Vision, ICCV, Barcelona, pp. 2564–2571 (2011)
8. Brown, M., Hua, G., Winder, S.: Discriminative learning of local image descriptors. IEEE Trans. Pattern Anal. Mach. Intell. **33**, 43–57 (2011)

9. Jain, P., Kulis, B., Davis, J.V., Dhillon, I.S.: Metric and kernel learning using a linear transformation. J. Mach. Learn. Res. **13**, 519–547 (2012)
10. Simonyan, K., Vedaldi, A., Zisserman, A.: Learning local feature descriptors using convex optimization. IEEE Trans. Pattern Anal. Mach. Intell. **36**, 1573–1585 (2014)
11. Han, X., Leung, T., Jia, Y., Sukthankar, R., Berg, A.C.: MatchNet: unifying feature and metric learning for patch-based matching. In: IEEE Conference on Computer Vision and Pattern Recognition, CVPR, Boston, pp. 3279–3286 (2015)
12. Zagoruyko, S., Komodakis, N.: Learning to compare image patches via convolutional neural networks. In: IEEE Conference on Computer Vision and Pattern Recognition, CVPR, Boston, pp. 4353–4361 (2015)
13. Zbontar, J., Lecun, Y.: Computing the stereo matching cost with a convolutional neural network. In: IEEE Conference on Computer Vision and Pattern Recognition, CVPR, Boston, pp. 1592–1599 (2015)
14. Altwaijry, H., Trulls, E., Hays, J., Fua, P., Belongie, S.: Learning to match aerial images with deep attentive architectures. In: IEEE Conference on Computer Vision and Pattern Recognition, CVPR, Las Vegas, pp. 4353–4361 (2015)
15. Yi, K.M., Trulls, E., Lepetit, V., Fua, P.: LIFT: learned invariant feature transform. In: Leibe, B., Matas, J., Sebe, N., Welling, M. (eds.) ECCV 2016. LNCS, vol. 9910, pp. 467–483. Springer, Cham (2016). https://doi.org/10.1007/978-3-319-46466-4_28
16. Altwaijry, H., Veit, A., Belongie, S.: Learning to detect and match keypoints with deep architectures. In: British Machine Vision Conference, BMVC, York, pp. 49.1–49.12 (2016)
17. Dana, K., Anandan, P.: Registration of visible and infrared images. In: Optical Engineering and Photonics in Aerospace Sensing International Society for Optics and Photonics, vol. 1957, pp. 2–13 (1993)
18. Coiras, E., Santamaria, J., Miravet, C.A.: Segment-based registration technique for visualir image. Opt. Eng. **39**, 282–289 (2000)
19. Kelman, A., Sofka, M., Stewart, C.: Keypoint descriptors for matching across multiple image modalities and non-linear intensity variations. In: IEEE Conference on Computer Vision and Pattern Recognition, CVPR, pp. 1–7 (2007)
20. Istenic, R., Heric, D., Ribaric, S., Zazula, D.: Thermal and visual image registration in Hough parameter space. In: The 14th International Conference on System, Signal and Image Processing, pp. 106–109 (2007)
21. Li, H., Zhou, Y.T.: Automatic visual/infrared image registration. Opt. Eng. **35**, 391–400 (1996)
22. Harris, C., Stephens, M.: A combined corner and edge detector. In: Proceedings of the Alvey Vision Conference, pp. 147–151 (1988)
23. Schroff, F., Kalenichenko, D., Philbin, J.: Facenet: a unified embedding for face recognition and clustering. In: IEEE Conference on Computer Vision and Pattern Recognition, CVPR, Boston, pp. 815–823 (2015)
24. Uijlings, J.R., Sande, K.E., Gevers, T., Smeulders, A.W.: Selective search for object recognition. Int. J. Comput. Vis. **104**, 154–171 (2013)
25. Fischler, M.A., Robert, C.: Bolles random sample consensus: a paradigm for model fitting with applications to image analysis and automated cartography. In: Communication of ACM, pp. 726–740. ACM, California (1981)
26. Rocco, I., Arandjelovic, R., Sivic, J.: Convolutional neural network architecture for geometric matching. In: IEEE Conference on Computer Vision and Pattern Recognition, CVPR, Hawaii (2017)

4 Collinear Points: Robust Point Set Registration Using Cross Ratio Invariance

Yiru Wang[1,2], Yinlong Liu[1,2], Zhijian Song[1,2(✉)], and Manning Wang[1,2(✉)]

[1] School of Basic Medical Science, Fudan University, Shanghai 200032, China
{yiruwang16,yinlongliu15,zjsong,mnwang}@fudan.edu.cn
[2] Shanghai Key Laboratory of Medical Imaging Computing and Computer Assisted Intervention, Digital Medical Research Center, Fudan University, Shanghai 200032, China

Abstract. In this paper, we present a robust point set registration method based on the cross ratio invariance of 4 collinear points, which is able to deal with point set registration problem under the most general linear transformation, projective transformation. On the basis of all combinations of 4 collinear points extracted by Hough transform, meaningful correspondences are identified by combining the cross ratio invariance of 4 collinear points and Randomized RANSAC. At the end, the underlying projective transformation matrix was estimate in a least square sense. It has been shown in the simulation experiments that the proposed approach remains robust in high level of degradations, including 45% outliers and 40% overlap ratio. Experiments with Oxford corridor sequence and ZuBuD wide baseline image database proved its usefulness in real application.

Keywords: Point set registration · Cross ratio invariance
Projective transformation

1 Introduction

Point set registration is a vital and fundamental problem in many computer vision and pattern recognition tasks, such as stereo matching, object recognition, medical image alignment and fusion [10]. Upon most occasions, the point samples are acquired from two or more different points of views [6]. The goal of point set registration is to identify the correspondences and estimate the transformation between the two sets of point samples.

In literature, a great deal of point set registration approaches have been developed in the past decades. Most of these registration methods aim to align the two point sets by estimating the transformations between the datasets. According to the process of searching for the approximate transformation, these methods can be divided into three categories: (1) Local optimization methods, which

The first two authors contributed equally to this work.

J. Yang et al. (Eds.): CCCV 2017, Part III, CCIS 773, pp. 540–550, 2017.
https://doi.org/10.1007/978-981-10-7305-2_46

gradually optimize the objective function from an initial coordinate transformation to the local optimum through the iterative method, including the Iterative Closest Point (ICP) algorithm [1], the Coherent Point Drift (CPD) algorithm [2,3], etc. However, these approaches are sensitive to the local minima in numerical optimization. (2) Global optimization methods, which can be divided into probabilistic global optimization method and deterministic global optimization method. These approaches are dedicated to find the global optimum of the objective function independent of the initial values [15]. However, the speed is slower in general, although they can obtain the global optimum determinately or with a high probability. (3) Hypothesis generation - evaluation methods, they are mostly based on the Random Sample Consensus (RANSAC) [7] framework. RANSAC is a randomized algorithm which is used to find the most probable hypothesis by repeating the process for picking up a subset of data randomly enough times. 4PCS [4], a typical example of such method, is based on a novel technique to extract all 4 coplanar points sets from a point set that match the same affine ratio with a given set of 4 coplanar points under affine transformation.

In practice, the point set registration has been a challenging problem with the degradations of datasets themselves: noise, outliers and limited amounts of overlap [5]. As 4PCS is still effective to 40% outliers, and down to 40% overlap with arbitrary initial positions as demonstrated in their experiments, it has become one of the most vogue approaches. 4PCS uses affine ratio invariance as a constraint to prune the correspondences and transformation, which makes it much faster than random sampling. However, this advantage is restricted to affine transformation. Although affine transformation is an important problem in the non-rigid registration, there is also a more general concept of linear transformation: perspective transform.

In this paper, we propose a novel point set registration algorithm that explores a similar idea with 4PCS but covers a more general form of geometry transformation: perspective transformation. The algorithm is based on the cross ratio invariance of 4 collinear points during the perspective transform. The main idea of the proposed algorithm is (a) to collect all combinations of 4 collinear points using Hough transform [12], (b) to identify candidate correspondences by combining the cross ratio invariance of 4 collinear points and Randomized RANSAC (R-RANSAC) [7,8,11], and (c) to estimate the underlying projective transformation matrix in a least square sense [13]. Moreover, we add a refine step based on Gaussian mixture models (GMM) [14] to further improve the accuracy. Experiments showed that our proposed registration algorithm could achieve great performances in the projective transformation with both robustness and effectiveness even under a large degree of degradations. Additionally, the experiment of ZuBuD image database on the wide-baseline feature matching demonstrates the strong capability of our method in constraining the search space for transformation and correspondences. The main contribution of this paper is that we propose a point set registration algorithm that can deal with the projective transformation, which is a more general form than affine and rigid transformation, and demonstrate its robustness under noise, outliers and low overlap ratio.

The paper is organized as follows: Sect. 2 details the proposed algorithm implementation. Section 3 lists our experimental results and analysis, and Sect. 4 comes to the conclusions.

2 Method: Point Set Registration Using Cross Ratio Invariance

2.1 Problem Formulation

Given a floating point set $P = \{(x_{pi}, y_{pi}) | (x_{pi}, y_{pi}) \in R^d, i = 1, \ldots, N_p\}$ and a reference point set $Q = \{(x_{qj}, y_{qj}) | (x_{qj}, y_{qj}) \in R^d, i = 1, \ldots, N_q\}$ as shown in Fig. 1, where N_p and N_q are the number of points in P and Q, respectively, we aim to find an optimal transformation T^* from P to Q. In our algorithm, T^* is a projective transformation [16], with rigid and affine transformation special cases of it. In a 2D to 2D registration, T^* can be represented as a matrix with eight degrees of freedom:

$$T^* = \begin{bmatrix} t_{11} & t_{12} & t_{13} \\ t_{21} & t_{22} & t_{23} \\ t_{31} & t_{32} & 1 \end{bmatrix} \tag{1}$$

In 3D to 3D registration, the transformation T^* is a 4 by 4 matrix with 15 degrees of freedom. For simplicity, we introduce our algorithm under the assumption of 2D to 2D registration, and it can be readily extended to 3D to 3D registration cases. Our main idea is to pick two random sets of 4 collinear points from the floating point set P as bases $B1$, $B2$, and then quickly find their correspondences in Q by searching in candidates with identical cross ratio.

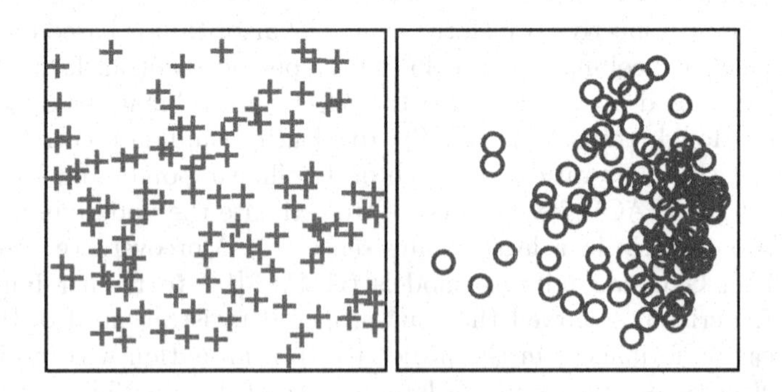

Fig. 1. Floating point set P (*left*) and reference point set Q (*right*)

In the following subsections, we first briefly introduce the cross ratio in perspective geometry [16] and then discuss how to efficiently extract corresponding 4 points in P and Q according to cross ratio, which is invariant during perspective transformation. At last, the complete algorithm will be given.

2.2 Cross Ratio

For completeness, we first introduce cross ratio in this section and details can be found in [16]. As shown in Fig. 2, 4 collinear points (A, B, C, D) in a 3D space are projected onto two planes and the projections on each plane $(A1, B1, C1, D1)$ and $(A2, B2, C2, D2)$ are also collinear. There is a projective transformation between the two 2D space defined on the two planes, that maps corresponding points on the two planes projected from the same 3D point from one to another. For any four collinear points on a plane, cross ratio is an invariance under projective transformation. We first give the definition of simple ratio for three collinear points (A, B, C) and cross ratio is based on it.

$$SR(A, B; C) = \frac{AC}{BC} \tag{2}$$

The simple ratio is determined by two directional line segments with the separation points and the base points. In Eq. (2), (A, B) are the base points and C is the separation point for the 3 collinear points (A, B, C). Then, the cross ratio can be computed based on a pair of simple ratios:

$$CR(A, B; C, D) = \frac{SR(A, B; C)}{SR(A, B; D)} = \frac{AC/BC}{AD/BD} \tag{3}$$

Cross ratio is invariant under projective transformation [16], and in Fig. 2, we have $CR(A, B; C, D) = CR(A1, B1; C1, D1) = CR(A2, B2; C2, D2)$.

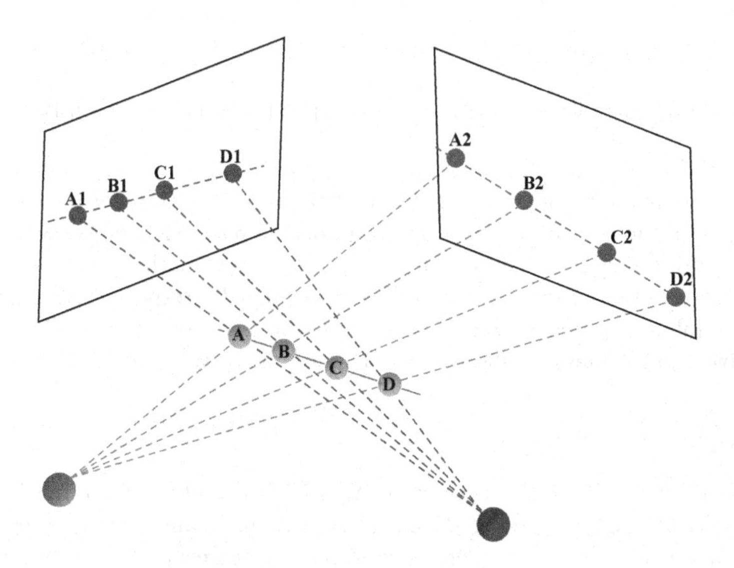

Fig. 2. The scene of perspective transformation

2.3 Extracting 4 Collinear Points and Finding Correspondence

We first detect lines in P and Q using Hough transform, since cross ratio is defined on collinear points. The definition of cross ratio in Formula (3) is dependent on the order of the four points. Therefore, given a set of four points, there are $A_4^4 = 24$ cross ratios corresponding to the full permutation of the four points. However, the collinear property makes things simpler, because we can restrict the ordering along the line direction and get only two permutations (A, B, C, D) and (D, C, B, A). Furthermore, according to Formula (3), $CR(A, B; C, D)$ equals $CR(D, C; B, A)$. Therefore, for any four collinear points, we can define one cross ratio, and we define these four points as a Base $B = \{(A, B, C, D), (D, C, B, A)\}$.

For a Base from P with cross ratio x_0, our goal is to find all potential corresponding Bases in Q with cross ratio $\{x_0 \pm u\}$, with a tolerance u. As shown in Fig. 3, there are four possible patterns of point correspondences in each corresponding Base pair.

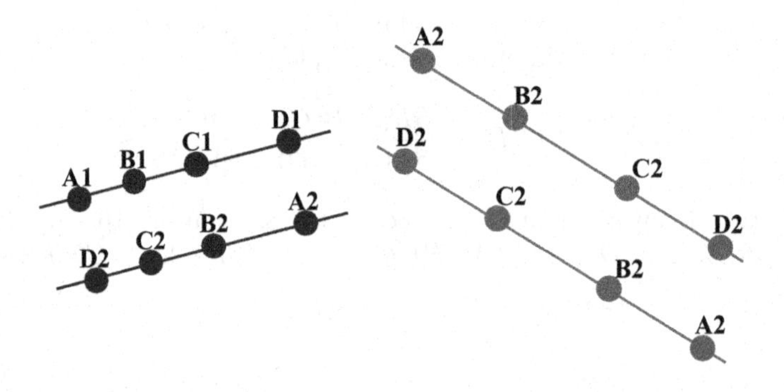

Fig. 3. Four possible combinations of correspondences between each Base pair

It is important to note that a 2D to 2D projective transformation can be determined by four points when they are not collinear. Hence, we need to find two corresponding pairs of Bases for calculating and evaluating a transformation. In practice, we randomly select two Bases $B1$ and $B2$, respectively, where m, n are the number of potential Bases. Therefore, for a pair of Bases in P, $m \cdot 2 \cdot n \cdot 2$ projective transformations should be calculated and evaluated.

2.4 Algorithm Framework

The algorithm we proposed is shown in Algorithm 1, and there are five inputs to the algorithm: P, Q, u, σ and R. P and Q are the two point sets to be registered. u is the allowable cross ratio difference in finding corresponding Bases and σ is the allowable residual error in evaluating a calculated projective transformation during the process of R-RANSAC. R depends on the overlap and outliers in P

and Q proposed in RANSAC framework [7]. R iterations are needed to guarantee the successful algorithm with the probability P_{sucess} :

$$R = \frac{\log(1 - P_{sucess})}{\log(1 - \gamma^N)} \tag{4}$$

where γ is the inlier or overlap ratio and N is the number of samples drawn each time.

Algorithm 1. Robust Point Set Registration Using Cross Ratio Invariance

1 PLine $\leftarrow$ Hough(**P**);
2 QLine $\leftarrow$ Hough(**Q**);
3 QLineCR $\leftarrow$ CalculateCR(QLine);
4 **for** $r = 1$ *to* $\boldsymbol{R}$ **do**
5     $B_{r1} \leftarrow$ PLine;
6   x_{pr1} = CalculateCR(B_{r1});
7     $B_{r2} \leftarrow$ PLine;
8   x_{pr2} = CalculateCR(B_{r2});
9     $U_r \leftarrow$ FindInQLine(x_{pr1}, QLineCR, QLine, $\boldsymbol{u}$);
10     $K_r \leftarrow$ FindInQLine(x_{pr2}, QLineCR, QLine, $\boldsymbol{u}$);
11     $UK_r \leftarrow$ Sort(U_r,K_r);
12   **for** *all* $UK_{ri} \in UK_r$ **do**
13         $T_{ri} \leftarrow$ best projective transform that matches $\{B_{r1}, B_{r2}\}$ and UK_{ri} in the Homogeneous Least Squares;
14   Distance = RMSDistance($T_{ri}(P_{partial})$, Q);
15   **if** *Distance* $\leq \boldsymbol{\sigma}$ **then**
16             $Erro_j$ = RMSDistance($T_{ri}(P)$, Q);
17     $Erro_k = \min\{Erro_j\}$;
18     $T_{result} = T_k$;
19     $T^* =$ GMMRefine(**P**, **Q**, T_{result})

First of all, we make use of the Hough transform to detect lines in P and Q. The points on each line stored in PLine and QLine are arranged in the ordering along the line direction. In order to improve the efficiency of the operation, we will complete all cross ratios of 4 collinear points at first, which would be saved as QLineCR. In the process of the r_{th} iteration, for the two Bases: B_{r1} and B_{r2} randomly selected in the PLine with cross ratios x_{pr1} and x_{pr2}, we can search for all potential corresponding Bases $U_r = \{U_{r1}, U_{r2}, \ldots\ldots, U_{rm}\}$ and $K_r = \{K_{r1}, K_{r2}, \ldots\ldots, K_{rn}\}$ in the QLineCR within allowable cross ratio difference of u. Next, as described in Sect. 2.3, there are $m \cdot 2 \cdot n \cdot 2$ kinds of correspondences situation of Bases B_{r1} and B_r2. Thus we can attain UK_r after sorting all elements of $U_r = \{U_{r1}, U_{r2}, \ldots\ldots, U_{rm}\}$ and $K_r = \{K_{r1}, K_{r2}, \ldots\ldots, K_{rn}\}$ in every possible combination, where UK_r is a set containing $m \cdot 2 \cdot n \cdot 2$ groups of 4 collinear points.

At present, for each element UK_{ri} in UK_r, we could use the Homogeneous Least Squares to calculate the best matching projective transform T_{ri}. To verify and evaluate T_{ri}, we first calculate the distance between the eight points in Base B_{r1} and B_{r2} and their corresponding points after registration. If the average Euclidean distance of these corresponding points is smaller than σ, the remaining points in model P will then be further verified. After repeating the operation for R times, we could select the best T_{result} and acquire the optimal T^* with the refine step of GMM.

In practice, we choose to use lines with 4–8 points after the Hough transform. In many experiments, we find that when the number of points is large in each line, the number of stored cross ratio increases rapidly. More importantly, these collinear points can always be detected because they are concentrated in a certain area, which is characterized by the distribution of relatively dense points. It is further understood that the cross ratio calculated by these points is not clearly distinguishable.

3 Experiments

To express concisely, our proposed robust point set registration using cross ratio invariance is abbreviated to CRI. The CRI was implemented with Matlab R2015b and experiments were done on an Intel Core E5 CPU 3.6 GHz with 32 GB RAM. Three kinds of point sets were used in our experiments: (1) random point sets, (2) Oxford corridor sequence [16], (3) extreme image pairs from the ZuBuD image database.

3.1 Robustness to Noise, Outliers and Partial-Overlap

First, we evaluated the robustness of CRI to noise, outliers and partial-overlap. We tested the algorithm on 100 random points as a floating point set P. Among them, some points were generated along 12 lines. Each point coordinate was a random variable between [0, 300] in every experiment. Applying a projective transformation to P, we could obtain a reference point cloud Q. In the registration with noise data, Q was under additive Gaussian noise with standard deviation ranging from 0–5. And then the registration accuracy with the increase of the disturbance was studied. In the experiment of outliers, we increased a certain number of points in P randomly to evaluate performance with outlier degradation. Similarly, we varied the amount of partial-overlap by deleting a certain number of points in P and Q respectively to evaluate the registration accuracy. Each experiment was tested on 20 randomized trials.

The registration results in Fig. 4 verify the method we proposed is robust to these degradations. The results of Hough transformation are in the left column. On this basis, we collect three examples of registration results under the noise, outliers and partial-overlap situations respectively. Estimation error was measured based on the standard RMS error. The existence of GMM enhanced the outcome especially with a large degree of noise level. Of course, another point we need to concern is that, our algorithm can also have an excellent performance when outliers percentage increases to 45% and the amounts of partial-overlap drops to 40% without the help of refine stage. In the outlier and partial-overlap experiments, the residual error came from the additive Gaussian noise with standard deviation of 1.

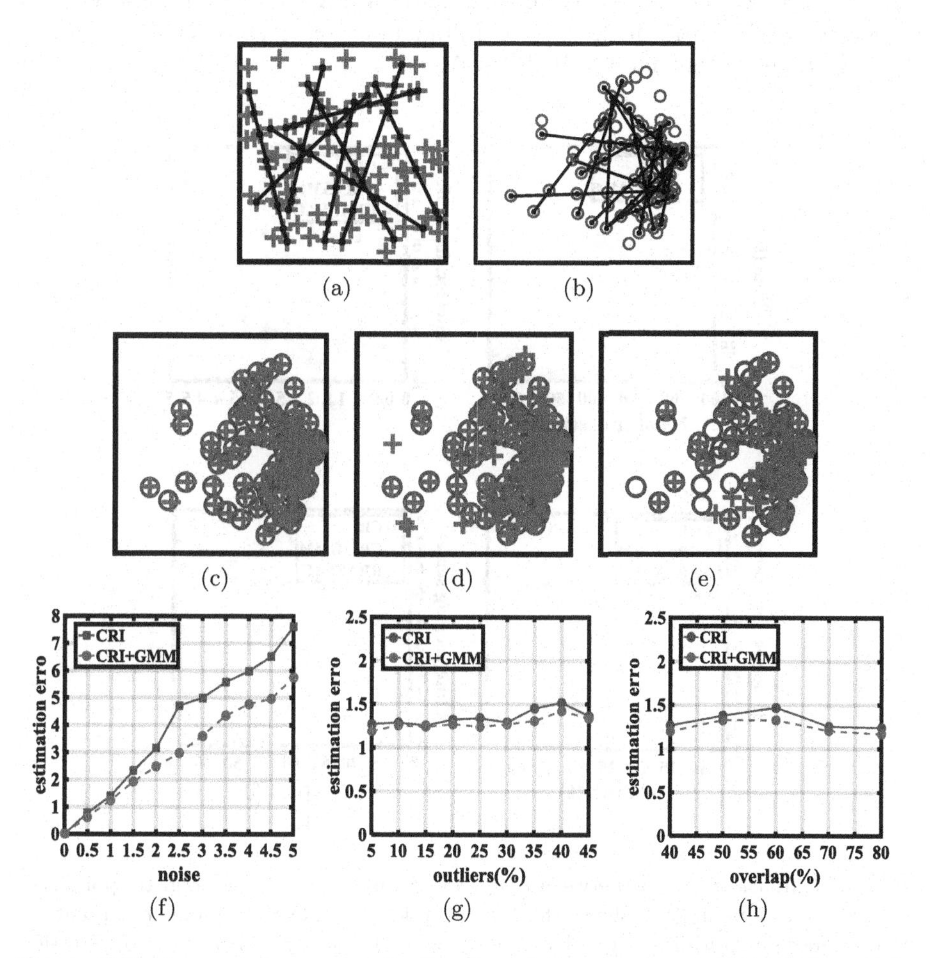

Fig. 4. Robustness and accuracy to the degradations: noise, outliers and partial-overlap. (a) and (b) are the results of Hough transformation in floating and reference point sets, respectively. (c), (d) and (e) are examples of registered points under the noise, outlier and partial-overlap, respectively. (f), (g) and (h) are the standard RMS error against noise level, outlier ratio and partial-overlap ratio.

3.2 Comparison to R-RANSAC

On this projective transformation context, it is difficult to provide the comparisons among the rare state of art algorithms. Hence, we chose the R-RANSAC algorithm to compare the speed of our algorithm when they were set the equal iterations R with the same degradations and dates described above. Figure 5(a) displays the time complexity curves of our algorithm with and without GMM. GMM costs a big proportion of the total time, especially when the number of points is small. When the number of samples is 500, the time of CRI is less than four seconds, however, the time of CRI with GMM increases around 86 s. As shown in Fig. 5(b)–(d), in all kinds of degradations, our method is several orders of magnitudes faster than the R-RANCAC.

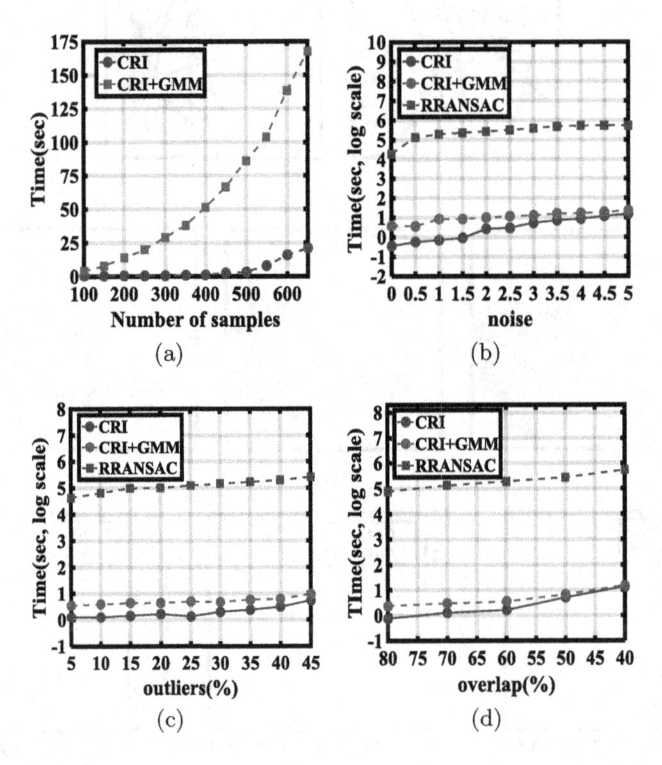

Fig. 5. Comparison of time between R-RANSAN and our method to register point sets on 100 random points. (a) shows the time complexity against the number of points of the proposed algorithm with and without GMM. (b), (c) and (d) are the registration time with noise, outliers and partial-overlap, respectively.

3.3 Real Application

We evaluated the performance of CRI in 2D real scene with both extreme image pair from the ZuBuD image database and local dense point sets pair from the Oxford corridor sequence. The ZuBuD image database is a kind of wide baseline stereo matching problem. For all the manually extracted points, the green 'o' points are matched in the real world. On the contrary, as outliers, the blue '+'

points should not be matched in the real world. After a strict Hough transformation, there were several lines detected and labeled with red color. Registration results of the ZuBuD image database are shown in the right image of the first row. The result shows that our proposed model could achieve great constraint force even in the extreme situation with a run time of around 3.5 s. For the two images from the Oxford corridor sequence, the left two images of the second row are frame 1 and frame 3 of the sequence, respectively. There were 175 corners (green '□') and 150 corners (red '+') detected based on the Harris detector. The result of registration shows the great performances in the process of the point set registration. However, the run time is around 60 s which is much longer than the registration of ZuBuD image pair because of the local dense points (Fig. 6).

Fig. 6. The registration results of our method for two different scenes (Color figure online)

4 Conclusions

In this paper, we propose a novel and simple point set registration framework based on the cross ratio invariance of 4 collinear points, which can recover the most general form of linear transformation: projective transformation. On the basis of all combinations of 4 collinear points collected by Hough transform, meaningful correspondences are identified by combining the cross ratio invariance of 4 collinear points and Randomized RANSAC. At the end, the underlying projective transformation matrix is estimate in a least square sense. Experiment with synthesis point sets show that the proposed approach is accurate and robust against noise, outliers and partial overlap. Preliminary experiments on ZuBuD and Oxford corridor image pairs show its potential in real application.

Acknowledgments. This study has been supported by: the Nation Natural Science Foundation of China (projects 60972102), the National Science and Technology Support Program (No. 2015BAK31B01), the National High Technology Research and Development Program (2015AA020507). This study is partly sponsored by Program of Shanghai Academic/Technology Research Leader project 16XD1424900 and project 15441905500, 17zr1401500 of Science and Technology Commission of Shanghai Municipality.

References

1. Besl, P.J., McKay, N.D.: A method for registration of 3-D shapes. IEEE Trans. Pattern Anal. Mach. Intell. **14**(2), 239–256 (1992)
2. Myronenko, A., Song, X., Carreira-Perpinan, M.: Non-rigid point set registration: coherent point drift. In: Advances in Neural Information Processing Systems, pp. 1009–1016 (2006)
3. Myronenko, A., Song, X.: Point set registration: coherent point drift. IEEE Trans. Pattern Anal. Mach. Intell. **32**(12), 2262–2275 (2010)
4. Aiger, D., Mitra, N.J., Cohen-Or, D.: 4-points congruent sets for robust pairwise surface registration. ACM Trans. Graph. (TOG) **27**(3), 85 (2008)
5. Tam, G.K., Cheng, Z.Q., Lai, Y.K., Langbein, F.C., Liu, Y., Marshall, D., et al.: Registration of 3D point clouds and meshes: a survey from rigid to nonrigid. IEEE Trans. Vis. Comput. Graph. **19**(7), 1199–1217 (2013)
6. Fitzgibbon, A.W.: Robust registration of 2D and 3D point sets. Image Vis. Comput. **21**(13), 1145–1153 (2003)
7. Fischler, M.A., Bolles, R.C.: Random sample consensus: a paradigm for model fitting with applications to image analysis and automated cartography. Commun. ACM **24**(6), 381–395 (1981)
8. Chum, O., Matas, J.: Randomized RANSAC with Td, d test. In: British Machine Vision Conferenc, vol. 2, pp. 448–457 (2002)
9. Oxford corridor sequence. http://www.robots.ox.ac.uk/~vgg/data/data-mview.html
10. Yan, J., Wang, J., Zha, H., et al.: Multi-view point registration via alternating optimization. In: AAAI Conference on Artificial Intelligence (AAAI) (2015)
11. Choi, S., Kim, T., Yu, W.: Performance evaluation of RANSAC family. In: BMVC (2009)
12. Duda, R.O., Hart, P.E.: Use of the Hough transformation to detect lines and curves in pictures. Commun. ACM **15**(1), 11–15 (1972)
13. Charnes, A., Frome, E.L., Yu, P.L.: The equivalence of generalized least squares and maximum likelihood estimates in the exponential family. J. Am. Stat. Assoc. **71**(353), 169–171 (1976)
14. Jian, B., Vemuri, B.C.: Robust point set registration using gaussian mixture models. IEEE Trans. Pattern Anal. Mach. Intell. **33**(8), 1633–1645 (2011)
15. Yan, J., Cho, M., et al.: Multi-graph matching via affinity optimization with graduated consistency regularization. IEEE Trans. Pattern Anal. Mach. Intell. (TPAMI) **38**(6), 1228–1242 (2016)
16. Hartley, R., Zisserman, A.: Multiple View Geometry in Computer Vision. Cambridge University Press, Cambridge (2003)

A Kind of Affine Weighted Moment Invariants

Hanlin Mo[1,2](✉), Shirui Li[1,2], You Hao[1,2], and Hua Li[1,2]

[1] Key Laboratory of Intelligent Information Processing,
Institute of Computing Technology, Chinese Academy of Sciences, Beijing, China
{mohanlin,lishirui,haoyou,lihua}@ict.ac.cn
[2] University of Chinese Academy of Sciences, Beijing, China

Abstract. A new kind of geometric invariants is proposed in this paper, which is called affine weighted moment invariant (AWMI). By combination of affine differential invariants and the framework of global integral, they can more effectively extract features of images, increase the number of low-order invariants. The experimental results show that AWMIs have good stability and distinguishability. Also, we find that the accuracy of image retrieval and classification has been improved by combining AWMIs with other traditional moment invariants.

Keywords: Differential invariants · Integral invariants
Weighted moment · Affine transform
Affine weighted moment invariants · Low-order

1 Introduction

Researchers have found that the geometric deformation, caused by the change of viewpoint, is an important factor leading to the object to be misidentified, as shown in Fig. 1. In order to solve this problem, various methods have been proposed to construct image features which are robust to geometric deformations. Moment and moment invariant are one of them.

The concepts of moment and moment invariant were first proposed by Hu [1]. He employed the theory of algebraic invariants, which was studied in 19th century [2], and defined geometric moment. Then he constructed seven geometric moment invariants which were invariant to the similarity transform. This set of invariants was widely used in various fields of pattern recognition, like [3]. But the similarity transform can't represent all geometric deformations. When the distance between the camera and the object is much larger than the size of the object itself, the geometric deformation of the object can be represented by the affine transform. The landmark work of affine moment invariants (AMIs) was proposed by Flusser and Suk [4]. They used geometric moments to construct several low-order and low-degree AMIs, which were more effective to practical applications, for example, image registration [5]. In order to obtain more AMIs, Suk and Flusser proposed the graph method which can generate AMIs of every order and degree [6]. Xu and Li [7] derived moment invariants in an intuitive

J. Yang et al. (Eds.): CCCV 2017, Part III, CCIS 773, pp. 551–564, 2017.
https://doi.org/10.1007/978-981-10-7305-2_47

way by multiple integrals of invariant geometric primitives, such as distance, area and volume. This method not only simplified the construction of AMIs, but also made them have a clear geometric meaning. Recently, Li *et al.* improved the method of geometric primitives [8]. They found a way to further simplify geometric primitives and used dot-product and cross-product of vectors to generate invariants. Additionally, previous studies have shown that low-order and low-degree moment invariants have better stability and less calculating cost than high-order and high-degree moment invariants. But the number of them was very limited. So, its very useful to get more low-order and low-degree moment invariants.

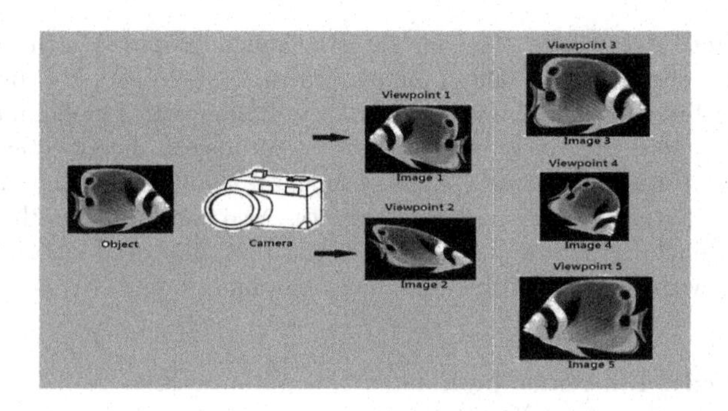

Fig. 1. The geometric deformation, caused by the change of viewpoint.

The studies of local differential invariants are another area, which should be concerned. Olver generalized the moving frame method and got differential invariants for general transformation groups [9]. He defined the affine gradient by using affine differential invariants [10]. Ge *et al.* [11] presented a local feature descriptor under color affine transformation by using the affine gradient. Wang *et al.* [12] proposed an effective method to derive a special type of affine differential invariants. However, they didn't explain how to use these local differential invariants in practical applications and how to improve the numerical accuracy of partial derivatives on discrete images. Recently, Li and Li [13] found the isomorphism between differential invariants and geometric moment invariants to general affine transformations. If affine moment invariants were known, relative affine differential invariants can be obtained by the substitution of moments by partial derivatives with the same order. This method made the construction of affine differential invariants very easily.

In this paper, we use the frame of geometric moments and partial derivatives to define a kind of weighted moments, which can be named differential moments (DMs). According to the definition of DMs and local affine differential invariants, affine weighted moment invariants (AWMIs) can be constructed easily, which use both global and local information. Some experimental results show that AWMIs have good stability and distinguishability. Also, they can improve the accuracy of image retrieval and classification.

2 Basic Definitions and Notations

In order to understand the structural frame of AWMIs more clearly, we first introduce some basic definitions and notations.

2.1 The Definition of Geometric Moment

The geometric moment of the image $f(x,y)$ is defined by

$$m_{pq} = \int_{\infty}^{-\infty}\int_{\infty}^{-\infty} x^p y^q f(x,y)dxdy \tag{1}$$

where $p,q \in \{0,1,2,\ldots\}$, $(p+q)$ is the order of m_{pq}. In order to eliminate the effect of translation, central geometric moments are usually used. The central moment is defined by

$$u_{pq} = \int_{\infty}^{-\infty}\int_{\infty}^{-\infty} (x-\bar{x})^p (y-\bar{y})^q f(x,y)dxdy \tag{2}$$

where

$$\bar{x} = \frac{m_{10}}{m_{00}}, \quad \bar{y} = \frac{m_{01}}{m_{00}} \tag{3}$$

2.2 Coordinate Transformation Under the Affine Transform

Suppose the image $f(x,y)$ is transformed into another image $g(u,v)$ by the affine transform A and the translation T. (u,v) is the corresponding point of (x,y). Then, there is a relation

$$\begin{pmatrix} u \\ v \end{pmatrix} = A \cdot \begin{pmatrix} x \\ y \end{pmatrix} + T = \begin{pmatrix} a_{11} & a_{12} \\ a_{21} & a_{22} \end{pmatrix} \cdot \begin{pmatrix} x \\ y \end{pmatrix} + \begin{pmatrix} t_1 \\ t_2 \end{pmatrix} \tag{4}$$

$$g(u,v) = f(x,y) \tag{5}$$

where A is a nonsingular matrix.

2.3 The Construction of Affine Moments Invariants

For the image $f(x,y)$, let (x_i,y_i) and (x_j,y_j) be two arbitrary points in the domain of $f(x,y)$. The geometric primitive proposed in [7] can be defined by

$$S^f(i,j) = \begin{vmatrix} (x_i-\bar{x}) & (x_j-\bar{x}) \\ (y_i-\bar{y}) & (y_j-\bar{y}) \end{vmatrix} \tag{6}$$

Suppose the image $f(x,y)$ is transformed into another image $g(u,v)$ by (4) and (5). (u_i,v_i), (u_j,v_j) in $g(u,v)$ are the corresponding points of (x_i,y_i), (x_j,y_j) in $f(x,y)$. Then, there is a relation

$$S^g(i,j) = |A| \cdot S^f(i,j) \tag{7}$$

where $|A|$ is the determinant of A. Therefore, using N points $(x_1, y_1), (x_2, y_2), \ldots, (x_N, y_N)$ in $f(x, y)$, $Core^f(N, m; d_1, d_2, \ldots, d_N)$ can be defined by

$$Core^f(N, m; d_1, d_2, \ldots, d_N) = \underbrace{S^f(1,2) \ldots S^f(k,l) \ldots S(r,N)}_{m} \tag{8}$$

where $k < l, r < N, k, l, r \in \{1, 2, \ldots, N\}$. d_i represents the number of times that the point (x_i, y_i) occurs in all geometric primitives, $i = 1, 2, \ldots, N$.

Let $(u_1, v_1), (u_2, v_2), \ldots, (u_N, v_N)$ in $g(u, v)$ be corresponding points of $(x_1, y_1), (x_2, y_2), \ldots, (x_N, y_N)$ in $f(x, y)$. It's obviously that

$$Core^g(N, m; d_1, d_2, \ldots, d_N) = |A|^m Core^f(N, m; d_1, d_2, \ldots, d_N) \tag{9}$$

Finally, using $Core^f(N, m; d_1, d_2, \ldots, d_N)$, AMI can be defined by

$$\begin{aligned} AMI &= \frac{I(Core^f(N, m; d_1, d_2, \ldots, d_N))}{(\iint f(x, y) dx dy)^{N+m}} \\ &= \frac{\iint \ldots \iint Core^f(N, m; d_1, d_2, \ldots, d_N)\, dx_1 dy_1 \ldots dx_N dy_N}{(\iint f(x, y) dx dy)^{N+m}} \end{aligned} \tag{10}$$

where $I(X)$ means multiple integrals. In [7], Xu and Li proved that (10) didn't change when the image was transformed by (4) and (5). (10) is the general form of AMI. In fact, this multiple integral can be expressed as the polynomial of central geometric moments.

$$\frac{I(Core^f(N, m; d_1, d_2, \ldots, d_N))}{(\iint f(x, y) dx dy)^{N+m}} = \frac{\sum_j a_j \cdot \prod_{i=1}^{N} u_{p_i q_i}}{(u_{00})^{N+m}} \tag{11}$$

where j represents the number of multiplicative items in the expansion, a_j is the coefficient of the j-th multiplicative item. In general, N is named as the degree of (10), $\max_i \{p_i + q_i\}$ is named as the order of (10).

2.4 Affine Differential Invariants

For the differentiable function $f(x, y)$, Olver [9] used the contact-invariant coframe to construct affine differential invariants. The differential invariants of $f(x, y)$ structured by using the first-order and second-order derivatives are listed in Table 1. $f(x, y)$ and $g(u, v)$ satisfy the relationship shown in (4) and (5). Note that we assume $x = x - \bar{x}, y = y - \bar{y}, u = u - \bar{u}$ and $v = v - \bar{v}$.

Obviously, ADI_4^f and ADI_5^f are pure differential invariants, which don't contain x or y. ADI_1^f and ADI_2^f are absolute differential invariants. ADI_3^f, ADI_4^f and ADI_5^f are relative differential invariants. In addition, Olver indicated that differential invariants shown in Table 1 were not independent [15]. The relationship of them can be represented by

$$(ADI_2^f(x, y))^2 - ADI_5^f(x, y) ADI_3^f(x, y) + (ADI_1^f(x, y))^2 ADI_4^f(x, y) = 0 \tag{12}$$

Table 1. ADIs (structured by using the first-order and second-order derivatives of $f(x.y)$)

No.	ADI	Relation
$ADI_1^f(x,y)$	$x\frac{\partial f}{\partial x}+y\frac{\partial f}{\partial y}$	$ADI_1^f(x,y)=ADI_1^g(u,v)$
$ADI_2^f(x,y)$	$x^2\frac{\partial f^2}{\partial^2 x}+2xy\frac{\partial f^2}{\partial x\partial y}+y^2\frac{\partial f^2}{\partial^2 y}$	$ADI_2^f(x,y)=ADI_2^g(u,v)$
$ADI_3^f(x,y)$	$x\frac{\partial f}{\partial y}\frac{\partial f^2}{\partial^2 x}+(y\frac{\partial f}{\partial y}-x\frac{\partial f}{\partial x})\frac{\partial f^2}{\partial x\partial y}-y\frac{\partial f}{\partial x}\frac{\partial f^2}{\partial^2 y}$	$ADI_3^f(x,y)=\frac{1}{\lvert A\rvert}ADI_3^g(u,v)$
$ADI_4^f(x,y)$	$\frac{\partial f^2}{\partial^2 x}\frac{\partial f^2}{\partial^2 y}-(\frac{\partial f^2}{\partial x\partial y})^2$	$ADI_4^f(x,y)=\frac{1}{\lvert A\rvert^2}ADI_4^g(u,v)$
$ADI_5^f(x,y)$	$(\frac{\partial f}{\partial y})^2\frac{\partial f^2}{\partial^2 x}-2\frac{\partial f}{\partial x}\frac{\partial f}{\partial y}\frac{\partial f^2}{\partial x\partial y}+(\frac{\partial f}{\partial x})^2\frac{\partial f^2}{\partial^2 y}$	$ADI_5^f(x,y)=\frac{1}{\lvert A\rvert^2}ADI_5^g(u,v)$

3 The Construction Frame of AWMIs

3.1 The Definition of DMs

Definition 1. Suppose that $f(x,y)$ is a differentiable function. Then, the first-order DM is defined by

$$d_{pqmn}=\int_{\infty}^{-\infty}\int_{\infty}^{-\infty}(x-\bar{x})^p(y-\bar{y})^q(\frac{\partial f}{\partial x})^m(\frac{\partial f}{\partial x})^n f(x,y)dxdy \tag{13}$$

where $p,q,m,n\in\{0,1,2,\ldots\}$.

The second-order differential moment is defined by:

$$\begin{aligned}d_{mnrst}^{pq}=&\int_{\infty}^{-\infty}\int_{\infty}^{-\infty}(x-\bar{x})^p(y-\bar{y})^q(\frac{\partial f}{\partial x})^m(\frac{\partial f}{\partial x})^n\\&(\frac{\partial f^2}{\partial^2 x})^r(\frac{\partial f^2}{\partial^2 y})^s(\frac{\partial f^2}{\partial x\partial y})^t f(x,y)dxdy\end{aligned} \tag{14}$$

where $p,q,m,n,r,s,t\in\{0,1,2,\ldots\}$.

Similarly, we can construct higher-order differential moments. But considering their convenience and calculation accuracy of partial derivatives, we only define two kinds of DM in this paper. Compared with (2), DMs are constructed by using polynomial functions and partial derivatives of $f(x,y)$. Thus, they can represent internal information of images better.

3.2 The Construction of AWMIs

Definition 2. Suppose that $f(x,y)$ is a differentiable function, $AWMI_1^f$ is defined by

$$\begin{aligned}AWMI_1^f&=\frac{I(DCore^f(N,m;d_1,d_2,\ldots,d_N,k_1,k_2,\ldots,k_N))}{(\iint f(x,y)dxdy)^{N+m}}\\&=\frac{\sum_j a_j\cdot\prod_{i=1}^{N}d_{p_iq_im_in_i}}{(d_{0000})^{N+m}}\end{aligned} \tag{15}$$

where

$$\begin{aligned}
&DCore^f(N,m;d_1,d_2,\ldots,d_N,k_1,k_2,\ldots,k_N)\\
&= Core^f(N,m;d_1,d_2,\ldots,d_N)(ADI_1^f(x_1,y_1))^{k_1}(ADI_1^f(x_2,y_2))^{k_2}\\
&\ldots(ADI_1^f(x_N,y_N))^{k_N}
\end{aligned} \tag{16}$$

Then, we can get the following theorem.

Theorem 1. Suppose the image $f(x,y)$ is transformed into another image $g(u,v)$ by (4) and (5). $(u_1,v_1),(u_2,v_2),\ldots,(u_N,v_N)$ in $g(u,v)$ are corresponding points of $(x_1,y_1),(x_2,y_2),\ldots,(x_N,y_N)$ in $f(x,y)$. The following equation is established.

$$\begin{aligned}
&\frac{I(DCore^f(N,m;d_1,d_2,\ldots,d_N,k_1,k_2,\ldots,k_N))}{(\iint f(x,y)dxdy)^{N+m}}\\
&=\frac{I(DCore^g(N,m;d_1,d_2,\ldots,d_N,k_1,k_2,\ldots,k_N))}{(\iint g(u,v)dudv)^{N+m}}
\end{aligned} \tag{17}$$

where

$$\begin{aligned}
&DCore^g(N,m;d_1,d_2,\ldots,d_N,k_1,k_2,\ldots,k_N)\\
&= Core^g(N,m;d_1,d_2,\ldots,d_N)(ADI_1^g(u_1,v_1))^{k_1}(ADI_1^g(u_2,v_2))^{k_2}\\
&\ldots(ADI_1^g(u_N,v_N))^{k_N}
\end{aligned} \tag{18}$$

The proof of (23) is the same as that of (10) proved in [7].

3.3 The Instances of AWMIs

In [6], Flusser and Suk proved that there were only 2 AMIs of which degrees $N \leqslant 3$ and orders $\max_i\{p_i+q_i\} \leqslant 3$. But now, we can use (15) to construct AWMIs. When $N \leqslant 3, \max_i\{p_i+q_i\} \leqslant 3$ and $\max_i\{m_i+n_i\} \leqslant 1$, there are 8 DCores which are listed in Table 2.

In Table 3, we list AWMIs constructed by Dcores in the Table 2. They are expressed by the polynomials of the first-order DM. It is worth noting that we have removed Dcores of which expansions are always 0 or contain d_{1000}, d_{0100}. In fact, using a similar definition to (15), we can obtain AWMIs constructed by using the second-order affine differential invariants. But here, we give a new definition. We want to point out that there are many different methods to construct AWMIs.

Definition 3. Suppose that $f(x,y)$ is a differentiable function, its AWMI which is constructed by the second-order affine differential invariants can be defined by

$$AWMI_2^f(1) = \frac{\iint ADI_4 f(x,y)dxdy}{\iint ADI_5 f(x,y)dxdy} \tag{19}$$

Table 2. DCores ($N \leqslant 3, \max_i \{p_i + q_i\} \leqslant 3, \max_i \{m_i + n_i\} \leqslant 1$)

No.	DCore
$DCore_1^f$	$(x_1y_2 - x_2y_1)^2(x_1\frac{\partial f}{\partial x_1} + y_1\frac{\partial f}{\partial y_1})$
$DCore_2^f$	$(x_1y_2 - x_2y_1)^2(x_1\frac{\partial f}{\partial x_1} + y_1\frac{\partial f}{\partial y_1})(x_2\frac{\partial f}{\partial x_2} + y_2\frac{\partial f}{\partial y_2})$
$DCore_3^f$	$(x_1y_2 - x_2y_1)(x_1y_3 - x_3y_1)(x_2\frac{\partial f}{\partial x_2} + y_2\frac{\partial f}{\partial y_2})(x_3\frac{\partial f}{\partial x_3} + y_3\frac{\partial f}{\partial y_3})$
$DCore_4^f$	$(x_1y_2 - x_2y_1)(x_1y_3 - x_3y_1)(x_1\frac{\partial f}{\partial x_1} + y_1\frac{\partial f}{\partial y_1})(x_2\frac{\partial f}{\partial x_2} + y_2\frac{\partial f}{\partial y_2})(x_3\frac{\partial f}{\partial x_3} + y_3\frac{\partial f}{\partial y_3})$
$DCore_5^f$	$(x_1y_2 - x_2y_1)(x_1y_3 - x_3y_1)^2(x_2\frac{\partial f}{\partial x_2} + y_2\frac{\partial f}{\partial y_2})$
$DCore_6^f$	$(x_1y_2 - x_2y_1)(x_1y_3 - x_3y_1)^2(x_3\frac{\partial f}{\partial x_3} + y_3\frac{\partial f}{\partial y_3})$
$DCore_7^f$	$(x_1y_2 - x_2y_1)(x_1y_3 - x_3y_1)^2(x_2\frac{\partial f}{\partial x_2} + y_2\frac{\partial f}{\partial y_2})(x_3\frac{\partial f}{\partial x_3} + y_3\frac{\partial f}{\partial y_3})$
$DCore_8^f$	$(x_1y_2 - x_2y_1)(x_1y_3 - x_3y_1)(x_2y_3 - x_3y_2)^2(x_1\frac{\partial f}{\partial x_1} + y_1\frac{\partial f}{\partial y_1})$

Table 3. AWMIs ($N \leqslant 3, \max_i \{p_i + q_i\} \leqslant 3, \max_i \{m_i + n_i\} \leqslant 1$)

No.	AWMI
$AWMI_1^f(1)$	$d_{0200}d_{2101} + d_{0200}d_{3010} + d_{0301}d_{2000} - 2d_{1100}d_{1201} - 2d_{1100}d_{2110} + d_{1210}d_{2000}$
$AWMI_1^f(2)$	$d_{0301}d_{2101} + d_{0301}d_{3010} - (d_{1201})^2 - 2d_{1201}d_{2110} + d_{1210}d_{2101} + d_{1210}d_{3010} - (d_{2110})^2$
$AWMI_1^f(3)$	$d_{0200}(d_{1101})^2 + 2d_{0200}d_{1101}d_{2010} + d_{0200}(d_{2010})^2 + (d_{0201})^2d_{2000} - 2d_{0201}d_{1100}d_{1101} - 2d_{0201}d_{1100}d_{2010} + 2d_{0201}d_{1110}d_{2000} - 2d_{1100}d_{1101}d_{1110} - 2d_{1100}d_{1110}d_{2010} + (d_{1110})^2d_{2000}$
$AWMI_1^f(4)$	$(d_{0201})^2d_{2101} + (d_{0201})^2d_{3010} - 2d_{0201}d_{1101}d_{1201} - 2d_{0201}d_{1101}d_{2110} + 2d_{0201}d_{1110}d_{2101} + 2d_{0201}d_{1110}d_{3010} - 2d_{0201}d_{1201}d_{2010} - 2d_{0201}d_{2010}d_{2110} + d_{0301}(d_{1101})^2 + 2d_{0301}d_{1101}d_{2010} + d_{0301}(d_{2010})^2 + (d_{1101})^2d_{1210} - 2d_{1101}d_{1110}d_{1201} - 2d_{1101}d_{1110}d_{2110} + 2d_{1101}d_{1210}d_{2010} + (d_{1110})^2d_{2101} + (d_{1110})^2d_{3010} - 2d_{1110}d_{1201}d_{2010} - 2d_{1110}d_{2010}d_{2110} + d_{1210}(d_{2010})^2$
$AWMI_1^f(5)$	$d_{0200}d_{0201}d_{3000} - d_{0200}d_{1101}d_{2100} + d_{0200}d_{1110}d_{3000} - d_{0200}d_{2010}d_{2100} - 2d_{0201}d_{1100}d_{2100} + d_{0201}d_{1200}d_{2000} - d_{0300}d_{1101}d_{2000} - d_{0300}d_{2000}d_{2010} + 2d_{1100}d_{1101}d_{1200} - 2d_{1100}d_{1110}d_{2100} + 2d_{1100}d_{1200}d_{2010} + d_{1110}d_{1200}d_{2000}$
$AWMI_1^f(6)$	$d_{0200}(d_{1101})^2 + 2d_{0200}d_{1101}d_{2010} + d_{0200}(d_{2010})^2 + (d_{0201})^2d_{2000} - 2d_{0201}d_{1100}d_{1101} - 2d_{0201}d_{1100}d_{2010} + 2d_{0201}d_{1110}d_{2000} - 2d_{1100}d_{1101}d_{1110} - 2d_{1100}d_{1110}d_{2010} + (d_{1110})^2d_{2000}$
$AWMI_1^f(7)$	$d_{0201}d_{0301}d_{3000} + d_{0201}d_{1200}d_{2101} + d_{0201}d_{1200}d_{3010} - 2d_{0201}d_{1201}d_{2100} + d_{0201}d_{1210}d_{3000} - 2d_{0201}d_{2100}d_{2110} - d_{0300}d_{1101}d_{2101} - d_{0300}d_{1101}d_{3010} - d_{0300}d_{2010}d_{2101} - d_{0300}d_{2010}d_{3010} - d_{0301}d_{1101}d_{2100} + d_{0301}d_{1110}d_{3000} - d_{0301}d_{2010}d_{2100} + 2d_{1101}d_{1200}d_{1201} + 2d_{1101}d_{1200}d_{2110} - d_{1101}d_{1210}d_{2100} + d_{1110}d_{1200}d_{2101} + d_{1110}d_{1200}d_{3010} - 2d_{1110}d_{1201}d_{2100} + d_{1110}d_{1210}d_{3000} - 2d_{1110}d_{2100}d_{2110} + 2d_{1200}d_{1201}d_{2010} + 2d_{1200}d_{2010}d_{2110} - d_{1210}d_{2010}d_{2100}$
$AWMI_1^f(8)$	$-2d_{0300}d_{1201}d_{3000} + 2d_{0300}d_{2100}d_{2101} + 2d_{0300}d_{2100}d_{3010} - 2d_{0300}d_{2110}d_{3000} + 2d_{0301}d_{1200}d_{3000} - 2d_{0301}(d_{2100})^2 - 2(d_{1200})^2d_{2101} - 2(d_{1200})^2d_{3010} + 2d_{1200}d_{1201}d_{2100} + 2d_{1200}d_{1210}d_{3000} + 2d_{1200}d_{2100}d_{2110} - 2d_{1210}(d_{2100})^2$

According to Table 1, we can prove that (19) won't change when $f(x, y)$ is transformed by (4) and (5) very easily. Its expansion is defined by

$$AWMI_2^f(1) = \frac{d_{0000110} - d_{0000002}}{d_{0002100} - 2d_{0011001} + d_{0020010}} \tag{20}$$

4 Experimental Results and Analysis

In this section, some experiments are conducted to evaluate the theoretical framework proposed in previous sections. Firstly, we introduce the method to calculate the partial derivatives of discrete images. Then, we test the stability and discernibility of AWMIs on the synthetic image database. Finally, we perform image retrieval and classification based on real databases. We find the accuracy of image retrieval and classification has been improved by combining AWMIs with traditional moment invariants as feature vectors.

4.1 The Partial Derivatives of Discrete Images

In the above, we assume that the function $f(x,y)$ is continuous and differentiable. Actually, general grayscale images are discrete two-dimensional functions. So we have to choose a way to calculate derivatives more accurately. Some researchers have confirmed that employing derivatives of the Gaussian function as filters to compute derivatives of discrete functions via convolution is a good way [14]. The 2D zero-mean Gaussian function and its the first-order and second-order partial derivatives are defined by

$$\begin{aligned} G(x,y) &= \frac{1}{2\pi\sigma^2}\mathrm{e}^{-\frac{x^2+y^2}{2\sigma^2}} & \frac{\partial G}{\partial x} &= -\frac{x}{2\pi\sigma^4}\mathrm{e}^{-\frac{x^2+y^2}{2\sigma^2}} \\ \frac{\partial G}{\partial y} &= -\frac{y}{2\pi\sigma^4}\mathrm{e}^{-\frac{x^2+y^2}{2\sigma^2}} & \frac{\partial^2 G}{\partial x^2} &= \frac{(x^2-\sigma^2)}{2\pi\sigma^6}\mathrm{e}^{-\frac{x^2+y^2}{2\sigma^2}} \\ \frac{\partial^2 G}{\partial x \partial y} &= \frac{xy}{2\pi\sigma^6}\mathrm{e}^{-\frac{x^2+y^2}{2\sigma^2}} & \frac{\partial^2 G}{\partial y^2} &= \frac{(y^2-\sigma^2)}{2\pi\sigma^6}\mathrm{e}^{-\frac{x^2+y^2}{2\sigma^2}} \end{aligned} \tag{21}$$

where σ is the standard deviation. Using (21) to convolve with the image function $f(x,y)$, we can get partial derivatives of $f(x,y)$. For example,

$$\frac{\partial f}{\partial x} = \frac{\partial G}{\partial x} \circledast f(x,y) \tag{22}$$

where $\circledast$ means convolution. For discrete images, we use (21) to convolve with the $N \times N$ neighborhood of (x,y) in the domain of $f(x,y)$. In general, N is odd and $\sigma = \frac{N-1}{6}$. In this paper, we make $N = 9$ and $\sigma = 1.33$.

4.2 Numerical Stability and Discernibility of AWMIs

We choose 5 kinds of fish pictures from Web page: https://www.igfa.org/Fish/Fish-Database.aspx. Original images are transformed by 5 different affine transforms and translations in Table 4. Thus, 30 images are obtained (512×512), which are shown in Fig. 2. They can be divided into 5 groups ($A \sim E$), each

Table 4. 5 Affine transforms

No.	a_{11}	a_{12}	a_{21}	a_{22}	t_1	t_2
1	0.69	−0.12	0.21	1.18	0	150
2	0.57	0.42	−0.42	0.42	160	280
3	0.60	−1.03	0.52	0.30	50	15
4	1.00	−1.00	0.00	1.00	100	50
5	1.50	0.00	0.00	0.80	30	10

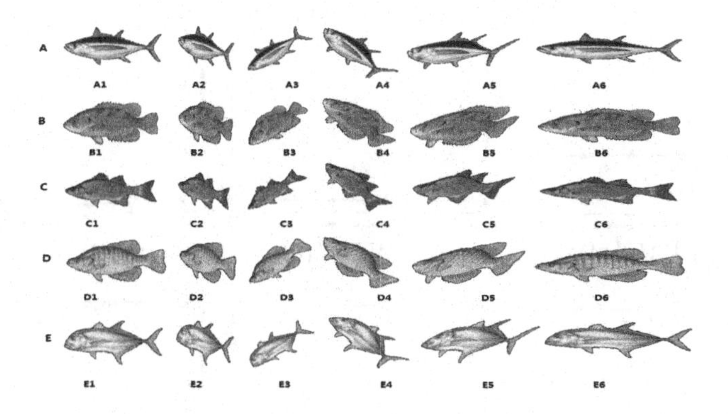

Fig. 2. The synthesis images, each line contains the original image and transformed versions.

group contains 6 images. AWMIs ($AWMI_1^f(1) \sim AWMI_1^f(8), AWMI_2^f(1)$) are computed for each image in Fig. 2, and the numerical values are shown in Fig. 3.

As shown in Fig. 4(a), we calculate the calculation errors of ($AWMI_1^f(f) \sim AWMI_1^f(8)$) which are defined by

$$Error = \frac{Max(intraclass\ invariants) - Min(intraclass\ invariants)}{|Max(intraclass\ invariants)| + |Min(intraclass\ invariants)|} \times 100\% \tag{23}$$

Obviously, AWMIs have good numerical stability, because all the calculation errors of AWMIs are less than 1%. Also, Chi-Square distance is used to calculate the feature distance between arbitrarily two images. Thus, as shown in Fig. 4(b), a 30×30 distance matrix is obtained. We can find the color of the area near the diagonal is lighter than those of other regions, indicating that feature distances of similar images are smaller, and vice versa.

According to this experimental result, we find that AWMIs have good stability and distinguishability. Therefore, it proved that the theoretical framework proposed in the previous sections is correct.

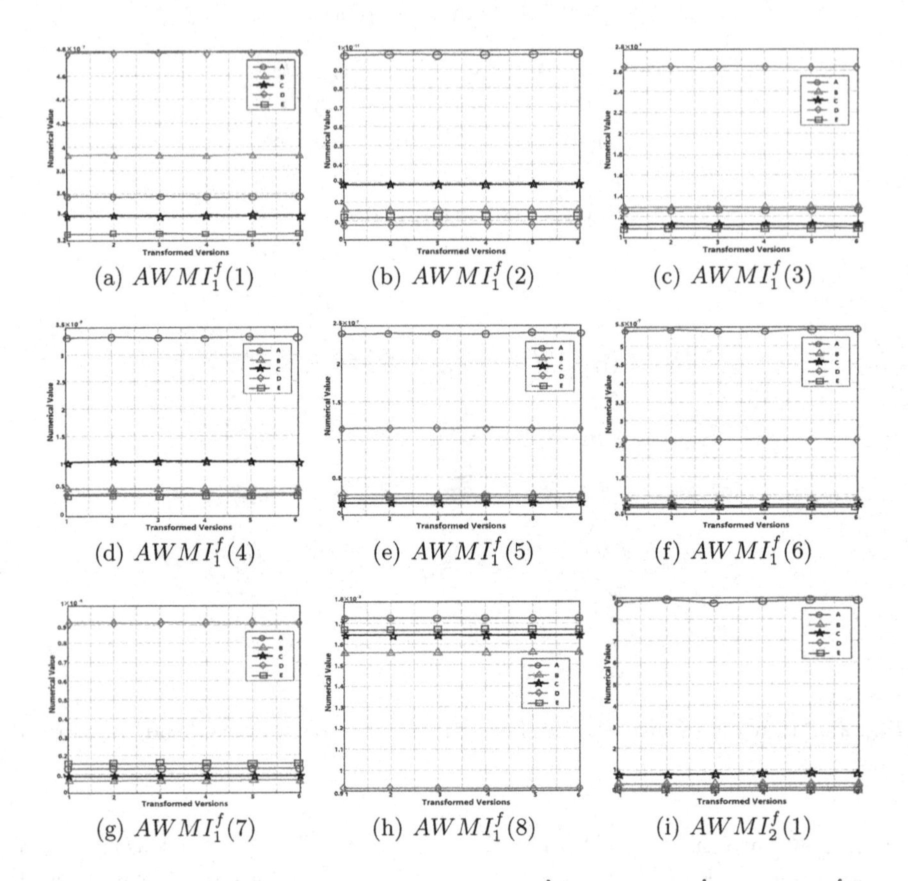

(a) $AWMI_1^f(1)$ (b) $AWMI_1^f(2)$ (c) $AWMI_1^f(3)$

(d) $AWMI_1^f(4)$ (e) $AWMI_1^f(5)$ (f) $AWMI_1^f(6)$

(g) $AWMI_1^f(7)$ (h) $AWMI_1^f(8)$ (i) $AWMI_2^f(1)$

Fig. 3. The numerical values of AWMIs ($AWMI_1^f(1) \sim AWMI_1^f(f), AWMI_2^f(8)$)

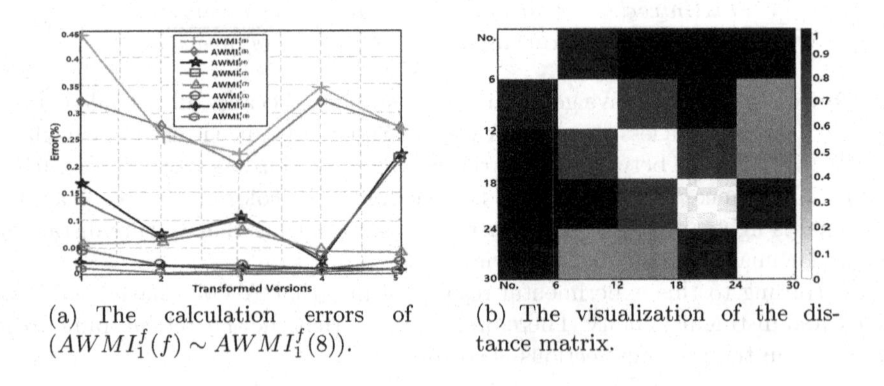

(a) The calculation errors of ($AWMI_1^f(f) \sim AWMI_1^f(8)$).

(b) The visualization of the distance matrix.

Fig. 4. The numerical stability and discernibility of AWMIs

4.3 Experiments on Real Image Databases

In order to verify the performance of AWMIs, we choose some real image databases for testing. COIL-20 in [15] which contains images of 20 objects taking from 72 different viewpoints, which are shown in Fig. 5. All images in this dataset have black background.

Fig. 5. 20 different classes of objects in COIL-20.

We choose 4 kinds of traditional moment invariants to combine with AWMIs.

1. AMIs: $(AMI_1, AMI_2, AMI_3, AMI_6, AMI_7, AMI_8, AMI_9)$, which were proposed in [6] and invariant to the affine transformation.
2. HMs: $(HM_1, HM_2, HM_3, HM_4, HM_5, HM_6, HM_7)$, which were proposed in [1] and invariant to the similarity transformation.
3. ZMs: $(Z_{11}, Z_{2,0}, Z_{2,2}, Z_{3,1}, Z_{3,3}, Z_{4,0}, Z_{4,2})$, which were proposed in [16] and invariant to the similarity transformation.
4. GHMs: $(\psi_1, \psi_2, \psi_3, \psi_4, \psi_5, \psi_6, \psi_7)$, which were proposed in [17] invariant to rotation and translation.

For image retrieval, we also adopt Chi-Square distance to measure the similarity of two feature vectors. Then we retrieval each image and draw 8 Precision-Recall curves of AWMIs+AMIs, AWMIs+HMs, AWMIs+ZMs, AWMIs+GHMs, AMIs, HMs, ZMs and GHMs in Fig. 6 to reflect their average levels of image retrieval.

Precision and Recall are defined by

$$Precision = \frac{|\{relevant\ images\} \cap \{retrieved\ images\}|}{|\{retrieved\ images\}|} \tag{24}$$

$$Recall = \frac{|\{relevant\ images\} \cap \{retrieved\ images\}|}{|\{relevant\ images\}|} \tag{25}$$

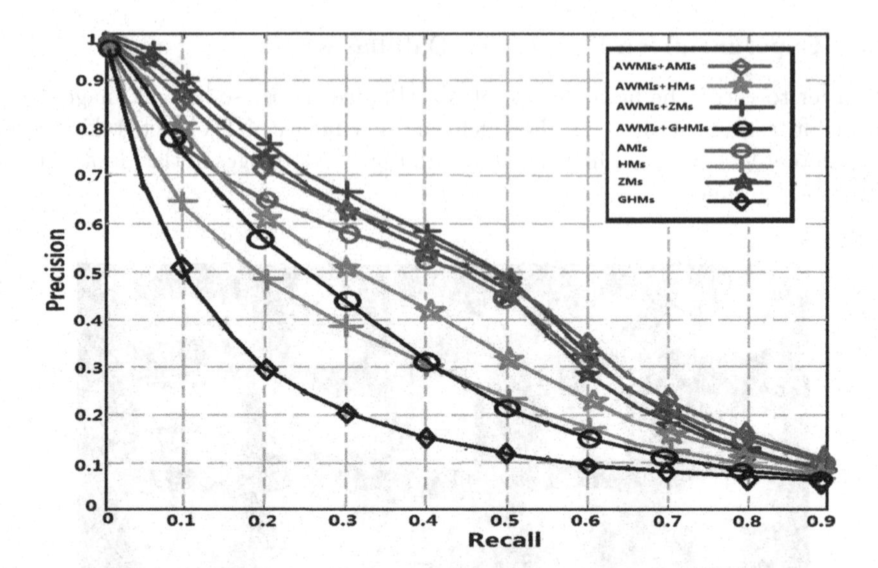

Fig. 6. Precision-Recall curves of AWMIs+AMIs, AWMIs+HMs, AWMIs+ZMs, AWMIs+GHMs, AMIs, HMs, ZMs and GHMs.

Then, image classification based on ALOI (Amsterdam Library of Object Images) [18] is carried out. Similar to COIL-20, ALOI contains images of 1000 objects taking from 72 different viewpoints. We choose 1000 images (size of 192 × 144) of 100 objects. Each object contains 10 images, which are shown in the Fig. 7.

20% images are selected randomly to be the training data and the rest 800 images make up the testing data. Also, we use the Nearest Neighbor classifier based on the Chi-Square distance to estimate the categories of the test images. The classification accuracy of ALOI is shown in Fig. 8.

Fig. 7. One object contains 12 images, which are taken from different viewpoints.

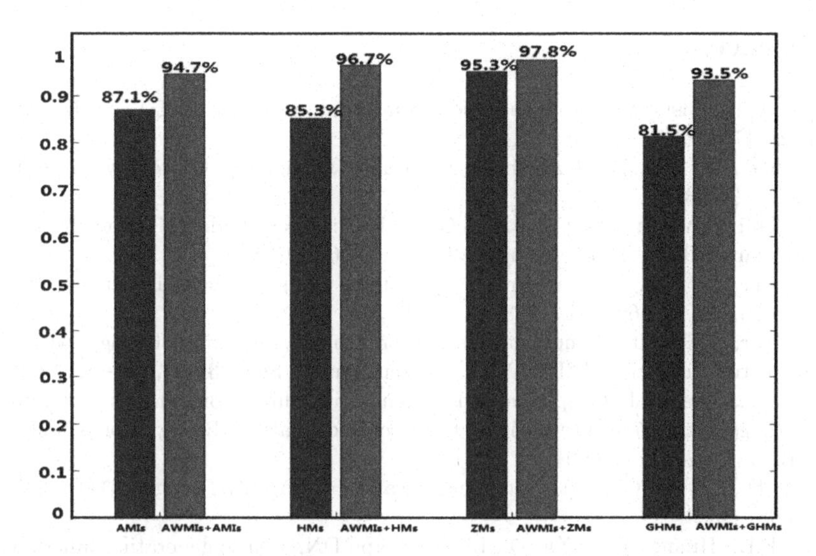

Fig. 8. The classification accuracy of ALOI, obtained by using AWMIs+AMIs, AWMIs+HMs, AWMIs+ZMs, AWMIs+GHMs, AMIs, HMs, ZMs and GHMs.

By using AWMIs, the number of low-order invariants is increased. According to Figs. 6 and 8, we can find the accuracy of image retrieval and classification is improved by combining AWMIs with traditional moment invariants as feature vectors. Thus, it makes sense of constructing AWMIs.

5 Conclusion

The contributions of this paper mainly include three aspects. Firstly, we extend the definition of moments, which is named as DMs. In theory, we can construct DMs by using arbitrary order partial derivatives. Secondly, by using local differential invariants and the structural framework of global integral invariants, we construct AWMIs. This approach greatly expands the number of low-order affine moments invariants. Meanwhile, it's important to note that there are many different ways to construct AWMIs. Thirdly, the final experimental results show that AWMIs have good stability and distinguishability. We find that the accuracy of image retrieval and classification can be improved by combining AWMIs with other traditional moment invariants.

In the future, we will design more structural formulas to expand the number of AWMIs. At the same time, it's also important to explore some methods of improving the accuracy of differential calculation, so that high-order differential moments can be used. Also, we want use AWMIs to construct new local descriptor of image, so that this kind of local feature has good invariance for geometric transformations.

Acknowledgments. This work has been funded by National Natural Science Foundation of China (Grant Nos. 60873164, 61227802 and 61379082).

References

1. Hu, M.K.: Visual pattern recognition by moment invariants. IRE Trans. Inf. Theory **8**(2), 179–187 (1962)
2. Hilbert, D.: Theory of Algebraic Invariants. Cambridge University Press, Cambridge (1993)
3. Dudani, S.A., Breeding, K.J., McGhee, R.B.: Aircraft identification by moment invariants. IEEE Trans. Comput. **26**(1), 39–46 (1977)
4. Flusser, J., Suk, T.: Pattern recognition by affine moment invariants. Pattern Recogn. **26**(1), 167–174 (1993)
5. Flusser, J., Suk, T.: A moment-based approach to registration of images with affine geometric distortion. IEEE Trans. Geosci. Remote Sens. **32**(2), 382–387 (1994)
6. Suk, T., Flusser, J.: Graph method for generating affine moment invariants. In Proceedings of the 17th International Conference on Pattern Recognition, Cambridge, UK, pp. 192–195 (2004)
7. Xu, D., Li, H.: Geometric moment invariants. Pattern Recogn. **41**(1), 240–249 (2008)
8. Li, E.B., Huang, Y.Z., Xu, D., Li, H.: Shape DNA: basic generating functions for geometric moment invariants (2017). https://arxiv.org/abs/1703.02242
9. Olver, P.J.: Equivalence, Invariants, and Symmetry. Cambridge University Press, Cambridge (1995)
10. Olver, P.J., Sapiro, G., Tannenbaum, A.: Affine invariants detection: edge maps, anisotropic diffusion, and active contours. Acta Applicandae Mathematicae **59**(1), 45–77 (1999)
11. Ge, J., Cao, W., Zhou, W., Gong, M., Liu, L., Li, H.: A local feature descriptor under color affine transformation. J. Comput.-Aided Des. Comput. Graph. **25**(1), 26–33 (2013). (in Chinese)
12. Wang, Y., Wang, X., Zhang, B.: Affine differential invariants of functions on the plane. J. Appl. Math. **2013**, 211–220 (2013)
13. Li, E.B., Li, H.: Isomorphism between differential and moment invariants under affine transform (2017). https://arxiv.org/abs/1705.08264
14. Schmid, C., Mohr, R.: Local grayvalue invariants for image retrieval. IEEE Trans. Pattern Anal. Mach. Intell. **19**(5), 530–535 (1997)
15. Nene, S.A., Nayar, S.K., Murase, H.: Columbia object image library (COIL-20). Technical report CUCS-005-96, February 1996
16. Khotanzad, A., Hong, Y.H.: Invariant image recognition by Zernike moments. IEEE Trans. Pattern Anal. Mach. Intell. **12**(5), 489–497 (2002)
17. Yong, B., Li, G.X., Zhang, H.L., Dai, M.: Rotation and translation invariants of Gaussian-Hermite moments. Pattern Recogn. Lett. **32**(9), 1283–1298 (2011)
18. Geusebroek, J.M., Burghouts, G.J., Smeulders, A.W.: The Amsterdam library of object images. Int. J. Comput. Vis. **61**(1), 103–112 (2005)

Statistical Methods and Learning

A Weighted Locally Linear KNN Model for Image Recognition

Yu-Lan Xu, Sibao Chen(✉), and Bin Luo

Key Lab of Intelligent Computing and Signal Processing of Ministry of Education, School of Computer Science and Technology, Anhui University, Hefei 230601, China
sbchen@ahu.edu.cn

Abstract. In this paper, we propose a weighted extension of Locally Linear KNN model (WLLKNN) model for image recognition. An iterative algorithm, with proof of convergence, is proposed for the weighted sparse minimization of the proposed model. Five different types of weighting forms in the WLLKNN model are investigated. The proposed WLLKNN model is evaluated on different tasks including action recognition, scenes recognition and face recognition. Experiments on several representative databases show the superiority of the proposed WLLKNN based classifier (WLLKNNC) and weighted Locally Linear Nearest Mean Classifier (WLLNMC).

Keywords: Sparse minimization · Weighting · Locally linear KNN Image recognition

1 Introduction

Image recognition, due to its wide applications, has received wide attention and research interests. During the past few decades, many methods of image recognition have been proposed in the literatures. The most successful methods are the subspace methods [9] and sparse representation (SR) methods [20].

Among the existing SR methods, SR-based Classification (SRC) [20] was first proposed for robust face recognition problems. Based on SR, some extension methods were proposed, such as local SR (LSR) [11], weighted SR (WSR) [13] and sparse subspace learning [6] for robust face recognition [4], partial occluded face recognition [15] and feature selection [5]. Then some discriminative dictionary learning methods have been proposed for sparse representation: Zhang and Li [23] proposed an objective function and applied a discriminative singular value decomposition (D-KSVD) method to learn an over-complete dictionary. Wang et al. [19] presented a Locality-constrained Linear Coding (LLC) which utilizes the locality constraints. Yang et al. [22] proposed the Fisher Discrimination Dictionary Learning (FDDL) method to learn a structured dictionary by exploits the discriminative information.

S.-B. Chen—This work was supported in part by National Natural Science Foundation of China under Grant 61472002, 61572030 and 61671018, and Collegiate Natural Science Fund of Anhui Province under Grant KJ2017A014.

J. Yang et al. (Eds.): CCCV 2017, Part III, CCIS 773, pp. 567–578, 2017.
https://doi.org/10.1007/978-981-10-7305-2_48

Since dictionary learning is very time consuming, Liu and Liu [12] recently proposed the locally linear KNN (LLKNN) model, which directly adopts training samples as dictionary atoms to circumvent dictionary learning. It is effectively a variant of local SRC [11] by adding a proximity term and using a coefficients trimming trick. It learns a sparse coefficients for each test sample by considering the construction, locality, and sparsity simultaneously. The representation coefficients of LLKNN model contain the grouping property of the nearest neighbors. Experiments showed that the LLKNN model has better performance.

In order to make the representation coefficients of LLKNN more sparse and improve the classification performance, we propose a more general framework of Locally Linear KNN model by weighting extension (WLLKNN) for image recognition. A new algorithm procedure of weighted sparse minimization is proposed for WLLKNN with proof of convergence. Five different types of weighting forms are investigated. Experimental results on action, scenes and face recognition show the proposed method outperforms some representative methods.

2 Weighted Locally Linear KNN Model

2.1 Formulation

Let $\{\mathbf{x}_1, \mathbf{x}_2, \cdots, \mathbf{x}_n\}$ be a training data set containing c classes, with $\mathbf{x}_i \in \mathbf{R}^m$ being an m-dimensional training sample vector belonging to one class. Let $\mathbf{y} \in \mathbf{R}^m$ be an m-dimensional test sample vector, which needs to be classified.

Theoretically, the proposed Weighted locally linear KNN (WLLKNN) model can be formulated as

$$\min_{\mathbf{v} \in \mathbb{R}^n} \|\mathbf{y} - \mathbf{B}\mathbf{v}\|^2 + \lambda \|\mathbf{W}\mathbf{v}\|_1 + \alpha \|\mathbf{v} - \beta \mathbf{d}\|^2, \tag{1}$$

where $\mathbf{B} = [\mathbf{x}_1, \mathbf{x}_2, \cdots, \mathbf{x}_n] \in \mathbf{R}^{m \times n}$ is the training sample matrix, $\mathbf{v} \in \mathbf{R}^n$ is an n-dimensional representation coefficient vector needing to be solved. L_1-norm $\|\mathbf{v}\|_1$ of vector $\mathbf{v} = (v_1, v_2, \cdots, v_n)^\top$ is the absolute sum of each elements, $\|\mathbf{v}\|_1 = \sum_{i=1}^n |v_i|$. Weighting matrix $\mathbf{W} \in \mathbf{R}^{n \times n}$ is a diagonal matrix, with diagonal elements $\mathbf{W}_{ii}$ being some type of weighting on the representation coefficients. The similarity vector $\mathbf{d} = (d_1, d_2, \cdots, d_n)^\top \in \mathbf{R}^n$. $d_i = \exp\{-\frac{1}{2\sigma^2}\|\mathbf{y} - \mathbf{x}_i\|^2\}$ is a kind of similarity between the i-th training sample $\mathbf{x}_i$ and the test sample $\mathbf{y}$.

The first term of our model in (1) represents the reconstruction error of the model, the second term in (1) is used to keep the sparse property of representation coefficients, and the third term in (1) is adopted to strengthen the local property of the model. The tuning parameters $\lambda > 0$, $\alpha > 0$ and $\beta > 0$ are taken to balance each term of the model.

2.2 Optimization Algorithm

The model (1) can be rewritten as

$$\min_{\mathbf{v} \in \mathbb{R}^n} \mathbf{v}^\top(\mathbf{B}^\top\mathbf{B} + \alpha \mathbf{I}_n)\mathbf{v} - 2\mathbf{v}^\top(\mathbf{B}^\top\mathbf{y} + \alpha\beta\mathbf{d}) + \mathbf{y}^\top\mathbf{y} + \alpha\beta^2\mathbf{d}^\top\mathbf{d} + \lambda\|\mathbf{W}\mathbf{v}\|_1, \tag{2}$$

where $\mathbf{I}_n$ is identity matrix of order n. Then the model can be denoted as:

$$\min_{\mathbf{v}\in\mathbb{R}^n} J(\mathbf{v}) \hat{=} f(\mathbf{v}) + \lambda\|\mathbf{W}\mathbf{v}\|_1 = \mathbf{v}^\top\mathbf{C}\mathbf{v} - 2\mathbf{v}^\top\mathbf{b} + \lambda\|\mathbf{W}\mathbf{v}\|_1, \tag{3}$$

where $f(\mathbf{v}) \hat{=} \mathbf{v}^\top\mathbf{C}\mathbf{v} - 2\mathbf{v}^\top\mathbf{b}$, $\mathbf{C} \hat{=} \mathbf{B}^\top\mathbf{B} + \alpha\mathbf{I}_n$, $\mathbf{b} \hat{=} \mathbf{B}^\top y + \alpha\beta\mathbf{d}$. Constant term $\mathbf{y}^\top\mathbf{y} + \alpha\beta^2\mathbf{d}^\top\mathbf{d}$ has no effect on the solution of (2) and is omitted.

Since $f(\mathbf{v})$ in (3) is a positive definite quadratic form, which is strictly convex with respect to $\mathbf{v}$, and L_1-norm $\|\mathbf{W}\mathbf{v}\|_1$ is also convex, weighted sparse minimization of (3) is strictly convex with respect to $\mathbf{v}$. Therefore, there exists an unique global minimum for weighted sparse minimization of (3). To obtain the global minimum, we propose an iterative algorithm, which is similar to the sparse minimization for positive definite quadratic form [1].

The whole optimization procedure of the proposed model (3) is summarized in Algorithm 1. In each iteration step, diagonal matrix $\mathbf{D}$ is calculated based on current $\mathbf{v}$ as in formula (4), then update the coefficient vector $\mathbf{v}$ with formula (5). The iteration procedure of the algorithm is repeated till meeting stop criterion.

Algorithm 1. Optimization procedure of Weighted Locally Linear KNN Model

1: **Input:** Training sample matrix $\mathbf{B} \in \mathbb{R}^{m\times n}$, test sample $\mathbf{y} \in \mathbb{R}^m$, similarity vector $\mathbf{d} \in \mathbb{R}^n$, diagonal weighting matrix $\mathbf{W} \in \mathbb{R}^{n\times n}$, initial non-zero solution $\mathbf{v}^{(0)} \in \mathbb{R}^n$, tuning parameters λ, α and $\beta > 0$, residual bound $\epsilon > 0$ or maximum number of iteration t_{max};

2: $\mathbf{C} = (\mathbf{B}^\top\mathbf{B} + \alpha\mathbf{I}_n)$, $\mathbf{b} = \mathbf{B}^\top\mathbf{y} + \alpha\beta\mathbf{d}$, set $t = 0$;

3: Update diagonal matrix,

$$\mathbf{D}^{(t)} = diag(\sqrt{|v_1^{(t)}|}, \sqrt{|v_2^{(t)}|}, \ldots, \sqrt{|v_n^{(t)}|}); \tag{4}$$

4: Update

$$\mathbf{v}^{(t+1)} = \mathbf{D}^{(t)}[\mathbf{D}^{(t)}\mathbf{C}\mathbf{D}^{(t)} + \frac{\lambda}{2}\mathbf{W}]^{-1}\mathbf{D}^{(t)}\mathbf{b}; \tag{5}$$

5: If $t > t_{max}$ or $|J(\mathbf{v}^{(t+1)}) - J(\mathbf{v}^{(t)})|/J(\mathbf{v}^{(t)}) < \epsilon$, go to step 6, otherwise, let $t = t+1$ and go to step 3;

6: **Output:** The optimal solution $\mathbf{v}^* = \mathbf{v}^{(t+1)}$.

2.3 Justification

In this subsection, we will show that the objective function value $J(\mathbf{v})$ in (3) does decrease monotonously along with the iteration of Algorithm 1, which is summarized in Theorem 1. Before we give the proof of Theorem 1, two lemmas are given and proved firstly.

Lemma 1. *Let $\mathcal{M}_t = \{i|v_i^{(t)} \neq 0, i = 1, 2, \cdots, n\}$ be the index set of non-zero coefficients at the t-th iteration. Define an auxiliary function*

$$F(\mathbf{v}, \mathbf{v}^{(t)}) = f(\mathbf{v}) + \lambda \sum_{i\in\mathcal{M}_t} \mathbf{W}_{ii}\frac{v_i^2}{2|v_i^{(t)}|}. \tag{6}$$

The following inequality holds with all the coefficient sequence $\{\mathbf{v}^{(t)}, t = 0, 1, 2, \cdots\}$ *obtained by Algorithm 1,*

$$F(\mathbf{v}^{(t+1)}, \mathbf{v}^{(t)}) \leq F(\mathbf{v}^{(t)}, \mathbf{v}^{(t)}). \tag{7}$$

The equality holds only at convergence, $\mathbf{v}^{(t+1)} = \mathbf{v}^{(t)}$.

Proof. The two terms of the auxiliary function $F(\mathbf{v}, \mathbf{v}^{(t)})$ in (6) are both positive definite quadratic form. It is easy to obtain the unique global optimal solution of minimizing $F(\mathbf{v}, \mathbf{v}^{(t)})$ by taking the derivatives and letting them equal to zero.

At the t-th iteration, $v_i^{(t)} = 0$ for subscript index $i \in \{1, \cdots, n\} - \mathcal{M}_t$, By the updating formulae (4) and (5) in Algorithm 1, we can get that $v_i^{(t+1)} = 0$, $i \in \{1, \cdots, n\} - \mathcal{M}_t$.

By using the notation $\mathbf{D}^{(t)}$ in (4), the auxiliary function $F(\mathbf{v}, \mathbf{v}^{(t)})$ can be rewritten as

$$F(\mathbf{v}, \mathbf{v}^{(t)}) = f(\mathbf{v}) + \frac{\lambda}{2} \sum_{i \in \mathcal{M}_t} \mathbf{W}_{ii} (\mathbf{D}_{ii}^{(t)})^{-2} v_i^2. \tag{8}$$

In order to obtain the global minimum of (8), we take its derivative with respect to v_i, $i \in \mathcal{M}_t$,

$$\frac{\partial F(\mathbf{v}, \mathbf{v}^{(t)})}{\partial v_i} = (2\mathbf{C}\mathbf{v} - 2\mathbf{b})_i + \frac{\lambda}{2} 2\mathbf{W}_{ii} (\mathbf{D}_{ii}^{(t)})^{-2} v_i, i \in \mathcal{M}_t. \tag{9}$$

By setting $\frac{\partial F(\mathbf{v}, \mathbf{v}^{(t)})}{\partial v_i} = 0, i \in \mathcal{M}_t$, and let $v_i = 0, i \in \{1, \cdots, n\} - \mathcal{M}_t$, we obtain the optimal solution of $F(\mathbf{v}, \mathbf{v}^{(t)})$,

$$\begin{aligned} \mathbf{v}^* &= \left[\mathbf{C} + \frac{\lambda}{2} (\mathbf{D}^{(t)})^{-2} \mathbf{W}_{ii} \right]^{-1} \mathbf{b} \\ &= \mathbf{D}^{(t)} \left[\mathbf{D}^{(t)} \mathbf{C} \mathbf{D}^{(t)} + \frac{\lambda}{2} \mathbf{W}_{ii} \right]^{-1} \mathbf{D}^{(t)} \mathbf{b}. \end{aligned} \tag{10}$$

We know that $\mathbf{v}^*$ in (10) is the global minimum of $F(\mathbf{v}, \mathbf{v}^{(t)})$. Thus $F(\mathbf{v}^*, \mathbf{v}^{(t)}) \leq F(\mathbf{v}, \mathbf{v}^{(t)})$ for any $\mathbf{v}$ satisfying $v_i = 0$, $i \in \{1, \cdots, n\} - \mathcal{M}_t$. Especially, $F(\mathbf{v}^*, \mathbf{v}^{(t)}) \leq F(\mathbf{v}^{(t)}, \mathbf{v}^{(t)})$. From formula (5), we know that $\mathbf{v}^{(t+1)} = \mathbf{v}^*$. Therefore, $F(\mathbf{v}^{(t+1)}, \mathbf{v}^{(t)}) \leq F(\mathbf{v}^{(t)}, \mathbf{v}^{(t)})$.

Furthermore, since the global minimum of $F(\mathbf{v}, \mathbf{v}^{(t)})$ is unique, therefore, $F(\mathbf{v}^{(t+1)}, \mathbf{v}^{(t)}) \leq F(\mathbf{v}^{(t)}, \mathbf{v}^{(t)})$. The equality $F(\mathbf{v}^{(t+1)}, \mathbf{v}^{(t)}) = F(\mathbf{v}^{(t)}, \mathbf{v}^{(t)})$ holds only at convergence $\mathbf{v}^{(t+1)} = \mathbf{v}^{(t)}$.

Lemma 2. *By alternately computing formulae (4) and (5) in Algorithm 1, the solution path* $\mathbf{v}^{(t)}$ *has the following property*

$$J(\mathbf{v}^{(t+1)}) - J(\mathbf{v}^{(t)}) \leq F(\mathbf{v}^{(t+1)}, \mathbf{v}^{(t)}) - F(\mathbf{v}^{(t)}, \mathbf{v}^{(t)}). \tag{11}$$

The equality holds only at convergence $\mathbf{v}^{(t+1)} = \mathbf{v}^{(t)}$.

Proof. Set $\kappa = (J(\mathbf{v}^{(t+1)}) - J(\mathbf{v}^{(t)})) - (F(\mathbf{v}^{(t+1)}, \mathbf{v}^{(t)}) - F(\mathbf{v}^{(t)}, \mathbf{v}^{(t)}))$. Note that $v_i^{(t+1)} = v_i^{(t)} = 0, \forall i \in \{1, \cdots, n\} - \mathcal{M}_t$.

$$\begin{aligned}
\kappa &= \lambda(\|\mathbf{W}\mathbf{v}^{(t+1)}\|_1 - \|\mathbf{W}\mathbf{v}^{(t)}\|_1) - \lambda\left(\sum_{i\in\mathcal{M}_t} \mathbf{W}_{ii}\frac{(v_i^{(t+1)})^2}{2|v_i^{(t)}|} - \sum_{i\in\mathcal{M}_t} \mathbf{W}_{ii}\frac{(v_i^{(t)})^2}{2|v_i^{(t)}|}\right) \\
&= -\frac{\lambda}{2}\sum_{i\in\mathcal{M}_t} \mathbf{W}_{ii}\frac{1}{|v_i^{(t)}|}\{-2|v_i^{(t+1)}||v_i^{(t)}| + 2|v_i^{(t)}|^2 + (v_i^{(t+1)})^2 - (v_i^{(t)})^2\} \\
&= -\frac{\lambda}{2}\sum_{i\in\mathcal{M}_t} \mathbf{W}_{ii}\frac{1}{|v_i^{(t)}|}\left(|v_i^{(t+1)}| - |v_i^{(t)}|\right)^2 \\
&\leq 0.
\end{aligned} \tag{12}$$

Therefore, $J(\mathbf{v}^{(t+1)}) - J(\mathbf{v}^{(t)}) \leq F(\mathbf{v}^{(t+1)}, \mathbf{v}^{(t)}) - F(\mathbf{v}^{(t)}, \mathbf{v}^{(t)})$. The equality $J(\mathbf{v}^{(t+1)}) - J(\mathbf{v}^{(t)}) = F(\mathbf{v}^{(t+1)}, \mathbf{v}^{(t)}) - F(\mathbf{v}^{(t)}, \mathbf{v}^{(t)})$ holds only at convergence $\mathbf{v}^{(t+1)} = \mathbf{v}^{(t)}$.

Theorem 1. *The objective function value $J(\mathbf{w})$ in (3) is decreasing monotonously, that is, at each iteration, we have $J(\mathbf{w}^{(t+1)}) \leq J(\mathbf{w}^{(t)})$. The equality holds only at convergence.*

Proof. From Lemmas 1 and 2, we know that

$$J(\mathbf{v}^{(t+1)}) - J(\mathbf{v}^{(t)}) \leq F(\mathbf{v}^{(t+1)}, \mathbf{v}^{(t)}) - F(\mathbf{v}^{(t)}, \mathbf{v}^{(t)}) \leq 0, \tag{13}$$

The equalities hold only at convergence $\mathbf{v}^{(t+1)} = \mathbf{v}^{(t)}$, that is,

$$J(\mathbf{v}^{(t+1)}) \leq J(\mathbf{v}^{(t)}). \tag{14}$$

The equality holds only at convergence. This completes the proof of Theorem 1.

2.4 Classification

In this paper, we use Weighted Locally Linear KNN based classifier (WLLKNNC) to classify the test sample $\mathbf{y}$, which is similar to LLKNNC in [12]. The classification rule is defined as follows:

$$i^* = \arg\max_i \sum_{(\mathbf{x}_j\in\mathbf{B}_i)\wedge(\mathbf{v}_j\in T_i(k))} \mathbf{v}_j, \tag{15}$$

where $i = 1, 2, \cdots, c$ is the class label, $\mathbf{B}_i$ is the set of training samples in the i-th class, $\mathbf{x}_j$ is the j-th training sample and $\mathbf{v}_j$ is its corresponding representation coefficient obtained by WLLKNN, $T_i(k)$ is the set of top k largest values of $\mathbf{v}_j$ for class i.

The effectiveness of the proposed WLLKNNC has the same property of LLKNNC in Theorem 3.2 of [12], which shows that the proposed WLLKNNC is related to Bayesian classifier in the view of kernel density.

We also propose weighted Locally Linear Nearest Mean Classifier (WLL-NMC) like the variant Locally Linear Nearest Mean Classifier (LLNMC) [12] based on class-specific reconstruction error minimization:

$$i^* = \arg\min_i \|\mathbf{y} - \sum_{(\mathbf{x}_j \in \mathbf{B}_i) \wedge (\mathbf{v}_j \in T_i(k))} \mathbf{v}_j \mathbf{x}_j\|_2^2. \tag{16}$$

3 Algorithm Analysis

In this section, we give the comprehensive analysis of the proposed Weighted LLKNN (WLLKNN) model. Especially, we evaluate our method on the following several critical issues: (1) the influence of different weighting forms, (2) the effectiveness of shifted power transformation (SPT) and the sensitiveness to the parameter λ_1 and λ_2 in SPT, and (3) the sensitiveness of tuning parameters λ, α, β and σ.

3.1 Effect of Different Weighting Forms

In this paper, we adopt five different types of weighting methods for the proposed Weighted LLKNN (WLLKNN) model. Different representation coefficients weighting forms may affect the effectiveness of iterative Algorithm 1. Five coefficient weighting forms are tested, whose diagonal elements are defined as the following:

(a) W_1: $\mathbf{W}_{ii} = a\|\mathbf{y} - \mathbf{x}_i\|_2^2/\Delta$, and let $\Delta = \max\{\|\mathbf{y} - \mathbf{x}_i\|_2^2, i = 1, 2, \cdots, n\}$.
(b) W_2: $\mathbf{W}_{ii} = \exp\{-a/(\|\mathbf{y} - \mathbf{x}_i\|_2^2/\Delta + \epsilon)\}$, and let $\Delta = \max\{\|\mathbf{y} - \mathbf{x}_i\|_2^2, i = 1, 2, \cdots, n\}$, where a small constant $\epsilon > 0$ is added to avoid dividing zero.
(c) W_3: $\mathbf{W}_{ii} = \frac{a}{1+|\mathbf{y}^\top \mathbf{x}_i|/\Delta}$, and let $\Delta = \max\{|\mathbf{y}^\top \mathbf{x}_i|, i = 1, 2, \cdots, n\}$.
(d) W_4: $\mathbf{W}_{ii} = \frac{a}{1+|\rho_i|}$, and let $\rho_i = \frac{\mathbf{y}^\top \mathbf{x}_i}{\|\mathbf{y}\|_2 \|\mathbf{x}_i\|_2}$.
(e) W_5: $\mathbf{W}_{ii} = \exp\{-|\rho_i^2|/a\}$, and let $\rho_i = \frac{\mathbf{y}^\top \mathbf{x}_i}{\|\mathbf{y}\|_2 \|\mathbf{x}_i\|_2}$.

In order to test the impact of different weighting forms, we test the proposed WLLKNNC and WLLNMC methods with the above five different weighting forms. Figure 1 shows the effect of five different weighting forms on the proposed WLLKNNC and WLLNMC methods under different weight parameter a on AR face database [14].

From the figure, we can see that the second type of weighting form (b) W_2 acquires higher classification accuracy and it is more stable with different weight parameter a setting. To simplify the following experiments, we just choose the second type of weighting form (b) W_2 in the following experiments.

3.2 Effect of Shifted Power Transformation

To test the effect of shifted power transformation (SPT) [12], we first test the sensitiveness of parameters λ_1 and λ_2 in SPT of features on the proposed WLLKNNC and WLLNMC methods, $\mathbf{y}' = |\mathbf{y} + \lambda_1 \mathbf{e}|^{\lambda_2} sign(\mathbf{y} + \lambda_1 \mathbf{e})$, $\lambda_1, \lambda_2 > 0$.

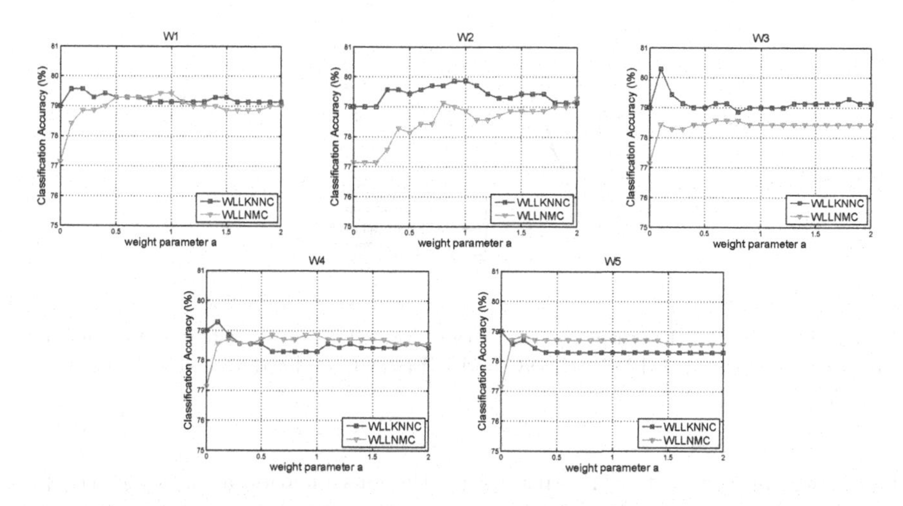

Fig. 1. The effect of different weighting methods along with parameter a on classification accuracy of the proposed WLLKNNC and WLLNMC on AR face database

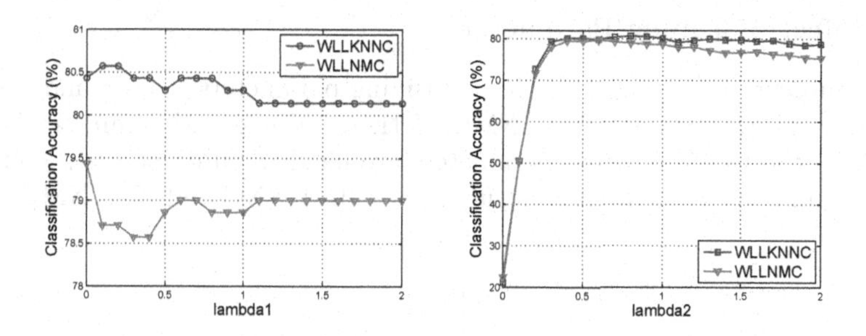

Fig. 2. The effect of different parameters λ_1 and λ_2 in SPT on the recognition accuracy of the proposed WLLKNNC and WLLNMC on AR face database

We evaluate the performance of our methods by setting different parameters λ_1 and λ_2. The values of all the others parameters are fixed. Figure 2 shows the variation of classification performance of WLLKNNC and WLLNMC under different parameter λ_1 and λ_2 on AR face database. It can be seen that the recognition accuracy of the proposed method are not changing significantly when the parameters λ_1 and λ_2 reach a certain range. We will choose one of the most suitable values of parameters λ_1 and λ_2 in the following experiments.

Then the effectiveness of SPT was evaluated on the proposed WLLKNNC and WLLNMC methods on AR face database. We compare the recognition accuracy of "WLLKNNC" (already with SPT), "WLLNMC" (already with SPT), "WLLKNNC without SPT" and "WLLNMC without SPT" when different value of KNN parameter k is set. Figure 3 shows the effect of shifted power transformation (SPT) on the classification accuracy variation of the proposed WLLKNNC and WLLNMC under different KNN parameter k on AR face database. From the

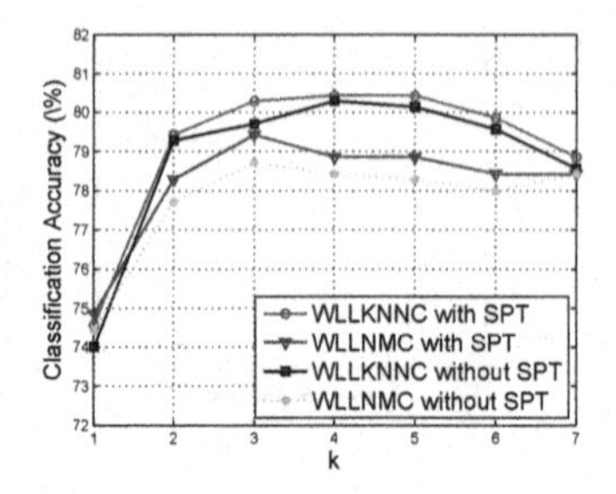

Fig. 3. The effect of shifted power transformation (SPT) on the classification accuracy variation of the proposed WLLKNNC and WLLNMC under different KNN parameter k on AR face database

figure, we can see that SPT can improve the classification accuracy of the proposed WLLKNNC and WLLNMC consistently with different KNN parameter k. Therefore, we will use the SPT on our methods in the following experiments.

3.3 Effect of Tuning Parameters

In this subsection, we test the effect of tuning parameters λ, α, β and σ on the proposed WLLKNNC and WLLNMC methods. When one parameter is being tested, all the values of other parameters are fixed. Figure 4 shows the effect of λ, α, β and σ on classification accuracy of WLLKNNC and WLLNMC on AR

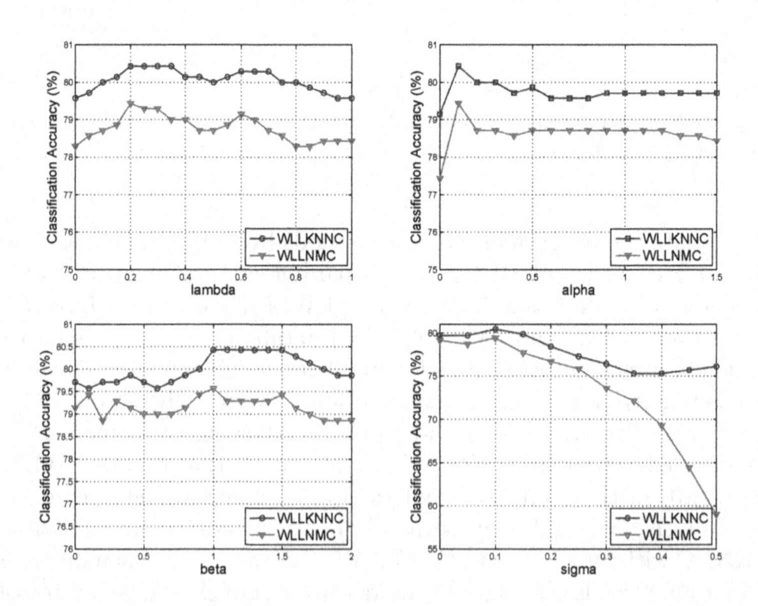

Fig. 4. The effect of different tuning parameters λ, α, β and σ on classification accuracy of the proposed WLLKNNC and WLLNMC on AR face database

face database. From all the four sub figures we can see that the tuning parameters have some effect on the proposed methods. However, it is easy to choose proper tuning parameters for better classification performance. For default setting, we could just set $\lambda = 0.2$, $\alpha = 0.1$, $\beta = 1.5$ and $\sigma = 0.1$. In the following experiments, we will list all the tuning parameters for each specific experiments.

3.4 Complexity of Algorithm 1

Given a test sample $\mathbf{y} \in \mathbb{R}^m$ and training sample matrix $\mathbf{B} \in \mathbb{R}^{m \times n}$, the computational complexity of the proposed Algorithm 1 is mainly dominated by the computation of formula (5) in loop. Note that diagonal matrix multiplication can be converted to element-wise multiplication. Therefore, the computation of formula (5) is dominated by matrix inverse operator, which is order $\mathrm{O}(n^3)$ for matrix of size $(n \times n)$. That is, we need $\mathrm{O}(n^3)$ to compute formula (5) in each loop of Algorithm 1. Let t be the number of iteration of Algorithm 1. Then the total computational complexity of Algorithm 1 of the proposed WLLKNN is $\mathrm{O}(tn^3)$.

4 Experiments

In this section, we evaluate the performance of the proposed WLLKNN model on the following recognition (classification) tasks, action recognition on UCF50 database [17]; scene recognition on 15 scenes database [10]; and face recognition on AR face database [14] and FERET database [16].

4.1 Action Recognition

We apply WLLKNN to action recognition on UCF50 database [17] in this subsection. The UCF50 database contains 6,676 videos from 50 action categories. The experimental setting in [18] is used, where it divided the database into five groups. In the following experiments, we randomly select one group for training and the remaining are used for testing. The action bank feature with dimension 14, 965 is provided by [18] and it's further reduced to 500 by PCA [9]. The parameters of WLLKNN are selected as $a = 0.1$, $\alpha = 0.01$, $\beta = 1.5$, $\lambda = 0.02$, and $\sigma = 1.5$. For the shift power transformation (SPT) [12], $\lambda_1 = 0.0$ and $\lambda_2 = 0.8$. For both WLLKNNC and WLLNMC, the value of KNN parameter $k = 20$.

Table 1 lists the recognition performance of the proposed WLLKNNC and WLLNMC on UCF50 database comparing with some popular methods SRC [20], LC-KSVD [8], FDDL [22], LLNMC and LLKNNC [12]. From the table we can see that the proposed methods are able to achieve better results than other representative methods.

4.2 Scene Recognition

We evaluate the proposed WLLKNN model to scene recognition problem in this subsection by using the 15 scenes database [10]. It contains 4, 485 images from

Table 1. Recognition performance of the proposed WLLKNNC and WLLNMC on UCF50 database comparing with some popular methods SRC [20], LC-KSVD [8], FDDL [22], LLNMC and LLKNNC [12].

Methods	SRC	LC-KSVD	FDDL	LLNMC	LLKNNC	WLLNMC	WLLKNNC
Accuracy (%)	30.93	42.5	46	49.44	57.4	**51.15**	**58.58**

Table 2. Recognition accuracy (%) of the proposed WLLKNNC and WLLNMC on the 15 scenes database comparing with some other popular methods.

Methods	KSPM [10]	ScSPM [21]	D-KSVD [23]	LC-KSVD [8]	LaplacianSC [2]	KC [3]
Accuracy	81.4 ± 0.5	80.3 ± 0.9	89.1	90.4	89.7	76.7 ± 1
Methods	TSR [7]	LLNMC [12]	LLKNNC [12]	WLLNMC	WLLKNNC	
Accuracy	87.1	97.5 ± 0.3	93.5 ± 0.5	$\mathbf{98.54} \pm 0.12$	$\mathbf{99.25} \pm 0.26$	

different scene categories. The experimental settings of the 15 scenes database can be found in [12]. First, for each class, 100 images are randomly selected for training and the remaining images are used for testing. Then we reduce the dimension of images to 500 by PCA [9]. The parameters of WLLKNN are selected as $a = 1.8$, $\alpha = 0.1$, $\beta = 1.5$, $\lambda = 0.2$, and $\sigma = 0.1$. For the shifted power transformation (SPT) [12], $\lambda_1 = 0.0$ and $\lambda_2 = 0.6$. For both WLLKNNC and WLLNMC, the value of KNN parameter $k = 26$.

In order to evaluate the effectiveness of the proposed WLLKNN, we compare the proposed methods with other popular methods, such as, KSPM [10], ScSPM [21], D-KSVD [23], LC-KSVD [8], LaplacianSC [2], KC [3], TSR [7], LLNMC and LLKNNC [12]. Table 2 lists the experimental results of the proposed WLLKNN and other comparative methods. From the table it can be seen that both our proposed methods WLLKNNC and WLLNMC can obtain better results comparing with other competing methods.

4.3 Face Recognition

In this subsection, we evaluate our algorithm to face recognition using AR face database [14] and FERET face database [16]. In order to evaluate the effectiveness of the proposed WLLKNN model, we compare the proposed methods with SRC [20], FDDL [22], LC-KSVD [8], TSR [7], LLKNNC and LLNMC [12].

The AR database [14] contains 3,120 frontal images from 120 individuals. For each individual, 26 pictures are taken from two separated sessions. We choose a subset consisting of 50 male subjects and 50 female subjects with dimension 50×40. For each subject, we select the images with variation of facial expression and illumination. The experiments settings in [12] is used: the seven images per subject from Session 1 are used for training and the other seven per subject from Session 2 are used for testing. Then the feature dimension is reduced to 180 by PCA [9]. For the model parameters, we set $a = 0.05$, $\alpha = 0.1$, $\beta = 1.5$, $\lambda = 0.2$, and $\sigma = 0.1$. For the shifted power transformation (SPT) [12], $\lambda_1 = 0.0$ and

$\lambda_2 = 0.85$. For WLLKNNC, the value of KNN parameter $k = 5$. For WLLNMC, the value of KNN parameter $k = 3$.

The left side of Table 3 shows the recognition performance of different methods on AR face database. From the table it can be seen that both the proposed WLLKNNC and WLLNMC outperforms the other competing methods.

Table 3. Recognition accuracy (%) of the proposed WLLKNNC and WLLNMC on AR and FERET face databases

Methods	AR face database	FERET face database
SRC [20]	75.86	73.07
LC-KSVD [8]	74.7	52.64
FDDL [22]	80.14	77.85
TSR [7]	70.43	63.79
LLNMC [12]	78.29	75.5
LLKNNC [12]	78.86	66.36
The proposed WLLNMC	**79.43**	**81.79**
The proposed WLLKNNC	**80.43**	**82.14**

The FERET database [16] has 14,126 images of 1,564 sets which consists of 1,199 individuals. We choose a subset includes 200 distinct individuals which each individual has 7 different images. 7-fold cross validation is used to evaluate the proposed methods. In each fold, we select one image per individual for testing and the remaining for training. Then the feature dimension is reduced to 120 by PCA [9]. For the shifted power transformation (SPT) [12], $\lambda_1 = 0.0$ and $\lambda_2 = 0.5$. The parameters of WLLKNN are selected as $a = 0.45$, $\alpha = 0.1$, $\beta = 1.5$, $\lambda = 0.55$, and $\sigma = 0.1$. For WLLKNNC, the value of KNN parameter $k = 4$. For WLLNMC, the value of KNN parameter $k = 6$.

The right side of Table 3 shows the recognition results of the proposed WLLKNNC, WLLNMC and other competing methods on FERET face database. From the table it can be seen that the proposed methods are able to obtain better results than other representative methods.

5 Conclusion

In this paper, we propose a Weighted Locally Linear KNN model (WLLKNN) model for image recognition. An iterative algorithm for weighted sparse minimization is proposed with proof of convergence. The effectiveness of the proposed WLLKNN model is evaluated on different tasks including action recognition, scenes recognition and face recognition. The superiority of the proposed WLLKNNC and WLLNMC is shown through the experimental results on several representative databases comparing with some representative methods.

References

1. Chen, S.B., Ding, C.H., Luo, B.: An algorithm framework of sparse minimization for positive definite quadratic forms. Neurocomputing **151**, 223–230 (2015)
2. Gao, S., Tsang, I.W.H., Chia, L.T.: Laplacian sparse coding, hypergraph laplacian sparse coding, and applications. IEEE Trans. PAMI **35**(1), 92–104 (2013)
3. van Gemert, J.C., Geusebroek, J.-M., Veenman, C.J., Smeulders, A.W.M.: Kernel codebooks for scene categorization. In: Forsyth, D., Torr, P., Zisserman, A. (eds.) ECCV 2008. LNCS, vol. 5304, pp. 696–709. Springer, Heidelberg (2008). https://doi.org/10.1007/978-3-540-88690-7_52
4. Gui, J., Liu, T., Tao, D., Sun, Z., Tan, T.: Representative vector machines: a unified framework for classical classifiers. IEEE Trans. Cybern. **46**(8), 1877–1888 (2016)
5. Gui, J., Sun, Z., Ji, S., Tao, D., Tan, T.: Feature selection based on structured sparsity: a comprehensive study. IEEE Trans. NNLS **28**(7), 1490–1507 (2017)
6. Gui, J., Sun, Z., Jia, W., Hu, R., Lei, Y., Ji, S.: Discriminant sparse neighborhood preserving embedding for face recognition. Pattern Recognit. **45**(8), 2884–2893 (2012)
7. He, R., Zheng, W.S., Hu, B.G., Kong, X.W.: Two-stage nonnegative sparse representation for large-scale face recognition. IEEE TNNLS **24**(1), 35–46 (2013)
8. Jiang, Z., Lin, Z., Davis, L.S.: Label consistent K-SVD: learning a discriminative dictionary for recognition. IEEE Trans. PAMI **35**(11), 2651–2664 (2013)
9. Jolliffe, I.: Principal Component Analysis. Springer, Heidelberg (1986). https://doi.org/10.1007/b98835
10. Lazebnik, S., Schmid, C., Ponce, J.: Beyond bags of features: spatial pyramid matching for recognizing natural scene categories. In: CVPR, pp. 2169–2178 (2006)
11. Li, C., Guo, J., Zhang, H.: Local sparse representation based classification. In: 20th International Conference on Pattern Recognition (ICPR), pp. 649–652 (2010)
12. Liu, Q., Liu, C.: A novel locally linear KNN model for visual recognition. In: CVPR, pp. 1329–1337 (2015)
13. Lu, C.Y., Min, H., Gui, J., Zhu, L., Lei, Y.K.: Face recognition via weighted sparse representation. J. Vis. Commun. Image Represent. **24**(2), 111–116 (2013)
14. Martínez, A.M., Kak, A.C.: PCA versus LDA. IEEE TPAMI **23**(2), 228–233 (2001)
15. Mi, J.X., Lei, D., Gui, J.: A novel method for recognizing face with partial occlusion via sparse representation. Optik - Int. J. Light Electron Opt. **124**(24), 6786–6789 (2013)
16. Phillips, P.J., Wechsler, H., et al.: The FERET database and evaluation procedure for face-recognition algorithms. Image Vis. Comput. **16**(5), 295–306 (1998)
17. Reddy, K.K., Shah, M.: Recognizing 50 human action categories of web videos. Mach. Vis. Appl. **24**(5), 971–981 (2013)
18. Sadanand, S., Corso, J.J.: Action bank: a high-level representation of activity in video. In: CVPR, pp. 1234–1241. IEEE (2012)
19. Wang, J., Yang, J., Yu, K., Lv, F., Huang, T., Gong, Y.: Locality-constrained linear coding for image classification. In: CVPR, pp. 3360–3367. IEEE (2010)
20. Wright, J., Yang, A.Y., Ganesh, A., Sastry, S.S., Ma, Y.: Robust face recognition via sparse representation. IEEE Trans. PAMI **31**(2), 210–227 (2009)
21. Yang, J., Yu, K., Gong, Y., Huang, T.S.: Linear spatial pyramid matching using sparse coding for image classification. In: CVPR, pp. 1794–1801 (2009)
22. Yang, M., Zhang, L., Feng, X., Zhang, D.: Sparse representation based Fisher discrimination dictionary learning for image classification. Int. J. Comput. Vis. **109**(3), 209–232 (2014)
23. Zhang, Q., Li, B.: Discriminative K-SVD for dictionary learning in face recognition. In: CVPR, pp. 2691–2698. IEEE (2010)

Analysis of Positioning Performance of the Algorithm of Time Sum of Arrival with RMSE

Fengxun Gong(✉) and Ma Yanqiu

College of Electronic Information and Automation,
Civil Aviation University of China, Tianjin 300300, China
gfxcauc@sina.com

Abstract. Multilateration location is widely used in airports surface and terminal areas with replying pulse signal. The theoretical positioning accuracy of Multilateration location based on TDOA is very high. However, there are some problems in practical applications. The literature presents that TSOA location algorithm can partially compensate for the shortcomings of the TDOA's in some complex scenarios. Here we analyze the location performance of TSOA algorithm using RMSE information in additive white Gaussian noise environment. The TSOA localization model is constructed. The distribution of location ambiguity region can be presented with 4 based stations pattern. And then, the location performance analysis is executed when the RMSE variation is calculated with the four base stations. Subsequently, the performance changes of TSOA location algorithm is simulated and anglicized using RMSE parameters, when the location parameters are changed in quantity of base stations, base station layout, baseline length and so on. So, the TSOA location characteristics are revealed. From the REMS state changing trend, the anti-noise performance and robustness of the TSOA localization algorithm are proved. The TSOA anti-noise performance will be used for reducing the blind-zone and the false location rate of MLAT systems.

Keywords: TSOA · Root Mean Square Error (RMSE)
Location performance analysis · Robustness · Passive location

1 Introduction

Multilateration (MLAT) positioning technology (also known as passive positioning) does not dependent on emission source, It has the advantages of simple process-ing technology, low price and so on. It is widely used in military, civil aviation and many fields. Especially used in airport surface and terminals surveillance, guidance, control for airport safety operations. The time of arrival (TOA) is the foundation of the time difference of arrival (TDOA) and the time sum of arrival (TSOA) positioning algorithm, they can be used in MLAT systems. Typically the targets position can be calculated with CHAN algorithm or Taylor

J. Yang et al. (Eds.): CCCV 2017, Part III, CCIS 773, pp. 579–591, 2017.
https://doi.org/10.1007/978-981-10-7305-2_49

algorithm based on TDOA. However, some issues have been presented in practical applications. Such as it has a lot of random blind areas, and mis-positioning rate is higher. So, the compensation methods for mis-positioning, and improving accuracy must be found. At the end of 1990s, the ellipse (time sum of arrival, or TSOA) algorithm started to be used for location and tracking targets by the American scholar at firstly. In 2005, the mobile targets coordinates are successfully calculated accurately using combination algorithm of TSOA and TDOA in the NLOS environment by Taiwan University of science and technology [1]. In 2009, the airport surface location by the TSOA algorithm was carry out successfully in MLAT by ERA company [2]. In 2010, it is preliminarily proved that targets location accuracy of the TSOA positioning algorithm is higher than that of the TDOA by Civil Aviation University of China (CAUC) [3–5]. But, none of the above literature tells us the accuracy of the TSOA localization algorithm. Generally positioning accuracy of MLAT system is determined mainly by two factors [6–12]. The one is geometric distribution of base stations in surveillance area, and the another is time measurement error between base station and transponder. In this paper, we construct the TSOA simulation positioning system according to the TSOA principle, with it, the influence factors about arrival time error and measurement error to the positioning accuracy are analyzed in detail. At same time, the performance variation of TSOA and TDOA in random noise environment are compared in different conditions. The preliminary conclusion is shown that location unambiguous performance of the TSOA algorithm is worse than the TDOA algorithm under the same positioning condition. But, if the quantity of base stations exceeds some specified number, the positioning performance of TSOA algorithm can also be greatly improved. The positioning performance of the TSOA algorithm is evaluated and analyzed using RMSE parameter with Matlab simulation. The simulation results indicated that the position accuracy of TSOA algorithm is better than that of TDOA's, under the same AGWN conditions with related system parameters changing. Therefore, the anti-error performance of TSOA algorithm in MLAT system is better than that of TDOA's under the same time measurement error location conditions.

2 The Positioning Principle Analysis of TSOA

Generally, the TSOA positioning algorithm is also known as ellipse location algorithm. The arrival time between the target replying pulse signals (or transponder responses) and base stations are detected by the ground base stations and collected by the master station, the arrival time sum can be calculated by the master station, then the target coordinate can be calculated when the equations of arrival time sum are solved. So, the target can be tracked according the continuous coordinates. In the two-dimensional plane, an ellipse is determined by two foci. When the base stations position coordinates are on the two foci, the ellipse will be determined by the arrival time sum of the targets signal. If there are three receiving base stations, the three ellipses will be formed. The three ellipses intersect each other to form only one common focus point. This point is the location coordinate of the target. The other intersections are called ambiguous points. These points must be removed using the operation information. At last the target position is determined. The basic positioning principle of TSOA algorithm is shown in the Fig. 1.

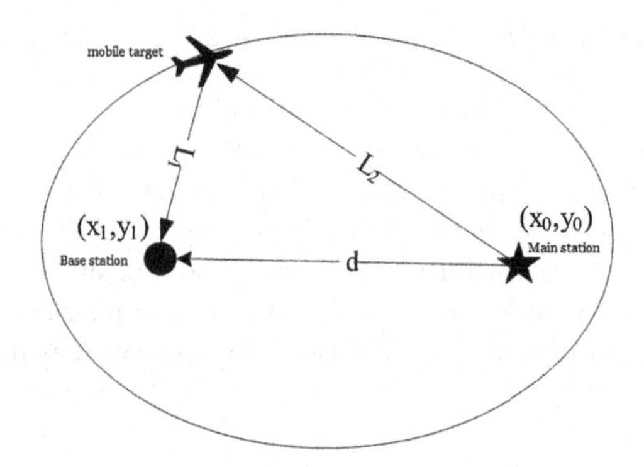

Fig. 1. The basic positioning principle of TSOA algorithm

The asterisk denotes the master station, and the black dot represents ground base stations on the two foci in Fig. 1. If (x, y) presents the target position coordinates to be estimated, and (X_i, Y_i) presents the known position coordinate of the No. i base station. Then, the distance be-tween the mobile target and the No. i base station is:

$$R_i = \sqrt{(X_i - x)^2 + (Y_i - y)^2} \tag{1}$$

$$R_i{}^2 = (X_i - x)^2 + (Y_i - y)^2 = K_i - 2X_i x - 2Y_i y + x^2 + y^2 \tag{2}$$

$$\Delta r = r_i + r_1 \tag{3}$$

One of them, $K_i = X_i^2 + Y_i^2$.

Let $R_{i,1}$ indicates the distance sum, which between the mobile target to the base station i, and the mobile target to the main station, then the distance sum can be ex-pressed as follow,

$$R_{i,1} = cT_{i,1} = R_i + R_1 = \sqrt{(X_i - x_1)^2 + (Y_i - y)^2} + \sqrt{(X_1 - x)^2 + (Y_1 - y)^2} \tag{4}$$

One of them, c is the speed of light. $cT_{i,1}$ is the time measured value of TSOA. Linearization of (4), we can obtain that

$$R_i^2 = (R_{i,1} - R_1)^2 \tag{5}$$

Then, (5) can be rewritten as,

$$R_{i,1}^2 - 2R_{i,1}R_1 + R_1^2 = K_i - 2X_i x - 2Y_i y + x^2 + y^2 \tag{6}$$

When i = 1, the (1), (2) and (3) can be simplified as:

$$R_1^2 = K_1 - 2X_i x - 2Y_i y + x^2 + y^2 \tag{7}$$

(6) minus (7), the result can be obtained:

$$R_{i,1}^2 - 2R_{i,1}R_1 = K_i - K_1 - 2X_{i,1}x - 2Y_{i,1}y \tag{8}$$

In the (8), $X_{i,1} = X_i - X_1, Y_{i,1} = Y_i - Y_1$. Consider x, y and R_1 as unknown parameters, consequently the nonlinear equations can be formed by the i (8). The solution of the equations can be used to calculate the position coordinates of mobile targets. According to the basic principles of the TDOA and TSOA algorithm, the stations distribution pattern(layout) of ambitious points can be obtained and shown as follow in Figs. 2 and 3. In the position simulation, there are four base stations, and the base stations spatial layout is in star distribution pattern.

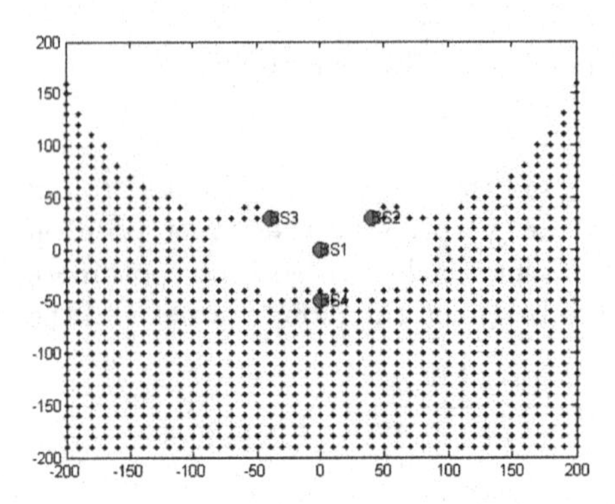

Fig. 2. Ambitious region of TDOA

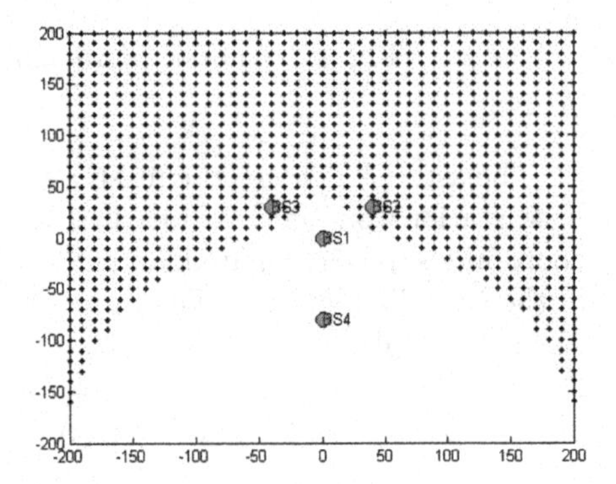

Fig. 3. Ambitious region of TSOA

Obviously, the distribution pattern of ambitious location points in the two algorithms is completely different shown as in Figs. 2 and 3. The TSOA positioning characteristics are very distinctness. And it is proved that the localization performance of the two algorithms can be complementary.

3 RMSE Analysis of TSOA Location Algorithm

In general, suppose the MLAT system with TSOA location algorithm operation environment includes random noise. Without loss of generality, the additional noise is additive Gaussian noises, which will be added to the measuring time, the Gauss white noise distribution model can be used on the time measurement error analysis between the base stations and target. In this paper, the variations of arrival time measurement error is described with the additive Gauss white noise(AWGN) model. The equations formed by formula (8) are used to evaluate the target localization accuracy of TSOA algorithm, and the error distribution between estimated position and ideal position is calculated and analyzed by using the root mean square error (RMSE) with AWGN. The formula of the RMSE can be expressed as follow.

$$RMSE = \sqrt{E\left[(x-\hat{x})^2 + (y-\hat{y})^2\right]} \tag{9}$$

3.1 RMSE Analysis in a Random Noise Environment

In order to analyze the RMSE of TSOA algorithm, the simulation positioning system based on TSOA algorithm is constructed with four base stations. That is to say, there are four base stations in basic positioning pattern. There is one main station and three auxiliary stations in the four base stations. The coordinate parameters of these base stations are shown as follow: the main station O(0, 0), the auxiliary station A(−6000, 3000), the auxiliary station B(−6000, 0), the auxiliary station C(0, 3000), and suppose the target coordinate is T(−2000, 4000). The distance unit is kilometer(km). The layout structure of the four base stations is rectangle model. The given TSOA time range is at 0–200 μs.

Under ideal condition, namely in light of sight (LOS) and no random noise, the RMSE of the TSOA algorithm gradually approaches a constant. Theoretically, it can close to zero. In fact, the targets response signal will be affected by various factors in the transmission, reception, recognition and processing, and those interference factors will be appended to the measurement arrival time in the form of AWGN.

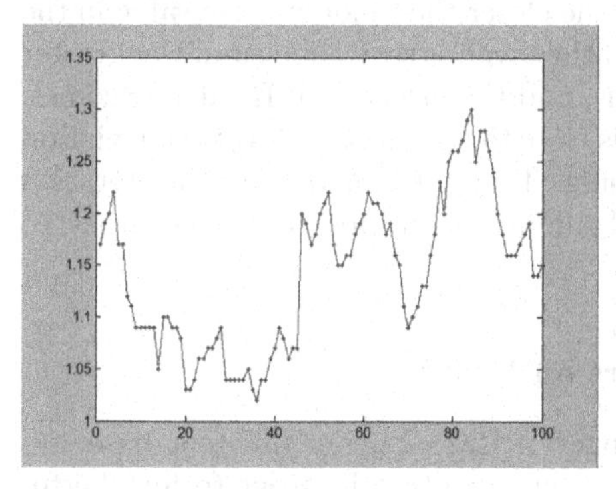

Fig. 4. Measurement error of TSOA 0.01 μs

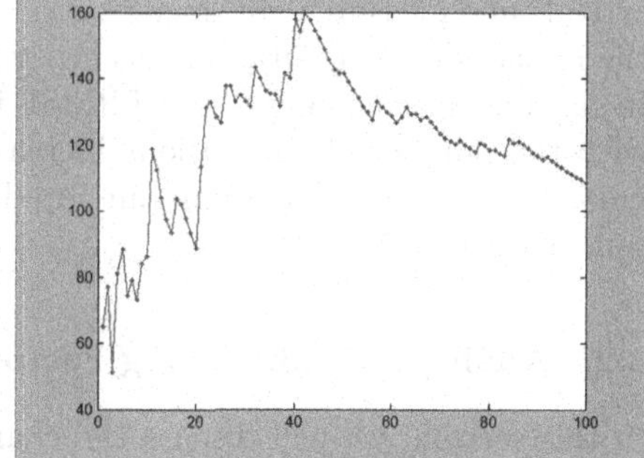

Fig. 5. Measurement error of TSOA 0.1 μs

The 100 random simulation numbers are selected as the time measurement error for arrival time sum of the TSOA algorithm(in other words, the AWGN is 100 random values, i.e. (0–1)∗(0.01 μs, 0.1 μs, 1.0 μs, 10.0 μs)). These simulating parameters are substituted into the TSOA algorithm. The variations of RMSE with time measurement errors are shown in the Figs. 4, 5 and 6. In the figures,

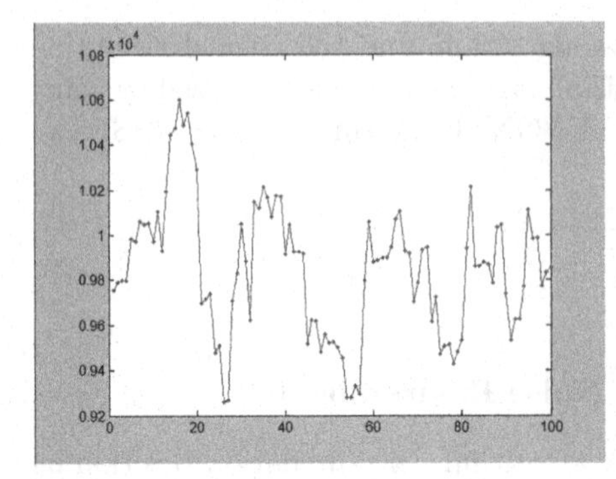

Fig. 6. Measurement error of TSOA 1.0 μs

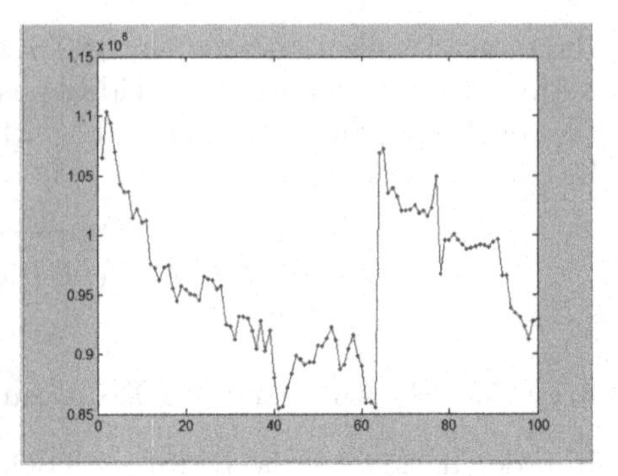

Fig. 7. Measurement error of TSOA 10 μs

the vertical axis unit is meters (m), and it describes the changes of the RMSE. The horizontal axis presents time of the simulated positioning algorithm.

In the LOS operation condition and specified simulation conditions, and at same time, the actual operation situation of an airport is referred when the simulation is carried out. Then the four measurement time will be obtained with AWGN. Subtracting the true time from them respectively, the measurement time deviations will be calculated. These time deviations leads to positioning error. So these time deviations must be extracted to evaluate location accuracy of the TSOA algorithm. Because of the AWGN in the measurement time, so the deviation between the calculated position and the real position of the target must be random with AWGN. Namely the pattern is shown in the Figs. 4, 5 and 6. And the simulating results present that, the closer the random noise value in the arrival time measurement deviation is to the given arrival time sum, the greater the impact on the positioning accuracy is, and the larger RMSE value becomes. In general, the changing trend of RMSE is close to the center of location deviation with the number of simulations increasing. Therefore, increasing the iteration number of TSOA algorithm can rapidly improve the positioning accuracy to some extent.

3.2 Analysis of Influencing Factors on RMSE

In this section, we will discuss the changes of RMSE under different base stations layout mode and different base stations quantity, in order to find better positioning models which the RMSE value is relatively small and positioning accuracy is better.

Analyzing RMSE changes under the same number of base stations and different station layout pattern. The horizontal axis represents arrival time measurement bias. Suppose, the 100 random points are selected in algorithm simulation. The vertical axis represents the location errors described with RMSE (m). And there are four stations which were selected in this simulation process. The simulation results are presented in the Figs. 7, 8 and 9. As shown in these figures, the simulation results can be seen that, the RMSE of the star pattern is minimum and the change is more stable than others.

Analyzing RMSE changes under the same station layout pattern and the different base stations quantity. The horizontal axis represents arrival time measurement bias. Suppose, the 100 random points are selected in algorithm simulation. The vertical axis represents RMSE(m) of the targets location. The base stations quantity is the independent variable in this simulation. The simulation results are presented in Figs. 10, 11 and 12. As shown in these figures, the simulation results can be seen that, directly increasing the quantity of base stations does not reduce the RMSE value, and also does not significantly improve the positioning accuracy. Therefore, it is not advisable to increase the positioning accuracy through increasing the number of base stations.

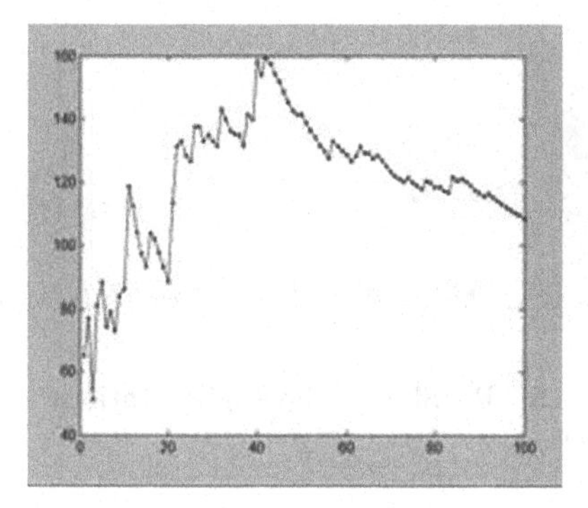

Fig. 8. RMSE of the rectangular layout

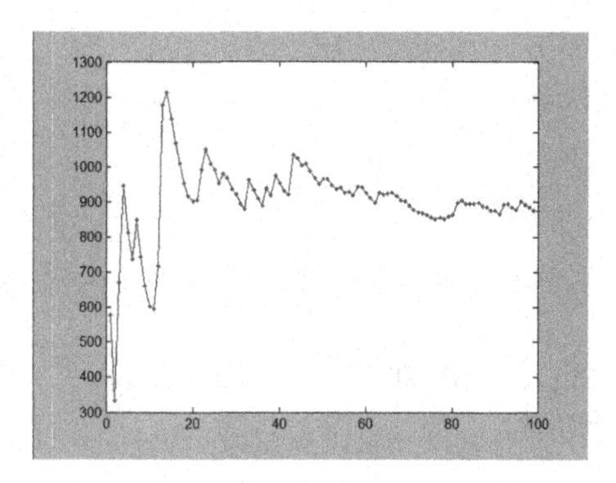

Fig. 9. RMSE of the parallelogram layout

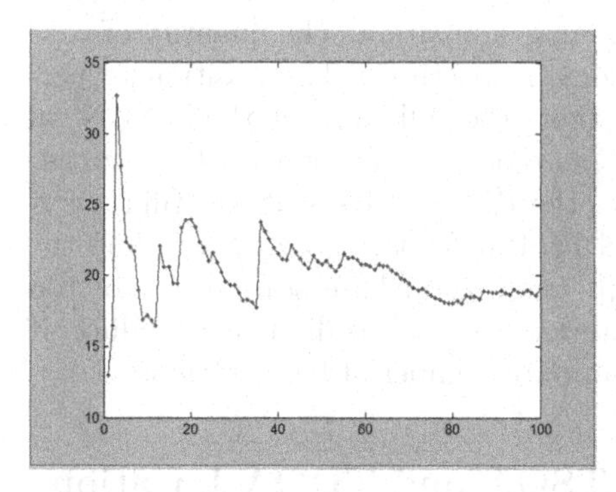

Fig. 10. RMSE of the star layout

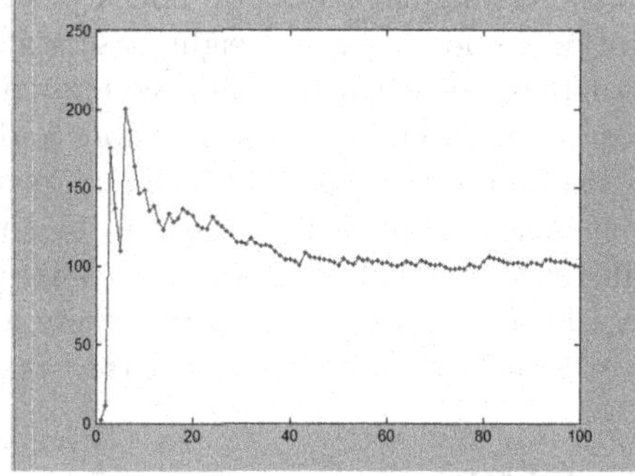

Fig. 11. RMSE of the four stations layout

3.3 Summarization of the TSOA Algorithm Location RMSE

From the simulations, some interesting phenomena have been presented, so we can deduce the conclusions. Different pattern of the base stations distribution has a greater impact on location accuracy using the given same number of base stations, which can be described with RMSE.

Under the same operation conditions, the anti-error performance of the TSOA algorithm gradually reduced in turn, which the base station layout pattern is the star, rec-tangle and parallelogram, and the simulation results are shown in the Figs. 7, 8 and 9. Therefore, the star layout pattern is first selected when the random noise is very difficult to reduce.

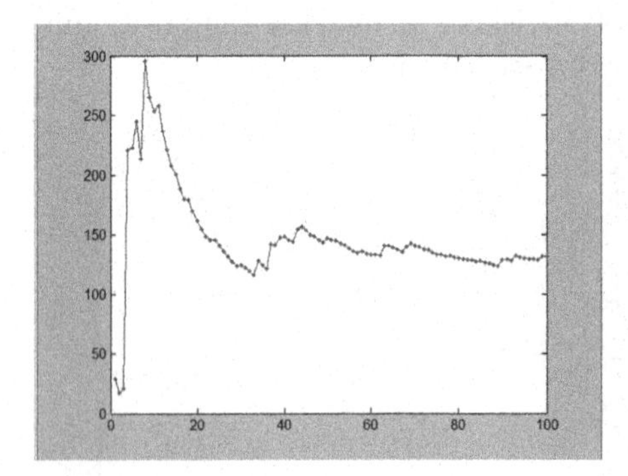

Fig. 12. RMSE of the six stations layout

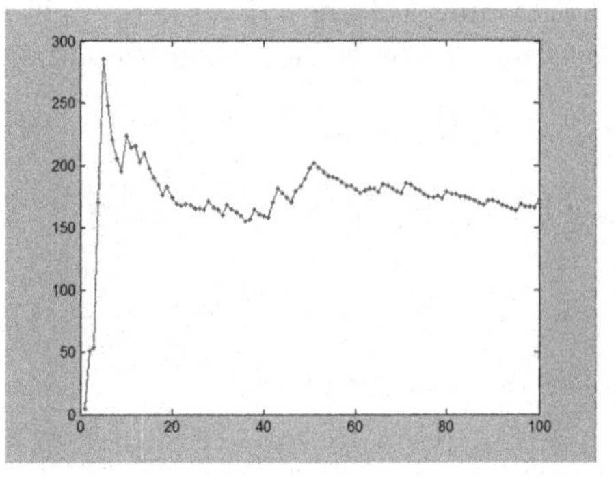

Fig. 13. RMSE of the eight stations layout

The quantity of base stations will have a impact on RMSE in the given same station layout pattern. But, directly increasing the number of base stations does not obtained greater performance improvements.

In the case of the same other operation patterns, and the base stations layout is rectangular mode for TSOA positioning simulation. The quantity of base stations is four, six and eight base stations, respectively. The positioning performance are calculated and compared that, the influence of the AWGN on positioning accuracy becomes more and more not significant with the increase of the number of base stations, and keeps the RMSE value stable within a certain range It is shown as Figs. 10, 11 and 12. But for the TSOA algorithm, the difference of the RMSE value of different quantity of base stations is not too big. Therefore, in environments where interference noise is difficult to reduce, it is extremely necessary to select the appropriate number of base stations.

4 RMSE Comparison of the TSOA and TDOA Location

The location simulation condition for the performance evaluation based on RMSE: The horizontal axis represents 100 random noise sampling values selected by simulation experiment, the vertical axis represents the location error expressed with the RMSE(m). And the location simulation is focus on the error change expressed with the RMSE with different stations layout pattern and stations quantity.

4.1 Different Layout Pattern with Four Stations

In the different layout pattern with four stations, the change of the RMSE is shown as in Figs. 13, 14, 15, 16, 17 and 18. The change analysis of the RMSE with four base stations in rectangular layout. The simulation results are shown as in Figs. 13 and 14. As shown, the deviation of the TDOA RMSE is greater than that of the TSOA. However, the deviation convergence of TDOA is faster than that of TSOA.

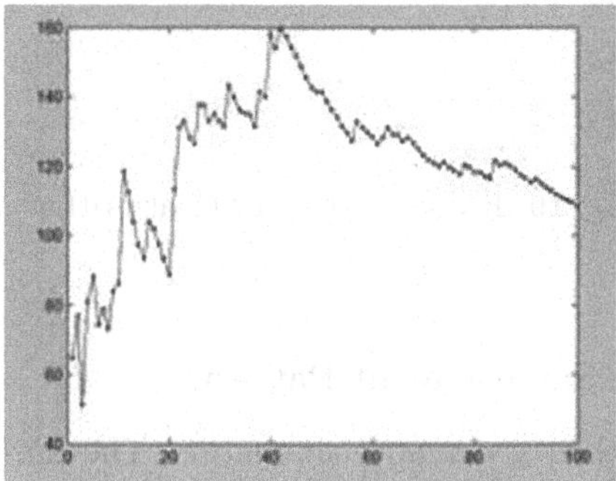

Fig. 14. RMSE of the TSOA algorithm

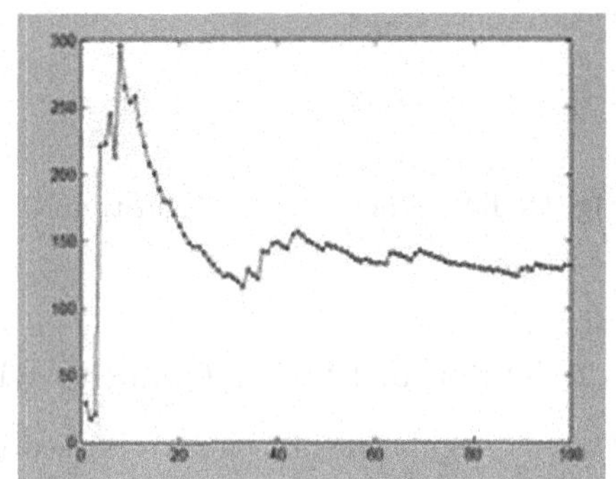

Fig. 15. RMSE of the TDOA algorithm

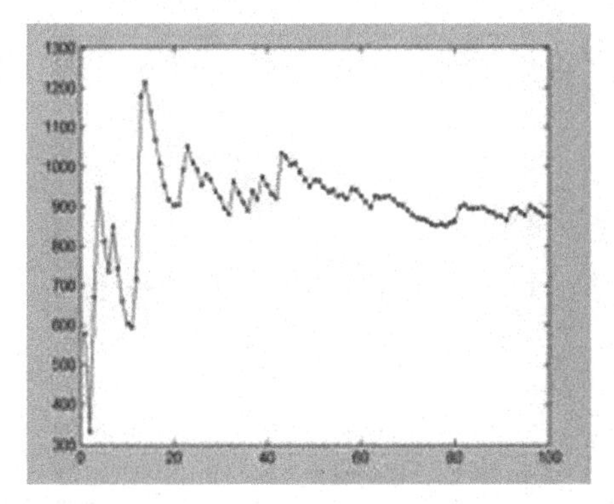

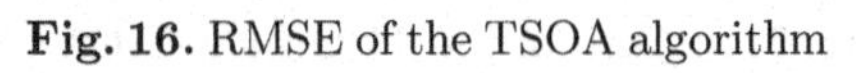

Fig. 16. RMSE of the TSOA algorithm

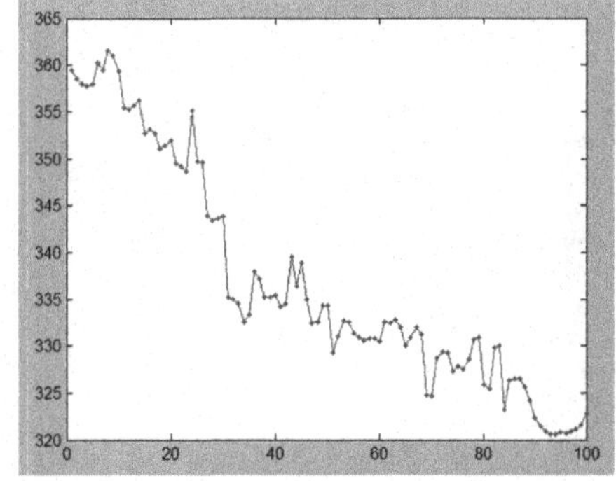

Fig. 17. RMSE of the TDOA algorithm

The change analysis of the RMSE with four stations in parallelogram layout. The simulation results are shown as in Figs. 15 and 16. As shown, the location deviation of the TSOA and TDOA are larger. But the TDOA is little smaller, and the TDOA algorithm deviation presents a rapid convergence trend.

The change analysis of the RMSE with four stations in star layout. The simulation results are shown as in Figs. 17 and 18. As shown, the location deviation of TSOA is obviously better than that of TDOA.

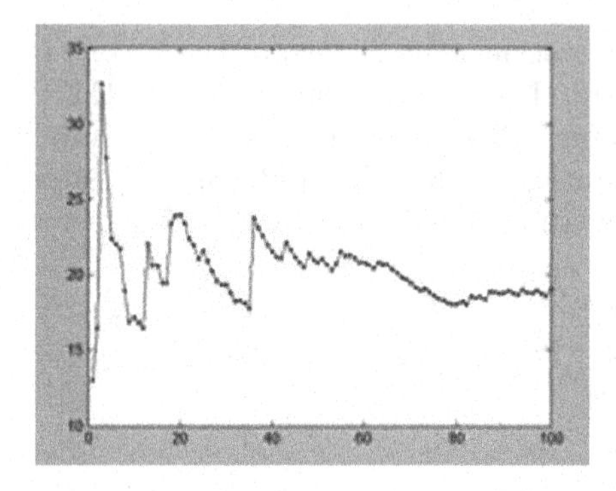

Fig. 18. RMSE of the TSOA algorithm

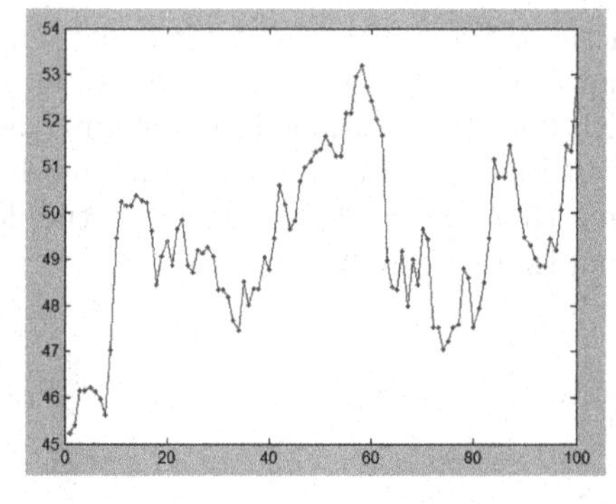

Fig. 19. RMSE of the TDOA algorithm

4.2 Different Station Quantity with Same Layout Pattern

The rectangular layout pattern is used in the performance simulation. The simulation results are shown as follow. The change of the RMSE with four stations. The simulation results are shown as in Figs. 19 and 20. As shown, the maximum localization deviation of the TSOA algorithm is less than that of the TDOA, but the convergence rate of the TDOA algorithm is more faster than the TSOA.

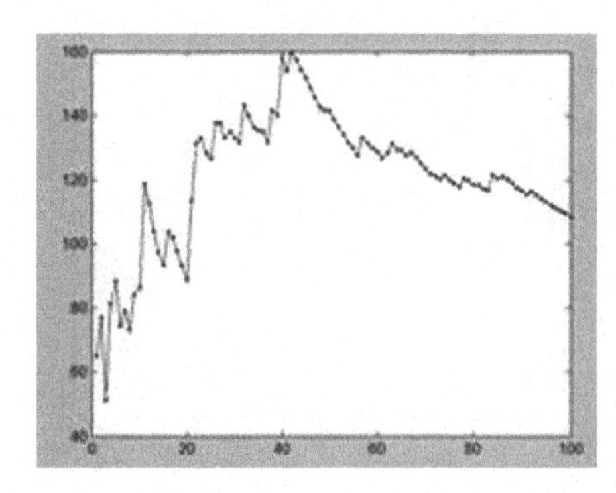

Fig. 20. RMSE of the TSOA algorithm

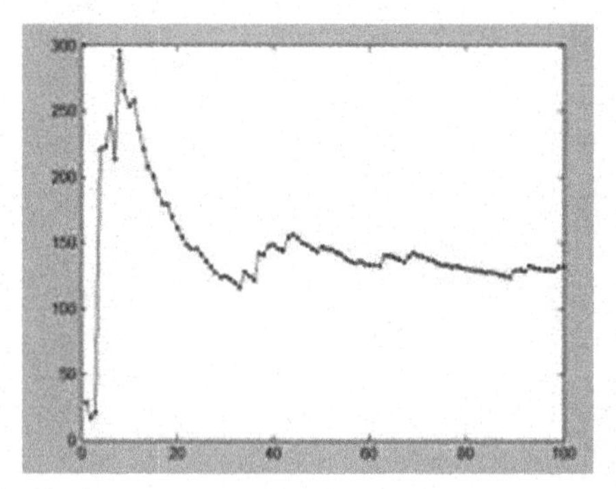

Fig. 21. RMSE of the TDOA algorithm

The change of the RMSE with six stations in the performance simulation. The simulation results are shown as in Figs. 21 and 22. As shown, the location deviation of TSOA algorithm is obviously smaller than that of TDOA. And the TDOA algorithm deviation presents a rapid divergence trend.

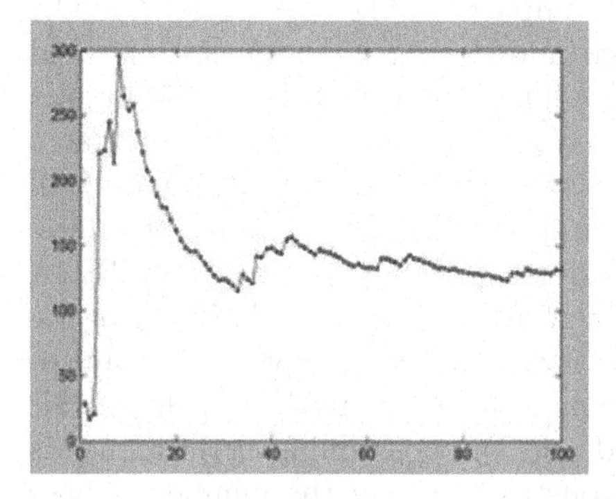

Fig. 22. RMSE of the TSOA algorithm

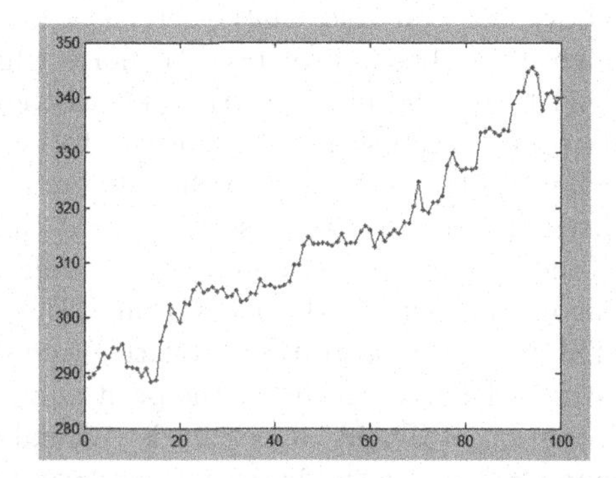

Fig. 23. RMSE of the TDOA algorithm

The change of the RMSE with eight stations in the performance simulation. The simulation results are shown as in Figs. 23, 24 and 25. As shown, the TSOA algorithm error presents a robust convergence trend, and the RMSE of TDOA algorithm fluctuates greatly and deviates from the central values. This means stability is poor.

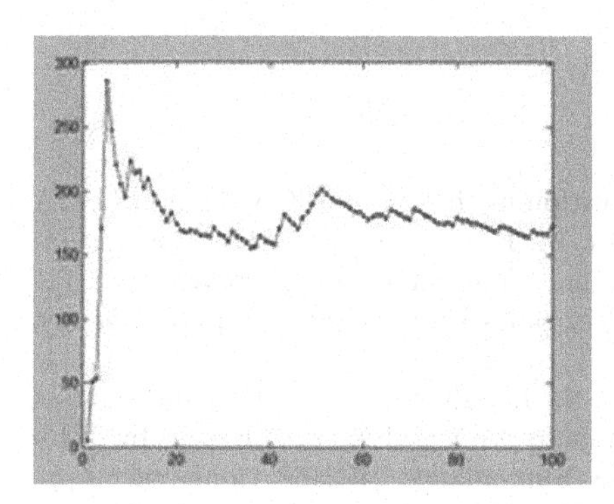

Fig. 24. RMSE of the TSOA algorithm

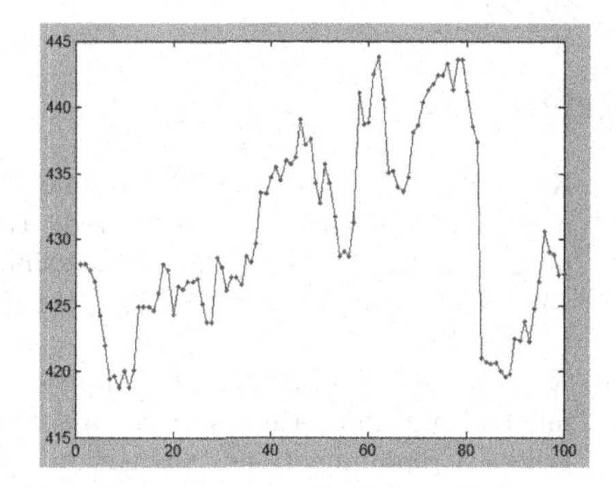

Fig. 25. RMSE of the TDOA algorithm

Therefore, by simulation and calculation results in details, some important conclusions can be drawn from the simulation results as follow: There is a great impact on the TDOA and TSOA algorithms because of random noise overlap-added on the measurement time, and it lead to positioning distortion and fluctuation of RMSE in target location. There is a great impact on the RMSE change trend because of different layout pattern of base stations. There is a great impact on the RMSE change trend because of base stations quantity. The robustness of the TSOA algorithm is affected by the change of relevant location parameters in layout.

5 Simulation Analysis and Conclusion

According to the location algorithm simulation results shown in the figures above, we can find that: there is a great impact on the TDOA and TSOA algorithms because of random noise which added on measurement time, and lead to positioning distortion and changing fluctuation of RMSE. The impact factors include stations quantity, stations layout pattern and so on. But, as a whole with the increase in the number of base stations, the impact of noise will become more and more stable, and it can keep the RMSE in a range of values. Under the same positioning performance, the anti noise performance of TSOA algorithm is relatively better than TDOA algorithm. And the robustness of the TSOA location algorithm is relatively good. Based on the principle of time sum of arrival(TSOA) algorithm, the positioning equations are derived. And the RMSE variation of TSOA algorithm is calculated and the positioning performance is analyzed, when the base stations layout construction and the number of base stations is different. Through comparison and analysis, it will be found, in the aspect of anti-noise performance, the TSOA algorithm index is better than the TDOA algorithm in most cases. Therefore, when the random noise on measurement time is difficult to reduce, the TSOA algorithm can be chosen, and the layout and number of base stations should be considered in detail.

Acknowledgment. This work is financially supported by the Joint civil aviation fund of National Natural Science Foundation of China (Grant No. U1533108 and No. U1233112).

References

1. Kuen-Tsiar, L., Wei-Kai, C.: Mobile positioning based on TOA/TSOA/TDOA measurements with NLOS error reduction. In: Proceedings of International Symposium on Intelligent Signal Processing and Communication Systems, pp. 545–548 (2005). Institute of Electrical and Electronics Engineering Computer Society, Hong Kong
2. Xu, N., Cassell, R., Evers, C., Hauswald, S., Langhans, W.: Performance assessment of multilateration systems a solution to NEXTGEN surveillance. In: Integrated Communications Navigation and Surveillance (ICNS) Conference, pp. D2-1-D2-8. IEEE Computer Society, Herndon, VA (2010)
3. Cheng, A.-H., Ji, Z.-H., Ge, B.-Z.: Novel hybrid ellipse-angle locating technique in NLOS environment. J. Syst. Simul. **21**, 3332–3337 (2009). (in Chinese)
4. Gong, J.: Research on TSOA/AOA positioning technology in CDMA Network. PLA Information Engineering University, Zhengzhou (2009). (in Chinese)
5. Gong, F., Lei, Y., Ma, Y.: Surface surveillance system of main station off-center and performance analysis. J. XIDIAN Univ. **38**, 1–7 (2011). (in Chinese)
6. Lee, H.B.: A novel procedure for assessing the accuracy of hyperbolic multilateration systems. IEEE Trans. AES **11**, 2–15 (1975)
7. Levanon, N.: Lowest GDOP in 2-D scenarios. In: IEE Proceedings of Radar Sonar Navigation, vol. 147, pp. 149–155 (2000)
8. Caffery Jr., J., Stuber, G.L.: Subscriber location in CDMA cellular networks. IEEE Trans. Veh. Technol **47**, 406–416 (1998)

9. Zheng, Z., Hua, J., Wu, Y., Wen, H., Meng, L.: Time of arrival and time sum of arrival based NLOS identification and localization. In: IEEE 14th International Conference on Communication Technology, pp. 1129–1133. Institute of Electrical and Electronics Engineers Inc., Chengdu (2012)
10. Thong-un, N.: Improvement of echolocation-based methods using ultrasonic wave by implementation of time difference of arrival and time sum of arrival for two reflectors. Acoust. Sci. Technol. **37**, 136–138 (2016)
11. Li, S., Hua, J., Zhong, G., Lu, W., Jiang, B.: A TSOA based localization algorithm in wireless networks. In: IEEE 11th Conference on Industrial Electronics and Applications, pp. 526–530. Institute of Electrical and Electronics Engineers Inc., Hefei (2016)
12. Zheng, X., Hua, J., Zheng, Z., Peng, H., Meng, L.: Wireless localization based on the time sum of arrival and taylor expansion. In: 19th IEEE International Conference on Networks, pp. 1–4. IEEE Computer Society, Singapore (2013)

A Novel Automatic Grouping Algorithm for Feature Selection

Qiulong Yuan and Yuchun Fang(✉)

School of Computer Engineering and Science, Shanghai University, Shanghai, China
ycfang@shu.edu.cn

Abstract. Feature selection is used in many application areas relevant to expert and intelligent system such machine learning, bioinformatics and image processing. Feature selection plays an important role in reducing the dimensionality of high-dimensional features. However, traditional feature selection methods are not able to intelligently learn intrinsic data structures. In this paper, we proposed a novel feature selection method, which can automatically learn grouping structure relation among features. Experiments are conducted on the selection of both raw features and statistically handled features. Experimental results demonstrate that the proposed method can identify important features by automatic grouping, and outperforms the other methods on several public data sets. Moreover, by using parallel computing, the training time consumed by our method is only 50% of that of the traditional methods.

Keywords: Feature selection · Mutual information
Automatic grouping

1 Introduction

Feature selection is used in many application areas relevant to expert and intelligent system such machine learning, bioinformatics and image processing. In computer vision, machine learning, and data mining, data are always represented by high-dimensional feature vectors. But high-dimensional data can increase the consumption of processing time and storage space during processing. Moreover, most of the existing machine learning methods, such as classification, regression, or other tasks, are mainly designed more adaptive to low-dimensional data. The computation of high-dimensional data can usually become much more complex and difficult. As a solution to this issue, feature selection (also known as variable selection) [2,13,20,21] is performed to choose a representative subset from the high dimensional features. The subset is expected to bear sufficient information about the original high-dimensional feature set for specific learning tasks. The purpose of feature selection is to find the relevant feature subset that best reflects the statistical properties of the pattern category to represent the original feature.

J. Yang et al. (Eds.): CCCV 2017, Part III, CCIS 773, pp. 592–603, 2017.
https://doi.org/10.1007/978-981-10-7305-2_50

According to different evaluation metrics, feature selection algorithms can roughly classified into three categories, i.e., filter, wrapper and embedded methods [13]. The filter methods rely on general characteristics of the data to evaluate and select feature subsets without considering the learning algorithm. Such as variance and Fisher score [11,14]. The wrapper model takes the performance of the learner to be used as the evaluation criterion. For embedded methods, the process of feature selection and learner training is integrated, both are completed in the same process, and in the process of training the learning algorithms automatic feature selection.

Recently, some new embedded feature selection methods that integrate the theory of sparse representation [3,8,24], compressed sensing, and feature selection [4,18,26] have been proposed. Sparsity-inducing feature selection methods have been widely used in face authentication [5,23], face detection, face attributes classification [9], and gene expression [25]. However, L_1 regulraization computation typically requires solving either NP-problem or an alternative problem that sill involves a costly iterative optimization [7].

In practical applications, the features have some essential structures. Integrating knowledge about the feature structures may help identify the important features. Ye and Liu [27] and Zhang et al. [28] proposed to use the group structure information of the data to carry on the feature selection. But the existence of these methods have a defect that the number of groups in the group structure is man-made, and the automatic grouping is not realized. On the one hand, the feature set F is artificially divided into k groups, and this grouping is usually carried out by experience, still do not achieve automatic grouping, on the other hand, the so-called grouping is to divide the adjacent m features into a group, the features that are adjacent to each other do not necessarily belong to the same group.

Taking into account these factors, in this paper, we propose a novel automatic grouping feature selection method, which use l_2 norm to ensure group effect [31] and use the mutual information measure to ensure the low redundancy within the group. The Laplacian regularization based on mutual information is used in the objective function of the method, so we call it MIL. Our experiments demonstrate the efficiency of the proposed method. The difference between our approach and the traditional group lasso method is shown in Fig. 1. As shown in Fig. 1, the traditional group lasso method is continuous and the number of groups is decided by people. In addition, the group is fixed once it is determined. The MIL method combines autonomously by judging the correlation between features, where each group of features can be discontinuous and the number of features of each group can be unequal.

The remainder of the paper is organized as follows. Sect. 2 presents different related methods and information theory used in our paper. Section 3 introduces our proposed method. Experimental evaluation is depicted in Sect. 4. At last, we conclude the paper in Sect. 5.

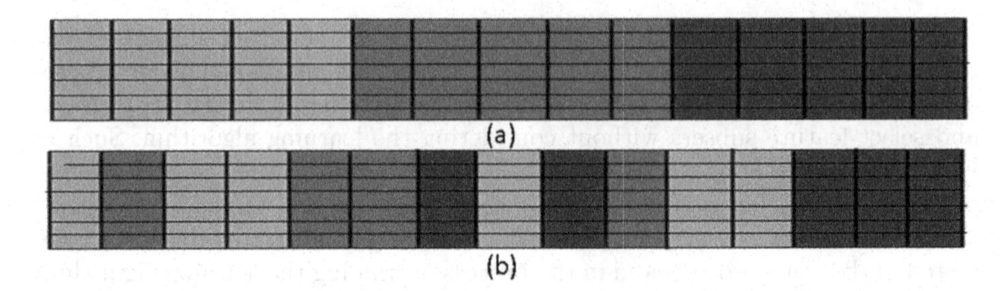

Fig. 1. Comparison between traditional group lasso and MIL methods. Suppose there is a data set $A \in R^{7\times 15}$, we use the 7×15 table to represent the A dataset, each row represents a sample, each column represents a feature, the same color indicates the same group, (a) traditional group lasso grouping result, (b) MIL grouping result. (Color figure online)

2 Related Work

In this section, we will introduce some of the existing feature selection algorithms, and then introduce the basics of the information theory, as we will use it in Sect. 3. Before the details are presented, we need to summarize the notations and the definitions used in this paper. Let $A = [a_1, a_2, \ldots, a_n] \in R^{n\times d}$ be n samples data in the d-dimensional space, where n is the number of sample and d is the number of features. Accordingly, denote label vector $y = [y_1, y_2, \cdots, y_n]^T \in R^{n\times 1}$, which $y_i \in \{+1, -1\}, i \in \{1, 2, \ldots, n\}$ if task is binary classification. Denote λ as the hyper parameter to balance the data misfit and the penalty. Denote $W = [w_1, w_2, \ldots, w_d]^T \in R^{d\times 1}$ as the unknown weight coefficient vector, which need we to be estimated.

2.1 Related Feature Selection Methods

As a kind of embedded method, regularization techniques based on L_1 norm has been widely used to cope with feature selection in machine learning tasks. According to compressive sensing theory, the minimum L_1 norm solution to an under determined system of linear equation is equivalent to the sparsest possible solution under general conditions. Destrero et al. [5,6] used lasso for feature selection in face detection and face authentication. Its objective function is:

$$\min_W \|y - WA\|_2^2 + \lambda\|W\|_1 \quad (1)$$

Lasso is suboptimal since it produces biased estimates for the large coefficients. Zou [30] found that Lasso uses the same degree of compression for all coefficients. And Lasso does not have the oracle properties. In order to improve the performance of Lasso, the adaptive Lasso [30] is proposed

$$\min_W \|y - WA\|_2^2 + \lambda\sum_{i=1}^{d} a_i\|w_i\|_1 \quad (2)$$

From Eq. 2, we know the only difference between Lasso and adaptive Lasso is the latter gives a weight coefficient for each. Different compression coefficients are used for different weights, let Lasso has oracle properties. It not only has good usability in prctice but also has an excellent character in theory. Obviously, when all coefficient is equal, the adaptive Lasso is equivalent to Lasso.

The fused Lasso introduced in [22] get a solution that has sparse both the coefficient and their successive differences. The objective function of fused Lasso can be represented as follows:

$$\min_{W} \|y - WA\|_2^2 + \|W\|_1 + \alpha \sum_{i=2}^{d} |w_i - w_{i+1}| \tag{3}$$

The bridge estimator [17] is defined as follows:

$$\min_{W} \|y - WA\|_2^2 + \lambda \sum_{i=1}^{d} |w_i|^\gamma \tag{4}$$

The bridge estimator has two important special cases. When $\gamma = 2$, it is popular ridge estimator. When $\gamma = 1$, it is the Lasso.

In many practical applications, some features often have a strong correlation. In this case, the lasso tends to select only one of the correlated features. To deal with feature with strong correlation, Zou and Hastie [31] proposed elastic net regularization as

$$\min_{W} \|y - WA\|_2^2 + \alpha \sum_{i=1}^{d} |w_i| + (1 - \alpha) \sum_{i=1}^{d} |w_i|_2 \tag{5}$$

where $|\cdot|_2$ is the L_2-norm. [31] show that L_2 regularization has the group effect and L_1 regularization does not have group effect Zou et al. [31] add a group effect to lasso by using L_2 regularization to handle feature with strong correlations. Furthermore, when α is equal to 1, the elastic net is lasso, and when α is equal to 0, the elastic net is ridge estimator [17].

The penalties introduced in Eqs. (1)–(5) are assume that features are independent and ignored the structures of features completely [27]. However, in practical application, the features have some essential structures, such as groups [27,28]. Suppose that features are divided into k groups. With the group structure, the W is rewritten as k groups $W = \{w_{G1}, w_{G2}, \ldots, w_{Gk}\}$, and the objective function of group lasso as follows:

$$\min_{W} \|y - WA\|_2^2 + \sum_{i=1}^{k} \beta_i \|w_{Gi}\|_q \tag{6}$$

where the $\|\cdot\|_q$ is indicate q-norm, and β_i is the weight coefficient of i-th group. There are different structured feature selection methods according to different q values or different constraints, such sparse group lasso [29]. The group structure provides good access to the structural property of the data. However, the number of groups in the group structure is man-made, and the automatic grouping is not realized. Moreover, the features that are adjacent to each other do not necessarily belong to the same group.

2.2 Information Theory

Given the two variables U, V, if their respective marginal probability distribution and joint probability distribution are respectively $p(u)$, $p(v)$ and $p(u,v)$, and then their mutual information $I(u,v)$ is defined as:

$$I(u,v) = \sum_{u,v} p(u,v) log \frac{p(u,v)}{p(u)\,p(v)} \tag{7}$$

When the variables U and V completely unrelated or independent of each other, the minimum mutual information, the result is zero, which means that there is no overlapping information between the two variables; on the other hand, the greater the interdependence, mutual information value will be greater.

3 MIL

In Sect. 2, we analyze some feature selection algorithms. On the one hand, L_1 regularization computation typically requires solving either NP-problem or an alternative problem that still involves a costly iterative optimization [7], on the other hand, Traditional methods based on structural features are not automatically group. Therefore, we proposed a novel feature selection model, MIL.

Assume given a set of training samples $A \in R^{n \times d}$ and the target labels $y \in R^{n \times 1}$ of the corresponding samples, the MIL uses the following criterion:

$$\min_W \|y - WA\|_2^2 + \alpha \|W\|_2 + \beta \sum_i \sum_j MI_{ij} (w_i - w_j)^2. \tag{8}$$

to find W, where α and β is the hyper parameter of the objective function. The first term of function (8) is the least-squares function. According to the least-squares method to seek the relationship of features and target labels. The second term of function (8) is l_2 norm. In this paper, the group effect of MIL is obtained via the L_2 norm, where the group is determined by the correlation between the features, and if the correlation between the two features is large, then we classify it as a group, and if the correlation is small, it is considered not a group. The third term of (8) is a manifold regularization where MI_{ij} is the correlation between the i-th and j-th feature. In this paper, the MI_{ij} is defined as the mutual information of the i-th and j-th feature (Mutual information can be used to measure the degree of interdependence between the two variables, and not limited to linear correlation, it also can be applied to nonlinear correlation). It is reasonable to require w_i and w_j close to each other if the i-th and j-th feature Similarity (redundancy) is low, which is the objective of the term of the third term of (8). In MIL, we use the third term of (8) to ensure that the redundancy in the group is minimal. In fact, the third term of (8) have the ability of l_1 penalty in feature selection, it can ensure the minimal redundancy of the feature subset. Denote MI as the similarity matrix constructed by all MI_{ij} and the diagonal matrix D where the element i-th of is the sum of the i-th row of the MI. Therefore, we can get the Laplacian matrix $L = D - MI$ [15], and the third term of (8) can be represented as $\beta W^T LW$ by simple algebra. The objective function can be rewritten as follows:

$$\min_{W} \|y - WA\|_2^2 + \alpha\|W\|_2 + \beta W^T LW. \tag{9}$$

Fortunately, (9) has an analytical solution as follows:

$$W = \left(A^T A + \alpha I + \beta L\right)^{-1} A^T y. \tag{10}$$

After obtain W, we can rank features according to $|w_i|$. The larger $|w_i|$ is, the more important this feature is [12]. We can either select a fixed number of the most important features or set a threshold and select the feature whose $|w_i|$ is larger than the value [16].

4 Experimental Evaluation

In the section, we present experimental results of our method. The performance of the newly proposed method in this paper, MIL, is mainly compared with five other methods: Lasso (L_1 regularization) [5], Ridge (L_2 regularization) [17], elastic-net [31], and group lasso [28]. These methods are chosen for the following reasons:(a) Like MIL approach, the choice of these methods based on regularization; (b) These methods are reported in the paper to provide good performance; (c) These methods contain structure and non-structure method, group effect and no group effect method, it helps to compare the performance of the algorithms. In the experiment, we use non-image and image data set.

The non-image data sets are Colon [1] and Leukemia [10], where Colon contains 2000 dimensions raw features and Leukemia contains 7029 dimensions raw features. For image data set, we select the CFW [19] face data set, which is a large collection of celebrity face images from the Internet. The data set contains 200,000 face images for 1,500 celebrities. We selected 8,000 face images (20 images $\times$ 400 people) from the CFW 60K to carry out attributes classification experiments, the 14 kinds of face attributes included in the selected pictures were gender, race, age and so on (more detail see Table 1). Then we use the ULBP to extracting low-level features. To be more specific, we scaling the face images to 140 $\times$ 160 pixels, and divided image into 7 $\times$ 8 cells, each of cell is 20 $\times$ 20 pixels. Then use the ULBP descriptor to extract 3,304-dimensional features as the raw features.

4.1 Performance Analysis with Colon and Leukemia Dataset

In the first experiment, we use the non-image data sets to evaluate our approach. We only selected the 50 features to observe the differences between the different methods. Figures 2 and 3 show the classification accuracy of the two datasets (Colon, Leukemia). As shown in Fig. 2, which illustrates the experiment with Colon, MIL reached stability with just 12 features and from the 4-th feature, MIL has always been far ahead of other methods. Compared to MIL, several other methods do not achieve better classification results. Without losing generality, the Fig. 3 shows similar results in Leukemia data set.

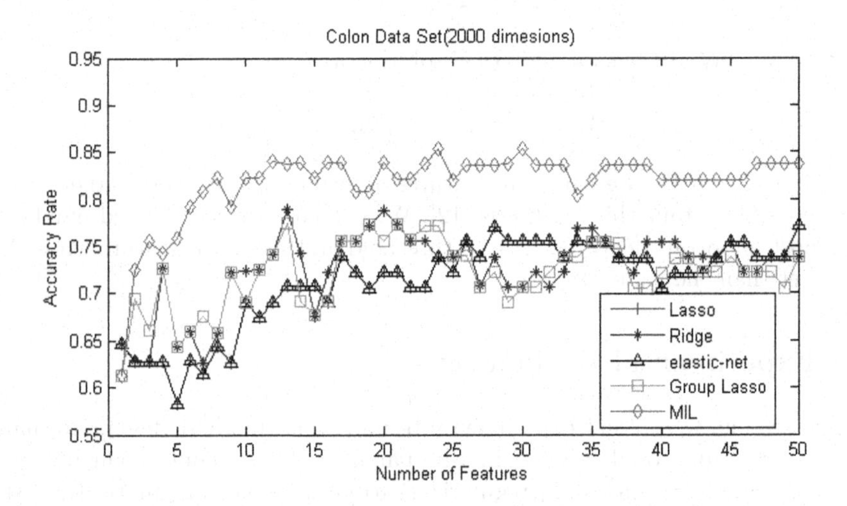

Fig. 2. Classification accuracy rate achieved with the Colon data set.

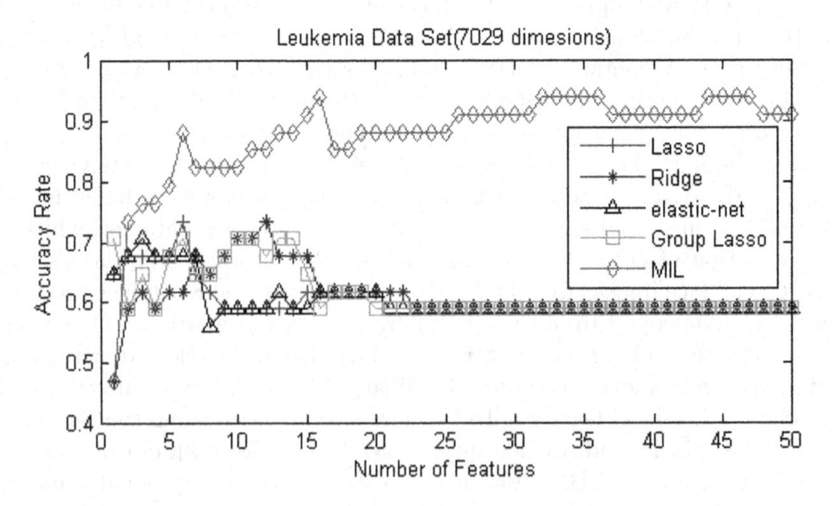

Fig. 3. Classification accuracy rate achieved with the Leukemia data set.

4.2 Performance Analysis on Face Dataset

In this experiment, we test MIL on image data set. In CFW dataset, we chose 14 face attributes (more detail see Table 1), and we used the MIL method to classify the 14 attributes. To observe the variation of accuracy with respect to a number of dimensions, the recognition rate is calculated from 10, 20, ..., to 100, and from 100, 200, ..., to 700. The results are shown in Table 1.

From the Table 1, we can find that our method has good performance. By calculating the average recognition rate of all dimensions, it can reflect the relationship between the recognition rate and the dimension. In general, if the average recognition rate is higher, then the number of dimensions required to achieve the same recognition rate will be less. However, the classification results obtained by the group sparse are not bad. We all know that most of the image feature description is based on statistical method, ULBP is the case, which itself counts as an area of information. In this article, the statistical size of the region is 20 × 20 pixels. Therefore, selecting continuous features as a set does not have much impact on the outcome of the experiment. The flaw in algorithms that cannot be grouped automatically is not so obvious.

Table 1. The average recognition accuracy rate (%) of five different methods, highest values are in blod, the average is calculated by the mean accuracy rate across a range of feature set size (from 10 to the maximum number (700) of selected features).

Attributes	Lasso	Ridge	Elastic-net	Group lasso	MIL
Man	79.52	89.11	89.57	89.40	**90.42**
Female	79.59	89.10	89.57	89.40	**90.31**
Asian	93.58	94.91	94.73	94.94	**95.16**
White	81.27	84.82	81.64	**84.92**	83.18
Black	90.88	93.54	**94.59**	93.60	93.12
Indian	93.52	95.12	94.59	95.14	**95.24**
Youth	66.28	71.71	64.33	**71.87**	69.88
Midddle age	67.12	69.11	67.25	**69.25**	66.88
Senior	90.79	93.20	92.37	93.26	**93.34**
No glasses	96.41	97.16	95.38	97.21	**97.60**
Eye glasses	97.43	97.50	96.66	97.54	**98.01**
Sun glasses	98.78	99.19	98.74	99.20	**99.20**
Positive expression	56.08	71.82	71.87	**71.92**	68.00
Neutral expression	55.01	71.64	71.87	**71.92**	68.14

4.3 Efficiency

Through the above analysis, we can say that our method is effective in feature selection. In this experiment, we test the efficiency of MIL compared to other

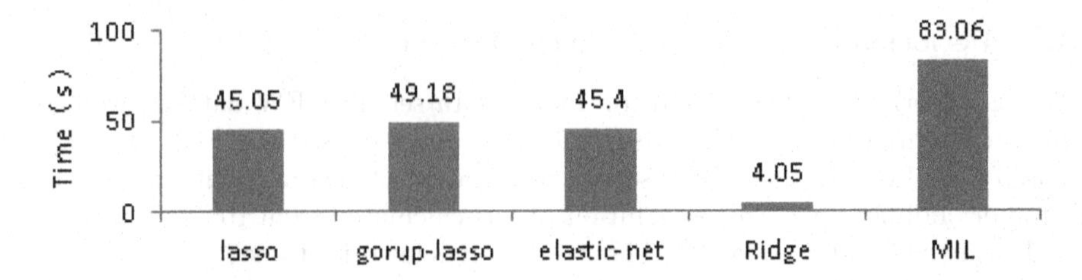

Fig. 4. Run time (s) comparison between MIL and other feature selection algorithms, measured by the time for computing W weight coefficient vector from gender attribute of the CFW data set.

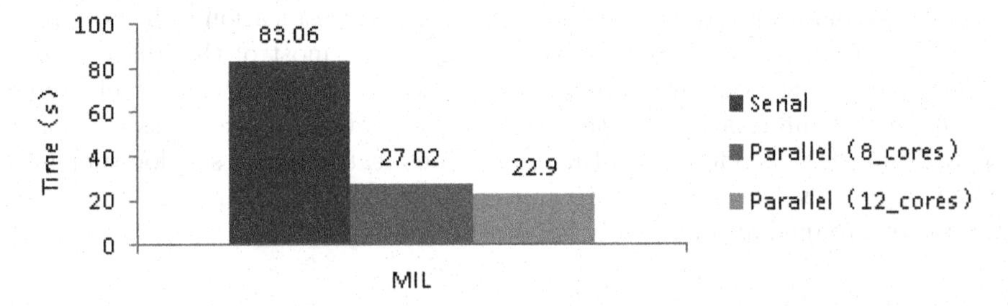

Fig. 5. Run time (s) comparison between serial and parallel MIL, measured by the time for computing W weight coefficient vector from gender attribute of the CFW data set.

approaches. However, we know the time complexity of MIL is $O(kNN)$, where N is defined as the feature numbers. MIL is thus generally more computationally demanding. In this paper, the raw features of face data set are 3304 dimensions. We use the time we calculate the weight coefficient W as a measure. In the experiment, which involves calculating L_1 regularization, we set the error tolerance to 10^{-6}. Taking face gender attribute as an example, the result is shown in Fig. 4.

As shown in Fig. 4, the proposed method takes 83.06 s to calculate CFW dataset with 3304 dimensions features. MIL is considerably expensive. Nevertheless, we believe that the time complexity should not be a major deterrent to the practicality of MIL. There are many applications where the data collecting time is far more than the time required for data mining tasks such as feature selection (e.g., days to months for data collection vs. hours for data mining). Commodity multi-core systems are common nowadays, and it is straightforward to parallelize MIL to harness this parallel processing power. In these cases, it is justifiable to spend significant amounts of time for data processing and the improved performance brought about by MIL will be worth the effort. Towards this end, we tested a parallel version of MIL. The effectiveness of parallelization can be clearly observed in Fig. 5.

As seen in Fig. 5, the parallel version of MIL can effectively reduce the computation time of W weights. When using a 8 core processor, the computational efficiency of the MIL is greatly improved, with a computation time of 27.02 s.

However, as the number of cores increases, the latter's promotion is not very obvious. After using parallel computing, the amount of time that MIL spends on computing the same dimensions training sample will be shortened to half of elastic-net and group-lasso.

5 Conclusions

In this paper, we proposed a novel feature selection method, which can realize automatic grouping. The proposed method is to use mutual information to achieve the minimum redundancy of each group. Because the amount of calculation of mutual information is increased by the increase in the number of features, we use parallel computing to reduce the cost of computing. It is worth mentioning that the calculation of L_1 regularization can not use parallel computing to speed up. We compare with other feature selection methods to evaluate the effectiveness of the proposed MIL. Experimental results show that the MIL can obtain high recognition rate with fewer feature dimensions, and outperforms the other methods on several public data sets.

Acknowledgement. The work is funded by the National Natural Science Foundation of China (*Nos*. 61371149, 61170155), Shanghai Innovation Action Plan Project (*No*. 16511 101200) and the Open Project Program of the National Laboratory of Pattern Recognition (*No*. 201600017)

References

1. Alon, U., Barkai, N., Notterman, D.A., Gish, K., Ybarra, S., Mack, D., Levine, A.J.: Broad patterns of gene expression revealed by clustering analysis of tumor and normal colon tissues probed by oligonucleotide arrays. Proc. Nat. Acad. Sci. **96**(12), 6745–6750 (1999)
2. Anne-Claire, H., Pierre, G., Jean-Philippe, V.: The influence of feature selection methods on accuracy, stability and interpretability of molecular signatures. PLoS One **6**(12), e28210 (2011)
3. Cheng, H., Liu, Z., Yang, L., Chen, X.: Sparse representation and learning in visual recognition: theory and applications. Sig. Process. **93**(6), 1408–1425 (2013)
4. Cong, Y., Wang, S., Liu, J., Cao, J., Yang, Y., Luo, J.: Deep sparse feature selection for computer aided endoscopy diagnosis. Pattern Recogn. **48**(3), 907–917 (2015)
5. Destrero, A., De Mol, C., Odone, F., Verri, A.: A sparsity-enforcing method for learning face features. IEEE Trans. Image Process. **18**(1), 188 (2009)
6. Destrero, A., De Mol, C., Odone, F., Verri, A.: A regularized approach to feature selection for face detection. In: Yagi, Y., Kang, S.B., Kweon, I.S., Zha, H. (eds.) ACCV 2007. LNCS, vol. 4844, pp. 881–890. Springer, Heidelberg (2007). https://doi.org/10.1007/978-3-540-76390-1_86
7. Donoho, D.L.: Compressed sensing. IEEE Trans. Inf. Theory **52**(4), 1289–1306 (2006)
8. Elad, M.: Sparse and Redundant Representations: From Theory to Applications in Signal and Image Processing. Springer, New York (2010). https://doi.org/10.1007/978-1-4419-7011-4

9. Fang, Y., Chang, L.: Multi-instance feature learning based on sparse representation for facial expression recognition. In: He, X., Luo, S., Tao, D., Xu, C., Yang, J., Hasan, M.A. (eds.) MMM 2015. LNCS, vol. 8935, pp. 224–233. Springer, Cham (2015). https://doi.org/10.1007/978-3-319-14445-0_20
10. Golub, T.R., Slonim, D.K., Tamayo, P., Huard, C., Gaasenbeek, M., Mesirov, J.P., Coller, H., Loh, M.L., Downing, J.R., Caligiuri, M.A., et al.: Molecular classification of cancer: class discovery and class prediction by gene expression monitoring. Science **286**(5439), 531–537 (1999)
11. Gu, Q., Li, Z., Han, J.: Generalized fisher score for feature selection, pp. 266–273 (2012)
12. Gui, J., Sun, Z., Ji, S., Tao, D., Tan, T.: Feature selection based on structured sparsity: a comprehensive study. IEEE Trans. Neural Netw. Learn. Syst. **PP**(99), 1–18 (2016)
13. Guyon, I., Elisseeff, A.: An introduction to variable and feature selection. J. Mach. Learn. Res. **3**(6), 1157–1182 (2003)
14. Guyon, I.: Pattern classification. Pattern Anal. Appl. **1**(2), 142–143 (1998)
15. He, X., Yan, S., Hu, Y., Niyogi, P., Zhang, H.J.: Face recognition using Laplacianfaces. IEEE Trans. Pattern Anal. Mach. Intell. **27**(3), 328–340 (2005)
16. Hou, C., Nie, F., Yi, D., Wu, Y.: Feature selection via joint embedding learning and sparse regression. In: Proceedings of the International Joint Conference on Artificial Intelligence, IJCAI 2011, Barcelona, Catalonia, Spain, July, pp. 1324–1329D (2011)
17. Huang, J., Horowitz, J.L., Ma, S.: Asymptotic properties of bridge estimators in sparse high-dimensional regression models. Ann. Stat. **36**(2), 587–613 (2008)
18. Kim, Y., Kim, J.: Gradient lasso for feature selection. In: International Conference on Machine Learning, p. 60 (2004)
19. Li, Y., Wang, R., Liu, H., Jiang, H.: Two birds, one stone: jointly learning binary code for large-scale face image retrieval and attributes prediction. In: IEEE International Conference on Computer Vision, pp. 3819–3827 (2015)
20. Saeys, Y., Inza, I., Larrañaga, P.: A review of feature selection techniques in bioinformatics. Bioinformatics **23**(19), 2507 (2007)
21. Tang, J., Alelyani, S., Liu, H.: Feature selection for classification: a review. In: Documentacin Administrativa, pp. 313–334 (2014)
22. Tibshirani, R., Saunders, M., Rosset, S., Zhu, J., Knight, K.: Sparsity and smoothness via the fused lasso. J. R. Stat. Soc. Ser. B (Stat. Methodol.) **67**(1), 91–108 (2005)
23. Vo, N., Moran, B., Challa, S.: Nonnegative-least-square classifier for face recognition. In: Yu, W., He, H., Zhang, N. (eds.) ISNN 2009. LNCS, vol. 5553, pp. 449–456. Springer, Heidelberg (2009). https://doi.org/10.1007/978-3-642-01513-7_49
24. Wright, J., Ma, Y., Mairal, J., Sapiro, G., Huang, T.S., Yan, S.: Sparse representation for computer vision and pattern recognition. Proc. IEEE **98**(6), 1031–1044 (2010)
25. Hang, X., Wu, F.X.: Sparse representation for classification of tumors using gene expression data. J. Biomed. Biotechnol. **2009**(1), 403689 (2009)
26. Yan, H., Yang, J.: Sparse discriminative feature selection. Pattern Recogn. **48**(5), 1827–1835 (2015)
27. Ye, J., Liu, J.: Sparse methods for biomedical data. ACM (2012)
28. Zhang, S., Huang, J., Li, H., Metaxas, D.N.: Automatic image annotation and retrieval using group sparsity. IEEE Trans. Syst. Man Cybern. Part B Cybern. **42**(3), 838–849 (2012). A Publication of the IEEE Systems Man & Cybernetics Society

29. Zhou, J., Liu, J., Narayan, V.A., Ye, J.: Modeling disease progression via fused sparse group lasso. In: ACM SIGKDD International Conference on Knowledge Discovery and Data Mining, pp. 1095–1103 (2012)
30. Zou, H.: The adaptive lasso and its oracle properties. J. Am. Stat. Assoc. **101**(476), 1418–1429 (2006)
31. Zou, H., Hastie, T.: Regularization and variable selection via the elastic net. J. R. Stat. Soc. **67**(2), 768–768 (2005)

Video Analysis and Event Recognition

Deep Key Frame Extraction for Sport Training

Meng Jian[1], Shijie Zhang[1], Xiangdong Wang[2](✉), Yudi He[1], and Lifang Wu[1]

[1] School of Information and Communication Engineering,
Beijing University of Technology, Beijing 100124, China
[2] Sports Science Research Institute of the State Sports General Administration,
Beijing 10000, China
908583913@qq.com

Abstract. For some professional sports, it is required to supervise and analyze the athletics pose in training of athletes. In order to facilitate the browse of training videos, it's necessary to extract the key frames from training videos. In this paper, we propose a deep key frame extraction method for analyzing sport training videos. To alleviate the bias from complex background, Fully Convolutional Networks (FCN) is employed firstly to extract the foreground region which contains the athlete and barbell. Then over the extracted foregrounds, Convolutional Neural Networks (CNN) are leveraged to estimate the pose probability of each frame and extract the key frames by the maximum probability on each pose. The experimental results demonstrate that the proposed method achieves good performance in key frame extraction of sport videos comparing method.

Keywords: Key frame extraction · Pose estimation · Sport video
Fully Convolutional Networks (FCN)
Convolutional Neural Networks (CNN)

1 Introduction

With the development of sports, effective analysis of sport training becomes increasingly important. In some sports, it's necessary to analyze athletes' pose accurately in continuous actions. For example, in weight lifting, there are four key poses: (1) the athlete picks up the barbell to knees in pose of squat. (2) the athlete lifts the barbell and extend the knees. (3) the barbell is lifted up rapidly. (4) the athletes squat and lift the barbell to the top. Figure 1 provides examples of the key poses. Just like weight lifting, every sport has the unique key poses. Moreover, even in the same sport, different athlete acts different to some degree

X. Wang—This work was supported in part by Beijing Municipal Education Commission Science and Technology Innovation KZ201610005012, in part by the China Postdoctoral Science Foundation funded project 2017M610026 and in part by Project 1602 supported by the Fundamental Research Founds for the China Institute of Sport Science.

J. Yang et al. (Eds.): CCCV 2017, Part III, CCIS 773, pp. 607–616, 2017.
https://doi.org/10.1007/978-981-10-7305-2_51

on the same pose. Therefore, analyzing the video scientifically is a promising way to improve training quality [9].

In this work, we aim to extract the key frames from the training videos for the coach who is professional in supervising athletes in sports. Automatic key frame extraction would facilitate the coaches work and assist athlete improvement in professional actions. Therefore the extracted key frames should contain the key pose of athlete in sport videos. However, the contents in sports videos are full of complex background with a big scope of variations.

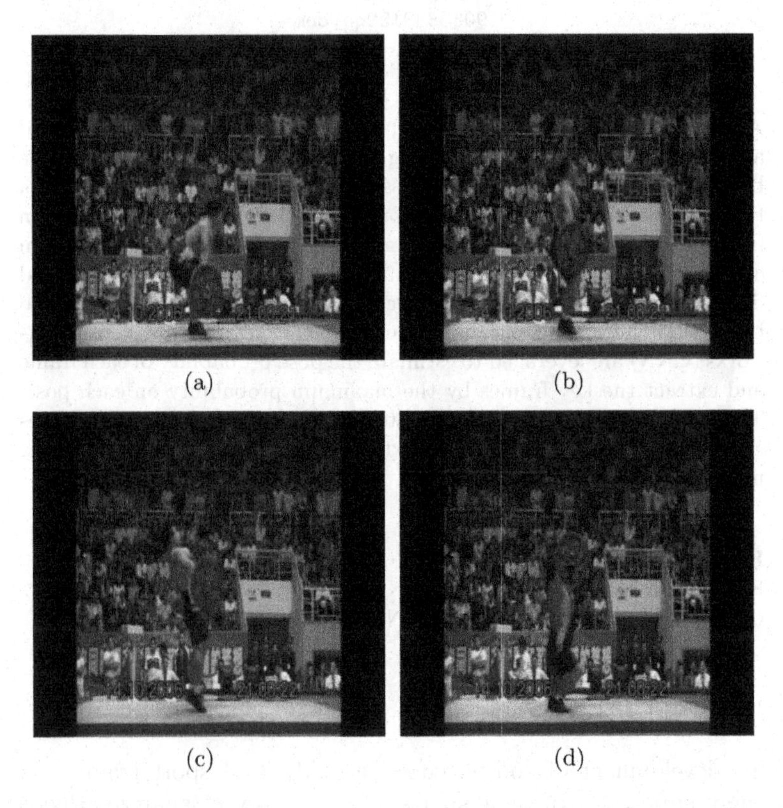

Fig. 1. Examples of key poses. (a) The athlete picks up the barbell to knees in pose of squat. (b) The athlete lifts the barbell and extend the knees. (c) The barbell is lifted up rapidly. (d) The athletes squat and lift the barbell to the top.

In some cases, the background of much audience becomes a big obstacle to focus on the characteristics of athlete in the video. In addition, there may exist some more barbells lying on the floor which also tends to make confusion to extract features of the barbell with the athlete. Figure 2(a) provides an example of complex background with many audiences and advertisement boards. Moreover, the other moving objects with various shifts also make confusion to analyze the key pose of the current frame. As an example, Fig. 2(b) illustrates a moving

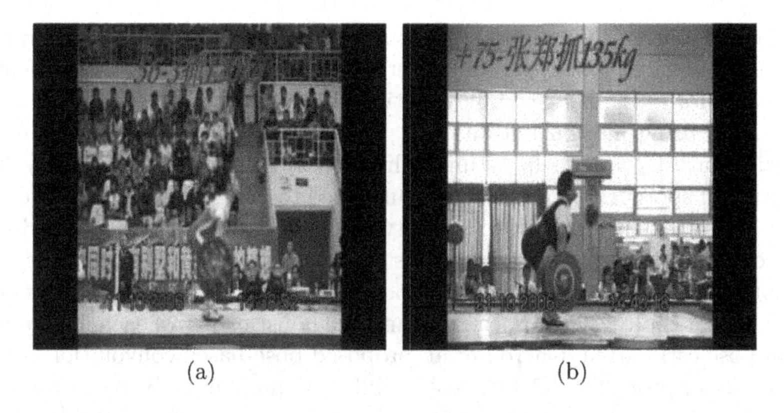

(a) (b)

Fig. 2. Representative examples of complex background in sport competition videos. (a) There are many audiences in the auditorium and advertisement boards in front of the auditorium. (b) The air fan is running, and the shift of the air fan between frames is larger than that of the athlete.

fan that takes larger shifts than the key athlete. It implies that optical flow based techniques [10] are not proper to capture the corresponding characteristics of the key athlete for pose analysis and key frame extraction. How to deal with the bias from complex background is challenging for key frame extraction methods. The neighboring frames with similar contents also make obstacle to extract the key frames for different poses as examples in Fig. 3.

(a) (b) (c) (d)

Fig. 3. Examples of neighboring frames with similar contents, especially similar key pose of athlete.

Some researchers employed motion information in analyzing actions of the videos [6–8]. Laptev et al. [6] present a method for video classification that builds upon and extends several recent ideas including local space-time features, space-time pyramids and multi-channel non-linear SVMs. The algorithm [7] is able to recognize and localize multiple actions in long and complex video sequences containing multiple motions. Schuldt et al. [8] demonstrate how such features can be used for recognizing complex motion patterns. However, the other moving objects also bring motion information. The motion information is not enough to distinct athlete and help estimate his poses. As a special case, key frame extraction of sport training videos is transferred to pose estimation and key pose extraction of frames. Recently, a variety of deep learning based methods [1–3] have been developed for human pose estimation. Ng et al. [1] fused raw frames and optical flow features to estimate poses of videos. Fan et al. [2] integrated both the local (body) part appearance and the holistic view of each local part for human pose estimation. Cheron et al. proposed pose-based convolutional neural network descriptor (P-CNN) [3] using different human body parts for action recognition. Inspired by these works, we employ deep neural networks of FCN and CNN to estimate poses of sport training videos, which aims to guide the training procedure and assist the coach to correct poses of athlete in professional sports.

In this work, we propose deep key frame extraction (DKFE) method for sport training videos. DKFE first extracts the foreground of athlete and barbell by FCN to alleviate the bias from complex background. Then we estimate key pose probabilities of each frame to extract the frames of key poses with the maximum probability. The main contributions of the proposed deep key frame extraction method are summarized as follows:

- Pose analysis of frames is constructed over foreground extraction of athlete by FCN for sport training video analysis. The extraction by FCN effectively avoids the bias from background.
- CNN is leveraged to estimate frame probability to each pose and extract key frames from athlete training videos with the maximum probability of each pose.
- The proposed deep key frame extraction method successfully applies the deep learning model of FCN and CNN to sport training video analysis.

2 Deep Key Frame Extraction for Sport Training Videos

In this section, we describe the details of the proposed deep key frame extraction (DKFE) method for sport training. Figure 4 illustrates the framework of the proposed method for key frame extraction of sport videos.

In Fig. 4, the proposed DKFE contains two parts: pose probability estimation of each frame and key frame extraction with the strategy of choosing the frame with relatively maximum probability located in the center of similar frames. DKFE focuses on investigating the pose of athlete in sport videos. Therefore, DKFE firstly investigates foreground of the first frame in sport video and employ FCN to extract the foreground region of the athlete and barbell, which effectively avoids the influence from complex background. Then all the frames of the video are cropped referring to the first frame. The CNN model is fine-tuned with training image set and leveraged to estimate key pose probabilities of the

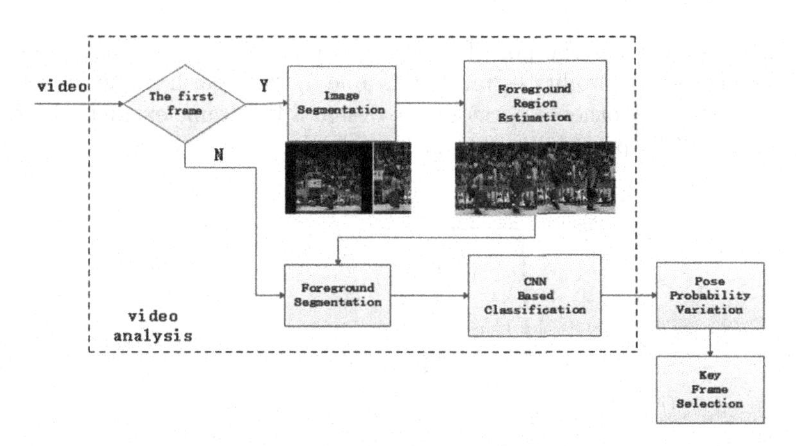

Fig. 4. The framework of the proposed deep key frame extraction method for sport videos.

frames. Finally, DKFE extracts the frames with relatively maximum key pose probabilities.

2.1 Foreground Extraction

In scenario of analyzing sport training videos, the athlete and barbell are considered as foreground region while the audiences, floor and the other objects are treated as background region. We randomly select 373 frames from all the videos labeled foreground and background regions to train a proper FCN model for foreground extraction.

2.2 CNN Based Pose Categorization

Then CNN model is trained over the extracted foreground regions without bias from background to distinct different key poses. We have 3062 key frames from 516 videos totally, where 1837 frames are selected to train a CNN model, 613 frames are collected to validate the model and 612 frames are remained to test the model to category different poses. The proposed DKFE employs CAFFE-NET to category the poses. Moreover, as CAFFE-NET has only 8 layers, it costs a short time to train a proper model. We fine-tune the CNN model in DKFE over the pre-train ImageNet model.

2.3 Pose Probability Estimation

Then the trained CNN model is conducted to estimate key pose probabilities of testing videos. Figure 5 provides the average estimated four key pose probabilities of videos, where x-axis denotes frame sequence of videos and y-axis represents the estimated key pose probabilities. In Fig. 5, four key poses denoted by different colors are investigated and the video plays poses one by one. The results in

Fig. 5 illustrate that every pose focuses on a quarter in order. It implies that CNN based pose probability estimation is able to distinguish frame sequences of the poses with each other. Therefore, we design a key frame extraction strategy based on the CNN probability estimation model.

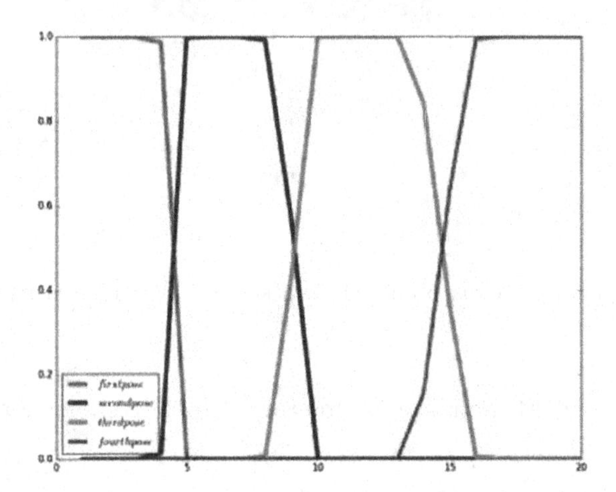

Fig. 5. Average estimated key pose probabilities of frames in sport videos. The four key poses are represented with four different colors, respectively. (Color figure online)

2.4 Key Frame Extraction Strategy

From a sport training video, the target of this work is to select the corresponding frames of key poses. Intuitively we attempt to extract the frame with maximum key pose probability. However as shown in Fig. 5, there exists many frames with the similar key pose probability which means they tend to contain similar foreground contents. Therefore, alternatively we select the center frame as the key frame when the probability difference between the neighboring frames is smaller than a tolerance as

$$F_i = \begin{cases} \arg\max\{p_i^j, i = 1, 2, 3, 4, j = 1, 2, \cdots, N\} & e_i \geqslant E \\ \arg\{p_i = p_i^{M/2}\} & e_i < E \end{cases} \tag{1}$$

where F_i represents the key frame of i-th pose, ${P_i}^j$ denotes the key pose probability of j-th frame to the i-th pose and M is the number of neighboring frames with similar probabilities. When the probability difference of adjacent frames e_i is larger than E, we select the frame with the maximum key pose probability as the key frame. If the e_i is less than E, we alternatively select the frame of ${p_i}^{M/2}$ as the corresponding key frame.

3 Experimental Results

We perform experiments and their corresponding analysis to verify the superiority of the proposed DKFE in key frame extraction of sport videos. In this section, we present the implementation details and compare the performance of the proposed method with some sport key frame extraction methods such as Wu's method [9] and deep learning based method [8]. We conduct experiments on a video database collected from general administration of sport of China, which contains 516 sport videos with various kinds of weightlifting competitions. All the videos are recorded from a side of the athlete.

3.1 Implementation Details

The proposed DKFE framework is built on the deep learning model of FCN [4] and CNN [5]. The whole network is fine-tuned with an initialization of the pre-trained CAFFE-NET. The parameters in FCN based foreground extraction network are set with mini-batch size of 1, base learning rate to 1×10^{-3}, momentum to 0.9, weight decay as 5×10^{-4}, and maximum number of training iterations to 2000. The parameters of CNN based pose probability estimation network are set as mini-batch size of 256, base learning rate to 1×10^{-4}, momentum to 0.9, weight decay to 5×10^{-4}, and maximum number of training iterations to 20000.

3.2 Experimental Analysis

For key frame extraction from sport training videos, the proposed DKFE leverages two deep models of FCN based foreground extraction and CNN based pose probability estimation modules to figure out the frames with key poses. In the following, the performance of foreground extraction with FCN and key pose probability estimation by CNN are presented respectively.

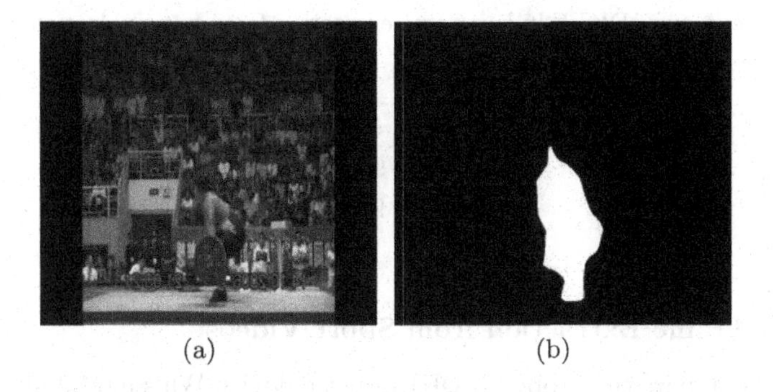

(a) (b)

Fig. 6. An example of foreground extraction result with FCN in DKFE.

Foreground Extraction with FCN. FCN model is trained with randomly selected 373 frames labeled foreground and background regions for athlete related foreground extraction. After a proper FCN model is trained, the corresponding threshold for FCN to separate foreground and background can be determined as 0.85 with exhaustive experiments on randomly selected testing frames. The pixels with the estimated foreground probabilities larger than the threshold are binarized to foreground, vice versa. Figure 6 provides an example frame of FCN based foreground extraction, (b) is the binary mask of foreground extraction by FCN on the frame in (a). The foreground extraction helps effectively prevent bias from the complex background in sport videos.

Key Pose Estimation by CNN. Then a CNN model is trained on the extracted foreground to estimate key pose probabilities of frames. For example, there are four key poses in weight lifting. The trained CNN model is employed to estimate likelihood probability of each frame to the four poses as Fig. 5. As the proportional size of training, validating and testing dataset in 3:1:1, we have 612 frames to test. Table 1 provides the accuracy of each pose estimated by CNN. The results indicate that the trained CNN model successfully estimates the poses of testing frames.

Table 1. Accuracy of each pose by CNN on testing frames.

	Total	Correct	Wrong	Accuracy
Pose1	169	165	4	97.6%
Pose2	130	123	7	94.6%
Pose3	155	152	3	98.1%
Pose4	158	156	2	98.7%

Table 2. Accuracy of DKFE in key frame extraction from sport videos compared with Wus method [9].

Method	Accuracy
Wu [9]	90.6%
DKFE	97.4%

3.3 Key Frame Extraction from Sport Videos

We further compare the proposed DKFE method with Wu's method [9] and deep learning based method [5]. CNN based method [5] did not employ techniques in avoiding bias from complex background. With the same settings to [9], DKFE is compared to [9] to extract key frames of sport videos in Table 2. The results illustrate DKFE outperforms Wus method [9] with almost 7% improvement.

Compared with the CNN method [5], Fig. 7 provides the estimated pose probability of DKFE for key frame extraction. In Fig. 7(a) the fourth pose did not distinct apparently to the others. Because in [5] the feature of the fourth pose is influenced by the complex background and the CNN model could not capture proper characteristics of the pose. However, Fig. 7(b) gives an estimation of the fourth pose as good as that of the other poses. It benefits from the foreground extraction by FCN for CNN based pose estimation for key frame extraction from sport videos. Therefore, we can conclude that the proposed DKFE is capable of extracting key frames from sport videos to aid the sport training.

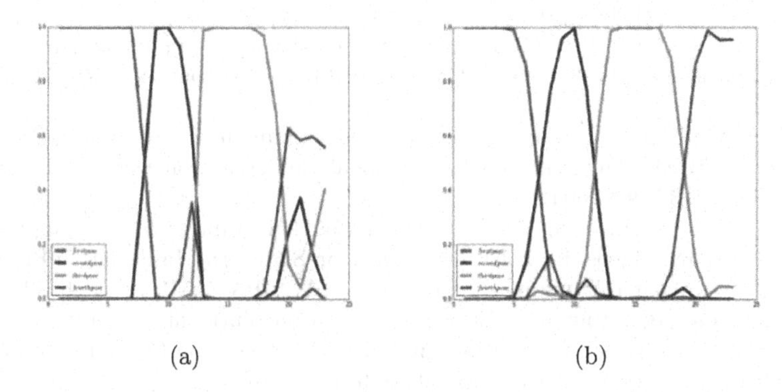

Fig. 7. The pose probability estimation of DKFE compared with CNN based method [5] for key frame extraction from sport videos.

4 Conclusion

In this paper, we propose deep key frame extraction (DKFE) for sport training videos. DKFE employs FCN based foreground extraction for the following pose estimation, which effectively prevents bias from complex background to the pose estimation of frames. Moreover, CNN model in DKFE performs to estimate key pose probability of frames in an video and extract the frames of key poses with relatively maximum probability centered at the frames of similar probability. The experimental results demonstrate that the proposed DKFE has a good ability in extracting key frames from sport training videos.

References

1. Ng, J.Y., Hausknecht, M., Vijayanarasimhan, S.: Beyond short snippets: deep networks for video classification. In: IEEE Conference on Computer Vision and Pattern Recognition (CVPR), pp. 4694–4702 (2015)
2. Fan, X., Zheng, K., Lin, Y., Wang, S.: Combining local appearance and holistic view: dual-source deep neural networks for human pose estimation. In: IEEE Conference on Computer Vision and Pattern Recognition (CVPR), pp. 1347–1355 (2015)

3. Cheron, G., Laptev, I., Schmid, C.: P-CNN: pose-based CNN features for action recognition. In: Proceedings of International conference on Computer Vision (ICCV), pp. 3218–3226 (2015)
4. Shelhamer, E., Long, J., Darrell, T.: Fully convolutional networks for semantic segmentation. In: IEEE Conference on Computer Vision and Pattern Recognition (CVPR), pp. 3431–3440 (2015)
5. Krizhevsky, A., Sutskever, I., Hinton, G.E.: ImageNet classification with deep convolutional neural networks. In: ACM Conference on Neural Information Processing Systems (NIPS), pp. 1097–1105 (2012)
6. Laptev, I., Marszalek, M., Schmid, C., Rozenfeld, B.: Learning realistic human actions from movies. IEEE Conference on Computer Vision and Pattern Recognition (CVPR), pp. 1–8 (2008)
7. Niebles, J.C., Wang, H., Li, F.F.: Unsupervised learning of human action categories using spatial-temporal words. ACM Trans. Int. J. Comput. Vis. **79**(3), 299–318 (2008)
8. Schuldt, C., Laptev, I., Caputo, B.: Recognizing human actions: a local SVM approach. In: IEEE Conference on International Conference on Pattern Recognition (ICPR), pp. 32–36 (2004)
9. Wu, L., Zhang, J., Yan, F.: A poselet based key frame searching approach in sports training videos. In: IEEE Conference on Signal and Information Processing Association Annual Summit and Conference (APSIPA ASC), pp. 1–4 (2012)
10. Feichtenhofer, C., Pinz, A., Zisserman, A.: Convolutional two-stream network fusion for video action recognition. In: IEEE Conference on Computer Vision and Pattern Recognition (CVPR), pp. 1933–1941 (2016)

Human Action Recognition Based on Sub-data Learning

Yang Chen[1], Tian Wang[1(✉)], Jiakun Li[1], Xiaowei Lv[2], and Hichem Snoussi[3]

[1] School of Automation Science and Electrical Engineering, Beihang University, Beijing, China
{youngcy1994,wangtian,lijiakun}@buaa.edu.cn

[2] Information Science Academy of China Electronics Technology Group Corporation, Beijing, China
lvxiaowei@cetc.com.cn

[3] Institut Charles Delaunay-LM2S-UMR STMR 6279 CNRS, University of Technology of Troyes, Troyes, France
hichem.snoussi@utt.fr

Abstract. Human action recognizing nowadays plays a key role in varieties of computer vision applications while at the same time it's quite challenging for the requirement of accuracy and robustness. Most current computer vision methods focus on algorithms designing classifiers with handcrafted features which are complex and inflexible. To automatically extract both spatial and temporal features, in this paper we propose a method of human action recognition based on sub-data learning which combines the proposed 3D convolutional neural network (3DCNN) with the One-versus-One (OvO) algorithm. We also employ effective data augmentation to reduce overfitting. We evaluate our method on the KTH and UCF Sports dataset and achieve promising results.

Keywords: Action recognition · 3DCNN · Sub-data learning

1 Introduction

The widely used surveillance cameras nowadays give rise to explosive demand on the recognition of objects in massive videos especially the human actions. The accurate human action recognition has found varieties applications in people's daily lives such as intelligent video surveillance, smart home, somatic gaming and so on. However, this recognition task is full of challenges due to obstacles like background clutters, scale variations, object occlusions, viewpoint shifts, etc. In the last decade, many efforts have been made in this area. However, most

T. Wang—This work is partially supported by the ANR AutoFerm project and the Platform CAPSEC funded by Région Champagne-Ardenne and FEDER, the Fundamental Research Funds for the Central Universities (YWF-14-RSC-102), the Aeronautical Science Foundation of China (2016ZC51022), the National Natural Science Foundation of China (U1435220, 61503017).

J. Yang et al. (Eds.): CCCV 2017, Part III, CCIS 773, pp. 617–626, 2017.
https://doi.org/10.1007/978-981-10-7305-2_52

of them focus on designing classifiers applying handcrafted feature extracting algorithms, which is inflexible when taking both accuracy and robustness into consideration. Schuldt *et al.* [18] constructed local space-time feature representations in videos and incorporated the support vector machine (SVM) method for action recognition. Scovanner *et al.* [19] introduced the 3-dimensional (3D) SIFT descriptor in action recognition and improved the performance a lot. To achieve better results, Wong and Cipolla [26] took the global information encoded in videos into consideration. They developed their research on the organization of pixels in the video sequences and proposed a detector utilizing a set of interest points. Besides, the combination of many effective feature descriptors has been widely used such as HOG (Histogram of Oriented Gradients) + HOF (Histogram of Optical Flow) + MBH (Motion Boundary Histogram) [22], DT (Dense Trajectories) + BOF (Bag of Features) [24] and so on.

The Convolutional Neural Networks (CNN) have shown its advantages in computer vision. Many tasks, such as image segmentation [16], object tracking [15], etc. have been handled successfully by CNN. The hierarchy of features are learnt by the CNN. In the human action recognition problem, the spatial-temporal features should be constructed. Ji *et al.* [6] developed a novel 3D CNN architecture which fused useful spatial-temporal features. While their model took gradients and optical flow feature as the input of the neural networks. Moreover, Karpathy *et al.* [8] and other researchers [7,11,25] proposed new architectures to recognize the action. Considerable amount of training data are needed for these complex architectures, but they tend to suffer from the overfitting problem while the train sample size is small.

In the deep learning models, many useful training strategies have been pointed out to improve the results, a representative among which is the data augmentation method. To prevent from overfitting, Krizhevsky et al. [9] employed translations, horizontal reflections and RGB intensity altering on the training images to generate more samples. Jung et al. [7] then added image rotation operation to obtain 14 times more training data. To further prevent from overfitting and improve the robustness, Molchanov et al. [11] creatively introduced two more data augmentation methods: spatial elastic deformation and image pixel drop-out.

In this paper, we are interested in the human action recognition on the KTH dataset [18] and the UCF Sports dataset [17]. We develop our model with the 3D convolutional neural networks. To reduce overfitting, we apply the effective data augmentation method on the input video volumes. To boost the model's performance to the best, we incorporate the One-versus-One (OvO) algorithm in our model which leads to an acceptable result as we wish.

2 Methodology

This part is organized as follows. We first briefly introduce the datasets used in our work in Sect. 2.1. We then provide some background information needed for our model in Sect. 2.2. Section 2.3 describes the details of the data preprocessing adopted in our model. Section 2.4 centers on the 3DCNN frameworks of our model. And Sect. 2.5 shows the details during training.

2.1 Datasets

We construct our model based on the KTH dataset and the UCF Sports dataset. The KTH action database contains six different types of human actions (walking, jogging, running, boxing, hand waving and hand clapping) and each of them is performed by 25 people in four different scenarios: outdoors, outdoors with scale variation, outdoors with different clothes and indoors. There are 100 videos for each of the human actions except the handclapping action which has only 99 videos. And all the videos have the frame rate of 25 fps and time length of four seconds shot by a static camera. To reduce the computation cost, the videos are all down-sampled to the resolution of 160 × 120 pixels.

The UCF Sports database consists of ten human action types (diving, golf swing, kicking, lifting, riding Horse, running, skateboarding, swing-bench, swing-side and walking). All these video sequences are collected from many sports scenes in broadcast television channels. There are 150 videos in total with the resolution of 720 × 480 pixels and they contain unconstrained background environments (Figs. 1 and 2).

Fig. 1. The KTH dataset: example video sequences with different people in four different scenarios.

Fig. 2. The UCF Sports dataset: different sports scenes.

2.2 Background

3D Convolution. The convolution operations are only applied to spatial dimensions in the typical 2D CNN, which cannot capture the temporal features useful for action recognition. Different from 2D convolutions, 3D convolutions extend the convolution operations to the temporal dimensions. By convolving 3D kernels on the give spatial-temporal video volumes, the network is capable of obtaining temporal dynamic information encoded in several adjacent frames, which is needed for action recognition.

Before the 3D convolutions, we firstly extract a few contiguous frames from the original video sequences and then stack them on the frame level to form a video volume of size $w \times h \times d$, which represent the width, height and depth (temporal length) separately. And then we apply 3D convolutional kernels with size of $w' \times h' \times d'$ across the volume to get numbers of feature maps. The calculation follows this way: the output value in the feature maps corresponding to the input value at position (x, y, z) is:

$$v_{xyz} = f(\sum_{i=0}^{w\prime-1}\sum_{j=0}^{h\prime-1}\sum_{k=0}^{d\prime-1} w_{ijk}k_{(x+i)(y+j)(z+k)} + b) \tag{1}$$

where f denotes the activation function, w_{ijk} denotes the weight value of the 3D kernel with index (i, j, k), $k_{(x+i)(y+j)(z+k)}$ denotes the input value at position $(x + i, y + j, z + k)$ and b denotes the bias value.

One-versus-One Algorithm. The OvO (One-versus-One) algorithm is one of the classic multiclass classification algorithms in machine learning [1]. The core idea of this algorithm is to reduce the problem of multiclass classification to multiple binary classification problems. In the reduction, we train $\frac{N\times(N-1)}{2}$ binary classifiers each of which receives the data samples from a pair of classes from the original training dataset. At the training stage, we focus on making each classifier learn to distinguish between the corresponding two classes. At prediction stage, we feed the test data to all these classifiers to get $\frac{N\times(N-1)}{2}$ results for each sample. With these results, the strategy we apply is the majority voting: the class that gets the highest votes is selected as the predicted class.

In our sub-data learning work, instead of directly designing a multiclass classification classifier, we train 15 and 45 binary classifiers on the two datasets according to the OvO algorithm. Finally we sum up all the validation results and produce the overall classification result.

2.3 Data Preprocessing

Target Area Segmentation. To achieve lower computational cost and make the result of higher precision, we don't feed our convolutional networks with the volumes made up of the original frames from the videos. Considering part of the video sequences contains scale variation and the area occupied by the person only accounts for a small percentage of the whole area in most of the frames, instead we adopt a human detector with the help of the HOG-SVM algorithm detailed described in [2]. After getting the detection results, we crop the area where the moving person always stay in the center out of the original frames and then resize them to the size of 40×60 pixels as shown in Fig. 3. When it comes to the temporal length of the volumes, the length should be big enough to catch a complete human motion, while at the same time as small as enough to gain reduction in computation. And according to our careful observation, the number 16 cannot be more appropriate for this very purpose. By this means, the extracted target area has a size of $40 \times 60 \times 16$ pixels.

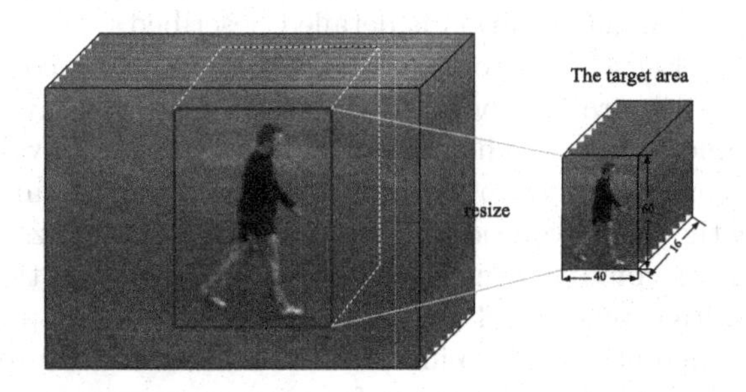

Fig. 3. Illustration of the target area segmentation method.

Data Augmentation. Note that the KTH dataset only contains 599 pieces of videos clips and the UCF Sports datasets just 150. It is far not enough to prevent from overfitting during training with this kind of sample size. As mentioned in [9], label-preserving transformation on the original dataset is the easiest way to reduce overfitting. Motivated by their work, we use two different data augmentation methods in our work to enlarge the training data while keeping the test data unchanged.

In our first method of data augmentation, we extract four 35×55 pixels patches from four corners of the 40×60 pixels target area cropped out according to the target area segmentation section and then do translations moving each of them in the diagonal direction (± 1 pixels along the x axis, and ± 1 pixels along the y axis). Then again we extract the center part with size of 35×55 pixels from the target area. So all above operations together make the number of patches for training increased by a factor of 9.

Our second method applies reverse operations along the temporal dimension to the output generated by the first augmentation method, with which we double the amount of the training data. For example, we create a new action that a man runs from left to right from the original action that the man runs from right to left. By this time, we increase the sample size by a factor of 18 in total.

Note that we only do the data augmentation scheme on the training data. Without the data augmentation our model suffers a lot from the overfitting problem during the experiments. And if the data augmentation is applied, the input size of the network will be smaller. Thus, at validation time we also extract 5 patches from the original test data (four corners and the center part) and then feed them to our model for validation.

2.4 Our Model Architecture

As mentioned above, our model combines the 3D convolutional neural networks (3DCNN) with the One-versus-One (OvO) algorithm. Thus, we divide the dataset into many sub-groups on each of which we train the 3DCNN which learns to distinguish between the corresponding two actions. In the following, the 3DCNN architecture that can effectively capture the temporal and spatial features useful for the classification is detailed described.

As depicted in Fig. 4, the proposed 3DCNN architecture contains six layers of which the first five are 3D convolutional or max-pooling layers and the last one is a fully connected layer. The input contains a number of volumes with size $35 \times 55 \times 16$, corresponding to picture size 35×55 and temporal length 16. Firstly we apply the 3D convolutional operation with a kernel size of $5 \times 5 \times 3$ (5×5 in the spatial dimension and 3 in the temporal dimension) on the input volume data. It's known that one kernel can only get one feature map from the input so we apply 16 different kernels to increase the number of feature maps. And to prevent the output size of the first layer from decreasing too sharply, we use the padding strategy that pads zeros around the image borders after convolution. Then with a $2 \times 2 \times 1$ 3D max pooling operation on each of the feature maps we get the 16 feature maps with reduced size $17 \times 27 \times 16$ in the layer S2. Subsequently, we further perform the 3D convolution on the feature maps of S2 with 32 kernels of size $6 \times 7 \times 3$, leading to 32 feature maps in C3. After that we apply the $3 \times 3 \times 1$ max pooling on the output of C3. By this time, the spatial size of the output is small (4×7) so at the next layer we only apply the 2D convolution operations. Then C4 is obtained by applying 2D convolutions with 64 kernels of size 4×7. After all the convolution and subsampling operations, we flatten the feature maps of C4 to concatenate all of them into a long $896D$ feature vector containing lots of useful motion information. Next we design the fully connected layer FC with 1024 nodes which is fully connected to each unit of the feature vector. Finally we set the number of the output to 2, which is as same as the number of types of human actions, and the two values represent the probability of each motion hypothesis separately with the help of the softmax regression function.

2.5 Details of Training

To train the network, we choose the average cross-entropy as the loss function to minimize it:

$$l = -\frac{1}{N}\sum_{i=0}^{N} P(x^i) \cdot log(Q(x^i)), \tag{2}$$

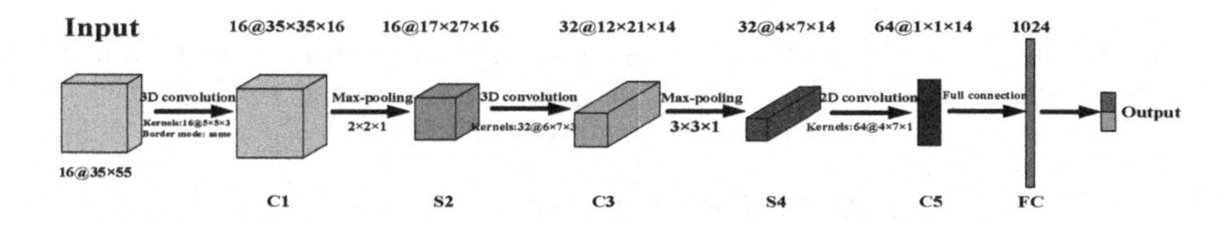

Fig. 4. The proposed 3DCNN structure in our model. It contains 3 convolutional layers, 2 max-pooling layers and 1 fully-connected layer.

where N is the total number of the samples of the data, x^i denotes the ith sample of the dataset, P and Q denote respectively the inherent probability distribution and the probability distribution of x predicted by the model.

In our experiments the weights in each layer are initialized from a truncated normal distribution centered on 0 with standard deviation $std = \sqrt{\frac{2}{n}}$ where n denotes the input or output connections at a layer. And we choose the ReLU activation function and set the biases for all the layers to 0 according to [4]. During the training stage, we apply drop-out strategy [21] with probability 0.5 after several layers and L2 regularization on the weights to overcome the possible overfitting problem. To accelerate the training, we also apply batch normalization [5] to the response of each layer.

3 Experimental Results

To evaluate the effectiveness of our model, we conduct the experiments on two benchmark datasets: the KTH dataset and the UCF Sports dataset. We first read the original video sequences frame by frame and down-sample them to the resolution of 40×60 pixels taking the memory and computation overhead into consideration. Then we extract and form the 35×55 pixels target areas according to the method mentioned in the Target Area Segmentation Section. Next, at train-test split, we randomly sample 10% from the reformed data as the validation part on the KTH dataset and the percentage is 33% on the UCF Sports dataset. After the data augmentation completes, for the KTH dataset, the training data contains 9702 video volumes with size $35 \times 55 \times 16$ pixels, and the test data contains 600 video volumes with the same size. And for the UCF Sports dataset, the corresponding two numbers are 1800 and 250, separately.

According to OvO algorithm, we generate 15 binary classifiers for the KTH dataset and 45 binary classifiers for the UCF Sports dataset. Each of this classifiers utilizes the 3DCNN architecture proposed by us and are fed by the subdata from the corresponding pair of classes. After rounds of training and the final fine-tune process, the classification results of these binary classifiers from the two datasets are reported in Tables 1 and 2. Then by summing up all these results and applying the majority voting strategy on the validation data, we finally get the overall validation accuracy: 94.0% on the KTH dataset and 95.6% on the UCF Sports dataset. The confusion matrices are shown in Figs. 5 and 6. And the comparison of our work to the peer work is demonstrated in Table 3.

Table 1. The 15 binary classifying results on the KTH dataset. A, B, C, D, E and F stands for boxing, handclapping, handwaving, jogging, running and walking separately. And the decimal numbers are the corresponding accuracies.

1	AB	0.96	6	BC	0.95	11	CE	0.99
2	AC	0.95	7	BD	0.98	12	CF	0.99
3	AD	0.99	8	BE	0.98	13	DE	0.96
4	AE	0.99	9	BF	0.98	14	DF	0.97
5	AF	0.99	10	CD	0.92	15	EF	0.97

Table 2. The 45 binary classifying results on the UCF Sports dataset. A, B, C, D, E, F, G, H, I and J stands for diving, golf swing, kicking, lifting, riding Horse, running, skateboarding, swing-bench, swing-side and walking separately. And the decimal numbers are the corresponding accuracies.

1	AB	1.00	10	BC	1.00	19	CE	1.00	28	DH	1.00	37	FH	0.91
2	AC	1.00	11	BD	1.00	20	CF	0.95	29	DI	1.00	38	FI	1.00
3	AD	1.00	12	BE	1.00	21	CG	1.00	30	DJ	1.00	39	FJ	0.94
4	AE	1.00	13	BF	1.00	22	CH	1.00	31	EF	1.00	40	GH	0.94
5	AF	1.00	14	BG	0.93	23	CI	1.00	32	EG	0.98	41	GI	1.00
6	AG	1.00	15	BH	1.00	24	CJ	1.00	33	EH	1.00	42	GJ	0.97
7	AH	0.92	16	BI	1.00	25	DE	1.00	34	EI	1.00	43	HI	1.00
8	AI	1.00	17	BJ	0.92	26	DF	1.00	35	EJ	0.91	44	HJ	1.00
9	AJ	1.00	18	CD	1.00	27	DG	1.00	36	FG	1.00	45	IJ	1.00

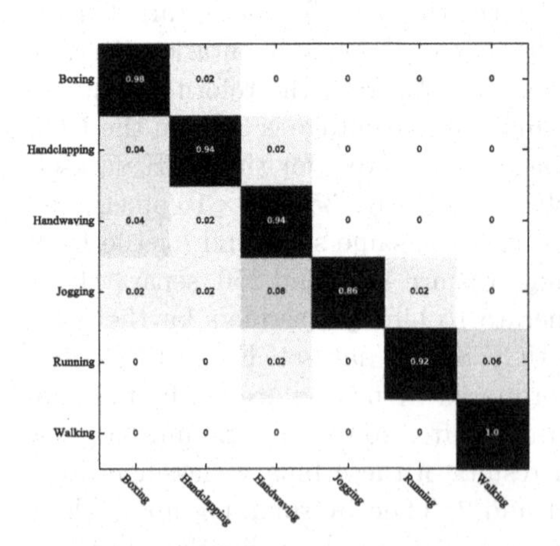

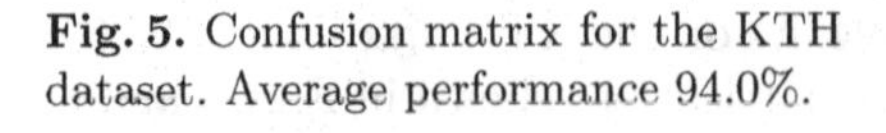

Fig. 5. Confusion matrix for the KTH dataset. Average performance 94.0%.

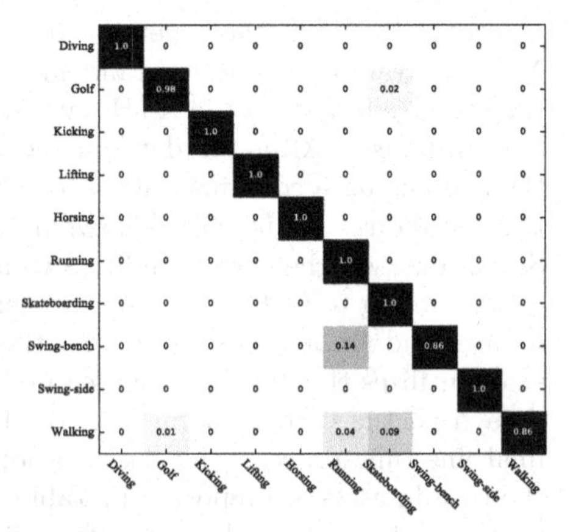

Fig. 6. Confusion matrix for the UCF sports dataset. Average performance 95.6%.

Table 3. Comparisons of our work to the peer work on the KTH and UCF Sports dataset.

Method	KTH	Method	UCF sports
Schuldt [18]	71.7	Wang [23]	85.5
Niebles [12]	83.3	Zhang [27]	87.3
Reddy [14]	89.8	OHara [13]	91.3
Ji [6]	90.2	Shao [20]	93.4
Wang [24]	94.2	Wang [22]	95.0
Liu [10]	95.5	Ghodrati [3]	95.7
Ours	94.0	Ours	95.6

4 Conclusions

In this paper, we focus on the action recognition problem on the KTH and UCF Sports dataset. Rather than capture the handcrafted features like most of the researchers do, we develope the 3D convolutional neural networks (3DCNN) to automatically caputure the useful spatial-temporal features. To boost our model's performance, we utilize the sub-data learning method that incorporate the One-versus-One (OvO) algorithm into our 3DCNN architecture. We achieve the high correct classification rate of 94.0% on the KTH dataset and 95.6% on teh UCF Sports dataset, which is quite competitive compared to the peer work.

References

1. Aly, M.: Survey on multiclass classification methods. Neural Netw. **19**, 1–9 (2005)
2. Dalal, N., Triggs, B.: Histograms of oriented gradients for human detection. In: 2005 IEEE Computer Society Conference on Computer Vision and Pattern Recognition (CVPR 2005) (2005)
3. Ghodrati, A., Diba, A., Pedersoli, M., Tuytelaars, T., Gool, L.V.: Deepproposals: hunting objects and actions by cascading deep convolutional layers. Int. J. Comput. Vis., pp. 1–17 (2017)
4. He, K., Zhang, X., Ren, S., Sun, J.: Delving deep into rectifiers: surpassing human-level performance on imagenet classification. In: Proceedings of the IEEE International conference on Computer Vision, pp. 1026–1034 (2015)
5. Ioffe, S., Szegedy, C.: Batch normalization: accelerating deep network training by reducing internal covariate shift. In: ICML (2015)
6. Ji, S., Xu, W., Yang, M., Yu, K.: 3D convolutional neural networks for human action recognition. IEEE Trans. Pattern Anal. Mach. Intell. **35**(1), 221–231 (2010)
7. Jung, H., Lee, S., Yim, J., Park, S., Kim, J.: Joint fine-tuning in deep neural networks for facial expression recognition. In: 2015 IEEE International Conference on Computer Vision (ICCV) (2015)
8. Karpathy, A., Toderici, G., Shetty, S., Leung, T., Sukthankar, R., Fei-Fei, L.: Large-scale video classification with convolutional neural networks. In: 2014 IEEE Conference on Computer Vision and Pattern Recognition (2014)

9. Krizhevsky, A., Sutskever, I., Hinton, G.E.: ImageNet classification with deep convolutional neural networks. In: NIPS (2012)
10. Liu, L., Shao, L., Rockett, P.: Boosted key-frame selection and correlated pyramidal motion-feature representation for human action recognition. Pattern Recogn. **46**, 1810–1818 (2013)
11. Molchanov, P., Gupta, S., Kim, K., Kautz, J.: Hand gesture recognition with 3D convolutional neural networks. In: 2015 IEEE Conference on Computer Vision and Pattern Recognition Workshops (CVPRW) (2015)
12. Niebles, J.C., Wang, H., Fei-Fei, L.: Unsupervised learning of human action categories using spatial-temporal words. Int. J. Comput. Vis. **79**, 299–318 (2006)
13. O'Hara, S., Draper, B.A.: Scalable action recognition with a subspace forest. In: 2012 IEEE Conference on Computer Vision and Pattern Recognition, pp. 1210–1217 (2012)
14. Reddy, K.K., Shah, M.: Recognizing 50 human action categories of web videos. Mach. Vis. Appl. **24**, 971–981 (2012)
15. Redmon, J., Divvala, S.K., Girshick, R.B., Farhadi, A.: You only look once: unified, real-time object detection. In: 2016 IEEE Conference on Computer Vision and Pattern Recognition (CVPR) (2016)
16. Ren, S., He, K., Girshick, R.B., Sun, J.: Faster R-CNN: towards real-time object detection with region proposal networks. In: NIPS (2015)
17. Rodriguez, M.D., Ahmed, J., Shah, M.: Action mach a spatio-temporal maximum average correlation height filter for action recognition. In: 2008 IEEE Conference on Computer Vision and Pattern Recognition, pp. 1–8 (2008)
18. Schuldt, C., Laptev, I., Caputo, B.: Recognizing human actions: a local svm approach. In: Proceedings of the 17th International Conference on Pattern Recognition 2004, ICPR 2004, vol. 3, pp. 32–36. IEEE (2004)
19. Scovanner, P., Ali, S., Shah, M.: A 3-dimensional sift descriptor and its application to action recognition. In: ACM Multimedia (2007)
20. Shao, L., Zhen, X., Tao, D., Li, X.: Spatio-temporal laplacian pyramid coding for action recognition. IEEE Trans. Cybern. **44**(6), 817–27 (2014)
21. Srivastava, N., Hinton, G.E., Krizhevsky, A., Sutskever, I., Salakhutdinov, R.: Dropout: a simple way to prevent neural networks from overfitting. J. Mach. Learn. Res. **15**, 1929–1958 (2014)
22. Wang, H., Kläser, A., Schmid, C., Liu, C.L.: Action recognition by dense trajectories. In: CVPR (2011)
23. Wang, H., Kläser, A., Schmid, C., Liu, C.L.: Dense trajectories and motion boundary descriptors for action recognition. Int. J. Comput. Vis. **103**, 60–79 (2012)
24. Wang, H., Schmid, C.: Action recognition with improved trajectories. In: 2013 IEEE International Conference on Computer Vision (2013)
25. Wang, L., Xiong, Y., Wang, Z., Qiao, Y.: Towards good practices for very deep two-stream convnets. CoRR abs/1507.02159 (2015)
26. Wong, S.F., Cipolla, R.: Extracting spatiotemporal interest points using global information. In: 2007 IEEE 11th International Conference on Computer Vision (2007)
27. Zhang, Z., Wang, C., Xiao, B., Zhou, W., Liu, S.: Action recognition using context-constrained linear coding. IEEE Signal Process. Lett. **19**, 439–442 (2012)

Hashing Based State Variation for Human Motion Segmentation

Yang Liu[1,2], Lin Feng[1,2], Muxin Sun[2,3], and Shenglan Liu[2,3](✉)

[1] Faculty of Electronic Information and Electrical Engineering, Dalian University of Technology, Dalian 116024, Liaoning, China
dlut_liuyang@mail.dlut.edu.cn, fenglin@dlut.edu.cn

[2] School of Innovation and Entrepreneurship, Dalian University of Technology, Dalian 116024, Liaoning, China
sunmuxin@neusoft.com, liusl@mail.dlut.edu.cn

[3] State Key Laboratory of Software Architecture (Neusoft Corporation) Shenyang, Shenyang, China

Abstract. Motion sequence segmentation is a fundamental work in human motion analysis, which can promote the deep understanding as well as wide application of human motion sequences. Mainstream human motion sequence segmentation methods only focus on the data characteristics of the sequence and neglect the physical characteristics of the motions. This paper proposes a hashing based state variation segmentation (HBSVS) method to realize human motion sequence segmentation by analyzing the changing characteristics of the motions on time series. To improve the computational efficiency and merge motions from the same class, HBSVS adopts hashing method to construct the state descriptor of each motion on the sequence. Experiments on CMU motion capture database and UT-Interaction standard dataset show that HBSVS outperforms several state-of-the-art human motion sequence segmentation methods.

Keywords: Human motion sequence segmentation
Motion change characteristics · State variation · Hashing

1 Introduction

Human motion analysis has been a hot research topic in multimedia research discipline, and is widely applied in athletic training, medical diagnostic test, video games and security monitoring, etc. [1–6]. Deep mining on human motion sequence is demanded in real world applications to meet actual needs (e.g. movement standardness evaluation in athletic training) [2,7]. However, since the temporality and complexity of human motion, analyzing the whole sequence directly is a tough issue. Moreover, the motions of different classes will influence each

Y. Liu—Student Author.

J. Yang et al. (Eds.): CCCV 2017, Part III, CCIS 773, pp. 627–638, 2017.
https://doi.org/10.1007/978-981-10-7305-2_53

other, which causes the decrease of analysis accuracy [8]. Hence, segmenting the human motion sequences can effectively facilitate the analysis of human motions and promote the analysis result of the sequences.

As a fundamental work in human motion sequence analysis, segmentation can facilitate the operations on human motion clips like clustering, classification, etc. [11,12]. Effective and accurate segmentations on human motion sequences are of great significance in understanding the semantic information of human motion sequences. Researches on human motion sequences have made great progress. Researchers segment the human motion sequences from different perspectives to analyze the temporal property of the sequence. For instance, considering the correlation of the same class motions, clustering methods are applied in human motion sequence segmentation [13,14]. On the other hand, based on the differences of motions from different classes, classification methods are introduced to human motions segmentations [15]. Some researchers segment the human motion sequences by detecting transition frames of the sequences. Dian et al. [16] proposed a kernel algorithm to detect the transition frames as well as the repetitive frames simultaneously of a motion sequence.

Apart from aforementioned segmentation methods, some methods realize human motions sequence segmentation by considering the property of each frame and analyze the physical properties [9,10] of human motions. For example, Principle Component Analysis (PCA) approach to human motion segmentation employs the intrinsic dimension to realize segmentation, and assigns a cut when the intrinsic dimensionality of a local model of the motion suddenly increases [17]. Probability Principle Component Analysis (PPCA) to human motion segmentation analyzes the distributions of motions of the sequence to detect the segmentation points. That is PPCA approach places a cut when the distribution of poses is observed to change [17]. Time series-Warp Metric Curvature Segmentation (TS-WMCS) constructs a curvature like descriptor to evaluate the changes of human motion sequence, and detects the transition point when the local curvature of a frame changes dramatically [8].

In general, the key issue in human motion segmentation is to detect the segmentation points in the sequences. Note that, human motions will change a lot near the segmentation points [8,17], this paper proposes a hashing based state variation segmentation (HBSVS) method for human motion sequence segmentation. HBSVS researches the human motion state variation by introducing the state machine theory to the human motion segmentation discipline and regarding the motion of each frame as a state. The states of motions from the same class will not change dramatically on the time series, but the variations of motion states will boom near the segmentation points for the appearance of new motion classes. Moreover, HBSVS employs a hashing representation method [18] to describe the state of each frame, and fully utilizes the ability of hashing methods in representing data similarities.

The main contributions of this paper can be summarized as follows.

Firstly, construct a state variation descriptor to represent the change of human motion.

Secondly, employ a data similarity based hashing method to describe the motion state. To our knowledge, it is the first time to utilize hashing method to realize human motion segmentation.

Thirdly, realize human motion segmentation from the perspective of motion state variation.

2 Related Works

2.1 PCA and PPCA Approaches to Human Motion Sequence Segmentation

PCA and PPCA approaches to human motion segmentation all focus on the motion poses, this section will briefly introduce the PCA and PPCA approaches. Given human motion sequence $X = [x_1, \cdots, x_n]^T \in R^{n \times D}$ with n frames and D dimensions. For each frame x_i, construct the neighborhood N_i for x_i with $k - NN$ method.

PCA based human motion segmentation method assumes that the error between neighborhood X_i of x_i and it's projection X_i' will not vary largely even if frames belong to different motion classes. However, the error will be large for the frames in the transition clips. The error e can be expressed as following Eq. (1).

$$e = \sum_{i=1}^{n} ||X_i - X_i'||^2 \tag{1}$$

The PCA approach to human motion sequence segmentation works well if motions sequences of different classes have clear transition clips. However, the error will not change significantly for the continuity of real world motions, which causes the incomplete detection of PCA approach in finding all transition clips [8].

PPCA based approach to human montion segmentation utilizes the Probability Principle Component Analysis to evaluate the distribution of motion sequences. PPCA approach extends the traditional PCA approach with Gaussian distribution, and constructs the correlation covariance matrix C to represent the relations of different motions. The correlation covariance matrix C can be utilized to compute the average Mahalanobis distance H of the neighborhood of frame K, which can evaluate how likely are motion frames $K + 1$ through $K + T$ to belong to the Gaussian distribution [17]. H can be caluculated by Eq.(2).

$$H = \frac{1}{T} \sum_{i=K+1}^{K+m} (x_i - \overline{x})^T C^{-1} (x_i - \overline{x}) \tag{2}$$

Thereinto, K is the number of the start frame and m is the length of a subsequence. PPCA based human motion sequence segmentation method captures the correlation in the motion of different joint angles as well as the variance of all joint angles [17]. However, the Gaussian distribution of the motions is too strong for real world applications, which will lower the segmentation accuracy of PPCA segmentation approach. Besides, the PCA and PPCA based human motion sequence segmentation methods are based on angles, which is hard to extend to real world applications [8].

2.2 Time Series-Warp Metric Curvature Segmentation (TS-WMCS) Algorithm

This section mainly introduces the Time series-Warp Metric Curvature Segmentation (TS-WMCS) algorithm to human motion segmentations. TS-WMCS utilizes a curvature like descriptor to depict the change degree of human motion sequences. Given human motion sequence $X = [x_1, \cdots, x_n]^T \in \mathbb{R}^{n \times D}$ with n frames. For each frame x_i, construct the neighborhood N_i of x_i with $k - NN$ method. Then, construct the angle between the samples and the tangent space in each neighborhood. The warp degree of the data can be described by the angle between x_j and its orthogonal projection, where x_j is the neighbor of x_i. $\alpha_{ij} = \alpha_j(Q_i)$, where $\alpha_{ij} \in [0, \pi/2]$. The local low dimensional space can be expanded by Q_i, and Q_i can be obtained by the optimization equation Eq. (3),

$$\arg\max_{Q_i} Tr\left(Q_i^T X_i Z X_i^T Q_i\right) \ s.t. \ Q_i^T Q_i = I \tag{3}$$

where X_i is the neighborhood of x_i and Z is the normalization matrix of $X_i X_i^T$. Then, the curvature-like descriptor can be expressed as Eq. (4).

$$c_i = \sum_{j \in N_i} \cos \alpha_{ij} \Big/ \sum_{j \in N_i} \|x_j\| \tag{4}$$

The transition points in the human motion sequences are those samples whose curvature-like descriptors are high. Apart from the curvature-like descriptor segmentation, TS-WMCS algorithm also reduces the dimension of original motion sequences, and utilizes the low dimensional temporal feature curves to assist the segmentation. For each neighborhood, compute the low dimensional embedding Θ_{k_i} of X_i, and then add affinity projection L_i for each Θ_{k_i} of X_i to map the low dimensional embedding to the global embedding $T = [\tau_1, \tau_2, \cdots, \tau_n]^T \in R^{n \times d}$. Hence, the low dimensional embedding T can be obtained by Eq. (5).

$$\min_{\tau_i, L_i} \sum_{i=1}^{n} c_i \left\| \left(T_{k_i} - \tau_i e_{k_i}^T\right) - L_i \Theta_{k_i} \right\|^2 \ s.t. \ TT^T = I_d \tag{5}$$

where I_d is a $d - by - d$ identity matrix, and e_{k_i} is a vector with all ones.

TS-WMCS algorithm utilizes the curvature-like descriptor to segment the human motion sequences, which can effectively depict the changes of human motions. However, a certain move in complex motion sequences (e.g. "white crane spreads its wing" in tai chi) will contain motions which are quite different from each other. TS-WMCS algorithm can not yield satisfying result on these kind of motion sequences [8]. Moreover, TS-WMCS algorithm employ the low dimensional temporal feature curves to assist the segmentation, which do not have clear physical significance.

3 Hashing Based State Variation Segmentation Approach for Human Motion Segmentation

In this section, we will introduce the proposed Hashing based state variation segmentation (HBSVS) approach to human motion sequence segmentation. HBSVS constructs the state variation to describe the changing degree of motions in the motion sequences. Note that, motions of the same class changes slightly while motions of different classes change greatly. Hence, we construct the state variation to fully reveal the change of motions, and details are in Sect. 3.2. Considering the outstanding ability of hashing methods in mapping similar samples to the same hashing codes (hash bucket), we utilize hashing representation of each frame to represent the state of the motion. Here, we employ the widely utilized hashing method Locality Sensitive Hashing(LSH) [18] to represent the status. Detailed introduction of LSH is in Sect. 3.1.

3.1 Local Sensitive Hashing (LSH)

As a typical hashing method, local sensitive hashing (LSH) is widely adopted in various applications (e.g. image retrieval, video surveillance, etc.) [18,19,23,24] with its high computational efficiency and satisfying performance.

LSH aims at constructing hashing functions to project similar samples to hamming space with smaller hamming distances. As for dissimilar samples, the constructed hashing functions should divide them apart in hamming space. After the hashing projection, the original data are mapped to a hashing table, and samples are aligned to different hashing buckets of the hashing table. Samples in the same bucket are more likely to be neighbors in the original space. In other words, original data set are divided into many subsets with the hashing functions, and similar samples or neighbors are mapped to the same subset. Hence, the original data are represented with a number of binary codes, and each binary representation can be regarded as a hashing bucket. Based on the binary representations, computation cost on the samples are reduced effectively.

Given samples x, y, and hashing function $h(\cdot)$. To realize the goals of LSH, the hashing functions should satisfy the following constraints.

$$\begin{cases} if\ d(x,y) \leq d_1,\ then\ P(h(x) = h(y)) \geq p_1 \\ if\ d(x,y) \geq d_2,\ then\ P(h(x) = h(y)) \leq p_2 \end{cases} \tag{6}$$

Thereinto, $d(\cdot)$ denotes the hamming distance of two binary representations and $d_1 < d_2$, $P(A)$ represents the probability of the occurrence of event A and $p_1 > p_2$. If $h(\cdot)$ satisfies the aforementioned constraints, then $h(\cdot)$ can be regarded as $(d_1, d_2, p_1, p_2)-$sensitive. To satisfy the constraints, we adopt the hashing functions adopted in [18], and utilize random projection to map the original data [18].

Random projection is powerful method in pattern recognition [19–21], which has been widely applied in dimensional reduction [20] and data clustering [21]. Based on random projection, similar data can be mapped closer in low-dimensional space, and will be encoded with similar hashing codes. On the other hand, dissimilar data will be projected apart either in the low-dimensional space or in hamming space. The low dimensional embedding $Y = [y_1, \cdots, y_n]^T \in \mathbb{R}^{n \times d}$ of X can be obtained as $Y = X \times P$, where P is a random projection matrix like [19–21] and d denotes the dimension of Y. In general, the binary codes $B = [b_1, \cdots, b_n]^T \in \{1, -1\}^{n \times d}$ of original data X can be obtained by LSH as follows:

$$B_{ij} = \begin{cases} 1, & y_{ij} > 0 \\ -1, & y_{ij} < 0 \end{cases} \tag{7}$$

where i is the index of the frame in a human motion sequence, j indicates the bit of binary code b_i.

3.2 Hashing Based State Variation Segmentation (HBSVS)

This section mainly introduces the way to construct state descriptor by LSH, and the way to evaluate the motion variation by HBSVS. Given human motion sequence X, for each frame x_i we can obtain the corresponding hashing representation $b_i \in \{1, 0\}^c$, and c is the length of binary representation.

The degree of motion changes reflects the correlation of human motions. The changes of motions from the same class will not vary largely on continuous frames of a sequence. However, the motions will change dramatically when motions of a new class appear. Based on this observation, we manage construct a motion change degree descriptor, which can effectively reflect the variation of human motions. Note that, hamming distance is always utilized to metric the similarity between two binary representations [18], and similar samples always have smaller hamming distance. According to the definition of hamming distance, binary representations of similar samples will have more identical binary bits. Hence, we utilize the binary bits to represent the states of human motions, which can maintain the fact that motions from the same class have more identical states.

Based on the hashing method, the state collection can be constructed as $S = \{s_1, s_2, \cdots, s_c\}$, and each state represents a bit of the binary representation. For the ith frame of the total motion sequence, we can construct state S_i for x_i, and $S_i = \{s_p | b_{ip} = 1, 0 < p < c\}$. Likely, the state for next frame x_{i+1} can be represented by $S_{i+1} = \{s_q | b_{(i+1)q} = 1, 0 < q < c\}$. From x_i to x_{i+1} on time series, states in $S_i = \{s_p | b_{ip} = 1, 0 < p < c\}$ are activated, and the state change of two adjacent frames can be represented by $V_i = \{s_p \rightarrow s_q | s_p \in S_i, s_q \in S_{i+1}\}$. Details can be captured from Fig. 1.

In human motion sequence, motions have the property of continuity as time goes by, especially for the motions from the same class. However, the initialized state collection S is not enough in representing the changing sate of a certain motion. Hence, we add the state change V_i to the state collection. At each moment, the referred original states $s_i \in S$ are activated as well as the changing state V_i.

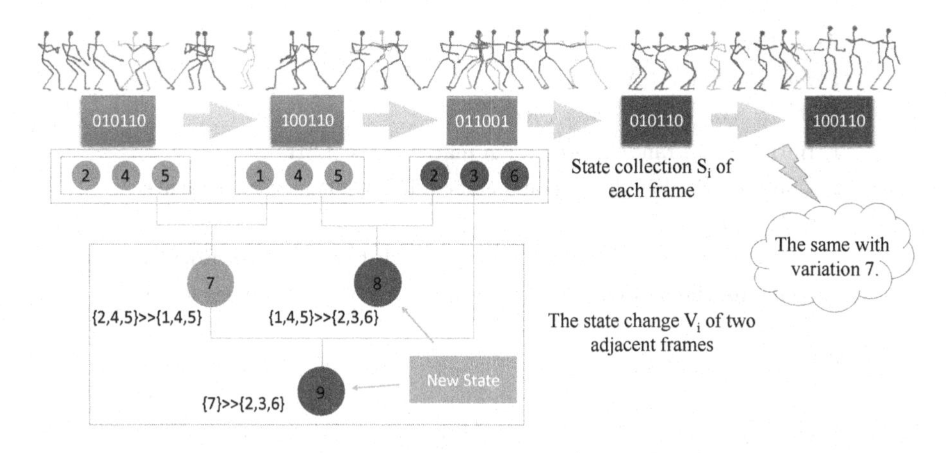

Fig. 1. The sketch of HBSVS in constructing the state variation

According to the state of next frame, we can construct new state changes. As for the repeated motions or similar motions in a sequence, the state changes have been constructed and will not construct much state changes. As is shown in Fig. 1, the change from '010110' to '100110' in gray rectangles has appeared, and HBSVS will not construct new state variation. Based on the mechanism, the state change variation can be measured by the following equation.

$$M_i = \frac{V_{i(new)}}{S_{i(activated)}} \tag{8}$$

where $V_{i(new)}$ is the newly constructed state variation, $S_{i(acativate)}$ is the activated state in the ith frame, including the states of $(i-1)th$ frame and the referred changing state, as is shown in Fig. 1. Based on the state change variation measurement M_i, we can effectively evaluate the change of motions on time series. As for the motions from the same class, since the state changes have been constructed at the start frames of a motion class. The number of newly constructed state variations will not be large. Nevertheless, when motions of a new class appears, many new state changes will be constructed. The main reason is that the states of motions from new class are quite different from that of the old class. Based on the state change variation M_i, we set a threshold ξ for to filter the significant state variation change. If $M_i > \xi$, add M_i to $Clips = \{c_1, \cdots, c_m\} \in \mathbb{R}^{1\times m}$. $Clips$ is the collection of transition clips.

4 Experiment Results

To fully validate the effectiveness of HBSVS method in segmenting human motion sequences, we conduct experiments on CMU motion capture database like many human motion segmentation methods [8,13,14,16,17]. Apart from the skeleton data of CMU motion capture database, we also conduct experiment on

Algorithm 4.1.

Input: Train data X

Output: Transition points $Clips$

1. Initialize the random projection matrix P for LSH.
2. Compute the low dimensional embedding Y for motion sequence X as $Y = X \times P$.
3. Encode the binary representation for each frame x_i with Eq. (7).
4. Generate the motion state variation M_i for the motion sequence.
5. ***If*** $M_i > \xi$

   Add M_i to $Clips$;

   End
6. Output $Clips$.

a real world video dataset UT-Interaction standard dataset [22]. Three methods are adopted as comparison methods to evaluate the segmentation performance of HBSVS, which are PCA, PPCA and TS-WMCS algorithms to human motion segmentation.

In order to verify the validity of the proposed algorithm in this paper, we employ two protocols utilized in [8,17] to evaluate the performances of the proposed algorithm and the comparison algorithms. The protocols include precision $P_p = Z/Q$ and recall $P_r = Z/N$, where Z is the number of the segment points located in the right segment clips computed by the algorithm, Q is the total segments detected by the algorithms, and N is the number of the right segment clips. To comprehensively evaluate the performance of the segmentation methods, we also adopt the F-Measure as one of our protocols. $F = P_p * P_r * 2/(P_p + P_r)$

4.1 Experiments on CMU Motion Capture Database

CMU motion capture database contains common human motions of 144 subjects, which are represented with skeletons of 31 joints. In our experiment, we randomly select the motion sequences of trail 1 to trail 8 in subject 15 to evaluate the segmentation performance of the proposed method. The selected sequences range from 5524 frames to 22948 frames, and mainly contain daily behaviors (e.g. wall, wander, and wash window, etc.). Like in [8], we unify the coordinate of 31 joints in the sequences to a global coordinate of 93 dimensions.

HBSVS utilizes the state variation to evaluate the changes of human motion sequences, which can effectively detect the changes in motion sequences and avoid detecting the inner changes of motions from the same class. As is shown in Table 1, HBSVS outperforms PCA and PPCA approach significantly in both precision and recall. When compared to TS-WMCS, the segmentation recall of HBSVS is lower. The main reason is that HBSVS aims to provide more accurate cuts for the human motion sequence segmentation, and tries to reduce the meaningless cuts. Note that the segmentation precision of HBSVS is much higher than TS-WMCS, and the total cuts of TS-WMCS are about 6 times over that of HBSVS. Taken together, F-Measure of HBSVS is better than the other three segmentation methods, which validates the effectiveness of HBSVS.

Figure 2 represents the segmentation results on the 12_04 motion sequence of the CMU database, which is a Tai Chi motion sequence. Motions of the same class in Tai Chi changes dramatically. However, we can find that HBSVS achieves satisfying segmentation results according to Fig. 2 and effectively reduces the meaningless cuts compared to other segmentation methods. The main reason is that HBSVS adpots hashing method to represent the state of human motions and focuses on the changing process, which can effectively reduce the influence of dramatic changes in a particular motion class. On the other hand, TS-WMCS, PCA and PPCA approach to human motion sequence segmentation only pay attentions to the change of a particular frame and neglect the influences of presented motions.

Table 1. Comparisons of precisions and recalls of 8 segmentation methods on CMU motion database

	N	Q	Z	P_p (%)	P_r (%)	F-Measure (%)
HBSVS	24	31	10	32.26	41.67	36.36
TS-WMCS	24	183	15	8.2	62.50	14.49
PCA	24	70	2	2.86	8.33	4.26
PPCA	24	95	8	8.42	33.33	13.45

Fig. 2. Segmentation results on the 12_04 motion sequence of the CMU Database (TaiChi)

4.2 Experiment on UT-Interaction Standard Dataset

The UT-Interaction standard dataset contains videos of continuous executions of 6 classes of human-human interactions, including shake-hands, point, hug, push, kick and punch. All the videos in UT-Interaction standard dataset are around 1 min. 8 videos of the subset1 are selected in our segmentaion experiments.

Note that the continuity of different motions are high in real world motions, which influences the accuracy in cutting the motion sequences. HBSVS utilizes the state variation to evaluate the changes of human motion sequences, which can effectively detect the changes in motion sequences. From Table 2 and Fig. 3 we can draw the conclusion that HBSVS outperforms the other segmentaion methods both in accuracy and recall.

TS-WMCS utilizes the curvature-like descriptor to realize motion sequence segmentation, which can not yield satisfying result in complex situations. PCA and PPCA only focuses on the motion data, which neglect the physical property of motion sequences. Hence, PCA and PPCA can not achieve satisfying result on real-world applications. As is shown in Table 2 and Fig. 3, HBSVS outperforms the other comparison methods.

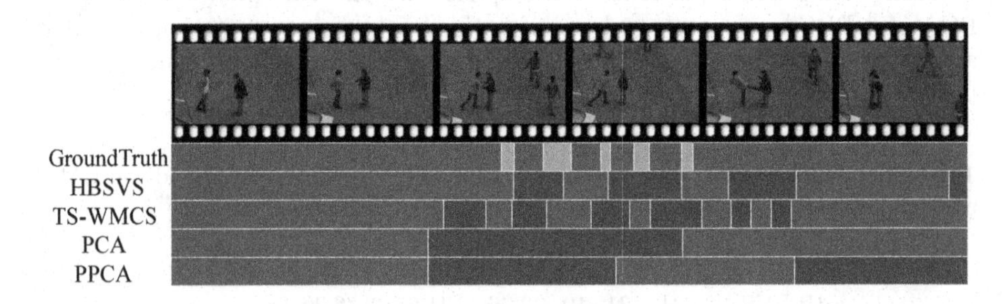

Fig. 3. Example of segmentation result on UT-Interaction standard video dataset

Table 2. Comparisons of precisions and recalls of 4 segmentation methods on UT-Interaction standard dataset

	N	Q	Z	P_p (%)	P_r (%)	F-Measure (%)
HBSVS	48	68	38	55.88	79.17	65.52
TS-WMCS	48	9	4	44.44	8.33	14.03
PCA	48	12	3	25	6.25	10
PPCA	48	26	9	34.62	18.75	24.33

5 Conclusion

This paper proposes a new algorithm for human motion segmentation named hashing based state variation segmentation (HBSVS), which analyzes the human motion sequences by the motion changing degree. The proposed method employs the hashing method to construct state descriptors to reveal the motion states in the sequences. Based on the motion state, HBSVS focuses on the changing process of the human motions and constructs the state variation evaluation criterion. The constructed evaluation criterion can effectively reveal the changing points in the human motion sequences and avoid meaningless cuts in the motion sequence segmentation tasks. Experimental results validate the effectiveness of the proposed method in detecting the changing points compared to several state-of-the-art methods.

Acknowledgement. This study was funded by National Natural Science Foundation of Peoples Republic of China (61173163, 61370200, 61672130, 61602082) and the Open Program of State Key Laboratory of Software Architecture, Item number SKL-SAOP1701. Yang Liu, Lin Feng, Muxin Sun and Shenglan Liu declare that they have no conflict of interest.

References

1. Chen, L., Wei, H., Ferryman, J.: A survey of human motion analysis using depth imagery. Pattern Recogn. Lett. **34**(15), 1995–2006 (2013)
2. Wang, L., Hu, W., Tan, T.: Recent developments in human motion analysis. Pattern Recogn. **36**(3), 585–601 (2003)
3. Simon, S.R.: Quantification of human motion: gait analysisbenefits and limitations to its application to clinical problems. J. Biomech. **37**(12), 1869–1880 (2004)
4. Carranza, J., Theobalt, C., Magnor, M.A., et al.: Free-viewpoint video of human actors. In: ACM Transactions on Graphics (TOG), vol. 22, no. 3, pp. 569–577. ACM (2003)
5. Lee, J., Chai, J., Reitsma, P.S.A., et al.: Interactive control of avatars animated with human motion data. In: ACM Transactions on Graphics (TOG), vol. 21, no. 3, pp. 491–500. ACM (2002)
6. Zhou, Z., Chen, X., Chung, Y.C., et al.: Activity analysis, summarization, and visualization for indoor human activity monitoring. IEEE Trans. Circuits Syst. Video Technol. **18**(11), 1489–1498 (2008)
7. Warren, M., Smith, C.A., Chimera, N.J.: Association of the functional movement screen with injuries in division I athletes. J. Sport Rehabil. **24**(2), 163–170 (2015)
8. Liu, S., Feng, L., Liu, Y., et al.: Manifold warp segmentation of human action. IEEE Trans. Neural Netw. Learn. Syst. (2017). https://doi.org/10.1109/TNNLS.2017.2672971
9. He, Q., Qiu, S., Fan, X., et al.: An interactive virtual lighting maintenance environment for human factors evaluation. Assembly Autom. **36**(1), 1–11 (2016)
10. Qiao, H., Li, C., Yin, P., et al.: Human-inspired motion model of upper-limb with fast response and learning ability-a promising direction for robot system and control. Assembly Autom. **36**(1), 97–107 (2016)
11. Sapienza, M., Cuzzolin, F., Torr, P.H.S.: Learning discriminative space-time action parts from weakly labelled videos. Int. J. Comput. Vis. **110**(1), 30–47 (2014)
12. Castrodad, A., Sapiro, G.: Sparse modeling of human actions from motion imagery. Int. J. Comput. Vis. **100**(1), 1–15 (2012)
13. Zhou, F., Torre, F., Hodgins, J.K.: Aligned cluster analysis for temporal segmentation of human motion. In: IEEE International Conference on Automatic Face and Gesture Recognition, pp. 1–7. IEEE (2008)
14. Zhou, F., De la Torre, F., Hodgins, J.K.: Hierarchical aligned cluster analysis for temporal clustering of human motion. IEEE Trans. Pattern Anal. Mach. Intell. **35**(3), 582–596 (2013)
15. Lin, J.F.S., Joukov, V., Kulic, D.: Human motion segmentation by data point classification. In: 2014 36th Annual International Conference of the IEEE Engineering in Medicine and Biology Society (EMBC), pp. 9–13. IEEE (2014)
16. Gong, D., Medioni, G., Zhao, X.: Structured time series analysis for human action segmentation and recognition. IEEE Trans. Pattern Anal. Mach. Intell. **36**(7), 1414–1427 (2014)

17. Barbi, J., Safonova A., et al.: Segmenting motion capture data into distinct behaviors. In: Graphics Interface, pp. 185–194. Canadian Human-Computer Communications Society (2004)
18. Datar, M., Immorlica, N., Indyk, P., et al.: Locality-sensitive hashing scheme based on p-stable distributions. In: Twentieth Symposium on Computational Geometry, pp. 253–262. ACM (2004)
19. Zhang, Y., Lu, H., Zhang, L., et al.: Video anomaly detection based on locality sensitive hashing filters. Pattern Recogn. **59**, 302–311 (2016)
20. Bingham, E., Mannila, H.: Random projection in dimensionality reduction: applications to image and text data. In: Proceedings of the Seventh ACM SIGKDD International Conference on Knowledge Discovery and Data Mining, pp. 245–250. ACM (2001)
21. Fern, X.Z., Brodley, C.E.: Random projection for high dimensional data clustering: a cluster ensemble approach. In: Proceedings of the 20th International Conference on Machine Learning (ICML2003), pp. 186–193 (2003)
22. Ryoo, M.S., Aggarwal, J.K.: UT-interaction dataset, ICPR contest on semantic description of human activities (SDHA). In: IEEE International Conference on Pattern Recognition Workshops, pp. 2–4 (2010)
23. Wang, J., Liu, W., Kumar, S., et al.: Learning to hash for indexing big data: a survey. Proc. IEEE **104**(1), 34–57 (2016)
24. Zhang, J., Peng, Y., Zhang, J.: SSDH: semi-supervised deep hashing for large scale image retrieval. arXiv preprint arXiv:1607.08477 (2016)

Improvement on Tracking Based on Motion Model and Model Updater

Tong Liu[1], Chao Xu[1](✉), Zhaopeng Meng[1], Wanli Xue[2], and Chao Li[3]

[1] School of Computer Software, Tianjin University, Tianjin 300354, China
{toddlt,xuchao,mengzp}@tju.edu.cn

[2] School of Computer Science and Technology, Tianjin University, Tianjin, China
xuewanli@tju.edu.cn

[3] College of Computer and Information Engineering, Tianjin Normal University, Tianjin, China
superlee@mail.tjnu.edu.cn

Abstract. Motion model and model updater are two important components for online visual tracking. On the one hand, an effective motion model needs to strike the right balance between target processing, to account for the target appearance and scene analysis, to describe stable background information. Most conventional trackers focus on one aspect out of the two and hence are not able to achieve the correct balance. On the other hand, the admirable model update needs to consider both the tracking speed and the model drift. Most tracking models are updated on every frame or fixed frames, so it cannot achieve the best state. In this paper, we approach the motion model problem by collaboratively using salient region detection and image segmentation. Particularly, the two methods are for different purposes. In the absence of prior knowledge, the former considers image attributes like color, gradient, edges and boundaries then forms a robust object; the latter aggregates individual pixels into meaningful atomic regions by using the prior knowledge of target and background in the video sequence. Taking advantage of their complementary roles, we construct a more reasonable confidence map. For model update problems, we dynamically update the model by analyzing scene with image similarity, which not only reduces the update frequency of the model but also suppresses the model drift. Finally, we integrate the two components into the pipeline of traditional tracker CT, and experiments demonstrate the effectiveness and robustness of the proposed components.

Keywords: Visual tracking · Motion model · Model updater
Image segmentation · Saliency detection · Image similarity

1 Introduction

Visual object tracking is part of the fundamental problems in computer vision [10,17,24,25]. It is a task of estimating the trajectory of a target in the video

J. Yang et al. (Eds.): CCCV 2017, Part III, CCIS 773, pp. 639–650, 2017.
https://doi.org/10.1007/978-981-10-7305-2_54

sequence [22]. The tracker has no prior knowledge of the object to be tracked such as category and shape. Despite extensive research on visual tracking, it remains challenging problems in handling complex object appearance changes caused by illumination, pose, occlusion [5] and motion [14]. According to the research of Wang [18], the tracker is composed of several modules: motion model, feature extractor, observation model, model updater, and ensemble post-processor. In particular, motion model and model updater contain many details that can affect the tracking result, but they are rarely concerned. Therefore, this paper will focus on these two components. More in detail, we try to combine salient region detection and image segmentation in the motion model and use image similarity in model updater.

Our approach is based on two major observations of previous work. First, as we all know, the motion model generates object proposals, it samples from the raw input image to forecast the possible candidate locations so as to confirm the scope of target searching. An effective sample selection mechanism can provide high-quality training samples which make the tracker recovers from failure and estimates appearance changes accurately. Hence, it is important to get more accurate samples in motion model. We develop a collaborative method based on image segmentation and salient region detection to analyze the appearance template consisting of the target object and its surroundings. This method differs significantly from existing motion model, such as the sliding window, which is prone to drifting in fast motion or large deformation video. Specifically, we use simple linear iterative clustering algorithm (SLIC [2]) for image segmentation in Sect. 3.1 and exploit frequency-tuned saliency analysis algorithm (FT [1]) for salient region detection in Sect. 3.2.

Second, it is critical to enhance the model updater of a tracker that adopts tracking-by-detection approach. Most of the tracking methods update the observation model in each frame, it reduces efficiency and more critical is that poor tracking results can cause the classifier to be contaminated, thus causing drift. Different from some other tracking paradigms that update model in a fixed manner such as updated every two frames, we formulates a simple and quick method to update observation model dynamically with image similarity. It not only improves the tracking speed, but also increases the accuracy. In Sect. 3.3, we will introduce a perceptual hash algorithm (pHash) in detail for image similarity.

The overview of our approach is illustrated in Fig. 1. The original image is subjected to image segmentation processing and salient region detection respectively, followed by cooperative learning, the result will be sent to CT [23] framework. Our main contribution of this work is to address the problems of tracking-by-detection trackers by effectively ameliorating the motion model and model updater. In order to make more rational use of the visual properties of the image, image segmentation is used to obtain more meaningful atomic regions in the field of color; Salient region detection is used to describe human's visual attention mechanism which involves distance, color, intensity, and texture. We use both methods to handle tracking scenes and targets in motion model thus achieving a more balanced appearance for visual tracking. In addition, we propose a novel

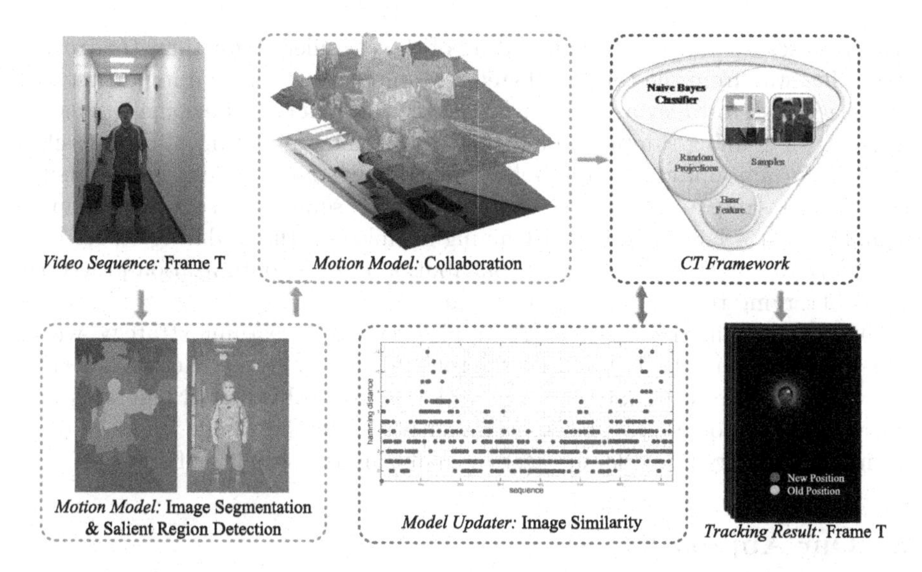

Fig. 1. Tracking pipeline

method to determine whether the estimated target is reliable in the time dimension and make a decision whether to update the observation model by using the hash of image similarity, it is simple and does reduce drift. We evaluate the proposed tracking algorithm on a large-scale benchmark with 50 challenging image sequences [20]. Experimental results show that our algorithm not only has good performance but also makes a significant improvement over the baseline tracker CT [23].

2 Related Work

Most trackers use statistical learning techniques to take charge of constructing robust object descriptors [12, 15] and building effective mathematical models for target identification [7, 21]. As estimated object position is converted into labeled samples, it is hard to give the accurate estimation of the object position. Hare [6] integrates the labeling positive and negative samples procedure into the learner by using online kernelized structured output support vector machine (Struck). And there are also many tracking algorithms [27] that focus on appearance and motion model definition to deal with the complex scene and avoid drifting. Compressed sensing theory is introduced into visual object tracking by Zhang [23] and he proposes compressive tracking algorithm. CT extracts Haar-like features in the compressed domain as the input characteristics to the classifier. It aims to design an effective appearance model and first compresses sample images of the foreground target and the background using the same sparse measurement matrix to efficiently extract the low-dimensional object descriptors.

In general, tracked results are chosen as the positive samples to update the classifier, noisy samples may often be included since they are not correct enough, which causes the failure of the updating of the classifier. After that, the tracker will drift away from the target. Therefore, sample selection is an important and necessary task for alleviating drift in the motion model. Additionally, massive amounts of training samples would hinder the online updating of the classifier without an appropriate sample selection strategy. Liu [11] designs a sparsity-constrained sample selection strategy to choose some representative support samples from a large number of training samples on the updating stage. It is necessary to integrate the samples contribution into the optimization procedure when observing the appearance of the target.

Most discriminative trackers [13] apply continuous learning strategy, where the observation model is updated rigorously in every frame. Research results show that excessive update strategy will lead to both lower frame-rates and degradation of robustness because of over-fitting in the recent frames. So we refine the strategy of model updater by analyzing the stability of scene.

3 Our Approach

To select high-quality samples, we construct a target-background confidence map according to the similarity of superpixels [19] in the surrounding region of the target between new frame and the first frame. Then it is refined by salient region detection result, the confidence map can facilitate tracker to distinguish the target and the background accurately. Finally, to accelerate the tracker, we control the model updater by judging the stability of scene, computing image similarity between frames.

3.1 Motion Model with Image Segmentation

Image segmentation process clusters pixels by the similarity of their feature and divides the raw image into several specific regions that may correspond to the tracked object. Superpixel is a kind of image segmentation algorithm, which provides a convenient primitive to compute local image features. There is a popular superpixel algorithm named SLIC (Simple Linear Iterative Clustering). SLIC [2] is fast, easy to use, and produces high-quality segmentations.

We segment the first frame into N superpixels. A color histogram is extracted as the feature vector f_i for each superpixel $sp(i)(i = 1, ..., N)$. We choose mean shift to cluster these superpixels. We can obtain n_{sp} ($n_{sp} < N$) different clusters and each cluster center $f_c(j)(j = 1, ..., n)$ by employing mean shift algorithm on the feature pool $F = \{f_i | i = 1, ..., N\}$.

Each cluster $clst(j)(j = 1, ..., n_{sp})$ is a set, and each cluster has its own members $\{f_i | f_i \in clst(j)\}$. Therefore, each cluster is represented by $f_c(j)$ and $clst(j)$ in the feature space. The members of each cluster $clst(j)$ corresponds to different superpixels in the image region. The weight $w(j)(j = 1, ..., n_{sp})$ is assigned to each cluster center $f_c(j)$ by exploiting a prior knowledge of the targets bounding box in the first frame, which indicates the likelihood that superpixel members of $clst(j)$ belong to the target area. We count two scores for each cluster $clst(j)$: $s^+(j)$ and $s^-(j)$. The former denotes the size of the overlapping area that all superpixel members of each cluster $clst(j)$ cover the bounding box, and accordingly the latter denotes the size of all superpixel members outside the target area. The weight is normalized between -1 and 1 and calculated as follows:

$$w(j) = \frac{s^+(j) - s^-(j)}{s^+(j) + s^-(j)}, \forall j = 1, ..., n_{sp} \quad (1)$$

where larger positive values indicate high confidence to assign the cluster center $f_c(j)$ to target and vice versa. To obtain a confidence map for the t-th frame, we first segment a surrounding region of the target into n_{sp} superpixels and then compute every superpixel confidence value. The surrounding region of the target is a square area, and its side length is $\eta\sqrt{S}$, where η is the constant parameter to control the size of this surrounding area and S is the area size of the target. The confidence value of a superpixel depends on two factors: the distance between this superpixel and the cluster center in the feature space, and the weight of the corresponding cluster center. $C_t(i)$ is the confidence value for superpixel i at the t-th frame. The confidence value of each superpixel is computed as:

$$C_t(i) = argmax_{1 \leq j \leq n}\{e^{||f_t(i) - f_c(j)||} \times w(j)\} \ \forall i = 1, ..., n_{sp} \quad (2)$$

where $f_t(i)$ and $f_c(j)$ denote the feature vector of the i-th superpixel in the t-th frame and the j-th cluster center in the first frame, respectively. Intuitively, the nearer the feature of a superpixel $f_t(i)$ is close to the targets cluster center $f_c(j)$, the more likely this superpixel belongs to the target area.

Each pixel in the i-th superpixel in the t-th frame shares the same confidence value $C_t(i)$. The surrounding area of the target is scanned with a sliding window that has the same size as the bounding box. At each position, the sum of the confidence value in this sliding window is computed, which demonstrates evidence for separating the target from the background. Then, the location of sliding window with the maximum value response will be selected as the new candidate location.

3.2 Motion Model with Salient Region Detection

Saliency is intentionally regarded as visual attention, it is determined as the local contrast of an image region with respect to its neighborhood at various scales, using one or more features of intensity, color, and orientation. The study of saliency detection comes from biological research. It is utilized to interpret complex scenes now. Scene analysis technique is integrated into visual tracking pipeline will significantly improve the performance, because it can separate the target from the background using high-quality saliency maps.

We use frequency-tuned saliency analysis algorithm (FT) to obtain the saliency map. This method can emphasize the largest salient objects and uniformly highlight whole salient regions. In order to have well-defined boundaries, the FT algorithm retains high frequencies from the original image. The frequency-tuned saliency analysis is formulated as

$$S_t(x, y) = ||I_\mu - I_{\omega hc}(x, y)|| \tag{3}$$

I_μ is the mean image feature vector of color and luminance, $I_{\omega hc}(x, y)$ is the corresponding image pixel vector value in the Gaussian blurred version (using a 5×5 separable binomial kernel) of the original image, and $|| \cdot ||$ is the L2 norm. Here we use the Lab color space, each pixel location is an $[L, a, b]^T$ vector, and the L2 norm is the Euclidean distance (Fig. 2).

Fig. 2. Saliency map

As we can see, the superpixels result is not stable. It only provides a coarse over-segmented image. To get the likelihood that superpixel members whether belong to the target area, we still need the prior knowledge of the targets bounding box in the first frame. The figure shows that salient region detection provides the probability of each pixel belonging to the foreground target, the result can be used to refine confidence map, it is formulated as:

$$Cmap_t = \phi C_t(i) + (1 - \phi) S_t \tag{4}$$

Here, each pixel of the i-th superpixel in the t-th frame shares the same confidence value $C_t(i)$, and each pixel has a sailency value in map S_t. We normalize the superpixel confidence value and FT saliency value to $[0, 1]$, and then fuse them to get the final confidence map. The parameter ϕ controls fusion of these two confidence value.

3.3 Model Updater with Image Similarity

We have integrated superpixels segmentation and salient region detection into CT, these procedure improves the performance of the base model. However, there is a computational overhead, it will slow down the base model. So we refine the strategy of model update to accelerate our tracker, the classifier will only be updated when the scene is not stable (background significantly changes).

We analysis the stability of scene by comparing similarity of incoming frame with previous frames. Here we use a perceptual hash algorithm pHash [16] to get the fingerprint of image, which has several properties: images can be scaled larger or smaller, have different aspect ratios, and even minor coloring differences (contrast, brightness, etc.) and they will still match similar images. The fingerprint result will not vary as long as the overall structure of the image remains the same. This can survive color histogram adjustments.

Algorithm 1. pHash Algorithm

1. Reduce size: Resize the input image to 32×32, pHash starts with a small image to simplify the DCT computation.
2. Reduce color: Reduce the image to grayscale to further simplify computation.
3. Compute the DCT: Separate the image into a collection of frequencies and scalars.
4. Reduce the DCT: Output of DCT is 32×32, just keep the top-left 8×8, which represents the lowest frequencies.
5. Further reduce the DCT: Compute the mean DCT value, and set the 64 hash bits to 0 or 1 depending on whether DCT values is above or below the average value.
6. Construct the hash: Set the 64 bits into a vector, it is the hash of image, we can compare the difference of images by computing hamming distance of their hash vector.

3.4 Tracking Framework

Compressive tracking (CT) aims to design an effective appearance model, which first compresses sample images of the foreground target and the background using the same sparse measurement matrix to efficiently extract the low-dimensional object descriptors [8], and then apply naive Bayes classifier with online update to classify the extracted features for object identification.

CT extracts haar-like features in the compressed domain as the input characteristics into classifier. The sparse measurement matrix facilitates efficient projection from the image feature space to a low-dimensional compressed subspace. The random measurement matrix is not a traditional Gaussian matrix, because it is dense with expensive memory space and computation. Finally, the maximal response classified by the classifier represents the tracking location in the current frame.

In order to locate the object of interest in the current frame, we introduce a confidence map based on scene analysis in the process of motion estimation and extend the coarse-to-fine sliding window search strategy. Firstly, the coarse-grained sliding window search is performed based on the previous object location within a large search radius γ_c to find the coarse location $\mathbf{l}_t$. Secondly, the sliding window obtains the tracking location $\hat{\mathbf{l}}_t$ with the maximal value response by shifting the window in the confidence map. Above two procedures not only consider the different shape or texture between the target and the background, but also take full advantage of discriminative color descriptors as a guidance. After that, the detected object location is close to the accurate object location. Thirdly, the fine-grained sliding window is carried out within a small search radius γ_f and then we can obtain the final object location $\mathbf{l}_t$ in the t-th frame. This strategy is more effective than previous work because of constructing the confidence map from color cues.

4 Experiment

In this section, we validate our tracking algorithm on OTB datasets, and compare our method with several trackers to demonstrate the effectiveness and robustness of our method. Our tracker is implemented in MATLAB on a 2.6 GHz Intel Core i5 CPU with 8 GB memory.

4.1 Qualitative Comparison

Comparison to the Baseline Tracker. To evaluate the impact of scene analysis, we compare our model with the standard CT on OTB [20], and run the One-Pass Evaluation (OPE) to verify the performance. As we can see, our method significantly ourperforms the baseline tracker CT in success rate and precision, the score has increased by about 50%. Therefore, the experiment results clearly demonstrate the importance and necessity of scene analysis. It helps the base tracker to handle target deformation and illumination change. The robustness of our method validates the important role of scene analysis component in visual tracking pipeline (Fig. 3).

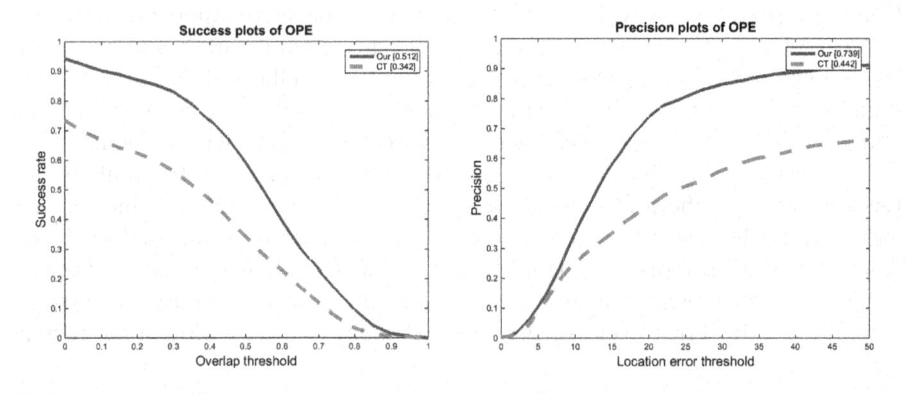

Fig. 3. OPE plot on OTB dataset, comparing to baseline CT

Comparison to the Basic Trackers. We compare our tracker with 7 tracking algorithms on dataset OTB over all 50 videos, these trackers are proposed almost the same period with CT. They are: CSK [4], SCM [26], Struck [6], ASLA [8], TLD [9], MIL [3] and CT. The results in Fig. 4 show that our method achieves almost the best performance using both the metrics. Our method is robust when the target deformation, illumination variatation and background clutter, as we can see, our method achieves higher score in the benchmark than other methods.

4.2 Quantitative Comparison

The robustness of our method is pretty obvious when there are target deformation, illumination variatation and background clutter. Figure 5 shows that our method has the advantages of dealing with complex scene, such as the squence Matrix and Cola. Most trackers are confused by background clutter because they don't have an efficient motion model to identify high-quality samples.

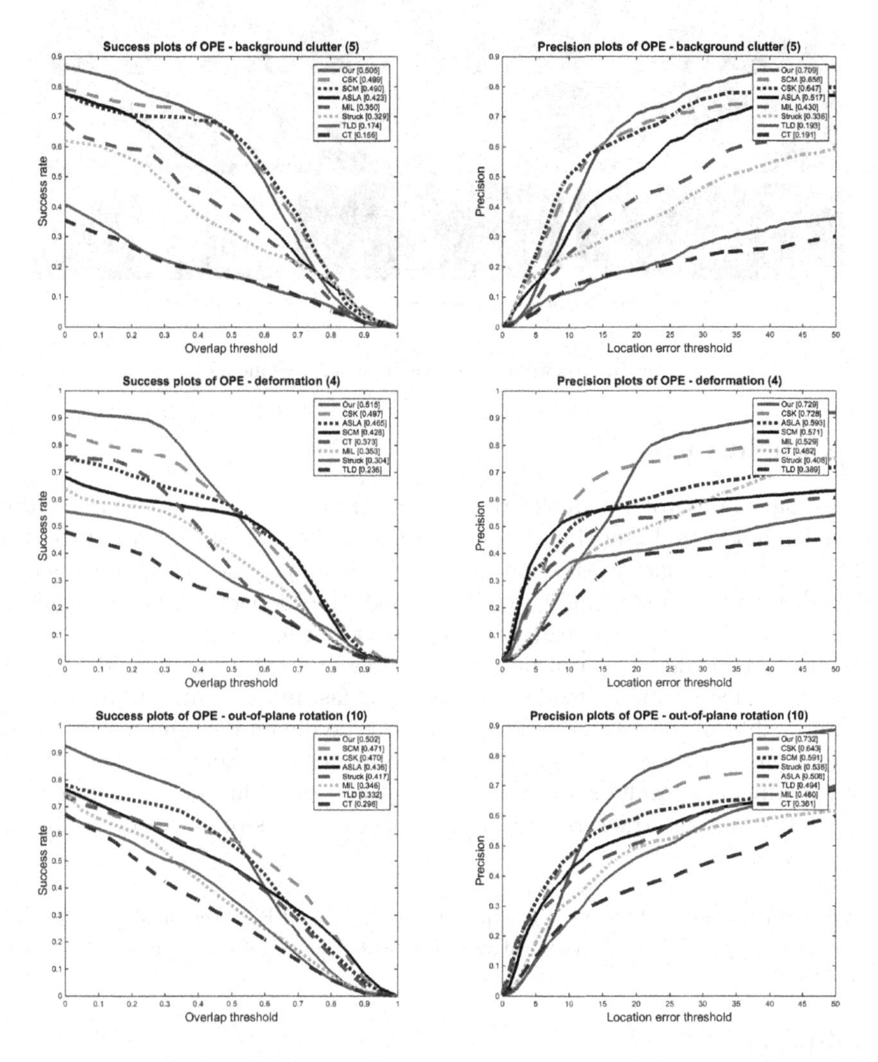

Fig. 4. OPE result with deformation, out-of-plane rotation and background clutter

Fig. 5. Tracking snapshot in several sequence

5 Conclusion

In this paper, we propose an effective algorithm for conventional visual tracking in motion model and model updater. Our method is more comprehensively considers the visual spatial attention factors in the appearance template, such as color, distance, intensity, and texture. Through the cooperation between salient region detection and image segmentation, we get an effective motion model which has the right balance between target processing and scene analysis. We further develop an effective online model updater using fast image similarity to measure the rationality of the estimated target in the time dimension, and it will reduce the frequency of the model update and improve the tracking accuracy. Extensive experimental results show that the proposed algorithm performs favorably against the baseline trackers and basic tracker in terms of efficiency, accuracy, and robustness.

Acknowledgments. This work is supported by National Key Technology R&D Program (No. 2015BAH52F00) and National Natural Science Foundation of China (No. 61304262).

References

1. Achanta, R., Hemami, S., Estrada, F., Susstrunk, S.: Frequency-tuned salient region detection, pp. 1597–1604. IEEE, June 2009
2. Achanta, R., Shaji, A., Smith, K., Lucchi, A., Fua, P., Süsstrunk, S.: Slic superpixels compared to state-of-the-art superpixel methods. IEEE Trans. Pattern Anal. Mach. Intell. **34**(11), 2274–2282 (2012)

3. Babenko, B., Yang, M.H., Belongie, S.: Visual tracking with online multiple instance learning. In: IEEE Conference on Computer Vision and Pattern Recognition, CVPR 2009, pp. 983–990. IEEE (2009)
4. Danelljan, M., Shahbaz Khan, F., Felsberg, M., Van de Weijer, J.: Adaptive color attributes for real-time visual tracking. In: Proceedings of the IEEE Conference on Computer Vision and Pattern Recognition, pp. 1090–1097 (2014)
5. Dong, X., Shen, J., Yu, D., Wang, W., Liu, J., Huang, H.: Occlusion-aware real-time object tracking. IEEE Trans. Multimedia **19**(4), 763–771 (2017)
6. Hare, S., Saffari, A., Torr, P.H.: Struck: structured output tracking with kernels. In: 2011 International Conference on Computer Vision, pp. 263–270. IEEE (2016)
7. Hong, X., Chang, H., Shan, S., Zhong, B., Chen, X., Gao, W.: Sigma set based implicit online learning for object tracking. IEEE Signal Process. Lett. **17**(9), 807–810 (2010)
8. Jia, X., Lu, H., Yang, M.H.: Visual tracking via adaptive structural local sparse appearance model. In: 2012 IEEE Conference on Computer Vision and Pattern Recognition (CVPR), pp. 1822–1829. IEEE (2012)
9. Kalal, Z., Mikolajczyk, K., Matas, J.: Tracking-learning-detection. IEEE Trans. Pattern Anal. Mach. Intell. **34**(7), 1409–1422 (2012)
10. Lin, L., Lu, Y., Li, C., Cheng, H., Zuo, W.: Detection-free multiobject tracking by reconfigurable inference with bundle representations. IEEE Trans. Cybern. **46**(11), 2447–2458 (2016)
11. Liu, Q., Ma, X., Ou, W., Zhou, Q.: Visual object tracking with online sample selection via lasso regularization. Signal Image Video Process. **11**(5), 1–8 (2017)
12. Ma, B., Hu, H., Shen, J., Liu, Y., Shao, L.: Generalized pooling for robust object tracking. IEEE Trans. Image Process. **25**(9), 4199–4208 (2016)
13. Ma, B., Huang, L., Shen, J., Shao, L.: Discriminative tracking using tensor pooling. IEEE Trans. Cybern. **46**(11), 2411–2422 (2016)
14. Ma, B., Huang, L., Shen, J., Shao, L., Yang, M.H., Porikli, F.: Visual tracking under motion blur. IEEE Transa. Image Process. **25**(12), 5867–5876 (2016)
15. Ma, B., Shen, J., Liu, Y., Hu, H., Shao, L., Li, X.: Visual tracking using strong classifier and structural local sparse descriptors. IEEE Trans. Multimedia **17**(10), 1818–1828 (2015)
16. Mıhçak, M.K., Venkatesan, R.: New iterative geometric methods for robust perceptual image hashing. In: Sander, T. (ed.) DRM 2001. LNCS, vol. 2320, pp. 13–21. Springer, Heidelberg (2002). https://doi.org/10.1007/3-540-47870-1_2
17. Tian, C., Gao, X., Wei, W., Zheng, H.: Visual tracking based on the adaptive color attention tuned sparse generative object model. IEEE Trans. Image Process. **24**(12), 5236–5248 (2015)
18. Wang, N., Shi, J., Yeung, D.Y., Jia, J.: Understanding and diagnosing visual tracking systems. In: Proceedings of the IEEE International Conference on Computer Vision, pp. 3101–3109 (2015)
19. Wang, S., Lu, H., Yang, F., Yang, M.H.: Superpixel tracking. In: 2011 IEEE International Conference on Computer Vision (ICCV), pp. 1323–1330. IEEE (2011)
20. Wu, Y., Lim, J., Yang, M.H.: Online object tracking: a benchmark. In: Proceedings of the IEEE Conference on Computer Vision and Pattern Recognition, pp. 2411–2418 (2013)
21. Xu, C., Tao, W., Meng, Z., Feng, Z.: Robust visual tracking via online multiple instance learning with fisher information. Pattern Recogn. **48**(12), 3917–3926 (2015)
22. Yang, H., Shao, L., Zheng, F., Wang, L., Song, Z.: Recent advances and trends in visual tracking: a review. Neurocomputing **74**(18), 3823–3831 (2011)

23. Zhang, K., Zhang, L., Yang, M.H.: Real-time compressive tracking. In: Fitzgibbon, A., Lazebnik, S., Perona, P., Sato, Y., Schmid, C. (eds.) ECCV 2012. LNCS, vol. 7574, pp. 864–877. Springer, Heidelberg (2012). https://doi.org/10.1007/978-3-642-33712-3_62
24. Zhao, L., Gao, X., Tao, D., Li, X.: Learning a tracking and estimation integrated graphical model for human pose tracking. IEEE Trans. Neural Netw. Learn. Syst. **26**(12), 3176–3186 (2015)
25. Zhao, L., Gao, X., Tao, D., Li, X.: Tracking human pose using max-margin markov models. IEEE Trans. Image Process. **24**(12), 5274–5287 (2015)
26. Zhong, W., Lu, H., Yang, M.H.: Robust object tracking via sparsity-based collaborative model. In: 2012 IEEE Conference on Computer Vision and Pattern Recognition (CVPR), pp. 1838–1845. IEEE (2012)
27. Zuo, W., Wu, X., Lin, L., Zhang, L., Yang, M.H.: Learning support correlation filters for visual tracking. arXiv preprint arXiv:1601.06032 (2016)

Learning Zeroth Class Dictionary for Human Action Recognition

Jiaxin Cai[1(✉)], Xin Tang[2], Lifang Zhang[3], and Guocan Feng[3]

[1] School of Applied Mathematics, Xiamen University of Technology, Xiamen 361024, China
caijiaxin@xmut.edu.cn

[2] College of Science, Huazhong Agricultural University, Wuhan 430070, China

[3] School of Applied Mathematics, Sun Yat-sen University, Guangzhou 510275, China
mcsfgc@mail.sysu.edu.cn

Abstract. In this paper, a discriminative two-phase dictionary learning framework is proposed for classifying human action by sparse shape representations, in which the first-phase dictionary is learned on the selected discriminative frames and the second-phase dictionary is built for recognition using reconstruction errors of the first-phase dictionary as input features. We propose a "zeroth class" trick for detecting undiscriminating frames of the test video and eliminating them before voting on the action categories. Experimental results on benchmarks demonstrate the effectiveness of our method.

Keywords: Human action recognition · Sparse coding
Dictionary learning · Fractional Fourier descriptor

1 Introduction

Recently, human action recognition has gained much interest for its great potential in many application areas such as video surveillance and human-computer interaction. The challenge of human action recognition usually results from the problem that different action classes often share some common motion patterns. Moreover, the action videos usually include many redundant frames indicating the background, large noise, clutter or small movements that are with limited help for recognition [4]. For action video classification, if the training and test videos contain some frames representing the common motion components among different action classes, or some redundant frames with large noise or useless clutters, then the discriminability of classifier learnt from training frames will be corrupted, and the recognition result of test video will also get corrupted when

J. Cai—This work is supported by Xiamen University of Technology High Level Talents Project (No. YKJ15018R) and the National Natural Science Foundation of P. R. China (No. 61602148).

J. Yang et al. (Eds.): CCCV 2017, Part III, CCIS 773, pp. 651–666, 2017.
https://doi.org/10.1007/978-981-10-7305-2_55

the labels of undiscriminating test frames are used to determine the class label of test video.

Sparse coding based recognition approaches have attracted much attention in the field of computer vision [14,32]. To learn a well-adapted dictionary for obtaining good reconstruction and recognition performance, many algorithms [15,27,31] for training dictionary with label information and discriminative criterion have been proposed. These algorithms can only work well for the cleaning training samples or training samples with small corruption [18]. So the efforts for learning low-rank and discriminative dictionary are made in [18,22] to solve this problem. However, the previous methods can not handle the problem resulted from the undiscriminating test frame samples for video recognition. Imagining the test video contains many frames representing the useless clutters or some common components among different video classes, the label of test video will get corrupted by these undiscriminating test frames. So it is necessary to develop a discriminative dictionary for the situation that both the training and test videos contain many corrupted or uninteresting frame samples.

In this paper, we attempt to learn a discriminative dictionary for action recognition to handle the situation that both the training and test videos contain undiscriminating frames with common components, redundant components, background, clutter or large noise. We propose a recognition framework for human action recognition in video by learning a discriminative dictionary called zeroth class dictionary. The "zeroth class" trick is proposed for detecting and filtering out the undiscriminating frames of the test video to eliminate the negative effect resulted from these frames during voting the action category of test video. The zeroth class dictionary method is a two-phase dictionary learning system [2,19] including three steps:

(1) Firstly, the discriminative frames of training videos are selected by Gentle Adaboost algorithm to learn the first-phase dictionary. The left undiscriminating frames are relabeled and assigned to the zeroth class. The zeroth class is a virtual class indicating undiscriminating frames which are with limited help for recognition, such as frames with common poses shared by different actions, frames with clutter or noise, and other redundant frames. Then the first-phase dictionary is learnt on the selected discriminative frames; the reconstruction errors of all frames corresponding to each dictionary atom are collected to build the new frame representations.
(2) Using the new frame representations, we learn the class-specific dictionary in which the sub-dictionary of each action class is learned on the corresponding selected discriminative frames and the zeroth class sub-dictionary is learnt on the undiscriminating frames. Then we obtain the preliminary labels of the test frames based on the learnt class-specific dictionary. The zeroth class is used for recognizing the redundant frames of test video, which are filtered out afterwards.
(3) After the undiscriminating frames of test video are excluded, the final action label is voted by all remained discriminative frames of the test video. Experimental results on benchmark data show that our method outperforms most state-of-the-art approaches.

The rest of the paper is organized as follows: Sect. 2 reviews the relative works. Section 3 presents the zeroth class dictionary learning framework for human action recognition. Experimental setup and results on benchmark datasets are presented in Sect. 4, and conclusions are given in Sect. 5.

2 Relative Works

Many efforts [12,29,33] have been devoted to studying action recognition by sparse representation and dictionary learning. Guha and Ward [12] provide a sparse representation for human action recognition by learning the over-complete bases on the local motion patterns. Zhang et al. [33] learn dictionary from spatiotemporal salient patches and use the sparse reconstruction coefficients of patches to represent image sequences of action videos. Lu and Peng [21] propose a structure sparsity ℓ_1-norm graph to learn the high-level visual-words for representing human action. Qiu et al. [26] build a compact and discriminative action attribute dictionary by minimizing the reconstruction error and Gaussian Process model. Wang et al. [29] propose a sparse model incorporating the similarity constrained term and the dictionary incoherence term for human action recognition. Recently, the deep learning technology has also been introduced to human action recognition. Wang et al. [30] propose a method using weighted hierarchical depth motion maps and deep convolutional neural networks for depth images based human action recognition. Liu et al. [20] propose a deep convolutional neural network to learn the spatio-temporal features for representing human action from depth videos. Li et al. [17] propose a method employing joint distance maps and convolutional neural networks to learn discriminative features for human action recognition.

Our work is also similar to the silhouette based action recognition approaches [3,5,6,8]. Chaaraoui et al. [5] develop a human action recognition method through extracting multi-view key poses sequences and handling variations in shape by dynamic time warping. Cheema et al. [6] propose a human action recognition method by extracting a scale invariant contour-based pose feature and clustering the features to construct distinctive key poses. Cai and Feng [3] present a human action recognition method by describing contour-based shape feature using fractional Fourier transform. Cheng et al. [8] propose a human action recognition approach based on human silhouettes by supervised temporal t-stochastic neighbor embedding and incremental learning via low-dimensional embedding.

3 Zeroth Class Dictionary Learning Based Action Recognition Framework

3.1 Feature Extraction

We use the fractional Fourier shape descriptor [3] to represent each frame of action videos. The fractional Fourier shape descriptor is built on the human pose represented by contour points of the binary silhouette. Given an image extracted from the action video, its binary silhouette is obtained from the segmented foreground region. Then the boundary of silhouette is extracted and the position of all points $\{(x(i), y(i))\}_{i=1}^{N}$ along the boundary is represented as a complex sequence $\{s(i)|s(i) = x(i) + jy(i)\}_{i=1}^{N}$, where $x(i)$ and $y(i)$ denote the horizontal and vertical coordinate of the ith point respectively. Here N is the total number of contour points, and j denotes the imaginary unit. Then we shift the base point of coordinate system to the center of mass (x_c, y_c) of contour points along the boundary.

$$\tilde{x}(i) = x(i) - x_c; \qquad \tilde{y}(i) = y(i) - y_c \tag{1}$$

After that, the length of sequence is normalized to a predetermined value L through down-sampling the contour. In our experiments, the normalized length L is set as 100.

$$\hat{x}(i) = \tilde{x}(\lceil i * \frac{N}{L} \rceil); \qquad \hat{y}(i) = \tilde{y}(\lceil i * \frac{N}{L} \rceil) \tag{2}$$

Afterwards, we compute the discrete fractional Fourier transform of the transformed contours $\{\hat{s}(i) : \hat{x}(i) + j\hat{y}(i)\}_{i=1}^{L}$, and get the response $\{S(i)\}_{i=1}^{L}$ in the fractional Fourier domain. For a continuous signal $\hat{s}(t)$, its p order continuous fractional Fourier transform is defined as:

$$S_p(u) = \begin{cases} B_\alpha \int_{-\infty}^{\infty} exp(j\frac{t^2+u^2}{2}cot\alpha - \frac{jtu}{sin\alpha})\hat{s}(t)dt, & \alpha \neq n\pi \\ \hat{s}(t), & \alpha = 2n\pi \\ \hat{s}(-t), & \alpha = (2n \pm 1)\pi \end{cases} \tag{3}$$

where $\alpha = p\pi/2$ is the rotation angle and $B_\alpha = \sqrt{\frac{1-jcot\alpha}{2\pi}}$. Here n denotes an integer. The order p is set as 0.9 in our experiments. For digital computation, we use a sampling-type discrete fractional Fourier transform proposed in [24] to calculate the response $\{S(i)\}_{i=1}^{L}$.

Then the amplitude of fractional response, $|S(i)|$, is calculated and normalized to obtain a scale invariant descriptor.

$$d(i) = \frac{|S(i)|^2}{\sum_{i=1}^{L} |S(i)|^2} \tag{4}$$

$\{d(i)\}_{i=1}^{L}$ consists the fractional Fourier shape descriptor of human pose contour. For each training video, we assign its action class label to its all affiliated frames.

3.2 First-Phase Dictionary Learning

Firstly, the Gentle AdaBoost algorithm is employed to select discriminative training frames. Gentle AdaBoost provides an approach for reweighting data points by updating weights of base classifiers and puts higher weights on undiscriminating data points than discriminative points [10]. Regression stump is used as the base classifier. The regression stump is a simple additive logistic regression based classifier, which classifies data points according to only one input dimension. For an input sample x whose kth dimensional feature is denote as $x(k)$, the output class label $f(x)$ of regression stump is defined by only four parameters (w, v, k, th), and represented as follows.

$$f(x) = w * sign(x(k) - th) + v \tag{5}$$

The "one-against-the-rest" technique is employed to extend the primary binary classification problem to multi-class case. A training frame with high weight imply that it contains common and undiscriminating patterns between different action categories. We select the frames with the lowest weights from the training frame set of each action class at a rate of R to build the discriminative subset for generating the first-phase dictionary, and the remained frames are pushed into a pool where they are relabeled as the zeroth class and will be used for detecting undiscriminating frames of the test video later.

After the discriminative subset of the training frames is selected out, we generate a dictionary D on this set. The aim is to learn a dictionary D so that the selected discriminative frames have a sparse representation B over the dictionary. It can be written as the following optimization problem [1]:

$$\begin{aligned} &min_{D,B}\|Y - DB\|_F^2 \\ &s.t.\ |b_i\|_0 \leq C,\ \forall i \end{aligned} \tag{6}$$

where Y is the selected discriminative subset of training frames represented by the fractional Fourier descriptor; D is the learned dictionary on the discriminative subset; b_i is the ith column of sparse coefficients matrix B, denoting the representing coefficient of ith frame. C is the parameter controlling the sparsity of coefficients. $\|\cdot\|_F$ denotes the Frobenius norm, and $\|\cdot\|_0$ is the l_0 norm enforcing the coefficients to be sparse.

Then for a frame y, the reconstruction error corresponding to the ith atom of the dictionary D is computed as [25]:

$$\begin{aligned} e_i(y) &= \|y - D\delta_i(\hat{\beta})\|^2 \\ \hat{\beta} &= argmin_{\beta}\|y - D\beta\|^2, \qquad s.t. \quad \|\beta\|_0 \leq C \end{aligned} \tag{7}$$

where the function $\delta_i(\beta)$ sets the jth dimension of β as 0 if $j \neq i$. Suppose m is the atom number of dictionary D, then the vector $[e_1(y), \ldots, e_m(y)]^T$ makes up a new feature of frame y, which would be used as the new frame feature in the next phase dictionary learning.

3.3 Second-Phase Dictionary Learning

After the new features of all frames in both training and test videos are computed, the class-specific dictionary learning [29] is performed. Suppose K is the number of action categories. Using the training frames belonging to the kth ($k = 0, 1, \ldots, K$) class (including the zeroth class), we learn the class-specific dictionary D_k using the new feature represented by reconstruction errors on the first-phase dictionary. The sub-dictionary D_k associated with the kth ($k = 1, \ldots, K$) nonzero action class is learnt on the corresponding selected discriminative frames of the kth class, and the zeroth class dictionary D_0 is learnt on the undiscriminating frames. Then the whole dictionary $\bar{D}$ is constructed by concatenating all the class-specific dictionaries, that is to say, $\bar{D} = [D_0|D_1|D_2|\ldots|D_K]$.

After the whole dictionary $\bar{D}$ is learned, the sparse representation a_i of a frame $\breve{x}_i$ of the test video can be estimated as follows.

$$a_i = argmin_a \|\breve{x}_i - \bar{D}a\|^2, \qquad s.t. \quad \|a\|_0 \leq C \tag{8}$$

The reconstruction error $r_k(\breve{x}_i)$ associated with the kth class can be defined as:

$$r_k(\breve{x}_i) = \|\breve{x}_i - \bar{D}\Theta_k(a_i)\|^2, \qquad k = 0, 1, \ldots, K \tag{9}$$

where $\Theta_k(a_i)$ produces a vector whose nonzero entries are coefficients of a_i associated with the kth class.

Then each frame of the test video is assigned to the class that corresponds to the minimum of reconstruction error with respect to each class (including the zeroth class). The estimated preliminary class $\breve{k}_i$ of the test frame $\breve{x}_i$ is given as:

$$\breve{k}_i = argmin_{k \in \{0,1,\ldots,K\}} r_k(\breve{x}_i) \tag{10}$$

Afterwards, we filtered out the undiscriminating frames in the test video which are labeled as the zeroth class. Then the max pooling or sum pooling criteria is used to vote the action label by the remained discriminative frames corresponding to nonzero classes of the test video. For max pooling policy, each frame of the test video is classified to the nonzero class that corresponds to the minimum of reconstruction error with respect to each non-zero class. Then the estimated action class $\hat{k}$ of the test video is given as:

$$\hat{k} = argmin_{k \in \{1,\ldots,K\}} min_{\hat{i} \in \{i|\breve{k}_i \neq 0\}} r_k(\breve{x}_{\hat{i}}) \tag{11}$$

For sum pooling policy, an overall residual is constructed by summing up the reconstruction errors corresponding to nonzero classes of each frames in the test video; then the test video is assigned to the nonzero class with respective to the minimum of overall error. The estimated action class $\hat{k}$ of the test video is given as:

$$\hat{k} = argmin_{k \in \{1,\ldots,K\}} \sum_{\hat{i} \in \{i|\breve{k}_i \neq 0\}} r_k(\breve{x}_{\hat{i}}) \tag{12}$$

4 Experimental Results

In order to evaluate the performance and practicability of the proposed approach, two human action recognition datasets, the Weizmann dataset [11] and

Table 1. Comparison of methods on Weizmann dataset

Method	Weizmann
Our method (sum pooling)	97.85%
Our method (max pooling)	95.70%
Class-specific Dictionary Learning (sum pooling)	91.40%
Class-specific Dictionary Learning (max pooling)	88.17%
Two-stage Dictionary Learning (sum pooling)	89.24%
Two-stage Dictionary Learning (max pooling)	89.24%
Chaaraoui et al. [5]	92.77%
Cheema et al. [6]	91.6%
Wang et al. [29]	96.7%
Cheng et al. [8]	94.44%
Cheng et al. [7]	93.18%
Cai and Feng [3]	93.55%
Zhao and Horace [34]	87.50%
Kumar and John [16]	95.69%
Guo et al. [13]	92.88%

the MuHAVi-MAS14 dataset [28], are used as benchmarks. For each class, we select frames with the lowest Gentle Adaboost weights as the discriminative subset at a rate of R. For $D_k(k = 1, 2, \ldots, K)$, the number of atoms equals LP proportion of that of the corresponding samples belonging to kth class. The leave-one-out cross validation strategy is employed to separate the training video set and test video set. All parameters are tuned by grid searching. The best recognition rates on Weizmann dataset is achieved as 97.85% when R is set as 0.2, the atom number of dictionary D is set as 90, the number of atoms employed in D_0 is set as 40, LP is set as 10%, and C is set as 15. The best recognition rates on MuHAVI-MAS14 dataset is achieved as 95.59% when R is 0.2, atom number of dictionary D is set as 150, the number of atoms employed in D_0 is set as 80, LP is 10%, and C is 15. Our method is compared to Class-specific Dictionary Learning and Two-stage Dictionary Learning whose parameters are turned within the same parameter search range as that of our method. The comparison results on Weizmann and MuHAVI-MAS14 dataset are presented in Tables 1 and 2 respectively. Experimental results on benchmarks show that our methods outperform recognition approaches through Class-specific Dictionary Learning and Two-layer Dictionary Learning performed on the same action feature. We also

Table 2. Comparison of methods on MuHAVI-MAS14 dataset

Method	MuHAVi-MAS14
Our method (sum pooling)	95.59%
Our method (max pooling)	95.59%
Class-specific Dictionary Learning (sum pooling)	88.97%
Class-specific Dictionary Learning (max pooling)	88.97%
Two-stage Dictionary Learning (sum pooling)	86.02%
Two-stage Dictionary Learning (max pooling)	86.02%
Chaaraoui et al. [5]	91.18%
Cheema et al. [6]	86.03%
Singh et al. [28]	82.35%
Eweiwi et al. [9]	91.90%
Murtaza et al. [23]	95.40%

compare the accuracy of our method to the reported accuracy of other state-of-the-art methods. Although the action features employed in most listed methods are different to ours, our method still shows a considerable performance and outperforms most listed methods on the benchmarks.

Analysis of the relation between recognition accuracy and parameters is carried out by changing one parameter when all other parameters are fixed as the default number. Figures 1 and 2 show the relation between accuracy and the rate R of selected discriminative frames in all frames on Weizmann and MuHAVI-MAS14 dataset respectively. Experiment results demonstrate that our Zeroth-class Dictionary Learning framework outperforms the traditional framework without zeroth class detection and filtering stages (when R is set as 1). However, if too many undiscriminating frames are selected into the discriminative training set for the first layer dictionary learning, the performance Zeroth-class Dictionary Learning framework of will decline and be worse than Two-layer Dictionary Learning framework. The experimental results demonstrate introducing the zeroth class is effective if a proper proration of zeroth class of the training set is set. Figures 3 and 4 show the relation between accuracy and the number of atoms of the first layer dictionary D learned on the selected discriminative training frames on Weizmann and MuHAVI-MAS14 dataset respectively. Figures 5 and 6 show the relation between accuracy of our method and the number of atoms belonging to the zeroth class dictionary D_0 on Weizmann and MuHAVI-MAS14 dataset respectively. Figures 7 and 8 show the relation between accuracy on Weizmann and MuHAVI-MAS14 dataset and the number of atoms belonging to each nonzero-class sub-dictionary in the second layer dictionary, which equal LP proportion of the number of samples belonging

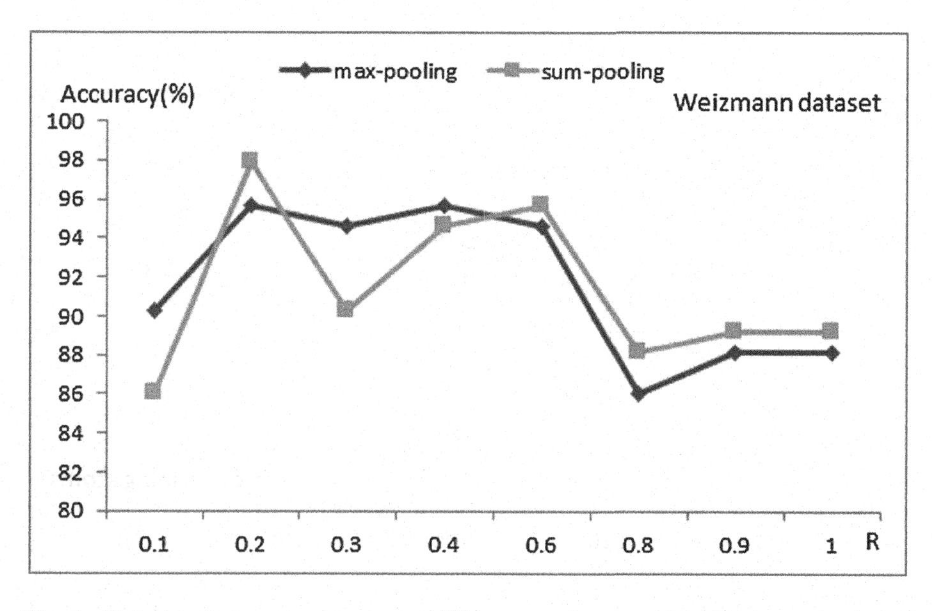

Fig. 1. Relation between accuracy on Weizmann dataset and the rate of selected discriminative frames

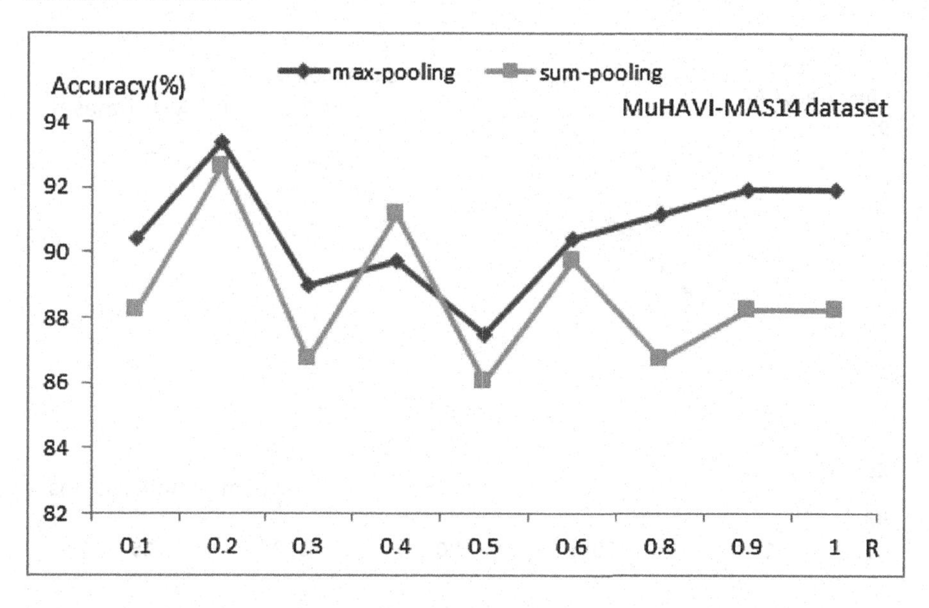

Fig. 2. Relation between accuracy on MuHAVI-MAS14 dataset and the rate of selected discriminative frames

to the corresponding class. Figures 9 and 10 show the relation between accuracy and parameter C on Weizmann and MuHAVI-MAS14 dataset respectively. The results demonstrate that it is easy to find proper parameters for achieving

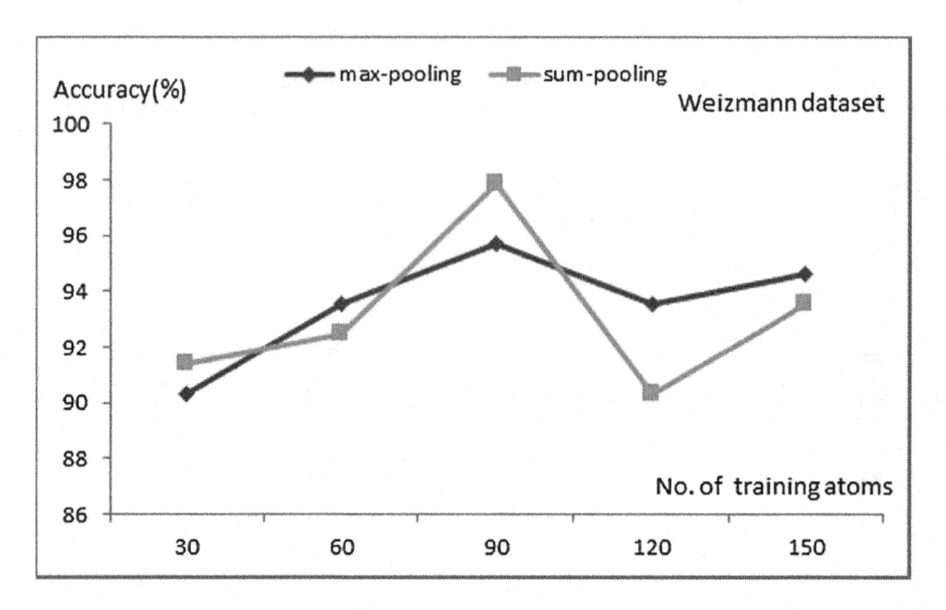

Fig. 3. Relation between accuracy on Weizmann dataset and the number of atoms of the first layer dictionary

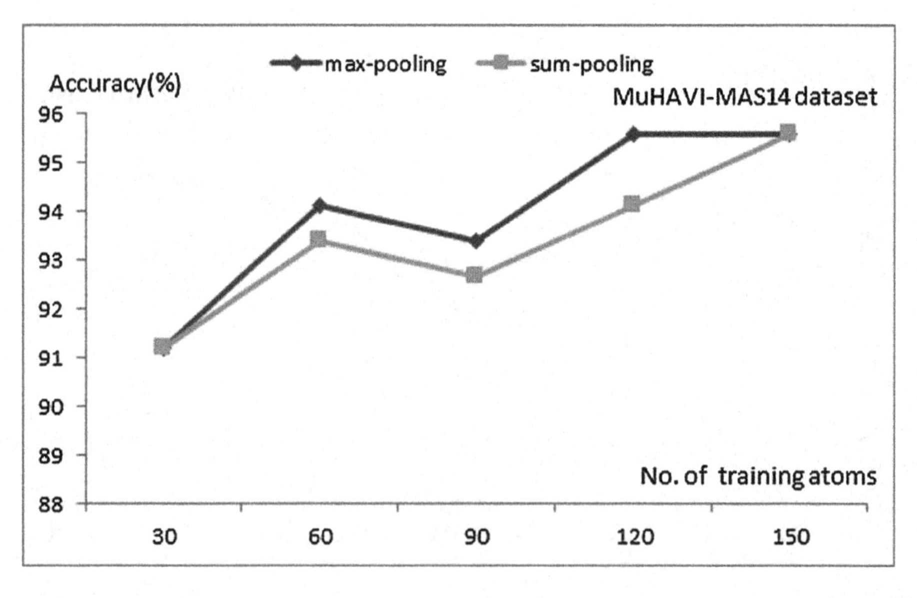

Fig. 4. Relation between accuracy on MuHAVI-MAS14 dataset and the number of atoms of the first layer dictionary

good classification performance through the zeroth class dictionary learning framework. Experiment results demonstrate that our method is effective in most cases.

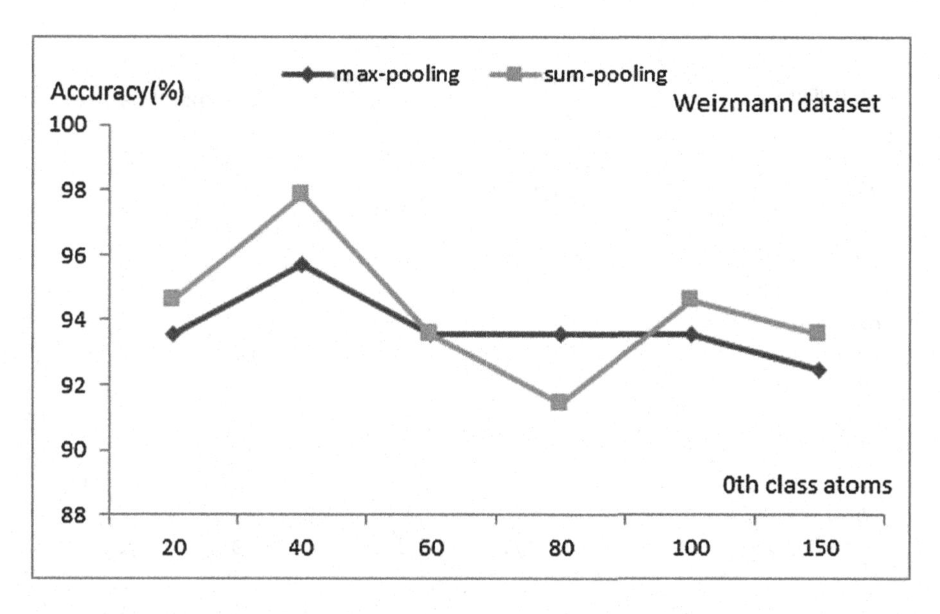

Fig. 5. Relation between accuracy on Weizmann dataset and the atom number of the zeroth class dictionary

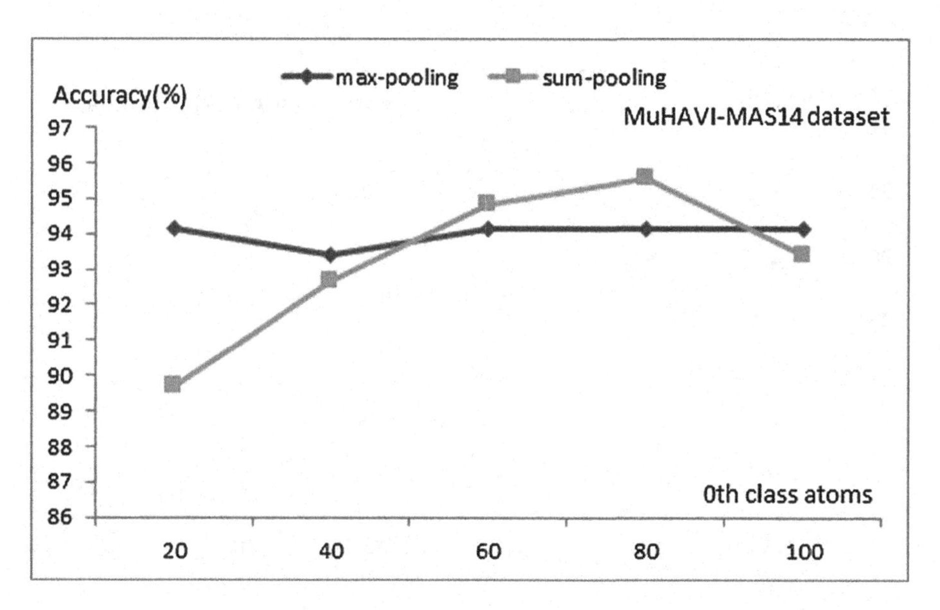

Fig. 6. Relation between accuracy on MuHAVI-MAS14 dataset and the atom number of the zeroth class dictionary

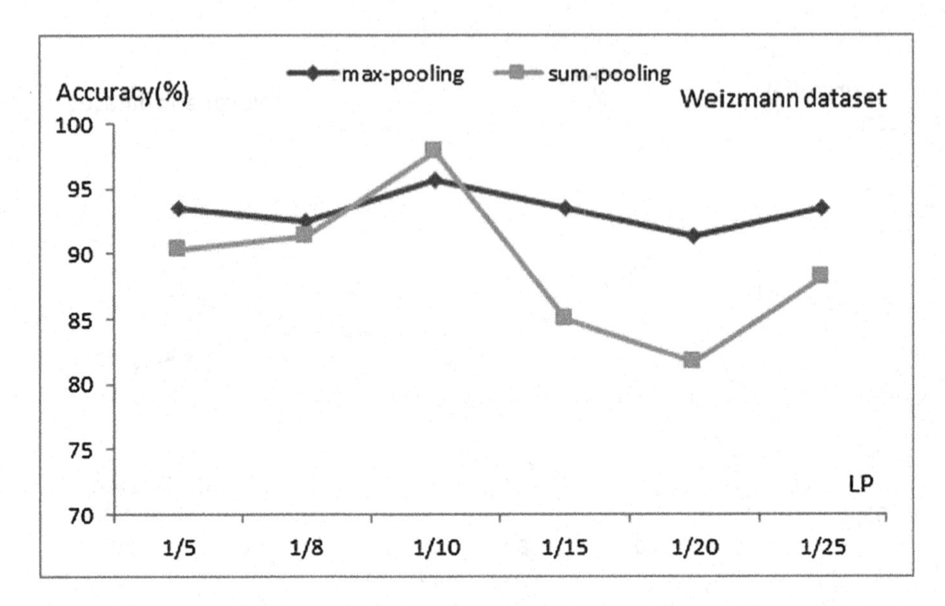

Fig. 7. Relation between accuracy on Weizmann dataset and the atom number of nonzero class dictionary

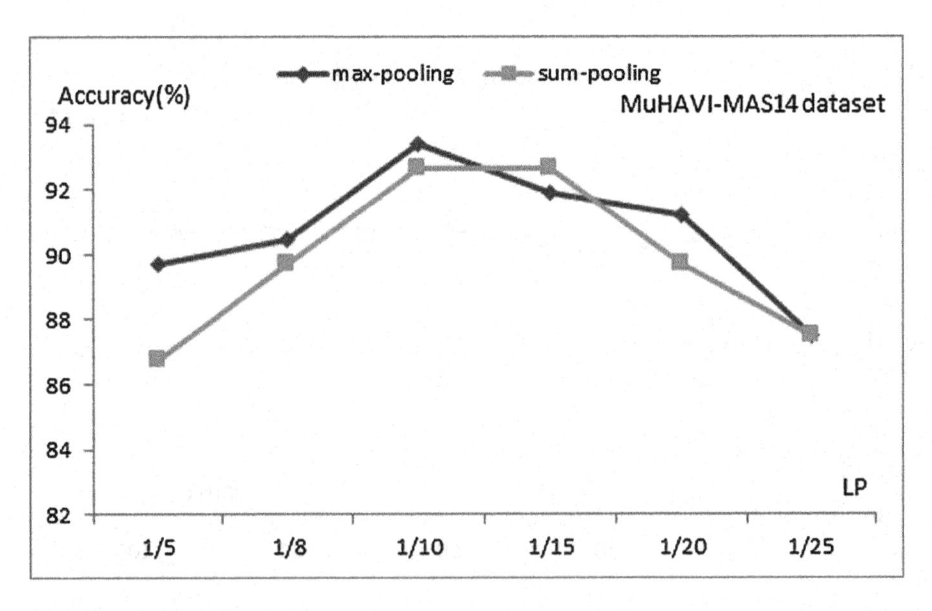

Fig. 8. Relation between accuracy on MuHAVI-MAS14 dataset and the atom number of nonzero class dictionary

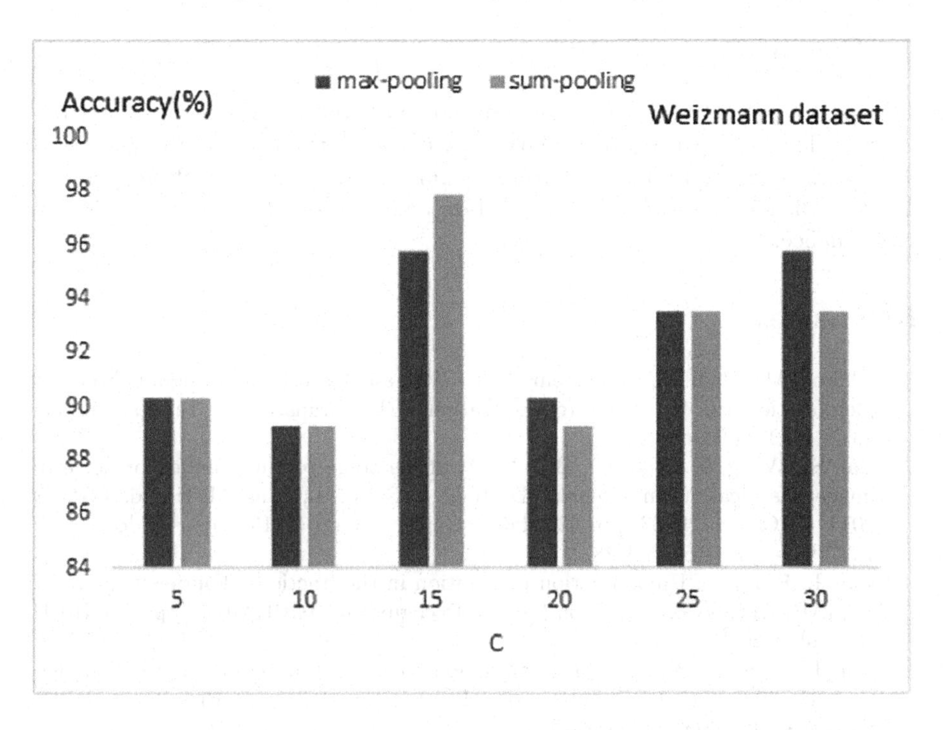

Fig. 9. Relation between accuracy on Weizmann dataset and parameter C

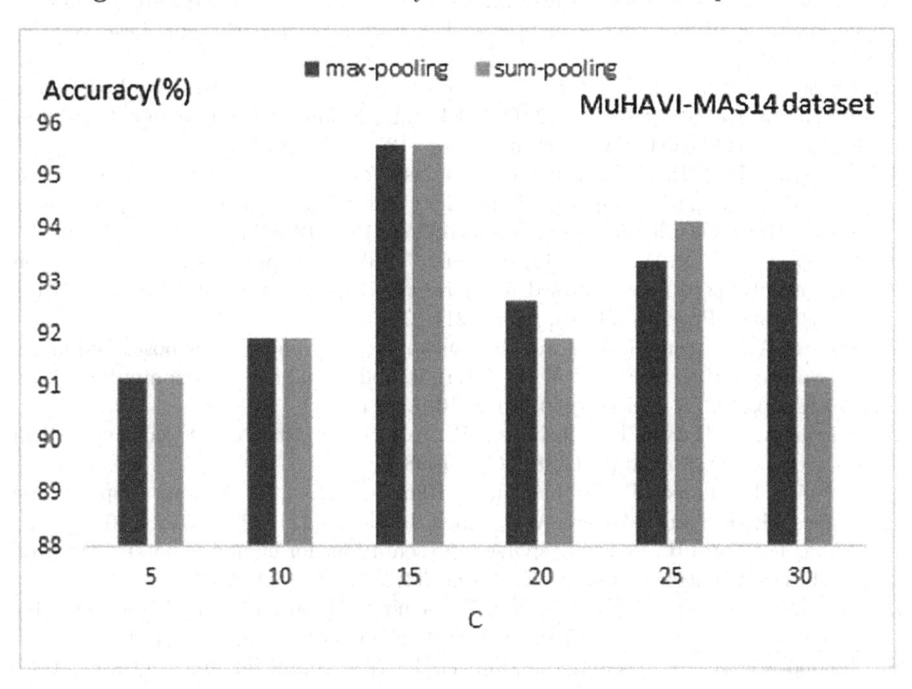

Fig. 10. Relation between accuracy on MuHAVI-MAS14 dataset and parameter C

5 Conclusion

This paper presents an action recognition method by using zeroth class dictionary. The zeroth class dictionary provides a method to detect and delete undiscriminating frames of test video for improving the classification accuracy. The recognition framework is validated on benchmarks, showing a considerable performance.

References

1. Aharon, M., Elad, M., Bruckstein, A.: K-SVD: an algorithm for designing overcomplete dictionaries for sparse representation. IEEE Trans. Signal Process. **54**(11), 4311–4322 (2006)
2. Bai, S., Wang, X., Yao, C., Bai, X.: Multiple stage residual model for accurate image classification. In: Cremers, D., Reid, I., Saito, H., Yang, M.-H. (eds.) ACCV 2014. LNCS, vol. 9003, pp. 430–445. Springer, Cham (2015). https://doi.org/10.1007/978-3-319-16865-4_28
3. Cai, J., Feng, G.: Human action recognition in the fractional Fourier domain. In: 3rd IAPR Asian Conference on Pattern Recognition (ACPR 2015), pp. 1–5. IEEE, November 2015
4. Cai, J., Feng, G., Tang, X.: Human action recognition using oriented holistic feature. In: 2013 20th IEEE International Conference on Image Processing (ICIP), pp. 2420–2424, September 2013
5. Chaaraoui, A.A., Climent-Prez, P., Flrez-Revuelta, F.: Silhouette-based human action recognition using sequences of key poses. Pattern Recogn. Lett. **34**(15), 1799–1807 (2013)
6. Cheema, S., Eweiwi, A., Thurau, C., Bauckhage, C.: Action recognition by learning discriminative key poses. In: 2011 IEEE International Conference on Computer Vision Workshops (ICCV Workshops), pp. 1302–1309 (2011)
7. Cheng, J., Liu, H., Li, H.: Silhouette analysis for human action recognition based on maximum spatio-temporal dissimilarity embedding. Mach. Vis. Appl. **25**(4), 1007–1018 (2014). https://doi.org/10.1007/s00138-013-0581-2
8. Cheng, J., Liu, H., Wang, F., Li, H., Zhu, C.: Silhouette analysis for human action recognition based on supervised temporal t-SNE and incremental learning. IEEE Trans. Image Process. **24**(10), 3203–3217 (2015)
9. Eweiwi, A., Cheema, S., Thurau, C., Bauckhage, C.: Temporal key poses for human action recognition. In: 2011 IEEE International Conference on Computer Vision Workshops (ICCV Workshops), pp. 1310–1317 (2011)
10. Friedman, J., Hastie, T., Tibshirani, R.: Additive logistic regression: a statistical view of boosting. Ann. Stat. **28**, 2000 (1998)
11. Gorelick, L., Blank, M., Shechtman, E., Irani, M., Basri, R.: Actions as space-time shapes. IEEE Trans. Pattern Anal. Mach. Intell. **29**(12), 2247–2253 (2007)
12. Guha, T., Ward, R.: Learning sparse representations for human action recognition. IEEE Trans. Pattern Anal. Mach. Intell. **34**(8), 1576–1588 (2012)
13. Guo, Z., Wang, X., Wang, B., Xie, Z.: A novel 3D gradient LBP descriptor for action recognition. IEICE Trans. Inf. Syst. **E100.D**(6), 1388–1392 (2017)
14. Huang, Y., Wu, Z., Wang, L., Tan, T.: Feature coding in image classification: a comprehensive study. IEEE Trans. Pattern Anal. Mach. Intell. **36**(3), 493–506 (2014)

15. Jiang, Z., Lin, Z., Davis, L.: Label consistent K-SVD: learning a discriminative dictionary for recognition. IEEE Trans. Pattern Anal. Mach. Intell. **35**(11), 2651–2664 (2013)
16. Kumar, S.S., John, M.: Human activity recognition using optical flow based feature set. In: IEEE International Carnahan Conference on Security Technology, pp. 1–5 (2017)
17. Li, C., Hou, Y., Wang, P., Li, W.: Joint distance maps based action recognition with convolutional neural networks. IEEE Signal Process. Lett. **24**(5), 624–628 (2017)
18. Li, L., Li, S., Fu, Y.: Learning low-rank and discriminative dictionary for image classification. Image Vis. Comput. **32**(10), 814–823 (2014)
19. Li, X., Song, Y., Lu, Y., Tian, Q.: Multi-layer orthogonal visual codebook for image classification. In: International Conference on Acoustics, Speech and Signal Processing, pp. 2312–2315 (2011)
20. Liu, Z., Zhang, C., Tian, Y.: 3D-based deep convolutional neural network for action recognition with depth sequences. Image Vis. Comput. **55**, 93–100 (2016)
21. Lu, Z., Peng, Y.: Latent semantic learning with structured sparse representation for human action recognition. Pattern Recogn. **46**(7), 1799–1809 (2013)
22. Ma, L., Wang, C., Xiao, B., Zhou, W.: Sparse representation for face recognition based on discriminative low-rank dictionary learning. In: 2014 IEEE Conference on Computer Vision and Pattern Recognition, pp. 2586–2593 (2012)
23. Murtaza, F., Yousaf, M.H., Velastin, S.A.: Multi-view human action recognition using 2D motion templates based on mhis and their hog description. IET Comput. Vis. **10**(7), 758–767 (2016)
24. Ozaktas, H., Erkaya, N., Kutay, M.: Effect of fractional fourier transformation on time-frequency distributions belonging to the Cohen class. IEEE Signal Process. Lett. **3**(2), 40–41 (1996)
25. Pati, Y.C., Rezaiifar, R., Krishnaprasad, P.S.: Orthogonal matching pursuit: recursive function approximation with applications to wavelet decomposition. In: Proceedings of the 27th Annual Asilomar Conference on Signals, Systems, and Computers, pp. 40–44 (1993)
26. Qiu, Q., Jiang, Z., Chellappa, R.: Sparse dictionary-based representation and recognition of action attributes. In: 2011 IEEE International Conference on Computer Vision (ICCV), pp. 707–714 (2011)
27. Shrivastava, A., Patel, V.M., Chellappa, R.: Non-linear dictionary learning with partially labeled data. Pattern Recogn. **48**(11), 3283–3292 (2015)
28. Singh, S., Velastin, S., Ragheb, H.: Muhavi: a multicamera human action video dataset for the evaluation of action recognition methods. In: 2010 Seventh IEEE International Conference on Advanced Video and Signal Based Surveillance (AVSS), pp. 48–55 (2010)
29. Wang, H., Yuan, C., Hu, W., Sun, C.: Supervised class-specific dictionary learning for sparse modeling in action recognition. Pattern Recogn. **45**(11), 3902–3911 (2012)
30. Wang, P., Li, W., Gao, Z., Zhang, J., Tang, C., Ogunbona, P.O.: Action recognition from depth maps using deep convolutional neural networks. IEEE Trans. Hum.-Mach. Syst. **46**(4), 498–509 (2016)
31. Yang, M., Zhang, L., Feng, X., Zhang, D.: Sparse representation based fisher discrimination dictionary learning for image classification. Int. J. Comput. Vis. **109**(3), 209–232 (2014)

32. Yu, Y.F., Dai, D.Q., Ren, C.X., Huang, K.K.: Discriminative multi-scale sparse coding for single-sample face recognition with occlusion. Pattern Recogn. **66**, 302–312 (2017)
33. Zhang, T., Xu, L., Yang, J., Shi, P., Jia, W.: Sparse coding-based spatiotemporal saliency for action recognition. In: 2015 IEEE International Conference on Image Processing (ICIP), pp. 2045–2049, September 2015
34. Zhao, Q., Horace, H.: Unsupervised approximate-semantic vocabulary learning for human action and video classification. Pattern Recogn. Lett. **34**(15), 1870–1878 (2013)

Global Motion Pattern Based Event Recognition in Multi-person Videos

Lifang Wu, Jiaoyu He, Meng Jian(✉), Siyuan Liu, and Yaowen Xu

Faculty of Information Technology, Beijing University of Technology, Beijing 100124, China
lfwu@bjut.edu.cn, hejiaoyu_1993@163.com, jianmeng648@163.com, liuliuliuliu0323@126.com, xuyao_wen@126.com

Abstract. Event analysis in multi-person videos is a great challengeable task, especially in the sports videos due to the intensive and fast movement of players with serious occlusions. In this paper, we propose a global motion pattern (GMP) based event recognition algorithm. In order to avoid obstacles from complex background noise, GMP of video frames in sports video is extracted by optical flow. In particular, both spatial and temporal features of GMP are devoted to event recognition by sequential CNN and LSTM. Experimental analysis demonstrates that the proposed algorithm is capable to take benefit from spatial and temporal characteristics of GMP for effective event recognition.

Keywords: Event recognition · Global motion pattern · Optical flow CNN · LSTM

1 Introduction

With the rapid development of multimedia technology, the sports videos are increasing with an explosion on broadcast and internet. In recent years, inspired by the huge amount of videos, it becomes urgent to analyze the content of sports videos automatically [1–6], especially for the complex content of basketball videos [1,2,5,6].

There are many research topics related to basketball videos analysis, including video segmentation, scoring detection, player detection, event recognition, etc. And the event recognition of basketball videos is one of the most challengeable problems. Basketball videos are generally broadcasted for long hours and attract a lot of audience. With the fast life rhythm, it becomes increasingly important to extract the main video clips from full-length videos and enhance effectiveness of the experience. For viewers, coaches or trainees, it is required to

This work was supported in part by the Beijing Municipal Education Commission Science and Technology Innovation Project(KZ201610005012) and in part by the China Postdoctoral Science Foundation funded project(2017M610026 & 2017M610027).

J. Yang et al. (Eds.): CCCV 2017, Part III, CCIS 773, pp. 667–676, 2017.
https://doi.org/10.1007/978-981-10-7305-2_56

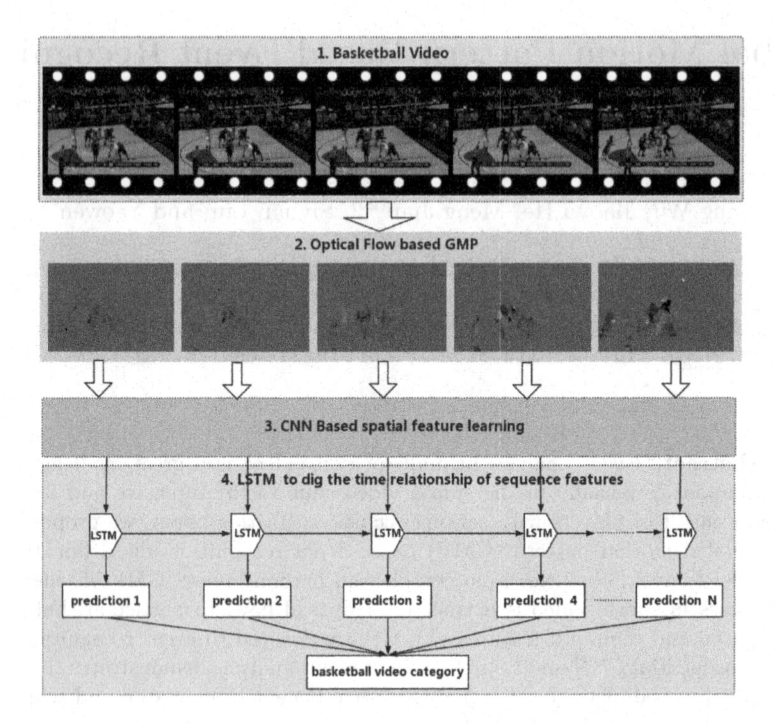

Fig. 1. An overview of our approach.

browse a particular type of basketball events for multiple times. In the past, most of the basketball broadcasters need the help of slow-motion replay scenes, text records and audiovisual features to distinguish the different events. Recently, a number of approaches have been proposed to automatically analyze the content of basketball videos. Tan et al. [2] used low-level information and combined it with domain knowledge of basketball to identify certain events and segment the given video into wide-angle and close-up shots. Tjondronegoro and Chen [5] used a definitive scope of detection and a universal set of audiovisual features for basketball. [1,6–8] analyzed the additional textual contents and described the key events based on the updated scoreboard displayed after a goal is created. However, character detection in the sports videos is very difficult, due to its small size, low resolution, complex background. The methods mentioned above were relied on low-level features of not only visual information but also text and audio information.

In recent years, with the development of deep learning, Donahue et al. [9] developed a novel recurrent convolutional architecture for large-scale visual learning, including basketball videos. However, the basketball videos used in [9] are the scenes of a single-person throwing rather than the actual multi-person videos. The obstacle of event recognition in multi-person videos lies in the dynamic mutual interference and interrelationship among players. A volume of researches [4,10,11] have explored models of group activities by concatenating every person's actions and their relationships. In fact, [4,10,11] have the ability to capture person-level information but in the course of the basketball game players occlusion problem is very serious, which tends to lose tracking the target person. Coupled with the complex background of basketball videos, it will lead to confusion and increase the amount of calculation.

In fact, multi-person videos in sports competitions can be seen as confrontation between two groups of players, and all the players collaborate to perform a specific event. Motivated by this phenomenon, we propose a GMP based event recognition algorithm. Figure 1 provides an overview of the proposed method. We combined CNN based spatial feature analysis and LSTM based temporal feature analysis of the GMP, which alleviates interference of background noises with the optical flow. The main contributions of this work are summarized as follows.

- The proposed method employs GMP to avoid considering occlusion among players in event recognition.
- We make use of optical flow to extract GMP and successfully alleviate bias of background to recognition procedure.
- The proposed method extracts spatial information by CNN from GMP and temporal context information by LSTM. Both of them guarantee the effectiveness of event recognition.

2 Event Recognition Based GMP

As illustrated in Fig. 1, the proposed method consists of three main conponents: (1) We extract the GMP of basketball videos based on the optical flow analysis; (2) Then CNN is introduced to learn the spatial features of GMP automatically; (3) Finally event recognition in multi-person videos is implemented on the temporal features of GMP by LSTM.

2.1 GMP Extraction

In this work, we employ optical flow images for event recognition instead of the original video frames, in order to preserve the motion information of players and GMP, while eliminating background interference to improve learning accuracy.

Optical flow [12,13] is a two-dimensional vector space which extracts characteristics of the object movement between two consecutive frames. We conduct optical flow as [13] and then convert the result to 3-channel images. The 3-channel optical flow images contain x-direction flow value, y-direction flow value and optical flow amplitude, where x, y directions and flow amplitude are normalized to 0–255. Images in part 2 of Fig. 1 are a set of optical flow images based GMP, which are constructed by difference between two consecutive frames. Compared to the original video frames, background noise is eliminated. Although in background there still exists a little noise, the players' profile is already very apparent. In part 2 of Fig. 1, optical flow preserves the location of the players fully and maintain GMP of all the players.

2.2 Spatial Feature Extraction of GMP Using CNN

Lots of works for sports video analysis are constructed on hand crafted features [14,15]. These low-level visual features lack of description on high-level semantic information. Inspired by the success of deep learning in extracting high-level features, in this work, we employ CNN [16,17] to extract spatial features of GMP, as illustrated in Fig. 2.

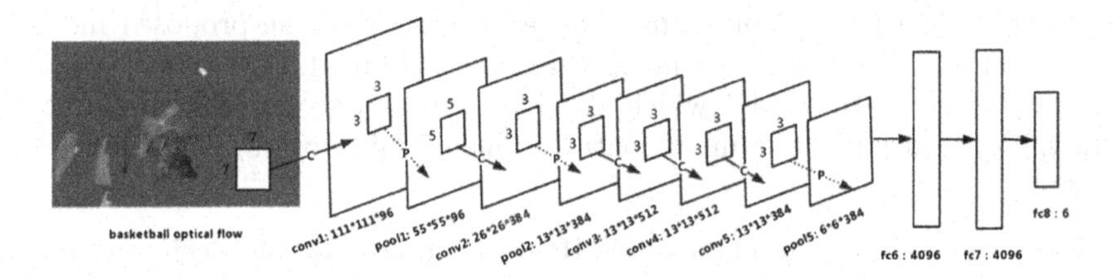

Fig. 2. The structure of CNN for spatial feature learning from GMP.

In the CNN diagram of Fig. 2, there are 5 convolutional layers, 3 max-pooling layers and 3 fully-connected layers. The input and output size of each layer have the following relationship

$$Y_i = (X_i + 2 * D_i - K_i)/S_i + 1 \tag{1}$$

where X_i and Y_i are respectively the size of input and output maps in the i-th layer, K_i is the size of the convolution kernel in the layer, S_i indicates the step size in convolution procedure, and D_i means the size of the pad, the default is 0. Taking the first convolutional layer *conv*1 as an example, the GMP based on optical flow is resized to $227 * 227$ as input, i.e. $X_1 = 227$. The *conv*1 layer contains 96 filters in size of $7 * 7$. With the step size of 2, that is $K_1 = 7$ and $S_1 = 2$. Therefore $Y_1 = 111$ and we could achieve 96 feature maps in size of $111 * 111$. Subsequently max-pooling *pool*1 is applied in $3 * 3$ local regions with step size of 2 resulting resized feature maps in $55 * 55$, where $Y_2 = 55$ with $X_2 = Y_1 = 111$, $K_2 = 3$ and $S_2 = 2$. The *conv*2 and *conv*5 layers are similar with the *conv*1 layer followed by a pooling layer, while differently the *conv*3 and *conv*4 layers are not followed by normalization and max-pooling layer. The following 3 fully-connected layers ($fc6, fc7, fc8$) play a role to classify the basketball video GMP based on optical flow. And the both hidden layers $fc6$ and $fc7$ have 4,096 neurons, while the last layer $fc8$ has 6 neurons corresponding to the six categories of basketball game events.

2.3 Event Recognition with Temporal Features of GMP

On event recognition, instead of relying on giving tracklets of multiple players [4,10,11], we adopt long short-term memory (LSTM) to investigate temporal features of the basketball game over GMP. This takes benefit of GMP by optical flow without background noise, which avoid analyzing the local occlusions among players and the efforts in tracking players. LSTM based temporal feature analysis and event prediction of sequential frames is given in part 4 of Fig. 1.

LSTM [18] has been used successfully for many sequential problems. LSTM unit consists of several cells with memory that stores information for a short temporal interval. The memory content of a LSTM makes it suitable for modeling complex temporal relationships that may span a long range. The content of the memory cell is made up of several gating units that control the flow of information in and out of the cells. This also helps in avoiding spurious gradient updates that can typically happen in training RNNs when the length of a temporal input is large. In this paper, we choose a single layer LSTM as part 4 of Fig. 1 to deal with temporal feature extraction and event prediction, as stated in [9] single layer LSTM is suitable in video recognition problems.

For a video clip with $T+1$ frames v_1, $v_2 \cdots v_{T+1}$, the optical flow v'_1, $v'_2 \cdots v'_T$ are calculated between consecutive frames to extract GMP hidden in visual features of frames. The GMP by optical flow is fed to CNN to capture a high-level compact expression x_1, $x_2 \cdots x_T$, which extract spatial features from each GMP image. Then LSTM is conducted over the CNN based spatial features of GMP to learn temporal representations of frames and perform event prediction for the given frames. Assume x_t is the input of the LSTM cell at time t, and w is the weight, which acts on the hidden layer output h_{t-1} at time $t-1$ and the input of x_t, and then a direct output z_t and hidden layer output h_t at time t are produced. Therefore, the processing of each video frame must be done in time, as shown in Fig. 1, from left to right in over to calculate, until h_T:

$$h_1 = F_w(x_1, h_0) = F_w(x_1, 0) \quad (2)$$

$$h_2 = F_w(x_2, h_1) \quad (3)$$

$$\vdots$$

$$h_T = F_w(x_T, h_{T-1}) \quad (4)$$

For the final event prediction, we use the softmax function to predict category of each frame. The event prediction process can be expressed as the following:

$$g_t = \frac{\sum_t^{t+T} s_m}{T} \quad (5)$$

where s_m is a score matrix of the m-th frame, g_t is the classification of the video clip where the m-th frame is located. The results of the T frames are averaged as inference to the video clip and the highest score is the predicted event category of the video clip.

3 Experiments

In this section, to verify the effectiveness we conduct experiments on the proposed GMP based event recognition in multi-person videos. The experiments are performed on an adjusted multi-person basketball video dataset [11]. We perform four comparison experiments including original frames based CNN, GMP based CNN, original frames based CNN with LSTM layer, and GMP based CNN with LSTM layer for event recognition.

3.1 Basketball Dataset

The NCAA basketball dataset was first presented in [11]. Different from the previous single-person shotting videos [9], this dataset is collected from YouTube in different venues and different age, which evolves real videos of multi-person games. The videos in the dataset take generally 1.5 h.

In this work, we aim at 6 key event types including 3-point, free-throw, layup, other 2-point, slam dunk and steal, respectively. The rules of the start and end of the event in this work follow [11]. And the average frames of video clips for 3-point, free-throw, layup, other 2-point, and steal event are 45, which provide enough temporal context to analyze the event. Especially the video clips of slam dunk contain an average of 15 frames, due to the transient nature of the event. It takes very few of context information to distinguish the event from the others. We select a subset of 250 games, and the videos were randomly divided into 200 for training and 50 for testing. From these videos, 9407 training and 2374 testing clips are split, each of which corresponds to a label of the 6 event categories.

3.2 Implementation Details

In the following, the implementation of the four comparison methods will be described in detail.

(1) **CNN:** In this work, it is required to pre-train a CNN model based on video frames. We implement CNN model using caffe [19] and the CNN is achieved by fine-tuning the AlexNet [20], which is elaborate in extracting high-level semantic features from images within a reasonable time and a reasonable amount of computation. As the data of slam dunk event is limited considerably, we randomly sample the other 5 categories in order to avoid the sample imbalance problem. The distribution of data for participating in both training and testing is shown in Table 1. During the CNN training and testing period, we randomly select 15443 video frames for training and 3341 video frames for testing. We take a batch size of 128, and a learning rate of 0.001 which is reduced by a factor of 0.95 every 5000 iterations. In order to demonstrate the ability of the CNN to extract the spatial features for event recognition, we perform event recognition by **CNN** with frames in 179 testing clips.

Table 1. Data distribution of the testing and training.

Event	CNN/frames		CNN + LSTM/clips	
	Train	Test	Train	Test
3-pointer	2574	557	150	30
Free-throw	2574	557	150	30
Layup	2574	557	150	30
Other 2-pointer	2574	557	150	30
Slam dunk	2573	556	149	29
Steal	2574	557	150	30

(2) **GMP based CNN:** We further perform experiments by GMP based CNN for event recognition to analyze the role of GMP based on optical flow. The training phase of this method is similar to the **CNN** training and the training parameters are the same as those of **CNN**. The difference is that the input frames are GMP by optical flow, which greatly reduces bias from background noise in the video frames.

(3) **CNN + LSTM:** To analyze the effect of LSTM in event recognition, we construct a **CNN+LSTM** network by fine-tuneing the trained **CNN**, which is CNN connected with a single-layer LSTM. There are 256 hidden nodes and 8 timesteps in the LSTM layer. Similar to the **CNN + LSTM** method for event recognition, a softmax layer is also deployed as the classification layer which corresponds to the 6 categories of video events. Considering the sample imbalance of the 6 event categories, during the training phase of the LSTM, we randomly sampled 899 clips for training and 179 clips for testing as shown in Table 1. In this work, we train the model of **CNN + LSTM** with consecutive 16 frames together of a video clip. If the last intercept exceeds the end of the video clip, we take the last 16 frames from the video clip. To verify the effectiveness of **CNN + LSTM** in event recognition, we use 179 basketball clips for testing the same as that in **CNN**.

(4) **GMP based CNN + LSTM:** The proposed method combines the benefit of both **CNN + LSTM** and **GMP based CNN**. The frames are GMP based on optical flow and fed to CNN and LSTM to extract spatial and temporal features respectively for event recognition. The training parameters are the same as **CNN + LSTM**. Similar to **CNN + LSTM**, the proposed **GMP based CNN + LSTM** is also constructed by fine-tuning the **CNN**. To verify the effectiveness of the proposed **GMP based CNN + LSTM** in event recognition, we use 179 basketball clips for testing the same as that in **CNN**.

3.3 Video Event Recognition

Table 2 provides accuracy of event recognition on the 179 testing video clips by the four comparisons. As shown in Table 2, compare **GMP based CNN**

Table 2. Accuracy of event recognition on testing video clips by the four comparisons.

Model	CNN	GMP based CNN	CNN + LSTM	GMP based CNN + LSTM
Accuracy	35.02	54.34	41.13	**60.28**

The bold numbers represent the best results (unit: % for accuracy).

Table 3. Confusion matrix of the classification.

	3-pointer	Free-throw	Layup	Other 2-pointer	Slam dunk	Steal
3-pointer	**17**	2	0	6	0	5
Free-throw	1	**27**	0	1	0	1
Layup	4	0	**10**	9	2	5
Other 2-pointer	6	2	2	**17**	1	2
Slam dunk	**12**	1	6	2	4	3
Steal	0	0	0	0	0	**30**
Accuracy	56.67	90	33.33	56.67	13.79	100

The bold numbers represent the best results in each measure (unit: % for accuracy).

with **CNN** the performance of **GMP based CNN** improves the accuracy of **CNN** by almost 20%. It implies the GMP from optical flow is able to alleviate obstacles from background noises and preserve effective motion information for event recognition. Moreover, it can be observed **CNN + LSTM** achieves better recognition rate of video clips and improves the accuracy of **CNN** by almost 6%. This indicates LSTM is good at extracting temporal characteristics of videos for event recognition. The results in Table 2 illustrate the proposed **GMP based CNN + LSTM** method achieves the best accuracy 60.28% in event recognition compared to the other three methods. It verifies the effectiveness of CNN and LSTM on GMP to predict events for videos.

We employ the **GMP based CNN + LSTM** model of best performance to infer events of the 179 testing video clips. Table 3 provides the performance distribution of every event category inferred by the proposed method. It can be observed that on free-throw and steal the proposed method achieves better accuracy, with almost all the inference correct. Because the GMP of these two events is more robust with relatively clear motion pattern. Due to some consecutive frames taking similar GMP, several 3-point and other 2-point video clips are confused with each other, and these two event categories achieve an accuracy more than 50%. And the performance of layup and slam dunk are slightly worse compared to the other categories. Because these two types of events happen in a short period which lacks of enough temporal features on GMP and makes the event recognition remain a great challenge. With analysis on the performance of the proposed CNN+LSTM on GMP, we can conclude that the event recognition based on GMP is promising to further spatial and temporal features extraction by CNN and LSTM, which devotes to the analysis in multi-person videos, especially sports videos.

4 Conclusion

We have proposed a event recognition approach based on GMP analysis in multi-person videos. The employed GMP effectively avoids the obstacles in tracking with human occlusion in multi-person videos. The proposed method investigates robust GPM with optical flow analysis, which effectively eliminates video background noise and improves the accuracy of event recognition by CNN and LSTM. Through a two-stage process, we extract spatial and temporal features of GMP using sequential CNN and LSTM. The experiments are conducted on a basketball dataset and the results illustrate the effectiveness of the proposed method by distinguishing 6 representative basketball events.

References

1. Zhang, Y., Xu, C., Rui, Y., Wang, J.: Semantic event extraction from basketball games using multi-modal analysis. In: IEEE International Conference on Multimedia and Expo, pp. 2190–2193 (2007)
2. Tan, Y.P., Saur, D.D., Kulkami, S.R., Ramadge, P.J.: Rapid estimation of camera motion from compressed video with application to video annotation. IEEE Trans. Circuits Syst. Video Technol. **10**, 133–146 (2002)
3. Jiang, H.H., Lu, Y., Xue, J.: Automatic soccer video event detection based on a deep neural network combined CNN and RNN. In: IEEE 28th International Conference on Tools with Artificial Intelligence (ICTAI), pp. 490–494 (2016)
4. Ibrahim, M.S., Muralidharan, S., Deng, Z., Vahdat, A., Mori, G.: A hierarchical deep temporal model for group activity recognition. In: Proceedings of the IEEE Conference on Computer Vision and Pattern Recognition, pp. 1971–1980 (2016)
5. Tjondronegoro, D.W., Chen, Y.P.P.: Knowledge-discounted event detection in sports video. IEEE Trans. Syst. Man Cybern. Part A Syst. Hum. **40**, 1009–1024 (2010)
6. Sato, T., Kanade, T., Hughes, E.K., Smith, M.A.: Video OCR for digital news archive. In: Proceeding IEEE International Workshop Content-Based Access Image Video Database, pp. 52–60 (1998)
7. Antani, S.C., Crandall, D., Kasturi, R.: Robust extraction of text in video. In: Proceeding 15th International Conference on Pattern Recognition, vol. 1, pp. 831–834 (2000)
8. Lienhart, R., Wernicke, A.: Localizing and segmenting text in images and videos. IEEE Trans. Circuits Syst. Video Technol. **12**(4), 256–268 (2002)
9. Donahue, J., Hendricks, L.A., Rohrbach, M., Venugopalan, S., Guadarrama, S., Saenko, K.: Long-term recurrent convolutional networks for visual recognition and description. In: AB Initto Calculation of the Structures and Properties of Molecules. Elsevier (2014)
10. Alahi, A., Goel, K., Ramanathan, V., Robicquet, A., Fei-Fei, L., Savarese, S.: Social lstm: human trajectory prediction in crowded spaces. In: Proceedings of the IEEE Conference on Computer Vision and Pattern Recognition, pp. 961–971 (2016)
11. Ramanathan, V., Huang, J., Abu-El-Haija, S., Gorban, A., Murphy, K., Fei-Fei, L.: Detecting events and key actors in multi-person videos. In: Proceedings of the IEEE Conference on Computer Vision and Pattern Recognition, pp. 3043–3053 (2016)

12. Bouguet, J.Y.: Pyramidal implementation of the affine lucas kanade feature tracker description of the algorithm. Intel Corporation. **4**, 1–10 (2001)
13. Brox, T., Bruhn, A., Papenberg, N., Weickert, J.: High accuracy optical flow estimation based on a theory for warping. In: Pajdla, T., Matas, J. (eds.) ECCV 2004. LNCS, vol. 3024, pp. 25–36. Springer, Heidelberg (2004). https://doi.org/10.1007/978-3-540-24673-2_3
14. Schuldt, C., Laptev, I., Caputo, B.: Recognizing human actions: a local SVM approach. Int. Conf. Pattern Recog. **3**, 32–36 (2004)
15. Wang, H., Klaser, A., Schmid, C., Liu, C.L.: Action recognition by dense trajectories. In: IEEE Conference on Computer Vision and Pattern Recognition, vol. 42, pp. 3169–3176 (2011)
16. Jia, Y., Shelhamer, E., Donahue, J.: Caffe: convolutional architecture for fast feature embedding. In: ACM International Conference on Multimedia, pp. 675–678. ACM (2014)
17. Zeiler, M.D., Fergus, R.: Visualizing and understanding convolutional networks. In: Fleet, D., Pajdla, T., Schiele, B., Tuytelaars, T. (eds.) ECCV 2014. LNCS, vol. 8689, pp. 818–833. Springer, Cham (2014). https://doi.org/10.1007/978-3-319-10590-1_53
18. Hochreiter, S., Schmidhuber, J.: Long short-term memory. Neural Comput. **9**, 1735–1780 (1997)
19. Jia, Y.: Caffe: an open source convolutional architecture or fast feature embedding (2013). http://caffe.berkeleyvision.org/
20. Krizhevsky, A., Sutskever, I., Hinton, G.E.: ImageNet classification with deep convolutional neural networks. In: International Conference on Neural Information Processing Systems, vol. 25, pp. 1097–1105. Curran Associates Inc. (2012)

Hierarchical Convolutional Features for Long-Term Correlation Tracking

Huizhi Chen(✉) and Baojie Fan

College of Automation, Nanjing University of Posts and Telecommunications, Nanjing 210046, China
machupicchuer@gmail.com, jobfbj@gmail.com

Abstract. Visual tracking is of great significance in computer vision. In this paper, we propose a long-term tracking method to deal with appearance variation caused by occlusion, fast motion, out-of-view, etc. Firstly, we extract CNNs features from both low and high depth of a pre-trained VGGNet and estimate target translation via correlation filters separately trained on features from three different layers. Then a HOG based scale correlation filter is applied to search target pyramids cropped around target position for optimal scale. In case of tracking failure, we train a SVM to re-detect the target. In addition, scale model is only updated when scale response is higher than a pre-defined threshold. Experimental results on OTB2013 show that our algorithm is of effectiveness and robustness in case of heavy appearance changes.

Keywords: Correlation filters · CNNs features · Long-term tracking
Scale estimation

1 Introduction

Visual Tracking is one of the most important area in computer vision with applications in surveillance, navigation, human computer interaction, robotics, etc. Visual object tracking can be described as: given an unknown target by a bounding-box on the first frame of a sequence and estimating states of the target on the following frames. After decades of development, tracking methods have been progressed a lot but there are still factors limiting tracking performance.

This paper is based on three observations upon prior works. Firstly, tracking methods suffer from large appearance changes caused by illumination changes, occlusion, abrupt motion, background clutter and deformation. Therefore, we apply an efficient strategy to update appearance templates, which is capable of preventing target models from being polluted by background information.

Secondly, convolutional networks have achieved significant success in object detection and image recognition regions. CNNs features, coming from large-scale vision datasets and efficient learning, prove to be more robust and discriminative than hand-crafted features. Thus we use deep CNNs features to cope with distractors.

J. Yang et al. (Eds.): CCCV 2017, Part III, CCIS 773, pp. 677–686, 2017.
https://doi.org/10.1007/978-981-10-7305-2_57

In addition, tracking methods based on hand-crafted features take advantage of efficient computation owing to low dimension. When it comes to deep features with hundreds of channels, tracking speed would be slowed down. To achieve computation efficiency, we make use of correlation filters and alleviate the sampling ambiguity at the same time.

2 Related Work

Lots of visual tracking methods have been proposed during decades of study. Tracking-by-detection is known as a popular discriminative framework for object tracking, which takes context information into consideration and classifies target and background by learning a classifier. Recently, different kinds of machine learning algorithms are applied into discriminative methods, for instance, multiple instance learning [1], boosting [3], structured support vector machine (SVM) [10], etc.

Apart from classification, target representations play an important role in visual tracking. CNNs features have promoted image recognition precious to a higher level than human. DLT [12] pre-trains a small network on tiny images dataset and then uses particle filters to localize targets. The idea of pre-trained model with offline tuning is inherited by many later methods. Wang et al. propose FCNT [11] based on fully convolutional networks, where features selected from conv4-3 and conv5-3 layer of VGGNet [6] are separately used to construct two special nets designed to catch position information and discriminate distractors. Ma et al. [7] apply features coarse-to-fine to learn three filters and the final response map is weighted sum of the three sub-maps, maximum value of which determines the location of targets. With the use of hierarchical features, FCNT and HCFT perform well with background clutter.

Correlation tracking has recently arisen much attention due to computational efficiency. Correlation tracking takes circular-shifted visions of features with Gaussian-weighted labels around target position as training samples, which alleviating the problem of sampling ambiguity. Nevertheless, circulant matrices enable correlation operations to be computed with high speed in frequency domain. Numerous extensions of correlation filters have been proposed to elevate tracking accuracy, including KCF [5] with HOG and Gaussian kernel and CSK [4] with linear kernel. Features like color-name or HOG adopted by these correlation filter based methods limit the robustness of tracker, leading to drifting under situations of severe deformation and occlusion.

3 Proposed Algorithm

This section consists of (1) correlation tracking, (2) online detection, and (3) model update. The following content will introduce these parts.

3.1 Correlation Tracking

As typical correlation trackers do, we learn a discriminative classifier and tracking targets by searching maximum value of correlation response map. The appearance of the target is modeled using correlation filter $w^{(l)}$. Feature vector $\boldsymbol{x}$ with size of $M \times N \times D$ are extract from searching window centered around target position, where M, N and D respectively indicate the width, height and depth of features. Training samples are all the circular shifts of $\boldsymbol{x}$ along the M and N dimensions, where each sample $x^{(l)}_{(m,n)}$, $m \in \{0, 1, \ldots, M-1\}$, $n \in \{0, 1, \ldots, N-1\}$ has a Gaussian function label $y^l(m,n)$ computed by: $y^{(1)}(m,n) = exp(-((m-M/2)^2 + (n-M/2)^2)/2\sigma^2)$, where σ is the kernel width. The correlation filter $w^{(l)}$ with the same size of feature $\boldsymbol{x}$ is trained by solving the ridge regression:

$$\min_{w^{(l)}} \sum_{m,n} \left| \Phi(x^{(l)}_{m,n} w^{(l)} - y^{(l)}(m,n)) \right|^2 + \left| w^{(l)} \right|^2 \tag{1}$$

where Φ denotes the mapping to a kernel space and λ is a non-negative regularization parameter. The learned filter $w^{(l)}$ can be expressed as

$$w^{(l)} = \sum_{m,n} a^{(l)}(m,n) \Phi(x^{(l)}_{m,n}) \tag{2}$$

where the coefficient a is defined as

$$A^{(l)} = \mathcal{F}(a^{(l)}) = \frac{\mathcal{F}(y^{(l)})}{\mathcal{F}(\Phi(x^{(l)} \cdot \Phi(x^{(l)}))) + \lambda} \tag{3}$$

In (3), $\mathcal{F}$ denotes the fast Fourier transformation operator and $\mathcal{F}(y^{(l)})$ is the Gaussian label. The response map in the new frame is computed on an image patch z within a $M \times N$ search window

$$\hat{y} = F^{-1}(A^{(l)} \odot F(\Phi(z) \cdot \Phi(\hat{x}))) \tag{4}$$

where $\hat{x}$ denotes the learned target appearance model and $\odot$ is the Hadamard product. Now, we get response maps for all layers denoted by $\hat{y}^{(l)}$. Then we sum the three maps with corresponding weights $\gamma^{(l)}$ and get summation $y_{sum} = \Sigma_l(\gamma^{(l)} \hat{y}^{(l)})$. Thus, the new target position is estimated by searching the maximum value of y_{sum}.

3.2 Online Detection

Obviously, re-detection step is significant to a long-term tracking algorithm in case of tracking failure. Our re-detection is performed on every frame. For computational efficiency and model robustness, we activate the re-detection module only when $\max(\hat{y}_scale) < T_{scale}$, where $\hat{y}_{scale}$ refers to scale response map and T_{scale} is a re-detection threshold.

To get scale response map, we construct a target pyramid around the estimated position $\arg_{m,n} y_{sum}$. The target size is assumed to be $W \times H$ in a test frame and K indicates the number of scales $s \in S$. For each scale in $S = \{a^k | k = -(K-1)/2, -(K-3)/2, \ldots, (K-1)/2\}$, we crop out an image patch with the size of $sW \times sH$ centered on the predicted position. Different from motion model, scale model is built on HOG features by solving the same ridge regression as (1). The scale most suitable for current target is

$$scale = \arg_j \max(\max(\hat{y}_1^s), \max(\hat{y}_2^s), \ldots, \max(\hat{y}_j^s)) \tag{5}$$

where $\hat{y}_j^s$ indicates scale correlation response and $j \in 1, 2, \ldots, K$ represent different scale levels. In addition, all patches are resized to $W \times H$ before doing correlation with scale filter.

For re-detection module, we use HOG features to train an online SVM classifier to re-detect targets. Training samples are patches with motion correlation response higher than confidence threshold $\mathcal{T}_{rd}$. The label y of a SVM train sample is determined by IOU value, which indicates the overlap ratios of the sample and target bounding box,

$$y = \begin{cases} +1, & \text{if } IOU > 0.9 \\ -1, & \text{if } IOU < 0.5 \end{cases} \tag{6}$$

SVM classifier is described in the form of $f(x) = w\Phi(x) + b$ learned from samples with labels $y \in -1, +1$. The object function is

$$\min_{w,b,\xi} \frac{1}{2} |w|^2 + C \sum_{i=1}^{N} \xi_i \tag{7}$$

subject to constrains

$$y_i(w\Phi(x) + b) - 1 + \xi_i \geq 0, \xi_i \geq 0 \tag{8}$$

Therefore, we learn a weight vector w by solving this quadratic convex optimization problem.

3.3 Model Update

In order to ensure the robustness of our tracker, it is necessary to update appearance models when targets undergo occlusion, deformation and abrupt motion. The motion model R_m is updated with a learning rate on every frame:

$$\hat{x}_t^{(l)} = (1 - \alpha)\hat{x}_{t-1}^{(l)} + \hat{x}_t^{(l)} \tag{9}$$

$$\hat{A}_t^{(l)} = (1 - \alpha)\hat{A}_{t-1}^{(l)} + \hat{A}_t^{(l)} \tag{10}$$

where t is the index of current frame, l indicates features layers.

However, the scale model R_s is only updated when scale correlation response $\max \hat{y}_{scale} \mathcal{T}_{update}$ and the learning rate is same as R_m.

4 Implementation

The main steps of our algorithm is presented in Algorithm 1 and more implementation details are described as follows.

Features. VGG-16 trained on ImageNet is adopted to extract hierarchical convolutional features in this work. Features used to train the motion model R_m are the output of conv5-4, conv4-4 and conv3-4 convolutional layer. However, the scale model R_s is constructed with HOG features. For the SVM detector, all different scale patches are resized to $W \times H$.

Kernel Selection. Both motion model and scale model adopt a Gaussian kernel $k(x, x') = e^{-|x-x'|^2/\sigma^2}$, which defined a mapping Φ as $k(x, x') = \Phi(x) \cdot \Phi(x')$.

Algorithm 1. Proposed tracking algorithm

Input: Initial target bounding box x_0
Output: Estimated object state $x_t = (\hat{x}_t, \hat{y}_t, \hat{s}_t)$, motion model R_m, scale model R_s, SVM classifier
Repeat
Crop out the searching window in frame t according to $(\hat{x}_{t-1}, \hat{y}_{t-1})$ and extract the features;
// *Translation estimation*
Compute the correlation map y_t using R_m and (4) to estimate the new position (x_t, y_t);
// *Scale estimation*
Build target pyramids around (x_t, y_t) and compute the correlation map y_s using R_s and (4);
Estimate the optimal scale using (5); $x_t = (\hat{x}_t, \hat{y}_t, \hat{s}_t)$;
// *Target re-detection*
if $\max(y_{\hat{s}}) < T_{scale}$ **then**
  Use SVM to perform re-detection and find the possible candidate states X;
  foreach *state* x'_i *in* X **do**
    computing confidence score y'_i using R_m and (4)
  end
  if $\max(y'_i) > T_{rd}$ **then**
    $x_t = x'_i$, where $i = \arg\max_i y'_i$
  end
end
// *Model update*
Update R_m using (9);
if $\max(y_{\hat{s}}) > T_{update}$ **then**
  Update R_s using (9);
end
Update SVM;
Untill end of video sequences;

SVM. The SVM classifier also applies a Gaussian kernel in order to discriminate non-linear samples. The re-detection step is implemented by scanning the whole frame with scanning windows with the size corresponding to the optimal scale obtain by correlation filter.

5 Experimental Results

Setups. We evaluate our algorithm LHCF on OTB2013 [13] that contains 50 different sequences and results of 29 trackers with comparisons to state-of-the-art methods by two metrics: distance precision and overlap success rate. Parameters for our test is listed in Table 1. Our algorithm is implemented in MATLAB on a 2.6 GHz Intel Core i5 CPU with 8 GB RAM.

Table 1. Parameters for implementation

Parameters	α	K	γ	$\mathcal{T}_{scale}$	$\mathcal{T}_{rd}$	$\mathcal{T}_{update}$
Value	0.01	21	[1 0.5 0.02]	0.25	0.5	0.5

Quantitative Evaluation. Our algorithm is evaluated on the benchmark with comparisons with (1) correlation based trackers: MEEM [15], KCF [5], CSK [4], DSST [9], (2) CNNs trackers: HCFT [7], DLT [12], FCNT [11] and (3) classifier based trackers: Struck [10], TLD [14], TGPR [2]. Figure 1 shows that our methods performs well against the other methods in overlap success rate of OPE (one pass evaluation) achieving an AUC score of 0.745 on OTB2013. Compared to HCFT and FCNT which designed for short term tracking, our algorithm achieves higher success rate. This result indicates that the re-detection step is of great importance to robustness. In terms of distance precision, LHCF (0.888) outperforms all methods based on hand-crafted features but performs a little inferior to HCFT (0.891). The precision achieved by our algorithm proves the fact that features extracted from convolution networks are more discriminative and robust than hand-crafted features in tracking as algorithms have done in other areas like image classification and detection.

Attribute-Based Evaluation. OTB2013 is annotated with 11 sequence attributes to describe different challenges in object tracking, including illumination variation (IV), out-of-plane rotation (OPR), scale variation (SV), occlusion (OCC), deformation (DEF), motion blur (MB), fast motion (FM), in-plane rotation (IPR), out-of-view (OV), background cluttered(BC) and low resolution (LR). Figure 2 shows distance precious plots on OPE for eight main attributes (fast motion, motion blur, occlusion, out-of-view, background clutter, deformation, scale variation). As it is shown in Fig. 2, our algorithm performs well in most attributes. It is known that sequences annotated with occlusion and out-of-view always contain targets which suffer temporal disappearance caused by long time occlusion or temporal out of plane. The proposed algorithm gets over this challenge depending on the SVM based re-detection step. Moreover, we can find out LHCF performs better than DSST and LCT both of which contain scale estimation which can be explained by the fact that deep layer features contain more semantic information than HOG. However, in terms of deformation, LHCF is unexpectedly defeated by HCFT which even has no re-detection module. Both algorithms apply similarly update model using a linear interpolation operation. When deformation happens and the response becomes smaller than the threshold, the re-detection step is activated but makes no contributes in this condition. This indicates that we have to optimize our model update solution in future works.

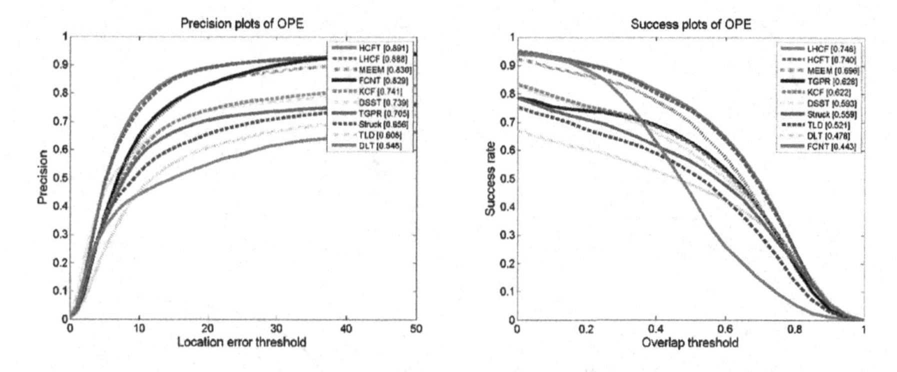

Fig. 1. The precision plots and success plots of OPE on OTB2013. The performance score for each tracker is shown in the legend. The performance score of precision plot is at error threshold of 20 pixels with the performance score of success plot is the AUC value.

Qualitative Evaluation. We display some tracking results of LHCF, compared with KCF [5], LCT [8], HCFT [7], DLT [12] and Struck [10] in Fig. 2. Struck and DLT are seriously affected by distractors (basketball). LHCF and FCNT track the rolling and fast moving motor perfectly due to robust appearance model based on CNNs features. However, when it comes to sequence lemming in which the target undergoes long time occlusion, only our algorithm performs well without drifting to the background. This result shows that the re-detection step is of great importance. Specially, LCT which also has a re-detection module loses the target as other short-term algorithms in this sequence. This again reveals the advantage of deep convolutional features (Fig. 3).

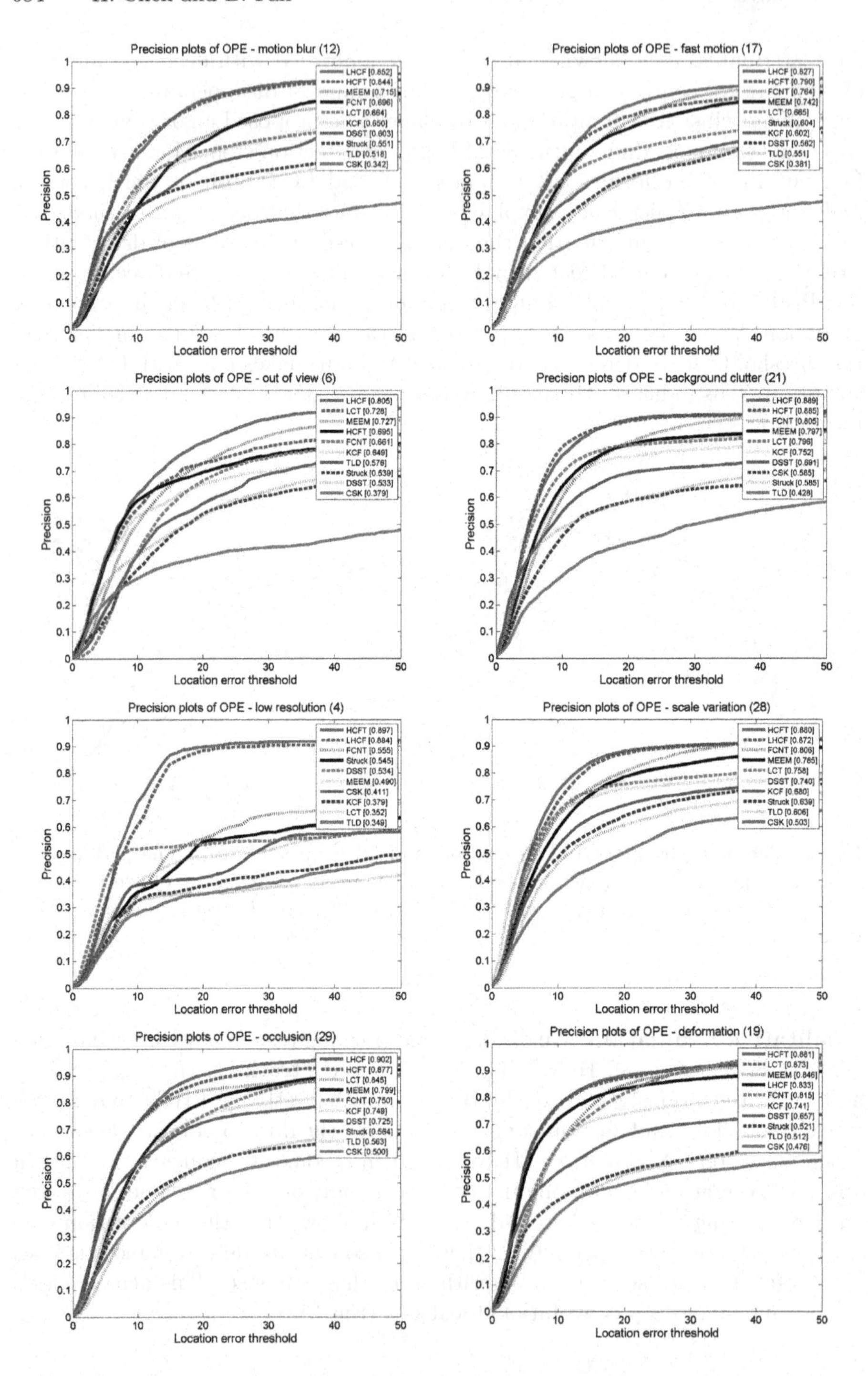

Fig. 2. The precision plots of OPE for 8 challenging attributes on OTB2013 dataset. The number of videos for each attribute is shown in parenthesis. Our algorithm LHCF almost perform best in all the attributes.

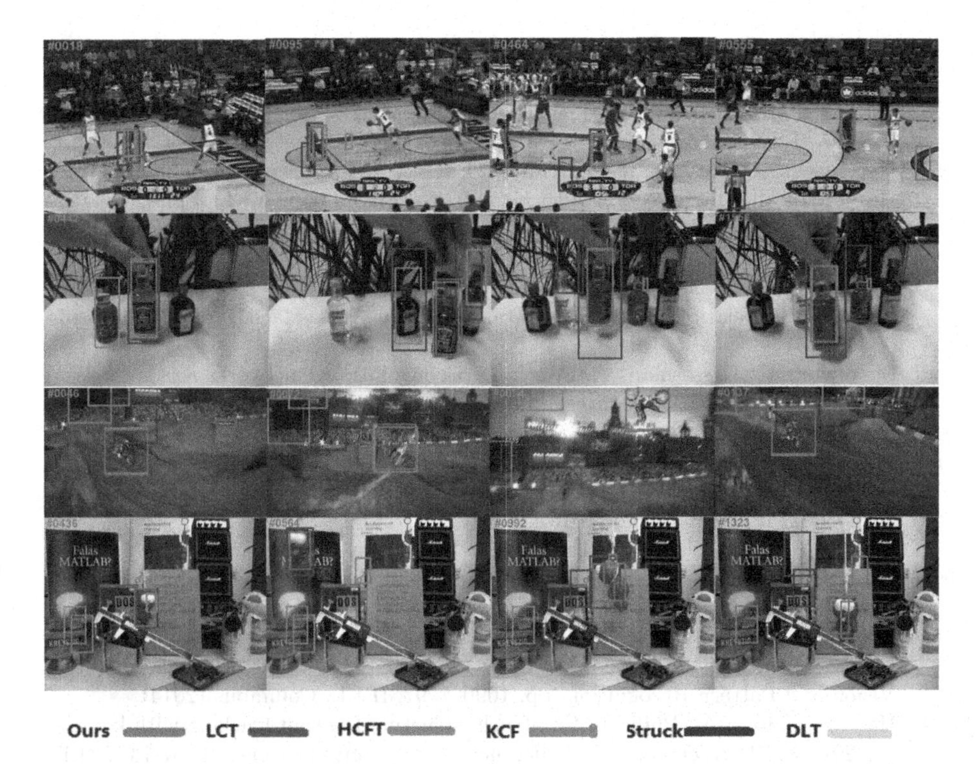

Fig. 3. Quality evaluation of proposed algorithm, LCT, HCFT, KCF, Struck and DLT on 4 challenging sequences (from top to bottom are Basketball, Liquor, MotorRolling and Lemming).

6 Conclusions

In this paper, we apply correlation filters and CNNs features into long-term visual tracking methods. We represent target with CNNs features and estimate position and scale with correlation filters. SVM is used as a re-detector to handle tracking failure. Experimental results show that our algorithm is of effectiveness and robustness.

References

1. Babenko, B., Yang, M.H., Belongie, S.: Robust object tracking with online multiple instance learning. IEEE Trans. Pattern Anal. Mach. Intell. **33**(8), 1619–1632 (2011)
2. Gao, J., Ling, H., Hu, W., Xing, J.: Transfer learning based visual tracking with Gaussian processes regression. In: Fleet, D., Pajdla, T., Schiele, B., Tuytelaars, T. (eds.) ECCV 2014. LNCS, vol. 8691, pp. 188–203. Springer, Cham (2014). https://doi.org/10.1007/978-3-319-10578-9_13

3. Grabner, H., Leistner, C., Bischof, H.: Semi-supervised on-line boosting for robust tracking. In: Forsyth, D., Torr, P., Zisserman, A. (eds.) ECCV 2008. LNCS, vol. 5302, pp. 234–247. Springer, Heidelberg (2008). https://doi.org/10.1007/978-3-540-88682-2_19
4. Henriques, J.F., Caseiro, R., Martins, P., Batista, J.: Exploiting the circulant structure of tracking-by-detection with Kernels. In: Fitzgibbon, A., Lazebnik, S., Perona, P., Sato, Y., Schmid, C. (eds.) ECCV 2012. LNCS, vol. 7575, pp. 702–715. Springer, Heidelberg (2012). https://doi.org/10.1007/978-3-642-33765-9_50
5. Henriques, J.F., Caseiro, R., Martins, P., Batista, J.: High speed tracking with kernelized correlation filters. IEEE Trans. Pattern Anal. Mach. Intell. **37**(3), 583–596 (2015)
6. Simonyan, K., Zisserman, A.: Very deep convolutional networks for large-scale image recognition. In: 3rd International Conference on Learning Representations, San Diego (2015)
7. Ma, C., Huang, J.B., Yang, X., Yang, M.H.: Hierarchical convolutional features for visual tracking. In: 2015 IEEE International Conference on Computer Vision, pp. 3074–3082. IEEE, Santiago (2015)
8. Ma, C., Yang, X., Zhang, C., Yang, M.H.: Long-term correlation tracking. In: 2015 IEEE Conference on Computer Vision and Pattern Recognition, pp. 5388–5396. IEEE, Boston (2015)
9. Danelljan, M., Shahbaz Khan, F., Felsberg, M., Van de Weijer, J.: Adaptive color attributes for real-time visual tracking. In: 2014 IEEE Conference on Computer Vision and Pattern Recognition, pp. 1090–1097. IEEE, Columbus (2014)
10. Hare, S., Saffari, A., Philip, H.S.: Struck: structured output tracking with kernels. In: 2011 IEEE International Conference on Computer Vision, pp. 6–13. IEEE, Bacelona (2011)
11. Wang, L., Ouyang, W., Wang, X., Lu, H.: Visual tracking with fully convolutional networks. In: 2015 IEEE International Conference on Computer Vision, pp. 3119–3127. IEEE, Boston (2015)
12. Wang, N., Yeung, D.Y.: Learning a deep compact image representation for visual tracking. In: Proceedings of 27th Annual Conference on Neural Information Processing Systems, Lake Tahoe (2013)
13. Wu, Y., Lim, J., Yang M.H.: Online object tracking: a benchmark. In: 2013 IEEE Conference on Computer Vision and Pattern Recognition, pp. 2411–2418. IEEE, Portland (2013)
14. Kalal, Z., Mikolajczyk, K., Matas, J.: Tracking-learning-detection. IEEE Trans. Pattern Anal. Mach. Intell. **34**(7), 1409–1422 (2012)
15. Zhang, J., Ma, S., Sclaroff, S.: MEEM: robust tracking via multiple experts using entropy minimization. In: Fleet, D., Pajdla, T., Schiele, B., Tuytelaars, T. (eds.) ECCV 2014. LNCS, vol. 8694, pp. 188–203. Springer, Cham (2014). https://doi.org/10.1007/978-3-319-10599-4_13

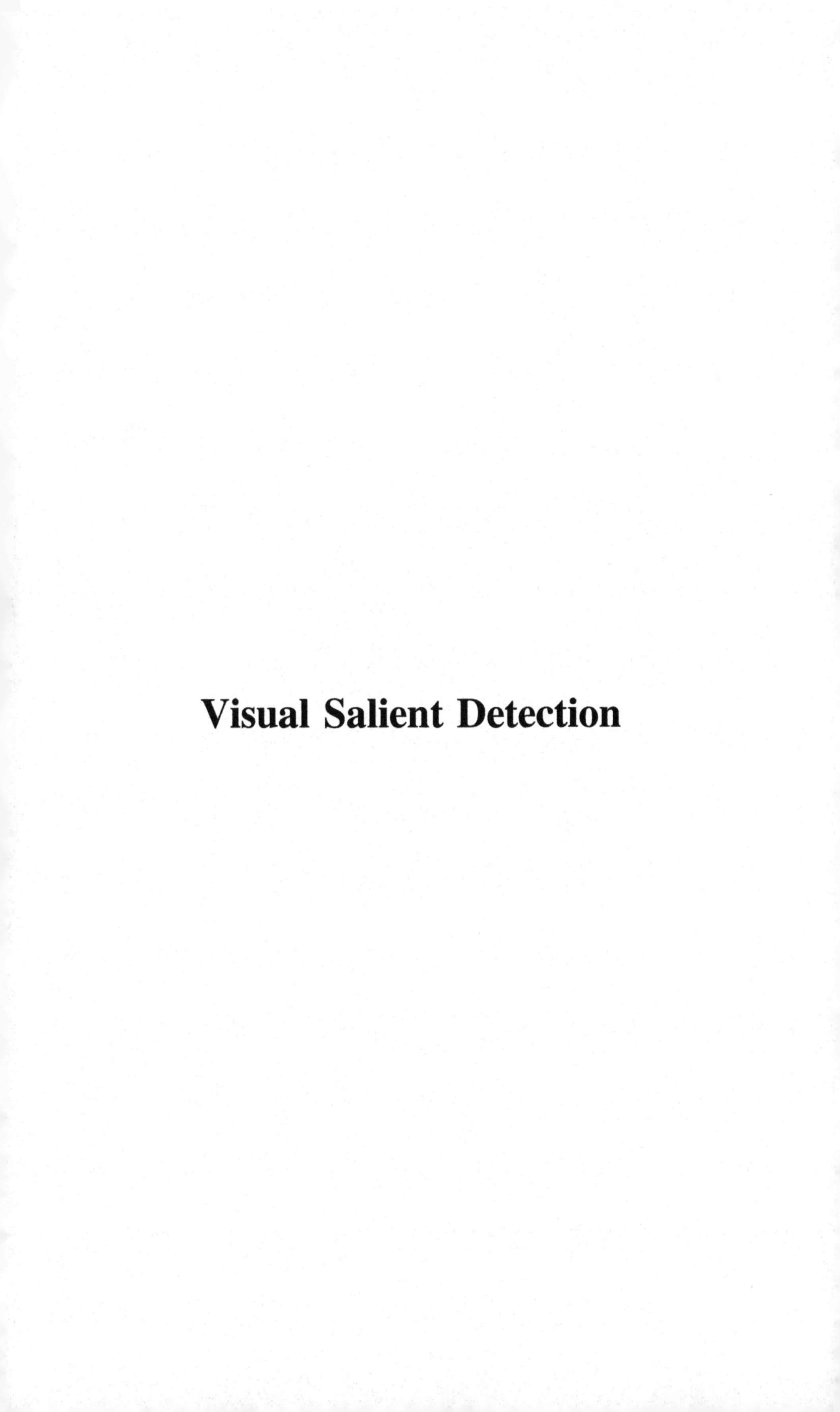

Visual Salient Detection

Saliency Detection Using Texture and Local Cues

Qiang Qi[1,2], Muwei Jian[1,2(✉)], Yilong Yin[1], Junyu Dong[2], Wenyin Zhang[3], and Hui Yu[4]

[1] School of Computer Science and Technology, Shandong University of Finance and Economics, Jinan, China
jianmuweihk@163.com
[2] Department of Computer Science and Technology, Ocean University of China, Qingdao, China
[3] School of Information Science and Engineering, Linyi University, Linyi, China
[4] School of Creative Technologies, University of Portsmouth, Portsmouth, UK

Abstract. In this paper, a simple but effective method is proposed for detecting salient objects by utilizing texture and local cues. In contrast to the existing saliency detection models, which mainly consider visual features such as orientation, color, and shape information, our proposed method takes the significant texture cue into consideration to guarantee the accuracy of the detected salient regions. Firstly, an effective method based on selective contrast (SC), which explores the most distinguishable component information in texture, is used to calculate the texture saliency map. Then, we detect local saliency by using a locality-constrained linear coding algorithm. Finally, the output saliency map is computed by integrating texture and local saliency cues simultaneously. Experimental results, based on a widely used and openly available database, demonstrate that the proposed method can produce competitive results and outperforms some existing popular methods.

1 Introduction

Saliency detection is the process of identifying the most informative location of objects in images, which is different to the traditional models of predicting human fixations [44]. In recent years, saliency detection has attracted wide attention of many researchers and become a very active topic in computer vision research. Since it plays a significant role in many computer-vision related applications such as: object detection [11,15,17,36], image/video resizing [3,35], image/video quality assessment [20,42], vision tracking [43] etc.

Generally, saliency detection methods can be traditionally summarized into two categories: the bottom-up and top-down approaches. Bottom-up saliency detection methods [37–39] mainly rely on visual-driven stimuli, which directly utilize low-level features for saliency detection. One of the pioneering computational models of bottom-up framework was proposed by Itti et al. in [13]. Later, in order to make salient object detection more precise, a novel graph-based method

J. Yang et al. (Eds.): CCCV 2017, Part III, CCIS 773, pp. 689–699, 2017.
https://doi.org/10.1007/978-981-10-7305-2_58

was proposed to estimate saliency map in [10]. While the graph-based method usually generates slow resolution saliency maps. Then, a spectral residual-based saliency detection method was pro-posed by Hou and Zhang in [12]. However, the early researches are hard to handle well in the cases with complex scenes. To solve this problem and acquire accurate salient regions, Ma and Zhang [21] presented a local contrast-based method for saliency detection, which increases the accuracy of saliency detection. From then on, many saliency detection models have been proposed and achieved excellent performance in various fields. In [7], a discriminant center-surround hypothesis model was pro-posed to detect salient objects. In [45], a bottom-up framework by using natural statistics was proposed to detect salient image regions. Achanta et al. [1] proposed a novel saliency detection method based on frequency tuned model to detect salient object. In [23], a new saliency measure was proposed by using a statistical frame-work and local feature contrast. Wang et al. [29] proposed a novel saliency detection model based on analyzing multiple cues to detect salient objects. In [14], a hierarchical graph saliency detection model was proposed via using concavity context to compute weights between nodes. Tavakoli et al. [25] designed a fast and efficient saliency detection model using sparse sampling and kernel density estimation. In [8], a context-aware based model, which aims at using the image regions to represent the scene, was proposed for saliency detection. Yang et al. [33] presented a new method based on foreground and background cues to achieve a final saliency map. In [32], a novel approach, which can combine contrast, center and smoothness priors, was proposed for salient detection. Ran et al. [24] introduced an effective framework by involving pattern and color distinctness to estimate the saliency value in an image. In [6], a cluster-based method was proposed for co-saliency detection. In [27], a novel saliency measure based on both global and local cues was proposed. Li et al. [19] proposed a visual saliency detection algorithm based on dense and sparse reconstruction. In [22], a novel Cellular Automata (CA) model was introduced to compute the saliency of the objects. In [14], a visual-attention-aware model was proposed for salient-object detection. Recently, a bottom-up saliency-detection method by integrating Quaternionic Distance Based Weber Descriptor (QDWD), center and color cues, was presented in [16]. In [2], a novel and effective deep neural network method incorporating low-level features was proposed for salient object detection.

Top-down based saliency detection models [40,41], which consider both visual information and prior knowledge, are generally task dependent or application oriented. Thus, the top-down model normally needs supervised learning and is lacking in extendibility. Compared with the bottom-up saliency detection model, not much work has been proposed. Kanan et al. [18] presented an appearance-based saliency model using natural statistics to estimate the saliency of an input image. In [4], a novel top-down framework, which incorporated a tightly coupled image classification module, was proposed for salient object detection. In [5], a weakly supervised top-down approach was proposed for saliency detection by using binary labels.

In this paper, a novel and efficient method based on a bottom-up mechanism, by integrating texture and local cues together, is proposed for saliency detection. In contrast to the existing saliency detection models, we compute the texture saliency map based on selective contrast (SC) method, which can guarantee the accuracy of the detected salient regions. In addition, we incorporated an improved locality-constrained linear coding algorithm (LLC) to detect the local saliency in an image. In order to evaluate the performance of our proposed method, we carried out experiments based on a widely used dataset. Experimental results, compared with other state-of-the-art saliency-detection algorithms, show that our approach is effective and efficient for saliency detection.

The remainder of the paper is organized as follows. In Sect. 2, we introduce the proposed saliency-detection algorithm in detail. In Sect. 3, we demonstrated our experimental results based on a widely used dataset and compared the results with other eight saliency detection methods. The paper closes with a conclusion and discussion in Sect. 4.

2 The Proposed Saliency Detection Method

This section presents the proposed saliency detection method by using selective contrast (SC) method for describing the texture saliency and locality-constrained method for estimating the local saliency. Texture saliency estimation method based on SC will first be described, followed by the locality-constrained method. All these different types of visual cues are fused to form a final saliency map.

2.1 Texture Saliency Detection Based on Selective Contrast

Texture is an important characteristic for human visual perception, which is caused by different physical-reflection properties on the surface of an object with the gray level or color changes. An image is not just a random collection of texture pixels, but a meaningful arrangement of them. Different arrangements of these pixels form different textures would provide us with important saliency information. As a basic property of image, texture also affects the similarity degree among regions, which is a useful cue in saliency detection, so we take the texture cue into consideration in our designed model.

In this section, we adopt selective contrast (SC) based algorithm [30] to estimate the texture saliency. Assume that the pixels in an input image are denoted as $X_n, n = 1, 2, 3, \ldots, N$, where N is the number of pixels in the image and X_n is a texture vector, which is achieved by using the uniform LBP [9]. Since the outputs of obtained textures span a very wide range of high dimensional spaces, we hope to express them in a more compact way. In order to solve this issue, we use $k - means$ to cluster these texture expressions and consider the cluster centers as the representative textures. After this transformation, each texture is denoted as its nearest texture prototype, this expression is called selective contrast (SC), which explores the most distinguishable component information in texture. Thus, each pixels texture saliency based on selective contrast (SC) can be written as follows [30]:

$$S_t(i, R_i) = \sum_{j \subset R_i} d(i, j) \tag{1}$$

where i is the examined pixel, R_i denotes the supporting region for defining the saliency of pixel i, and $d(i, j)$ is the distance of texture descriptors between i and j. The l_2 norm can be used to define the distance measure of textures descriptors:

$$d(i, j) = \|k_i - k_j\| \tag{2}$$

where k_i and k_j are the transformed textures expressions of pixels i and j by $k - means$.

In order to further improve the efficiency by reducing the computational complexity, a limited number of textons are trained from a set of images and the textures can be quantized to M textons. By this means, the computation is reduced to looking up a distance dictionary D_t of $M * M$ dimensions. Thus, the Eq. (1) can be rewritten as:

$$S_t(i, R_i) = \sum_{\psi(j)} f(\psi(j)) D_t(\psi(i), \psi(j)) \tag{3}$$

where i, R_i have a similar meaning in Eq. (1), $\psi(i)$ is the function mapping pixel i to its corresponding prototype texture and $f(\psi(j))$ is the frequency of $\psi(j)$ in region R_i. Readers can be referred to the [30] for more details.

Figure 1(b) shows some texture saliency maps of the input images. From the results we can see that the texture saliency maps can exclude most of the background pixels and detect almost the whole salient objects in images, but it leads to missing some homogeneous regions which have the similar texture appearances.

2.2 Local Saliency Detection Using Locality-Constrained Method

The motivation of local estimation is the local outliers, which are standing out from their neighbors with different colors or textures and tend to attract human attention. In order to detect local outliers and get acceptable performance, local coordinate coding method [34], which described the locality is more essential than sparsity, has been used in saliency detection. Furthermore, the proposed SC model in Sect. 2.1 only takes the texture information into consideration, which misses some local in-formation. Thus, we employ an approximated algorithm based on locality-constrained linear coding (LLC) [28] to estimate the local saliency. For a given image, we first over-segmented the image into N regions, $r_i, i = 1, 2, \ldots, N$. For each region r_i, let X be a set of 64-dimensions local descriptors extracted from the image, and $X = [x_i^0, x_i^1, \ldots, x_i^{63}]^T, i = 1, 2, \ldots, N$. Therefore, the original function of LLC method is written as follows [28]:

$$\min_B \sum_{i=1}^{N} (\|x_i - Db_i\|^2 + w\|dr_i. * b_i\|^2), s.t. 1^T b_i = 1, \forall_i \tag{4}$$

where $B = [b_1, b_2, \ldots, b_N]$ is the set of codes for X, $D = [d_1, d_2, \ldots, d_M]$ is the codebook with M entries, and the parameter ω is used to balance the weight between the penalty term and regularization term. The constraint $1^T b_i = 1$ follows the shift-invariant requirements of the LLC coding and .* denotes an element-wise multiplication. Here, the dr_i is the locality adaptor that gives different freedom for each codebook vector based on its similarity to the input descriptor x_i, and is defined:

$$dr_i = exp(\frac{dist(x_i, D)}{\lambda}) \tag{5}$$

where $dist(x_i, D) = [dist(x_i, d_1), dist(x_i, d_2), \ldots, dist(x_i, d_M)]^T$, $dist(x_i, d_i)$ denotes the Euclidean distance between x_i and the code book vector d_i, λ is used to adjust the weight decay speed for the locality adaptor, and M is the number of elements in the codebook. More details about LLC can be referred to [28,34].

In this paper, we adopted an approximated LLC algorithm [27] to detect the local cue. We consider the K nearest neighbors in spatial as the local basis D_i owing to the vector b_i in Eq. (4) with a few non-zero values, which means that it is sparse in some extent. It should be noted that the K is smaller than the size of the original codebook M. Thus, the Eq. (4) can be rewritten as follows:

$$\min_B(\sum_{i=1}^{N} \|x_i - Db_i\|^2), s.t. 1^T b_i = 1, \forall_i \tag{6}$$

where D_i denotes the new codebook for each region r_i, $i = 1, 2, \ldots, N$ and K is the size of the new codebook and is empirically set at $K = 2M/3$.

Unlike the traditional LLC algorithm, solving the improved LLC algorithm is simple and the solution can be derived analytically by

$$b_i = \frac{1}{(C_i + \omega * dig(C_i))}, \tag{7}$$

$$\widetilde{b}_i = \frac{b_i}{1^T b_i}, \tag{8}$$

where $C_i = (D_i - 1x_i^T)(D_i - 1x_i^T)^T$ represents the covariance matrix of the feature and ω is a regularization parameter, which is set to be 0.1 in the proposed algorithm. As the solution of the improved LLC method is simple and fast, therefore, the local saliency value of the region r_i can be defined as follows [27]:

$$S(r_i) = \|x_i - D_i \widetilde{b}_i\|^2, \tag{9}$$

where $\widetilde{b}_i$ is the solution of Eq. (6), which is achieved by Eqs. (7) and (8).

Figure 1(c) shows some local saliency maps. We can see that the local saliency maps achieve more reliable local detailed information owing to the locality-constrained coding model.

Fig. 1. The proposed method for saliency detection. (a) Input images, (b) texture saliency maps, (c) local saliency maps, (d) the final saliency maps.

2.3 Final Saliency Fusion

The texture saliency map and the local saliency map are linearly combined with adaptive weights to define the final saliency map. Therefore, the final saliency map can be defined as follows:

$$Sal = \alpha S_t + \beta S_l, \tag{10}$$

where S_t and S_l are the texture saliency and the local saliency, respectively. α and β are the weights of the texture saliency and the local saliency, accordingly. In this paper, we introduced a more effective and logical fusion method to adjust the weights between the different feature maps adaptively. We used the *DOS* (degree-of-scattering) of saliency map to determine the weighting parameters and set the $\alpha = 1 - DOS^t$, $\beta = 1 - DOS^t$ according to the method in [26].

3 Experimental Results

We perform saliency detection experiments based on a widely used dataset: ECSSD [31], which included 1000 images acquired from the internet, and compare our method with other eight state-of-the-art methods including the Spectral Residual (SR) [12], Saliency detection using Natural Statistics (SUN) [45], Segmenting salient objects (SEG) [23], Context-aware (CA) saliency detection [8],

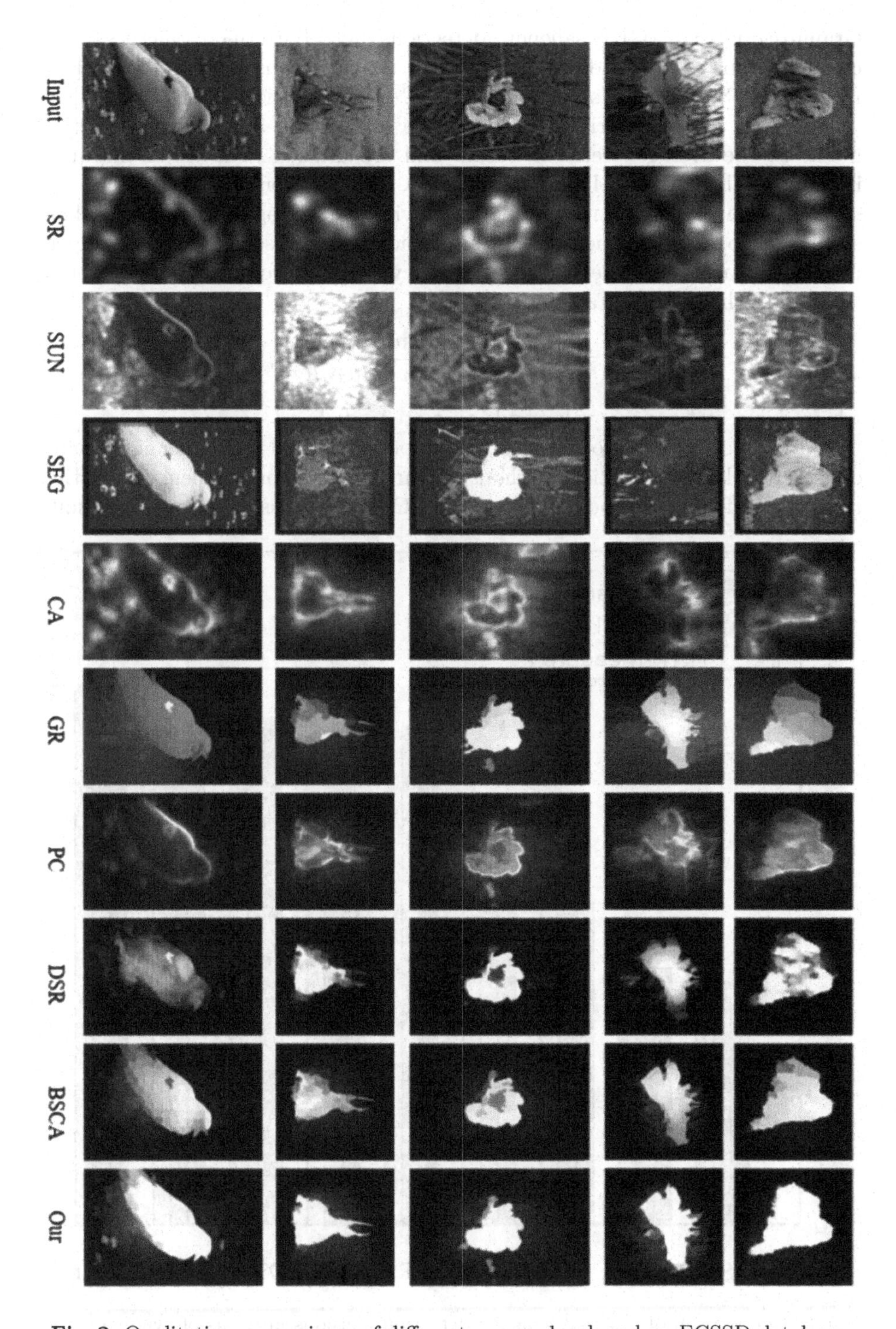

Fig. 2. Qualitative comparisons of different approaches based on ECSSD database.

Graph-regularized (GR) saliency detection [32], Pattern distinctness and color (PC) based method [24], Dense and sparse reconstruction (DSR) [19], Background-Single Cellular Automata (BSCA) based saliency detection [22].

In order to quantitatively compare the state-of-the-art saliency-detection methods, the average precision, recall, and are utilized to measure the quality of the saliency maps based on setting a segmentation threshold for binary segmentation. The adaptive threshold is twice the average value of the whole saliency map to get the accurate results. Each image is segmented with superpixels and masked out when the mean saliency values are lower than the adaptive threshold. The is defined as follows:

$$F_\beta = \frac{(1+\beta) * Precision * Recall}{\beta * Precision + Recall}, \quad (11)$$

where β is a real positive value and is set at $\beta = 0.3$.

We compared the performance of the proposed method with other eight state-of-the-art saliency detection methods. Figure 2 shows some saliency detection results of different methods based on the ECSSD database. The output results

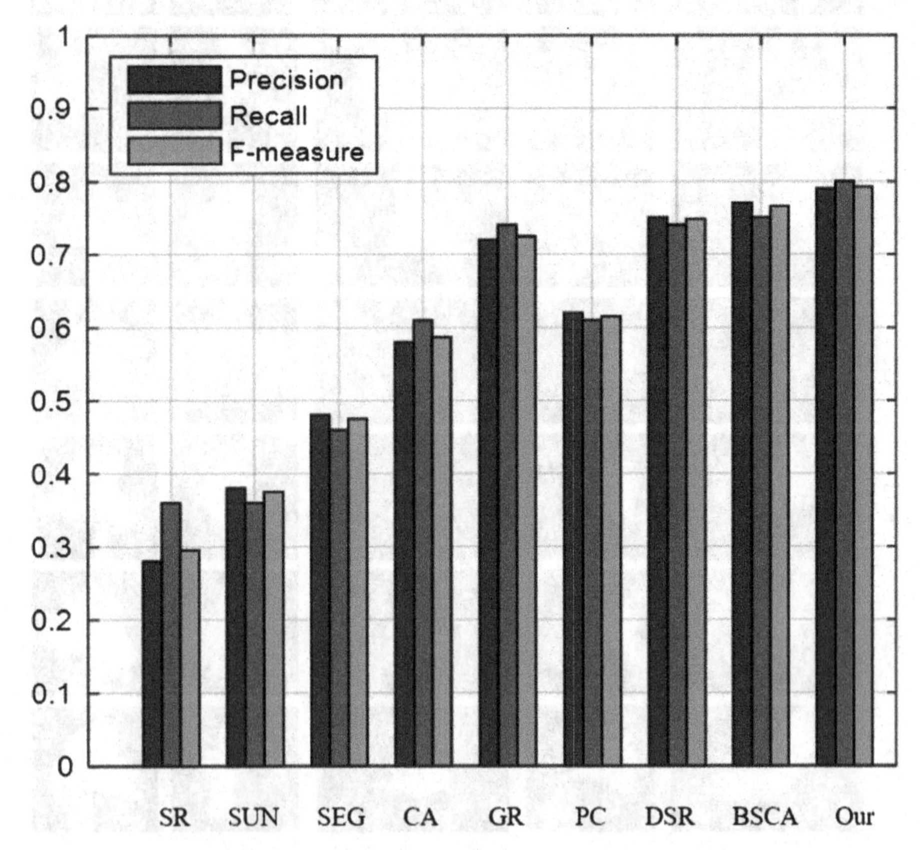

Fig. 3. Comparison of different saliency-detection methods in terms of average precision, recall, and based on ECSSD database.

show that our final saliency maps can accurately detect almost entire salient objects and preserve the salient objects contours more clearly.

We also used the precision, recall and the $F-measure$ to evaluate the performance of different methods objectively. Figure 3 shows the comparisons of different methods under different evaluation criterions. As can be seen from Fig. 3, our pro-posed method outperforms the other eight methods in terms of detection accuracy and the proposed method achieves the best overall saliency-detection performance (with precision = 79.0%, recall = 80.0%), and the $F-measure$ is 79.2%. The experiment results show that the proposed model is efficient and effective.

4 Conclusion and Discussion

This paper proposes a novel bottom-up method for efficient and accurate image saliency detection. This proposed approach integrated texture and local cues to estimate the final saliency map. The texture saliency maps are computed based on the selective contrast (SC) method, which explores the most distinguishable component information in texture. The local saliency maps are achieved by utilizing a locality-constrained method. We also evaluated our method based on a publicly available dataset and compared our proposed approach with other eight different state-of-the-art methods. The experimental results show that our algorithm can produce promising results compared to the other state-of-the-art saliency-detection models.

Acknowledgments. This work was supported by National Natural Science Foundation of China (NSFC) (61601427); Natural Science Foundation of Shandong Province (ZR2015FQ011); Applied Basic Research Project of Qingdao (16-5-1-4-jch); China Postdoctoral Science Foundation funded project (2016M590659); Postdoctoral Science Foundation of Shandong Province (201603045); Qingdao Postdoctoral Science Foundation funded project (861605040008) and The Fundamental Research Funds for the Central Universities (201511008, 30020084851).

References

1. Achanta, R., Hemami, S., Estrada, F., Susstrunk, S.: Frequency-tuned salient region detection. In: IEEE Conference on Computer Vision and Pattern Recognition, CVPR 2009, pp. 1597–1604 (2009)
2. Chen, J., Chen, J., Lu, H., Chi, Z.: CNN for saliency detection with low-level feature integration. Neurocomputing **226**(C), 212–220 (2017)
3. Cheng, M.-M., Liu, Y., Hou, Q., Bian, J., Torr, P., Hu, S.-M., Tu, Z.: HFS: hierarchical feature selection for efficient image segmentation. In: Leibe, B., Matas, J., Sebe, N., Welling, M. (eds.) ECCV 2016. LNCS, vol. 9907, pp. 867–882. Springer, Cham (2016). https://doi.org/10.1007/978-3-319-46487-9_53
4. Cholakkal, H., Johnson, J., Rajan, D.: A classifier-guided approach for top-down salient object detection. Sig. Process. Image Commun. **45**(C), 24–40 (2016)
5. Cholakkal, H., Johnson, J., Rajan, D.: Weakly supervised top-down salient object detection (2016)

6. Fu, H., Cao, X., Tu, Z.: Cluster-based co-saliency detection. IEEE Trans. Image Process. **22**(10), 3766 (2013). A Publication of the IEEE Signal Processing Society
7. Gao, D., Mahadevan, V., Vasconcelos, N.: The discriminant center-surround hypothesis for bottom-up saliency. In: Advances in Neural Information Processing Systems, vol. 20, pp. 497–504 (2007)
8. Goferman, S., Zelnikmanor, L., Tal, A.: Context-aware saliency detection. IEEE Trans. Pattern Anal. Mach. Intell. **34**(10), 1915–1926 (2012)
9. Guo, Z., Zhang, L., Zhang, D.: Rotation invariant texture classification using LBP variance (LBPV) with global matching. Pattern Recogn. **43**(3), 706–719 (2010)
10. Harel, J., Koch, C., Perona, P.: Graph-based visual saliency. In: International Conference on Neural Information Processing Systems, pp. 545–552 (2006)
11. Hou, Q., Cheng, M.M., Hu, X.W., Borji, A., Tu, Z., Torr, P.: Deeply supervised salient object detection with short connections (2017)
12. Hou, X., Zhang, L.: Saliency detection: a spectral residual approach. In: IEEE Conference on Computer Vision and Pattern Recognition, CVPR 2007, pp. 1–8 (2007)
13. Itti, L., Koch, C., Niebur, E.: A model of saliency-based visual attention for rapid scene analysis. IEEE Trans. Pattern Anal. Mach. Intell. **20**(11), 1254–1259 (1998). IEEE Computer Society
14. Jian, M., Lam, K.M., Dong, J., Shen, L.: Visual-patch-attention-aware saliency detection. IEEE Trans. Cybern. **45**(8), 1575–1586 (2015)
15. Jian, M., Qi, Q., Dong, J., Sun, X., Sun, Y., Lam, K.-M.: The OUC-vision large-scale underwater image database (2017)
16. Jian, M., Qi, Q., Dong, J., Sun, X., Sun, Y., Lam, K.M.: Saliency detection using quatemionic distance based weber descriptor and object cues. In: Signal and Information Processing Association Summit and Conference, pp. 1–4 (2017)
17. Jian, M., Qi, Q., Dong, J., Sun, X., Sun, Y., Lam, K.-M.: Saliency detection using quaternionic distance based weber local descriptor and level priors. Multimed. Tools Appl. (2017). https://doi.org/10.1007/s11042-017-5032-z
18. Kanan, C., Tong, M.H., Zhang, L., Cottrell, G.W.: SUN: top-down saliency using natural statistics. Vis. Cogn. **17**(6–7), 979 (2009)
19. Li, X., Lu, H., Zhang, L., Xiang, R., Yang, M.H.: Saliency detection via dense and sparse reconstruction. In: IEEE International Conference on Computer Vision, pp. 2976–2983 (2013)
20. Ma, Q.: New strategy for image and video quality assessment. J. Electron. Imaging **19**(1), 011019 (2010)
21. Ma, Y.F., Zhang, H.J.: Contrast-based image attention analysis by using fuzzy growing. In: Eleventh ACM International Conference on Multimedia, pp. 374–381 (2003)
22. Qin, Y., Lu, H., Xu, Y., Wang, H.: Saliency detection via cellular automata. In: Computer Vision and Pattern Recognition, pp. 110–119 (2015)
23. Rahtu, E., Kannala, J., Salo, M., Heikkilä, J.: Segmenting salient objects from images and videos. In: Daniilidis, K., Maragos, P., Paragios, N. (eds.) ECCV 2010. LNCS, vol. 6315, pp. 366–379. Springer, Heidelberg (2010). https://doi.org/10.1007/978-3-642-15555-0_27
24. Ran, M., Tal, A., Zelnikmanor, L.: What makes a patch distinct? In: Computer Vision and Pattern Recognition, pp. 1139–1146 (2013)
25. Rezazadegan Tavakoli, H., Rahtu, E., Heikkilä, J.: Fast and efficient saliency detection using sparse sampling and kernel density estimation. In: Heyden, A., Kahl, F. (eds.) SCIA 2011. LNCS, vol. 6688, pp. 666–675. Springer, Heidelberg (2011). https://doi.org/10.1007/978-3-642-21227-7_62

26. Tian, H., Fang, Y., Zhao, Y., Lin, W.: Salient region detection by fusing bottom-up and top-down features extracted from a single image. IEEE Trans. Image Process. **23**(10), 4389–4398 (2014). A Publication of the IEEE Signal Processing Society
27. Tong, N., Lu, H., Zhang, Y., Xiang, R.: Salient object detection via global and local cues. Pattern Recogn. **48**(10), 3258–3267 (2015)
28. Wang, J., Yang, J., Yu, K., Lv, F., Huang, T., Gong, Y.: Locality-constrained linear coding for image classification, vol. 119, no. 5, pp. 3360–3367 (2010)
29. Wang, L., Xue, J., Zheng, N., Hua, G.: Automatic salient object extraction with contextual cue, vol. 23, no. 5, pp. 105–112 (2011)
30. Wang, Q., Yuan, Y., Yan, P.: Visual saliency by selective contrast. IEEE Trans. Circ. Syst. Video Technol. **23**(7), 1150–1155 (2013)
31. Yan, Q., Xu, L., Shi, J., Jia, J.: Hierarchical saliency detection. In: Computer Vision and Pattern Recognition, pp. 1155–1162 (2013)
32. Yang, C., Zhang, L., Lu, H.: Graph-regularized saliency detection with convex-hull-based center prior. IEEE Signal Process. Lett. **20**(7), 637–640 (2013)
33. Yang, C., Zhang, L., Lu, H., Ruan, X., Yang, M.H.: Saliency detection via graph-based manifold ranking. In: Computer Vision and Pattern Recognition, pp. 3166–3173 (2013)
34. Yu, K., Zhang, T., Gong, Y.: Nonlinear learning using local coordinate coding. In: International Conference on Neural Information Processing Systems, pp. 2223–2231 (2009)
35. Zhang, J.: Seam carving for content-aware image resizing. ACM Trans. Graph. **26**(3), 10 (2007)
36. Song, H., Liu, Z., Du, H., et al.: Depth-aware salient object detection and segmentation via multiscale discriminative saliency fusion and bootstrap learning. IEEE Trans. Image Process. **26**(9), 4204–4216 (2017)
37. Guan, Y., Jiang, B., Xiao, Y., et al.: A new graph ranking model for image saliency detection problem. In: Software Engineering Research, Management and Applications (SERA), pp. 151–156 (2017)
38. He, Z., Jiang, B., Xiao, Y., Ding, C., Luo, B.: Saliency detection via a graph based diffusion model. In: Foggia, P., Liu, C.-L., Vento, M. (eds.) GbRPR 2017. LNCS, vol. 10310, pp. 3–12. Springer, Cham (2017). https://doi.org/10.1007/978-3-319-58961-9_1
39. Peng, H., Li, B., Ling, H., et al.: Salient object detection via structured matrix decomposition. IEEE Trans. Pattern Anal. Mach. Intell. **39**(4), 818–832 (2017)
40. Yang, J., Yang, M.H.: Top-down visual saliency via joint CRF and dictionary learning. IEEE Trans. Pattern Anal. Mach. Intell. **39**(3), 576–588 (2017)
41. Deng, T., Yang, K., Li, Y., et al.: Where does the driver look? top-down-based saliency detection in a traffic driving environment. IEEE Trans. Intell. Transp. Syst. **17**(7), 2051–2062 (2016)
42. Liu, Y., Yang, J., Meng, Q., et al.: Stereoscopic image quality assessment method based on binocular combination saliency model. Signal Process. **125**, 237–248 (2016)
43. Zhang, K., Liu, Q., Wu, Y.: Robust visual tracking via convolutional networks without training. IEEE Trans. Image Process. **25**(4), 1779–1792 (2016)
44. Lee, S.H., Kang, J.W., Kim, C.S.: Compressed domain video using global and local spatiotemporal features. J. Vis. Commun. Image Represent. **35**, 169–183 (2016)
45. Zhang, L., Tong, M.H., Marks, T.K., Shan, H., Cottrell, G.W.: Sun: a Bayesian framework for saliency using natural statistics. J. Vis. **8**(7), 32 (2008)

An Improved Saliency Detection Method Using Hypergraphs on Adaptive Multiscales

Feilin Han[2], Aili Han[1(✉)], and Jing Hao[1]

[1] Department of Computer Science and Technology, Shandong University, Weihai, China
hanal@sdu.edu.cn

[2] College of Computer Science and Technology, Zhejiang University, Hangzhou, China

Abstract. We present an improved saliency detection method by means of hypergraphs on adaptive multiscales (HAM). An input image is charaterized by hypergraphs in which hyperedges are used to capture contextual properties of regions. Thus, the saliency detection problem is transformed into that of finding salient vertices and hyperedges in hypergraphs. The HAM method first adjusts adaptively the ranges of pixel-values in R, G, B channels in an input image and uses the ranges to determine adaptive scales. And then, it models the image as a hypergraph for each scale in which hyperedges are clustered by means of agglomerative mean-shift. The HAM method can get more single-scale hypergraphs and thus has higher accuracy than the previous ones because each hypergraph is on an adaptive scale instead of a fixed scale. Extensive experiments on three benchmark datasets demonstrate that the HAM method improves the performance of saliency detection, especially for the images with narrow ranges of pixel-values.

Keywords: Saliency detection · Adaptive multiscale · Hypergraph

1 Introduction

Saliency detection plays an important role in the field of computer vision, which aims to identify the most attractive region in an image. Its important applications include image segmentation [12], target recognition [10] and image retrieval [13]. In order to extract salient object from input image, many visual attention models have been proposed [4,6–9,11].

Since Itti et al. [7] proposed a visual attention system inspired by animal visual system, the researches on saliency detection have been developed rapidly. Ma and Zhang [9] designed a saliency detection method based on local contrast. Harel et al. [11] gave a graph-based visual saliency model. Cheng et al. [4,6] proposed a saliency detection method based on global contrast and space coherence. Li et al. [8] employed hypergraphs to capture contextual attributes for saliency detection.

J. Yang et al. (Eds.): CCCV 2017, Part III, CCIS 773, pp. 700–710, 2017.
https://doi.org/10.1007/978-981-10-7305-2_59

The contextual hypergraph modeling method for saliency detection (CHMS) [8] has good performance of capturing salient objects in most images, which has higher accuracy than many previous methods. Through the experiments, we find that, for the images with wide ranges of pixel-values (e.g. covering almost the whole range of $[0, 255]$), the CHMS method always has good performance, as shown in the top three lines in Fig. 1. However, for the images with narrow ranges of pixel-values (e.g. covering only the first or middle or last part of the range $[0, 255]$), the CHMS method usually cannot get very good performance, as shown in the bottom three lines in Fig. 1. Further experiments show that hypergraphs on different scales affect experimental results directly. For example, the images in the buttom three lines in Fig. 1 have no sufficient valid single-scale hypergraphs, so the experiments cannot give good performance by the CHMS method.

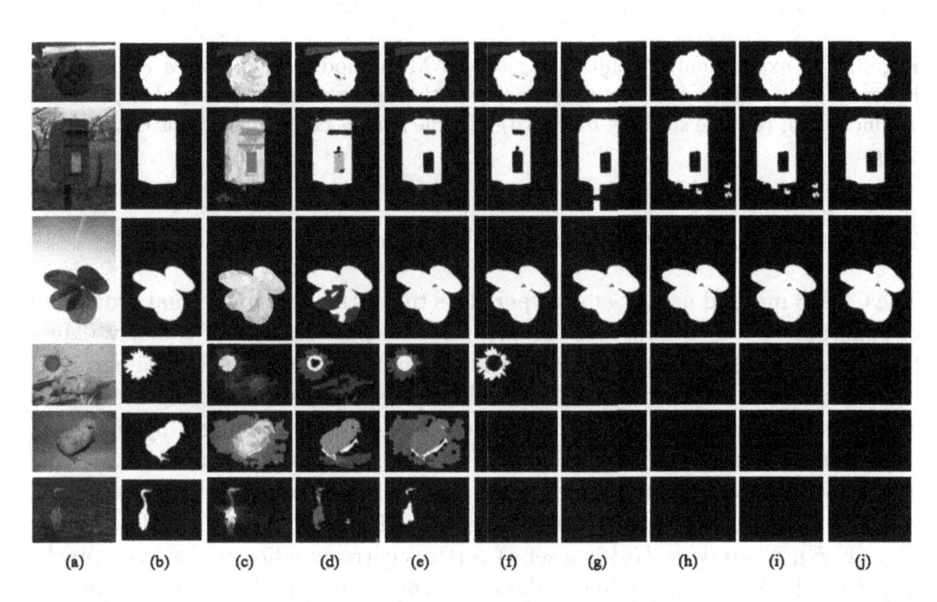

Fig. 1. The experimental results by the CHMS method. The top three lines are for the images with wide ranges of pixel-values, and the bottom three lines are for the images with narrow ranges of pixel-values. (a) The input images; (b) The ground truth; (c) The saliency maps by the CHMS method; (d)–(j) The hypergraphs on a fixed scale in $[0.15, 0.25, 0.35, 0.45, 0.55, 0.65, 0.75]$. From the experimental results, we can find that the CHMS method has good performance for the images in the top three lines and bad performance for the images in the bottom three lines.

Inspired by this, we propose an improved saliency detection method using hypergraphs on adaptive multiscales (HAM), which is an improvement on the CHMS method. The HAM method adaptively adjusts the ranges of pixel-values in R, G, B channels and detects salient objects on adaptive multiscales. As shown in Fig. 2, the HAM method provides more valid single-scale hypergraphs and better final saliency maps. The experimental results show that the HAM method improves the performance of saliency detection, especially for the images with narrow ranges of pixel-values. The basic idea of adjusting adaptively the ranges of pixel-values in an image can be widely used in other applications in computer vision or artificial intelligence.

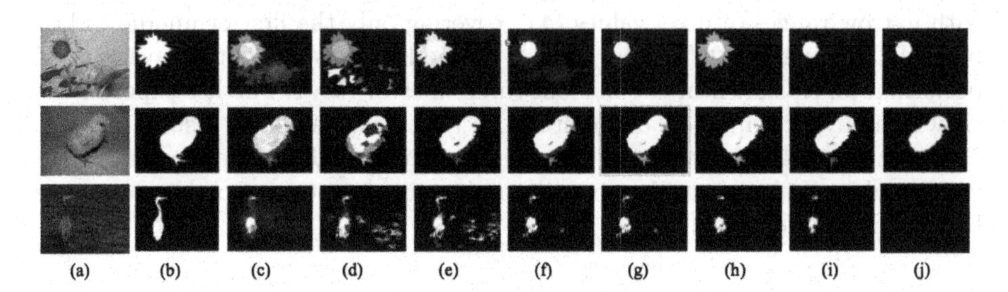

Fig. 2. The experimental results using the HAM method for the images with narrow ranges of pixel-values. (a) The input images with narrow range of pixel-values; (b) The ground truth; (c) The saliency maps; (d)–(j) The single-scale hypergraphs on adpative scales.

2 Contextual Hypergraph Modeling

The CHMS method uses a set of hyperedges to capture the contextual properties of superpixels, which improves significantly the performance of saliency detection. The CHMS method firstly segments an input image into some superpixels by means of the SLIC method [2], which applies k-means clustering method to generate superpixels efficiently. And then, it clusters superpixels by means of agglomerative mean-shift [1] on a set of fixed scales, which is based on an iterative query set compression mechanism and runs in linear time. Thus, an input image I is converted to a set of hypergraphs. Each hypergraph is denoted by $G = (V, E)$, where $V = \{v_i\}$ is a set of vertices corresponding to the superpixels, and $E = \{e_j\}$ is a set of hyperedges (a hyperedge is a clique of superpixels) that satisfy $\bigcup_{e_j \in E} e_j = V$. Thus, the saliency detection problem is converted to that of finding salient vertices and hyperedges in hypergraphs on adaptive multiscales.

The saliency of a hyperedge can be determined by the gradient magnitudes of the superpixels within a narrow band along the boundary of the hyperedge. For any hyperedge e_j, the saliency score of e_j is defined as follows.

$$\Gamma(e_j) = \omega_{e_j}(\|I_g^* \bullet M_g(e_j)\|_1 - \rho(e_j)), \tag{1}$$

where ω_{e_j} is the weight of e_j, I_g^* is the binary gradient map, $M_g(e_j)$ is a binary mask indicating the superpixels within a narrow band along the boundary of the hyperedge e_j, $\bullet$ is the elementwise dot product operator, $\| \; \|_1$ is the 1-norm, and $\rho(e_j)$ is a penalty factor that is equal to the number of the superpixels in the intersection of the hyperedge e_j and the boundary superpixels of the input image.

The saliency of a vertex (i.e. a superpixel) is associated to the superpixel and its contexts. For any hyperpixel v_i in a hypergraph, the saliency score of v_i is defined as follows.

$$HSa(v_i) = \sum_{e_j \in E} \Gamma(e_j) h(v_i, e_j), \tag{2}$$

where $\Gamma(e_j)$ is the saliency score of e_j, and $h(v_i, e_j)$ is the element value in the ith row and j^{th} column in the incidence matrix H. If $v_i \in e_j$, then $h(v_i, e_j) = 1$; otherwise $h(v_i, e_j) = 0$.

3 Saliency Detection Using Hypergraphs on Adaptive Multiscales

The CHMS method uses a set of empirical values [0.15, 0.25, 0.35, 0.45, 0.55, 0.65, 0.75] as the fixed scales to detect salient objects in an image, which has good performance for most images. It is widely acecepted that there are great differences between different images. If all the pixel-values in an image lie in the first or middle or last part of the range $[0, 255]$, there may be no hyperedges for some scales which result in bad performance of saliency detection, as shown in the last three lines in Fig. 1.

We propose an improved saliency detection method using hypergraphs on adaptive multiscales (HAM). Different from the CHMS method, our HAM method uses a set of adaptive scales instead of fixed scales. Thus, it can get more single-scale hypergraphs than the previous ones for an input image, which results in higher accuracy.

3.1 Adaptive Scaling of Pixel-Values

In order to adaptively adjust the ranges of pixel-values in R, G, B channels, we count for the pixel-values in each channel by means of histogram. The statistical results are used to determine the range of pixel-values covering more than 95% pixels in each channel. Selecting the range of covering more than 95% pixels is to avoid the influence of outliers [6]. The pixel-values outside the range are replaced by the nearest pixel-values. And then, all the pixel-values are normalized and remapped to the range $[0, 255]$. For each channel i, $i \in \{R, G, B\}$, the formula for normalization and remapping is as follows.

$$\left(\left(\frac{I_i - low_i}{high_i - low_i}\right)^{\gamma} \bullet (high_{i,out} - low_{i,out})\right) + low_{i,out}, \tag{3}$$

where I_i is a pixel value in channel i in the input image I, low_i is the lower bound of pixel-values in channel i, $high_i$ is the upper bounder of pixel-values in channel i, $low_{i,out}$ is the lower bound of the remapped values in channel i, $high_{i,out}$ is the upper bound of the remapped values in channel i, and γ indicates the shape of curve. Here, $low_{i,out} = 0$, $high_{i,out} = 255$, and $\gamma = 1$ representing a linear mapping.

Algorithm 1. Adaptive scaling of pixel-values.

Input:

an RGB image I.

Output:

the image I' after adaptive scaling.

1: Count for the pixel-values in R, G, B channels, respectively, to generate the histograms of R, G, B channels;

2: Judge whether a pixel-value belongs to the range of covering more than 95% pixels in the descending order of the number of pixel-values in R, G, B channels, respectively. For R channel, the minimal pixel-value belonging to the range of covering more than 95% pixels is used as the lower bound low_r, and the maximal is used as the upper bound $high_r$. Same for G and B channels. Thus, we obtain the three ranges of pixel-values, i.e. $[low_r..high_r]$, $[low_g..high_g]$, and $[low_b..high_b]$;

3: Normalize the three ranges of pixel-values, and then remap the normalized ranges of pixel-values to $[0, 255]$, respectively. For each channel i, $i \in \{R, G, B\}$, the normalization and remmapping are done according to the equation (3).

3.2 Construction of Hypergraphs on Adaptive Scales

The CHMS method constructs hypergraphs on a set of fixed scales by means of the Alggo-MS method. For the details, refer to [14]. Different from the CHMS method, we use a set of adaptive scales instead of fixed scales. We first adjust adaptively the ranges of pixel-values in R, G, B channels in an image to get three remapped ranges of pixel-values, and then combine the remapped pixel-values with a set of fixed scales. The results of the two operations are equivalent to that of the original image with a set of adaptive scales. Thus, in the HAM method, we use hypergraphs on a set of adaptive scales to detect salient objects. The distance between any pixels x_i and x_j on an adaptive scale μ is computed as follows.

$$dist = \left(\frac{x_i - x_j}{\mu}\right)^2. \tag{4}$$

The adaptive scaling of pixel-values in an image can result in more single-scale hypergraphs for the images with narrow ranges of R, G, B pixel-values (covering only the first or middle or last part of the range $[0, 255]$). Thus, the performance of saliency detection has been improved significantly by means of the HAM method. Take the three images shown in the 1st column of Fig. 2 as examples. When using the HAM method to detect salient objects, the number of single-scale hypergraphs is $7, 7, 6$, respectively, as shown in Fig. 2; when using the CHMS method, the number of single-scale hypergraphs is $3, 2, 2$, respectively, as shown in the last three lines in Fig. 1. By comparing the results in Fig. 2 with the last three lines in Fig. 1, it can be concluded that the HAM method has better performance than the CHMS method for the images with narrow ranges of pixels values.

From more experimental results, we find that, for the images with narrow ranges of pixel-values (all the pixel-values lie in the first or middle or last part of the range [0, 255]), the number of single-scale hypergraphs obtained by the HAM method is always greater than the number of single-scale hypergraphs obtained by the CHMS method; for the images with wide ranges of pixel-values, the two methods always get similar number of single-scale hypergraphs.

3.3 Saliency Detection Using Hypergraph on Adaptive Multiscales

The CHMS method uses hypergraphs on a set of fixed scales, no considering the color difference of different input images. For any input image, it takes the empirical values [0.15, 0.25, 0.35, 0.45, 0.55, 0.65, 0.75] as the fixed scales. In order to detect salient objects according to the color difference in an image [15], we make an adaptive scaling of pixel-values and use hypergraphs on a set of adaptive scales to detect salient objects. The HAM method improves signifycantly the performance of saliency detection for the images with narrow ranges of R, G, B pixel-values.

Algorithm 2. Saliency detection using hypergraphs on adaptive multiscales.

Input:
an RGB image I.

Output:
the saliency map S of image I.

1: Apply Algorithm 1 to I to get the ranges of pixel-values in R, G, B channels, which are normalized and remapped to [0,255] to get a remapped image;
2: Employ the SLIC algorithm to oversegment the remapped image into some superpixels to get a superpixel image $I^{'}$;
3: Combine the superpixel image with a set of fixed scales, which is equivalent to that of the original input image with a set of adaptive scales. For each scale, use the Agglo-MS method to cluster superpixels into hyperedges to get a hypergraph;
4: Apply the Gaussian operator to the remapped image to get a gradient map I_g, and then use I_g to compute the saliency scores of hyperedges and superpixels to get a saliency map S_{sg};
5: Minimize a cost-sensitive classification function to get a LS-SVM based saliency map S_{svm};
6: Linearly combine the saliency maps S_{sg} and S_{svm} to get the final saliency map S.

In step 5, the saliency map based on LS-SVM is obtained from one alternative method in the CHMS method [8]. In order to achieve a fair comparison, we also use this method. For further details, refer to [8].

The HAM method is an improvement on the CHMS method. The comparisons between the HAM method and the CHMS method are illustrated in Fig. 3 (More experimental results are shown in Sect. 4). Consider the images in the 1st line in Fig. 3, there are less noises in the saliency map obtained by the HAM method than that by the CHMS method, and the boundaries of the CAUTION banner are smoother in the saliency map obtained by the HAM method than that by the CHMS method. For the images in the 2nd line in Fig. 3, the background around the dog is usually marked as a part of salient object by the previous methods including the CHMS method because the dog has the similar color features with the background. The HAM method can detect the salient object better than the previous ones. For the images in the 3rd line in Fig. 3, the remarkable object is a flower, but the green leaves are marked as salient object by the CHMS method due to their large size; while the HAM method highlights the flower and darkens the background which achieves better performance than the CHMS method. The experimental results show that the HAM method can better capture salient objects and improve the performance of saliency detection.

Fig. 3. Three examples for the comparisons between the CHMS method and the HAM method. (a) The input images; (b) The saliency map obtained by the CHMS method; (c) The saliency map obtained by the HAM method.

4 Experimental Results and Analysis

The HAM method is compared with six state-of-the-art saliency detection methods including CHMS [8], GC [5], RC [15], MSS [3], FT [1], LC [16] on the public datasets MSRA-1000 and SED-100, which are used as standard benchmarks. In addition, we use a new dataset, marked as IMNR, to evaluate the HAM method.

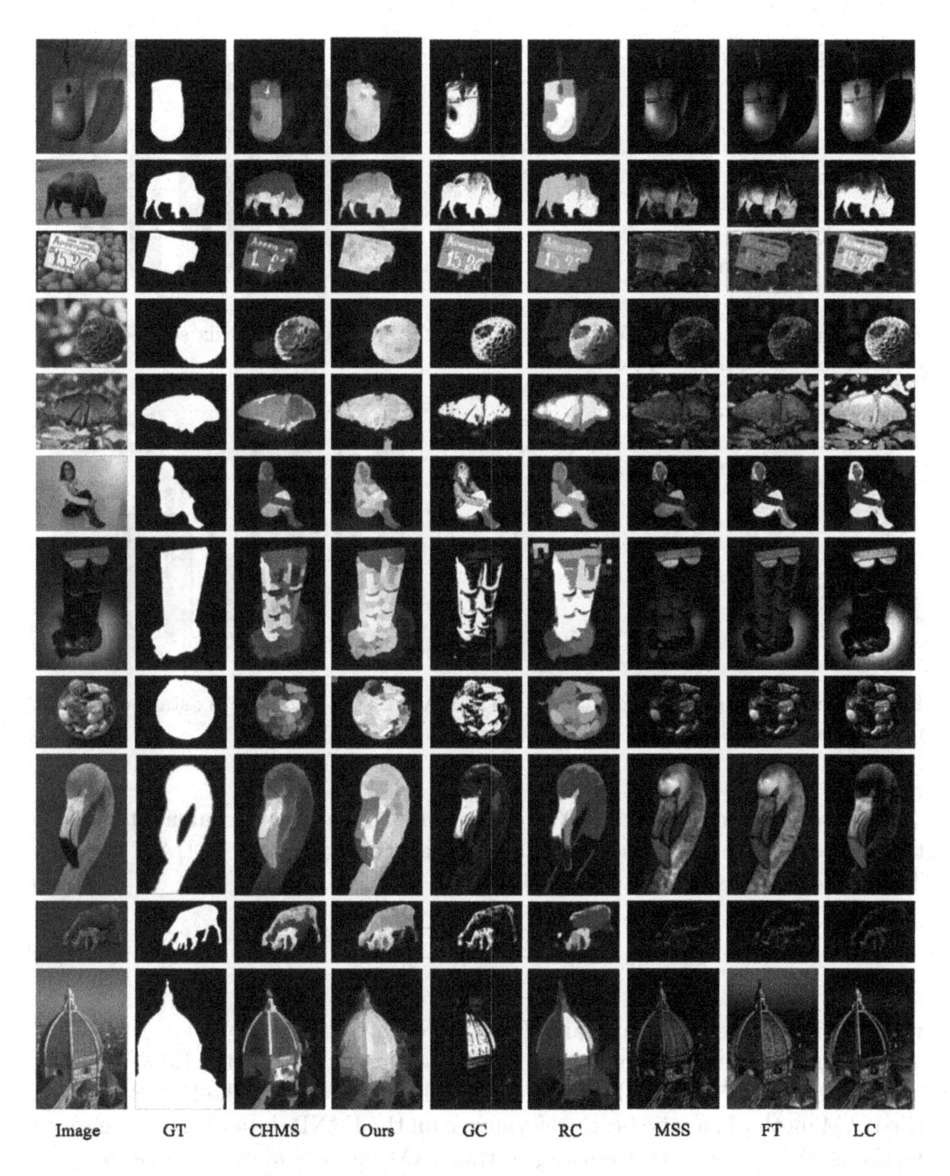

Fig. 4. Comparisons between our HAM method and the six state-of-the-art methods includeing CHMS [8], GC [5], RC [15], MSS [3], FT [1] and LC [16] on the three datasets: MSRA-1000 (top six rows), SED-100 (middle two rows) and IMNR (bottom two rows).

Images in the IMNR dataset is with narrow ranges of R, G, B pixel-values. Some experimental results are shown in Fig. 4. The experimental results show that the HAM method is the most competitive one in all the seven methods.

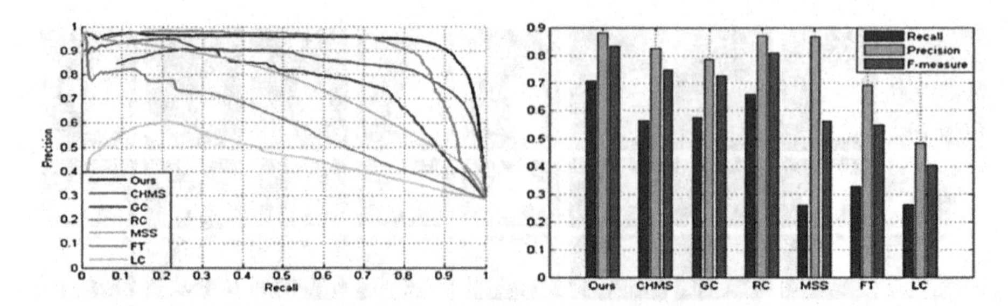

Fig. 5. The quantitative performance of the HAM method and the six state-of-the-art methods on the IMNR dataset.

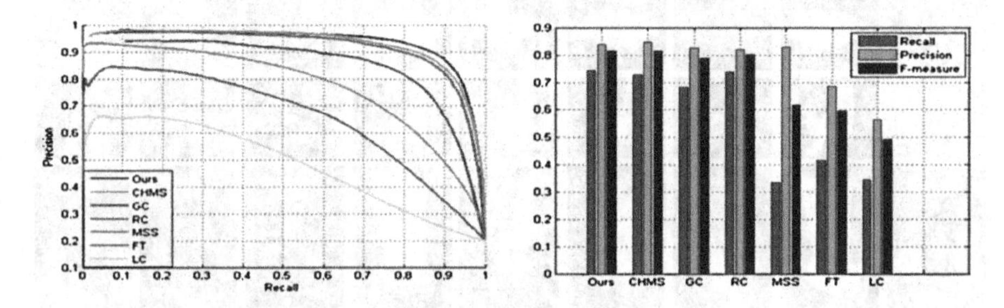

Fig. 6. The quantitative performances of the HAM method and the six state-of-the-art methods on the MSRA-1000 dataset.

We use the precision-recall (PR) curves and F-measures to evaluate the quantitative performances of the six state-of-the-art methods and the HAM method. The F-measures is computed as follows.

$$F_{\beta} = \frac{(1 + \beta^2)P \times R}{\beta^2 P + R}, \tag{5}$$

where P is the precision rate, and R is the recall rate. Let $\beta^2 = 0.3$.

The quantitative performances of all the seven methods on the IMNR dataset are shown in Fig. 5. From the PR curves and F-measures on the IMNR dataset, the HAM method has the best performance on the IMNR dataset in all the seven methods. As you can see from Fig. 5, the HAM method always has better performance than the CHMS method. When the recall rate is greater than 0.7, the HAM method is better than all the six state-of-the-art methods. Furthermore, the saliency map obtained by the HAM method is more smooth and robust than other six state-of-the-art methods.

The quantitative performances of all the seven methods on the MSRA-1000 dataset are shown in Fig. 6. From the PR curves and F-measures on MSRA-1000, the HAM method also has the best performance in all the seven methods but its advantages are not obvious since the most images in MSRA-1000 are with wide ranges of pixel-values and seldom of them are with narrow ranges of pixel-values.

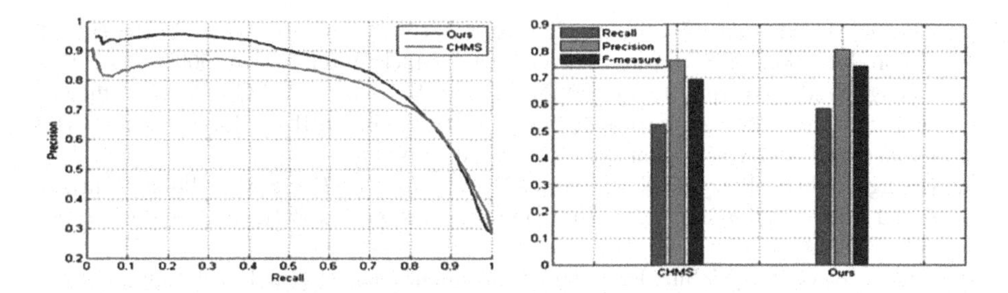

Fig. 7. The quantitative performance of the CHMS and HAM methods on the SED-100 dataset.

The quantitative performances of the CHMS and HAM methods on the SED-100 dataset are shown in Fig. 7. From the PR curves and F-measures, the HAM method consistently performs better than the CHMS method.

5 Conclusion

The previous saliency detection methods based on hypergraphs uses a set of fixed scales, no considering the color difference of different input images. In order to detect salient objects according to the color difference in an image, we propose an improved saliency detection method baserd on hypergraphs on adaptive multiscales (HAM). Our HAM method adaptively adjusts the ranges of pixel-values in R, G, B channels, and uses hypergraphs on a set of adaptive scales to detect salient objects. The experimental results show that our HAM method improves significantly the performance of saliency detection for the images with narrow ranges of R, G, B pixel-values. For the images with wide ranges of pixel-values, the performance is also improved to some extent. The basic idea of our HAM method can be widely used in other applications in computer vision or artificial intelligence.

Acknowledgement. This work is supported by the Shandong Provincial Natural Science Foundation of China under Grant No. ZR2016FM20.

References

1. Achanta, R., Hemami, S., Estrada, F., Susstrunk, S.: Frequency-tuned salient region detection. In: IEEE Conference on Computer Vision and Pattern Recognition, CVPR 2009, pp. 1597–1604 (2009)
2. Achanta, R., Shaji, A., Smith, K., Lucchi, A., Fua, P., SüSstrunk, S.: SLIC superpixels compared to state-of-the-art superpixel methods. IEEE Trans. Pattern Anal. Mach. Intell. **34**(11), 2274–2282 (2012)
3. Achanta, R., Süsstrunk, S.: Saliency detection using maximum symmetric surround. In: IEEE International Conference on Image Processing, pp. 2653–2656 (2010)

4. Cheng, M., Mitra, N.J., Huang, X., Torr, P.H.S.: Global contrast based salient region detection. IEEE Trans. Pattern Anal. Mach. Intell. **37**(3), 569–582 (2015)
5. Cheng, M.M., Warrell, J., Lin, W.Y., Zheng, S., Vineet, V., Crook, N.: Efficient salient region detection with soft image abstraction. In: IEEE International Conference on Computer Vision, pp. 1529–1536 (2013)
6. Cheng, M.M., Zhang, G.X., Mitra, N.J., Huang, X., Hu, S.M.: Global contrast based salient region detection. In: Computer Vision and Pattern Recognition, pp. 409–416 (2011)
7. Itti, L., Koch, C., Niebur, E.: A model of saliency-based visual attention for rapid scene analysis. IEEE Trans. Pattern Anal. Mach. Intell. **20**(11), 1254–1259 (1998)
8. Li, X., Li, Y., Shen, C., Dick, A., Hengel, A.V.D.: Contextual hypergraph modeling for salient object detection. In: IEEE International Conference on Computer Vision, pp. 3328–3335 (2013)
9. Ma, Y.F., Zhang, H.J.: Contrast-based image attention analysis by using fuzzy growing. In: Eleventh ACM International Conference on Multimedia, pp. 374–381 (2003)
10. Navalpakkam, V., Itti, L.: An integrated model of top-down and bottom-up attention for optimizing detection speed. In: IEEE Computer Society Conference on Computer Vision and Pattern Recognition, pp. 2049–2056 (2006)
11. Harel, J., Koch, C., Perona, P.: Graph-based visual saliency. In: Advances in Neural Information Processing Systems, vol. 19, pp. 545–552 (2007)
12. Wang, L., Xue, J., Zheng, N., Hua, G.: Automatic salient object extraction with contextual cue. In: IEEE International Conference on Computer Vision, pp. 105–112 (2011)
13. Wang, X.J., Ma, W.Y., Li, X.: Data-driven approach for bridging the cognitive gap in image retrieval (2004)
14. Yuan, X., Hu, B.G., He, R.: Agglomerative mean-shift clustering via query set compression. IEEE Trans. Knowl. Data Eng. **24**(2), 209–219 (2011)
15. Yuan, Y., Han, A., Han, F.: Saliency detection based on non-uniform quantification for RGB channels and weights for lab channels. In: Zha, H., Chen, X., Wang, L., Miao, Q. (eds.) CCCV 2015. CCIS, vol. 546, pp. 258–266. Springer, Heidelberg (2015). https://doi.org/10.1007/978-3-662-48558-3_26
16. Zhai, Y., Shah, M.: Visual attention detection in video sequences using spatiotemporal cues. In: ACM International Conference on Multimedia, pp. 815–824 (2006)

Visual Saliency Detection via Prior Regularized Manifold Ranking

Yun Xiao, Bo Jiang, Zhengzheng Tu, and Jin Tang(✉)

School of Computer Science and Technology, Anhui University,
Jiulong Road No. 111, Hefei 230601, China
{xiaoyun,jiangbo}@ahu.edu.cn, zhengzhengahu@163.com, ahu_tj@163.com

Abstract. Bottom-up saliency detection has been widely used in many applications, such as image retrieval, object recognition, image compression and so on. Manifold ranking (MR) model can identify the most salient and important area from an image efficiently. One limitation of the MR model is that it fails to consider the prior information in its ranking process. To overcome this limitation, we propose a new manifold ranking model, called prior regularized manifold ranking (RegMR), which uses the prior calculating by boundary connectivity and employs the foreground possibility in the first stage and background possibility in the second stage. We compare our model with fifteen state-of-the-art methods. Experiments show that our model performs well than all other methods on four public databases on PR-curves, F-measure and so on.

Keywords: Manifold ranking · Boundary connectivity
Background possibility · Foreground possibility

1 Introduction

The human vision system can identify the most salient and important area from an image. In order to simulate this ability in computer vision, more and more researchers pay attention to saliency detection. It has been widely used in many applications, such as image retrieval [1–4], object recognition [5–8], image compression [9–12] and so on. We focus on bottom-up methods [13–27], which rely on the assumptions of the background and foreground instead of the high-level knowledge in top-down methods [28–35]. Some state-of-the-art saliency detection methods are presented in [36]. Boundary prior, contrast prior, boundary connectivity and so on are widespread to use in many models [13,21,24,25,37]. As a pioneer, Itti [13] et al. propose a model to get saliency map by fuse the color, direction and gray features. Then, a large number of contrast-based models have been developed. Classically, Wei et al. [21] exploit geodesic saliency by using boundary and connectivity priors. Zhu et al. [25] describe the boundary connectivity and present a general energy optimization framework to optimize the final results. By all these models, we can learn that the prior is useful to the saliency detection, which can lead to raise a better performance.

J. Yang et al. (Eds.): CCCV 2017, Part III, CCIS 773, pp. 711–722, 2017.
https://doi.org/10.1007/978-981-10-7305-2_60

By using of the graph structure to explain the intrinsic relationship in image. More and more graph based methods are appeared. Harel et al. [14] propose a graph-based visual saliency detection model. Gopalakrishnan et al. [19] construct two kinds of graphs by image patch and regard the saliency detection as Markov random walk on graphs. Yang et al. [24] use a manifold ranking function to calculate the saliency value by the relationships of all super-pixels.

In actual conditions, we can get many prior information from kinds of ways. The traditional manifold ranking model [24] is not considering the existing prior information. For better to use them, we propose an efficient model to obtain the saliency map. The results are showed as Fig. 1. The main contributions of this work are enunciated as follows: Firstly, we make better use of the existing prior information. Secondly, we can get the close-form solution and obtain the final results of the function. At last but not the least, we get more efficient results by many experiments.

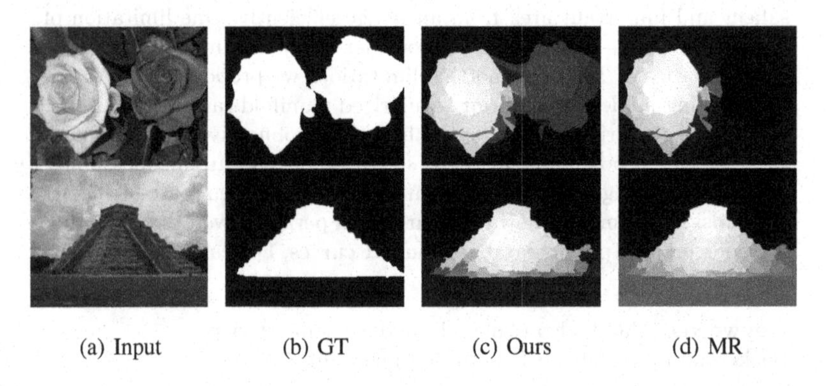

(a) Input (b) GT (c) Ours (d) MR

Fig. 1. (a) Input image; (b) The ground truth; (c) The results by using our model; (d) The results by using MR method.

2 Brief Review of Manifold Ranking

For an image, manifold ranking(MR) method [24] uses simple linear iterative clustering (SLIC) algorithm [38] to gain n super-pixels.

Formally, let $X = \{x_1, \ldots, x_q, x_{q+1}, \ldots, x_n\}$ be the set of super-pixels, where $X_q = \{x_1, \ldots, x_q\}$ are labeled queries and the rest $X_u = \{x_{q+1}, \ldots, x_n\}$ are unlabeled super-pixels. The aim of MR is to assign a ranking value $\mathbf{r}_i$ for each unlabeled region $x_i \in X_u$ according to its relevance to the labeled queries X_q. In order to do this, a graph $G = (V, E)$ is first constructed, where nodes V represent the super-pixels X and edges E denote the affinities $\mathbf{W}$ between pairs of super-pixels. Let $\mathbf{q} = (\mathbf{q}_1, \mathbf{q}_2, \ldots, \mathbf{q}_n)$ be the indication vector of queries, where $\mathbf{q}_i = 1$ if $x_i \in X_q$, otherwise, $\mathbf{q}_i = 0$.

Then, MR computes the optimal ranking $\mathbf{r}$ by solving

$$\min_{\mathbf{r}} \quad J_{\text{MR}} = \frac{1}{2}\sum_{i=1}^{n}\sum_{j=1}^{n}\mathbf{W}_{ij}(\frac{\mathbf{r}_i}{\sqrt{\mathbf{d}_i}} - \frac{\mathbf{r}_j}{\sqrt{\mathbf{d}_j}})^2 + \mu\sum_{i=1}^{n}(\mathbf{r}_i - \mathbf{q}_i)^2,$$

where $\mathbf{d}_i = \sum_{j=1}^{n} \mathbf{W}_{ij}$. It is known that the above MR model has a closed-from solution and the optimal solution [24] $\mathbf{r}^*$ is given by

$$\mathbf{r}^* = (\mathbf{I} - \frac{1}{1+\mu}\mathbf{S})^{-1}\mathbf{q} \tag{1}$$

where $\mathbf{S} = \mathbf{D}^{-\frac{1}{2}}\mathbf{W}\mathbf{D}^{-\frac{1}{2}}, \mathbf{D} = \text{diag}(d_1, d_2, \cdots d_n)$ and $\mathbf{I}$ is an identity matrix.

To get more effective result, MR model obtains another ranking function [24] by using the unormalized Laplacian matrix as,

$$\mathbf{r}^* = (\mathbf{D} - \frac{1}{1+\mu}\mathbf{W})^{-1}\mathbf{q} \tag{2}$$

3 Prior Regularized Graph Ranking

One limitation of the above MR model [24] is that it fails to consider the prior (background or foreground possibility) information in its ranking process. For saliency detection tasks, the prior information has been shown importantly in saliency computation problem. Our aim in this section is to propose a new manifold ranking model, called prior regularized manifold ranking (RegMR). The aim of RegMR is to consider the prior information in ranking process.

Model formulation. Formally, let $X = \{x_1, \ldots x_q, x_{q+1}, \ldots x_n\}$ be the set of super-pixels, $\mathbf{W}$ is an affinity matrix of relationship between pairs of super-pixels. Let $\mathbf{q} = (\mathbf{q}_1, \mathbf{q}_2, \ldots, \mathbf{q}_n)$ be the indication vector of queries, where $\mathbf{q}_i = 1$ if x_i is a query node, otherwise, $\mathbf{q}_i = 0$. Let $\mathbf{p}_i$ be the prior of the node x_i, the larger is $\mathbf{p}_i$, the less reference of the node x_i ranking value. The aim of MR is to assign a ranking value $\mathbf{r}_i$ for each unlabeled region $x_i \in X_u$ according to its relevance to the labeled queries X_q. Formally, by incorporating the prior regularization $\Psi(\mathbf{r}_i, \mathbf{p}_i)$ in MR, our RegMR can be formulated as

$$\min_{\mathbf{r}} J_{\text{RegMR}} = \frac{1}{2}\sum_{i=1}^{n}\sum_{j=1}^{n}\mathbf{W}_{ij}(\mathbf{r}_i - \mathbf{r}_j)^2 + \mu\sum_{i=1}^{n}\mathbf{d}_i(\mathbf{r}_i - \frac{\mathbf{q}_i}{\sqrt{\mathbf{d}_i}})^2 + \lambda\sum_{i=1}^{n}\Psi(\mathbf{r}_i, \mathbf{p}_i), \tag{3}$$

where $\mathbf{p}_i$ denotes some prior information. Many regularization functions can be used here. In this paper, we set $\Psi(\mathbf{r}_i, \mathbf{p}_i) = \mathbf{p}_i\mathbf{r}_i^2$ and propose a kind of RegMR as follows,

$$\min_{\mathbf{r}} J_{\text{RegMR}} = \frac{1}{2}\sum_{i=1}^{n}\sum_{j=1}^{n}\mathbf{W}_{ij}(\mathbf{r}_i - \mathbf{r}_j)^2 + \mu\sum_{i=1}^{n}\mathbf{d}_i(\mathbf{r}_i - \frac{\mathbf{q}_i}{\sqrt{\mathbf{d}_i}})^2 + \lambda\sum_{i=1}^{n}\mathbf{p}_i\mathbf{r}_i^2, \tag{4}$$

Optimization. Our RegMR model is convex and the global optimal solution can be computed. Using vector representation, problem Eq.(4) is equivalently formulated as

$$\min_{\mathbf{r}} J_{\text{RegMR}} = \mathbf{r}^T\mathbf{D}\mathbf{r} - \mathbf{r}^T\mathbf{W}\mathbf{r} + \mu(\mathbf{r}^T\mathbf{D}\mathbf{r} - 2\mathbf{r}^T\mathbf{D}^{\frac{1}{2}}\mathbf{q} + \mathbf{q}^T\mathbf{q}) + \lambda\mathbf{r}^T\mathbf{P}\mathbf{r}, \tag{5}$$

The optimal solution is computed by setting the first derivative of the above function $J_{\text{RegMR}}(\mathbf{r})$ w.r.t $\mathbf{r}$ to be zero, i.e.,

$$\frac{\partial J_{\text{RegMR}}(\mathbf{r})}{\partial \mathbf{r}} = \mathbf{Dr} - \mathbf{Wr} + \mu(\mathbf{Dr} - \mathbf{D}^{\frac{1}{2}}\mathbf{q}) + \lambda \mathbf{Pr} = \mathbf{0} \tag{6}$$

Thus, we obtain the result,

$$\mathbf{r}^* = \left(\mathbf{D} - \frac{1}{1+\mu}\mathbf{W} + \frac{\lambda}{1+\mu}\mathbf{P}\right)^{-1}\mathbf{q} \tag{7}$$

4 Saliency Detection by the Proposed Model

We describe the details of the saliency detection progress. At first, we calculate the prior vector by using the boundary connectivity. Additionally, we construct a novel graph model. At last, we describe the saliency detection progress based on the prior regularized graph model.

4.1 The Prior by Boundary Connectivity

The prior in this model can be calculated by any method. We introduce the boundary connectivity here to indicate the background and foreground prior value. The connotation of the boundary connectivity [25] is the proportion of the region occupied the boundary comparing with the square root of the whole area. If boundary connectivity is low, the foreground probability is high, the background probability is low. Then, the foreground probability and the background probability by boundary connectivity [25] are calculated as,

$$\mathbf{p}_i^{fg} = exp(-\frac{BndCon^2(x_i)}{2\sigma_1^2}), \tag{8}$$

$$\mathbf{p}_i^{bg} = 1 - exp(-\frac{BndCon^2(x_i)}{2\sigma_1^2}), \tag{9}$$

where $BndCon(x_i)$ is the boundary connectivity of super-pixel x_i and σ_1 is a parameter.

4.2 The Graph Construction

We construct the graph of our model in this section. The graph is constructed as Fig. 2. The detail is explained as follows.

(1) Each node is connected to those neighbor nodes and also connected to the 2-hop neighbor nodes, that is to say, each node x_i is connected to both its directly neighbors x_j and 2-hop neighbors x_k.

Fig. 2. Graph construction: Supe-rpixels are gained by SLIC algorithm on the original image. The blue dots are the directly neighbors and 2-hop neighbors of the yellow dot. The light blue dots are the extending nodes. The black dots are the boundary nodes. (Color figure online)

(2) For extending the node connection, we continue to search the adjacent nodes x_l of the 2-hop node neighbors x_k. If the distance between x_l and x_k is low, the super-pixel x_l is close to the super-pixel x_k, then, we add the edge between x_i and x_l.
(3) We constraint the graph as a close-loop graph, it means that all nodes on four boundaries are adjacent.

If the super-pixel x_i and x_j are contiguous, $a_{ij} = 1$, otherwise, $a_{ij} = 0$. Then, the edge weight $\mathbf{W}_{ij}$ between super-pixel x_i and x_j can be calculated as,

$$\mathbf{W}_{ij} = e^{-\frac{\|x_i - x_j\|^2}{\sigma_2^2}}, \tag{10}$$

where x_i and x_j are feature vectors that are extracted from LAB color value. σ_2 is a parameter.

4.3 Saliency Detection via Prior Regularized Graph Ranking

We use two stages ranking method to get the final saliency maps. In the first stage, boundary super-pixels are clustered into K clusters by k-means clustering method. We separately use the super-pixels in the K clusters as background queries. Let $\mathbf{q}_1^k (k = 1, 2, \ldots, K)$ contain the first stage queries for the K boundary clusters, respectively. $\mathbf{p}_i^{fg}$ is the prior of the foreground probability. We can get the result as,

$$\mathbf{r}_1^k = \left(\mathbf{D} - \frac{1}{1+\mu}\mathbf{W} + \frac{\lambda}{1+\mu}\mathbf{P}^{fg}\right)^{-1}\mathbf{q}_1^k, k = 1, 2, \cdots, K, \tag{11}$$

where $\mathbf{P}^{fg} = \text{diag}(\mathbf{p}_i^{fg})$. After get the K ranking results, we calculate the most different cluster k_{max} ranking vector and remove it by the Euclidean distance with other ranking vectors. Then, we combine the $K-1$ ranking results,

$$\mathbf{s}_{1i} = \prod_{i=1,\cdots,n;k=1,\cdots,K,k\neq k_{max}} \mathbf{s}_{1i}^k \tag{12}$$

where

$$\mathbf{s}_{1i}^k = 1 - \mathbf{r}_{1i}^k, k = 1, 2, \cdots, K. \tag{13}$$

In the second stage, we use the first stage result as the second stage query, if $\mathbf{s}_{1i}^k \geq mean(\mathbf{s}_{1i}^k)$, then $\mathbf{q}_{2i} = 1$, else $\mathbf{q}_{2i} = 0$, $\mathbf{p}_i^{bg}$ is the prior of the background probability. Then, we compute the refinement saliency map by,

$$\mathbf{s} = \mathbf{r}_2 = \left(\mathbf{D} - \frac{1}{1+\mu}\mathbf{W} + \frac{\lambda}{1+\mu}\mathbf{P}^{bg}\right)^{-1} \mathbf{q}_2, \tag{14}$$

where $\mathbf{P}^{bg} = \text{diag}(\mathbf{p}_i^{bg})$. Then, we get the final saliency value.

5 Experiments

We set the number of super-pixels $n = 200$ in all experiments. We set the boundary super-pixels clustering number $K = 4$ by experience. The parameter σ_1 in prior is set to 1. The parameter σ_2^2 in graph construction is set to 10. Besides, there are two controlling parameters, $\alpha = \frac{1}{1+\mu} = 0.99$ and $\beta = \frac{\lambda}{1+\mu} = 10$.

Our proposed method is tested on four public datasets: ECSSD [39], DUT-OMRON [24], SED [40] and SOD [21,41]. ECSSD dataset contains 1000 images. DUT-OMRON dataset contains 5168 complex images, in which the salient objects have different sizes. SED contains 200 images, 100 images have only one salient object, the other 100 images have two salient objects. SOD contains 300 challenging images, in which the salient objects are multiple and the backgrounds are more complicated.

For verifying the effectiveness of our method, we compare with fifteen state-of-the-art saliency object detection methods including CA [18], FT [17], SEG [42], BM [43], SWD [44], SF [20], GCHC [45], LMLC [46], PCA [47], MC [23], MR [24], MS [48], RBD [25], RR [26] and MST [49].

We use precision-recall curves (PR-curves), F-measure and mean absolute error (MAE) to evaluate all methods with our model. PR-curves are constructed by comparing the saliency map with ground truth by thresholds from 0 to 255. Precision is the ratio of the number of correctly salient pixels to the number of all ground truth salient pixels. Recall is the percentage of all selected salient pixels number to the number of all ground truth pixels. F-measure is an overall performance, which is calculated as,

$$F = \frac{(1+\gamma)Precision \cdot Recall}{\gamma Precision + Recall}, \tag{15}$$

referred to [17,24], we set $\gamma = 0.3$ in our experiments.

MAE is an additionally measurement. We normalize both of saliency map and ground truth in the range [0, 1], and then calculate the average difference between them. MAE score as [49],

$$MAE = \frac{1}{H \times W} \sum_{i=1}^{H} \sum_{j=1}^{W} |S(i,j) - GT(i,j)|, \tag{16}$$

where H, W denote the height and width of image, respectively. $S(i,j)$ is the saliency value of pixel level, $GT(i,j)$ is the ground truth of pixel level.

Figure 3 displays the results of our RegMR comparing with other methods on ECSSD database. Figures 4, 5 and 6 show the comparison results of PR-curves and F-measure separately on DUT-OMRON, SED and SOD datasets.

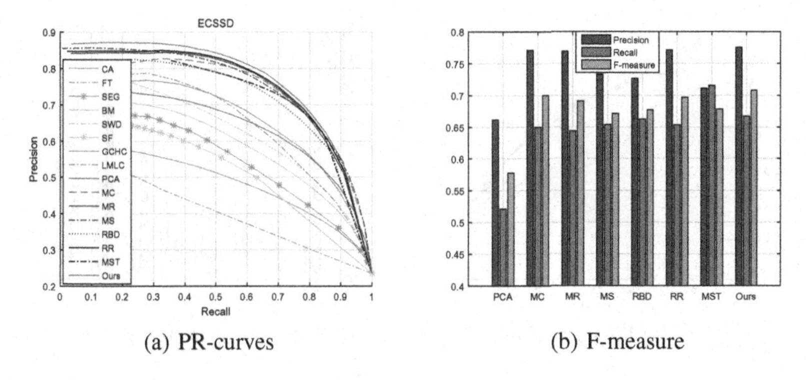

(a) PR-curves (b) F-measure

Fig. 3. Precision-recall curves and F-measures comparing with different methods on ECSSD database.

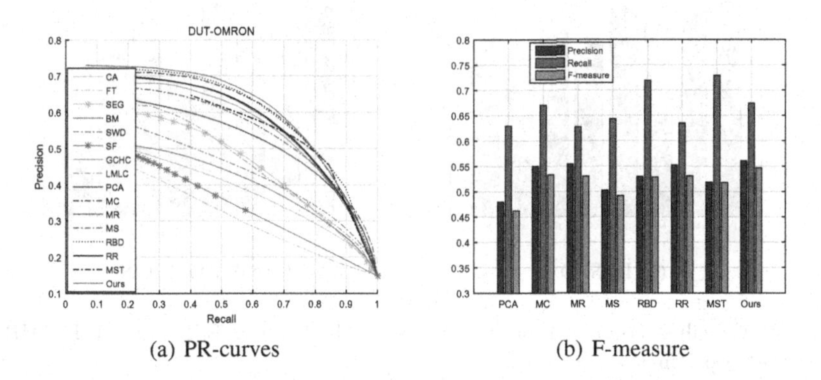

(a) PR-curves (b) F-measure

Fig. 4. Precision-recall curves and F-measures comparing with different methods on DUT-OMRON database.

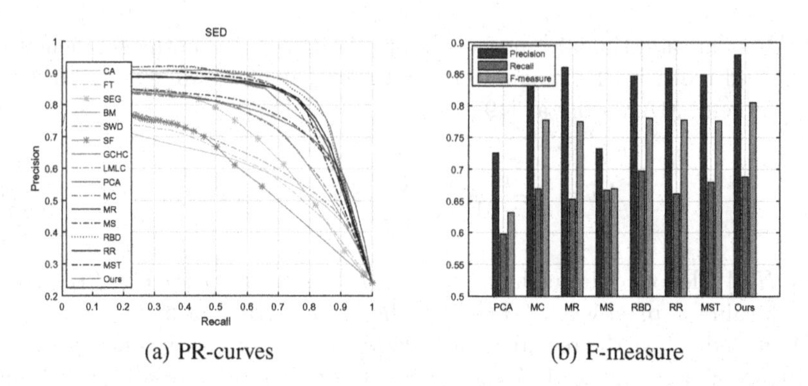

(a) PR-curves (b) F-measure

Fig. 5. Precision-recall curves and F-measures comparing with different methods on SED database.

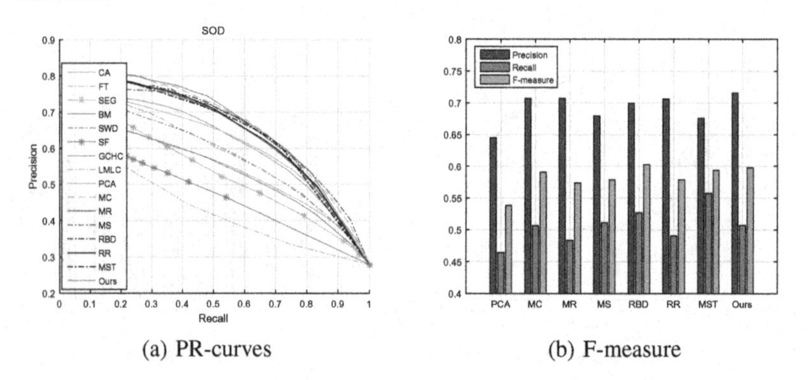

(a) PR-curves (b) F-measure

Fig. 6. Precision-recall curves and F-measures comparing with different methods on SOD database.

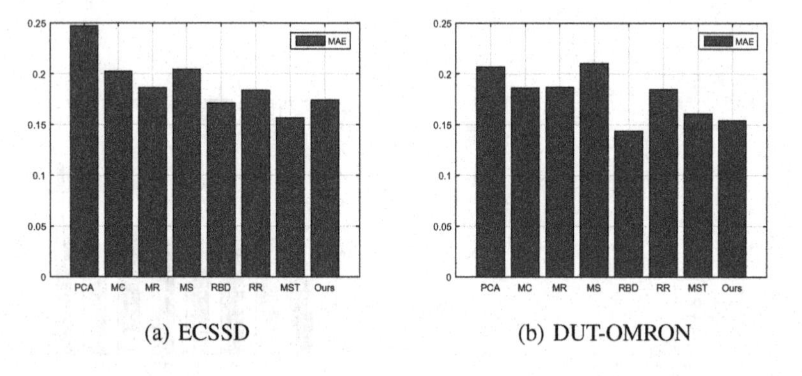

(a) ECSSD (b) DUT-OMRON

Fig. 7. MAE values comparing with different methods on ECSSD and DUT-OMRON (we choose top 7 methods to display).

From above evaluation, our method better than fifteen state-of-the-art methods by measuring PR-curves, which also performances well comparing with others on F-measures except RBD algorithm on SOD database. The MAE value of the proposed method is the top 3 on all four databases which can be see in Figs. 7

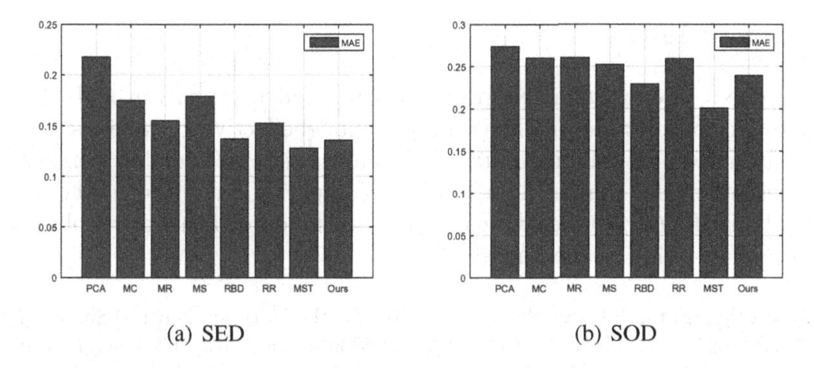

Fig. 8. MAE values comparing with different methods on SED and SOD (we choose top 7 methods to display).

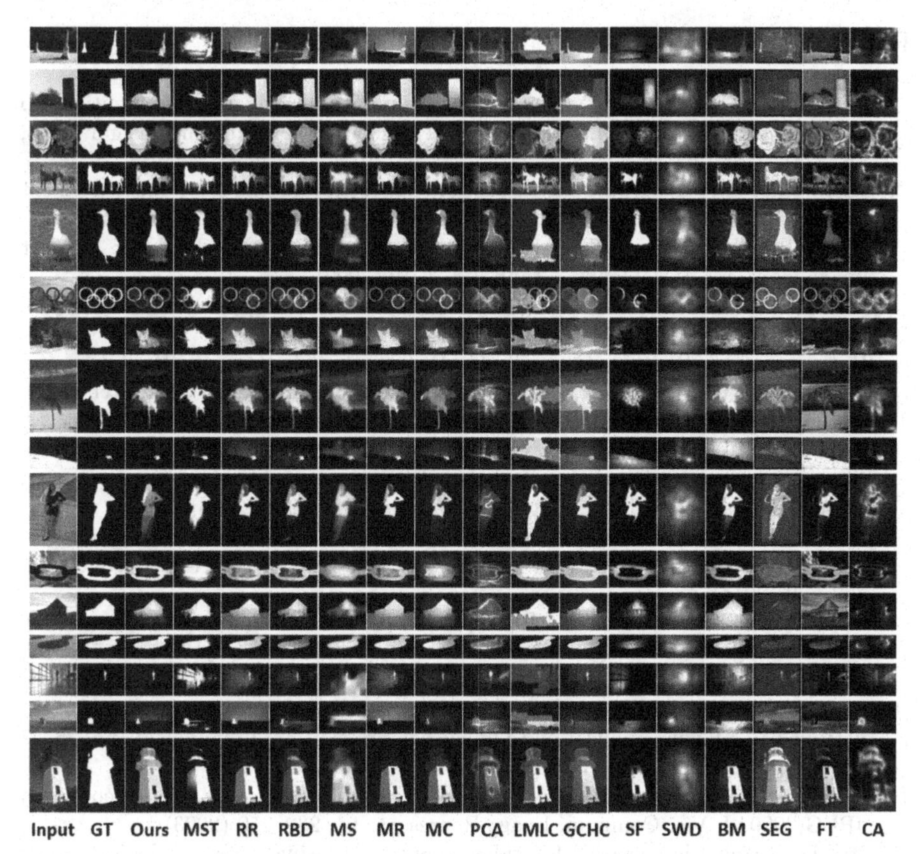

Fig. 9. Examples of output saliency maps results using different algorithms on the ECSSD, DUT-OMRON, SED and SOD datasets.

and 8. Figure 9 shows some sample saliency maps from four datasets. Our model can get more accurately saliency maps intuitively. On the whole, the proposed method better than other fifteen methods.

6 Conclusion

We propose a prior regularized graph ranking model, which can make full use of existing prior information. The proposed model effectively combines the prior information with the manifold structure information. We further demonstrate that the new model achieve higher results comparing with fifteen state-of-the-art methods. In the future, we will use more kinds of features to obtain the intrinsic relationship in the images.

Acknowledgments. This work was sponsored by the National Natural Science Foundation of China (Nos. 61472002, 61602001, 61602006), the Natural Science Foundation of Anhui Higher Education Institution of China (KJ2015A110) and Co-Innovation Center for Information Supply & Assurance Technology, Anhui University.

References

1. Wang, X.J., Ma, W.Y., Li, X.: Data-driven approach for bridging the cognitive gap in image retrieval. In: IEEE International Conference on Multimedia and Expo, pp. 2231–2234 (2004)
2. Hiremath, P.S., Pujari, J.: Content based image retrieval using color boosted salient points and shape features of an image. Int. J. Image Process. **2**, 10–17 (2008)
3. Chen, T., Cheng, M.M., Tan, P., Shamir, A., Hu, S.M.: Sketch2photo: internet image montage. ACM Trans. Graph. **28**, 1–10 (2009)
4. Cheng, M.M., Mitra, N.J., Huang, X., Hu, S.M.: Salientshape: group saliency in image collections. Vis. Comput. **30**, 443–453 (2014)
5. Rutishauser, U., Walther, D., Koch, C., Perona, P.: Is bottom-up attention useful for object recognition? In: IEEE Computer Society Conference on Computer Vision and Pattern Recognition, pp. 37–44 (2004)
6. Navalpakkam, V., Itti, L.: An integrated model of top-down and bottom-up attention for optimizing detection speed. In: IEEE Computer Society Conference on Computer Vision and Pattern Recognition, pp. 2049–2056 (2006)
7. Ren, Z., Gao, S., Chia, L.T., Tsang, I.W.H.: Region-based saliency detection and its application in object recognition. IEEE Trans. Circ. Syst. Video Technol. **24**, 769–779 (2014)
8. Tao, D., Lin, X., Jin, L., Li, X.: Principal component 2-D long short-term memory for font recognition on single Chinese characters. IEEE Trans. Cybern. **46**, 756–765 (2016)
9. Stentiford, F., Park, A., Heath, M.: An estimator for vival attention through competitive novelty with application to image compression. In: Immunology, pp. 25–27 (2001)
10. Bradley, A.P., Stentiford, F.W.M.: Visual attention for region of interest coding in JPEG 2000. J. Vis. Commun. Image Represent. **14**, 232–250 (2003)
11. Itti, L.: Automatic foveation for video compression using a neurobiological model of visual attention. IEEE Trans. Image Process. **13**, 1304–1318 (2004)
12. Fang, Y., Chen, Z., Lin, W., Lin, C.W.: Saliency detection in the compressed domain for adaptive image retargeting. IEEE Trans. Image Process. **21**, 3888–3901 (2012). A Publication of the IEEE Signal Processing Society
13. Itti, L., Koch, C., Niebur, E.: A model of saliency-based visual attention for rapid scene analysis. IEEE Trans. Pattern Anal. Mach. Intell. **20**, 1254–1259 (1998)

14. Harel, J., Koch, C., Perona, P.: Graph-based visual saliency. Adv. Neural Inf. Process. Syst. **1**, 5 (2006)
15. Hou, X., Zhang, L.: Saliency detection: a spectral residual approach. In: IEEE Conference on Computer Vision and Pattern Recognition, pp. 1–8 (2007)
16. Gao, D., Mahadevan, V., Vasconcelos, N.: The discriminant center-surround hypothesis for bottom-up saliency. Adv. Neural Inf. Process. Syst. **20**, 497–504 (2007)
17. Achanta, R., Hemami, S., Estrada, F., Susstrunk, S.: Frequency-tuned salient region detection. In: IEEE Conference on Computer Vision and Pattern Recognition, pp. 1597–1604 (2009)
18. Goferman, S., Zelnik-Manor, L., Tal, A.: Context-aware saliency detection. In: IEEE Conference on Computer Vision and Pattern Recognition, pp. 2376–2383 (2010)
19. Gopalakrishnan, V., Hu, Y., Rajan, D.: Random walks on graphs for salient object detection in images. IEEE Trans. Image Process. **19**, 3232–3242 (2010)
20. Perazzi, F., Krähenbühl, P., Pritch, Y., Hornung, A.: Saliency filters: contrast based filtering for salient region detection. In: IEEE Conference on Computer Vision and Pattern Recognition, pp. 733–740 (2012)
21. Wei, Y., Wen, F., Zhu, W., Sun, J.: Geodesic saliency using background priors. In: Fitzgibbon, A., Lazebnik, S., Perona, P., Sato, Y., Schmid, C. (eds.) ECCV 2012. LNCS, vol. 7574, pp. 29–42. Springer, Heidelberg (2012). https://doi.org/10.1007/978-3-642-33712-3_3
22. Li, X., Lu, H., Zhang, L., Ruan, X., Yang, M.H.: Saliency detection via dense and sparse reconstruction. In: IEEE International Conference on Computer Vision, pp. 2976–2983 (2013)
23. Jiang, B., Zhang, L., Lu, H., Yang, C., Yang, M.H.: Saliency detection via absorbing Markov chain. In: IEEE International Conference on Computer Vision, pp. 1665–1672 (2013)
24. Yang, C., Zhang, L., Lu, H., Ruan, X., Yang, M.H.: Saliency detection via graph-based manifold ranking. In: IEEE Conference on Computer Vision and Pattern Recognition, pp. 3166–3173 (2013)
25. Zhu, W., Liang, S., Wei, Y., Sun, J.: Saliency optimization from robust background detection. In: IEEE Conference on Computer Vision and Pattern Recognition, pp. 2814–2821 (2014)
26. Li, C., Yuan, Y., Cai, W., Xia, Y., Feng, D.D.: Robust saliency detection via regularized random walks ranking. In: IEEE Conference on Computer Vision and Pattern Recognition, pp. 2710–2717 (2015)
27. Xia, C., Li, J., Chen, X., Zheng, A., Zhang, Y.: What is and what is not a salient object? learning salient object detector by ensembling linear exemplar regressors. In: IEEE Conference on Computer Vision and Pattern Recognition, pp. 4321–4329 (2017)
28. Liu, T., Sun, J., Zheng, N.N., Tang, X., Shum, H.Y.: Learning to detect a salient object. In: IEEE Conference on Computer Vision and Pattern Recognition, pp. 1–8 (2007)
29. Gao, D., Han, S., Vasconcelos, N.: Discriminant saliency, the detection of suspicious coincidences, and applications to visual recognition. IEEE Trans. Pattern Anal. Mach. Intell. **31**, 989–1005 (2009)
30. Khuwuthyakorn, P., Robles-Kelly, A., Zhou, J.: Object of interest detection by saliency learning. In: Daniilidis, K., Maragos, P., Paragios, N. (eds.) ECCV 2010. LNCS, vol. 6312, pp. 636–649. Springer, Heidelberg (2010). https://doi.org/10.1007/978-3-642-15552-9_46

31. Mehrani, P., Veksler, O.: Saliency segmentation based on learning and graph cut refinement. In: BMVC, pp. 1–12 (2010)
32. Yang, J., Yang, M.H.: Top-down visual saliency via joint CRF and dictionary learning. In: IEEE Conference on Computer Vision and Pattern Recognition, pp. 2296–2303 (2012)
33. Jiang, H., Wang, J., Yuan, Z., Wu, Y., Zheng, N., Li, S.: Salient object detection: a discriminative regional feature integration approach. In: IEEE Conference on Computer Vision and Pattern Recognition, pp. 2083–2090 (2013)
34. Lu, S., Mahadevan, V., Vasconcelos, N.: Learning optimal seeds for diffusion-based salient object detection. In: IEEE Conference on Computer Vision and Pattern Recognition, pp. 2790–2797 (2014)
35. Kim, J., Han, D., Tai, Y.W., Kim, J.: Salient region detection via high-dimensional color transform. IEEE Trans. Image Process. **25**, 9–23 (2016)
36. Borji, A., Cheng, M.M., Jiang, H., Li, J.: Salient object detection: a benchmark. IEEE Trans. Image Process. **24**, 5706–5722 (2015)
37. Tian, Y., Li, J., Yu, S., Huang, T.: Learning complementary saliency priors for foreground object segmentation in complex scenes. Int. J. Comput. Vis. **111**(2), 153–170 (2015)
38. Achanta, R., Shaji, A., Smith, K., Lucchi, A., Fua, P., Süsstrunk, S.: Slic superpixels. Technical report 149300, EPFL (2010)
39. Yan, Q., Xu, L., Shi, J., Jia, J.: Hierarchical saliency detection. In: IEEE Conference on Computer Vision and Pattern Recognition, pp. 1155–1162 (2013)
40. Alpert, S., Galun, M., Basri, R., Brandt, A.: Image segmentation by probabilistic bottom-up aggregation and cue integration. In: IEEE Conference on Computer Vision and Pattern Recognition, pp. 1–8 (2007)
41. Movahedi, V., Elder, J.H.: Design and perceptual validation of performance measures for salient object segmentation. In: Computer Vision and Pattern Recognition Workshops (CVPRW), pp. 49–56 (2010)
42. Rahtu, E., Kannala, J., Salo, M., Heikkilä, J.: Segmenting salient objects from images and videos. In: Daniilidis, K., Maragos, P., Paragios, N. (eds.) ECCV 2010. LNCS, vol. 6315, pp. 366–379. Springer, Heidelberg (2010). https://doi.org/10.1007/978-3-642-15555-0_27
43. Xie, Y., Lu, H.: Visual saliency detection based on Bayesian model. In: IEEE International Conference on Image Processing, pp. 645–648 (2011)
44. Duan, L., Wu, C., Miao, J., Qing, L., Fu, Y.: Visual saliency detection by spatially weighted dissimilarity. In: IEEE Conference on Computer Vision and Pattern Recognition, pp. 473–480 (2011)
45. Yang, C., Zhang, L., Lu, H.: Graph-regularized saliency detection with convex-hull-based center prior. IEEE Sig. Process. Lett. **20**, 637–640 (2013)
46. Xie, Y., Lu, H., Yang, M.H.: Bayesian saliency via low and mid level cue. IEEE Trans. Image Process. **22**, 1689 (2013). A Publication of the IEEE Signal Processing Society
47. Margolin, R., Tal, A., Zelnik-Manor, L.: What makes a patch distinct? In: IEEE Conference on Computer Vision and Pattern Recognition, pp. 1139–1146 (2013)
48. Tong, N., Lu, H., Zhang, L., Ruan, X.: Saliency detection with multi-scale superpixels. IEEE Sig. Process. Lett. **21**, 1035–1039 (2014)
49. Tu, W.C., He, S., Yang, Q., Chien, S.Y.: Real-time salient object detection with a minimum spanning tree. In: IEEE Conference on Computer Vision and Pattern Recognition, pp. 2334–2342 (2016)

Author Index

Printed in the United States
By Bookmasters